HEALTH SCIENCES LIBRARY
LUTHERAN COLLEGE
3024 FAIRFIELD AVE.
FORT WAYNE IN 46807-1697

Calcium Channels:
Their Properties, Functions, Regulation, and Clinical Relevance

Editors

Leon Hurwitz, Ph.D.
Professor Emeritus
Department of Pharmacology
University of New Mexico
School of Medicine
Albuquerque, New Mexico

L. Donald Partridge, Ph.D.
Associate Professor
Department of Physiology
University of New Mexico
School of Medicine
Albuquerque, New Mexico

John K. Leach, M.D.
Professor Emeritus
Departments of Pharmacology and Medicine
University of New Mexico
School of Medicine
Albuquerque, New Mexico

CRC Press
Boca Raton Ann Arbor Boston London

Library of Congress Cataloging-in-Publication Data

Calcium channels: their properties, functions, regulation, and
    clinical relevance / editors, Leon Hurwitz, L. Donald Partridge,
    John K. Leach.
            p.   cm.
        Includes bibliographical references and index.
        ISBN 0-8493-8807-4
        1. Calcium channels. 2. Calcium channel blockers. I. Hurwitz,
    Leon, 1921–   . II. Partridge, L. Donald. III. Leach, John K.
    [DNLM: 1. Calcium Channels. QH 601 C1444]
    QP535.C2C2627   1991
    612.7′4—dc20
    DNLM/DLC
        for Library of Congress   90-15171
                                                                                         CIP

Developed by: Telford Press

This book represents information obtained from authentic and highly regarded sources. Reprinted material is quoted with permission, and sources are indicated. A wide variety of references are listed. Every reasonable effort has been made to give reliable data and information, but the author and the publisher cannot assume responsibility for the validity of all materials or for the consequences of their use.

Neither this book nor any part may be reproduced or transmitted in any form or by any means, electronic or mechanical, including photocopying, microfilming, and recording, or by any information storage and retrieval system, without permission in writing from the publisher.

Direct all inquiries to CRC Press, Inc., 2000 Corporate Blvd., N. W., Boca Raton, Florida, 33431.

© 1991 by CRC Press, Inc.

International Standard Book Number 0-8493-8807-4

Library of Congress Card Number 90-15171
Printed in the United States   2 3 4 5 6 7 8 9 0

Printed on acid-free paper

PREFACE

Calcium channels are a common component of the membranes of a wide range of excitable cells and their presence is crucial to the functioning of these cells. As a consequence, workers in diverse areas of life science ranging from the cell biologist to the physician have an overlapping interest regarding the operation of these channels. The last decade has seen a vast expansion of the knowledge about calcium channels emanating from workers with widely different research interests. The information covered by this research spans the gamut from the protein structure of these channels to the effectiveness of specific channel blockers in the treatment of clinical diseases. In this book, an attempt was made to bring together the vast body of factual and conceptual material that has been gathered about calcium channels into an organized compendium-like presentation. Other more specialized books deal effectively with this channel but nowhere else, we believe, have such diverse topics been brought together in a single volume.

The book is divided into four sections that provide a current review of the biophysics, physiology, pharmacology, and clinical role of calcium channels. Although the preparation of individual chapters in each of the four sections of the book necessarily relied on the expertise of specialists in their respective areas of research, every effort was made to interrelate the considerable amount of information covered by focusing on a central theme, namely the characterization and modulation of calcium channels. Such an effort, it was felt, would be of interest to a broad scientific readership ranging from the biophysicist to the physician. The biophysics section covers the operation of the single calcium channel. It includes chapters on ion permeation, channel activation, channel inactivation, and second messenger modulation. The physiology section deals with the role of calcium channels in the function of a number of cell types. The chapters in this section cover excitation-contraction coupling, excitation-secretion coupling, sensory transduction, regulation of electrical activity, and the regulation of cell growth and development. The pharmacology section discloses the manner in which chemical agents affect calcium channels. This section includes chapters that describe the effects of permeant and inhibitory inorganic ions, blocking and activating effects of organic ions, and the regulatory effects of naturally occurring compounds. The final section explores the role of calcium channels in clinical medicine. Its chapters deal with the modulation of calcium channels in the treatment of ischemic heart disease, cardiomyopathies, hypertension, cardiac arrhythmias, peripheral vascular diseases, platelet-related disorders, neurological disorders, and psychiatric disorders.

It is our hope that a unified presentation of this broad spectrum of material will serve to enhance the understanding of this complex subject and to further the utilization of this information.

EDITORS

Leon Hurwitz, Ph.D., is a professor emeritus in the Pharmacology Department at the University of New Mexico School of Medicine in Albuquerque.

Dr. Hurwitz received his B.S. degree from Cornell University in 1947 and his Ph.D. degree from the University of Rochester in 1953. In 1954, he acquired a position as pharmacologist in the Pharmacology Division of the Federal Food and Drug Administration. After 18 months, he became Chief of the Acute Toxicity Branch in that division. In 1957, he accepted a faculty position in the Pharmacology Department at Vanderbilt University School of Medicine. During his 15 year tenure at Vanderbilt University he rose to the rank of professor. In 1972, he was awarded the position of Chairman of the Pharmacology Department at the University of New Mexico, School of Medicine. He remained in that position until 1986, when he retired from active teaching and research.

His primary research work focuses on the disposition and functional role of inorganic ions in mammalian smooth muscle. He has served on the editorial board or as a reviewer for a number of scientific journals and on national committees associated with the National Institutes of Health.

L. Donald Partridge, Ph.D., is an associate professor in the Department of Physiology at the University of New Mexico School of Medicine in Albuquerque.

Dr. Partridge received his B.S. degree in biology from M.I.T. in 1967 and his Ph.D. degree in physiology from the University of Washington in 1973. He carried out postdoctoral study at the University of Bristol as a Welcome Research Fellow and at the Friday Harbor Laboratories as a N.I.H. postdoctoral fellow. He was awarded a Fulbright Grant as an investigator at the Max Planck Institute of Psychiatry in Munich.

Dr. Partridge's research interests center on the slow ionic currents that underlie neuronal firing patterns in neurons. Most recently he has pursued collaborative studies supported by stipendiums from the Max Planck Institutes in Munich and Göttingen.

John K. Leach, M.D., is a professor emeritus in the Departments of Pharmacology and Medicine at the University of New Mexico School of Medicine in Albuquerque.

Dr. Leach received his B.S. degree in biology from Baldwin–Wallace College in 1943 and his M.D. from Albany Medical College in 1947. He completed a residency in internal medicine at the University of Colorado Medical Center and fellowships in cardiology at Albany Medical College and Massachusetts General Hospital. He was awarded research fellowships at Albany Medical College and U.C.L.A. Medical Center. He is certified by the American Board of Internal Medicine and Cardiovascular Disease and is a Fellow of the American College of Physicians and the American College of Cardiology.

Dr. Leach's research interests include cardiac muscle mechanics, excitation-contraction coupling, and hemodynamics.

CONTRIBUTORS

Yeon S. Ahn, M.D.
Professor
Department of Internal Medicine
University of Miami School of Medicine
Miami, Florida

Cristina R. Artalejo
Assistant Professor
Department of Pharmacological and
 Physiological Sciences
University of Chicago
Chicago, Illinois

George J. Augustine, Ph.D.
Associate Professor
Department of Biological Sciences
University of Southern California
Los Angeles, California

Philip M. Best, Ph.D.
Associate Professor
Department of Physiology and Biophysics
University of Illinois at Champaign-
 Urbana
Urbana, Illinois

Ghassan Bkaily, Ph.D.
Associate Professor
Department of Physiology and Biophysics
University of Sherbrooke
Sherbrooke, Quebec, Canada

Emilio Carbone, Ph.D.
Professor
Department of Anatomy and Physiology
University of Torino
Torino, Italy

Matthias Dose, M.D.
Vice Director
District Hospital
Department of Psychiatry
Ansbach, Germany

Hinderk M. Emrich, M.D.
Professor
Max-Planck-Institute for Psychiatry
Munich, Germany

Aaron P. Fox, Ph.D.
Assistant Professor
Department of Pharmacological and
 Physiological Sciences
University of Chicago
Chicago, Illinois

William F. Graettinger, M.D.
Assistant Professor
Department of Medicine
University of California, Irvine
V.A. Medical Center
Long Beach, California

Stephen D. Hess, Ph.D.
Research Associate
Department of Biological Sciences
University of Southern California
Los Angeles, California

Lane D. Hirning, Ph.D.
Research Fellow
Department of Pharmacological and
 Physiological Sciences
University of Chicago
Chicago, Illinois

Janet Holliday, Ph.D.
Research Associate
Department of Neuropharmacology,
 BCRI
Scripps Clinic and Research Foundation
La Jolla, California

Leon Hurwitz, Ph.D.
Professor Emeritus
Department of Pharmacology
University of New Mexico School of
 Medicine
Albuquerque, New Mexico

Ronald A. Janis, Ph.D.
Principal Staff Scientist
Miles Institute for Preclinical
 Pharmacology
West Haven, Connecticut

Gaétan Jasmin, M.D., Ph.D.
Professor
Department of Pathology
University of Montreal
Montreal, Quebec, Canada

Wenche Jy, Ph.D.
Assistant Professor
Department of Medicine
Center for Blood Diseases
University of Miami
Miami, Florida

Juan I. Korenbrot, Ph.D.
Professor
Department of Physiology
University of California
San Francisco, California

Wai-Meng Kwok
Post-doctoral Fellow
Department of Physiology
University of Rochester
Rochester, New York

Nallanna Lakshminarayanaiah, Ph.D.
Honorary Research Professor
Department of Pharmacology
Jefferson Medical College
Philadelphia, Pennsylvania

John K. Leach, M.D.
Professor Emeritus
Departments of Pharmacology and
  Medicine
University of New Mexico School of
  Medicine
Albuquerque, New Mexico

Hans-Dieter Lux, M.D.
Professor
Neurophysiology Department
Max-Planck-Institute for Psychiatry
Martinsried, Germany

Amy MacDermott, Ph.D.
Assistant Professor
Department of Physiology and Cellular
  Biophysics
Columbia University
New York, New York

Andres Villu Maricq, M.D., Ph.D.
Fellow
Department of Pharmacology
University of California
San Francisco, California

Richard Miles, Ph.D.
Assistant Professor
Department of Neurology
Columbia University
New York, New York

Richard J. Miller
Professor
Department of Pharmacological and
  Physiological Sciences
University of Chicago
Chicago, Illinois

David J. Mogul
Research Fellow
Department of Pharmacological and
  Physiological Sciences
University of Chicago
Chicago, Illinois

Lionel H. Opie, M.D.
Professor
Heart Research Unit
Department of Medicine
University of Cape Town Medical School
Cape Town, South Africa

L. Donald Partridge, Ph.D.
Associate Professor
Department of Physiology
University of New Mexico School of
  Medicine
Albuquerque, New Mexico

André Pasternac, M.D.
Associate Clinical Professor
Montreal Heart Institute
Montreal, Quebec, Canada

Nicholas J. Penington
Research Fellow
Department of Pharmacological and
  Physiological Sciences
University of Chicago
Chicago, Illinois

Libuse Proschek
Researcher
Department of Pathology
University of Montreal
Montreal, Quebec, Canada

Gary A. Rosenberg, M.D.
Professor and Chairman
Department of Neurology
University of New Mexico School of
 Medicine
Neurology Service Veteran's Medical
 Center
Albuquerque, New Mexico

Reese S. Scroggs
Research Fellow
Department of Pharmacological and
 Physiological Sciences
University of Chicago
Chicago, Illinois

Bramah N. Singh, M.D.
Professor
Department of Medicine
UCLA School of Medicine
Wadsworth V.A. Medical Center
Los Angeles, California

Nicholas Spitzer, Ph.D.
Professor
Department of Biology and Center for
 Molecular Genetics
University of California, San Diego
La Jolla, California

Dieter Swandulla, M.D., Ph.D.
Max-Planck-Institute for Biophysical
 Chemistry
Göttingen, Germany

David J. Triggle, Ph.D.
Dean
School of Pharmacy
State University of New York
Buffalo, New York

Michael A. Weber, M.D.
Professor
Department of Medicine
University of California, Irvine
V.A. Medical Center
Long Beach, California

Xiaoping Xu
Research Assistant
Department of Physiology and Biophysics
University of Illinois at Urbana-
 Champagne
Urbana, Illinois

TABLE OF CONTENTS

Section I: Biophysics of Calcium Channels
Chapter 1
Biophysics of Calcium Channels: An Overview 3
L. Donald Partridge

Chapter 2
Calcium Channel Permeability and Ionic Block 9
Hans-Dieter Lux

Chapter 3
Activation and Deactivation of Neuronal High Voltage-Activated,
Fast Deactivating Calcium Channels 21
Dieter Swandulla

Chapter 4
Calcium Channel Inactivation 35
Emilio Carbone and Dieter Swandulla

Section II: The Physiology of Calcium Channels
Chapter 5
The Physiology of Calcium Channels: An Overview 63
L. Donald Partridge

Chapter 6
Calcium Channels in Excitation-Contraction Coupling 69
Philip M. Best, Wai-Meng Kwok, and Xiaoping Xu

Chapter 7
Calcium Channels as Triggers of Secretion 87
George J. Augustine and Stephen D. Hess

Chapter 8
Calcium Channels in Sensory Transduction: A Case Study in Photoreceptors 107
Juan I. Korenbrot and Andres V. Maricq

Chapter 9
Calcium Channels and Patterns of Electrical Activity in Neurons 125
Amy MacDermott and Richard Miles

Chapter 10
Calcium Channels in the Regulation of Cell Development and
Cellular Interactions 137
Janet Holliday and Nicholas Spitzer

Section III: Pharmacology of Calcium Channels
Chapter 11
The Effects of Chemical Agents on Calcium Channels: An Overview 159
Leon Hurwitz

Chapter 12
Effects of Inorganic Ions on Calcium Channels 163
Nallanna Lakshminarayanaiah

Chapter 13
Drugs Acting on Calcium Channels 195
Ronald A. Janis and David J. Triggle

Chapter 14
Modulation of Calcium Channels by Neurotransmitters, Hormones and Second
Messengers 251
**Aaron P. Fox, Lane D. Hirning, David J. Mogul, Cristina R. Artalejo,
Nicholas J. Penington, Reese S. Scroggs, and Richard J. Miller**

Section IV: Modulation of Calcium Channels in Clinical Medicine
Chapter 15
Modulation of Calcium Channels in Clinical Medicine: An Overview 267
John K. Leach

Chapter 16
Modulation of Calcium Channels in the Treatment of Ischemic Heart Disease 273
Lionel H. Opie

Chapter 17
Modulation of Calcium Channels in the Management of Cardiomyopathies 295
Gaétan Jasmin, André Pasternac, Ghassan Bkaily, and Libuse Proschek

Chapter 18
Experiences with Calcium Channel Blockers in the Treatment of Hypertension 309
Michael A. Weber and William F. Graettinger

Chapter 19
Control of Cardiac Arrhythmias by Modulation of the Slow Myocardial Channel 327
Bramah N. Singh

Chapter 20
The Effects of Calcium Channel Blockers on Platelets and their Application
in the Management of Vascular Diseases 363
Wenche Jy and Yeon S. Ahn

Chapter 21
Calcium Channel Blockers in Neurological Disorders 377
Gary A. Rosenberg

Chapter 22
Use of Calcium Channel Blockers in Psychiatric Disorders 385
Matthias Dose and Hinderk M. Emrich

References 397

Index 481

Section I
Biophysics of Calcium Channels

Chapter 1

Biophysics of Calcium Channels: An Overview

L. Donald Partridge

Selective permeability of membranes to small ions is a property that is crucial to the functioning of cells across the broad range of living organisms. Excitable cells regulate the selective permeability of their membranes in response to a variety of circumstances, including: (1) input transduction, such as light-dependent changes of sodium permeability of the rod outer segment; (2) integration, such as postsynaptic transmitter-dependent changes of membrane permeability in a neuron soma; (3) signaling, such as the classic changes of sodium and potassium permeability in a nerve axon; and (4) coupling of output phenomena, such as the changes in calcium permeability that underlie excitation-contraction coupling in muscles. Considerable advances have been made in recent years in our understanding of the mechanisms by which excitable cells regulate their selective membrane permeability. These advances, which are still very much in progress, are the result of the work of numerous biophysicists, molecular biologists, geneticists, and biochemists.

At the beginning of this century, Julius Bernstein (1902) hypothesized correctly that excitable cells maintain an intracellular potential as a result of their membranes being selectively permeable to potassium ions. He further proposed that excitation occurs as a result of the transient loss of this selective permeability. Although this latter concept proved to be wrong, it prompted the experiments of Alan Hodgkin, Andrew Huxley, Kenneth Cole, and Howard Curtis (Cole and Curtis, 1938; Hodgkin and Huxley, 1952b) during the middle decades of this century, which, in turn, set the stage for our current understanding of the changes in selective membrane permeability that occur in excitable cells. Although their work dealt with the membrane permeability of the squid axon to Na^+ and K^+, their findings have been extended to include a multitude of ion permeabilities in a host of diverse cell types.

While the recent advances in our understanding of selective membrane permeability have branched out in several directions, all center around the one important issue of ion channels. Selective membrane permeability exists because of intrinsic protein macromolecules that span the membranes of cells and thereby form aqueous pores through which ions can permeate the hydrophobic lipid membrane (Hille, 1984). Erwin Neher and Bert Sakmann

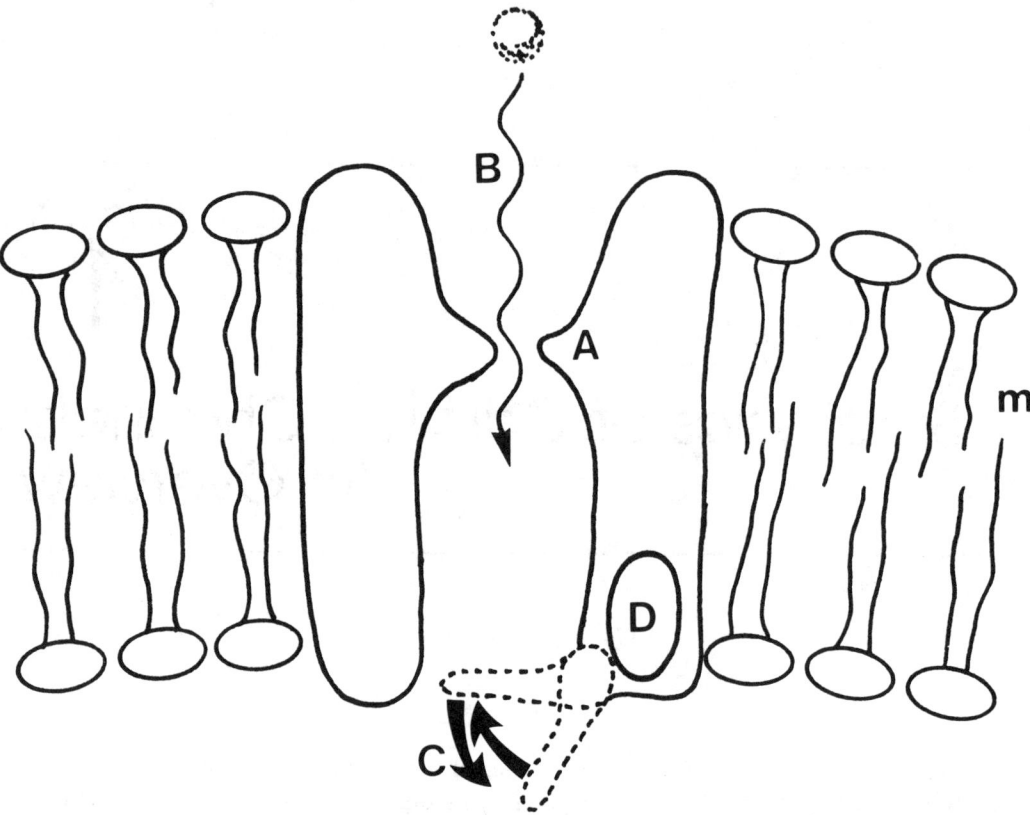

Figure 1. Schematic drawing of a generic ion channel. This channel is shown as a protein molecule that spans the lipid bilayer of the cell membrane (m). The labeled elements of this channel are: (A) the selectivity filter that determines which ions can pass through the channel; (B) the electrical conductance of the channel that indicates the ease with which selected ions actually transit the channel; (C) the channel gate(s) that can move between open and closed positions; and (D) the sensor that determines the position of the gate in response to the electrical field across the membrane or the presence of ions or transmitters.

(Hamill et al., 1981) have developed a technique called patch clamping that permits the measurement of ionic currents through single membrane channels. Laboratories such as that of Lawrence Salkoff (Salkoff and Tanouye, 1986) have made considerable advances in the understanding of the genetic regulation of ion channels. Groups, such as that headed by S. Numa (Noda, et al., 1984), have made enormous strides in providing the amino acid sequence and probable tertiary configuration of various ion channel proteins.

From these various sources of information we have developed a rather general understanding of the nature of ion channels. Figure 1 is a simplified schematic of a generic channel. This schematic emphasizes four characteristics that are common to most ion channels: (1) ion selectivity; (2) electrical conductance; (3) gating kinetics; and (4) chemical or electrical field sensing. A single excitable cell may contain 5 to 10 different channel types that are distinguishable by these four characteristics. Because of these properties of ion channels they can serve their unique functions in the cell. These characteristics are ultimately a result of the genetically determined amino acid sequence of the channel protein.

Selectivity is the property of ion channels by which they are usually named; a sodium channel selects for Na^+ ions, while a calcium channel selects for Ca^{2+} ions. This ability to select for a specific species of ion is generally rather good, so that preferred ions are passed through the channel at least a hundred times more easily than rejected ions. In the simplest form, selectivity is accomplished by steric means (Hille, 1975). Thus, the channel has a narrow portion where ions larger than a certain size are excluded and those smaller than this size are passed. Additional means by which channels accomplish ion selectivity have to do with the relative binding strength of the selected ion to sites within the channel compared to the binding strength of that ion to the surrounding water molecules in the aqueous medium.

Ca^{2+} is somewhat unique among permeant ions in that Ca^{2+} ions themselves serve crucial functions within the cell in addition to merely carrying electrical charge. Chapter 2, Section I will consider the important issues regarding the selectivity of calcium channels. While steric effects certainly exist, binding of Ca^{2+} ions within the channels is important to the establishment of channel selectivity. These binding sites provide the interesting additional possibility that the binding of other ions can influence the permeation of Ca^{2+} ions through calcium channels. This topic will also be introduced in this chapter.

The conductance of a channel is a measure of the ease with which ions pass through that channel when it is in a permissive state. Conductance is an electrical measurement determined as the ratio of the current flowing through the channel to the electrical driving force on the ions carrying the current:

$$g_s = I_s/(E - E_s)$$

where g_s is the conductance to ion S, I_s is the current produced by ion S moving through its selective channels, E is the potential difference across the membrane, and E_s is the potential at which ion S is at electrochemical equilibrium.

Single channels have conductances of picosemens (10^{-12} ohm^{-1}) and pass picoamperes (10^{-12} amps) of current, this is equivalent to tens of millions of ions passing through the channel per second. The flux of Na^+ ions down their transmembrane electrochemical gradient through the population of conducting sodium channels can have a rapid and profound effect upon the transmembrane potential but with a fairly insignificant effect upon intracellular Na^+ concentration. Calcium channels, on the other hand, control the flux of a charged ion that is maintained at an exceedingly low intracellular concentration. Thus, the Ca^{2+} flux through the population of conducting calcium channels, while having an effect on transmembrane potential, may also alter the intracellular Ca^{2+} concentration by as much as two orders of magnitude.

Channel gating is a determination of the state of permissiveness of a channel for ion conduction. To a first approximation, channels have only two states—open and closed. Generally the gates move from one state to the other state in response to an appropriate stimulus and then return to the original state with kinetics intrinsic to the channel protein. Two general categories of gates can be identified. Those gates that open in response to a stimulus are referred to as activation gates, while gates that close the channel following a stimulus are referred to as inactivation gates.

The action of activation and inactivation gates warrants some discussion as a result, in part, of the vast amount of classical biophysical literature devoted to these processes. The movement of individual gates cannot be observed physically in a single protein molecule but single channel recordings provide the opportunity to directly observe the effects of gate movements. Much of the older, but important and insightful literature is based on measurements of currents flowing across large areas of cell membranes utilizing the voltage clamp technique. This corpus of experimental work delineates three distinguishable (gating) processes: (1) activation; (2) deactivation; and (3) inactivation. The brief description of these processes presented here will make use of the schematic gates shown in Figure 2, although the original literature treats these processes more empirically. Since, in these experiments, the cells are *voltage* clamped, the voltage is under the experimenter's direct control. Normally, membrane potential is clamped to a steady "holding" potential, E_h, and then suddenly changed to some variable "clamp" potential, E_c, during which time the ionic current flow

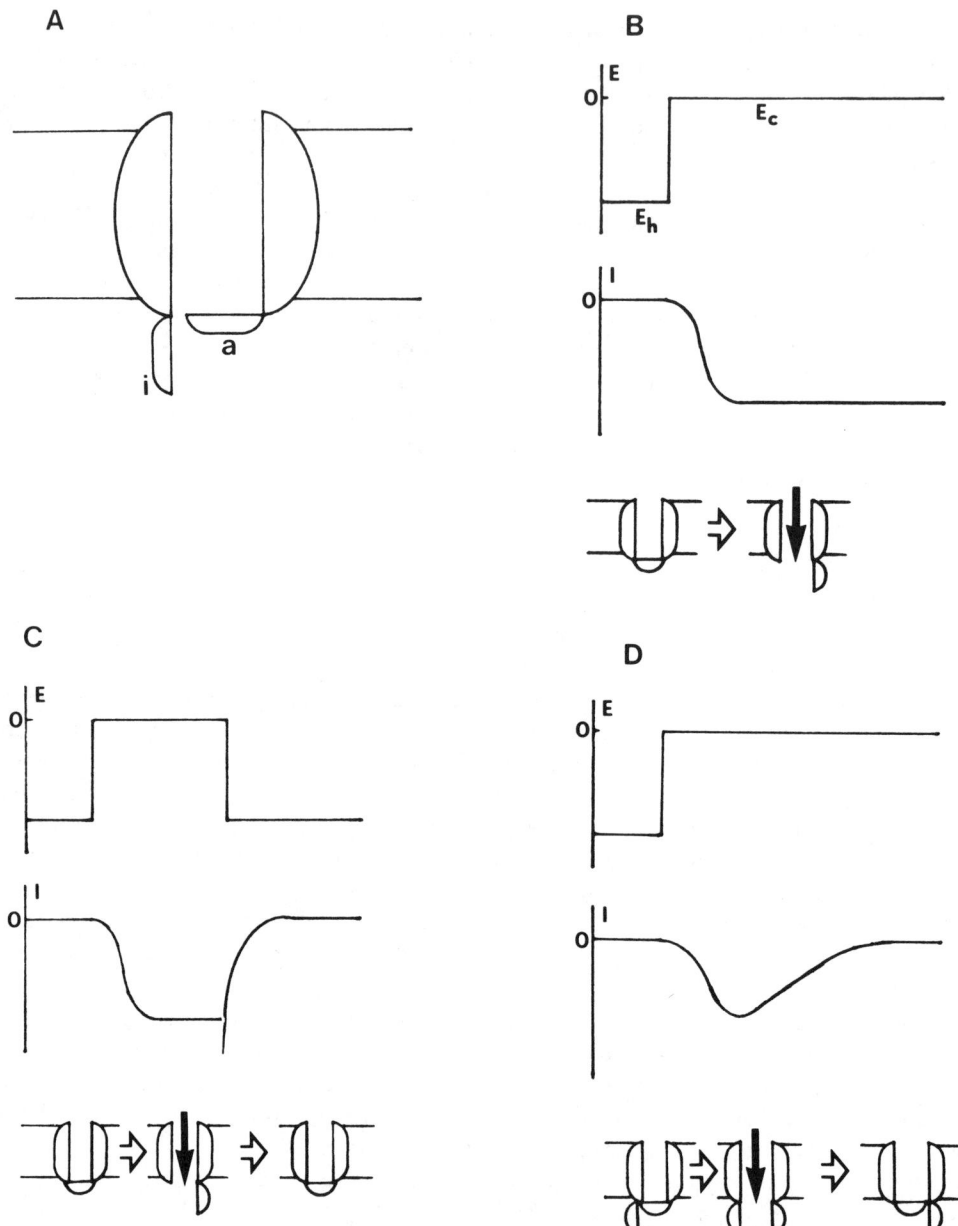

Figure 2. Activation, deactivation, and inactivation of ion channels. (A) Simplified representation of an ion channel with an activation gate (a) and an inactivation gate (i). This shows the typical resting condition with the activation gate closed and the inactivation gate open. (B) Activation; the membrane potential in a voltage clamped cell is changed suddenly from a negative value (E_h) to a depolarized one (E_c) and as a result an ionic current begins to flow. This current flow can occur because the depolarization causes activation gates to open. In records of voltage clamp experiments, inward currents are displayed as negative values. Furthermore, because the interesting results are usually in the current records, voltage traces are often omitted with values for E_h and E_c simply being listed. (C) Activation and deactivation; if the membrane potential is returned to a negative value after the activation gates are open, these gates will close. The resultant current tail is the result of deactivation. (D) Activation and inactivation; many channels possess activation and inactivation gates. In this example, the depolarization causes the activation gates to open and, more slowly, the inactivation gates to close. The resultant inward current rises to a maximum because of activation and subsequently falls because of inactivation.

is measured. Figure 2B shows the process of activation in which a depolarization of the membrane potential opens the activation gate and initiates a maintained current flow through the channel. In Figure 2C, the membrane potential is returned to its holding potential following a brief depolarization. The current flow in this case first increases due to the opening of activation gates and then decreases due to closure of these same activation gates (deactivation). This decreasing ionic current is often referred to as the current "tail". Figure 2D shows a channel with two distinct types of gates, the activation gate and an additional inactivation gate that closes slowly in response to depolarization. In this case, a maintained depolarization leads to the sequence of first increasing ion current due to activation followed by decreasing ion current due to inactivation. Deactivation can also be observed in a channel with inactivation gates but the return of the membrane potential to the holding potential must be accomplished before significant inactivation has had time to occur.

Chapter 3, Section I, will discuss both activation and the related process of deactivation in calcium channel gating. In particular, this chapter will address the voltage dependence and kinetics of calcium channel activation gates. It will further consider means by which ions or drugs can affect the action of activation gates.

Chapter 4 in this section will deal with the mechanism of action of inactivation gates on calcium channels. Again, the voltage dependency and kinetics of inactivation gates will be considered along with modulation of the inactivation process. Calcium channel inactivation is of additional interest because it provides an important means with which to distinguish between different classes of Ca^{2+}-selective channels.

The fourth channel property that we consider here is that of sensing. Gating sensors can be divided into two general groups, although some channels may possess properties of both groups. These two major groups were originally delineated in early work on the sodium channel of the squid giant axon, which has a voltage-sensitive gating mechanism (Hodgkin and Huxley, 1952b) and in the similar early work on the acetylcholine-activated channel of the neuromuscular junction, which has a ligand-sensitive gating mechanism (Takeuchi and Takeuchi, 1959). Voltage-sensitive channels open in response to the often rather large changes in electrical field that can exist across cell membranes. The transmembrane potential is an effective stimulus not only in opening and closing a channel's activation gates, but also in opening and closing the channel's inactivation gates. The other type of sensor is one that causes a change in channel gating when a specific molecule is present at a receptor site associated with that channel. Channels with this type of sensor include those that are affected by neurotransmitters or neuromodulators that have access to the extracellular end of the channel and also to those that are sensitive to various second messengers and modulatory ions at the cytoplasmic end of the channel. The effect of intracellular Ca^{2+} ions in the regulation of calcium channel inactivation will also be considered in Chapter 4, Section I.

Calcium channels are a ubiquitous feature of excitable cells. They are an important component of the palette of membrane channels of: invertebrate muscle; vertebrate smooth, cardiac and embryonic skeletal muscles; neuron somata, presynaptic terminals and embryonic axons; various sensory receptors; and a wide range of cells that secrete hormones, neuromodulators, neurotransmitters, or mucus. In most of these cells, the flux of Ca^{2+} through these channels provides the means by which changes in membrane potential are transduced into nonelectrical activities. Calcium channels are not as highly conserved as, for instance, sodium channels but vary rather widely among these several tissues in such properties as their conductance and inactivation mechanisms.

There is, of course, no equivalent to a fossil record from which we can deduce the course of evolution of calcium channels. We can glean, however, some information from the genetics and molecular biology of channels that exist today (Hille, 1984). Ion channels seem to be some of the most highly conserved protein molecules found among living organisms. The first primitive ancestral cation channels probably evolved before the time of transition from prokaryotes to eukaryotes. These channels became more specific in eukaryotic organisms, eventually becoming specialized as primitive potassium and calcium channels. Ca^{2+} ions have long played a dual role, on the one hand in the regulation of cellular functions, and on the other hand in the regulation of cellular excitability. Calcium channels, then, are capable of both significantly modifying membrane potential and of regulating intracellular Ca^{2+} concentration. It is probable that the sodium channel evolved as a modification of the calcium channel when it became advantageous to separate the excitability of a cell from the multitude of coupling functions that intracellular Ca^{2+} ions had acquired. Calcium channels are found today in those cells or portions of cells where excitation is tied to regulation of other cellular functions.

Our concept of selective membrane permeability has been significantly altered with the recent advances in understanding of ion channels. Before the last 15 years it could have been reasonably argued that selective permeability was a distributed membrane property. We now understand it to be very much a discrete property that results from the action of specific protein macromolecules within the lipid matrix of the membrane. The opening of individual channels is a probabilistic function, with the open channel probability being a function of the activation state of a gating sensor. One can, however, predict with a fair degree of accuracy, the percentage of time that a channel will be open given the state of the sensor. The authors in this section of the book will discuss aspects of the selective membrane permeability of excitable cells on the basis of the current understanding of the biophysics of calcium channels.

Chapter 2

Calcium Channel Permeability and Ionic Block

H. D. Lux

TABLE OF CONTENTS

I.	Introduction	10
II.	Divalent Permeant and Blocking Ions	10
III.	An Apparent Paradox	12
IV.	Block by Mg^{2+}	13
V.	Mg^{2+}-Blockage of Currents Carried by Monovalent Ions Through Calcium Channels	14
VI.	Ca^{2+}-Block of Sodium Currents Through LVA Calcium Channels	15
VII.	Voltage-Dependent Block	17
VIII.	Conclusions	18

I. INTRODUCTION

Most voltage-activated membrane channels transport specific ions to the exclusion of others. Their ionic selectivity together with the driving force produced by the electrochemical gradient across the membrane provides for directed ion flux. This is necessary for electrical signaling and, as it applies to calcium channels, for the chemical signaling that it is necessary for diverse calcium-dependent cellular functions. The selectivity of ion channels is generally limited, but calcium channels seem unique in this regard when compared to other ionic channels. The sodium channel is known to be permeant to Ca^{2+} ions and K^+ ions at a ratio of 0.05 to 0.1 of its Na^+ permeability (Hille, 1984). Calcium channels would not be very useful if the permeability of these channels to Na^+ was similar to the Ca^{2+} permeability of the sodium channel. Poorly selective calcium channels operating in a physiological medium could not reliably select for Ca^{2+} ions against the vast majority of Na^+ ions and against other cations such as K^+, and Mg^{2+}.

This chapter will review the basic properties of the calcium channel that relate to its unique ion selectivity. It is argued that this channel's selectivity is due to ionic affinities for a binding site rather than to the sieving of the pore against undesired ions. This will be demonstrated by observations on the blockage (by Ca^{2+} ions themselves and by Mg^{2+} ions) of nonspecific (Na^+) currents flowing through calcium channels.

II. DIVALENT PERMEANT AND BLOCKING IONS

Ionic selectivity is, by definition, derived from the zero current or reversal potential that is measured when two permeant ion species, at specified concentration ratios, are separated by the membrane. Even with only millimolar Ca^{2+} concentration on the outside and much larger ($>0.1\ M$) monovalent cation concentration on the inside of the membrane, the currents reverse at positive membrane potentials (Reuter and Scholz, 1977; Fenwick et al., 1982; Lee and Tsien, 1982; Fukushima and Hagiwara, 1985). The permeability of Ca^{2+} ions must therefore be dominant, since a positive membrane potential that generates an increased driving force for the internal monovalent cations is necessary to balance the inward calcium current by an outward flow of the monovalent ions. These experiments confirm that calcium channels greatly prefer Ca^{2+}, Ba^{2+}, and Sr^{2+} over all alkaline ions for which a permeant ratio as low as 0.0003 (for Cs^+) and 0.001 (Na^+) are usually calculated.

The interpretation of results with reversal potentials can be ambiguous, however. This theory is based on the principle of independent ion movement. This implies that a more mobile ion of the same charge that carries the greater current should predominate in determining the zero current potential. This expectation is not met if inward currents carried by Ca^{2+}, Ba^{2+}, Sr^{2+}, and Mn^{2+} ions are compared, by means of their reversal potentials, in their efficacy to balance outward sodium currents (Fukushima and Hagiwara, 1985). Ba^{2+} ions, the most effective current carriers, were found to produce a zero current potential quite similar to that of the rather poor current carrying Mn^{2+} ions, see Figure 1. Ca^{2+} ions, when compared with other divalent ions, should have been the most permeant by this measure, but their currents about equaled that of Ba^{2+} ions. In the case of the calcium channels, the permeability as determined by zero current potential measurements, poorly addresses ionic conduction, it reflects instead the affinity of permeant and even blocking ions for binding sites of the channel.

Calcium channels of different types (see also Chapter 4, Section I, Chapter 7, Section II and Bean, 1989) share comparable permeabilities for alkaline earth ions with the exception that Mg^{2+} ion is nearly impermeant. The low-voltage-activated (LVA or T) channel is equally permeable to Ca^{2+} and Ba^{2+} (Bean, 1985; Fedulova et al., 1985; Fox et al., 1987; Nilius et al., 1985; Hagiwara et al., 1988), shows some small preference for Sr^{2+} (Carbone

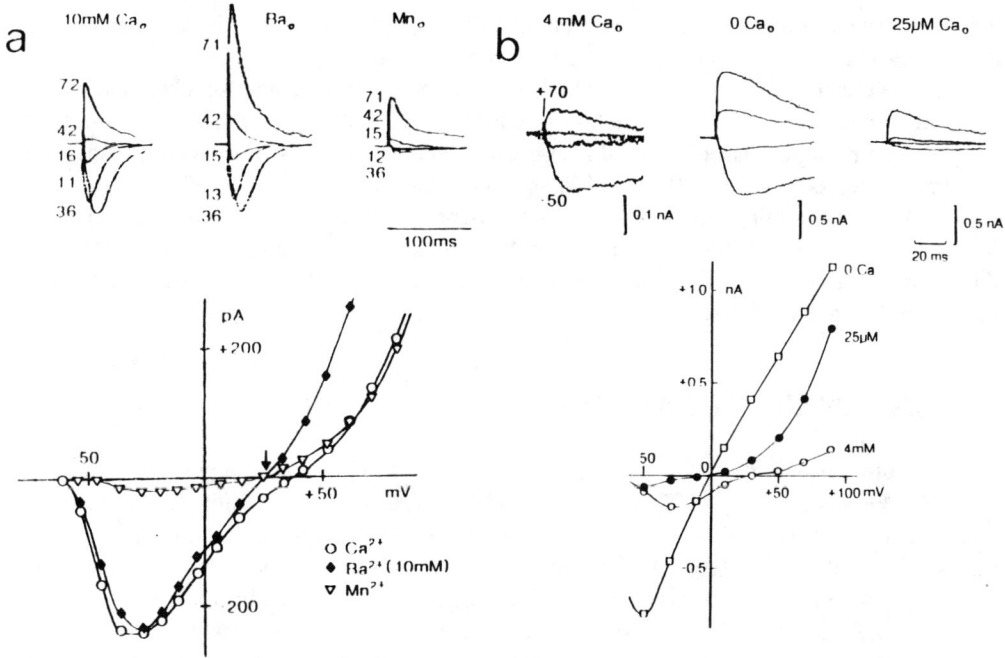

Figure 1. Effects of different permeable cation species on the membrane current. The cells were perfused by Na^+-containing solutions; 155 mM in (a) and 120 mM in (b). The external solutions contained Na^+ ions at similar concentrations. (a) Upper panel: superimposed samples of whole-cell currents for Ca^{2+}, Ba^{2+}, and Mn^{2+} ions, recorded at indicated membrane potentials. Lower panel: current-voltage relation at current peaks. Arrow indicates the current reversal potential of Ba^{2+}- and Mn^{2+}-inward currents flowing against the Na^+-outward currents. Modified from results of Fukushima and Hagiwara, 1985, on the calcium channel in mouse neoplastic B lymphocytes. (b) Upper panel: superimposed samples of currents through LVA calcium channels in dorsal root ganglionic cells. Left: currents with 4 mM Ca^{2+} in the external solution recorded at -50, -10, $+30$, and $+70$ mV membrane potential, holding potential -110 mV. The zero current potential is near $+30$ mV. Middle: sodium currents of the cell in the presence of 10 mM EGTA with no divalent cations, current reversal near zero mV. Right: the external solution contained 25 μM free Ca^{2+}. Note the pronounced block of Na^+ inward currents. Lower panel: current-voltage relations at current peaks for the different Ca^{2+} concentrations.

and Lux, 1987a), is especially sensitive to block by Ni^{2+} ions, and is somewhat resistant to block by Cd^{2+} ions. In contrast, the high voltage-activated (HVA N,L) types of calcium channels carry Ba^{2+} nearly twice as effectively as they do Ca^{2+} (Bean, 1985).

The relative potency of calcium channel block by divalent metal ions was well demonstrated in a study on mouse neuroblastoma cells (Narahashi et al., 1987, Chapter 12, Section 3). The blocking efficacy of the LVA- or T-type calcium channel was shown to follow the order of dissociation constants: Ni^{2+}, 47 μM > Cd^{2+}, 160 μM = Co^{2+}, 160 μM. The series for the high-voltage-activated calcium channel was: Cd^{2+}, 7 μM ≫ Ni^{2+}, 280 μM > Co^{2+}, 560 μM. The blocking actions were compared at -20 mV and $+10$ mV membrane potentials, respectively. These series indicate both a rule with exceptions, and some quantitative differences between different preparations (Bean, 1989; Pelzer, 1990). Moreover, the intensity of the blockade may depend on the concentration of permeant ions, on the voltage, or on the presence of Mg^{2+}. Whether or not block is a function of voltage, is not precisely known. Such a possibility is suggested by the findings of Lansman et al. (1987) who demonstrated that membrane hyperpolarization results in an increase in the unblocking rate (i.e., relief from block) of Cd^{2+}-induced blockade of unitary barium currents in heart myocytes. The block of barium currents by Cd^{2+} was previously found to be less severe at negative membrane potentials in snail neurons (Byerly et al., 1985). It would appear from these data that the blocking Cd^{2+} ions can be ejected from a channel into the cytoplasm by a sufficiently strong electric field. While it is likely that blocking ions are expelled at very positive membrane potentials, this is difficult to investigate because of the decreasing amplitudes of the current signals (see below for Mg^{2+}).

It is interesting that Mn^{2+} ions, which are potent calcium channel blockers (see Figure 1), comparable to Co^{2+} in frog skeletal muscle fibers, (Palade and Almers, 1985) were also shown to carry significant currents through Ca^{2+} channels in muscular tissue (Ochi, 1970; see for review Pelzer et al., 1990) and in snail neurons (Akaike et al., 1983). This behavior is explained by assuming a higher affinity of a channel site for Mn^{2+} than for Ca^{2+} or Ba^{2+}. Its occupation by Mn^{2+} thus interferes with the permeation of alkaline earth ions. However, the apparent affinity for Mn^{2+} is still not high enough to prevent Mn^{2+} fluxes at appropriately large Mn^{2+} concentrations.

III. AN APPARENT PARADOX

Calcium currents that flow during channel openings (Lux and Nagy, 1981; Reuter et al. 1982; Fenwick et al., 1982; Hagiwara and Ohmori, 1983; see Tsien et al., 1987), as observed in isolated membrane patches, reveal fluxes of several thousand Ca^{2+} ions per millisecond and calcium current amplitudes appear to be invariant (when measured in a solution containing more than a minimum Ca^{2+} concentration) even in the face of changing amounts of external alkaline cations (Isenberg, 1982; Matsuda and Noma, 1984). Thus calcium channels are both very selective and highly permeable. The assumption that calcium channels are selective simply because they repel other ions, such as Na^+, is inappropriate, however, since large currents of Na^+ ions (see Figure 1b) or of other alkaline ions are observed to flow through calcium channels in different preparations in the absence of Ca^{2+} ions (Rougier et al., 1969; Kostyuk and Krishtal, 1977; Prosser et al., 1977; Yamamoto and Washio, 1979; see Tsien et al., 1987, for review). These currents are known to disappear in the presence of micromolar amounts of Ca^{2+} ions (see Figure 1b and Figure 3) with an apparent K_d of the block of 0.7 to 2 μM in various neuronal and muscle preparations (Kostyuk et al., 1983; Hess and Tsien, 1984; Almers and McCleskey, 1984; Fukushima and Hagiwara, 1985; Carbone and Lux, 1987b; see also Tsien et al., 1987).

A quantitative evaluation of these findings illustrates the problem that is encountered in explaining the selective permeability of the calcium channel. In a simple blocking-unblocking scheme, the rate by which the blocking molecule exists from its binding site is given by the product of the dissociation constant, the K_d of the block ($\sim 10^{-6} M$), and the rate with which the molecule enters the block. To obtain a high estimate, the entry or blocking rate may be set equal to the diffusion-limited rate ($\sim 10^5 - 10^6 M^{-1}ms^{-1}$) by which a Ca^{2+} ion hits a target that geometrically could represent the mouth of the Ca^{2+} channel (Almers and McCleskey, 1984; Almers et al., 1985). This rate complies well with actual transfer rates of Ca^{2+} ions as observed with single channel Ca^{2+} fluxes at millimolar Ca^{2+} concentrations. An entry rate of the blocking molecule on the order of $10^5 M^{-1}ms^{-1}$ is also substantiated by the more direct experimental evidence that is shown later. Thus the exit of a blocking Ca^{2+} ion could hardly occur more frequently than once per millisecond. The large discrepancy between this number and measured rates of Ca^{2+} flux seems incompatible with the otherwise reasonable supposition that in each case Ca^{2+} ions bind to, and then leave, the same Ca^{2+}-specific sites of the channel. The selective permeability of calcium channels may thus demand rather specific affinity properties for the binding sites of the channel. Assuming separate calcium-specific and nonspecific sites does not remove the problem of how channel block and channel passage can be reconciled in a single file process.

IV. BLOCK BY Mg^{2+}

The impermeant Mg^{2+} ion is a constituent of physiological media and is known for its interference with ion transport through calcium channels. Even if Mg^{2+} ions hardly pass, it is still possible that they enter the channel to exert their block. The question of whether it must enter the channel or whether it blocks by binding to an external site arises because of the large hydration shell of this relatively small alkaline earth ion and, especially, because of the low rates of water substitution at its inner hydration shell. This occurs three to four orders of magnitude more slowly than for the other alkaline earth ions (Diebler et al., 1969) as shown for various Mg^{2+} water complexes. Both Ni^{2+} and Co^{2+} ions are somewhat similar in these aspects.

An inhibitory action of Mg^{2+} on inward calcium currents was uncovered early on by demonstrating $[Mg^{2+}]_o$-induced changes of the maximum rise of the calcium action potential in barnacle muscle fibers (Hagiwara and Takahashi, 1967). In the presence of 100 mM Mg^{2+}, the apparent K_d of this Ca^{2+}-channel site complex was found to be between 25 and 40 mM $[Ca^{2+}]_o$, but was calculated to be below 15 mM Ca^{2+} in the absence of $[Mg^{2+}]_o$. The strength of the block was 1:46 of that of Co^{2+} ions, which are potent calcium current blockers at and below millimolar concentrations. These results pointed to a weak Mg^{2+}-induced block of calcium currents.

Inward calcium currents in snail neurons, measured at +25 mV membrane potential were found to be of the same magnitude both in the absence and in the presence of 15 mM $[Mg^{2+}]_o$ (Akaike et al., 1978). The relatively strong depolarization used in these experiments could have offset the action of Mg^{2+} ions, if the block were voltage dependent in the manner suggested by the observation made on the LVA calcium channel (see below). A considerable block of barium currents by Mg^{2+} was indeed observed at low levels of depolarization (Wilson et al., 1983) but was attributed to the binding of Mg^{2+} to a specific site at the channel entrance.

The rate of actual entry of Mg^{2+} ions into cardiac calcium channels has recently been derived from the analysis of single channel activity (Lansman et al., 1986). The terms entry and exit imply here, and in the forthcoming, the intermediate binding of a blocking ion to a site of the channel. To optimally visualize the single channel events, relatively high Ba^{2+} concentrations (20 to 110 mM) were used by these authors. In addition, mean open times of the channel were increased by introducing an activator dihydropyridine (BAY K 8644). These steps facilitated the analysis of the Mg^{2+}-induced effects. The blockage by Mg^{2+} was manifested by channel closures that reduced the mean open times. The rate at which these closures occurred was measured as the reciprocal of the mean open time and was evaluated as a function of $[Mg^{2+}]_o$. This relationship showed a slope of $1.9 \times 10^5 M^{-1}s^{-1}$. This value most likely represents a lower limit since the Ba^{2+} present in the external medium would compete with Mg^{2+} for the channel binding site, assuming that a single channel binding site can interact with either one of these divalent ions. Extrapolating to low Ba^{2+} concentrations probably would increase this estimate to $1 \times 10^6 M^{-1}s^{-1}$. The Mg^{2+} exit rates, which were taken as the inverse of the mean closed times (times to unblock of the channel), ranged between 1 and $3.5 \times 10^3 s^{-1}$. The ratio of the exit and entry values corresponds to the K_d of the Mg^{2+}-induced block in the millimolar range. The patch-to-patch variation in the Mg^{2+}-induced block probably prohibited the resolution of the effects of $[Mg^{2+}]_o$ lower than 1 mM. It should be noted that the values were determined at a fixed membrane potential (0 mV), and the block was not investigated for a possible voltage dependence.

V. Mg^{2+}-BLOCKAGE OF CURRENTS CARRIED BY MONOVALENT IONS THROUGH CALCIUM CHANNELS

In the absence of divalent cations, calcium channels are known to carry alkaline cations with high efficacy (Kostyuk and Krishtal, 1977; Tsien et al., 1987) but low selectivity for the different monovalent ions. Unitary lithium currents through cardiac calcium channels were found to be blocked by $[Mg^{2+}]_o$ (Lansman et al., 1986), but to produce significant effects, a minimum of 1 mM Mg^{2+} had to be applied. This complies with results of Almers et al. (1984) on similar nonspecific currents carried by Na^+ or other monovalent ions through calcium channels in skeletal muscle fibers. A stronger block by Mg^{2+} of sodium currents through cardiac calcium channels was observed by Matsuda (1986). With a K_d of 60 μM at -33 mV, the block by Mg^{2+} was still only 2% of that by Ca^{2+}. In the LVA calcium channel of chick sensory neurons (Lux et al., 1990) as well as that of neoplastic lymphocytes (Fukushima and Hagiwara, 1985), the K_d of the Mg^{2+}-induced block appeared even smaller than that found in heart muscle (see below).

Similar to observations in heart cells (Hess et al., 1986), the LVA calcium channel activity was considerably altered in the presence of 120 mM Na^+ or Li^+ when the external free Ca^{2+} concentration was reduced to submicromolar values. The free divalent ion (Mg^{2+} and Ca^{2+}) concentrations were adjusted by using appropriate divalent ion chelators (5 or 10 mM EGTA, HEDTA, or NTA) for the desired range of divalent ion concentration. Under these conditions the single channel and whole cell currents were about five times larger, but their activation-inactivation time courses and their sensitivity to holding potentials were similar to LVA calcium currents (Carbone and Lux, 1987b). The currents were effectively blocked by the addition of metal ions (Ni^{2+} or Cd^{2+}) to the bath, and were not observed in the absence of alkali ions. All this indicated that these large-amplitude currents were carried by Na^+ or Li^+ passing through LVA channels. The holding potential was appropriately increased in order to compensate for the observed shift of about -30 to -35 mV of the I-V relationship following reduction of Ca^{2+} concentration from 10 mM to submicromolar values (Carbone and Lux, 1987b).

At Ca^{2+} concentrations below micromolar values, the single channel sodium currents of LVA channels lasted an average 0.26 ± 0.06 ms. The open times are thus considerably shorter than those of unitary calcium currents, which are distributed around millisecond values (Carbone and Lux, 1984b; 1987b). Openings occur in bursts, with intermediate closures of similar mean duration. In this respect, the unitary sodium currents through the LVA calcium channel resemble calcium or barium currents. The average sodium current amplitude in single channels during depolarizations to -40 mV was 1.85 ± 0.35 pA, which was about five times larger than that recorded with 20 mM Ca^{2+} concentration at corresponding potentials. Activation and inactivation properties do not appear to depend on the current carrying ion if the shift in activation voltage is accounted for.

At membrane potentials around -40 mV, the activity of unitary sodium currents through the LVA calcium channel in isolated membrane patches (Figure 2) was found to be reduced by Mg^{2+} concentrations as low as 25 μM, while 1 mM Mg^{2+} suppressed the activity to a low percentage of the control. This effect is manifested by channel closures that are prolonged when compared with spontaneous closures of controls in the absence of divalent ions. The single channel conductance appears unaltered at low Mg^{2+} concentrations. The rare openings observed at millimolar $[Mg^{2+}]_o$ are shortened on the average and the apparent reduction in unitary current amplitudes is attributable to the reduced resolution of the short-lasting events. Thus a primary effect of Mg^{2+} on the conductivity of the permeation pathway seems less likely. Instead, it appears that Mg^{2+} ions intermittently occupy the channel for some period interdicting the passage of the permeant ions. The features are similar to those observed with Ca^{2+} ions as blockers (see below). The latter, however, show a K_d of the block near 2 μM as compared to the K_d for Mg^{2+} of about 40 μM at this potential.

2. Calcium Channel Permeability

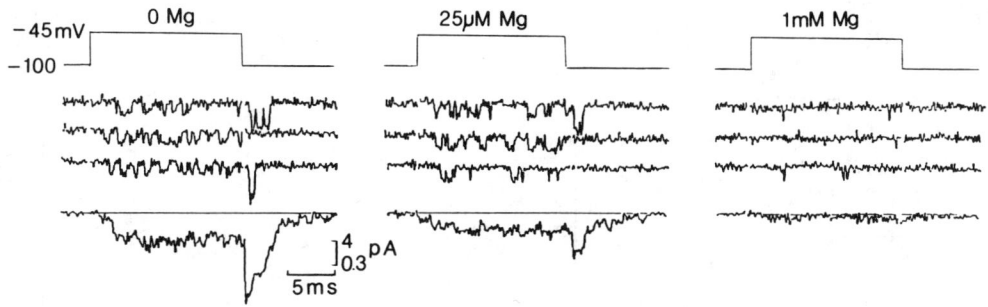

Figure 2. Mg^{2+} blockage of unitary sodium currents through a LVA calcium channel. Single channel sodium currents in an outside-out patch from a chick dorsal root ganglionic cell were recorded during successive depolarizations to -45 mV in (left) bivalent ion free solution (10 mM EGTA) with 25 μM free Mg^{2+} (middle) and with 1 mM free Mg^{2+} (right). Bottom traces (with zero current line) are sample averages taken from more than 20 trials at each concentration. External and internal solutions contained 120 mM NaCl, holding potential -110 mV, cut-off frequency 3.6 kHz.

The prolongation of mean closed times (Figure 2) can be attributed to times during which the channel is blocked after the entry of Mg^{2+} ions. The difference of several ms in mean closed times from those of the control can be used for an estimate of the exit rate of the blocking ion. Together with actual Mg^{2+} concentration values (near the apparent K_d), this results in entry rates on the order of $10^6 - 10^7 \, M^{-1} s^{-1}$. The lowest estimate is obtained by assuming that the channel is also available for Mg^{2+} entry and Mg^{2+} block during the period when the channel is closed. Repeated block during channel closures is most consistent with an increasing contribution of Mg^{2+} to long-lasting closed times as the concentration of Mg^{2+} is increased.

The difference in entry rates of channel blockers in neuronal LVA and cardiac calcium channels may in part arise from the different conditions used in the studies performed. With increasing concentration of blocker, the block could approach saturation without being complete. There is necessary competition between blocking ions and the probe and the transported ion species for occupation of channel sites. An underestimation of the blocking or entry rates could result if an empty binding site, a precondition for the block, was not always available. However, a comparison of the data where it is possible (≈ 1 mM Mg^{2+}) suggests that the block is somewhat stronger in the neuronal LVA channel than in cardiac channels. This conclusion pertains even if a potential-dependent K_d is assumed to hold for the data on the cardiac calcium channel as is suggested by the observations on the LVA calcium channel (Figure 4).

VI. Ca^{2+}-BLOCK OF SODIUM CURRENTS THROUGH LVA CALCIUM CHANNELS

The amplitude of whole cell sodium currents through LVA calcium channels and of sample averages of single channel recordings becomes strongly reduced (see Figures 1b, 3, and 4) when Ca^{2+} concentration is increased from submicromolar to 10 μM values (Carbone and Lux, 1987b). However, similar to the data with Mg^{2+} concentrations, the amplitude of single channel sodium currents remains unchanged (Figure 3). The main effect is an increase of closed times, thus the number of openings per activating pulse decreases in proportion to the reduction of averaged currents (Carbone and Lux, 1987c). To significantly decrease the mean open times, larger Ca^{2+} concentrations are necessary. The channel life times are halved at a Ca^{2+} concentration of 50 μM, that is about 20 times larger than the concentration

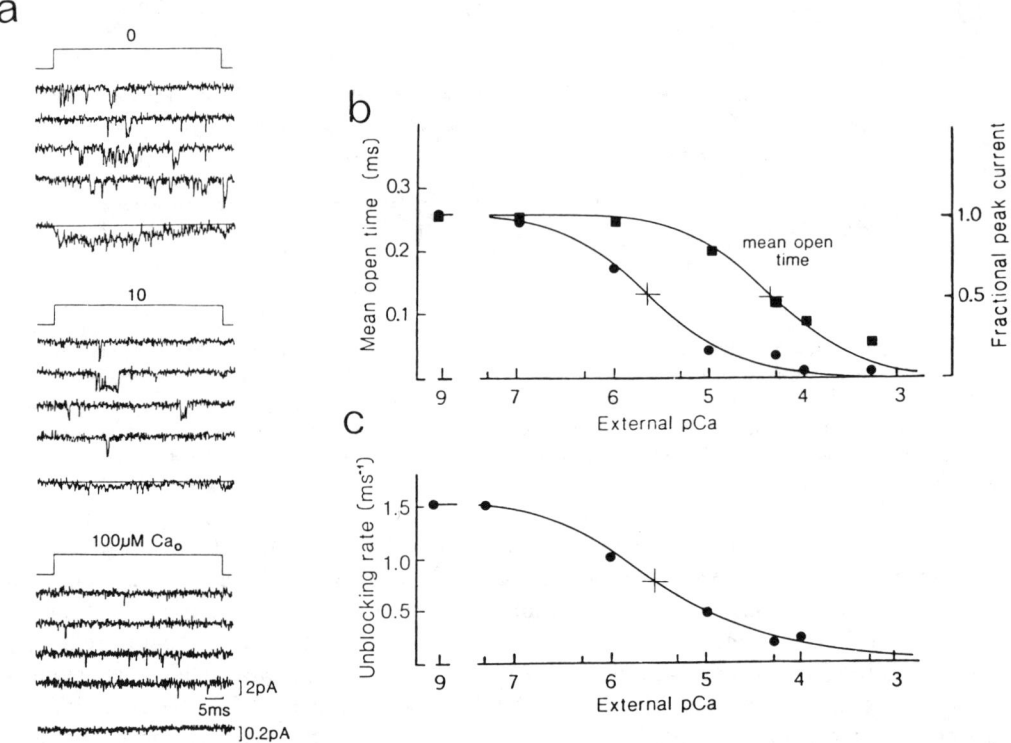

Figure 3. Ca^{2+} blockade of unitary sodium currents through a LVA calcium channel. (a) Examples of single channel sodium currents recorded during successive depolarizations to -40 mV in the presence of nominally zero (upper panel), 1×10^{-5} (middle), and 1×10^{-4} M $[Ca^{2+}]_o$ (lower panel), holding potential -110 mV. Bottom traces (with zero current lines) are sample averages of the currents recorded during 25 (0 $[Ca^{2+}]_o$) to 73 (100 μM $[Ca^{2+}]_o$) trials. Note the large increase in silent periods already at 10 μM $[Ca^{2+}]_o$ with little effect on open durations. Openings are rarely observed with 100 μM $[Ca^{2+}]_o$, and traces with activity were selected. Other conditions as in Figure 2. (b) Mean values (4 to 7 experiments each) of averaged currents normalized at zero Ca^{2+} (●) and open times (■) as a function of $-\log [Ca^{2+}]_o$. Half value of block of averaged currents at 2 μM and half value of mean open times near 50 μM $[Ca^{2+}]_o$ are indicated by crosses. (c) Mean values of (fast) unblocking rates calculated from measured closed times of experiments similar to those in (a).

required (2.3 μM) to produce a half-block of the averaged single channel current and of whole cell currents at membrane potentials around -40 mV. It appears reasonable to assume that the half reduction of the average open time at about 50 μM Ca^{2+} represents the balance between spontaneous channel closures and block of the open channel after entry of a Ca^{2+} ion. The actual value of the average open time of 130 μs at this Ca^{2+} concentration then provides an estimate of the blocking or apparent entry rate of Ca^{2+} ions of $(50 \times 10^{-6} M \times 0.13 \times 10^{-3} s)^{-1} = 1.5 \times 10^8 M^{-1} s^{-1}$. This value compares well with earlier estimates on Ca^{2+} entry rates, that apply to Ca^{2+} flux, based on probable geometric target sizes of the channel entry and diffusional characteristics of the Ca^{2+} ion (Almers and McCleskey, 1984). This estimate would suggest that access to the channel for Ca^{2+} ions is not altered when Na^+ ions are the current carriers. Thus the blocking rate for the open channel agrees with supposed Ca^{2+} entry rates, but the half-blocking Ca^{2+} concentration in these single channel records gives a different value than the apparent K_d of the Ca^{2+}-induced block of whole cell or averaged currents. This result also speaks against open channel block as a main factor. With Ca^{2+} concentrations of 10^{-4} M, closures frequently outlast the pulse duration, and "blanks" (records with no channel openings) are common. Thus reduction of averaged single channel currents as well as of whole-cell currents with increased Ca^{2+}

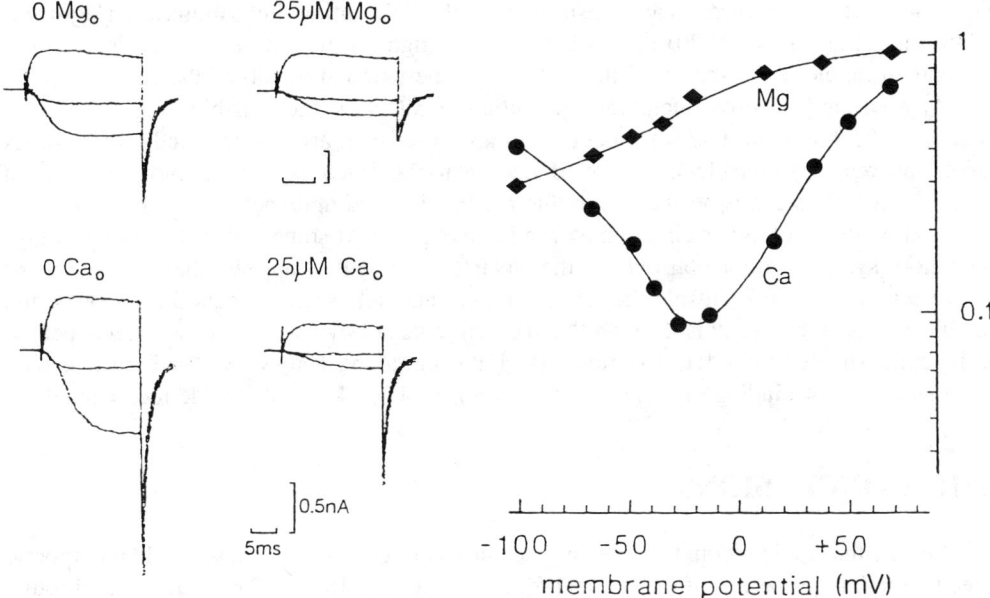

Figure 4. Voltage dependencies of the suppression of sodium currents through LVA calcium channels by external Ca^{2+} and Mg^{2+} ions. Left: examples of sodium inward and outward currents in nominally zero concentrations of divalent cations and with 25 μM Ca^{2+} or Mg^{2+}. Current recording (Mg^{2+}) at -50, -20, and $+50$ mV and, in a different cell (Ca^{2+}), at -50, -10, and $+50$ mV membrane potential. Maximum current suppression in the case of Mg^{2+} is seen on the deactivation (tail current) on return to the holding potential. With Ca^{2+}, the suppression is most pronounced in the recording at -10 mV. Right: ratios of peak current amplitudes at 25 μM and zero divalent concentrations are plotted against the membrane potential on semilogarithmic scale. Note the relief of the sodium current block by Ca^{2+} at positive and negative potentials with Mg^{2+} attaining the blocking efficacy of Ca^{2+} at negative membrane potential.

concentration can be attributed largely to the increase of the periods during which channels are unavailable for sodium permeation. It is obvious that a reduction of single channel currents will be recorded, if the entry of the blocking ion occurs at an average frequency that is above experimental resolution. This is probably the case when investigating the blocking effect of Ca^{2+} above concentrations of 0.1 mM.

Ba^{2+} ions were found to act similar to Ca^{2+} ions at similar concentrations. However, this comparison is limited to the 100 μM concentration range because of the difficulty in buffering residual Ca^{2+} impurities in the applied salts against Ba^{2+} ions. Outward sodium currents of a similar average time course could be recorded in inside-out patches (Hamill et al., 1981). When exposing the former cytoplastic site of the membrane to Na^+ solutions containing micromolar free Ca^{2+}, the Ca^{2+} ions were found to be much less effective in blocking the sodium currents than was the external application of Ca^{2+} (Carbone and Lux, 1987b). As in external Ca^{2+} application, however, the blocking action was manifested primarily by prolonged closed times.

VII. VOLTAGE-DEPENDENT BLOCK

To either reach or leave its blocking site within the channel, the ion experiences the electric field applied across the pore. Because a permeant ion can exit the pore in either direction, the block exerted by Ca^{2+} on Na^+ passage should be relieved by either hyperpolarization or depolarization of the membrane. For an impermanent ion such as Mg^{2+}, the block will be relieved if the inside of the cell is made more positive. The results shown in

Figure 4 on whole cell currents are consistent with the assumption (Fukushima and Hagiwara, 1985; Lux et al., 1989, 1990) that in both cases a high affinity site is located deep within the pore at an electric distance of about 0.6 from the external mouth of the channel.

At a strongly positive membrane potential ($+80$ mV), the half-blocking [Ca^{2+}] for sodium and lithium currents was about 20 times larger than at -20 mV and a three times larger apparent K_d of the block was found to apply to the deactivation (tail) currents recorded at -100 mV. There was no indication that the block could approach a constant level with increased voltages of either direction as would be expected if some part of the high affinity, regulatory system were acting outside the electric field. It thus appears that the process of ion selection takes place inside the LVA calcium channel. At fixed membrane potentials, the block of sodium currents through this calcium channel by Ca^{2+} concentration appeared to be approximated by a 1:1 stoichiometry (Fukushima and Hagiwara, 1985) in line with the assumption of binding of a single divalent ion to a blocking site inside the channel.

VIII. CONCLUSIONS

Models that could account for the ionic selectivity of calcium channels should incorporate specific binding properties of the channel for the ion to be selected. Ca^{2+} selectivity through ion sieving by the pore structure involving rigid energy barriers appears unlikely for reasons stemming from the physical properties of alkaline earth ions as compared with those of sodium ion and other alkaline ions. It is, in particular, inconsistent with the significant passage of monovalent ions in the absence of the preferred divalent ions.

Selection by affinity can only account for the large preference of the calcium channel for divalent ions as manifested by the block of alkaline ion passage. The channel conductivity for divalent ions as observed in the high (millimolar) concentration range remains unexplained and seems difficult to reconcile with high affinity binding of these ions.

The very different half saturating Ca^{2+} concentrations (K_ds) that apply to calcium flux and to block by Ca^{2+} ions of nonspecific (Na^+) cation flux through calcium channels suggest that different strengths of Ca^{2+}-binding sites prevail in the two situations. Removal of the block and, simultaneously, effective Ca^{2+} ion flow at higher [Ca^{2+}], could be expected if an increased probability of simultaneous Ca^{2+}-occupation of multiple binding sites was assumed (Almers and McCleskey, 1984; Hess and Tsien, 1984) in a rational extension of the blocking hypothesis. The occupation of double high affinity sites could enact repulsive forces between the ions that would offset strong individual ion binding. Thus mobilization of the Ca^{2+} ion is equivalent to a decreased affinity, and an increased Ca^{2+} ion flux would result. The model can well account for both, channel block by the Ca^{2+} ion and Ca^{2+} flux, but is in its present form not very applicable to the results on sodium current block in the LVA calcium channel. With two fixed binding sites in place of a central one, the electric distances between the two energy wells and the barriers decrease resulting in a weaker voltage dependence of the block (Almers and McCleskey, 1984; Lux et al., 1989), as actually observed. In multiple binding site models, the transitions from the Ca^{2+} blocked forms of the channel toward Na^+ conductive forms arise from mixed occupations of the sites and subsequent Ca^{2+}-Na^+ repulsion. Thus, the unblocking time courses should depend on [Ca^{2+}] while the unblocking rate, as well as the blocking rate, should increase with increased [Ca^{2+}]. The latter effect is not displayed by the LVA channel and a modification of the model would be necessary to account for the time-dependent features of the block of sodium current by Ca^{2+}.

Alternatively, ion-channel interactions could be envisaged. Ca^{2+} ions and other divalent ions with high affinity could be strongly bound initially but force a ligand in the channel to attain a conformation with weaker electrostatic attraction allowing for Ca^{2+} ion passage.

2. Calcium Channel Permeability

The transformation from strongly to weakly binding states (Lux et al., 1990) is time dependent in the manner encountered by the block of unitary nonspecific (Na^+) currents. The selection of Ca^{2+} ions over Na^+ ions could as well utilize ligand-field or ligand-polarization effects for the differently charged ions and their very large differences in free energies of solvation (hydratization). A barrier that screens against monovalent ions should be feasible on this basis. The supposed ion-channel interactions could be formally described by a "fluctuating barrier" model (Läuger, 1983). Such models are, in principle, able to account for all known relevant permeability features of the calcium channel, but a physical picture of the reactions underlying affinity changes is still awaited.

Chapter 3

Activation and Deactivation of Neuronal High Voltage-Activated, Fast Deactivating Calcium Channels

D. Swandulla

TABLE OF CONTENTS

I. Introduction 22

II. Activation Kinetics of HVA Calcium Channel Currents 22

III. Are Calcium Channels Occupied by Ions when Gated? 26

IV. Models for Activation 27
    A. Activation Kinetics 30
    B. Deactivation Kinetics 30

V. Modulation of Calcium Channel Activation 31

VI. Conclusions and Perspectives 31

I. INTRODUCTION

Calcium channels are known to exist in muscle, neuronal, and secretory cells. Voltage-sensitive calcium channels play important roles in excitation and signal transduction. They, like other ionic channels, are membrane-spanning proteins that consist of several transmembrane subunits. A principal subunit forms the ion-conducting pore and is expressed with a variable number of associated subunits in different cell types (Campbell et al., 1988; Catterall, 1988). Voltage-gated calcium channels, as other channels, must have charged structures or dipoles that sense the membrane electric field and reorient in response to changes in the field.

In biophysical studies the membrane of intact cells is voltage clamped and macroscopic currents, resulting from ion fluxes through a large number of channels in response to quick voltage changes, are measured. The patch clamp and planar bilayer techniques allow the study of currents through single channels. The analysis of both whole cell and single channel current kinetics have provided profound, although limited, insight into channel gating. Using this approach the voltage and time dependence of channel gating can be modeled as a series of transitions among conducting and nonconducting states of the channel controlled by charges that move in response to changes of the electric field in the membrane.

Recently multiple types of calcium channels have been found or suggested in a variety of excitable cells that differ in their kinetic behavior and pharmacological properties (see Chapter 4, Section I). In an attempt to describe principles of calcium channel activation by voltage, I will concentrate in this report on one type of calcium channel that is widely distributed in neuronal tissue and that has been extensively studied over the last years (for reviews see Tsien et al., 1988; Carbone and Swandulla, 1989; Bean, 1989a; Swandulla et al., 1991). These channels exhibit a high threshold of activation (> -50 mV) and are therefore termed high voltage-activated (HVA) calcium channels. They co-exist with other types of calcium channels in many neuronal membranes that require strong hyperpolarization—low voltage-activated (LVA) calcium channels—before they can be activated. For details of LVA calcium channel kinetics see Chapter 4, Section I.

II. ACTIVATION KINETICS OF HVA CALCIUM CHANNEL CURRENTS

Following the step-like onset of a depolarizing change in membrane potential applied to a voltage-clamped neuron, calcium channels open and the calcium current turns on or activates. In the course of an activating step, which lasts several tens of milliseconds, the current reaches a peak and then declines as channels inactivate, i.e., enter a nonconducting state in which they are not available for activation (see Chapter 4, Section I). On return to the initial membrane potential, open channels close and the calcium current turns off or deactivates. Figure 1 shows the voltage-dependent inward currents carried by Ca^{2+} and Ba^{2+} ions through HVA calcium channels in a chick dorsal root ganglion (DRG) neuron. During depolarizations from the holding potential of -80 mV at which all of the channels are available for activation, i.e., not in an inactivated state, whole cell calcium currents activate with a sigmoidal time course and reach their maximal amplitude within several milliseconds. The time to peak amplitude depends on the applied voltage and decreases with more positive test pulse potentials. The inward currents are preceded by a rather large initial transient of outward current that probably represents the calcium channel gating current (Kostyuk et al., 1977; Adams and Gage, 1977) that results from charge movements in the channel protein induced by the step change in membrane voltage. On repolarization, calcium channels deactivate, or close. When returning to a more negative potential following the activating

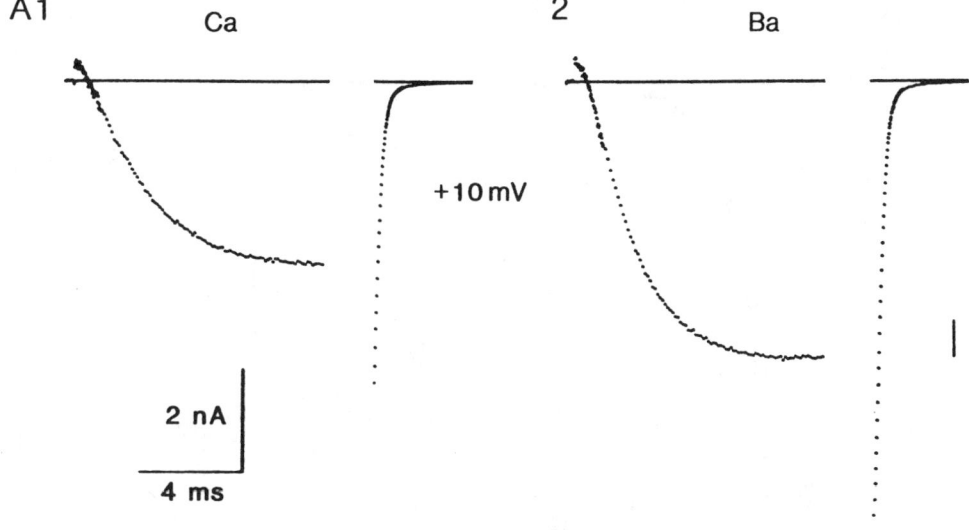

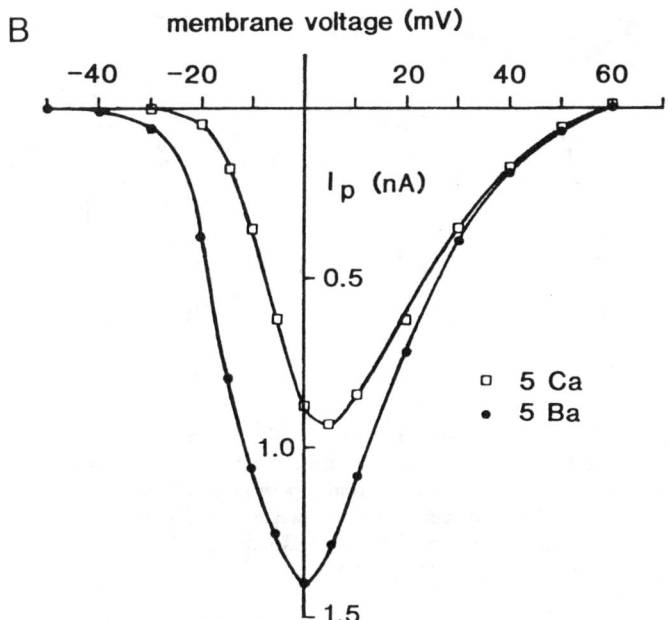

Figure 1. A: whole-cell calcium (1) and barium (2) currents recorded at +10 mV (single cell). Test pulses (10 ms) were from −80 mV. External solutions contained 10 mM Ca^{2+} and Ba^{2+}, respectively. Note different current scale for tail currents (indicated in A2) (20°C). B: current to voltage relationship for calcium (5 mM) and barium (5 mM) currents through calcium channels. Peak amplitudes were plotted against membrane potential (same cell). Test pulses (15 ms) were from −80 mV holding potential (15°C). (Modified from Swandulla and Armstrong, 1988.)

pulse, the current jumps in magnitude because of the increased driving force on Ca^{2+} and Ba^{2+} ions at these potentials. Subsequently, the channels deactivate, giving rise to a "tail" current that decays in magnitude as the channels close. This is illustrated in Figure 1A for a step from +10 to −80 mV with Ca^{2+} and Ba^{2+} as charge carriers. The current-to-voltage relationship for both, Ca^{2+} and Ba^{2+} ions, (Figure 1B), demonstrates that the peak current amplitude depends strongly on voltage. It rises steeply as more channels are activated with

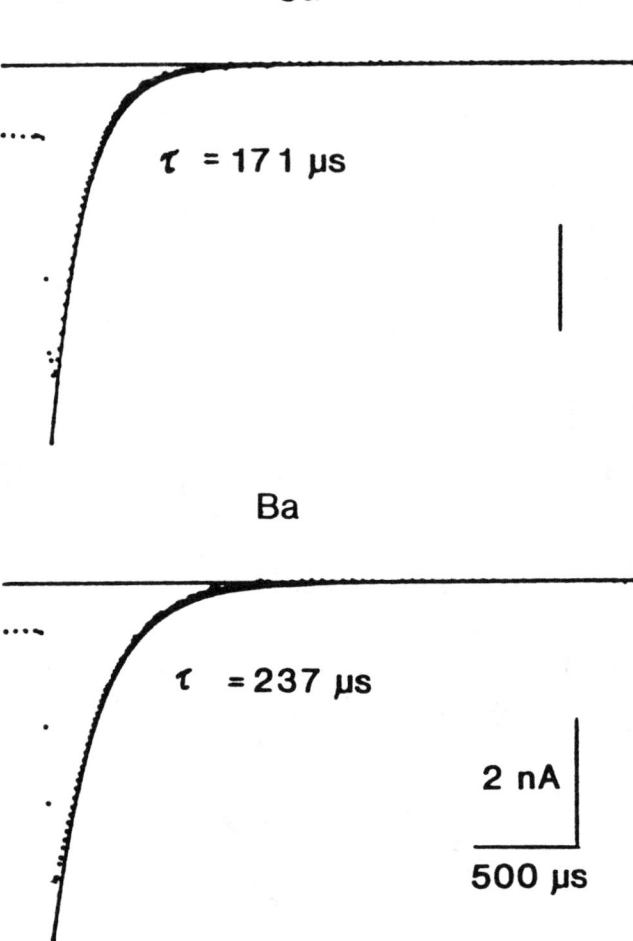

Figure 2. Calcium channel tail currents carried by Ca^{2+} (20 mM, upper trace) and Ba^{2+} ions (10 mM, lower trace). Tail currents were recorded on repolarization to -60 mV. Activating pulses (15 ms) were from -80 to $+20$ mV. The continuous curves are single exponentials fitted to the tail current decay. Time constants were as indicated (20°C). (Modified from Swandulla and Armstrong, 1988.)

increasing depolarization and reaches a maximum at 0 mV for Ba^{2+} and $+5$ mV for Ca^{2+}. Even though many channels are opened with further depolarization (see below), the calcium channel current becomes smaller again, since the driving force on Ca^{2+} and Ba^{2+} ions is less.

When Ba^{2+} is the charge carrier substituting for Ca^{2+}, the maximum amplitude of this current is larger by a factor of ~ 1.5, as is the amplitude of the tail current (Figure 2). Tail currents in Ca^{2+} as well as Ba^{2+} can be very well fitted by a single exponential expression (Swandulla and Armstrong, 1988). In Ca^{2+} solutions, the time constant of the current decay at -80 mV is near 170 μs at 20°C whereas in Ba^{2+} solutions it is slightly slower (Figure 2).

The tail current decay becomes progressively faster with more negative repolarization potentials, as illustrated in Figure 3, whereas variations of the magnitude of the activating pulse do not produce changes in the tail current decay. The amplitude of the tail currents

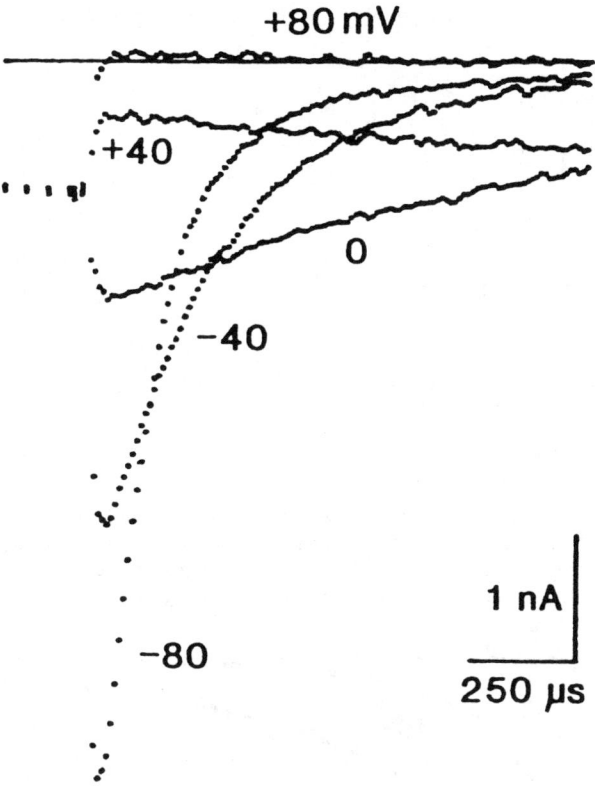

Figure 3. Tail currents at different membrane potentials. Tail currents were recorded at a variety of return potentials (as indicated) following activation of barium currents with 10 ms pulses from −80 to +20 mV. 5 mM Ba^{2+} (20°C). (Modified from Swandulla and Armstrong, 1988.)

at a given return potential should be proportional to the fraction of the channels, p_o (V), that are activated by a certain potential (V). Figure 4A shows a plot of tail amplitudes measured at −80 mV following activating pulses to various membrane voltages. This conductance-voltage relationship illustrates that p_o reaches a maximum at about +40 mV and is constant for more positive voltages, indicating saturation of the p_o (V) curve at high potentials (see also Llinas et al., 1981a; Fenwick et al., 1982).

The open channel or instantaneous I-V-relationship for calcium channels can be determined by varying the return potential that is reached after a test pulse. The tail current amplitude is measured as soon as possible (''instantaneously'') and is plotted as a function of membrane potential (Figure 4B). The resulting curve is nonlinear and its slope decreases progressively for voltages above −10 mV. This behavior is expected for the highly asymmetric distribution of permeant ions inside and outside the cell (for Ca^{2+}, the internal concentration is normally ≈10^6 times lower than the external concentration) and the curved open-channel I-V-relation can be well approximated using the Goldman (1943), Hodgkin and Katz (1949) equation (see below).

At lower temperatures, activation and deactivation kinetics are slowed down. Figure 5 illustrates this slowing of the tail current decay. The time constant of the current decline decreases by a factor of about 2.5 when the temperature is changed by 10°C. Similar observations are made for the time course of activation, which is slowed by a factor of about 3.5 for the same temperature range. Figure 6 illustrates the activation time course of barium currents measured at different membrane potentials at 10°C. These values differ significantly

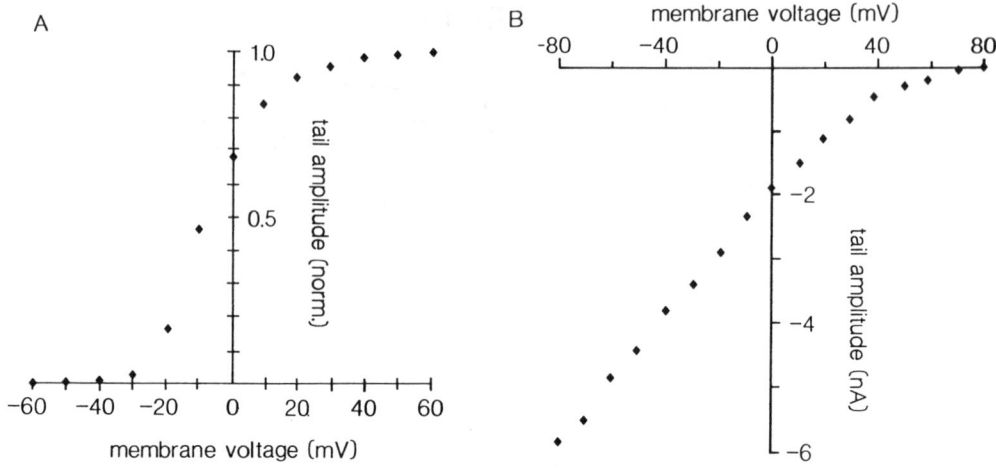

Figure 4. A: conductance to voltage relationship. Tail currents were recorded at −80 mV following activation of calcium channels by 10 ms pulses to a variety of potentials. Normalized tail current amplitudes were plotted as a function of the activating voltage. 2 mM Ba^{2+} (15°C). B: instantaneous current (open channel) to voltage relationship. Tail amplitudes from Figure 3 were plotted as a function of the return potential level.

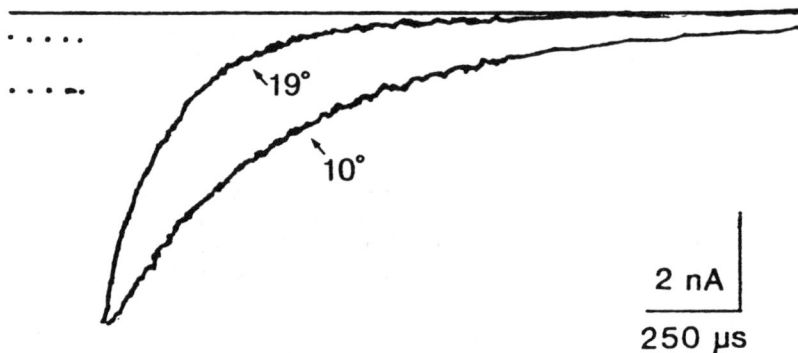

Figure 5. Temperature effect on tail currents. Tail currents were recorded at −60 mV following activating pulses from −80 to 0 (19°C) and +20 mV (10°C). At the temperatures indicated. 5 mM Ba^{2+}. (Modified from Swandulla and Armstrong, 1988.)

from those reported for low voltage-activated calcium channels found in neurons (Nobile et al., 1990) indicating different pathways of activation for these two channels (see also Carbone and Lux, 1987).

III. ARE CALCIUM CHANNELS OCCUPIED BY IONS WHEN GATED?

Current models of alkaline ion permeation through calcium channels assume sites located in the conducting pore to which the different ions bind with different strength in the course of their passage (see Chapter 2, Section I). Ions that bind particularly strongly such as Cd^{2+}, Co^{2+}, or Ni^{2+} are potent channel blockers. Divalent ions can reach these binding sites when calcium channels are in the closed state at rest (see for example, Swandulla and Armstrong, 1989). Thus, an interesting question is whether channel gating may be influenced by the occupying ion. To address this question, we have used Cd^{2+} ions as a probe for calcium channels and their gating apparatus (Swandulla and Armstrong, 1989).

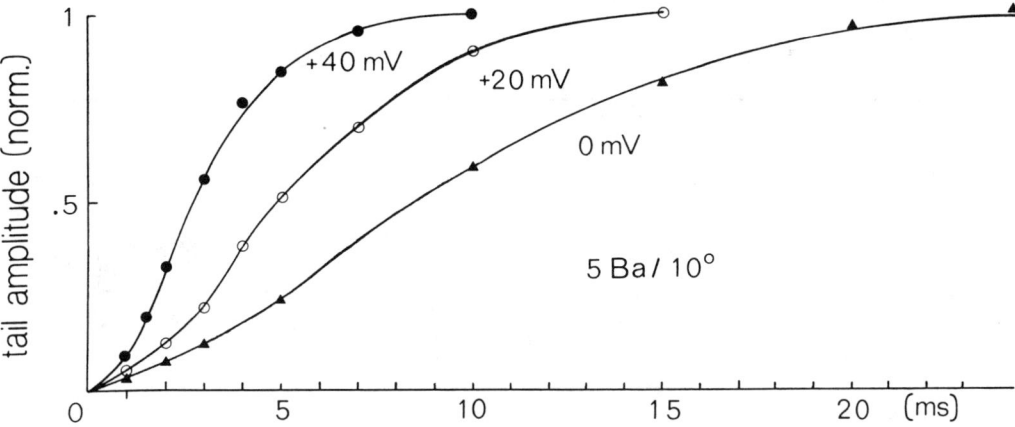

Figure 6. Time course of activation of barium currents through calcium channels determined by tail current measurements. Tail amplitudes at −60 mV were measured following activating pulses of varied duration from −80 mV to the potentials indicated. Tail current amplitudes, which are a measure of activation during the preceding test pulse, were plotted normalized as a function of pulse duration.

As shown in Figure 7, Cd^{2+} is an effective blocker of HVA calcium channels when applied in micromolar concentrations to the outside of the membrane. The control trace illustrates the sigmoidal activation time course of barium current through calcium channels, and the deactivation of the channels on repolarization to the holding potential. Addition of Cd^{2+} to the bathing solution (Figure 7B) blocks calcium current during the activating pulse almost completely, but there is a substantial tail current when V_m is returned to −80 mV. Closer analysis of tail current kinetics at different return potentials showed that tail current amplitudes in Cd^{2+} are not reached instantaneously as in control but gradually increase as Cd^{2+} comes out of the blocked channels, and then decrease as the channels deactivate (see also Taylor, 1988b). As illustrated in Figure 7C and 7D, the decline from the current peak had the same kinetics in control and in Cd^{2+}. This indicates that Cd^{2+} does not have to leave the channels before they close but that channels can close when Cd^{2+}-occupied. In case channels had to be cleared of Cd^{2+} before closing, one would have expected a measurable slowing of the tail current decay. Do these Cd^{2+}-occupied channels gate normally, i.e., do they activate at a normal rate, and with normal voltage-dependence? Since Cd^{2+}-occupied channels do not conduct during activating pulses, we used tail current amplitudes, recorded at the end of activating pulses, with increasing duration to study activation. Figure 8A shows a plot of maximum tail current amplitudes following activating pulses for the times indicated. It is obvious from the figure that the time course of activation in control and Cd^{2+} are the same. Similarly, tail current amplitudes were measured after activating pulses to a range of voltages to study the effect of Cd^{2+} on the voltage dependence of activation. As illustrated in Figure 8B, the presence of Cd^{2+} in the channel does not affect the channel's ability to open on depolarization. These findings lead to the conclusion that calcium channels can gate normally when occupied by a divalent cation. They strongly suggest that occupancy of the channel by a divalent cation, normally Ca, may be an essential part of the gating mechanism (see also Saimi and Kung, 1982), similar to what has been reported for potassium channels (Armstrong and Taylor, 1980; Armstrong et al., 1982; Armstrong and Matteson, 1986).

IV. MODELS FOR ACTIVATION

The kinetics of calcium channels have been explored in a variety of preparations. Kinetic

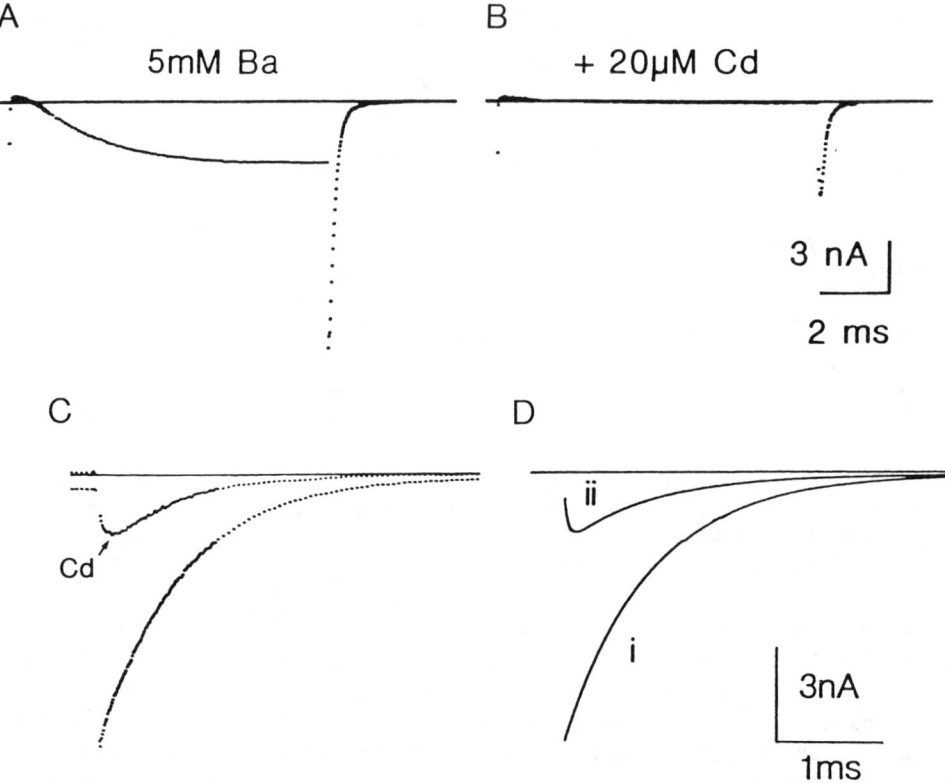

Figure 7. A, B: tail currents through calcium channels were not effectively blocked by Cd^{2+}. Barium currents were activated by pulses from -80 to $+20$ mV. While pulse currents were almost completely blocked by adding 20 μM Cd^{2+} to the external medium, substantial tail currents were recorded on repolarization to -80 mV. 20°C. Cd^{2+} block does not slow closing of calcium channels. C: tail currents with (upper trace) and without 20 μM Cd^{2+} (lower trace) were recorded on return to -60 mV following activating pulses (10 ms) to $+20$ mV. 5 mM Ba^{2+} (15°C). D: the time course of the tail currents was modeled for Cd^{2+} free (i), and Cd^{2+}-blocked (ii) channels, assuming that the channels are able to close with Cd^{2+} ion inside using the following reaction scheme:

$$\begin{array}{ccc}
\text{gate open} & b = 6 \text{ ms}^{-1} & \text{gate open} \\
Cd^{2+}\text{-blocked} & \rightleftharpoons & \text{conducting} \\
 & c = 17 \text{ ms}^{-1} & \\
\Updownarrow d = 1.24 \text{ ms}^{-1} & & \Updownarrow a = 1.24 \text{ ms}^{-1} \\
\text{gate closed} & \rightleftharpoons & \text{gate closed} \\
Cd^{2+}\text{-blocked} & &
\end{array}$$

(Modified from Swandulla and Armstrong, 1989.)

analysis of the activation process, as has been achieved for sodium channel activation, was hampered for a long time by the time resolution of the voltage-clamp systems available for measuring calcium channel currents. Hodgkin and Huxley (1952b) modeled sodium current activation using a variable, m, raised to the third power, which is determined by a first-order process.

Many investigators followed this approach by using the Hodgkin and Huxley formalism for calcium current (I_{Ca}) activation. The basic assumption is that the current can be described by:

$$I_{Ca} = g_{Ca}(V)m^q(V,t) \tag{1}$$

3. Activation and Deactivation

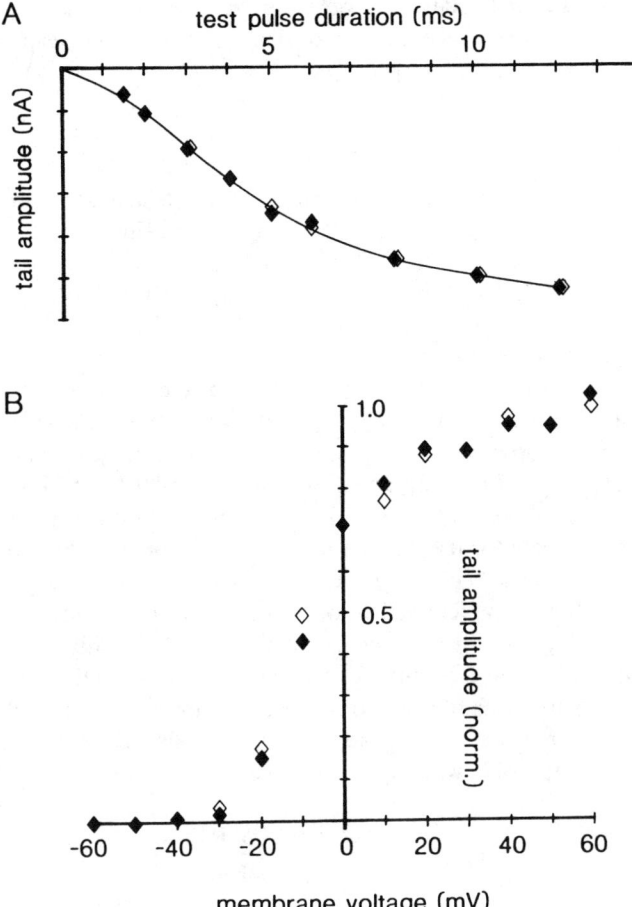

Figure 8. Cd^{2+} does not affect activation kinetics (A) or voltage dependence of activation (B) of calcium channels. A: tail currents with (♦) and without (◊) 20 μM Cd^{2+} were recorded at −80 mV following activating pulses to +20 mV of varying duration. Tail current amplitudes were plotted as a function of pulse duration. Amplitudes in Cd^{2+} have been scaled up by a factor of 6.12. B: tail currents with (♦) and without (◊) Cd^{2+} were recorded following activation of calcium channels by pulses (20 ms) to a variety of potentials. Normalized tail currents were plotted as a function of the activating pulse potential. 2 mM Ba^{2+} (15°C). (Modified from Swandulla and Armstrong, 1989.)

where $g_{Ca}(V)$ is the conductance at different membrane voltages, q a constant, and m(V,t) a continuous variable that has values from 0 to 1 indicating the probability that the channel gates are in a permissive state. The variable m is determined by the kinetic relationship:

$$1 - m \underset{\beta_m}{\overset{\alpha_m}{\rightleftharpoons}} m$$

that yields the equation:

$$dm(V,t)/dt = \alpha_m(1 - m) - \beta_m m = (m_\infty - m)/\tau_m \qquad (2)$$

where α_m and β_m are the first order rate constants that determine the opening and closing kinetics of the channel, respectively. m_∞ ($= \alpha_m/[\alpha_m + \beta_m]$) is the steady-state value of m and the time constant $\tau_m = 1/(\alpha_m + \beta_m)$.

A. Activation Kinetics

At the onset of a depolarizing pulse, the calcium currents activate with a sigmoidal time course that can be fitted to solutions of the step response of Equation 2.

$$m^q(V,t) = \{m_\infty - (m_\infty - m_o)\exp[-t/\tau_m(V)]\}^q \qquad (3)$$

where m_o is the value of m at time t = 0.

The values of q used to describe turn-on kinetics of the calcium current range from 1 to 5 (see for example, Akaike et al., 1978; Kostyuk et al., 1977, 1981; Llinas et al., 1981 [m = 5]; Byerly and Hagiwara, 1982; Hagiwara and Ohmori, 1982; Eckert and Ewald, 1983; Kay and Wong, 1987 [m = 2]; Jones and Marks, 1989 [m = 1]). This wide range may reflect space clamp problems in some of these studies. However, differences in activation kinetics of calcium channels for the various preparations cannot be excluded.

Steady state activation (m_∞^q) of the calcium current can be estimated from the amplitude of tail currents that follow complete activation of the calcium current.

From Equation 1 it follows that the open channel I-V-relationship, g(V), can be derived from the ratio of the measured I-V-curve to the steady state activation curve. As expected from the asymmetric distribution of Ca^{2+} ions, the curve approaches the voltage axis asymptotically (see Figure 4B), which is consistent with the constant-field equation (Goldman, 1943; Hodgkin and Katz, 1949) with zero Ca^{2+} inside the cell:

$$g(V) = \frac{PV \exp(2FV/RT)}{[1 - \exp(-2FV/RT)]}$$

where P is a constant that depends on the internal Ca^{2+} concentration.

B. Deactivation Kinetics

The Hodgkin-Huxley model predicts a single exponential for the tail current at potentials where activation is zero. For m^2 kinetics, which have been used in several studies (see above) to model calcium current activation, tail currents should thus decay with a time course that is a single exponential with a time constant of 0.5 τ_m, if at the holding potential the steady state value of m is zero.

Tail currents evoked by stepping to voltages within the range of activation of calcium currents are expected to be composed of the sum of two exponentials, with time constants of τ_m and 0.5 τ_m.

While activation kinetics of calcium currents have been well described in Hodgkin-Huxley terms, this model does not sufficiently describe deactivation behavior in many cases. Recently, clear inconsistencies with this model have been found, namely that deactivation or turning off of calcium tail currents has a component with a time constant that is much smaller than the activation time constants determined from the Hodgkin-Huxley model (see for example, Byerly and Hagiwara, 1982; Fenwick et al., 1982; Brown et al., 1983).

These indications of the presence of a faster process in calcium current activation have been supported by results from studies of single calcium channels in snail neurons (Lux and Nagy, 1981), cardiac cells (Reuter et al., 1982), chromaffin cells (Fenwick et al., 1982), and pituitary cells (Hagiwara and Ohmori, 1982). The single-channel calcium current has been shown to flicker on and off with a time course that is considerably faster than that of the activation of the macroscopic current. In simultaneous recordings from whole cell and cell-attached patches Lux and Brown (1984b, 1985) have found that activation of whole cell calcium currents and membrane patch calcium currents proceed along identical lines. These results exclude a literal Hodgkin-Huxley model.

As an alternative, linear transition-state models with voltage-dependent rate constants have been proposed to describe activation and deactivation kinetics. Fenwick et al. (1982) were able to fit most of their data by using a linear sequential model of the form:

$$C_1 \rightleftharpoons C_2 \rightleftharpoons 0$$

with two closed and one open state of the channel. Other authors have needed at least three closed states to describe activation and deactivation kinetics (Hagiwara and Ohmori, 1982, see also Brown et al., 1983; Byerly et al., 1984; Brown et al., 1984; Lux and Brown, 1984; Taylor, 1988a).

V. MODULATION OF CALCIUM CHANNEL ACTIVATION

Modulation of calcium channel kinetics is an important mechanism to control Ca^{2+} influx through the channels. Calcium channels can be modulated by transmitters, enzymes, and drugs (for reviews see Reuter, 1983; Tsien et al., 1988; Bean, 1989a; Carbone and Swandulla, 1989; Swandulla et al., 1991). In neurons and neurosecretory cells there is evidence that certain transmitters and peptides slow down activation of HVA calcium channels and thus reduce Ca^{2+} entry. Examples are: dopamine (Marchetti et al., 1986), acetylcholine (Wanke et al., 1987), GABA (Deisz and Lux, 1985; Grassi and Lux, 1989), its agonist baclofen (Dolphin and Scott, 1987), noradrenaline (Forscher and Oxford, 1985; Marchetti et al., 1986; Bean, 1989b; Docherty and McFadzean, 1989; Lipscombe et al., 1989), somatostatin (Lewis et al., 1986; Luini et al., 1986; Tsunoo et al., 1986; Ikeda and Schofield, 1989), leucine-encephalin (Tsunoo et al., 1986), and the neuromodulator adenosine (Dolphin et al., 1986; Kasai and Aosaki, 1989). This effect on activation kinetics has been attributed to a change of the voltage dependence of calcium channel activation (Marchetti et al., 1986; Bean, 1989b), and the involvement of G-proteins in this process has been suggested (Dolphin and Scott, 1987, 1989; Bean, 1989b; Grassi and Lux, 1989; Kasai and Aosaki, 1989; Marchetti and Robello, 1989). Recently Grassi and Lux (1989) have provided evidence that voltage dependence of G-protein coupling to the channel may be the essential mechanism for transmitter-induced slowing of the activation process (see also Marchetti and Robello, 1989). When depolarizing the membrane for a rather short period with a prepulse, calcium currents in the presence of GABA, which were activated during a subsequent test pulse, show much faster activation kinetics than those recorded without prior depolarization (Figure 9). The GABA effect clearly depends on the activation state of a G-protein since the GABA action is mimicked when adding to the pipette solution GTP-γ-S, which activates G-proteins. The exact mechanism of G-protein action remains, however, to be determined.

VI. CONCLUSIONS AND PERSPECTIVES

The primary structure of the cardiac dihydropyridine-sensitive calcium channels have been recently analyzed (Mikami et al., 1989) and the α_1-subunit of the dihydropyridine-sensitive receptor of skeletal muscle has been shown to function as a calcium channel (Perez-

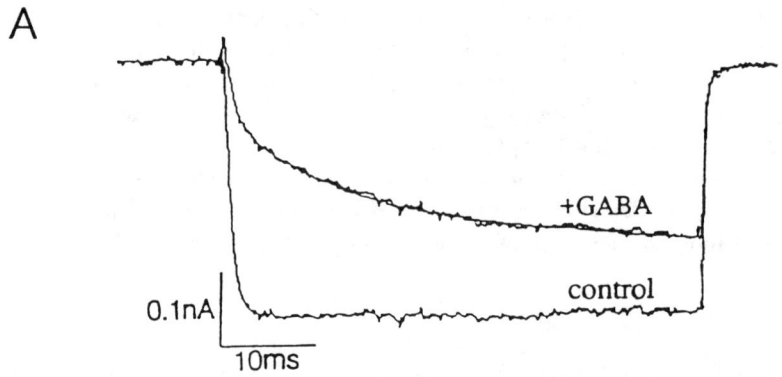

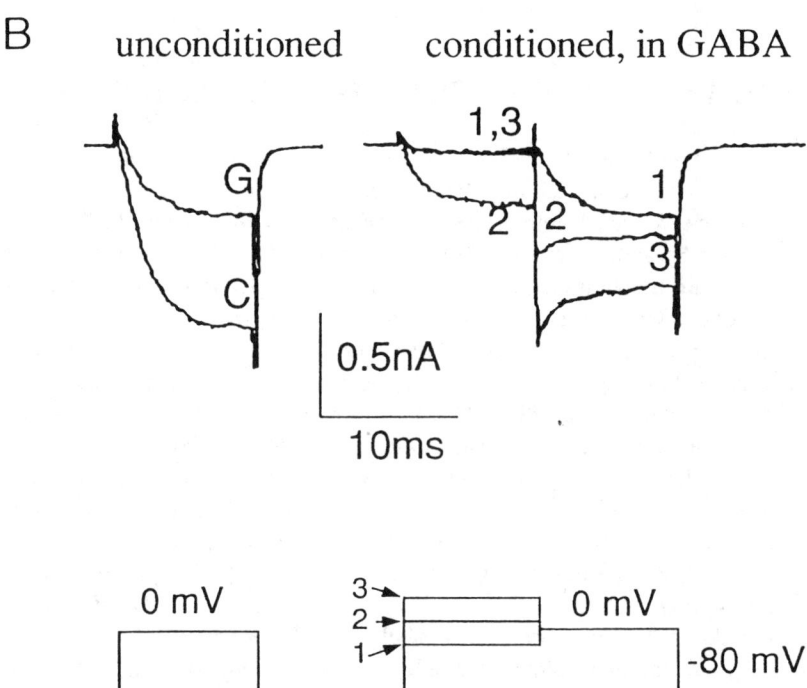

Figure 9. Slowing of calcium current activation by GABA. (A) illustrates the effect of 50 μM external GABA on calcium currents in chick DRG neurons. Currents were activated by voltage steps from −80 mV to 0 mV. B, left traces: unconditioned currents in control (C) and after application of GABA (G) right traces show currents in the presence of GABA recorded following conditioning prepulses to −20 (1), +10 (2) and +40 mV (3). 2 mM Ca^{2+} (23°C). (Modified with permission from Grassi and Lux, 1989.)

Reyes et al., 1989). However, the structural parts of the channel protein involved in voltage-dependent channel gating, unlike those of sodium channels, have not been elucidated for calcium channels. Recently it has been shown that point mutations introduced into a particular region of sodium channels, whereby positively charged amino acid residues are replaced by neutral or negatively charged residues, leads to a reduction in voltage-dependence of activation (Stühmer et al., 1989). This, for the first time, provided evidence for the involvement of positive charges in a particular region of the channel protein as the voltage-sensing device for activation.

3. Activation and Deactivation

It can be hypothesized, that strong similarities with regard to the voltage sensing part of the sodium channel protein exist for calcium and potassium channels, and it is most likely that a mechanism similar to that found in sodium channels leads to opening and closing of calcium channels in response to changes in the voltage across the membrane.

However, as long as structural mutations for the calcium channel are not available, biophysical experiments remain the sole approach to analyze the gating mechanisms of calcium channels.

Chapter 4

Calcium Channel Inactivation

Emilio Carbone and Dieter Swandulla

TABLE OF CONTENTS

I.	Introduction	36
II.	Ion Channel Inactivation	36
	A. The Concept of Ion Channel Inactivation	36
	B. Inactivation Processes Controlled by Voltage	37
	C. Inactivation Viewed Through Gating Currents Measurements	38
	D. Ca^{2+}-Dependent Inactivation	39
	E. Inactivation Viewed Through Single Channel Measurements	39
III.	Mechanisms of Calcium Channel Inactivation	41
	A. Calcium Channel Types	41
	B. The Low-Threshold Channel	42
	1. Voltage-Dependent Inactivation	42
	2. Voltage-Dependent Recovery from Inactivation	43
	3. Single Low-Threshold Channels	44
	4. Kinetic Models of the Low-Threshold Channel	45
	C. The High-Threshold Channels	47
	1. Voltage- and Ca^{2+}-Dependent Inactivation	47
	2. A Model for the HVA Channel Inactivation	50
	3. Ca^{2+} Buffering and Inactivation Gating	53
	4. Functional Implications of Voltage- and Ca^{2+}-Dependent Inactivation	53
	5. Single High-Threshold Calcium Channels	54
	D. Do Inactivation Kinetics Distinguish Between Different Types of HVA Calcium Channels?	55
IV.	Modulation of Calcium Channel Inactivation	55

	A.	HVA Channel Inactivation Induced by External Menthol and Bay K 8644 56
	B.	Enzymatic Modulation of Inactivation of Calcium Channels 58
	C.	Intracellular Agents Affecting HVA Channel Inactivation 59
V.	Summary and Conclusions 59	

Acknowledgments 60

I. INTRODUCTION

On a short time scale, calcium channel activation and inactivation are probably the most effective mechanisms governing Ca^{2+} influx through plasma membranes. While activation can be easily identified with a sequence of molecular events leading to channel opening, the nature of calcium channel inactivation has long been debated and has not yet been fully clarified (Brown et al., 1981; Plant and Standen, 1981; Eckert and Chad, 1984; Lux and Brown, 1984a; Gutnick et al., 1989). The elucidation of calcium channel inactivation has become particularly important, following the discovery of multiple types of voltage-sensitive calcium channels in excitable tissues (for recent reviews see Tsien et al., 1988; Bean, 1989a). Together with the voltage range of activation, pharmacological sensitivity, and single channel properties, inactivation kinetics are one of the most commonly used parameters to identify calcium channel types.

In this report we will describe basic principles governing calcium channel inactivation, focusing on its physiological significance and modulation by endogenous agents. We will also attempt a critical overview of the criteria adopted to identify inactivation processes of different calcium channel types.

II. ION CHANNEL INACTIVATION

A. The Concept of Ion Channel Inactivation

The number of ions that flow through a channel is set by the constraints of the conducting pore, the driving force acting on the ions, and the kinetics of channel opening and closing. The probability of finding a channel open or closed is controlled either by membrane voltage (voltage-operated-channel, VOC) or by ligand binding to the channel protein (receptor-operated-channel, ROC). In both cases, however, the rapid equilibrium between channel openings and closings can be modulated by a comparatively slowly developing process that leads the channel to a temporarily inactive (nonconductive) form that differs from the closed state of the channel. This inactivation process may depend on membrane voltage, intracellular Ca^{2+}, or other factors and may be linked to some degree to channel activation (see below).

Although there are no clear definitions, channel inactivation is usually indicated as "fast" or "slow" depending on its inactivation time constant (τ_h). Typical examples of well-separated fast and slow inactivation are reported for the voltage-dependent sodium

4. Calcium Channel Inactivation

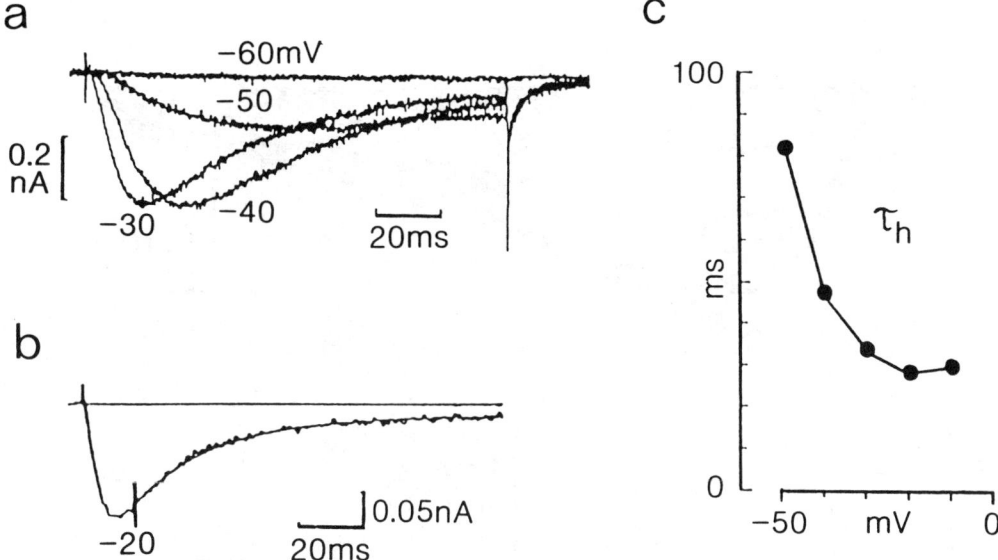

Figure 1. (a) Whole-cell clamp LVA calcium currents recorded from an embryonic rat DRG at the potentials indicated. Holding potential (V_h) −80 mV. Step repolarizations to V_h. External bath (mM): 120 cholineCl, 5 CaCl$_2$, 2 glucose, 10 NaHEPES (pH 7.3), 3 μM TTX. Pipette filling solution (mM): 130 CsCl, 20 TEACl, 0.25 CaCl$_2$, 5 EGTA-OH, 10 glucose, 10 CsHEPES (pH 7.3). Temp. 12°C. (b) An LVA current record at −20 mV in 5 mM Ca^{2+}. The inactivating phase of the current is fitted with a single exponential with a time constant of 27 ms (τ_h). The beginning of the fitting is indicated by a vertical bar. (c) Plot of τ_h vs. voltage for the records of panel b. (From Carbone and Lux, 1986b; 1987a. With permission.)

channel of peripheral neurons (Hodgkin and Huxley, 1952b; Chandler and Meves, 1970) and for the low threshold calcium channel of vertebrate sensory neurons (Carbone and Lux, 1984a). For the fast and slow sodium channel inactivation, τ_h varies between 1 and 10 ms and between 0.2 and 10 s, respectively, depending on membrane voltage, temperature, and type of cell (see Table 3 in Carbone and Lux, 1986a). Values about one order of magnitude slower have been observed for the corresponding inactivation of the low threshold calcium channel (Bossu and Feltz, 1986; Carbone and Lux, 1987a). Fast and slow inactivation can also refer to gating mechanisms that develop on a much closer time scale, as for the neuronal high-threshold calcium channels whose τ_{slow} can be only 2 to 10 times longer than τ_{fast}.

B. Inactivation Processes Controlled by Voltage

In general, voltage-dependent inactivation implies that the rate of channel inactivation ($1/\tau_h$) is a steep function of membrane voltage (Figure 1). In other words, linear plots of τ_h vs. voltage yield bell shaped curves that peak at rather negative membrane potentials (−60 to −50 mV), with steep voltage dependency above and below these values. τ_h vs. voltage is usually determined by fitting with one or two exponentials the decaying phase of whole-cell current recordings at potentials where ion currents can be measured directly on step depolarizations (Figure 1c). A marked decrease of τ_h with membrane depolarization from −50 to −30 mV is then taken as a good indication of a voltage-dependent inactivation process.

Inactivation processes are also identified by S-shaped curves that describe their "steady-state" voltage dependence and indicate the fraction of channels still available for activation after inactivation has fully developed (Figure 2b and 2c). Curves for fast (h_∞) or slow (s_∞) inactivation are derived by plotting the normalized current values measured during a brief depolarization (V_t) following conditioning prepulses (V_c) of variable amplitude but fixed

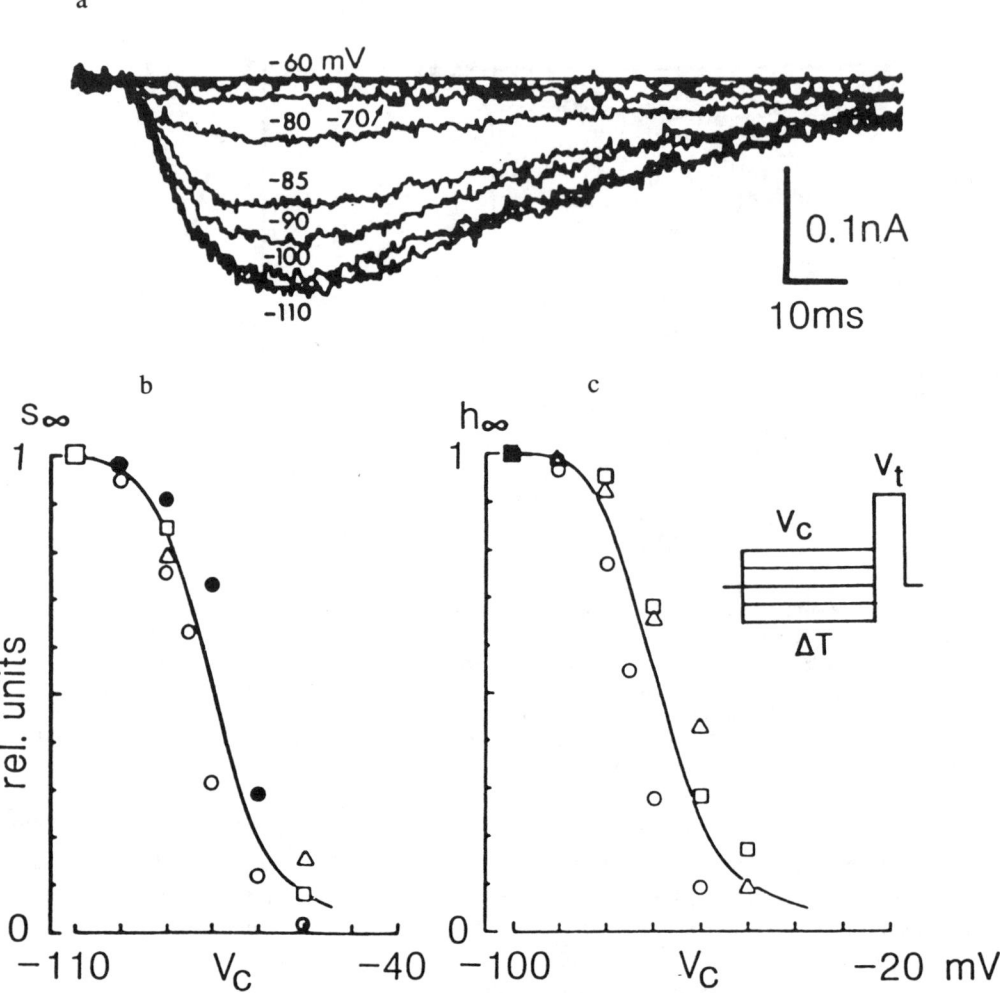

Figure 2. Slow and fast inactivation of the low-threshold calcium channel of chick sensory neurons. (a) Superimposed LVA calcium currents recorded during step depolarization to -40 mV from varied preconditioning prepulses between -60 and -110 mV. The cell was maintained for 90 s (ΔT) at the potentials indicated. (b) The voltage dependence of slow LVA channel inactivation. The data are normalized peak LVA current plotted vs. V_c (external Ca^{2+} concentration, 5 mM). Different symbols indicate different cells. The line was drawn by visual approximation. (c) The voltage-dependent fast inactivation process of LVA calcium currents (h_∞). The data are normalized values of peak LVA currents at -40 mV (V_t) plotted vs. the potential level of the conditioning prepulse (V_c), that lasted 150 ms (ΔT). The inset indicates the pulse protocol used for the determination of s_∞ and h_∞. (From Carbone and Lux, 1984a. With permission.)

duration (see inset in Figure 2c). The duration (ΔT) of the conditioning prepulses sets the type of inactivation tested. Fast inactivation usually requires ΔT of 50 to 200 ms while slow inactivation necessitates ΔT of several seconds.

C. Inactivation Viewed Through Gating Currents Measurements

At the molecular level, voltage-dependent inactivation implies that charged groups ("gates") associated with the channel protein move in response to the membrane electric field, preventing further ion passage through the open channel. As a consequence of the gating-charge displacements, one should expect to observe sizeable "gating currents" with amplitude and time course depending on the number of elementary charges per gate, the density of the channels, and the on-off rate of the gate. However, detailed studies have shown that only gating currents associated with channel activation can be measured accurately (Armstrong and Bezanilla, 1973; Adams and Gage, 1977; Kostyuk et al., 1977). Inactivation processes are apparently too slow or bear too little charge to produce detectable gating currents (Armstrong and Bezanilla, 1977). In spite of this, channel inactivation can cause gating charge "immobilization" that is revealed by a loss of displaced charges during channel deactivation. Thus, full development of sodium channel inactivation in squid giant axons immobilizes about two thirds of the gating charges displaced during channel activation (Armstrong and Bezanilla, 1977). Remarkably, agents that prevent sodium channel inactivation, such as pronase and high pH_i, cause removal of gating charge immobilization (Swenson, 1980; Wanke et al., 1983), supporting the view that ion channel inactivation can also be viewed through gating current recordings. Presently, there are no similar reports for calcium channel inactivation.

D. Ca^{2+}-Dependent Inactivation

Early evidence for Ca^{2+}-driven inactivation came from calcium current measurements in ciliates (Brehm and Eckert, 1978) and molluscan neurons (Tillotson, 1979; Eckert and Tillotson, 1981; Plant and Standen, 1981), and quickly extended to other cell preparations including insect muscle (Ashcroft and Stanfield, 1982b) and heart cells (Fischmeister et al., 1981; Marban and Tsien, 1982). A more complete list of these reports has been given in excellent reviews on this topic (see for example, Eckert and Chad, 1984; Chad, 1989). Therefore, in this section we will focus only on the basic criteria used to identify the Ca^{2+} dependence of calcium channel inactivation.

Evidence for Ca^{2+}-driven inactivation comes from the following observations:

1. The rate of calcium current inactivation depends on current amplitude. It increases with increasing amplitude of calcium currents (Figure 3).
2. The relationship of inactivation to membrane voltage appears "U-shaped", suggesting that decreased Ca^{2+}-entry at high membrane potentials reduces calcium channel inactivation (see Eckert and Chad, 1984).
3. Barium or sodium currents through calcium channels inactivate less than do calcium currents (Figure 3b).
4. Inactivation is strongly relieved in the presence of internal calcium chelators.
5. Increasing internal Ca^{2+} concentration by microinjections of Ca^{2+} ions into the cells or by photo-releasing Ca^{2+} ions from photosensitive Ca^{2+} buffers introduced intracellularly causes increased calcium channel inactivation (see e.g., Eckert and Chad, 1984; Morad et al., 1988b).

The above criteria allow a rather sharp separation between Ca^{2+} and voltage-dependent inactivation processes among different types of calcium channels. In fact, voltage-dependent inactivation preserves its properties independently of the current carrying ion (Ca^{2+}, Ba^{2+}, or Na^+), the size of the current, or the amount of internal Ca^{2+} buffering (see below). On the other hand, Ca^{2+}-dependent inactivation develops with kinetics that usually show an opposite voltage dependency to that expected for a strictly voltage-sensitive inactivation process.

E. Inactivation Viewed Through Single Channel Measurements

Single channel current measurements utilizing the patch-clamp technique offer an independent way to study the kinetics of ion channel inactivation (Sakmann and Neher, 1984). Recent studies using various membrane patch configurations (Hamill et al., 1981) have furnished rather detailed information about this phenomenon.

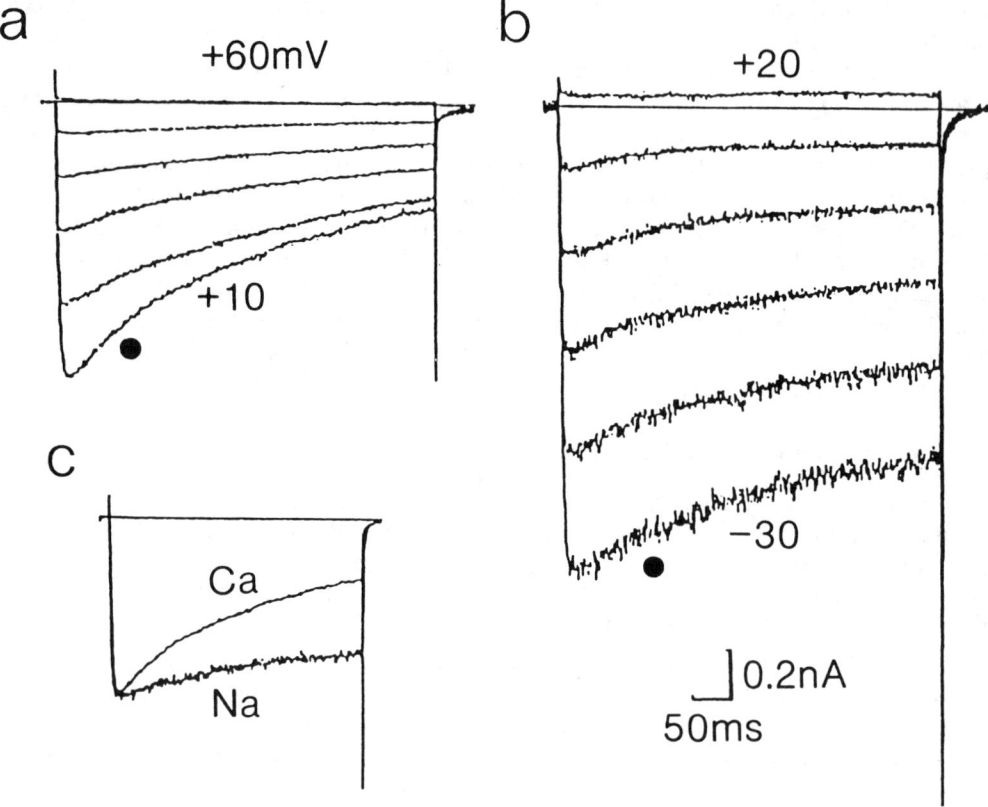

Figure 3. Time course of calcium (a) and sodium currents (b) through high-threshold calcium channels in chick DRG neurons. The bath contained (mM): 20 Ca^{2+}, 120 Na^+, 2 Mg^{2+} in (a) and 140 Na^+, 4 Ca^{2+}, 5 EGTA to give 0.1 μM free Ca^{2+} in (b). Step depolarizations were as indicated. Holding potential, -90 mV. (c) Comparison of HVA current inactivation when Ca^{2+} or Na^+ are the current carrying ions. The two overlapped traces were taken from (a) and (b) (●) and normalized to the peak amplitude. (From Carbone and Lux, 1988a. With permission.)

Usually, at the single channel level, inactivation is viewed as prolonged interruptions of channel activity. This implies that the inactivated state of the channel is "absorbing" and follows the open state, i.e., once inactivated the channel does not return readily to the open configuration. This possibility is represented, in a simple case, by the following scheme:

$$C \rightleftarrows O \nearrow\!\!\!\!\!\searrow I \qquad (1)$$

where C, O, and I indicate the closed, open, and inactivated states of the channel. Under these conditions, channel inactivation is viewed either as a final interruption of channel openings (transition O → I) or as a lack of elementary events during membrane depolarizations (transition C → I). If transitions O → I and C → I are voltage-dependent, the interruptions of channel activity will be observed rarely at low voltages (-90 to -50 mV), but more often during membrane depolarizations. Figure 4 shows examples of unitary sodium channel events at various potentials resolved in an outside-out membrane patch of a chick sensory neuron (Carbone and Lux, 1986a). At -90 mV, there is no sign of channel inactivation. Channels open occasionally or not at all during pulses of 160 ms. Channel inactivation becomes evident at potentials less negative than -50 mV, at which openings appear

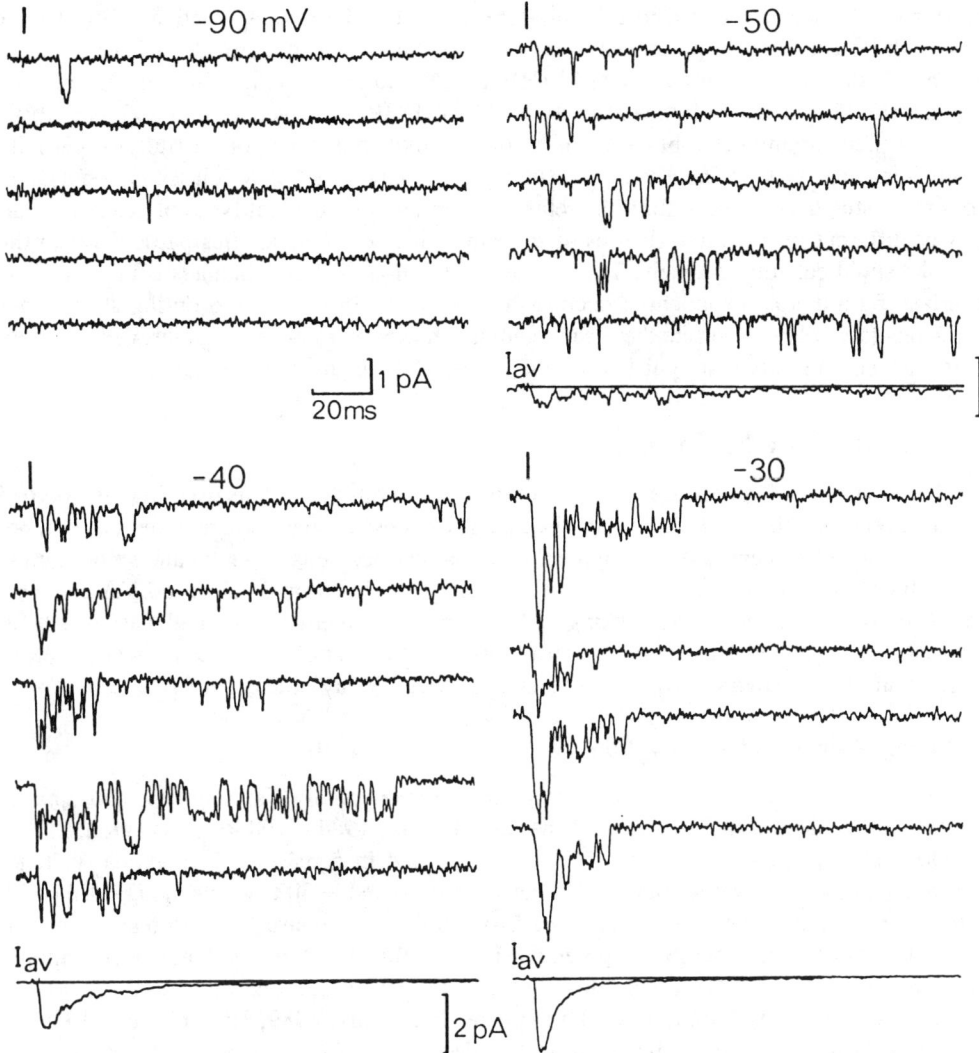

Figure 4. Single sodium channel currents in an outside-out patch recorded at the potential indicated. Bars indicate the onset of the depolarizing voltage steps. The holding potential was −90 mV. The patch apparently contained 7 to 8 channels as estimated from the maximum number of simultaneous openings in response to large depolarizations. Out (mM): 120 Na$^+$, 7 Mg^{2+}. In (mM): 130 Cs$^+$. (From Carbone and Lux, 1986a. With permission.)

more frequently shortly after the onset of the step depolarization than at later times. At −30 mV, multiple channel openings occur immediately after the beginning of a pulse and activity ceases within about 30 ms. The time course of sample averages (I_{av}) compares well with that of fast inactivating sodium currents recorded in whole-cell conditions, suggesting a close correlation between single channel events and macroscopic current inactivation.

III. MECHANISMS OF CALCIUM CHANNEL INACTIVATION

A. Calcium Channel Types

Calcium channels are commonly distinguished on the basis of their voltage range of activation (see Chapter 3, Section I). Channels that activate at potentials more positive than

−50 mV are usually referred to as "low-threshold" or "low-voltage activated" (LVA or T) and those activated at potentials more positive than −10 mV are identified as "high-threshold" or "high-voltage activated" (HVA or N and L). Calcium channels can also be distinguished by their level of sensitivity to drugs (DHP-sensitive, ω-Conotoxin-sensitive, Cd^{2+}-blocked, or amiloride-blocked) or by their inactivation time course (fully or partially inactivating, fast or slow inactivating). Although convenient, the latter classification usually does not distinguish between multiple relaxations reflecting different types of calcium channels or different inactivation kinetics of the same channel. This applies particularly to the high-threshold calcium channels. The rate of inactivation of these channels is usually characterized by a double exponential decline that cannot be fully resolved during short depolarizations (see below). Despite these shortcomings, however, low- and high-threshold calcium channels can be easily distinguished by the nature of their inactivation gate.

B. The Low-Threshold Channel

LVA calcium channels are a unique class of calcium channels possessing stereotyped inactivation properties across a variety of cells. Their presence has been reported in a number of tissues including central and peripheral neurons, cardiac cells, skeletal and smooth muscles, fibroblasts, osteoblasts, and oocytes. In cardiac and neuronal cells the LVA channels are likely to be involved in pacemaking and bursting phenomena (Llinas and Yarom, 1981a; Bean, 1985). Detailed lists of the many reports on this matter have been given in recent reviews and books (Bean, 1989a; Morad et al., 1988a; Wray et al., 1989a).

1. Voltage-Dependent Inactivation

Inactivation of LVA channels develops monoexponentially, is complete, and appears strictly voltage dependent (Figure 1) (Carbone and Lux, 1984a, 1984b; Bossu et al., 1985; Fedulova et al., 1985). In chick DRG neurons bathed in 5 mM Ca^{2+}, maximal voltage-sensitivity of the inactivation rate occurs between −40 and −30 mV, changing e-fold with a 14 mV potential variation (Carbone and Lux, 1987a). In Figure 1, τ_h decreases from 80 to 30 ms by raising the membrane potential from −50 to −10 mV and attains a minimum of 25 ms at −20 mV. Neuronal (Bossu and Feltz, 1986; Carbone and Lux, 1987a) and cardiac LVA channels (Bean, 1985; Droogmans and Nilius, 1989; Hirano et al., 1989) of mammals show a nearly two-fold faster inactivation time course than avian calcium channels suggesting tissue-specific kinetic properties.

The voltage-dependent nature of LVA channel inactivation is also indicated by the observation that its rate is independent of the amount of Ca^{2+} entry into the cell and of the type of the current carrying ion. The former is best demonstrated by comparing the time course of LVA currents recorded on step depolarizations from increasingly negative holding potentials (−60 to −100 mV) (Figure 2a). Despite the different size, the corresponding LVA currents have nearly identical inactivation time courses. The latter becomes evident when replacing external Ca^{2+} by Ba^{2+} or Na^+. Replacement of Ca^{2+} with Na^+ causes a nearly four-fold increase of the low-threshold current size with little effect on its activation and inactivation kinetics (Figure 5) (see Carbone and Lux, 1988b). Similar observations were reported for the low-threshold calcium channel of mouse neoplastic B lymphocytes (Fukushima and Hagiwara, 1985). Activation and inactivation kinetics of this channel are unaffected when Ca^{2+} is replaced with Na^+ and the size of the corresponding sodium currents is depressed by micromolar additions of external Ca^{2+} or Mg^{2+}.

The similarity of LVA channel inactivation with that of other voltage-dependent ion channels extends to its temperature sensitivity. In chick sensory neurons, a temperature change from 20 to 30°C causes a 2.2-fold acceleration of the LVA inactivation time course (Q_{10} 2.2) and a comparable acceleration of the activation process (Nobile et al., 1990). Somewhat larger values have been reported for similar LVA channels of a mouse neuro-

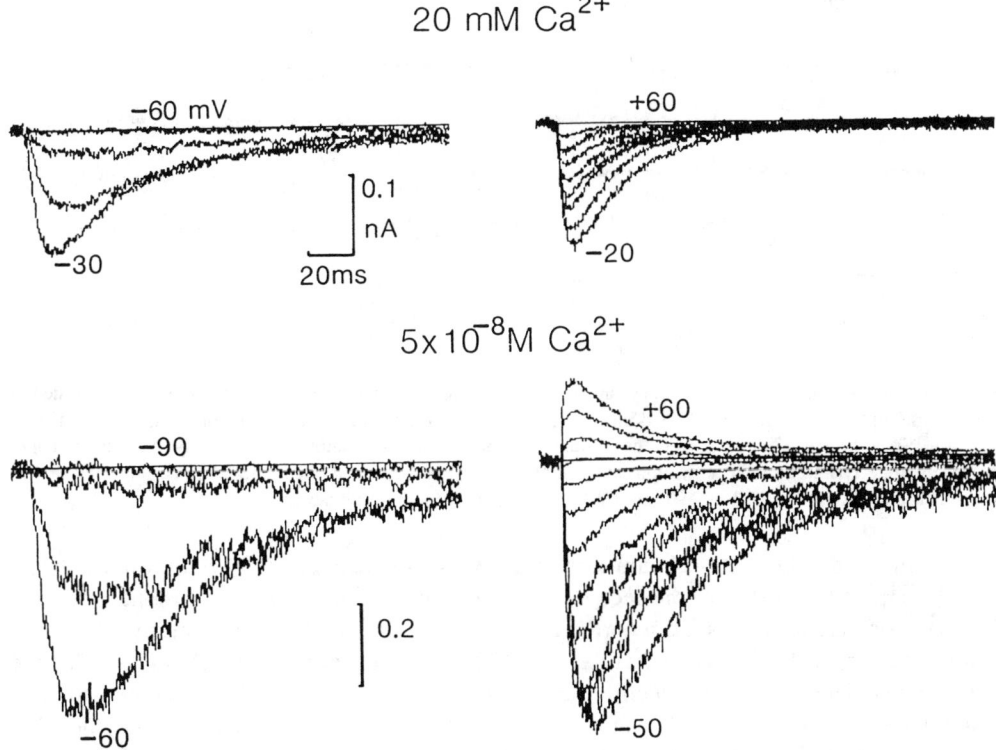

Figure 5. Calcium and sodium currents through LVA channels in a chick DRG neuron. The calcium currents (top) were recorded in 20 mM Ca^{2+} at the membrane potential indicated. Depolarizations were separated by steps of 10 mV. The sodium currents (bottom) were recorded in Na-EGTA (pCa 8), starting from -90 mV. The cell was not treated with any drug and apparently possessed very few HVA channels. V_h -100 mV. (From Carbone and Lux, 1988b. With permission.)

blastoma cell line (Q_{10} 2.5, Narahashi et al., 1987) and rat thalamic relay neurons (Q_{10} 2.8, Coulter et al., 1989c) for the same range of temperatures. These values compare well with those of sodium channel inactivation (Q_{10} 2.3 to 3) in peripheral neurons (Chiu et al., 1979; Kimura and Meves, 1979) and cardiac cells (Colatsky, 1980), but are markedly smaller than those for HVA channel inactivation ($Q_{10} > 9$, Nobile et al., 1990). Thus, temperature reveals structural differences between HVA and LVA inactivation gating, suggesting similar activation energies for similarly voltage-dependent inactivation processes.

2. Voltage-Dependent Recovery from Inactivation

LVA currents can be completely inactivated after prolonged depolarizations to potentials still below resting values (Figure 2a). However, LVA channels recover from inactivation, as do HVA channels, provided that the membrane is hyperpolarized to negative membrane potentials (-80 to -100 mV) for sufficiently long periods of time (Figure 2a). For instance, hyperpolarizations of 200 ms to -100 mV following steady depolarizations to -30 mV are sufficient to elicit sizeable LVA currents when the membrane is once again depolarized. Increasing the duration of the hyperpolarization causes a progressive increase of LVA current amplitudes to maximal values with hyperpolarizations of about 6 s (see Figure 6 in Carbone and Lux, 1987a).

As with sodium channels, recovery from inactivation of LVA channels in vertebrate neurons is time and voltage dependent. The time course of recovery from inactivation was measured by holding the cell at -80 mV for various lengths of time between two depolarizing

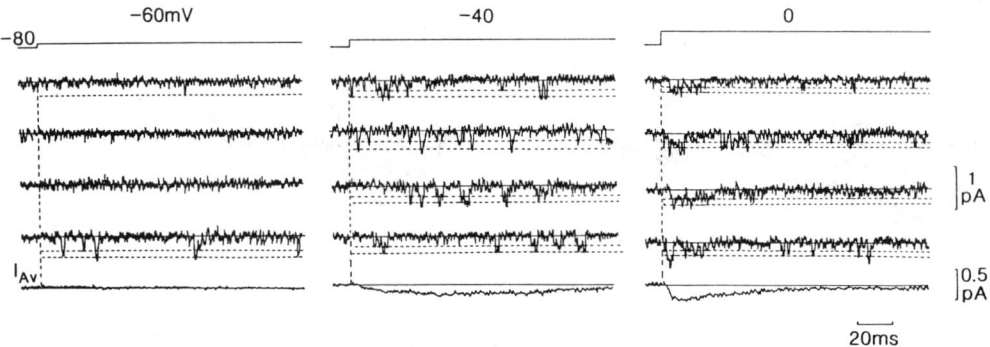

Figure 6. Calcium channel currents in an outside-out patch from a chick DRG cell in 50 mM Ca^{2+} recorded at the potential indicated. Horizontal dashed lines represent the mean amplitudes of channel openings. The last trace (I_{av}) at -60, -40 and 0 mV represents the average over 30, 68, and 41 samples, respectively. Note the voltage-dependent time course of their activation and inactivation. I_{av} resembles strongly the LVA current recorded in whole-cell conditions. V_h -80 mV. (From Carbone and Lux, 1987b. With permission.)

pulses. The amplitude of the calcium current recovers monoexponentially with a time constant of 1.4 s (τ_r). At -120 mV, the recovery is faster and τ_r falls to 0.4 s (Carbone and Lux, 1987a). Comparable values are reported for the low-threshold channels of rat hypothalamic and thalamocortical neurons (Akaike et al., 1989a; Coulter et al., 1989c) and cells of the lateral geniculate nucleus (Crunelli et al., 1989). Detailed studies in rat cranial ganglia neurons (Bossu and Feltz, 1986) have shown that recovery from inactivation may develop with one or two time constants depending on whether a short (several hundreds of milliseconds) or a long conditioning depolarization (tens of seconds) is applied (see also Akaike et al., 1989b). Fast recovery was better observed after large, brief depolarizations and slow recovery after small, long duration inactivating prepulses. Bossu and Feltz (1986) concluded that the two components of recovery are independent processes and that the slow one has characteristics of a Ca^{2+}-mediated process. Recent studies on LVA channels of canine Purkinje cells, however, show that recovery from inactivation of cardiac LVA channels follows complex kinetics that depend strictly on voltage but not on Ca^{2+} (Hirano et al., 1989).

3. Single Low-Threshold Channels

Single LVA calcium channels were first resolved in excised patches of chick sensory neurons and guinea-pig ventricular myocytes using high extracellular Ca^{2+} or Ba^{2+} concentrations (40 to 100 mM) (Carbone and Lux, 1984b; Nilius et al., 1985). These ionic conditions were needed to resolve the unitary LVA current events that were particularly small with Ca^{2+} as the current-carrying ion. Despite the low signal-to-noise ratio of those recordings, single LVA channel events could be clearly identified by their sensitivity to extracellular Ca^{2+}, their voltage-dependent activation and inactivation kinetics, and their sensitivity to holding potential. Their prolonged activity in excised membrane patches suggested that, contrary to HVA channels, there was no particular metabolic requirement for their functioning.

LVA channel activity started at membrane potentials of about -60 mV showing discrete inward going current events with virtually no sign of channel inactivation during pulses of 200 ms duration (Figure 6). At -40 mV, channel activity and the average number of openings per trace increased considerably. Events were separated by short- and long-lasting closures that indicated fast and slow bursting activity of the channel. Time-dependent inactivation became evident at more positive membrane voltages (> -20 mV) where the probability of channel openings was higher shortly after the onset of the step than it was at

4. Calcium Channel Inactivation

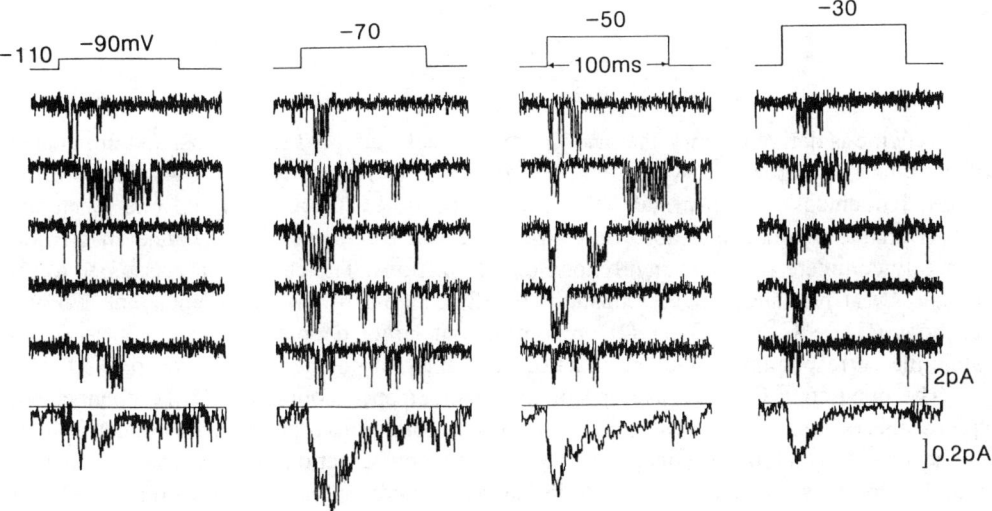

Figure 7. Unitary LVA sodium currents in an outside-out patch from a chick DRG cell recorded at the potential indicated. The bath contained 3×10^{-8} Ca^{2+}, 140 mM Na^+, 10 mM Na-HEPES, and 3 µM TTX. The last trace of each panel represents the current average over 60 (-90 mV), 65 (-70 mV), 58 (-50 mV) and 30 (-30 mV) samples. 3 kHz low-pass filter. Sampling rate: 36 µs per point. Note the increased unit sizes of the events and the rapid bursting kinetic during openings. (From Carbone and Lux, 1987b. With permission.)

later times. There was a close correlation between single channel and macroscopic inactivation kinetics.

LVA channel inactivation kinetics are also preserved when the external free Ca^{2+} concentration is reduced to submicromolar values, and Na^+ or Li^+ are the major current-carrying ions (Figure 7). Unitary sodium currents are larger but are similar to calcium currents in their activation and inactivation time courses and sensitivity to holding potentials. They are blocked by additions of 5 mM Ni^{2+} or 1 mM Cd^{2+} to the bath and are hardly measurable in the absence of alkali metal ions. This indicates that monovalent and divalent ions cross LVA chennels without interfering with the voltage-sensitive activation and inactivation gating of the channel.

LVA channel activity can also be measured in cell-attached patches using pipette solutions containing 40 mM Ca^{2+} or 95 mM Ba^{2+} while the bath contains normal saline. Under these conditions, step depolarizations of 50 to 60 mV from the holding potential elicit openings of about -0.4 pA with mean open times and inactivation kinetics similar to those observed in outside-out patches.

4. Kinetic Models of the Low-Threshold Channel

As shown above, activation and inactivation time courses of isolated LVA currents appear to be strictly voltage-dependent. Activation develops sigmoidally (see Chapter 3, Section I) while inactivation appears to follow a single exponential and is much slower than activation. Single-channel data from excised membrane patches containing few channels provide a more direct insight into the kinetic features of this channel and help to discriminate among kinetic models (Carbone and Lux, 1987b). Particularly, they indicate whether activation and inactivation of the channel reflect two independent voltage-operated processes (weak coupling) or whether inactivation derives its voltage-dependency directly from the activation process that might be slow and rate limiting (strong coupling).

Single-channel and whole-cell LVA currents of chick DRG neurons can be well described by the following kinetic scheme (Carbone and Lux, 1987b):

$$C_1 \rightleftarrows C_2 \overset{\nearrow \uparrow \nwarrow}{\rightleftarrows} O \quad \overset{I}{} \tag{2}$$

This scheme is derived from "the basic kinetic model" adopted for the fast sodium channel of neurons (Bezanilla and Armstrong, 1977; Vandenberg and Horn, 1984; Kunze et al., 1985). It includes one open state (O), two closed states (C_1 and C_2) and one inactivated state (I) that is assumed to be absorbing (Aldrich et al., 1983) since LVA channel inactivation is largely complete after sufficiently long depolarizations. The channels are assumed to stay in state C_1 at rest and the rate constants of all reactions are voltage dependent. Forward reactions ($C_1 \rightarrow C_2$ and $C_2 \rightarrow O$) predominate at larger depolarizations (-30 to 0 mV) while the corresponding backward reactions dominate at potentials negative to -30 mV.

The open state O is the average of two distinct open conditions of the channel with comparable occupational probability. The two open conditions represent subconductive levels of the channel that may reflect two stable open configurations of the channel in rapid equilibrium. This seems to be a common feature of calcium channels (Lacerda and Brown, 1989; Plummer et al., 1989), rather than being the result of unresolved short openings due to a limited measurement frequency resolution. We have observed maximal slope conductances of 5.2 and 3.7 pS in outside-out patches of chick DRG in 40 mM external Ca^{2+} (Carbone and Lux, 1987b), that compare well with those of the cardiac LVA channel in cell-attached patches: 6.8 and 3.4 pS in 110 mM external Ca^{2+} (Droogmans and Nilius, 1989).

Neuronal (Carbone and Lux, 1984b; 1987b; Nowycky et al., 1985a) and cardiac LVA calcium channels (Nilius et al., 1985; Droogmans and Nilius, 1989) exhibit rather similar kinetic features that can be summarized as follows:

1. The channels possess two subconductive open states with mean lifetimes (1 to 3 ms) that are weakly voltage dependent.
2. Openings occur in bursts (i.e., with a high probability of reopenings), and first latency of opening is long-lasting and strongly voltage dependent. As a consequence, channel activation derives its voltage dependency mainly from the strong voltage dependency of the transition $C_1 \rightarrow C_2$.
3. The channel reopens several times before it inactivates, suggesting that macroscopic inactivation is not a reflection of appropriately delayed first openings but it occurs independently of channel activation. Inactivation may proceed also from closed states attained before or after the open state. In other words, the channel can inactivate without opening.
4. In chick DRG, inactivation of LVA channels is weakly coupled to activation and derives its voltage dependency from its inherent voltage-dependent gate, very likely, associated with some charged groups of the channel protein subunit responsible for inactivation.
5. Macroscopic inactivation and mean burst duration of LVA channels in avian neurons last longer than in mammalian DRG and cardiac cells, supporting the view that in mammals inactivation of LVA channels is at least partially controlled by microscopic activation. Thus it resembles that of fast sodium channels (Aldrich et al., 1983; Vandenberg and Horn, 1984).

Data from cardiac Purkinje cells suggest that Scheme 2 requires an additional inactivated state (I_2) reached only from state I (Hirano et al., 1989). This would account for the lag of about 50 ms in the onset of recovery from inactivation observed after prolonged depolarizations. An effect that may have been overlooked in vertebrate neurons (Bossu and Feltz,

1986; Carbone and Lux, 1987a). In conclusion, with the appropriate modifications, Scheme 2 accounts sufficiently well for the voltage-dependent activation and inactivation characteristics of the low threshold channels in a variety of tissues.

C. The High-Threshold Channels

The HVA channels form a broad class of calcium channels that are widely distributed in neuronal, muscular, and secretory cells. Besides their high threshold of activation, they can easily be identified by their unique type of inactivation that shares voltage- and Ca^{2+}-dependent features. Obviously, this dual nature of inactivation introduces some complexity in the overall behavior of HVA channels but also provides a basis for the understanding of the regulatory action of these channels in various cellular functions.

1. Voltage- and Ca^{2+}-Dependent Inactivation

An example of voltage- as well as Ca^{2+}-dependent inactivation has been recently studied in detail for the HVA calcium channel of snail neurons (Gutnick et al., 1989). Inactivation of the calcium current is evident in the decay from its peak during a sustained depolarizing step. The current relaxation exhibits an initial fast phase followed by a slower phase (Figure 8a). Inactivation kinetics can be approximated by a double exponential function with time constants of about 30 and 300 ms, respectively, with activating pulses to +30 mV at 30°C. Replacing Ca^{2+} with Ba^{2+} as the charge carrier results in a slowing of inactivation kinetics, which is a clear indication of a Ca^{2+}-dependent inactivation process (Figure 8b). Inactivation can also be visualized when comparing the currents associated with two sequential activating voltage pulses. If the interval between the pulses is short and does not allow for substantial recovery of inactivated channels, inactivation is evident as a decrease of the peak current amplitude during the second activating pulse. By changing the amplitude of the first activating pulse (prepulse), Ca^{2+}-entry during this period can be varied in a systematic fashion and its effect on the peak current amplitude associated with the second activating pulse (test) can be studied. This procedure is illustrated in Figure 9, showing the effect of varying the potential of a 60 ms prepulse (P_1) on the peak calcium current evoked 10 ms later by a test pulse to +30 mV (P_2) (see inset in Figure 9). The degree of inactivation (filled triangles) increases continuously, reaches a maximum at +60 mV (88%), and then declines slightly at more positive potentials. At +90 mV, however, inactivation is still substantial (70%) even though Ca^{2+} entry is also zero, as can be seen from the corresponding I-V plot that shows the relationship between peak calcium current induced by the prepulses and the prepulse potential (filled circles). Similar results were obtained with Ba^{2+} as the charge carrier.

These findings deviate somewhat from those reported in other molluscan neurons mainly with regard to the higher degree of inactivation observed in snail neurons at very positive potentials (+50 to +90 mV) where calcium currents are small in amplitude (Tillotson, 1979; Plant and Standen, 1981; Eckert and Chad, 1984). Such discrepancy could be due to the different durations of prepulse and interpulse intervals used in these experiments. Prolonged interpulses (100 to 400 ms) and short prepulses (50 to 300 ms) favor a substantial recovery of channels that may result in a more U-shaped inactivation curve (Brown et al., 1981; Gutnick et al., 1989). The different observations could also arise from a different Ca^{2+}-sensitivity of HVA channel inactivation in mollusks. This might also be the case for cardiac and vertebrate neurons (Fischmeister et al., 1981; Marban and Tsien, 1982; Kasai and Aosaki, 1988). In summary, these findings indicate that inactivation of HVA channels is not simply due to a build up of internal Ca^{2+} during channel activity but depends on Ca^{2+} *and* voltage in a rather complex manner.

The most convincing evidence for the voltage dependence of the inactivation process comes from the finding that recovery of inactivated channels (repriming) is sensitive to

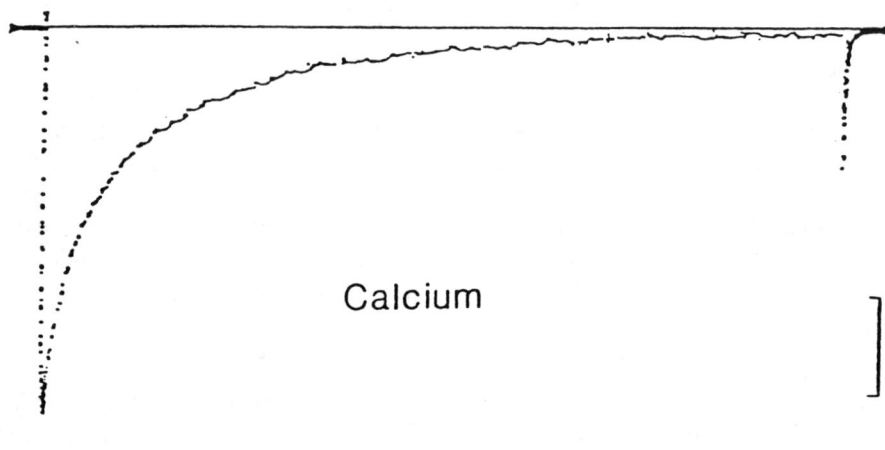

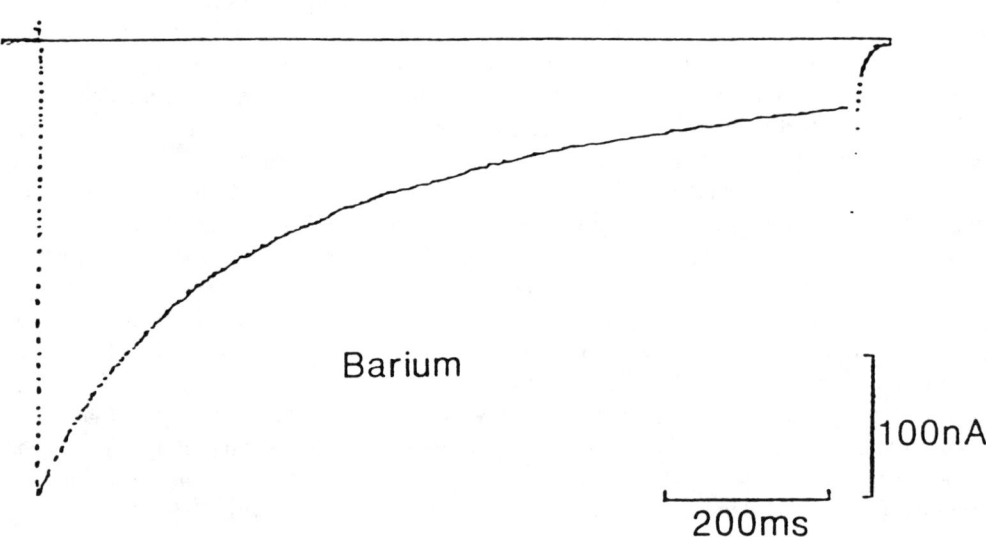

Figure 8. Inactivation of whole cell calcium channel current in an *Helix pomatia* neuron. Activating pulses were from V_h −50 mV to +30 (a) and +20 mV (b). The current in (b) was recorded 10 min after substituting Ca^{2+} with 40 mM Ba^{2+} isosmotically. Linear components of leakage and capacitance were removed by subtracting currents obtained with hyperpolarizing pulses of the same magnitude. 30°C.

membrane potential (Yatani et al., 1983). The repriming effect of even brief hyperpolarizations is illustrated in Figure 10a. In this experiment, the current induced by a sustained depolarizing step to −50 to +30 mV is compared to the current generated by a train of depolarizing pulses that was separated by 4 ms hyperpolarizations to −110 mV. The calcium current induced by the pulse train inactivated substantially less even though Ca^{2+} entry, as inferred from the time integral of the current, was about twice that occurring during the sustained pulse. From these results it is obvious that hyperpolarization enhances recovery from inactivation and that Ca^{2+} entry is not the sole mechanism that determines inactivation kinetics. The role of Ca^{2+} in the inactivation process becomes evident when determining

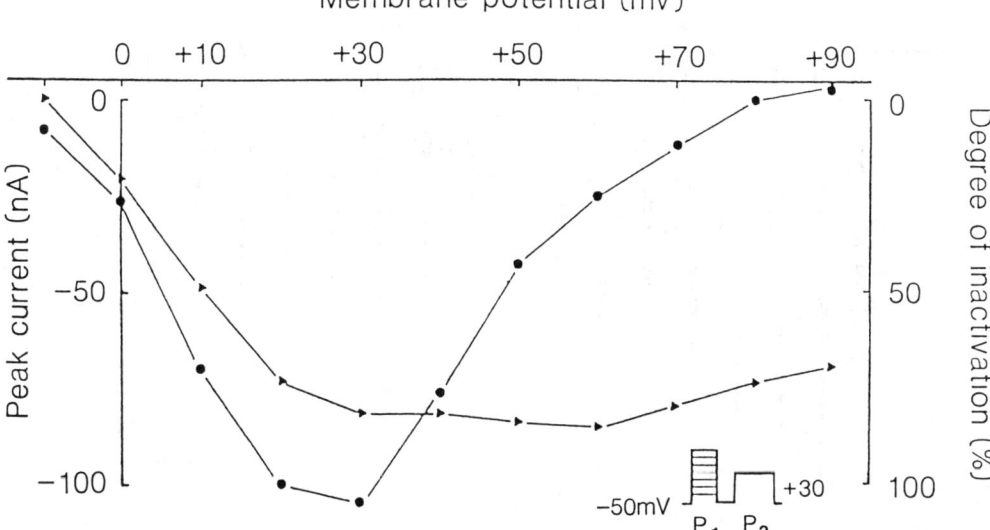

Figure 9. I-V relationship (dots) for peak calcium currents (40 mM Ca^{2+}). Activating pulses (400 ms) were from V_h −50 mV. Degree of inactivation as a function of prepulse amplitude (triangles). Prepulses (P_1, 60 ms) to various potentials were followed by test pulses (P_2) to −30 mV. Between the pulses the membrane was repolarized for 10 ms to −50 mV (see inset). Peak currents recorded during the test pulses (I_2) were normalized to peak current of an unconditioned test pulse (I_1) and plotted as (1-I_2/I_1) against prepulse voltage. 30°C.

the relationship between the internal Ca^{2+} concentration and the time course of HVA channel inactivation. The relationship between free intracellular Ca^{2+} concentration and inactivation can be determined by introducing Ca^{2+} buffers such as EGTA, HEDTA, and BAPTA intracellularly. As shown in Figure 10b, injections of Ca^{2+} buffers yielding free levels of Ca^{2+} in the nanomolar range slow down inactivation most likely by reducing the fast phase of calcium current decay. Inactivation is also less complete even with longer depolarizations.

Fitting double exponentials to the decay of the current reveals that the time constant of the fast component decreases slightly with decreasing free [Ca^{2+}]$_i$, while the slow time constant is not affected (Figure 11a). A dramatic effect, however, is observed in the relative amplitudes of the two components of the total current. The relative contribution of the two components (fast/slow) decreases markedly with decreasing [Ca^{2+}]$_i$ below 1×10^{-7} M (Figure 11b). At 1×10^{-9} M the relative contribution of the fast phase is almost zero.

Recovery from inactivation is also dependent on the free internal Ca^{2+} concentration (Yatani et al., 1983; Gutnick et al., 1989). As illustrated in Figure 12a the ability of hyperpolarizing pulses to restore inactivated calcium currents is strongly enhanced in cells that had been massively injected with Ca^{2+} buffers to keep free [Ca^{2+}]$_i$ low. Under these conditions, the speed and the effectiveness of repriming is considerably increased (Figure 12a). A plot of the maximal amplitude of reprimed currents vs. [Ca^{2+}]$_i$ shows that the effectiveness of the hyperpolarizing interval increases steeply when [Ca^{2+}]$_i$ is less than 1×10^{-7} M.

From the above it is obvious that HVA channel inactivation does not depend simply on the accumulation of Ca^{2+} near the inner membrane surface but is rather influenced by preexisting intracellular levels of Ca^{2+}. These findings are in line with observations on single calcium channels in the same preparation. Single calcium current recordings have shown that inactivation of whole cell calcium currents is associated with a decrease in the probability of channel opening and that inactivation of single channel currents is only weakly related to prior Ca^{2+} entry through the channels (Lux and Brown, 1984a).

a

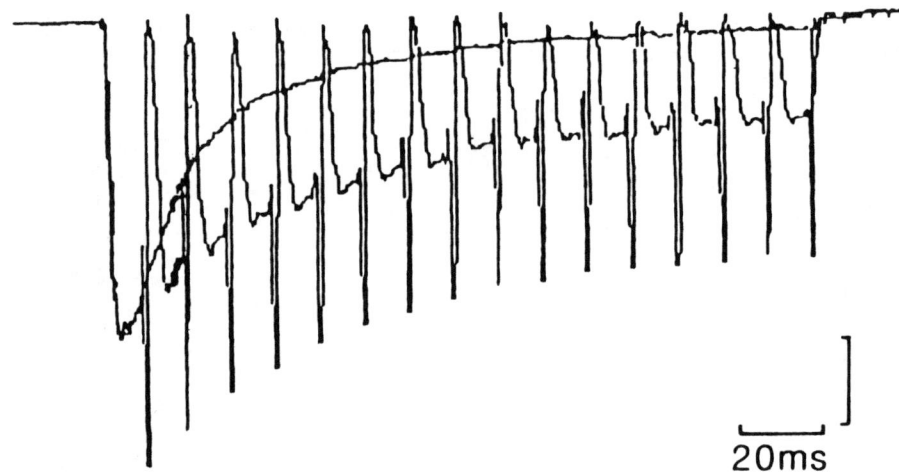

b

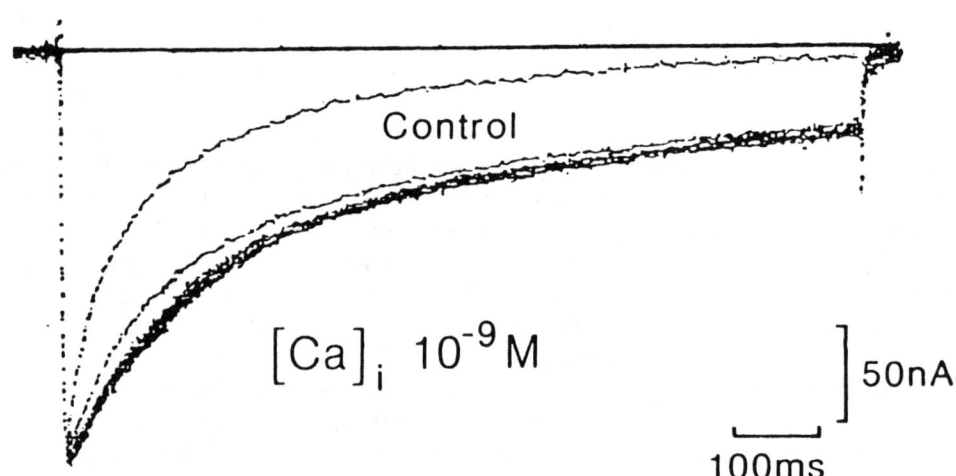

Figure 10. (a) Recovery from HVA channel inactivation is voltage dependent. Current induced by a sustained depolarizing step from -50 to $+30$ mV was superimposed on the current generated by a depolarizing step that was frequently interrupted by 4 ms intervals of hyperpolarization to -110 mV. 18°C. (b) Effect of internal calcium buffering on inactivation. The cell was injected with EGTA-Ca^{2+} buffer to fix internal Ca concentration at the level indicated. Inactivation was slowed with subsequent injections. 20°C. (From Gutnick et al., 1989. With permission.)

2. A Model for the HVA Channel Inactivation

The complex kinetics of calcium channel activation can be modeled by assuming that inactivation starts from an active state (A) reaching two inactivated states (I_V and I_{Ca}), as illustrated below:

$$\begin{array}{c} A \\ \swarrow \quad \searrow \\ I_V \leftrightarrow I_{Ca} \end{array} \qquad (3)$$

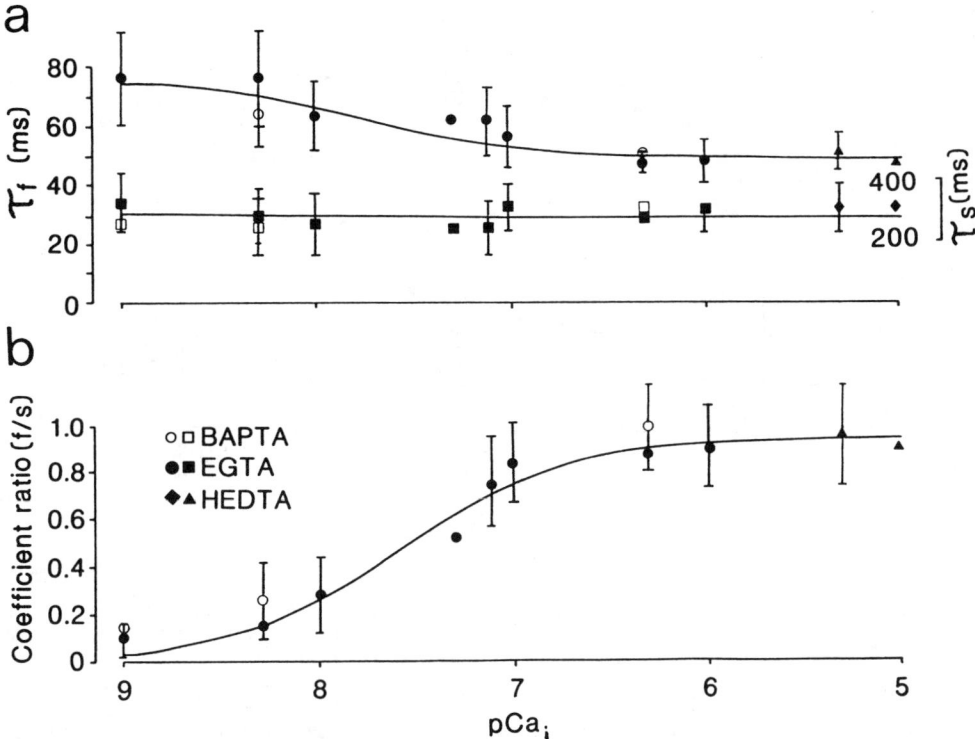

Figure 11. (a) Relationship between fast (τ_f, upper curve) and slow (τ_s, lower curve) time constant of inactivation and internal Ca^{2+} concentration. (b) The ratio (f/s) of the amplitudes of the fast (f) and slow (s) component of inactivation is plotted as a function of the negative logarithm of $Ca^{2+}{}_i$ (pCa_i). Internal Ca^{2+} chelators were as indicated. Vertical bars indicate standard deviations from the mean (n = 4). Continuous lines represent the predictions obtained from computer simulations using the three state model described in the text. (From Gutnick et al., 1989. With permission.)

In Scheme 3 the active state (A) includes open and closed states of the channel that are lumped together because of their rapid equilibrium reactions. Transitions leading to state I_{Ca} (A → I_{Ca} or I_V → I_{Ca}) are assumed to be Ca^{2+} and voltage dependent while the other transitions are simply voltage dependent. Channels in state A are at rest and relax to states I_{Ca} and I_V depending on voltage and on $[Ca^{2+}]_i$. The probability that a channel reaches state I_{Ca} is determined by the Ca^{2+} bound to a receptor site at or near the channel according to a 1:1 equilibrium reaction. Thus, probability of occupying state I_{Ca} increases with $[Ca^{2+}]_i$ while occupation of state I_V is favored by low $[Ca^{2+}]_i$ and prolonged depolarizations. Appropriate values for the six rate constants allow one to fit the time constants and amplitude ratio of the fast and slow inactivating components vs. $Ca^{2+}{}_i$ (Figure 11). Besides that, the model also accounts for the voltage and Ca^{2+} dependence of recovery from inactivation if transitions $I_V \rightleftharpoons A$ and $I_V \rightleftharpoons I_{Ca}$ are assumed to be steeply voltage-dependent at very negative membrane potentials. In low $[Ca^{2+}]_i$, occupation of state I_V is favored and repriming is faster than in high $[Ca^{2+}]_i$, where I_{Ca} predominates and channel recovery develops more slowly (Figure 12b).

While Scheme 3 simplifies the complex behavior of HVA channel gating, it accounts fairly well for the biphasic inactivation time course, the Ca^{2+} dependence of the inactivation rate, and the voltage and the Ca^{2+} dependence of repriming that play an important role in the regulation of firing patterns in bursting neurons (see below). With the appropriate modifications, the model may also account for similar complex mechanisms of inactivation described in heart cells (Kass and Sanguinetti, 1984; Lee et al., 1985; Hadley and Hume,

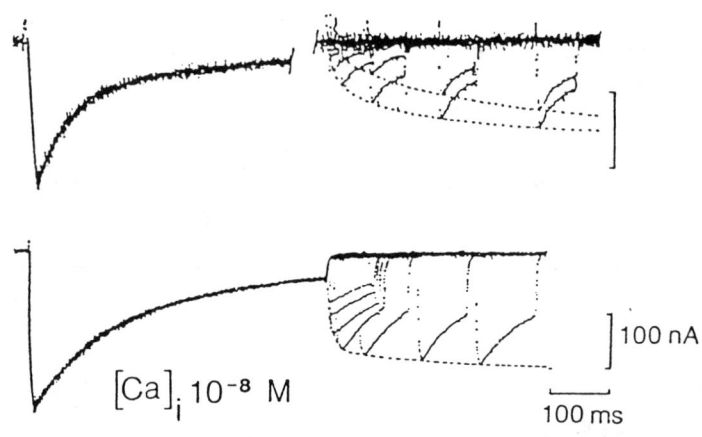
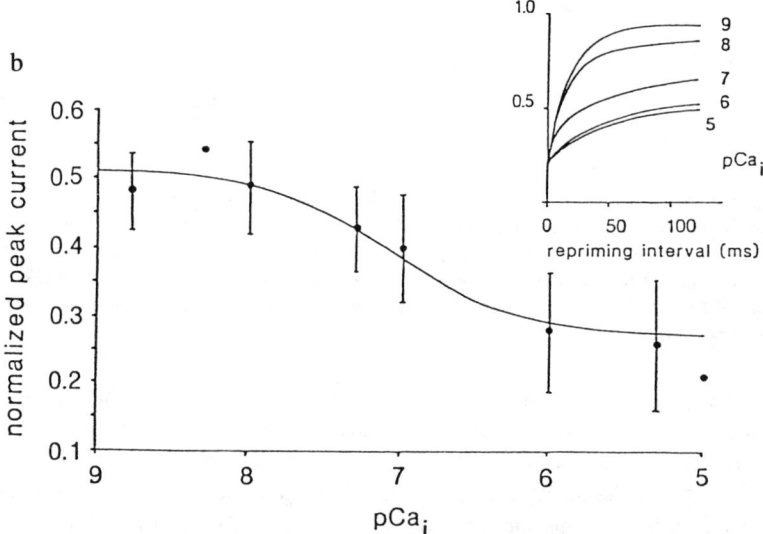

Figure 12. (a) Recovery from inactivation depends on the internal Ca^{2+} concentration. Upper traces: currents recorded during 1 s prepulses (shown for the first 0.5 s) followed by test pulses to +30 mV after repolarizing pulses of various duration were superimposed. Dashed lines indicate the time course of recovery from inactivation for interpulses to −50 mV (upper curve) and −130 mV (lower curve). Lower trace: internal Ca^{2+} concentration was set to the level indicated by injecting EGTA-Ca^{2+} buffer into the neuron. Interpulses were to −130 mV. Note that in this case recovery was faster and that the interpulses caused more effective recovery of HVA channels; 20°C. (b) Relationship between internal Ca^{2+} concentration and degree of recovery from inactivation. Activating pulses from −50 to +30 mV were separated by 20 ms interpulses to −130 mV. Peak currents during the test pulse were normalized to peak currents during the prepulse and plotted against pCa_i. Vertical bars indicate standard deviation from the mean (n = 4). Continuous line illustrates predictions of the model (see also Figure 11). Inset shows time course of the recovery process as modeled for various internal Ca^{2+} concentrations. (From Gutnick et al., 1989. With permission.)

1987; Argibay et al., 1988; Campbell et al., 1988a), vertebrate and mollusk neurons (Brown et al., 1981; Yatani et al., 1983; Akaike et al., 1988; Kasai and Aosaki, 1988; Jones and Marks, 1989), and secretory cells (Satin and Cook, 1989). All these reports agree on inactivation being jointly dependent on Ca^{2+} and voltage although the exact mechanisms by which Ca^{2+} exerts its action remain to be elucidated.

3. Ca^{2+} Buffering and Inactivation Gating

To gain more insight into the mechanisms of HVA channel inactivation, Gutnick et al. (1989) have injected large amounts of the Ca^{2+} buffer EGTA (up to 100 mM) into molluscan neurons to properly control internal Ca^{2+} levels. They have also applied BAPTA, a Ca^{2+} chelator that is far more effective than EGTA in buffering fast changes in the intracellular Ca^{2+} concentration (for a comparison see Neher, 1986). Under these conditions, increases in the internal Ca^{2+} concentration due to Ca^{2+} influx through channels are restricted to the small volume surrounding the inner mouth of a single channel. Since chelation of the Ca^{2+} ions by the buffer is fast (about 1 µs) compared to the time course of inactivation and even channel opening and closing kinetics (see Lux and Brown, 1984b), the spatial concentration profile that develops during Ca^{2+} entry in the area surrounding a channel is stationary and, following channel closure, the Ca^{2+} concentration returns quickly to the level set by the buffer. Even in areas of high channel density with up to 10 channels per μm^2 (Lux and Brown, 1984b) a spatial accumulation of Ca^{2+} should not occur. As can be inferred from Figure 1 in Gutnick et al. (1989), Ca^{2+} buffering to the levels set by the BAPTA$-Ca^{2+}$ buffers is already reached 10 to 20 nm away from the inner channel mouth, while EGTA-Ca^{2+} buffers become effective only at distances about 4 times longer. Inactivation, however, does not significantly differ in either BAPTA or EGTA (see Figure 11). These findings strongly suggest that the regulatory site for Ca^{2+}-dependent inactivation is in an area where the internal Ca^{2+} concentration is under the control of the buffer, i.e., in a location that can be reached by Ca^{2+} only by diffusion over a distance of about 100 nm or more.

The assumption that the regulatory site is at the channel mouth where the Ca^{2+} concentration changes drastically during Ca^{2+} influx (see Figure 1, Gutnick et al., 1989) is difficult to reconcile with the high sensitivity of inactivation kinetics to Ca^{2+} concentration (see Figure 11). In this case, one would expect that even brief openings of the channel would lead to maximal inactivation. This, however, cannot be observed in single channel recordings (see Lux and Brown, 1984a).

4. Functional Implications of Voltage- and Ca^{2+}-Dependent Inactivation

Two types of processes are known to limit Ca^{2+} entry in cells under physiological conditions. These types are the activation of an outward current, generally carried by K^+ ions, and the inactivation of voltage-gated calcium channels. Since inactivation is sensitive to internal Ca^{2+} concentration, Ca^{2+} itself controls further Ca^{2+} entry via a negative feedback loop. If intracellular Ca^{2+} accumulation were the only cause for inactivation, high internal Ca^{2+} levels would lead to complete calcium channel inactivation and recovery from inactivation would be largely determined by internal Ca^{2+} buffering and sequestration. In repetitively firing neurons, such as bursting pacemaker cells, Ca^{2+} accumulation during even a short train of action potentials leads to a considerable increase in the intracellular Ca^{2+} concentration that can reach values one order of magnitude higher than resting levels (Gorman and Thomas, 1978). Removal of internal Ca^{2+} is slow on the time scale of a single action potential and thus calcium channels would soon become unavailable for activation. A solution to this problem is given by the joint regulation of calcium channel inactivation by internal Ca^{2+} and membrane potential. As demonstrated for inactivation of the calcium channel in molluscan neurons, even short duration hyperpolarizations, as occur during action potential repolarization, can significantly remove inactivation despite considerable intracellular Ca^{2+} accumulation. Thus, even under conditions of relatively high sustained intracellular Ca^{2+} levels, voltage-dependent calcium channels will become readily available for activation.

5. Single High-Threshold Calcium Channels

Single HVA calcium channels were first resolved in cell attached patches of neuronal (Lux and Nagy, 1981; Brown et al., 1982; Hagiwara and Ohmori, 1982), secretory (Fenwick et al., 1982), and cardiac cells (Reuter et al., 1982; Cavaliè et al., 1983). These early reports suggested the existence of a single calcium channel type with activation-inactivation kinetics and sensitivity to Ca^{2+} and Ba^{2+} similar to that of the macroscopic HVA calcium currents previously described. Microscopic inactivation of this channel was never complete, even after prolonged depolarizations, and showed fast and slow phases that varied from cell to cell (Cavaliè et al., 1983; Lux and Brown, 1984b). There are also indications of a large single channel conductance in high $[Ca^{2+}]_i$ (9 to 12 pS in 50 mM Ca^{2+}) that nearly doubled in isotonic internal Ba^{2+} (20 to 25 pS in 110 mM Ba^{2+}). The channel was blocked by inorganic calcium channel-blockers (Cd^{2+}, Ni^{2+}, La^{2+}, etc.) or, as in cardiac tissues, effectively activated by the 1,4-dihydropyridine Bay K 8644 (Hess et al., 1984; Brown et al., 1984b) and modulated by β-adrenergic stimulation (Reuter, 1983; Bean et al., 1984).

Subsequent reports in vertebrate sensory neurons indicated the existence of a second type of HVA channel (N-type) that differed from the L type mainly by its greater sensitivity to holding potentials, lower single channel conductance, and faster and strictly voltage-dependent inactivation kinetics (Nowycky et al., 1985a; 1985b). There is, however, a poor correlation between single channel and macroscopic calcium current recordings (Fox et al., 1987a; 1987b) that impeded a general acceptance of L and N type as distinct Ca^{2+} channel types (Carbone and Lux, 1987a; Swandulla and Armstrong; 1988). In addition, the N channel proved to be pharmacologically inseparable from the L-type channel. The two channels showed equal sensitivity to ω-conotoxin (ω-CgTx) (McCleskey et al., 1987b; but see below) and Cd^{2+} (Fox et al., 1987a), and displayed a common resistance to LVA channel blockers (Carbone et al., 1987b). Both channels also proved to be highly unstable in excised membrane patches (Carbone and Lux, 1987b; Fox et al., 1987b). In addition, recent studies have shown that the inactivation kinetics of the N channel may change greatly from neuron to neuron leading to the proposal that its inactivation may be Ca^{2+} and voltage dependent (Hirning et al., 1988). As a consequence, in some neurons N-channel inactivation is not at all complete and is hardly distinguishable from that of the L-type (DHP-sensitive) channel (Kongsamut et al., 1989). Other reports argue against significantly different single channel conductances (Plummer et al., 1989) and in favor of similar sensitivities to holding potentials between the two HVA channels (Aosaki and Kasai, 1989; Usowicz et al., 1989; Carbone et al., 1990b).

Although there are still contrasting results at the single channel level, due to the heterogeneity and diversity of neuronal HVA channels, recent findings seem to agree on some important points:

1. ω-CgTx identifies a class of HVA channel that gives rise to a slowly and incompletely inactivating macroscopic current (ω-CgTx-sensitive, N-type?).
2. Residual ω-CgTX-resistant HVA channels show a high sensitivity to DHP. Channels that respond to DHP (DHP-sensitive) are clearly ω-CgTX-resistant but the contrary may not be true.
3. ω-CgTX-sensitive (N-type?) channels are probably more sensitive to holding potentials than DHP-sensitive (L-type) channels. However, they can be partly activated even at very low holding potentials (-30 mV, see Plummer et al., 1989).
4. In most peripheral neurons of vertebrates, the contribution of ω-CgTx-sensitive

channels predominates over that of DHP-sensitive channels. The same may not be true for central neurons (Sah et al., 1989).

D. Do Inactivation Kinetics Distinguish Between Different Types of HVA Calcium Channels?

Consistent with the above arguments, various groups have adopted a more pharmacological approach to identifying HVA calcium channels on the basis of their sensitivity to ω-conotoxin (ω-CgTx) and DHP derivatives (Kasai et al., 1987; Sher et al., 1988; Aosaki and Kasai, 1989, Plummer et al., 1989; Carbone et al., 1990a). A rationale for this comes from the following observations. First, saturating concentrations of ω-CgTX (3 to 6 μM) only partly depress the HVA calcium currents in a variety of neurons. In sympathetic neurons (Plummer et al., 1989; Jones and Marks, 1989), sensory ganglia (Aosaki and Kasai, 1989; Carbone et al., 1990a), and human neuroblastoma (Carbone et al., 1990b), ω-CgTx blocks 70 to 90% of the total current (Figure 13), while in central neurons the toxin seems to spare a larger fraction of the current (50 to 60%) (Sah et al., 1989). Second, DHP derivatives can either block or modulate the HVA calcium currents. Their action, however, is more marked in ω-CgTx preincubated neurons. Particularly evident is the activating action of the 1,4-dihydropyridine, Bay K 8644. This DHP causes a maximal two- to three-fold increase of HVA current amplitudes as well as a drastic prolongation of HVA channel deactivation in ω-CgTx-treated neurons but more reduced effects in normal cells (Figure 13b) (Carbone et al., 1990b).

These findings suggest some interesting conclusions. First, the ω-CgTx-sensitive component contributes to most of the HVA current in peripheral neurons. In rat sympathetic neurons, chick sensory ganglia, and human neuroblastoma, the block of HVA currents is complete in 50% of the cells while in the remaining half the toxin blocks 90% of the currents. Thus, ω-CgTx-sensitive currents predominate in peripheral neurons. This supports recent observations on single exponential deactivation kinetics of HVA calcium currents in chick DRG that suggest a predominance of one type of HVA channel that is insensitive to DHP (Swandulla and Armstrong, 1988). Second, the inactivation time course of ω-CgTx-sensitive currents is slow and incomplete. Thus, the association of these currents with N-type calcium channels as defined by Nowycky et al., 1985a (with fast, voltage-dependent inactivation) would require a basic redefinition of the N-channel inactivation kinetics. This process will not be necessarily fast, but slow and incomplete for pulses of 200 ms duration. The rate of inactivation would be dependent on voltage, $[Ca^{2+}]_i$, or other factors rather than being strictly voltage dependent. Third, HVA channel inactivation is probably the most unreliable parameter for identifying HVA channel subtypes. The inactivation time course of HVA currents in Figure 8 would hardly suggest the existence of multiple HVA current components. They did, however, show different pharmacological sensitivities to ω-CgTX and DHP derivatives. These data suggest that association of fast or slow inactivating current components with ω-CgTx- or DHP-sensitive HVA currents should be made with caution, if there is no control made available by pharmacological assays. This is crucial for a number of observations concerning the role and function of HVA channels in neurons (see Dunlap et al., 1987; Bean, 1989a; Dolphin and Scott, 1989) and to the relative contribution of HVA channel subtypes to the release of peptidergic neurotransmitters (Miller, 1987).

IV. MODULATION OF CALCIUM CHANNEL INACTIVATION

As is the case with other ion channels, HVA channel inactivation can be modulated by various compounds. Most of them act intracellularly, but some seem to act preferentially from the extracellular side of the channel, suggesting that the inactivation gate spans the

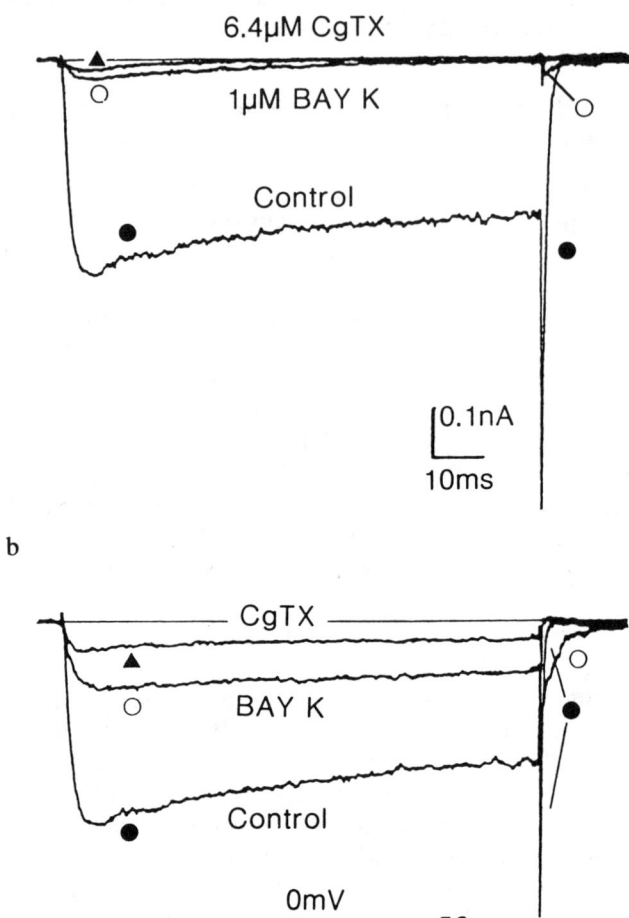

Figure 13. (a,b) Time course of ω-CgTx- and DHP-sensitive HVA Ba^{2+} currents from two IMR32 differentiated cells. The current traces were recorded before (●), after 2 min bath application of 6.4 μM ω-CgTx (▲) and in the presence of 1 μM Bay K 8644 (○) in 10 mM external Ba^{2+}. Notice the fast deactivation kinetics at control (●) and the prolongation of the tail with 1 μM Bay K 8644 (○). (Modified from Carbone et al., 1990b. With permission.)

plasma membrane and probably is a substantial part of the channel protein. Surprisingly enough, there are no reports of endogenous modulatory effects on the fast LVA channel inactivation. In the following sections we will review the most well-documented modulatory actions on HVA channel inactivation.

A. HVA Channel Inactivation Induced by External Menthol and Bay K 8644

Recent studies have shown that menthol (a cyclic alcohol derived from peppermint oil; Figure 14a) exerts a specific action on low- and high-threshold calcium channels in vertebrate neurons (Swandulla et al., 1986; 1987). At 0.5 mM, menthol increases three- to six-fold the rate of HVA channel inactivation and depresses by about 70% the size of LVA currents

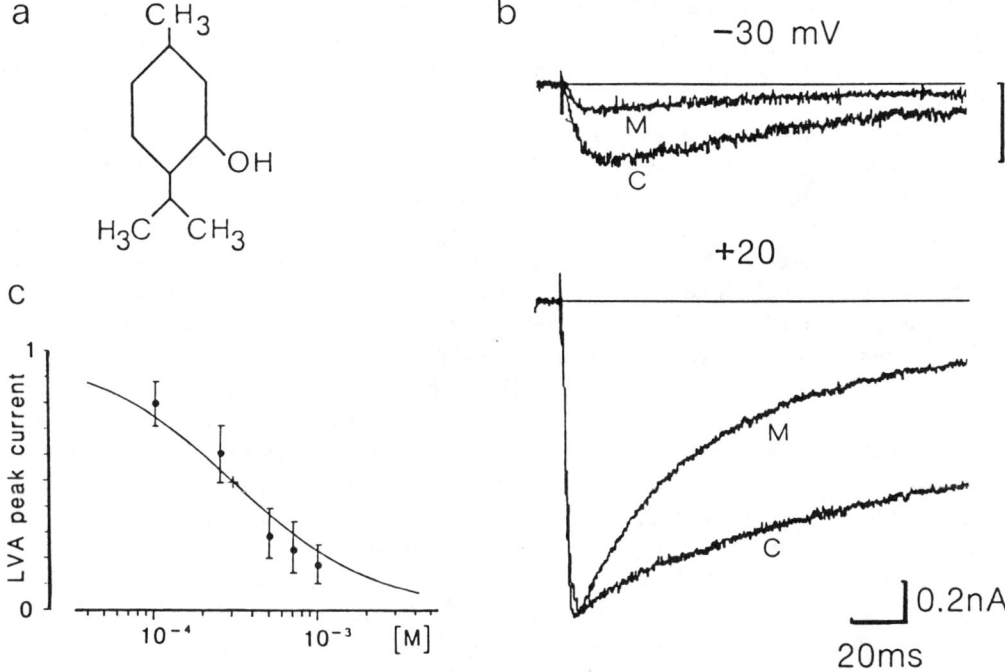

Figure 14. Action of menthol on calcium currents in a chick DRG cell. (a) Chemical structure of menthol. (b) LVA and HVA currents recorded before (C) and during application (M) of 0.5 mM menthol at the potentials indicated. The currents recovered completely after five minutes of washing. V_h −80 mV. The bath contained (mM): 120 NaCl, 20 CaCl$_2$, 2 MgCl$_2$, 10 Na-HEPES (pH 7.3) plus 0.3 μM TTX. (c) Dose-response curve of normalized peak LVA current vs. menthol concentration. The continuous curve was drawn assuming that menthol interacts with a receptor site following a first order chemical reaction. (Modified from Swandulla et al., 1987. With permission.)

with little change in their time course (Figure 14b). The action of menthol on LVA and HVA currents is dose dependent, fully reversible, and independent of voltage (Figure 14c) (Swandulla et al., 1987).

Menthol action on HVA current inactivation has some interesting features. Menthol accelerates and enhances HVA current inactivation (Figure 14b, bottom). The time constant of inactivation decreases from 120 to 40 ms and becomes rather insensitive to calcium current amplitude, current carrying ion, and internal buffering (Swandulla et al., 1986, 1987). As previously shown (Swandulla et al., 1986), menthol facilitates the inactivation gating of HVA channels, thereby most likely suppressing the inactivating action of intracellular Ca^{2+}. On the other hand, preliminary experiments have shown that the action of menthol is strongly reduced when Na^+ ions flow through open HVA channels (Carbone, Swandulla and Lux, unpublished observations). Taken together, these findings suggest that the action of menthol may depend on the permeability state of the channel being much less pronounced when the channel is in the Na^+-mode as compared to the Ca^{2+}-mode (for details of channel modes see Lux et al., 1989).

In all cases, the action of menthol seems unrelated to the presence of the fast inactivating current observed in neurons (N-type) and could not be due to a transformation of slowly inactivating (L-type) into rapidly inactivating (N-type) HVA channels. This was concluded from the observation that: (1) menthol also accelerates HVA current inactivation in cells with weakly inactivating calcium currents; (2) the action of menthol on calcium current inactivation was identical if the holding potential was changed from −80 to −50 mV; and (3) in menthol-treated cells (1 mM), HVA currents showed no sign of voltage-dependent inactivation positive to +10 mV (see Figure 6 and 7 in Swandulla et al., 1987).

Since in chick DRG neurons the majority of high-threshold calcium currents are carried through ω-CgTx sensitive channels it is most likely that menthol mainly accelerates the inactivation process of these channels without affecting their activation and deactivation kinetics.

The Ca^{2+} agonist Bay K 8644 has also been shown to be an effective modulator of HVA channel inactivation. In heart cells, Bay K 8644 preferentially acts by increasing the size of high-threshold calcium currents, prolonging their deactivation kinetics, and accelerating their inactivation time course (Hess et al., 1984; Brown et al., 1984; Sanguinetti et al., 1986; Markwardt and Nilius, 1988; Lacerda and Brown, 1989). In neurons, the action of Bay K 8644 is more variable. In some cases, 1 μM Bay K 8644 shows only slight or no activating action on HVA calcium channels (Scott and Dolphin, 1988; Jones and Marks, 1989). In other cases, Bay K 8644 acts as an antagonist of HVA calcium currents (Boll and Lux, 1985). These ambiguities very likely reflect the heterogeneity of neuronal HVA channels and may be overcome by assuming that neuronal DHP-sensitive channels contribute to only a minor fraction of the total HVA current (see above). Recent studies in peripheral neurons have shown that Bay K 8644 has a marked activating action when the majority (70 to 90%) of HVA channels are persistently blocked by ω-CgTx (Plummer et al., 1989; Aosaki and Kasai, 1989; Carbone et al., 1990b). Particularly, in adult rat DRG with HVA currents strongly depressed by ω-CgTx, the activating action of Bay K 8644 is accompanied by a sharp acceleration of the HVA channel inactivation. Like menthol, the inactivation rate increases nearly three-fold in the presence of Bay K 8644 and is apparently independent of membrane potential and current size (Carbone et al., 1990a). Unlike menthol, however, the Bay K-induced inactivation causes a marked current amplitude increase and tail current prolongation. This suggests a combined action of the dihydropyridine on the activation and inactivation processes of the same channel that may derive from a prolongation of the mean open time of the channel in the presence of the Ca^{2+} channel agonist. It is also likely that the two compounds act preferentially on different HVA channel types: Bay K 8644 being more selective for cardiac and neuronal DHP-sensitive (L-type) HVA channels and menthol acting on neuronal ω-CgTx-sensitive channels.

B. Enzymatic Modulation of Inactivation of Calcium Channels

A considerable body of evidence has been accumulated over the last years showing that calcium channels can be modulated by internal enzymatic reactions (for a recent review see Hartzell, 1988). Early evidence for this came from the observation that, in dialyzed cells and excised membrane patches, HVA calcium channels rapidly disappear (Kostyuk et al., 1981; Byerly and Hagiwara, 1982; Fenwick et al., 1982; Cavalié et al., 1983; Carbone and Lux, 1984a; Nilius et al., 1985) and that this "wash-out" could be slowed or partially prevented by agents that promote enzymatic phosphorylation such as cAMP, ATP, Mg^{2+}, and/or the catalytic subunit of cAMP-dependent protein kinase (Doroshenko et al., 1982; Reuter, 1983; Trautwein and Pelzer, 1985; Armstrong and Eckert, 1987). In cardiac cells it was demonstrated that β-adrenergic stimulation induces phosphorylation of calcium channels via cAMP (Reuter, 1974; Kameyama et al., 1985, 1986). This leads to an increase in the number of channels that can be activated by voltage (Cachelin et al., 1983; Bean et al., 1984; Brum et al., 1984). Recently it has been shown that the rate of calcium current inactivation decreases when the intracellular cAMP production is stimulated maximally by forskolin in GH3 cells (Armstrong and Kalman, 1988). Thus phosphorylation of the calcium channel appears to be important for channel functioning in several tissues. It has been hypothesized that dephosphorylation of these channels may be the basic mechanism of inactivation of calcium channels regulated by cAMP-dependent phosphorylation (Eckert and Chad, 1984; Chad and Eckert, 1986; Chad, 1989). According to this hypothesis Ca^{2+} entering the cell during channel activity activates a protein phosphatase such as calcineurin that subsequently dephosphorylates calcium channels, a process that finally leads to channel inactivation. Rephosphorylation of the channel is then necessary to restore its ability to open (Armstrong and Eckert, 1987).

Recently, Hescheler and Trautwein (1988) have presented evidence that the high-threshold L-type calcium channel of guinea pig ventricular myocytes is strongly affected by internal application of the endopeptidase trypsin. Besides increasing the current amplitude by about three-fold, the enzyme was particularly effective in prolonging the time course of calcium channel inactivation. Prolongation of HVA channel inactivation by trypsin also occurred during induced β-adrenergic stimulation and could not be suppressed by effective blockers of the cAMP-dependent phosphorylation. Their results were interpreted as a direct effect of trypsin on either the voltage- or the Ca^{2+}-dependent inactivation of this channel that was independent of cAMP or the stimulation of a cAMP-dependent protein kinase.

C. Intracellular Agents Affecting HVA Channel Inactivation

N-methyl-D-glucamine (NMG) is known to be a good substitute for Cs^+ ions as a potassium channel blocker (Fernandez et al., 1984). For this reason NMG is often used to study pure calcium currents in a variety of cells. A recent report has shown that NMG may affect the kinetics and voltage dependence of calcium currents in guinea pig ventricular cells (Malécot et al., 1988), suggesting that some caution should be observed in the use of this compound as an ion channel blocker. The rationale for this relies on the observation that NMG increases the half-time of calcium current activation and prolongs the slow phase of inactivation, causing a -30 mV shift of τ_h at positive potentials. The removal of HVA channel inactivation is particularly evident when Ba^{2+} is the main current-carrying ion. Thus, besides blocking potassium channels, NMG seems to affect the gating properties of cardiac L-type calcium channels. However, as was suggested by these authors, other mechanisms of action, such as an increased intracellular Ca^{2+} buffering capability in the presence of NMG, cannot be ruled out.

The fast inactivating phase of HVA calcium currents in guinea pig atrial myocytes can also be prevented when the internal perfusate contains citric or dipicolinic acid (DPA) as Ca^{2+} chelators in place of EGTA (Bechem and Pott, 1985). The effects of these compounds are already visible soon after establishing the whole-cell recording conditions but become more prominent when the size of the current is reduced due to HVA channel "rundown". Inactivation of reduced HVA currents is nearly absent under these conditions even during prolonged depolarizations (1 s). These observations, however, seem to be limited to cardiac cells.

Hartzell and White (1989) have recently reported that changing free Mg^{2+} in isolated cardiac myocytes by internal perfusion has dramatic effects on HVA current inactivation. Increasing $[Mg^{2+}]_i$ causes a marked increase in the rate and extent of voltage-dependent inactivation of calcium channels, suggesting that internal Mg^{2+} can block cardiac HVA channels in a time- and voltage-dependent fashion and that this block can be partly responsible for the "voltage-dependent" inactivation.

V. SUMMARY AND CONCLUSIONS

The recent discovery of multiple types of voltage-activated calcium channels with distinct pharmacology has greatly enhanced the multidisciplinary approach to determining their role in Ca^{2+}-triggered events. As Ca^{2+} is involved in the regulation of a variety of cellular activities, the identification and characterization of calcium channels has become a prerequisite for understanding in detail the mechanisms underlying these processes. Calcium channel inactivation is of particular significance. Although important points still await clarification, many groups have contributed to highlight the basics of this phenomenon. Accepted features are the strictly voltage-dependent inactivation of the LVA calcium channel and the joint Ca^{2+} and voltage dependence of HVA channel inactivation. Molecular details, such as the proposal of a phosphorylation-dephosphorylation mechanism controlling the rate of HVA channel inactivation, require further experimental support and refinement to gain general acceptance. Most of the future progress will rely on the availability of new compounds that selectively down- or up-modulate the inactivation gating. Complications, however, can always be expected when trying to attribute the modulatory effects to either the Ca^{2+} or voltage-dependent sensitivity of channel inactivation. The same holds true for the usage of inactivation kinetics as parameters to distinguish subclasses of HVA calcium channels. This trend will certainly change as more pharmacological tools and molecular manipulations of calcium channels become available in the near future.

Since voltage-activated calcium channels form a broad family of membrane proteins with close activation kinetics, recovery from inactivation, and Ca^{2+}/Na^+ selectivity, it would not be surprising if strong analogies between their amino acid sequences and quaternary structures are soon discovered. Major molecular differences are expected between the high- and low-threshold calcium channels. Genetic mutations in this case should account for the different voltage range of activation, inactivation mechanisms, Ca^{2+}/Ba^{2+} selectivity, single channel conductance, and drug sensitivity. Minor differences are expected among the high-threshold calcium channels whose heterogeneity seems to reflect a different pharmacological sensitivity to Ca^{2+}-channel agonists and antagonists rather than diverse biophysical properties.

ACKNOWLEDGMENTS

We wish to thank Professor H. D. Lux for stimulating discussions. This work was partly supported by NATO (Grant No. 0576/87) and the Consiglio Nazionale delle Ricerche (Grant No. 88.00458.04).

Section II
The Physiology of Calcium Channels

Chapter 5

The Physiology of Calcium Channels: An Overview

L. Donald Partridge

The ground-breaking observations of Luigi Galvani in *De Viribus Electricitatis In Motu Musculari Commentarius* published at the end of the 18th century set the stage for the study of electricity in excitable cells. Each subsequent advance in the technology of measurement has advanced our understanding of membrane currents and voltages in excitable cells. Since the nervous system depends on the action of excitable cells, it is easy to overlook the fact that this system is not primarily an electrical device. With only a few exceptions, neither the input to nor the output from the nervous system is electrical. Input information is transduced into an electrical signal that, in turn, is used solely for internal signaling. The electrical signal must be transduced again in order to accomplish any one of a great number of nonelectrical output actions. Ca^{2+} ions are well suited to this transduction function for several reasons. Ca^{2+} is a highly diffusible ion capable of moving throughout the cytoplasm. Ca^{2+} ions bind readily to the large anions that are found either associated with the plasma membrane, within the cytoplasm, or within specialized intracellular organelles and, once bound, can have important steric effects on these molecules. Intracellular free Ca^{2+} ions are vigorously controlled so that repeated rapid changes in concentration are possible. Thus calcium channels, by controlling Ca^{2+} flux into the cell, play a crucial role in this transduction process. It is not an entirely unreasonable simplification to state that the role of the electrical processes in the nervous system is to regulate membrane Ca^{2+} flux.

A multitude of cellular processes, summarized in Figure 1, work in concert to regulate the intracellular free, physiologically active, Ca^{2+} concentration at its extremely low level of 10 to 300 nm. Ca^{2+} is transported outward across the cell membrane against its electrochemical gradient either in exchange for Na^+ movement down its gradient or directly by an energy requiring pump. Ca^{2+} is pumped into mitochondria in exchange for H^+ ions as part of the electron transport process. Specialized structures, such as the sarcoplasmic reticulum in striated muscle and other intracellular membrane-enclosed organelles, concentrate Ca^{2+} and thereby remove it from the pool of free cytoplasmic Ca^{2+}. Finally, proteins and other cytoplasmic buffering systems, as well as plasma membranes, bind much of the remaining free intracellular Ca^{2+}. Because of the low levels brought about by these mech-

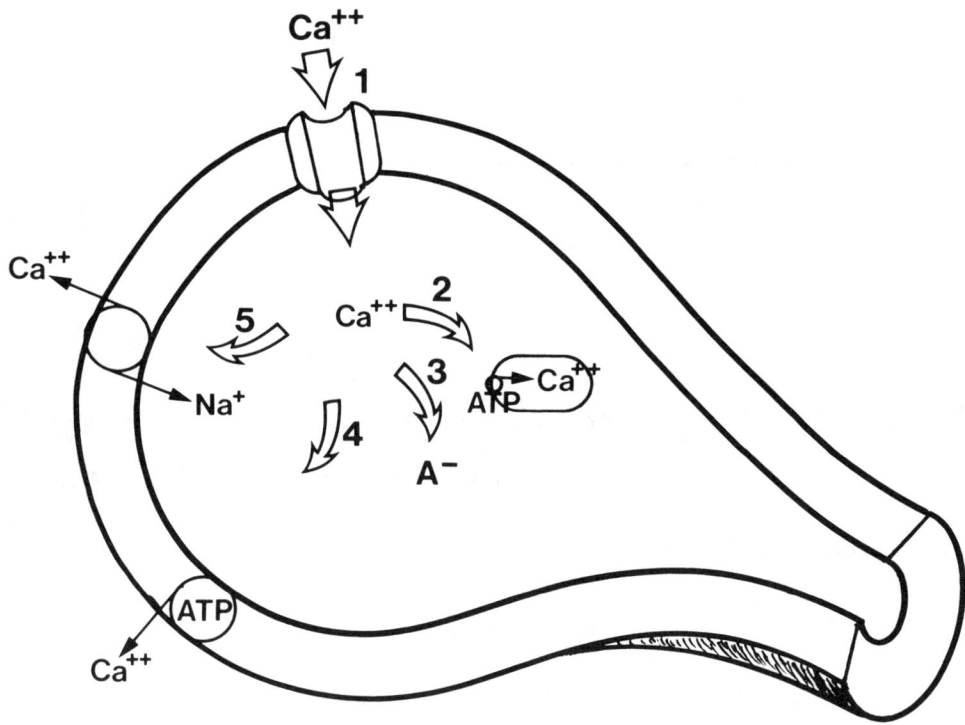

Figure 1. Regulation of intracellular Ca^{2+}. (1) Ca^{2+} ions enter through gated calcium channels and create a concentration gradient from the channel into the cytoplasm. Cytoplasmic Ca^{2+} is (2) pumped into mitochondria in exchange for H^+ and into other organelles by means of ATP-requiring pumps, (3) buffered by cytoplasmic proteins and by the plasma membrane, (4) pumped across the plasma membrane by ATP-dependent pumps, and (5) co-transported in exchange for the movment of Na^+ into the cell. The result of the action of these mechanisms on intracellular Ca^{2+} is that free Ca^{2+} in the cytoplasm is maintained in the nanomolar range.

anisms, even a brief transmembrane flux of Ca^{2+}, following the opening of calcium channels, can cause a transient rise in the concentration of cytoplasmic free Ca^{2+} by perhaps 20 times.

The buffering, pumping, and sequestering mechanisms continue to operate while Ca^{2+} enters a cell. These mechanisms, in addition to the transient nature of the influx, have the result that the concentration increase is slowed as it rises and is subsequently terminated. Moreover, these same mechanisms are responsible for the restoration of Ca^{2+} concentration to its steady-state level once the influx is finished. Not only the temporal, but the spatial profile of Ca^{2+} concentration within the cell are determined by these processes. The concentration rises more rapidly and to a higher level just adjacent to the open calcium channels than it does further away (Smith and Augustine, 1988). Apparently calcium channels are preferentially located in those specific areas of cells where a local rise in Ca^{2+} concentration would be expected to be most effective. Alterations of the temporal and spatial qualities of the Ca^{2+} signal result whenever the cellular homeostatic mechanisms for Ca^{2+} are altered.

It is not our object here to discuss mechanisms of cellular Ca^{2+} homeostasis. We are concerned, however, with the role that calcium channels play in transducing cellular events by affecting a change in intracellular Ca^{2+} concentration. Some of these events are summarized in Figure 2. Calcium channels are frequently associated with three important cellular processes in which electrical signals are transduced into a cellular action. These are contraction, secretion, and gating. The role of Ca^{2+} in each case is one of coupling the electrical event to the cellular action. We will also consider in this section two other important Ca^{2+}-regulated processes, sensory transduction and cell growth.

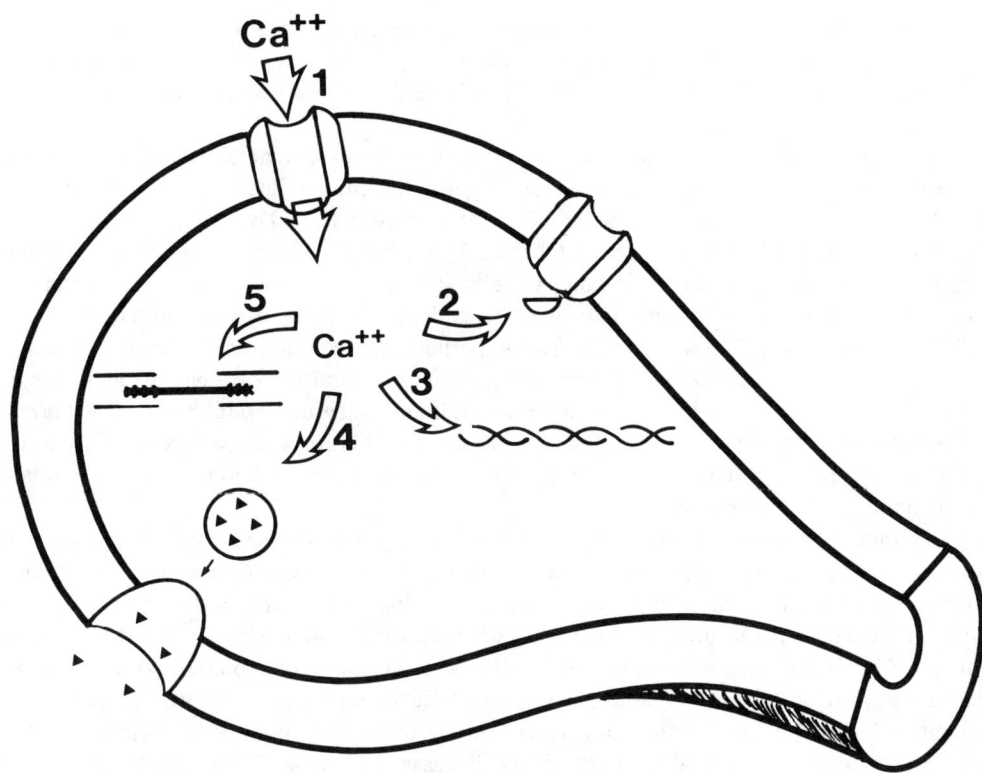

Figure 2. Role of Ca^{2+} in the regulation of cellular processes. (1) Ca^{2+} ions enter through gated calcium channels and transiently raise the concentration of cytoplasmic free Ca^{2+}. Depending on the cell type, this Ca^{2+} can (2) control the gating of ion channels, (3) activate or regulate protein synthesis or enzyme action, (4) initiate the exocytotic release of secretory products, (5) initiate the events of cell motility. Ca^{2+} ions are probably the most important means available to the nervous system to transduce electrical events into effector actions.

An obvious and important example by which electrical events are transduced into a useful action is muscle contraction. The movement of joints by skeletal muscles is an obvious end result of such a process. Other equally important end results are the contraction of cardiac muscle that produces a pressure pulse with a consequent blood flow, and the contraction of visceral, vascular, and other smooth muscles that, with few exceptions, regulates movement of substances through hollow structures. In each case, the interaction of contractile proteins is coupled to a membrane signal by means of a change in Ca^{2+} concentration (Ebashi et al., 1969). As a matter of fact, calcium channels mediate a multitude of forms of cell motility. These processes are, in general, referred to as excitation-contraction coupling and will be considered in Chapter 6, Section II. Skeletal muscle possesses three types of calcium channels, two in the surface membrane and a third in the sarcoplasmic reticulum membrane. Since the vast majority of Ca^{2+} for excitation-contraction coupling in skeletal muscle is released from the sarcoplasmic reticulum, the calcium channels in the sarcoplasmic reticulum are of special interest. These channels are especially unique because their gates respond to a potential change, not across the sarcoplasmic reticulum membrane in which the channels reside, but across the adjacent cell surface membrane. Cardiac muscle also contains at least two types of calcium channels that exhibit voltage-sensitive gating and can be further modulated by transmitters. As in skeletal muscle, the gating mechanism of cardiac sarcoplasmic reticulum calcium channels is of interest. Two mechanisms, one involving a linkage between surface membrane sensors and sarcoplasmic reticulum channels, and a second that utilizes Ca^{2+} itself as a second messenger have been invoked to explain this process. Finally, smooth muscle presents more of a *terra incognita* regarding calcium channel mechanisms. Certainly Ca^{2+} influx across the surface membrane is more crucial. The membrane channels that are responsible for this influx fall into both the voltage-dependent and the receptor-operated varieties.

Another important cellular process that is regulated by electrical signals is the secretion of various agents from intracellular stores. Again, the intracellular Ca^{2+} concentration, regulated by calcium channels, is the coupler in this transduction. The process of excitation-secretion coupling will be addressed in Chapter 7, Section II. Excitation-secretion coupling is central to the function of a diverse range of cells. Two major sources of Ca^{2+} for excitation-secretion coupling are inositol trisphosphate-dependent release from intracellular stores and Ca^{2+} influx through calcium channels. The importance to secretion of Ca^{2+} influx through calcium channels is central to our understanding of neurotransmitter release (Katz and Miledi, 1965). Similar Ca^{2+}-dependent mechanisms of excitation-secretion coupling are now known to exist in such other diverse cells as adrenal medullary cells, pancreatic β cells, mast cells, lymphocytes, and parathyroid cells. In each of these examples, calcium channels regulate the secretory action of the cell.

Calcium channels have an important action in input transduction in neural sensory systems. A clear example of this is in the single cellular organism *paramecium*. Mechano-sensitive calcium channels in the anterior portion of this animal respond when the animal comes in contact with an obstacle. This contact causes the channels to open and initiate an influx of Ca^{2+} ions. As a result, the cell membrane of the animal depolarizes and ultimately causes the direction of ciliary beating to be reversed (Eckert and Brehm, 1979). More complex organisms have separate neurons for sensory and motor function but Ca^{2+} continues to play a central role in sensory transduction as will be discussed in Chapter 8, Section II. In sensory cells, Ca^{2+} acts as a second messenger between transduction molecules and ion channels (e.g., in taste receptor cells), it is involved in information processing within the sensory cells (e.g., in photoreceptors), and it is essential for synaptic transmission to second order sensory neurons. In photoreceptors, Ca^{2+} has a modulatory role in the cGMP-dependent change in membrane conductance that is initiated by light. Electrical tuning of auditory hair cells depends on the interaction of voltage-dependent calcium channels and Ca^{2+}-activated potassium channels.

A fourth important Ca^{2+}-dependent process is the coupling of the gating of various ion channels to the influx of Ca^{2+}. By this means, calcium channels modulate the function of other channels (Meech, 1974). This topic will be considered in Chapter 9, Section II. The influx of Ca^{2+} through calcium channels provides a depolarizing drive that is important in determining the shape of the action potential in many neurons and muscle cells. Intracellular Ca^{2+} has the important second messenger function of modulating potassium channels, non-selective cation channels, and calcium channels. Through this second messenger function of Ca^{2+}, calcium channels are important in determining the firing patterns of neurons.

The fifth role to be considered here is more general and is covered in Chapter 10, Section II. Some of the earliest studies of calcium currents were in egg cells (Hagiwara and Jaffe, 1979). Calcium channels in some eggs are responsible for the action potential-like response that occurs upon fertilization and in the subsequent activation of the egg cell. In other cells, calcium action potentials are common early in development only to be replaced by sodium action potentials later. The implication is that intracellular Ca^{2+} plays a distinct role in cells at certain stages of their development. Some examples of such a role are in synaptic development, in the control of neurite outgrowth, in regulation of transmitter phenotype, in regulation of transmitter sensitivity, in the establishment of neuromuscular interactions, and in determining the extent and strength of gap junction coupling.

The topics considered in this section have far-reaching implications in the interactions carried out by cells and organisms. Much of this work is quite recent and new examples of biological roles played by calcium channels are to be found in new issues of scientific journals. The objective of this section of the book is to provide a representative sample and demonstrate the importance of calcium channels to the successful operation of biological tissues.

Chapter 6

Calcium Channels in Excitation-Contraction Coupling

Philip M. Best, Wai-Meng Kwok, and Xiaoping Xu

TABLE OF CONTENTS

I.	Introduction	70
II.	Skeletal Muscle	71
	A. Two Types of Surface Membrane Calcium Currents	71
	B. Inward Calcium Current is not Necessary for Twitch Activation	72
	C. Calcium Channel Blockers can Prevent Muscle Contractures	72
	D. The Dihydropyridine Receptor in the T. Tubule Membrane is the Voltage Sensor for Excitation-Contraction Coupling	73
	E. The S.R. Calcium Channel has a Large Cytoplasmic Domain that is the Foot Structure Linking the T. Tubule and S.R. Membranes	74
	F. E-C Coupling in Skeletal Muscle	76
III.	Cardiac Muscle	77
	A. Two Types of Surface Membrane Calcium Channels in Cardiac Muscle	77
	B. Evidence that L-Type Calcium Current Rather than a Charge-Coupled Mechanism Gates S.R. Ca^{2+} Release in Cardiac Muscle	78
	C. Calcium- and Voltage-Dependent Inactivation of L-Type Calcium Current	79
	D. Calcium Channel Modulators Affect Contraction	80
	E. Role of Calcium Current in Replenishing S.R. Ca^{2+} Stores	80
	F. The Calcium Release Channel of the S.R. is Ca^{2+} Activated	80
	G. Ryanodine Binds to the Cardiac S.R. Calcium Channel	81
	H. Relative Contribution of S.R. Ca^{2+} Release to Contraction	81

IV. Vascular Smooth Muscle 81
    A. Surface Membrane Calcium Channels 82
        1. Voltage-Dependent Channels 82
        2. Receptor-Operated Channels 83
        3. Leak Channels 84
    B. Calcium Release from Intracellular Stores 84

V. Summary 84

I. INTRODUCTION

In smooth, skeletal and cardiac muscle, cytoplasmic calcium ions regulate contractile protein activation and tension generation. The cellular mechanisms that link surface membrane excitation with an increase in intracellular calcium and thus with the initiation of contraction are termed excitation-contraction coupling. Contractile activation in striated (skeletal and cardiac) muscle is initiated by membrane depolarization (electromechanical coupling). In smooth muscle contraction can be initiated by both electrical stimulation and by binding of agonists to receptor sites on the surface membrane (pharmacomechanical coupling). An intracellular organelle called the sarcoplasmic reticulum (s.r.) acts as a reservoir for Ca^{2+} ion within muscle cells (Endo, 1985; Fabiato, 1985; Somlyo, 1988). Following membrane excitation, Ca^{2+} is released from the s.r. and activates the contractile proteins. In some cases (i.e., frog ventricle and some smooth muscle) Ca^{2+} entry through surface membrane channels is sufficient to activate the contractile proteins directly without amplification by internal release.

A number of hypotheses have been suggested to account for the coupling of membrane excitation to the initiation of s.r. Ca^{2+} release. In general, the proposed mechanisms fall into two classes. The first of these, which we will refer to as charge coupling, assumes a direct structural linkage between voltage sensing proteins in the surface membrane and calcium channels in the s.r. This idea was first proposed by Chandler, Rakowski, and Schneider (1976b) based on work done with skeletal muscle. A second class of mechanisms involves the release of a second messenger following excitation of the plasma membrane. Both Ca^{2+} ion (calcium-induced Ca^{2+} release) and inositol 1,4,5-triphosphate (IP_3) have been proposed as second messengers in striated as well as smooth muscle.

In both striated and smooth muscle, specialized junctions exist between the plasmalemma and the s.r. Invaginations of the surface membrane, the transverse tubules (t. tubules), come into close proximity (about 15 nm) with the s.r. membrane in skeletal muscle. Electron dense structures called "feet" (Franzini-Armstrong, 1970) are seen in electron micrographs to bridge the gap between the two membrane systems. Similar bridging structures have been identified between junctional s.r. and the plasmalemma of both cardiac and smooth muscle. Thus a structural element that might link the surface membrane with the s.r. exists in many muscle cells. Despite similarities in structure, the physiological mechanisms coupling surface membrane excitation to s.r. calcium release seem to vary quite significantly between the different types of muscle. During pharmacomechanical coupling in vascular smooth muscle the evidence is compelling that IP_3 acts as a second messenger. The cellular processes underlying electromechanical coupling in skeletal, cardiac, and smooth muscle are not as well understood and involve different mechanisms in different types of muscle. Calcium-induced Ca^{2+} release seems to play a major role in activation of mammalian cardiac muscle. However, in skeletal muscle, a charge coupled mechanism involving a direct molecular linkage between the t. membrane and the s.r. has considerable support, although evidence implicating a role for IP_3 and perhaps Ca^{2+} in the coupling process also exists.

In the discussion to follow we will focus on the contributions of calcium channels to excitation-contraction coupling in skeletal, cardiac, and vascular smooth muscle. The importance of the s.r. calcium channel in supporting the rise in intracellular Ca^{2+} that accompanies contraction is obvious. The role of surface membrane calcium channels is not always as clear, but may include (depending on muscle type) contributing to surface membrane excitability, acting as the voltage sensor for surface membrane depolarization, acting as a source of trigger Ca^{2+} that activates s.r. Ca^{2+} release, acting as a source of activator Ca^{2+} that binds directly to the contractile proteins, and acting as a source of Ca^{2+} ions to replenish intracellular stores.

II. SKELETAL MUSCLE

There are at least three types of calcium channels in skeletal muscle including two types found in the surface membrane and one found in the s.r. The surface membrane calcium channels are associated with two distinct types of macroscopic currents, one that is rapidly activated and another that is slowly activated (see Avila-Sakar et al., 1986 for review). Calcium channel types are discussed in detail in Chapters 3 and 4 of Section I of this book.

A. Two Types of Surface Membrane Calcium Currents

Calcium influx during muscle activation has been measured by determining the rate of entry of ^{45}Ca (Bianchi and Shanes, 1959; Curtis, 1966). Following trains of action potentials or prolonged depolarization, the Ca^{2+} influx increases substantially compared to that at rest. Electrophysiological methods have confirmed that there is a movement of Ca^{2+} across the surface membrane in depolarized muscle fibers (Beaty and Stefani, 1976; Sanchez and Stefani, 1978; Almers and Palade, 1981). In skeletal muscle from frog, a slowly activating inward calcium current with a peak current time of hundreds of milliseconds has been identified. The activation kinetics of this current are much slower than the time course of a twitch in these muscles and roughly ten times slower than the activation of the L-type calcium currents in cardiac and smooth muscle that share similar pharmacology and voltage dependency. These slow currents are activated at approximately -40 mV (Stanfield, 1977; Sanchez and Stefani, 1978) which is near the mechanical threshold of muscle cells (Hodgkin and Horowicz, 1960). The channels through which the slowly activated currents flow are sensitive to block by dihydropyridines and D600 but are insensitive to blocking agents specific for potassium or sodium channels. A similar current has been identified in mammalian muscle (Donaldson and Beam, 1983).

Localization of the channels responsible for the slowly activating current was accomplished using fibers treated with glycerol shock (Nicola Siri, et al., 1980; Potreau and Raymond, 1980a,b). The glycerol treatment uncouples the t. tubules electrically from the surface membrane of the muscle fibers (Fujino et al., 1961). These detubulated fibers propagate action potentials with normal sodium and potassium currents but with calcium currents that are either greatly reduced or abolished. The amount of residual calcium current is linearly correlated with the fraction of tubular capacitance remaining (Nicola Siri et al., 1980). Thus the effect of detubulation on the slowly activating calcium currents in skeletal muscle suggests that the calcium channels are found predominantly in the t. tubules rather than on the surface membranes. A similar conclusion was reached from an investigation of the decline of calcium current that followed a depletion of Ca^{2+} in the t. tubular lumen (Almers et al., 1981).

A fast activating calcium channel has also been reported from frog muscle fibers (Cota and Stefani, 1986). It is a low threshold (-60 mV), rapidly activating current that is insensitive to dihydropyridine block much like the T-type calcium currents seen in other cell types. However, unlike other T-type currents it does not inactivate appreciably. The kinetics of this current are such that it should be activated during the action potential. Therefore, it could account for the extra ^{45}Ca entry measured during the twitch compared to the resting state.

B. Inward Calcium Current is not Necessary for Twitch Activation

One of the earliest mechanisms proposed to account for the initiation of contraction in skeletal muscle was that an influx of Ca^{2+} across the surface membrane acts as the primary activating signal for contraction (Sandow, 1952). Stimulation of s.r. Ca^{2+} release by an elevation in myoplasmic free Ca^{2+} (calcium-induced Ca^{2+} release) was first described in skeletal muscle fibers (Endo et al., 1970; Ford and Podolsky, 1970). However, a major physiological role for calcium-induced Ca^{2+} release in skeletal muscle seems unlikely based on the conditions necessary to produce the response (Endo, 1985; but see Rios and Pizarro, 1988). For instance, calcium-induced Ca^{2+} release is substantially inhibited at concentrations of Mg^{2+} ion that are considered physiological.

Calcium entry through the surface membrane of skeletal muscle is, in fact, not necessary for twitch activation. It has already been mentioned that the slow inward calcium current in frog muscle reaches its peak amplitude after hundreds of milliseconds. Since the mechanical transients initiated by action potentials in frog muscles last only tens of milliseconds, the slowly activating inward calcium current could not be a significant factor in excitation-contraction coupling or, for that matter, in the generation of the action potential in frog muscle. In contrast, the slow currents in mammalian fibers activate on a time scale similar to the mechanical response and thus may allow for significant Ca^{2+} entry during a twitch (Donaldson and Beam, 1983). However, skeletal muscle fibers are capable of sustained contractile activity in the absence of external Ca^{2+} indicating that Ca^{2+} influx via either fast or slow channels is not necessary to activate contraction (Armstrong et al., 1972; Luttgau and Spiecker, 1979; Cota and Stefani, 1981). This notion is supported by the observation that concentrations of the calcium channel blocking agent diltiazem that are sufficient to block the slow inward calcium current do not inhibit contraction (Gonzalez-Serratos et al., 1982). In addition, voltage clamp pulses to above the reversal potential for calcium (where calcium current should be outward) produce s.r. Ca^{2+} release (Brum et al., 1988a). Thus inward calcium currents are not necessary for the activation of contraction in skeletal muscle.

C. Calcium Channel Blockers can Prevent Muscle Contractures

Although Ca^{2+} influx via surface membrane calcium channels is not a prerequisite for activation of a muscle twitch, certain calcium channel blockers such as D600 and a variety of dihydropyridines (DHP) block contractures resulting from prolonged membrane depolarization (Eisenberg et al., 1983; Berwe et al., 1987; Caputo and Bolanos, 1987; Dulhunty and Gage, 1988; Fill and Best, 1989).

The tension response elicited by prolonged depolarization produced either by voltage clamp or by elevated extracellular potassium rises rapidly to a peak and then declines slowly (a few seconds) to baseline levels. As long as the fiber remains depolarized, a second contracture cannot be elicited. During this period of unresponsiveness, the fiber is said to be in the inactivated state. The kinetics of inactivation are slow. Therefore, fibers stimulated by action potentials rather than by prolonged depolarization do not become appreciably inactivated. Following inactivation, a subsequent contracture can be initiated if the fiber is allowed to recover at normal resting potential. This recovery process is known as repriming.

In the presence of micromolar amounts of the calcium channel blocking agent D600, single fibers exposed to elevated potassium concentration produce a single contracture. After the membrane potential is returned to the resting level no further response can be elicited, either by an electrical stimulus or by further application of potassium (Eisenberg et al., 1987). Following block of activation by D600, fibers have normal resting and action potentials, and can still contract in response to caffeine (a drug that directly activates s.r. Ca^{2+} release) (Rousseau, 1988). These results suggest that the normal coupling between the t. tubule membrane and s.r. Ca^{2+} release is disrupted by calcium channel blockers. In fact, the drugs seem to stabilize fibers in the inactivated state (Berwe et al., 1987; Caputo and Bolanos, 1987). The paralyzing effects of D600 and dihydropyridines have also been reported in skinned (plasmalemma removed) fiber preparations (Donaldson et al., 1984; Fill and Best, 1989). Block seems to arise from the binding of drug to the cytoplasmic face of the t. tubule membrane (Donaldson et al., 1984; Fill and Best, 1989). Blockers bind with higher affinity to the inactivated state (Fill and Best, 1989; Rios and Brum, 1987) and prevent repriming thereby accounting for the lack of an effect of these drugs on muscle twitches stimulated by action potentials.

The ability of calcium channel blockers to prevent contracture is explained by their effect on intramembrane charge movement. Charge movements are nonlinear capacity currents that are thought to be a manifestation of an early event in excitation-contraction coupling (Schneider and Chandler, 1973; Huang, 1988). Presumably, they arise from the movement of dipoles in the t. tubule membrane in response to changes in membrane potential. Thus they may correspond to changes in orientation of proteins in the t. tubule membrane that act as voltage sensors for excitation-contraction coupling. Charge becomes immobilized following prolonged depolarization explaining the inactivation of contracture (Chandler, Schneider, and Rakowski, 1976b). Both D600 (Hui et al., 1984, 1987) and dihydropyridines (Lamb, 1986; Rios and Brum, 1987) block charge movement in muscle. Dihydropyridines prevent the remobilization of charge following a return to normal resting potentials. This accounts for the ability of the drugs to block contractures (but not twitches) and suggests a role for the DHP receptor in excitation-contraction coupling.

D. The Dihydropyridine Receptor in the T. Tubule Membrane is the Voltage Sensor for Excitation-Contraction Coupling

The ability of calcium channel blockers to block contractures implies a functional role for the DHP receptor in excitation-contraction coupling. Although the receptor can act as a calcium channel (see below), inward calcium currents have been shown to be unnecessary for contractile activation. Instead, it appears likely that the critical function played by the DHP receptor is that it acts as the voltage sensor that responds to changes in t. tubule membrane potential.

The t. tubules of skeletal muscle are a rich source of receptors for 1,4 dihydropyridines, a family of calcium channel blocking drugs. The DHP receptor has been isolated and purified (Curtis and Catterall, 1984; Borsotto et al., 1985; Flockerzi et al., 1986; Leung et al., 1987; Nakayama et al., 1987) and has been shown to be a functional calcium channel when reconstituted into lipid bilayers (Flockerzi et al., 1986). Flockerzi and co-workers isolated a purified nitrendipine receptor complex which retained both its biochemical and pharmacological properties. The purified receptor was then incorporated into a lipid bilayer, and spontaneous channel openings with a single channel conductance of approximately 20 pS were recorded when a constant potential gradient was applied across the membrane. Phosphorylation of the receptor complex by cAMP-dependent protein kinase resulted in prolonged open-channel lifetimes and shortened closed intervals between channel openings. These characteristics are similar to those of the L-type calcium channel.

The primary structure of the alpha 1 subunit of the DHP receptor shows extensive similarities to the voltage-dependent sodium channel (Tanabe et al., 1987). Each of the four internal repeated units has five hydrophobic segments and one positively charged segment. These six segments share characteristic structural features with the corresponding segments of the sodium channel. The positively charged segment contains 5 or 6 arginine or lysine residues at every third position with mostly nonpolar residues intervening between the basic residues. The corresponding positively charged segment of the sodium channel is highly conserved in all sodium channels of known sequence. The positive charges in this segment of the DHP receptor, many of which presumably form dipoles, might act as a voltage sensor.

It is not certain whether all of the DHP receptors in muscle act as functional calcium channels (Schwartz et al., 1985). Comparison of DHP binding with voltage-clamp measurements of calcium current suggests that there are 35 to 50 times more DHP binding sites than there are voltage-dependent calcium channels that can be activated to pass current. A possible role for these "extra" DHP receptors has been suggested by Rios and Brum (1987). They have shown that low concentrations of DHP inhibit, in a quantitatively similar way, both nonlinear capacity current (charge movement) and calcium release from the s.r. They propose that the DHP receptor is the molecular structure that generates charge movement. Consistent with this idea, Rios and co-workers have also shown (Brum et al., 1988b) that either monovalent or divalent metal ions are needed to support charge movements, and that the selectivity sequence of the ions for the support of charge movements is identical to the selectivity sequence of the L-type calcium channel. Thus the DHP receptor may act as the voltage sensor that links surface membrane depolarization to s.r. Ca^{2+} release.

Confirmation of the critical role played by the alpha 1 subunit of the DHP receptor in linking excitation to contraction in skeletal muscle has come from work with mice suffering from a lethal recessive mutation called muscular dysgenesis. Skeletal muscle from dysgenic mice lacks normal excitation-contraction coupling (Powell and Fambrough, 1973). Since the muscle cells conduct normal action potentials and contract when exposed to caffeine, the defect seems to involve steps that link surface membrane voltage changes to the stimulation of s.r. Ca^{2+} release (Powell and Fambrough, 1973; Bournaud and Mallart, 1987). Cultured myotubes from dysgenic mice lack the DHP-sensitive slow calcium current (Beam et al., 1986). Dysgenic muscles specifically lack the alpha 1 subunit of the DHP receptor (Knudson et al., 1989) while retaining the other protein components thought to play important roles in excitation-contraction coupling. Microinjection of an expression plasmid carrying cDNA for the alpha 1 subunit of the DHP receptor into cultured myotubes from dysgenic mice restored both the slow inward calcium current and excitation-contraction coupling (Tanabe et al., 1988). Spontaneous and electrically stimulated contractions that were absent in the dysgenic cells returned after the cDNA encoding the DHP receptor was microinjected. These results confirm the essential role played by the DHP receptor in excitation-contraction coupling in skeletal muscle. It remains unclear whether a given DHP receptor can act simultaneously as a calcium channel and as the voltage sensor for excitation-contraction coupling. It is conceivable that two populations of receptors act independently of one another.

E. The S.R. Calcium Channel has a Large Cytoplasmic Domain that is the Foot Structure Linking the T. Tubule and S.R. Membranes

The final step in excitation-contraction coupling in skeletal muscles is the release of Ca^{2+} from the s.r. and the binding of Ca^{2+} to the contractile proteins thereby resulting in contraction. The calcium release channels situated in the s.r. are pharmacologically different from those in the t. tubule membrane (McCleskey, 1985). Structurally, the s.r. can be divided into two sections, the terminal cisternae and the longitudinal cisternae. The terminal cisternae are closely apposed to the t. membrane. Electron dense structures known as feet span the gap between the two membrane systems (Franzini-Armstrong, 1970). The feet have a characteristic and unique tetragonal symmetry (Ferguson et al., 1984).

Calcium release channels are concentrated in heavy s.r. vesicles that are thought to be derived from the terminal cisternae. Calcium efflux from s.r. vesicles is activated by caffeine, Ca^{2+}, and adenine nucleotides, and is inhibited by Mg^{2+} (Kim et al., 1983; Meissner, 1986). The release channels from the heavy s.r. vesicles have been incorporated into planar lipid bilayers (Smith et al., 1985) to allow single channel recordings. With such a preparation, conductance measurements of the s.r. calcium channels were made in the presence of Ca^{2+}, Mg^{2+}, and adenine nucleotides (Smith et al., 1986). A high unit conductance of 100 pS was observed in 53 mM Ca^{2+}. Similar to the results obtained from the vesicle studies, Ca^{2+} and adenine nucleotide were found to be activators of the s.r. calcium channel while Mg^{2+} was an inhibitor.

Ryanodine, a plant alkaloid, binds to the calcium release channels in the terminal cisternae (Fleischer et al., 1985). The ryanodine receptor has been isolated (Campbell et al., 1987; Inui et al., 1987; Lai et al., 1987). Ryanodine binding is dependent on Ca^{2+}, stimulated by ATP, and inhibited by ruthenium red, all properties common to Ca^{2+} release in studies of isolated s.r. vesicles (Campbell et al., 1987).

The ryanodine receptor was identified as the s.r. calcium release channel by several groups who incorporated the purified receptor into lipid bilayers (Imagawa et al., 1987; Lai et al., 1988; Hymel et al., 1988; Smith et al., 1988). The receptor protein acts as a calcium channel with properties similar to those found earlier in vesicle studies. Namely, submicromolar Ca^{2+} and millimolar ATP concentrations activated the channel, and millimolar Mg^{2+} and micromolar ruthenium red concentrations inhibited channel activity. This calcium channel exhibited rather low ion selectivity and high conductance with at least two conducting states of approximately 45 to 50 pS and 90 to 110 pS. The effect of ryanodine was such that it locked the channel into a new state with lower conductance.

The dimensions and shape of the purified ryanodine receptor corresponded to those of the foot structure that bridges the gap between the s.r. and t. tubule membranes (Inui et al., 1987). From electron micrographs, the structure of the ryanodine receptor was shown to be a tetramer, consisting of four individual subunits approximately 400 kDa in weight with a central cavity (Lai et al., 1988; Saito et al., 1988). The receptor has been cloned and its primary structure determined from sequencing the cDNA (Takeshima et al., 1989). The predicted secondary structure suggests that the calcium channel activity of the protein comes from the C-terminal region that inserts into the s.r. membrane. The rest of the molecule is cytoplasmic and lies between the t. tubule and s.r. membranes and thus is the foot region seen in electron micrographs (see Figure 1). Evidence of a direct physical interaction between the ryanodine receptor in the s.r. and the DHP receptor protein in the t. tubule membrane has been provided by structural analysis of the two membrane systems (Franzini-Armstrong and Nunzi, 1983; Block et al., 1988). Particles in the t. tubule membrane form square shaped clusters that align with the subunits of the foot protein. The implication is that there is a large protein complex that bridges the gap between the two membrane systems. Assuming that the particles in the t. tubule membrane are DHP receptors, this suggests that a voltage-sensing protein in the t. tubule membrane is in direct contact with the s.r. calcium channel protein and can regulate its function. Thus, the ryanodine receptor protein has two important functions with respect to excitation-contraction coupling. First, it is the s.r. calcium release channel. In addition, based on striking structural similarities, it has been identified as the foot protein that bridges the gap between the t. tubule and s.r. membranes.

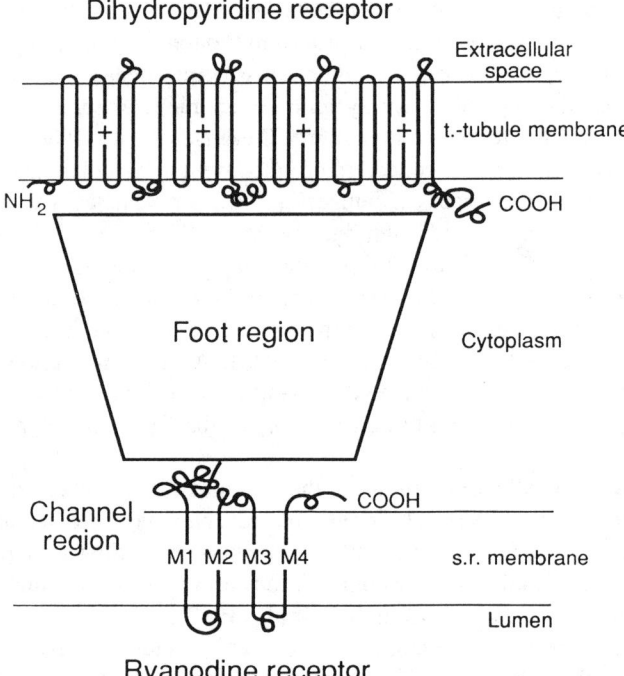

Figure 1. Diagram (modified by Takeshima et al., 1989) showing the molecular arrangement of the dihydropyridine (DHP) receptor protein and ryanodine receptor protein in skeletal muscle. DHP receptors are integral membrane proteins found in the t. membrane. They presumably act as both the voltage sensor for excitation-contraction coupling and as calcium channels. The ryanodine receptor protein is the calcium release channel of the sarcoplasmic reticulum. It contains a large cytoplasmic domain that corresponds to the "foot" structure seen in electron micrographs to bridge the space between t. membrane and the s.r. It is thought that interaction between these two proteins couples t. membrane voltage changes to s.r. Ca^{2+} release.

F. E-C Coupling in Skeletal Muscle

Despite detailed knowledge of the structures located at the t. tubule-s.r. junction, the fundamental question concerning excitation-contraction coupling in skeletal muscle remains unanswered. How is information concerning t. tubule membrane voltage conveyed to the s.r. release channels? Structurally there is good evidence for a direct protein linkage between the two membranes; this could allow direct molecular interaction between the DHP and ryanodine receptor proteins (a charge-coupled mechanism) as first suggested by Chandler et al. (1976a). Another possibility is that the foot structures merely anchor the two membrane systems together, while the actual transfer of information is dependent on a second messenger. For example, inositol (1,4,5)-triphosphate has been implicated in modulating Ca^{2+} release in skeletal muscle (Vergara et al., 1985; Suarez-Isla et al., 1988; Donaldson et al., 1987) although significant negative evidence arguing against a direct role for inositol (1,4,5)-triphosphate also exists (Somlyo et al., 1988). Hybrid schemes, assuming that not all s.r. calcium channels are gated identically, are also possible. Block et al. (1988) have shown that not all of the foot structures line up with putative DHP molecules in the t. tubule membrane. It is possible that the calcium release channels that are not directly associated with DHP receptors are gated by a mechanism different from those that are. One proposed scheme envisions Ca^{2+} released from the s.r. via charge-coupled channels activating adjacent calcium-dependent channels (Rios and Pizarro, 1988).

III. CARDIAC MUSCLE

The importance of extracellular Ca^{2+} for the activation of cardiac muscle has been recognized since Ringer's discovery in 1883 that the isolated heart ceases to beat in Ca^{2+}-free bathing solution. About 25 years after Ringer's first observation, Mines (1913) found that, though perfusion with a Ca^{2+}-free solution stopped the contraction, the surface electrical activity of the heart remained. These early experiments revealed the essential role of extracellular Ca^{2+} in coupling cardiac muscle excitation and contraction. More recent experiments have confirmed these observations. In thin bundles of guinea pig ventricular trabeculae, both calcium current and contractile force could be enhanced within a few seconds by increasing the extracellular Ca^{2+} concentration (Kitazawa, 1984). Similar results have been obtained in single rabbit and rat myocytes where removal of extracellular Ca^{2+} inhibits the next contraction while adding Ca^{2+} back to the bathing solution restores contraction without delay (Rich et al., 1988; Nabauer et al., 1989). This behavior is quite different from that observed in skeletal muscle where contractions continue after Ca^{2+} removal.

The most widely held theory of excitation-contraction coupling in cardiac muscle states that s.r. Ca^{2+} release is triggered by a rise in intracellular Ca^{2+} that results from the activation of surface membrane calcium channels (Fabiato, 1983, 1985a,b; Beuckelmann and Wier, 1988; Callewaert et al., 1988). This mechanism is referred to as calcium-induced Ca^{2+} release. The essential role of surface membrane voltage in this model is the modulation of plasmalemmal calcium permeability via voltage-gated calcium channels. An alternate model assumes a charge-coupled mechanism similar to that proposed for skeletal muscle in which a voltage sensing protein in the surface membrane gates the s.r. calcium channel directly. A discussion of calcium channels and cardiac arrhythmias can be found in Chapter 20, Section IV.

A. Two Types of Surface Membrane Calcium Channels in Cardiac Muscle

Early information about calcium current in cardiac muscle came from the studies of action potentials (Noble, 1984). Following a rapid inward current carried by sodium ions, a "slow inward current" contributes to the plateau phase of the cardiac action potential. The slow inward current was identified as being carried predominantly by Ca^{2+} ion since, unlike the fast inward current, it was a function of the extracellular Ca^{2+} concentration and could not be blocked by tetrodotoxin. The inward calcium current could be dissected into an earlier, rapidly inactivating component and a later, sustained component (Noble, 1984).

Early studies to characterize cardiac calcium channels were hampered by the complexities of multicellular preparations in which cells are electrically coupled, and by the difficulty of separating calcium current from other overlapping current components. During the past few years, these difficulties have been largely overcome by using single myocytes, which can be internally perfused, and by the use of patch-clamp techniques to record single calcium channel activity. Both whole cell, and single patch studies reveal that cardiac tissue from most species contains two different types of calcium channels including a transient (T-type) calcium channel activated by small depolarization (-60 mV) from resting membrane potential and long-lasting (L-type) calcium channel activated by larger depolarization (-30 mV) (Bean, 1985; Nilius et al., 1985; Mitra and Morad, 1986; Bonvallet, 1987; Tsien et al., 1987; Hess, 1988). The T-type calcium channel is almost completely inactivated at holding potentials more depolarized than -50 mV and is DHP resistant, while the L-type calcium channel is only slightly inactivated when the cell is held at -40 mV but is blocked by DHP (Bean, 1985; Hagiwara et al., 1988; Hess, 1988).

The relative current density of the two channel types differs between atrial and ventricular myocytes and also varies among species. Generally, the density of the L-type calcium current is larger than that of the T-type component (Bean, 1985; Hess, 1988). This is especially true for ventricular myocytes, where T-type current has been reported to be absent or too small to be detected in cells from frog, adult rabbit, and rat (Tsien et al., 1987; Gonzalez-Rudo et al., 1989; Richard et al., 1989).

The physiological function of the T-type calcium channel in cardiac tissue remains poorly understood. Because it is activated at relatively hyperpolarized membrane potentials, the channel might help set the firing threshold in myocardial cells and Purkinje fibers (Hess, 1988). In rabbit SA node cells, T-type current is important for the generation of the slow diastolic pacemaker potential (Hagiwara et al., 1988). Whether T-type current has a direct role in stimulating s.r. Ca^{2+} release is not clear. The low density, fast inactivation and negative voltage range of activation and inactivation of T-type channels do not suggest an important role in E-C coupling under physiological conditions.

B. Evidence that L-Type Calcium Current Rather than a Charge-Coupled Mechanism Gates S.R. Ca^{2+} Release in Cardiac Muscle

The relationship between L-type calcium current, the extent of contraction, and membrane potential has been studied in isolated myocytes as well as in multicellular preparations using voltage clamp techniques. Both calcium current and contraction depend on membrane potential in a similar bell-shaped manner (New and Trautwein, 1972; London and Krueger, 1986). This contrasts with the behavior of skeletal muscle where both charge movement and contraction are saturating, sigmoidal functions of voltage (Chandler et al., 1976a). Cardiac calcium current and contraction have similar thresholds and both are maximally activated at membrane potentials of 0 to 10 mV. As the membrane potential becomes more positive and approaches the calcium equilibrium potential, calcium current and contraction decrease in parallel. In single myocytes, a quick shortening of variable amplitude was observed to follow repolarization from very positive membrane potentials, a maneuver that causes an inward calcium tail current as the driving force on the ion is increased (London and Krueger, 1986). In contrast, repolarization does not elicit s.r. Ca^{2+} release in skeletal muscle (Miledi et al., 1983), where repolarization would be expected to return gating charge to its resting state. Thus, the bell-shaped voltage dependency of calcium current and contraction and the brief contractions caused by repolarization are unique characteristics of cardiac muscle and differ significantly from the behavior of skeletal muscle. These responses are consistent with the idea that Ca^{2+} flux through the L-type calcium channel controls the activation of contraction in cardiac muscle.

Both calcium current and myoplasmic Ca^{2+} transients have been simultaneously recorded in voltage clamped single ventricular myocytes dialyzed internally with the fluorescent Ca^{2+} indicator Fura-2. Peak calcium current and the peak intracellular Ca^{2+} transient have very similar voltage dependency in both guinea pig and rat ventricular cells (Beuckelmann and Wier, 1988; Callewaert et al., 1988). A rapid Ca^{2+} transient accompanying the inward calcium tail current is elicited upon repolarization from membrane potentials greater than 30 mV (Cannell et al., 1987; Beuckelmann and Wier, 1988; Callewaert et al., 1988). Measurements of intracellular Ca^{2+} transients and contraction confirm that an influx of Ca^{2+} is mandatory for cardiac muscle activation. The substitution of Ba^{2+} or Na^+ as charge carriers through the calcium channel does not lead to activation of cardiac muscle even though these ions support intramembrane charge movements in skeletal muscle (Nabauer et al., 1989). These results are consistent with the idea that it is the inward Ca^{2+} flux rather than the membrane potential per se that grades the Ca^{2+} transient and contraction in cardiac muscle. In mammalian cardiac cells, the influx of Ca^{2+} through the surface membrane is not sufficient to activate the contractile proteins directly. Blocking s.r. Ca^{2+} release by ryanodine significantly depresses the Ca^{2+} transient and contraction (Wier et al., 1985; Callewaert et al., 1988). These experiments suggest that calcium-induced Ca^{2+} release triggered by Ca^{2+} flux through L-type channels plays a major role in activating contraction in mammalian cardiac cells.

There is some evidence, however, that part of the s.r. Ca^{2+} release in mammalian cardiac cells may be modulated directly by surface membrane potential rather than by calcium current. Membrane repolarization during the rising phase of the cytoplasmic Ca^{2+} transient rapidly terminates Ca^{2+} release (Cannell et al., 1987; Beuckelmann and Wier, 1988). During the repolarization, the s.r. Ca^{2+} flux is much larger than that resulting from the plasmalemma calcium current so the deactivation of surface membrane calcium channels should not affect the intracellular Ca^{2+} transient significantly. Why, then, is the release rapidly terminated? Is calcium entering the cell from the outside able to control release while that released from the s.r. is not? In skeletal muscle, repolarization under these circumstances would be expected to return membrane charge to the resting state and thus terminate release. Asymmetrical charge movement can be recorded from ventricular cells. Its properties are complex, but can be resolved into two components associated with sodium and calcium channel gating (Field et al., 1988; Robert and Lederer, 1989; Bean and Rios, 1988). The D600 blockable charge movement is larger than expected if it were associated only with complete opening of calcium channels (Bean and Rios, 1989). It is possible that, like skeletal muscle, one component of charge movement in cardiac muscle is produced by voltage sensors for excitation-contraction coupling (Cohen and Lederer, 1989; Bean and Rios, 1989).

C. Calcium- and Voltage-Dependent Inactivation of L-Type Calcium Current

The inactivation properties of the L-type calcium channel are also important in the modulation of contraction in cardiac tissue. There is considerable evidence suggesting that the L-type calcium channel can be inactivated both by Ca^{2+} and by voltage. The amount of inactivation induced by a prepulse appears to be related to calcium current magnitude. The dependence of inactivation on the prepulse potential is not a smooth curve, as would be expected for a solely voltage-dependent mechanism, but instead is partially U-shaped (Lee et al., 1985; Hadley and Hume, 1987). The rate of inactivation of calcium current is increased by high extracellular Ca^{2+}, and decreased by the substitution of Sr^{2+} or Ba^{2+} for Ca^{2+} (Kass et al., 1984; Lee et al., 1985). On the other hand, inactivation by a prepulse of the nonspecific cation current through calcium channels is directly related to the prepulse potential. Therefore, the inactivation of L-type calcium current is both Ca^{2+}- and voltage-dependent. Voltage-dependent inactivation accounts for much of the total inactivation, with Ca^{2+}-dependent inactivation making a strong contribution over the plateau range of potentials (Hadley and Hume, 1987). Calcium-dependent inactivation provides an important negative feedback mechanism for regulation of Ca^{2+} ion entry and of mechanical output in contracting heart cells (Fabiato, 1983; Lipp et al., 1987). This mechanism functions on a beat-to-beat basis since the Ca^{2+} transient due to the Ca^{2+} release from the s.r. rises quickly enough to provide the signal for shutting off Ca^{2+} influx during the plateau (Boyett et al., 1988). When ryanodine is used to inhibit s.r. Ca^{2+} release, the inactivation of calcium current is reduced (Boyett et al., 1988; Callewaert et al., 1988; Cohen and Lederer, 1988). Voltage-dependent inactivation may prevent a secondary rise in Ca^{2+} concentration when myoplasmic Ca^{2+} falls during maintained depolarization of cardiac cells.

D. Calcium Channel Modulators Affect Contraction

Calcium channel blockers like dihydropyridines, verapamil, and D600 can greatly decrease the myoplasmic Ca^{2+} transient during activation and thereby depress or completely abolish the twitch (Wier and Isenberg, 1982; Morgan et al., 1983; Morad et al., 1983). The DHP agonist Bay K 8644 can increase inward calcium current and contraction more or less in parallel (Sanguinetti and Kass, 1984). Bay K 8644 does not affect Ca^{2+} sensitivity or Ca^{2+} release of skinned cardiac fibers, suggesting that Bay K 8644 exerts its positive inotropic effect on cardiac muscle mainly by directly enhancing calcium current (Thomas et al., 1985).

Beta-adrenergic agonists, such as isoprendine, adrenaline and noradrenaline, have positive inotropic effects on cardiac muscle contraction (Reuter, 1983). Reuter and Scholz (1977) ascribed this effect to an enhancement of inward calcium current. Isoprendine was found to increase both calcium current and the force of contraction by roughly 5 times in guinea pig ventricular cells and by 3 times in rabbit ventricular cells (Hescheler et al., 1988). The effect of isoprendine on the calcium current can be mimicked by introducing cAMP or the catalytic subunit of cAMP-dependent protein kinase into single ventricular cells using intracellular dialysis. The effects of cAMP and C subunit on calcium current and those of isoprendine do not sum linearly suggesting that the substances affect the same site on the calcium channel (Kameyama et al., 1985). Brum et al. (1983) also found that injecting cAMP or active cAMP-dependent protein kinase into cardiac muscle can increase the duration of the action potential plateau and hence, the calcium current. Myocytes injected with cAMP or cAMP-dependent protein kinase contract more forcefully, similar to the effect of adrenaline or noradrenaline. Therefore, it was proposed that agonists binding to the beta-receptor activate adenylate cyclase to produce cAMP which in turn actives the catalytic subunit of cAMP-dependent protein kinase. This leads to the phosphorylation of the calcium channel or of a protein closely associated with the channel (Reuter, 1983; Kameyama et al., 1985). Electrophysiological studies show that the phosphorylation increases the probability of channel opening (Tsien et al., 1986). As a result, macroscopic calcium current is increased during membrane depolarization.

E. Role of Calcium Current in Replenishing S.R. Ca^{2+} Stores

In addition to inducing Ca^{2+} release from the s.r., the sarcolemmal Ca^{2+} influx also serves to replenish the s.r. Ca^{2+} store, thus maintaining or enhancing Ca^{2+} release in subsequent contractions (Fabiato, 1985b; Mitchell et al., 1985). Any change in net Ca^{2+} entry or extrusion can affect tension generation. For example, prolongation of the action potential by injecting current does not increase the strength of the accompanying twitch, but potentiates the following contraction (Wood et al., 1969). This would be expected if the Ca^{2+} entering during the plateau phase serves to fill the s.r. Ca^{2+} store, so that a greater amount of Ca^{2+} is available for release in the next contraction. Another well-known example is the effect of high stimulus frequency on adult mammalian cardiac muscle (Morgan and Blinks, 1982; Wohlfart and Noble, 1982). When stimulus frequency increases, the time-averaged Ca^{2+} influx increases because cells spend more time depolarized. The increased Ca^{2+} influx potentiates the Ca^{2+} filling of the s.r. over several cycles. Consequently, the contractile response increases gradually and reaches a new steady state.

F. The Calcium Release Channel of the S.R. is Ca^{2+} Activated

The s.r. calcium release channel of cardiac muscle is activated by cytoplasmic Ca^{2+}. Calcium-induced Ca^{2+} release was first described in skeletal muscle (Endo et al., 1970; Ford and Podolsky, 1970). Fabiato (1983, 1985a,b) has studied the response in skinned (sarcolemma removed) cardiac myocytes and was the first to argue strongly that it was the physiological mechanism initiating contraction in the heart. He showed that a sudden increase of the free Ca^{2+} from 0.1 μM to 0.2 to 0.5 μM elicited a large, transient contraction in skinned myocytes with functional s.r. but no contraction if the s.r. was destroyed by detergent (Fabiato, 1983, 1985a). It was suggested that the rise in free Ca^{2+} released stored Ca^{2+} from the s.r. in a sufficient amount to activate the myofilaments (Ca^{2+}-induced Ca^{2+} release). It was shown that Ca^{2+}-induced Ca^{2+} release is not only graded with the peak level of free Ca^{2+} used as a trigger, but also with the rate of change of free Ca^{2+}. In addition, the s.r. calcium release channel seems to be slowly inactivated by Ca^{2+}. This suggests that released Ca^{2+} could act via negative feedback to modulate Ca^{2+} release once it has been initiated.

G. Ryanodine Binds to the Cardiac S.R. Calcium Channel

As in skeletal muscle, ryanodine binds with high affinity to the s.r. calcium channel in cardiac muscle. The ryanodine receptor has been purified from cardiac s.r. and found to be an oligomer composed of a 340 kDa polypeptide. It is the major component of the foot structure that bridges the gap between the junctional s.r. and the surface membrane (Inui et al., 1987; Hymel et al., 1988). The protein has a dense central core enclosed by a frame having a pinwheel appearance (Saito et al., 1989). The ryanodine receptor isolated from the cardiac muscle have been incorporated into lipid bilayers and found to have the characteristics of the calcium release channel in s.r. (Hymel et al., 1988). It was found that the calcium conducting channel formed in the bilayer by ryanodine receptors was activated by Ca^{2+}, ATP, and ryanodine and was inhibited by ruthenium red and Mg^{2+}. This behavior is very similar to that found for the calcium release channel incorporated into bilayers from the purified cardiac s.r. vesicles (Rousseau et al., 1986). The reconstituted channel shows time-correlated subconductance states from 4 to 60 pS which are integral multiples of the smallest observed stable conductance levels, 4 and 8 pS. This is suggestive evidence for an oligomeric channel structure.

H. Relative Contribution of S.R. Ca^{2+} Release to Contraction

The relative importance of Ca^{2+} entry across the surface membrane and s.r. Ca^{2+} release to contractile activation in cardiac muscle varies from species to species. Frog ventricular cells have very small diameters (about 5 μm) and very sparse s.r. Calcium ions entering the cell via the surface membrane can reach the contractile proteins rapidly by diffusion. The action potential of frog cardiac muscle has a very long plateau, which allows a considerable amount of Ca^{2+} to enter the myoplasm and directly elicit contraction (Fabiato, 1983; Morad et al., 1983). Contraction of frog ventricle is not significantly affected by ryanodine. Thus in frog ventricle, entry of extracellular Ca^{2+} through surface membrane calcium channels leads to direct activation of the contractile proteins. In contrast, rat cardiac cells depend heavily on s.r. Ca^{2+} release to activate contraction (Bers, 1985). In rat ventricular cells, the action potential exhibits only a very brief plateau phase, so that the extent of Ca^{2+} influx is small (Mitchell et al., 1985). Rat ventricular cells are more sensitive to ryanodine depression than are ventricular cells from other mammalian species (Sutko et al., 1985; Rich et al., 1988). The importance of s.r. Ca^{2+} release in mammalian cardiac muscle also changes with development (Cohen and Lederer, 1988; Orchard and Lakatta, 1985).

IV. VASCULAR SMOOTH MUSCLE

As in skeletal and cardiac muscle, cytoplasmic Ca^{2+} ion regulates contractile activation in smooth muscle. The cellular mechanisms that control intracellular Ca^{2+} concentration are complicated and not completely understood. There is great heterogeneity in the behavior of smooth muscles from different sources. However, in most tissues, depolarization of the surface membrane either in the form of action potentials or graded depolarization can lead to an increase in tension via a mechanism that has been called, like the voltage-dependent activation of striated muscle, electromechanical coupling (Johansson and Somlyo, 1980). In addition, a second mechanism termed pharmacomechanical coupling (Somlyo and Somlyo, 1968) has been suggested to explain activation of contraction by drugs and transmitters independent of changes in membrane potential. This discussion will be limited to the role of calcium channels in vascular smooth muscle. Other smooth muscle types differ significantly in structure (i.e., they may contain less s.r.) and physiological behavior.

Vascular smooth muscle contains well-developed s.r. that forms specialized junctions with the surface membrane. These junctions contain bridging structures that may link the s.r. and surface membranes and are similar to the "feet" seen in striated muscle (Somlyo and Franzini-Armstrong, 1985). At least for pharmacomechanical coupling, inositol-1,4,5-triphosphate has been implicated as the chemical messenger that is responsible for triggering s.r. Ca^{2+} release (Somlyo et al., 1988). The steps that couple surface membrane depolarization to a rise in free intracellular Ca^{2+} are less well understood but also seem to involve, at least in some tissues, release of Ca^{2+} from intracellular stores. In arterial smooth muscle, the cell membrane potential is closely correlated with tension generation. Furthermore, changes in membrane potential can modulate the subsequent effects of applied agonists (Harder, 1982).

A. Surface Membrane Calcium Channels

Studies of ion fluxes and electrophysiological measurements from multicellular preparations have led different investigators to suggest that vascular smooth muscle may contain three types of channels for calcium entry. These include voltage-dependent channels, receptor-operated channels, and leak channels (Loutzerhiser et al., 1985; Benham and Tsien, 1986).

1. Voltage-Dependent Channels

Many vascular smooth muscles generate action potentials or local membrane responses that persist in sodium-free or tetrodotoxin-containing solutions. Electrical activity is, however, generally sensitive to the removal of Ca^{2+} ion from the external media suggesting that inward calcium flux through voltage-dependent channels is responsible for membrane depolarization in these cells. Thus calcium channels play a critical role in electromechanical coupling in vascular smooth muscle by maintaining cell excitability. Whether calcium influx across the surface membrane is predominantly involved in activation of muscle proteins or in initiating release of Ca^{2+} from the s.r. is not clear. The amount of Ca^{2+} entry may not be sufficient to activate contraction directly in all tissues. The importance of Ca^{2+} as a trigger of s.r. Ca^{2+} release has yet to be clearly established.

Evidence for two types of surface membrane calcium currents has been obtained from voltage clamp experiments of small segments of rat cerebral artery. Under conditions where outward potassium currents were inhibited, Hirst et al. (1986) found two kinetic components to the regenerative response of this tissue that were Ca^{2+}-dependent and were not blocked by tetrodotoxin. Under voltage clamp conditions, two kinetically distinct inward calcium currents with different voltage dependency were revealed. One was activated by small depolarizations and did not inactivate. The other was activated at larger depolarizations and inactivated rapidly. The relationship of these currents found in a multicellular preparation to those seen in isolated cells is, as yet, unclear.

Studies using isolated cells from arteries and veins have, in most cases, revealed two types of calcium channels that are similar to the L-type and T-type channels found in cardiac myocytes. The high-threshold current (L-type) activated by large depolarizations shows relatively slow inactivation and can be evoked from depolarized holding potentials. (Bean et al., 1986; Benham et al., 1987; Loirand et al., 1986; Sturek and Hermsmeyer, 1986; Yatani et al., 1987; Wang et al., 1989). Current amplitude is increased when Ba^{2+} is the charge carrier and the current is sensitive to 1,4-dihydropyridine derivatives. Single channel recordings (Benham et al., 1987; Yatani et al., 1987) indicate conductances of 20 to 25 pS (110 mM Ba^{2+}) for the L-type channel. A low-threshold (T-type) current that shows rapid inactivation and low conductance (7 pS in 100 mM Ba^{2+}) has also been described by many of these authors. To suggest that all calcium currents in smooth muscle adhere to this simple classification may be unwarranted (Aaronson et al., 1988). Worley et al. (1986) have described a low conductance channel that is DHP sensitive. Similarly, Loirand et al. (1989) have described a calcium current from cultured myocytes from rat portal vein with many of the activation characteristics of a T-type current that is, however, blocked by dihydropyridines.

The physiological role that these calcium currents play in muscle activation is problematical. The L-type current would be activated during action potentials and would be expected to contribute significantly to calcium influx during sustained depolarization. Their high sensitivity to dihydropyridines indicates that they are the site of action of these potent pharmacological agents. Dihydropyridine blocking compounds cause vasodilation and hypotension. The T-type currents may be involved in maintaining cell excitability and generation of the action potential.

Noradrenaline is a potent activator of vascular smooth muscle contraction that can cause an increase in intracellular Ca^{2+} without changing membrane potential (Droogmans, et al., 1977). This suggests either the existence of receptor operated channels or Ca^{2+} release from intracellular stores (see next sections). In addition, recent evidence shows clearly that L-type calcium channels in the plasma membrane of arteries are modulated by noradrenaline (Droogmans et al., 1987; Benham and Tsien, 1988; Nelson et al., 1988). Whole cell patch clamp recordings from rabbit ear arterial cells disclose a significant enhancement of macroscopic L-type current with micromolar concentrations of noradrenaline (Benham and Tsien, 1988). The voltage dependency of activation and inactivation of these whole cell currents is unchanged by the drug. However, the open state probability of single, voltage-dependent calcium channels from smooth muscle cells taken from mesenteric arteries is increased by noradrenaline (Nelson et al., 1988). Thus, under conditions in which smooth muscle cell membranes are depolarized, noradrenaline will increase Ca^{2+} entry via voltage-dependent channels. This will presumably affect cell excitability and contractile activation.

2. Receptor-Operated Channels

The observation that noradrenaline at certain concentrations can activate contraction of smooth muscle cells without changing membrane potential or in cells that are fully depolarized leads to the suggestion that receptor-operated calcium channels exist in the plasma membranes of these cells (Somlyo, 1968; Bolton, 1979; van Breemen et al., 1979). Activation without depolarization is dependent on agonist concentration (Large and Bolton, 1986). At high bath concentrations cells generally depolarize, whereas lower concentrations cause contraction without membrane depolarization. It has been suggested that two classes of adrenergic receptors exist in smooth muscle and only those associated with nerve endings cause cell depolarization (Hirst and Neild, 1980, 1981). The concept of receptor-operated channels is somewhat controversial (Rink, 1988; van Breeman and Saida, 1989). Agonist activated, voltage-insensitive channels have not, until recently, been found using patch-clamp techniques (Droogmans, 1987). In addition, alternate explanations such as agonist-dependent release of Ca^{2+} from intracellular stores with subsequent activation of Ca^{2+}-dependent outward currents (Benham and Bolton, 1986) and the influence of agonists on voltage-dependent channels (see above) could account equally well for many of the observed phenomena that first lead to the suggestion that receptor-operated channels exist. The only unequivocal demonstration of a receptor-operated calcium channel in isolated smooth muscle cells is provided by the work of Benham and Tsien (1987). They showed that ATP, a sympathetic neurotransmitter, directly activated a calcium-selective channel (3:1 Ca^{2+} over Na^+) in arterial smooth muscle. The channel could be activated at voltages near the resting potential and was clearly distinct from either the L-type or T-type voltage-dependent channels seen in a variety of cells. Second messengers are apparently not involved in the activation of the channel, suggesting that channel activation is directly coupled to ATP binding. Presumably the channels are involved in nonadrenergic activation of smooth muscle and produce excitatory junction potentials.

3. Leak Channels

There is a significant inward flux of Ca^{2+} into quiescent smooth muscle cells (van Breeman et al., 1979). It has been postulated that voltage-insensitive leak channels may account for this Ca^{2+} entry (Cauvin et al., 1983), although other pathways could be involved. Benham and Tsien (1986) have recorded single calcium channel currents that are active at very negative potentials and only weakly voltage sensitive. It is possible that they could account for the inward movement of Ca^{2+} in unstimulated cells.

B. Calcium Release from Intracellular Stores

The s.r. is the intracellular source of Ca^{2+} released during pharmacomechanical and electromechanical coupling in vascular smooth muscle (Itoh et al., 1983; Bond et al., 1984; Kowarski et al., 1985; Kobayashi et al., 1986; van Breeman and Saida, 1989). The calcium release channel from smooth muscle s.r. has not been isolated. However, both ryanodine and caffeine, agents known to affect gating of the s.r. calcium channel in skeletal muscle, stimulate release of Ca^{2+} from intracellular stores in smooth muscle (Hwang and van Breeman, 1987; Hisayama and Takayanagim, 1988). Thus, at least with regard to these two drugs, the pharmacological profile of the s.r. calcium channel from smooth muscle is similar to that seen in striated muscle.

It seems highly likely that agonist-dependent activation of vascular smooth muscle (pharmacomechanical coupling) is mediated by second messenger stimulation of s.r. calcium release channels. Inositol 1,4,5-triphosphate (IP_3) is clearly an important physiological activator of s.r. Ca^{2+} release in smooth muscle (Suematsu et al., 1984; Somlyo et al., 1985) whereas its role in striated muscle is still controversial (Somlyo et al., 1988). IP_3 directly activates Ca^{2+} release from the s.r. of smooth muscle at physiological rates (Walker et al., 1987). The appearance of IP_3 in the cytoplasm may be linked to surface membrane receptors by G protein activation of membrane lipases (Kanaide et al., 1986; Kobayashi et al., 1988). In addition, GTP can apparently activate release independently of IP_3 (Kobayashi et al., 1988). Pharmacomechanical coupling persists in Ca^{2+}-free media suggesting that an influx of Ca^{2+} is not necessary for activation (Bond et al., 1984). Procaine does not block IP_3 stimulation of Ca^{2+} release (Kobayashi et al., 1988) although it is a potent inhibitor of caffeine induced release in skeletal muscle.

V. SUMMARY

The contributions of calcium channels to excitation-contraction coupling in muscle are diverse. The roles played by surface membrane calcium channels in the activation of contraction vary considerably with muscle type. In skeletal muscle, calcium currents do not contribute significantly to the generation of the action potential and transmembrane flux of Ca^{2+} ion is not necessary for contractile protein activation. However, t. membrane proteins that bind dihydropyridines apparently function as voltage sensors that link surface membrane activation to release of Ca^{2+} from the s.r. In contrast, calcium current through L-type channels in the surface membranes of cardiac cells contributes significantly to action potential shape and calcium influx is a critical step linking membrane depolarization to muscle activation. In vascular smooth muscle, surface calcium channels contribute significantly to cell excitability and act as a pathway through which extracellular Ca^{2+} can enter the cytoplasm leading to muscle activation. Finally, calcium channels in the s.r. membrane regulate the release of Ca^{2+} from this intracellular organelle.

Chapter 7

Calcium Channels as Triggers of Secretion

George J. Augustine and Stephen D. Hess

TABLE OF CONTENTS

I.   A General Overview of Cellular Secretion 88

II. Sources of Trigger Calcium 89

III. Influx of Ca^{2+} Through Voltage-Gated Channels as a Trigger for
     Exocytosis 90
     A.   Neuronal Secretion as a Case Study 90
     B.   General Criteria for Identifying Calcium-Dependent Secretion 95
     C.   Calcium-Dependent Secretion is a Widespread Process 95
          1.   Catecholamine Secretion from the Adrenal Medulla 96
          2.   Pancreatic Beta Cells 96
          3.   Other Examples of Ca^{2+}-Dependent Hormone Secretion 98
          4.   Discharge of Trichocysts from *Paramecium* 99
          5.   Cortical Granule Discharge During Egg Fertilization 100
          6.   Histamine Secretion from Mast Cells 100
          7.   Antibody Secretion from B Lymphocytes 101
          8.   Parathyroid Hormone Secretion 102

IV. Conclusions 103
     A.   Calcium is an Important Secretory Trigger 103
     B.   Modes of Calcium Triggering in Secretory Cells 104
     C.   Diversity of Voltage-Gated Calcium Channels 105

Acknowledgments 105

I. A GENERAL OVERVIEW OF CELLULAR SECRETION

Secretion is the preferential discharge of selected molecules from cells. It is a very general property of cells. In some cases secretion is a highly visible cellular activity that is readily detected simply by examining the anatomical organization of a tissue. Examples include the cells of exocrine glands—whose functions are to secrete tears, milk, mucus, saliva and other digestive juices, etc. into well-defined secretory ducts—and endocrine cells, which secrete hormones into the bloodstream. In other cases it is more difficult to visually identify secretion as an important cellular activity. One such example is the slow trickle of glycoproteins that is responsible for constructing the basal lamina surrounding most cells. Whether obvious or subtle, secretion is a primary means of communication between all cells in multicellular organisms. Even single-celled organisms, such as protozoans and yeast, have well-characterized secretory activities. In sum, it is likely that some form of secretion occurs in all cells.

Cellular secretion can be divided into two categories distinguished by the rate-limiting step in its control (Kelly, 1985). One form of secretion is *constitutive*, meaning that it occurs constantly and is limited only by the rate of synthesis of the secreted molecule. An example of this form of secretion is the secretion of the molecules of the basal lamina (Burgess and Kelly, 1987). The other form of secretion, termed *regulated*, is more complex. Regulated secretion requires the storage of secretory products within the cytoplasm of the secretory cell, often concentrated in membrane-bound organelles called vesicles or granules. Secretion of these stored molecules occurs in response to a physiological signal, so that this physiological signal (or the reaction(s) that it triggers) limits the rate of secretion. Most of the examples listed in the previous paragraph are the result of regulated secretion. Numerous distinctions between, and properties of, these two forms of secretion have been discussed by Burgess and Kelly (1987). The present chapter will focus on regulated secretion, particularly on secretion that is regulated by Ca^{2+}.

Regulated secretion is mediated by at least two general classes of mechanisms. Selective permeation through the plasma membrane is one mechanism for molecules to be secreted from cells. Permeation may occur due to the opening of specific ion channels; for example, fluid and electrolyte secretion from exocrine gland cells seems to be mediated by Ca^{2+}-activated chloride and potassium channels (Marty et al., 1984; Iwatsuki et al., 1985; Cook et al., 1988). Other substances are translocated by active membrane transport: one example is the secretion of bicarbonate ions by epithelial cells (Ammar et al., 1987).

The second, and more prominent, mechanism of regulated cellular secretion is *exocytosis*. Exocytosis is a complex process that involves cytoplasmic vesicles containing stored secretory products. During exocytosis these vesicles fuse with the plasma membrane of the secretory cells, resulting in the ejection of the contents of the vesicle into the extracellular medium. Because all of the examples of cellular secretion that we will consider later in this chapter result from exocytosis, we will briefly summarize several of the steps that are involved in exocytosis.

Before exocytosis occurs, a number of preparatory steps must take place. These steps may include (1) the formation of vesicles and specific plasma membrane sites of exocytosis, (2) the synthesis of the secreted molecules and the storage of these molecules in the vesicles, (3) the intracellular transport of vesicles to specific sites of secretion, and (4) the docking of vesicles at these sites. Burgess and Kelly (1987) and Kelly (1988) summarize our limited understanding of these preparatory processes. The exocytosis event itself involves several other steps, potentially including (5) initiation by the binding of a second messenger (often Ca^{2+}) to an intracellular receptor molecule, (6) an effector action that leads to fusion of vesicle and plasma membranes, and (7) the dispersion of the contents of the vesicle into the extracellular medium. Although much has been written about these steps (e.g., Burgoyne,

1984; Baker and Knight, 1987; Augustine et al., 1987; Zimmerberg, 1987), a detailed understanding of them is still lacking.

In this chapter we will concentrate on step number five. In particular, we will concentrate on the role of Ca^{2+} in acting as a second messenger to trigger exocytotic secretion from many types of cells. While Ca^{2+} is not the only second messenger involved in regulating exocytosis (Gomperts, 1986; Baker and Knight, 1987; Penner and Neher, 1988a), much more is understood about its role in secretion than about any other regulatory molecule. Perhaps elucidation of the specific molecular means by which Ca^{2+} triggers exocytosis will help illuminate the means by which other messenger signals regulate secretion.

II. SOURCES OF TRIGGER CALCIUM

Consideration of the triggering of secretion by Ca^{2+} must begin with an examination of cellular sources of Ca^{2+}. Cellular Ca^{2+} can come from two places: (1) the extracellular medium; or (2) internal stores. Wherever the Ca^{2+} originates, it ultimately triggers secretion by finding its way to the sites of secretion that are located in the most superficial region of the cell interior (e.g., Plate 1).*

Cells bind or sequester significant quantities of Ca^{2+} within their interiors. For example, although the total concentration of Ca^{2+} in the cytoplasm of a giant nerve cell from squid is several mM, the cytoplasmic concentration of free Ca^{2+} ions, $[Ca]_i$, is roughly $10^{-7}M$ (Baker, 1972; Brinley, 1978). Thus, more than 99.9% of the cytoplasmic Ca^{2+} in this cell is bound to molecules or stored in organelles and presumably this is also true for other cells. This stored cytoplasmic Ca^{2+} is potentially available for use as an intracellular signal to trigger secretion and other physiological events.

The means by which intracellularly stored Ca^{2+} can be released has been a subject of intensive investigation in recent years. Attention has focused on the inositol lipids, in particular inositol 1,4,5-triphosphate (IP_3), as second messengers that release Ca^{2+} from cellular storage sites (Berridge and Irvine, 1984). Without attempting a detailed summary of this hyperdynamic topic (Berridge and Irvine, 1989; Williams and Monck, 1989), we will say that it is clear that elevation of the intracellular concentration of IP_3, produced by activation of phospholipase C, can produce a rise in $[Ca]_i$. This Ca appears to be released from endoplasmic reticulum and/or other intracellular organelles (Berridge and Irvine, 1984; Meldolesi et al., 1988). IP_3 appears to release Ca^{2+} from these organelles by activating calcium-permeable channels in the organellar membranes (Ehrlich and Watras, 1988).

IP_3-induced release of Ca^{2+} has been shown to trigger secretion in some types of cells. In sea urchin eggs, intracellular microinjection of IP_3 produces an increase in $[Ca]_i$ that promotes the exocytosis of proteins (Whitaker and Irvine, 1984; Turner et al., 1986; Swann and Whitaker, 1986). This and other evidence suggests that IP_3-induced release of Ca^{2+} may be an important event during fertilization of this egg (reviewed in Jaffe, 1988). Antigenic stimulation of mast cells causes these cells to secrete histamine via exocytosis; one of the early events in the secretory response of these cells is a transient rise in $[Ca]_i$ that is mimicked by intracellular injection of IP_3 (Neher, 1988; Penner et al., 1988). Chromaffin cells of the adrenal gland have IP_3-sensitive Ca^{2+} stores that may be involved in regulating catecholamine secretion (Stoehr et al., 1986). Other endocrine cells also have been found to have IP_3-induced Ca^{2+} release (Ronning et al., 1982; Prentki and Matchinsky, 1987). Thus, it is clear that IP_3 can regulate secretion by releasing Ca^{2+} from internal storage organelles. It is possible that still other intracellular messengers, including Ca^{2+} itself (Fabiato, 1985), may play similar roles in mobilizing Ca^{2+} from intracellular sites.

In addition to release from internal stores, entry of Ca^{2+} from the extracellular medium

---

* Plate 1 appears after page 94.

is a very important source of Ca^{2+} for triggering secretion. Because passive leakage of Ca^{2+} into cells is normally small, it appears that the opening and closing of specific, calcium-permeable channels is the means by which Ca^{2+} enters cells. The opening and closing, or gating, of these channels can be regulated by either the cellular membrane potential (voltage-gated calcium channels; see Chapter 3, Section I) or the presence of extracellular or intracellular chemical signals (ligand-gated calcium channels; see Chapter 14, Section III). Relatively little is known about ligand-gated calcium channels, in secretory or other cells (Meldolesi and Pozzan, 1987; Tsien et al., 1988). In lymphocytes a calcium-permeable channel that is activated by intracellular IP_3 has been reported (Kuno and Gardner, 1987), but its possible role in secretion is unclear. An IP_3-sensitive component of Ca^{2+} influx appears to influence secretion from mast cells, but whether this influx is caused by a calcium channel is not yet known (Penner et al., 1988). Mast cells also contain a second messenger-gated, cation-permeable channel that may play a role in elevating $[Ca]_i$ and regulating secretion (Penner et al., 1988). Identification of other ligand-gated calcium channels, and elucidation of their role in secretion, will be an important area for future investigation.

Because the influx of Ca^{2+} through voltage-gated calcium channels appears to be a more widespread means of triggering secretion, the remainder of this chapter will concentrate on this form of secretory control.

III. INFLUX OF Ca^{2+} THROUGH VOLTAGE-GATED CHANNELS AS A TRIGGER FOR EXOCYTOSIS

A. Neuronal Secretion as a Case Study

The best case for voltage-gated calcium channels acting as a source of Ca^{2+} for triggering exocytosis comes from studies of transmitter secretion from neurons. The bulk of available evidence indicates that neuronal secretion occurs via exocytosis (Heuser et al., 1979; Torri-Tarelli et al., 1985) and most often occurs at the presynaptic terminals, although somatic and dendritic secretion has also been reported (Nieoullon et al., 1977). Individual neurons secrete one or more of the dozens of known transmitter substances; there may be a variety of mechanisms involved in secreting these transmitters, depending upon whether the transmitter is a classical, fast-acting molecule (such as acetylcholine, glutamate, GABA, or glycine) or a more slowly acting peptide or amine (DeCamilli and Navone, 1987).

The idea that calcium channels trigger transmitter secretion was first set forth in the *Calcium Hypothesis* of Katz (1969). This hypothesis proposed that transmitter secretion occurs as the result of the following sequence of events:

1. An action potential in the presynaptic neuron produces a transient depolarization of the membrane potential.
2. This depolarization opens voltage-gated calcium channels present in the terminal of the presynaptic neuron.
3. Opening of the calcium channels leads to an influx of Ca^{2+} into the terminal and an accumulation of Ca^{2+} within the presynaptic cytoplasm.
4. This elevation of intracellular Ca^{2+} concentration causes transmitter-laden synaptic vesicles to fuse with the presynaptic plasma membrane and exocytotically secrete the transmitter into the extracellular space between the pre- and postsynaptic cells.

The calcium hypothesis of transmitter secretion is so well established that it now is generally regarded as fact (Katz, 1969; Rahamimoff et al., 1978a; Augustine et al., 1987). However, there is at least one well-documented example of depolarization-activated trans-

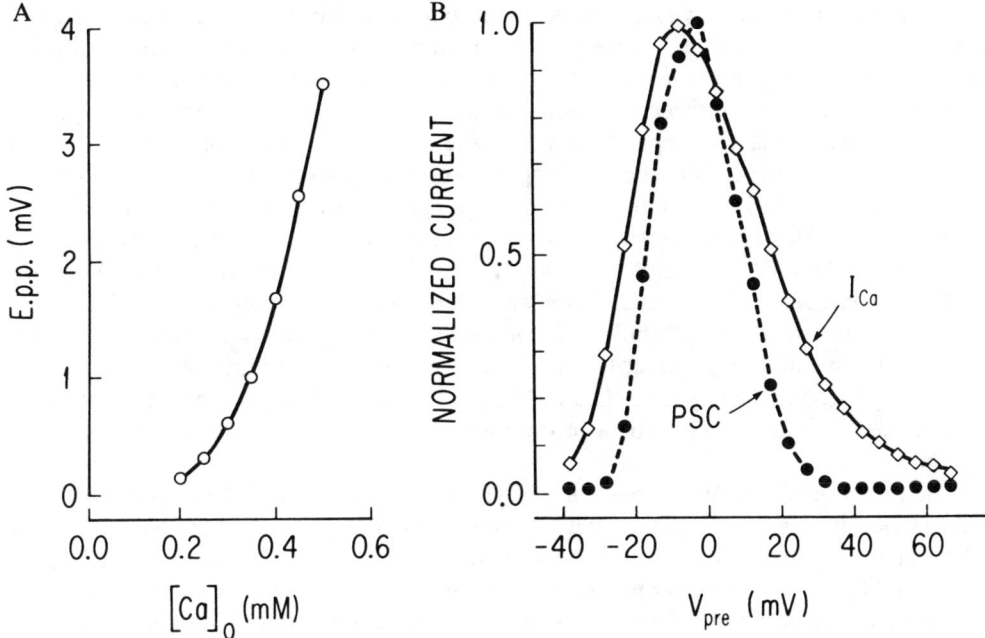

Figure 1. Dependence of neurotransmitter secretion upon external Ca^{2+} concentration and membrane potential. (A) The amplitude of the end-plate potential (E.p.p.) recorded at frog neuromuscular synapses varies with the external Ca^{2+} concentration ($[Ca]_o$). Under the conditions of this experiment, end-plate potential amplitude is proportional to the amount of transmitter secreted, so that this experiment reveals that transmitter secretion is a sensitive function of $[Ca]_o$. From Dodge and Rahamimoff (1967), with permission. (B) The amplitudes of both postsynaptic current (PSC), a measure of transmitter secretion and presynaptic calcium current (I_{Ca}), and a measure of Ca^{2+} influx, exhibit a bell-shaped dependence on presynaptic membrane potential (V_{pre}) at a voltage-clamped giant synapse of squid. From Smith et al. (1985).

mitter secretion that is calcium independent (Schwartz, 1987). In this section we will summarize some of the many lines of evidence that have helped establish the calcium hypothesis of neurotransmitter secretion. Our goal is to use neuronal secretion as an exemplary "case study" of how calcium channels trigger secretion. This summary will not be comprehensive, but instead will attempt to identify the most compelling evidence—more or less in chronological sequence—with reference to a few relevant contemporary findings.

The earliest evidence suggesting the involvement of Ca^{2+} in transmitter secretion came from studies of the dependence of transmitter release upon extracellular Ca^{2+}. Modern analysis of this problem began with the study of del Castillo and Katz (1954) demonstrating that the amount of neurotransmitter secreted from a motor nerve terminal in response to an action potential depended upon the extracellular Ca^{2+} concentration, $[Ca]_o$. This evoked transmitter release increased as $[Ca]_o$ was raised and was eliminated when there was no Ca^{2+} in the extracellular medium. Dodge and Rahamimoff (1967) examined the quantitative relationship between $[Ca]_o$ and rate of transmitter release; they found that release was exceedingly sensitive to $[Ca]_o$ and could be described as a fourth-order function of $[Ca]_o$ (Figure 1A). Similar findings have been found at other synapses (reviewed in Martin, 1977 and Silinsky, 1985) and are consistent with a model that multiple Ca^{2+} ions act "cooperatively" to trigger the secretion of transmitter quanta (Dodge and Rahamimoff, 1967; Augustine and Charlton, 1986). In summary, secretion of transmitter from presynaptic terminals requires the presence of Ca^{2+} in the extracellular environment.

The original evidence that extracellular Ca^{2+} enters presynaptic terminals via calcium channels was rather indirect, because most presynaptic terminals are very small and thus inaccessible to direct scrutiny. Firm evidence in support of this notion only emerged when attention turned to the "giant" nerve terminal of squid, a structure several orders of magnitude larger than most other nerve terminals (Llinas, 1982). In this terminal, it was found that transmitter release is initiated by an action potential in the presynaptic terminal (Hagiwara and Tasaki, 1958). Pharmacological methods that eliminate the presynaptic action potential permit graded depolarization of the presynaptic membrane potential and revealed that transmitter release is a bell-shaped function of membrane potential (Figure 1B), reaching a maximum near 0 mV and being suppressed at potentials more positive than +60 to +100 mV (Katz and Miledi, 1967; Kusano et al., 1967; Llinas et al., 1981b; Augustine et al., 1985b). This peculiar voltage dependence was interpreted as resulting from the dual effect of membrane potential on Ca^{2+} influx: depolarization opens more calcium channels, while increasing the membrane potential beyond 0 mV decreases the driving force on Ca^{2+}, by approaching E_{Ca}, and thereby decreases Ca^{2+} entry. Thus, the voltage dependence of transmitter release was as expected for a process that depended upon influx of Ca^{2+} through calcium channels.

The presence of calcium channels in presynaptic terminals was more directly indicated by the demonstration that squid terminals generate calcium-dependent action potentials after sodium and potassium channel currents are pharmacologically attenuated (Figure 2A; Katz and Miledi, 1969). This not only proved that calcium channels were present presynaptically, but also indicated that their density was sufficient to support action potential generation (under some circumstances). Similar results have been obtained at other nerve terminals with extracellular electrical recording (e.g., Gundersen et al., 1982; Brigant and Mallart, 1982; Penner and Dreyer, 1986; Lindgren and Moore, 1989) or optical methods (Salzberg and Obaid, 1988; Obaid et al., 1989).

Detailed characterization of the gating of squid presynaptic calcium channels awaited application of the voltage clamp method. Such studies required pharmacological elimination of sodium and potassium channel currents, to unmask the smaller current flowing through the calcium channels (Figure 2B). The presynaptic calcium current was measured with this method for the first time by Llinas et al. (1976) and was found to have the voltage dependence expected from earlier studies of transmitter release (Figure 1B). More recent work has characterized the biophysical properties of presynaptic calcium channels, in nerve terminals of squid (Llinas et al., 1981a; Augustine et al., 1985a) as well as crab and mammalian neurosecretory terminals (Lemos et al., 1986; Lemos and Nowycky, 1989). Measurements of Ca^{2+} influx into synaptosomes, which are pinched-off nerve terminals, also have provided a means of documenting and characterizing presynaptic Ca^{2+} entry (Nachshen and Blaustein, 1982). In general, the properties of these channels are not remarkably different from those of calcium channels found in extrasynaptic locations (reviewed in Augustine et al., 1989; see also Chapters 2 to 4).

Pharmacological studies also have showed that transmitter release is blocked by several inorganic ions that block calcium channels (see Chapter 12, Section III). These ions include Mn^{2+}, Co^{2+}, Cd^{2+}, and others, including Mg^{2+} (Figure 2B; reviewed in Silinsky, 1985). More recently, a number of organic compounds, such as dihydropyridines and omega-*Conus* toxin, were found to block transmitter release in some nerve terminals (Miller, 1987; Augustine et al., 1987; Smith and Augustine, 1988). Some organic calcium channel agonists, such as BAY K 8644 (Lindgren and Moore, 1989), leptinotarsin (McClure et al., 1980) and maitotoxin (Takahashi et al., 1982), also were found to selectively increase release from certain nerve terminals. These results not only indicate that calcium channels are present in nerve terminals but also suggest that a variety of pharmacologically distinct types of these channels are found in different terminals (see Chapter 13, Section III).

A final piece of evidence supporting the idea that transmitter release is triggered by presynaptic calcium channels came from studies of the relationship between presynaptic

7. Calcium Channels as Triggers of Secretion

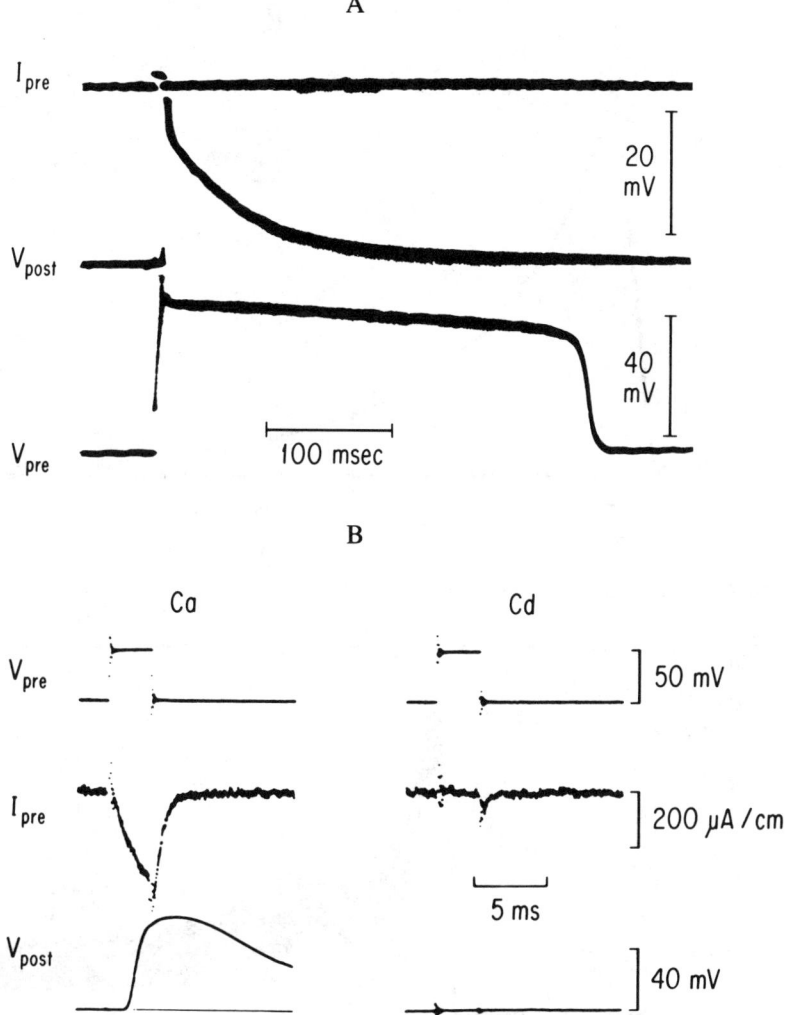

Figure 2. Properties of presynaptic calcium channels in the squid giant nerve terminal. (A) A brief pulse of depolarizing current (I_{pre}) elicits a Ca^{2+}-dependent action potential in the presynaptic terminal (V_{pre}). This action potential produces secretion of neurotransmitter, as evidenced by a depolarizing response in the postsynaptic neuron (V_{post}). From Katz and Miledi (1969), with permission. (B) Left: depolarization of the membrane potential (V_{pre}) of a voltage clamped presynaptic terminal, in the presence of pharmacological agents that block the much larger (circa 1 mA/cm²) sodium and potassium channel currents, reveals an inward current (I_{pre}) flowing through presynaptic calcium channels. This presynaptic Ca^{2+} influx then triggers transmitter secretion, indicated by a change in the postsynaptic membrane potential (V_{post}). Right: treatment with the calcium channel blocker, Cd^{2+} (2 mM), blocks both the presynaptic calcium current and postsynaptic response evoked by an identical presynaptic depolarization. From Augustine and Eckert (1984).

calcium current and transmitter release at the synapse formed by squid giant terminals. These studies found that the timing of transmitter release depends upon the time course of the presynaptic calcium current (Llinas et al., 1981b; Augustine et al., 1985b). More critically, the magnitude of this current was also closely correlated with transmitter release, with this relationship described by a linear (Llinas et al., 1976, 1981b) or exponential (Augustine et al., 1985b; Augustine and Charlton, 1986) function. These correlations leave no doubt that influx of calcium through channels is responsible for triggering release.

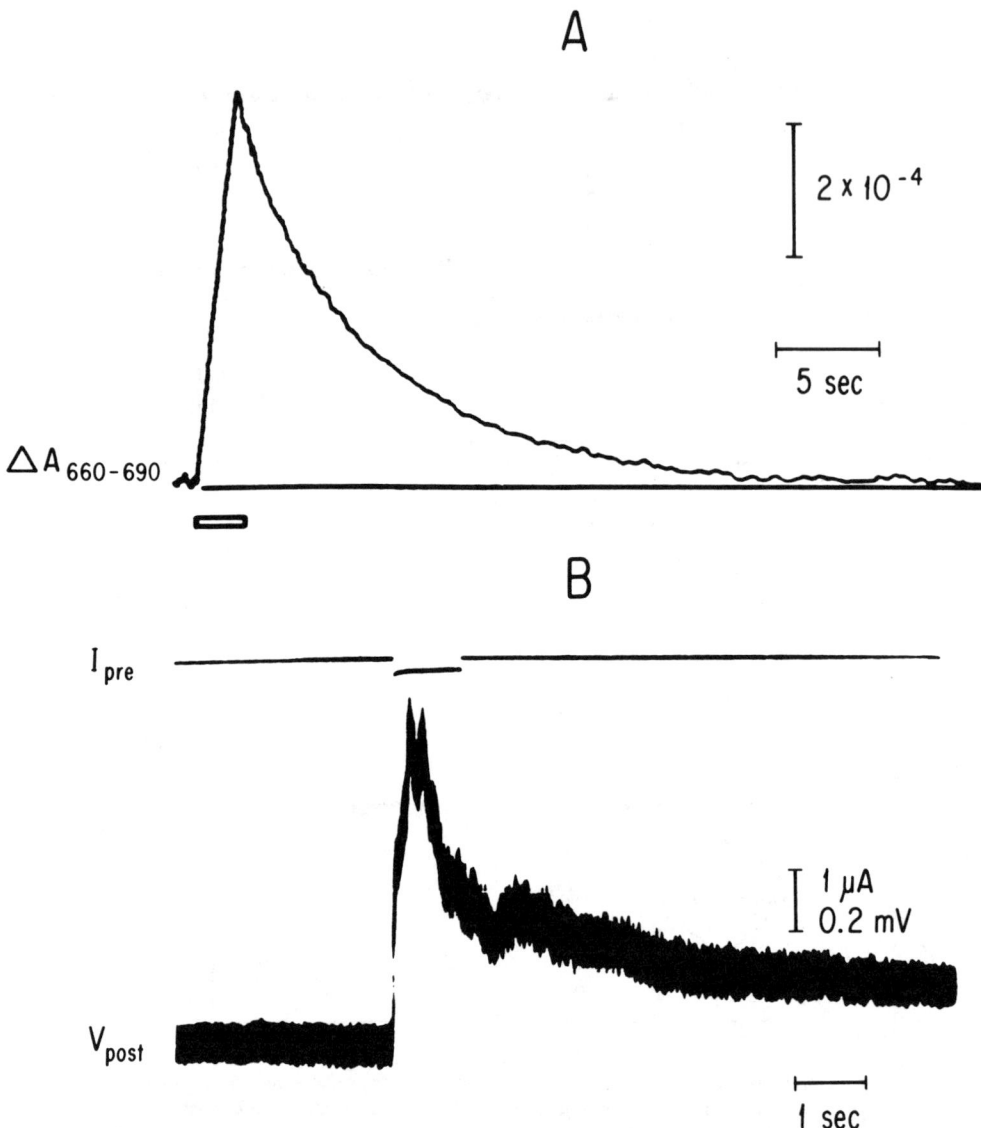

Figure 3. Intracellular Ca^{2+} and transmitter release. (A) A train of sixty-six action potentials in a squid nerve terminal injected with the calcium indicator, Arsenazo III, causes a change in dye absorbance indicative of a rise in presynaptic $[Ca]_i$. The vertical calibration bar corresponds to a change in average presynaptic $[Ca]_i$ of approximately 0.55 μM. From Charlton et al. (1982), with permission. (B) Iontophoretic injection of Ca^{2+} into a squid presynaptic terminal, at the time indicated by the current pulse (I_{pre}), causes a depolarization of the postsynaptic neuron (V_{post}). This postsynaptic response is presumably due to the massive and prolonged secretion of transmitter as a consequence of the elevation of presynaptic $[Ca]_i$. From Miledi (1973), with permission.

The next increment in the understanding of the role of Ca^{2+} in triggering transmitter secretion came with appreciation that it is the concentration of Ca^{2+} within the presynaptic terminal, $[Ca]_i$, that determines the rate of release. The first direct indication of this came with the observation that $[Ca]_i$ is elevated when squid giant terminals are stimulated (Llinas et al., 1972; Llinas and Nicholson, 1975). Measurements of $[Ca]_i$ changes during presynaptic activity have been refined by subsequent application of improved techniques (Figure 3A; e.g., Miledi and Parker, 1981; Charlton et al., 1982; Stockbridge and Ross, 1984; Llinas, 1984; Augustine et al., 1985a) and it is now possible to demonstrate the diffusion of Ca^{2+} away from sites of entry within the presynaptic terminal (Augustine et al., 1989).

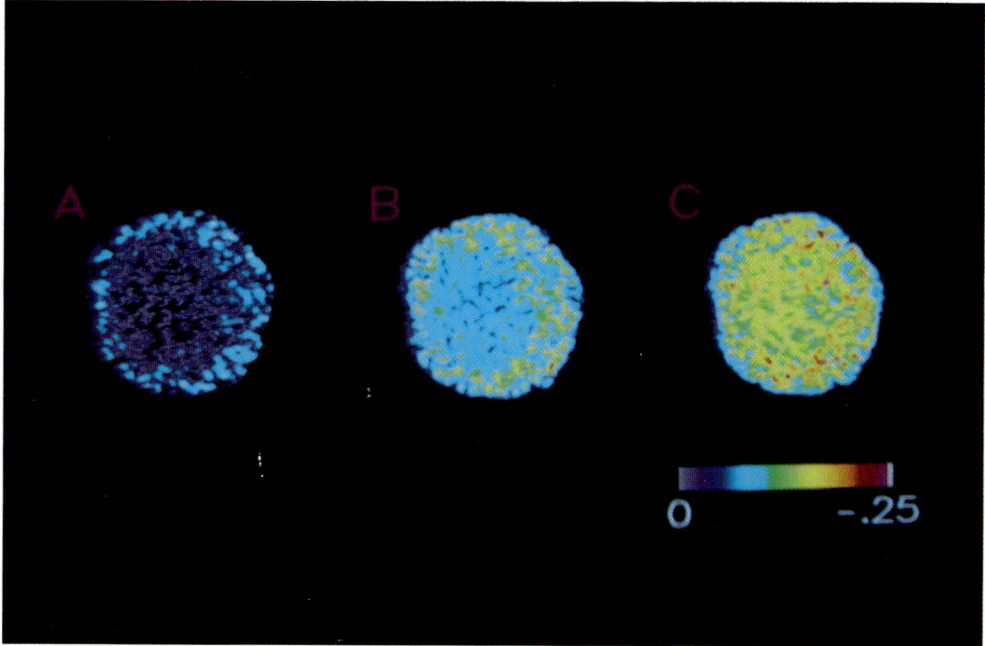

PLATE 1. Digital imaging of depolarization-induced rises in [Ca]$_i$ in a bovine chromaffin cell. The cell was voltage clamped with a patch pipette and depolarized to +30 mV for 500 ms. The colored circles are single-frame images of the changes in fluorescence of Fura-2, a calcium indicator dye that was introduced into the cell cytoplasm via the recording pipette. These images were recorded approximately 50 ms (A) and 300 ms (B) after the onset of the depolarization, and 1 s after the depolarization was terminated (C). The magnitude of the [Ca]$_i$ changes induced by depolarization were estimated from the changes in Fura-2 fluorescence, measured with digital imaging techniques, and encoded into the pseudocolor scale shown below. The maximum change in fluorescence (−0.25) corresponds roughly to a 200 nM rise in [Ca]$_i$. (From G. Augustine and E. Neher, unpublished results.)

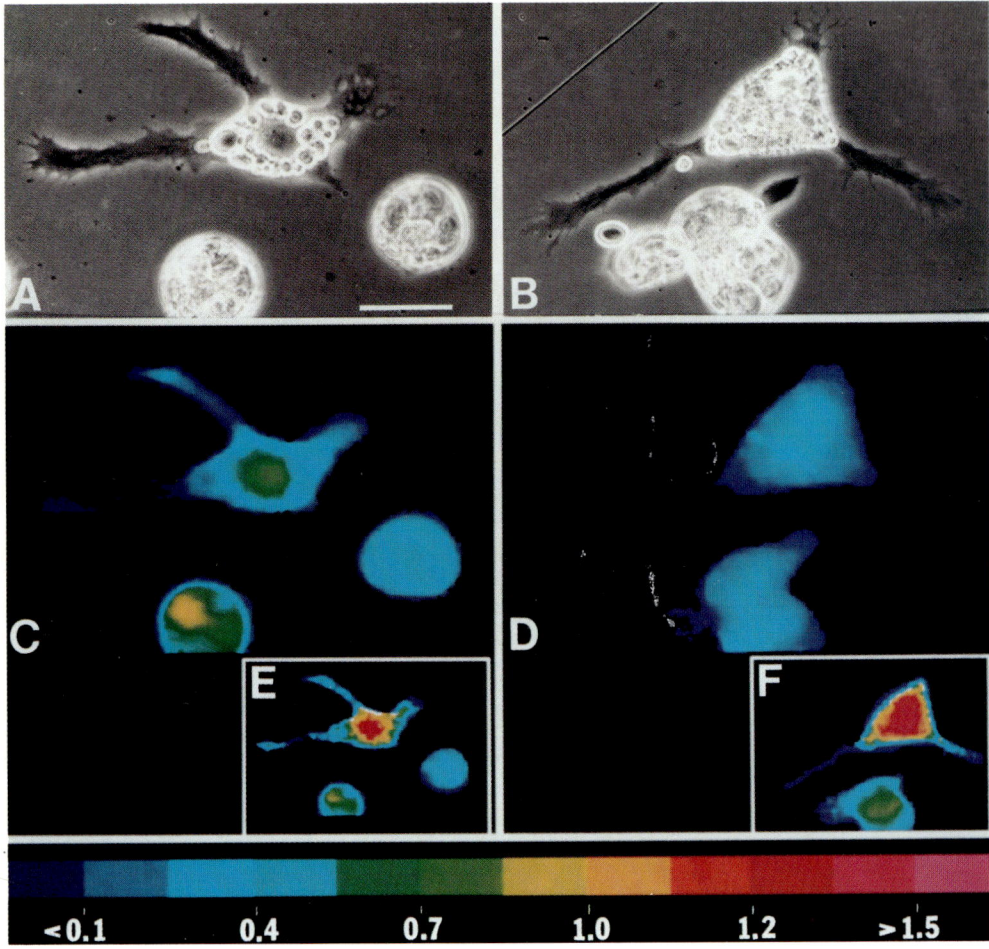

PLATE 2. Intracellular Ca^{2+} levels of resting and depolarized neurons at a developmental stage when Ca^{2+}-dependent action potentials are expressed. Neurons cultured from embryonic *Xenopus* neural plate begin to extend processes in either standard medium (A, 10 mM Ca^{2+}) or Ca^{2+}-free medium (B, 10mM Mg^{2+} and 1 mM EGTA) at 6 h in culture. The levels of Ca^{2+} are displayed in pseudocolor (C-F). Neurons growing in standard culture medium (A, C) often have different apparent concentrations of Ca^{2+} in separate processes extending from the same cell soma and elevated concentrations associated with their nucleus. This nuclear elevation is observed in many undifferentiated cells as well. Neurons growing in the absence of extracellular Ca^{2+} (B, D) maintain levels in the cell soma that are close to those seen in its presence, although the Ca^{2+} concentration in their processes is below detectable levels under these measurement conditions. Elevations of Ca^{2+} associated with the nucleus are generally small or absent (D). Depolarization of cells by the addition of K^+ (to 50 mM) in the presence of extracellular Ca^{2+} (5 to 10 mM) causes an increase in intracellular Ca^{2+} in neurons within 1 min (E, F). Increases occur in the soma, nuclear region and in processes while only small changes occur in most undifferentiated cells. Cells 7 h in culture. Calibration bar in A is 20 μM; the color scale refers to Ca^{2+} concentration in μM. (From Holliday and Spitzer, 1990. With permission.)

That this elevation of $[Ca]_i$ triggers transmitter release has been demonstrated by several experiments that manipulated presynaptic $[Ca]_i$. Miledi (1973) performed the most critical experiment by showing that microinjection of Ca^{2+} into squid presynaptic terminals was able to evoke release (Figure 3B). Independently, Glagoleva et al. (1970) and Alnaes and Rahamimoff (1975) found that treatment with metabolic inhibitors, which should elevate $[Ca]_i$ by inhibiting Ca^{2+} removal from the presynaptic terminal, caused dramatic increases in transmitter release. Further, elevating $[Ca]_i$ by ionophores (Kita and Van der Kloot, 1976) or liposomes (Rahamimoff et al., 1978b; Kharasch et al., 1981) also increases transmitter secretion. Finally, microinjection of Ca^{2+} chelators into squid presynaptic terminals attenuates evoked transmitter release (Adler et al., 1991). In summary, elevating $[Ca]_i$ increases release and decreasing $[Ca]_i$ reduces release. This proves that an increase in $[Ca]_i$ couples Ca^{2+} influx through the voltage-gated calcium channels to triggering of transmitter secretion.

B. General Criteria for Identifying Calcium-Dependent Secretion

These studies of neurotransmitter secretion make it possible to define general criteria for identifying whether secretion in a given system is triggered by calcium entry through voltage-gated calcium channels. Important criteria for proving such a scheme include:

1. A requirement for external Ca^{2+}:
   a. Eliminating external Ca^{2+} eliminates secretion
   b. There is a proportionality (which may be nonlinear) between $[Ca]_o$ and rate of secretion
2. A correlation with Ca^{2+} influx:
   a. Action potentials are present and capable of evoking secretion
   b. Direct depolarization of the membrane potential also causes secretion, but depolarizations to very positive potentials (in excess of $+60$ mV) should not cause secretion
   c. Calcium-dependent action potentials and/or voltage-dependent calcium currents are present
   d. Calcium channel antagonists and agonists alter secretion in ways consistent with their actions upon calcium channels
   e. There is a close correspondence between calcium entry and rate of secretion
3. A dependence upon $[Ca]_i$:
   a. $[Ca]_i$ rises during secretion
   b. Elevation of $[Ca]_i$ increases secretion and reduction of $[Ca]_i$ reduces secretion

At present there are few secretory systems in which all of these criteria have been established, because technical limitations often make it difficult to fulfill all criteria in a single experimental system. Further, additional criteria might be required in the future. For example, it presently is not clear how an increase in $[Ca]_i$ leads to exocytosis (Reichardt and Kelly, 1983; Augustine et al., 1987). Once the relevant effector molecules have been identified, it may be possible to use the presence of these molecules as an additional criterion for demonstrating calcium-dependent secretion.

C. Calcium-Dependent Secretion is a Widespread Process

We next will use the above criteria as a framework for reviewing some of the available information on the triggering of secretion in various other cells and tissues. Our goal will be to determine how well the calcium hypothesis is able to account for secretion in these other systems. The discussion first considers several systems that appear to adhere quite closely to this scheme and then examines some other systems in order of increasing deviation from the predictions of the calcium hypothesis.

1. Catecholamine Secretion from the Adrenal Medulla

Catecholamine secretion from the chromaffin cells of the adrenal medulla normally occurs in response to the release of acetylcholine from sympathetic neurons. We will briefly review only a very small number of the entries in the vast literature on this form of endocrine secretion. The reviews of Burgoyne (1984), Winkler et al. (1986), and Knight (1988) are good points of access to the rest of the literature on this subject.

The dependence of adrenal medullary catecholamine secretion upon $[Ca]_o$ was first established by Douglas and Rubin (1961). They demonstrated that secretion elicited both by acetylcholine and direct depolarization (via elevated external K) was abolished by elimination of external Ca^{2+}. They also demonstrated a correlation between $[Ca]_o$ and the rate of catecholamine secretion.

Several of the criteria required for documenting Ca^{2+} influx into chromaffin cells also have been satisfied. First, action potentials have been recorded from chromaffin cells (Kidokoro and Ritchie, 1980). Additionally, as mentioned above, direct depolarization of the chromaffin cell elicits secretion, and there appears to be a bell-shaped relationship between membrane potential and the rate of secretion from single chromaffin cells (Clapham and Neher, 1984). Completely convincing evidence for voltage-gated calcium channels is available, including direct recordings of macroscopic and unitary calcium channel currents (Fenwick et al., 1982; Clapham and Neher, 1984; Hoshi and Smith, 1987) and less-direct tracer measurements of voltage-dependent Ca^{2+} influx (Artalejo et al., 1987). Pharmacological agents that interact with voltage-gated calcium channels also have the predicted effects upon catecholamine secretion (e.g., Garcia et al., 1984; Gandia et al., 1987). Technical difficulties thus far have hindered analysis of the quantitative relationship between Ca^{2+} entry and catecholamine secretion from single chromaffin cells, but efforts are underway (Bookman and Schweizer, 1988).

The correlation between the $[Ca]_i$ of chromaffin cells and catecholamine secretion is even better established than in neurons. A rise in $[Ca]_i$ is evident during secretion (Plate 1, Knight and Kesteven, 1983; O'Sullivan et al., 1989). One important additional line of evidence comes from studies of chromaffin cells whose membranes have been rendered permeable to Ca^{2+} (Baker and Knight, 1984). In such preparations it has been possible to measure the quantitative relationship between $[Ca]_i$ and rate of secretion (e.g., Knight and Baker, 1982; Dunn and Holz, 1983; Wilson and Kirshner, 1983), with the general observation being that the secretory apparatus is half-activated at a $[Ca]_i$ of approximately 1 μM (Figure 4). Although detailed comparison between permeabilized and intact chromaffin cells is not yet possible, these studies on permeabilized cells provide a completely unambiguous demonstration of a $[Ca]_i$ requirement for secretion.

In summary, secretion from chromaffin cells appears to adhere quite closely to the model developed for neurotransmitter secretion. Indeed, many of the central tenets of the calcium hypothesis of neurotransmitter secretion were developed in parallel by Douglas (1968) as a consequence of his classic work on adrenal catecholamine secretion. The coherence of the pictures of "excitation-secretion" coupling resulting from studies of these two systems has provided much of the motivation for attempts to implicate Ca^{2+} in triggering of secretion from other cells.

2. Pancreatic Beta Cells

Beta cells are endocrine secretory cells that are dispersed in small "islets" throughout the exocrine tissue of the pancreas. The secretion of insulin from these cells also appears

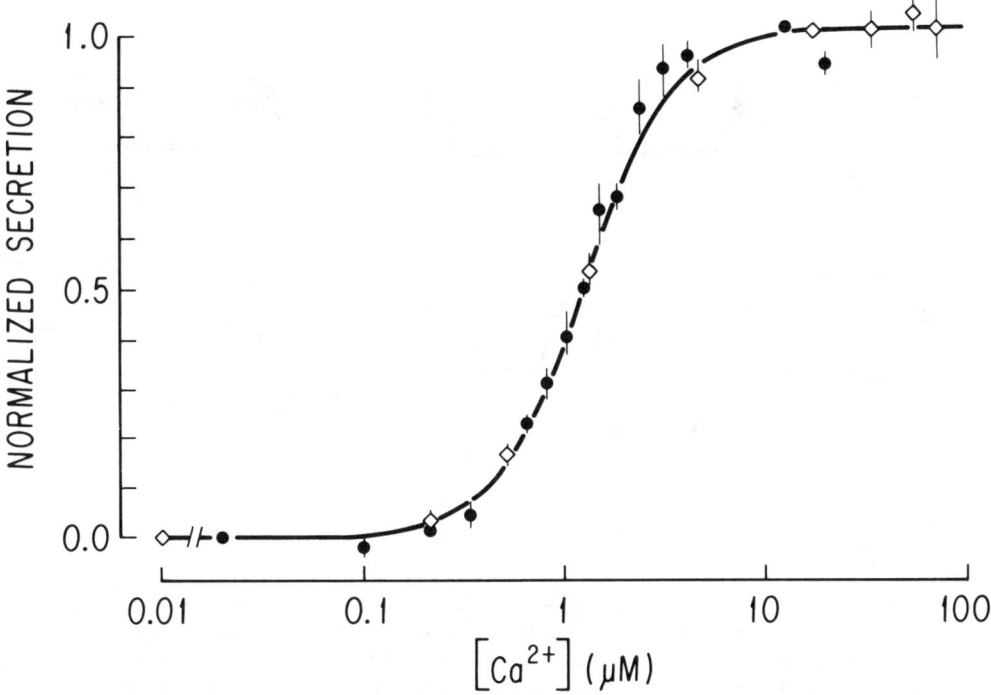

Figure 4. Relationship between [Ca]$_i$ and rate of catecholamine secretion from permeabilized chromaffin cells. The symbols indicate measurements made under different conditions, with the vertical lines indicating ±1 S.E.M., and the solid line is a function based on saturable binding of Ca^{2+}, with an affinity constant of 1.8 μM and a Hill coefficient of 2.2. From Knight and Baker (1982), with permission.

to be a calcium-dependent process very similar to that described for neurons and chromaffin cells. Several relevant reviews are available (Hellman and Gylfe, 1986; Peterson and Findlay, 1987; Prentki and Matschinsky, 1987).

The dependence of insulin secretion on extracellular Ca^{2+} has been well established. Physiological triggers of insulin secretion, such as glucose, require the presence of external Ca^{2+} (Curry et al., 1968; Wollheim and Sharp, 1981). Many studies have shown that beta cells produce calcium-dependent action potentials (Atwater and Beigelman, 1976; Ribalet and Beigelman, 1981; reviewed in Petersen and Findlay, 1987). These action potentials are generated in glucose-regulated bursts of activity and are correlated with similar bursts of insulin secretion (Figure 5). The oscillations in membrane potential that underlie the bursts apparently are controlled by ATP-sensitive K^+ channels (Rorsman and Trube, 1985; Peterson and Findlay, 1987).

Since quantification of insulin release from single cells has only just begun (e.g., Penner and Neher, 1988a), most evidence correlating secretion to the electrical activity of single cells has been indirect (Rosario et al., 1986; Prentki and Matschinsky, 1987). In the cell line RINm5F, which is able to secrete insulin in response to agents other than glucose, Wollheim and Pozzan (1984) used voltage sensitive dyes to correlate glyceraldehyde-stimulated changes in cellular membrane potential with secretion. Such studies indicate that membrane depolarization precedes secretion and more firmly implicate the glucose-sensitive bursts of action potentials in initiating secretion.

Patch clamp studies have established that one or more types of voltage-gated calcium channels are present in beta cells (Satin and Cook, 1985; Rorsman and Trube, 1986; Trube and Rorsman, 1987; Plant, 1988; Rorsman et al., 1988; Satin and Cook, 1988) and RINm5F

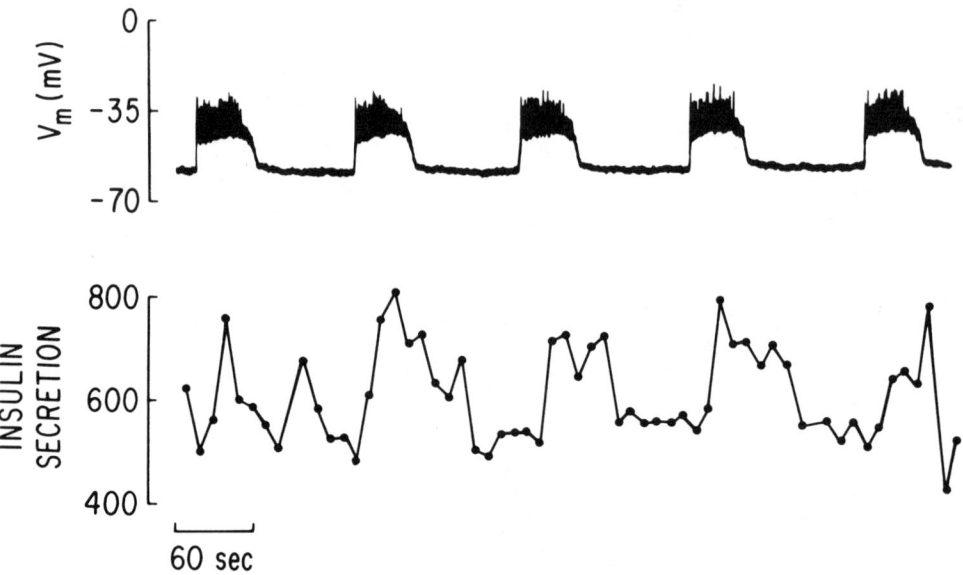

Figure 5. Correlation between electrical activity and insulin secretion in a pancreatic islet. Top, electrical activity of a single β-cell (V_m), measured with a microelectrode. Bottom, insulin secretion, measured with an immunoassay. Units are insulin-like immunoreactivity, expressed as pg/islet/min. Each burst of β-cell action potentials is associated with an increased rate of insulin secretion from the islet. From Rosario et al. (1986), with permission.

cells (Findlay and Dunne, 1985; Velasco et al., 1988). Because the calcium channels are activated over the voltage range traversed by the action potential, it is likely that these channels serve as the triggers for insulin release. Further, agents that block calcium channels—such as nifedipine (Malaisse-Lagae et al., 1984), verapamil (Wollheim and Pozzan, 1984) and D600 (Abrahamsson et al., 1985)—block insulin secretion (see also Chapter 13, Section III). Malaisse-Lagae et al. (1984) also found that the calcium channel agonist BAY K 8644 stimulated $^{45}Ca^{2+}$ uptake and insulin secretion from isolated rat pancreatic islets. Although influx of external Ca^{2+} through voltage-gated channels seems to serve as the trigger for glucose-stimulated insulin release, epinephrine and carbamylcholine can also mobilize internal calcium stores (Wollheim and Biden, 1986). Such release of internal Ca^{2+} might serve to amplify the secretory response that is triggered by influx through voltage-gated calcium channels (Prentki and Matchinsky, 1987).

Insulin secretion also appears to depend upon $[Ca]_i$. Measurements of $[Ca]_i$ in beta cells (Abrahamsson et al., 1985) or in RINm5F cells (Wollheim and Pozzan, 1984; Wollheim et al., 1984) show that intracellular Ca^{2+} levels rise two- to three-fold following stimulation. Using electropermeabilized beta cells, Jones et al. (1985) examined the dependence of insulin release on $[Ca]_i$ and found that maximum rates of secretion were observed with 1 to 10 μM Ca^{2+}.

Owing to the small size of beta cells, the difficulties in measuring release from single cells, and electrical coupling between cells in islets, the evidence that release of insulin from beta cells fits the calcium hypothesis is less complete than desired. Nonetheless, the hypothesis is presently capable of accounting for the process of action-potential mediated secretion of insulin from these cells. There is also evidence for a second, calcium-independent form of secretion in these cells (Wollheim et al., 1987).

3. Other Examples of Ca^{2+}-Dependent Hormone Secretion

In the interest of brevity, we will group together information obtained from other kinds of endocrine cells, both in intact tissue and in culture, to summarize the evidence for the general applicability of the calcium hypothesis to hormone secretion.

With the exception of the parathyroid gland, which will be described in a separate section, secretion from most other types of endocrine tissues appears to depend upon $[Ca]_o$ (e.g., Nordmann, 1983). It also has been known for some time that action potentials initiate secretion from endocrine tissues. Among the many demonstrations of action potentials in endocrine cells, those of Kidokoro (1976), Taraskevitch and Douglas (1977), and Salzberg and Obaid (1988) are particularly noteworthy. One important feature of hormone secretion from many endocrine cells is that it occurs in response to patterned trains of action potentials, with the specific form of these patterns exerting an important influence on the amount (and even the identity) of hormone secreted (Cazalis et al., 1985; Gainer et al., 1986; Bicknell, 1988). The causes of this pattern-sensitivity are incompletely understood, but it appears to be at least partially due to frequency-dependent changes in the action potential waveform (Bicknell, 1988).

Action potentials appear to trigger hormone secretion by opening voltage-gated calcium channels. Calcium channels have been shown to be present in many types of hormone-secreting cells, including those of the adrenal glomerulosa (Durroux et al., 1988), thyroid parafollicular cells (Kawa, 1988), anterior pituitary (Hagiwara and Ohmori, 1982; Dubinsky and Oxford, 1984; Matteson and Armstrong, 1986; DeRiemer and Sakmann, 1986; Mason et al., 1988; Leong, 1988), pituitary intermediate lobe (McBurney and Kehl, 1988) and the neurosecretory cells of the posterior pituitary (Cazalais et al., 1987a; Obaid et al., 1989; Lemos and Nowycky, 1989). There are as many as three types of calcium channels found in single endocrine cells (Durroux et al., 1988). Although at least some of these calcium channels should be involved in triggering hormone secretion, the reasons for such an abundance of calcium channel types are not yet clear. One possibility is that certain calcium channels are used to generate action potentials and/or regulate action potential frequency, while other types of channels serve as the site of entry for the Ca^{2+} influx that triggers secretion.

Hormone secretion from permeabilized neurosecretory terminals is a sensitive function of $[Ca]_i$, with half-maximal secretion occurring at levels of roughly 2 μM Ca^{2+} (Cazalis et al., 1987b). Presumably, Ca^{2+} influx through the voltage-gated calcium channel is the primary means of elevating $[Ca]_i$ and triggering hormone secretion (Brethes et al., 1987; Connor et al., 1987; Schegel et al., 1987). However, some excitatory hormones appear to augment endocrine secretion by releasing Ca^{2+} from internal stores (Ronning et al., 1982; Gershengorn, 1989). Thus, while secretion from endocrine tissues seems to correspond quite closely to the criteria of the calcium hypothesis, it is possible that both Ca^{2+} entry through voltage-gated calcium channels and release of Ca^{2+} from internal stores initiate hormone secretion.

4. Discharge of Trichocysts from *Paramecium*

Paramecium cells discharge the contents of membrane-bound intracellular organelles, known as trichocysts, in response to a variety of external stimuli. The result of this form of secretion is the appearance of needle-like trichocyst matrices, which represent the fully expanded contents of trichocyst vesicles (Satir et al., 1988). At least some of the criteria of the calcium hypothesis apply to trichocyst discharge.

Picric acid-stimulated secretion of the trichocyst matrix, as well as the expansion of isolated trichocysts (Bilinski et al., 1981), is absolutely dependent on $[Ca]_o$ (Satir and Oberg, 1978; Gilligan and Satir, 1983). The role of membrane depolarization in initiating secretion remains unclear. While *Paramecium* displays Ca^{2+}-dependent action potentials (Naitoh et al., 1972) and voltage-gated calcium channels (Eckert and Brehm, 1979; Ehrlich et al.,

1984), direct electrophysiological evidence relating a given calcium channel type to trichocyst secretion is lacking. Satir et al. (1988) found that verapamil and Cd^{2+}, but not nifedipine, inhibit secretion induced by trinitrophenol. Because it is thought that voltage-gated calcium channels are present only on the cilia of *Paramecium* (Ogura and Takahashi, 1976; Dunlap, 1977), the observation that *Paramecium* mutants lacking functional ciliary calcium channels are able to secrete when stimulated with trinitrophenol (Satir et al., 1988) casts doubt on the role of these calcium channels in triggering trichocyst discharge.

The calcium ionophore A23187 stimulates trichocyst discharge (Satir and Oberg, 1978; Gilligan and Satir, 1983), indicating a role for $[Ca]_i$ in regulating secretion. Using isolated trichocysts, Garofalo and Satir (1984) examined the relationship between $[Ca]_i$ and matrix expansion and found that expansion increased sharply with $[Ca]_i$ above 1 μM. Vilmart-Seuwen et al. (1986) have also found that elevating $[Ca]_i$ in permeabilized *Paramecium* leads to trichocyst discharge. Thus, $[Ca]_i$ probably is the primary regulator of trichocyst discharge.

To summarize, Ca^{2+} from external sources serves as the trigger for secretion of trichocysts and may enter through calcium channels, if such channels are present on the nonciliary membrane of *Paramecium*. More work is needed to correlate the role of membrane electrical events to the process of trichocyst secretion in *Paramecium*. An increase in $[Ca]_i$ appears to be the ultimate signal for trichocyst discharge, although such a signal has not yet been demonstrated during this form of secretion.

5. Cortical Granule Discharge During Egg Fertilization

As mentioned in a previous section, the exocytotic event associated with fertilization of sea urchin eggs, and presumably other types of eggs, is calcium dependent. This event is termed the cortical reaction, because the secretory vesicles involved are called cortical granules and secretion takes place from the periphery, or cortex, of the egg. Numerous experiments have demonstrated that $[Ca]_i$ rises during this cortical reaction (Steinhardt et al., 1977; Eisen et al., 1984), treatment with calcium ionophores (Steinhardt and Epel, 1974; Chambers et al., 1974) or microinjection of Ca^{2+} into egg cytoplasm (Hamaguchi and Hiramo, 1981; Turner et al., 1986) causes the reaction to take place, and microinjection of Ca^{2+} chelators blocks the reaction (Zucker and Steinhardt, 1978; Hamaguchi and Hiramoto, 1981). In addition, isolated cortical "lawns", consisting of isolated cortical granule-plasma membrane complexes (Vacquier, 1975), undergo secretion when exposed to Ca^{2+} concentrations of 1 μM or higher (Steinhardt and Alderton, 1982; Moy et al., 1983; Whitaker and Baker, 1983). Thus, the involvement of $[Ca]_i$ in the cortical reaction is based on convincing evidence.

However, the cortical reaction deviates from the calcium hypothesis because this form of secretion does not require the presence of Ca^{2+} in the extracellular medium (Steinhardt et al., 1977). Thus, Ca^{2+} influx from the extracellular medium does not serve as a primary trigger of exocytosis. Instead, Ca^{2+} is released from intracellular stores, presumably as a consequence of IP_3 production during fertilization (Whitaker and Irvine, 1984; Jaffe, 1988).

6. Histamine Secretion from Mast Cells

The secretion of histamine from mast cells, a process important in the production of allergic responses, deviates from the expectations of the calcium hypothesis in more radical ways. First, removal of extracellular Ca^{2+} does not completely eliminate this secretion (Foreman and Mongar, 1972). Second, there is no evidence for voltage-gated calcium channels in these cells (Lindau and Fernandez, 1986). There are, however, at least two types of ligand-gated calcium influx pathways that are activated during stimulation (Penner et al., 1988). Thus, although removal of external Ca^{2+} does not eliminate secretion, influx from the outside normally appears to be important in the secretory activity of this cell.

7. Calcium Channels as Triggers of Secretion

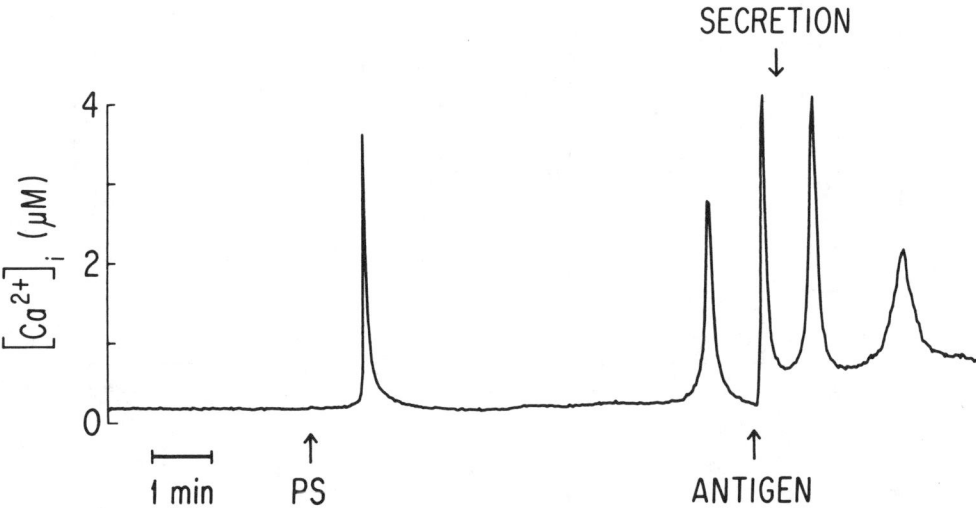

Figure 6. Secretion and [Ca]$_i$ in a single mast cell. The cell was loaded with Fura-2, to measure [Ca]$_i$, and was treated with phosphatidyl serine (PS) and antigen at the times indicated by the lower arrows. PS treatment produced two transient rises in [Ca]$_i$ that did not initiate secretion. Subsequent exposure to antigen produced a third transient rise in [Ca]$_i$ that was then followed by secretion (at the time indicated by the upper arrow) and further [Ca]$_i$ transients. From Neher and Almers (1986), with permission.

The [Ca]$_i$-dependence of mast cell secretion is more perplexing. [Ca]$_i$ rises dramatically in secreting mast cells, but this rise is not temporally linked to secretion (Figure 6; Neher and Almers, 1986; Neher, 1988). As discussed above, this rise in [Ca]$_i$ appears to be due to both release of Ca^{2+} from internal stores and influx from the external medium. Elevation of [Ca]$_i$ with ionophores triggers secretion (Foreman et al., 1973; Cochrane and Douglas, 1974) and secretion is activated by tonic elevation of [Ca]$_i$ above 1 μM (Bennett et al., 1981; Penner and Neher, 1988b). However, secretion can take place without a detectable rise in [Ca]$_i$ in cells loaded with the calcium buffer, EGTA (Neher, 1988). In summary, Ca^{2+} both enters and is released from internal stores during mast cell secretion, but secretion is not tightly coupled to these [Ca]$_i$ changes. It appears that secretion in these cells is accelerated by rises in [Ca]$_i$, but some other messenger molecule may be the critical trigger for secretion.

7. Antibody Secretion from B Lymphocytes

Secretion of immunoglobulins, in response to the presence of an appropriate stimulatory antigen, is the primary function of B lymphocytes. This response to antigen consists of several steps: antigenic stimulation initially causes a differentiation of B lymphocytes to form plasma cells and these plasma cells then perform the function of antibody secretion. Studies of antibody secretion have been hindered by the complexity of this response to antigen and further hindered by the fact that it has been difficult to identify and study the plasma cells directly.

With the discovery of voltage-gated calcium channels in an antibody-secreting, lymphocyte-derived cell line (Fukushima and Hagiwara, 1985), it was natural to suppose that these channels trigger antibody secretion from the plasma cells. Studies of these cultured cells revealed a good correlation between calcium channel density and the mean amount of antibody secreted per culture (Fukushima et al., 1984). Further, stimulation of native B lymphocytes with IgM can, under certain conditions, depolarize their membrane potential (MacDougall et al., 1988). However, depolarization of B lymphocytes, via elevated external K^+, does not raise Ca$_i$ (Grinstein et al., 1989). Further, patch clamp recordings provide no

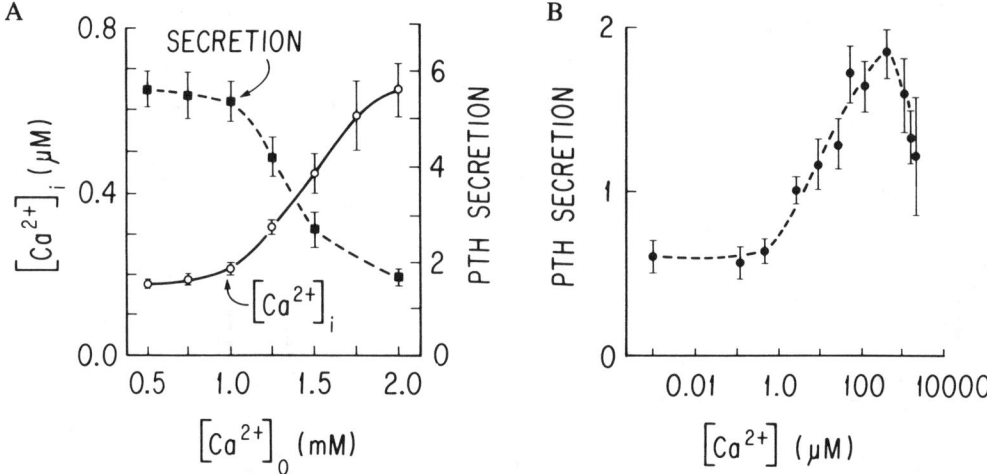

Figure 7. Ca^{2+}-dependence of parathyroid hormone (PTH) secretion. (A) Elevation of $[Ca]_o$ causes a modest rise in $[Ca]_i$ (open symbols, left axis) but a decline in PTH secretion (closed symbols, right axis). PTH secretion is expressed in units of ng PTH/10^5 cells/h. From Shoback et al. (1984), with permission. (B) PTH secretion is enhanced in permeabilized parathyroid cells when $[Ca]_i$ is raised above 1 μM. PTH secretion is expressed in units of ng PTH/10^5 cells/15 min. From Oetting et al. (1987), with permission.

evidence for voltage-gated calcium channels in native B cells (Lewis and Cahalan, 1990). This suggests that depolarization-initiated opening of voltage-gated calcium channels is not the stimulus for antibody secretion in plasma cells, although it has not yet been possible to make measurements of calcium signals directly from plasma cells. The ability of octanol to block both the voltage-dependent channels and proliferation of the antibody-secreting cell line suggests that these channels instead may be necessary for cell growth (Bosma and Sidell, 1988), a possible explanation for the correlation described by Fukushima et al. (1984).

Exposure of native B lymphocytes to anti-IgM causes a rise in Ca^{2+} (MacDougall et al., 1988). IP_3-regulated internal Ca^{2+} stores appear to be part of the source of this signal (reviewed by Cambier and Ransom, 1987). However, lymphocytes loaded with the calcium buffer, BAPTA, to chelate Ca^{2+} released from internal stores still produce a rise in Ca^{2+} in response to anti-IgM stimulation (MacDougall et al., 1988). These results suggest that B lymphocytes also have ligand-gated calcium channels, perhaps similar to those described for mast cells (Penner et al., 1988) and T lymphocytes (Lewis and Cahalan, 1989). Thus, antigen stimulation can produce a rise in $[Ca]_i$ in native B lymphocytes without requiring voltage-gated calcium channels.

The role of this $[Ca]_i$ signal in triggering antibody secretion remains unclear. It is conceivable that antibody secretion is not directly regulated by Ca^{2+}, because antibody secretion is constitutive (Tartakoff and Vassalli, 1977; Henkart et al., 1987). Perhaps the rise in Ca^{2+} produced by ligand-gated calcium channels and IP_3-mediated release of Ca^{2+} from internal stores instead is important in regulating the differentiation of B lymphocytes to plasma cells. Clearly, more data are needed—particularly from plasma cells—to clarify the role of calcium channels and $[Ca]_i$ changes in antibody secretion. For the present we suggest that the triggering of antibody secretion from plasma cells is not explained by the calcium hypothesis, a conclusion not unexpected for a constitutive form of secretion.

8. Parathyroid Hormone Secretion

The most radical deviation from the calcium hypothesis appears to be found in the secretion of parathyroid hormone from endocrine cells of the parathyroid gland. These cells secrete their calcium-mobilizing hormone in response to a drop in $[Ca]_o$, while a rise in $[Ca]_o$ inhibits hormone secretion (dashed curve in Figure 7A; Care et al., 1966). Because parathyroid hormone increases blood Ca^{2+} levels, such $[Ca]_o$-dependence is essential for this hormone to be able to keep Ca^{2+} levels in the blood constant. However, the $[Ca]_o$-dependence of parathyroid hormone secretion clearly is opposite to that found in other secretory cells. Further, raising $[Ca]_o$ depolarizes these cells (Lopez-Barneo and Armstrong, 1983), which would be expected to augment secretion if it were mediated by conventional, voltage-gated calcium channels.

The picture becomes even less simple when examining the relationship between $[Ca]_i$ and secretion of parathyroid hormone. Raising $[Ca]_o$ raises $[Ca]_i$, to levels near 1 μM, and reductions in $[Ca]_o$ have the opposite effect (solid curve in Figure 7A; Shoback et al., 1984; Nemeth and Scarpa, 1987). Elevating $[Ca]_i$ with a calcium ionophore also decreases parathyroid hormone secretion (Shoback et al., 1984). These observations could suggest that secretion in these cells has a reverse dependence upon $[Ca]_i$, with an increased $[Ca]_i$ leading to a decreased rate of secretion. However, secretion from permeabilized parathyroid cells is *enhanced* when $[Ca]_i$ is raised above 1 μM (Figure 7B; Oetting et al., 1987). Thus, it appears that parathyroid cells have a secretory mechanism that is sensitive to $[Ca]_i$, in the same way that is found in the more conventional secretory cells. However, the physiological signal for parathyroid hormone secretion, fluctuations in $[Ca]_o$, causes changes in $[Ca]_i$ that are too small to activate this mechanism.

There appears to be some other intracellular signal that triggers secretion from parathyroid cells. It has been proposed that parathyroid cells have a receptor in their plasma membrane that binds external calcium (Lopez-Barneo and Armstrong, 1983; Nemeth and Scarpa, 1987). Upon binding external calcium, this receptor may generate intracellular second messengers, such as cyclic AMP (Brown et al., 1978), diacylglycerol (Oetting et al., 1987) and/or inositol phospholipids (Shoback et al., 1988), that could regulate secretion independently of $[Ca]_i$. Whatever the nature of the signal, it is clear that parathyroid hormone secretion cannot be explained by the calcium hypothesis. Presumably an alternative strategy for secretory control was developed because of the unique requirement that this hormone be secreted in response to a reduction in blood Ca^{2+} concentration.

IV. CONCLUSIONS

We have summarized a great deal of information regarding the triggering of secretion from various secretory cell types. In this section we will attempt to extract some generalizations from this collection of facts.

A. Calcium is an Important Secretory Trigger

There are several, and perhaps many, second messenger systems capable of regulating secretion. However, Ca^{2+} appears to be the most prevalent intracellular regulator of cellular secretion. This is supported by the observation that most (and perhaps all) of the cells that we have considered secrete when their cytoplasmic Ca^{2+} concentration is elevated within a certain (cell-specific) range. Many of these cells take advantage of these capabilities by using an intracellular Ca^{2+} transient as a physiological trigger for secretion; parathyroid cells are a glaring exception to this generalization. It has been proposed that secretion from cells that possess voltage-gated calcium channels, termed electrically excitable cells, may be predominantly regulated by Ca^{2+}, while secretion from nonexcitable cells may be under the regulatory control of a wider spectrum of intracellular signals (Penner and Neher, 1988a). This appears to be a useful distinction, as long as it is appreciated that Ca^{2+} is capable of playing *some* regulatory role in every secretory cell in which $[Ca]_i$ has been manipulated.

B. Modes of Calcium Triggering in Secretory Cells

Among those cells that use Ca^{2+} as a secretory trigger, only a subset conform to the expectations of the calcium hypothesis. In particular, most deviations result from the fact that multiple strategies have developed to deliver Ca^{2+} to cytoplasmic secretory sites. While voltage-gated calcium channels are a common means of elevating $[Ca]_i$ during secretion, they are not the only means available because ligand-gated calcium channels and release of Ca^{2+} from internal storage depots also contribute to the regulation of secretion in some cases.

Selection of the mode of Ca^{2+} delivery appears to be governed by several factors. One important consideration is the nature of the *physiological signal* that initiates secretion. Electrically excitable cells, such as neurons and many endocrine cells, normally respond to rapidly acting, extracellular chemical signals by producing action potentials. These action potentials trigger secretion by opening voltage-gated calcium channels. Other cells, such as egg and mast cells, appear to respond primarily to slowly developing, extracellular chemical signals. Such cells often are incapable of producing conventional action potentials and instead utilize IP_3 and other ligands to open calcium channels that yield intracellular Ca^{2+} mobilization and/or a modest rate of Ca^{2+} entry from the external medium. Some endocrine cells appear to utilize both classes of Ca^{2+} sources, having both voltage-gated calcium channels and well-developed release of Ca^{2+} from internal stores. This may reflect a need to secrete in response to both action potentials and slower-acting extracellular ligands.

The mode of Ca^{2+} delivery also places important constraints on the *kinetics* of the secretory response. Excitable cells probably employ voltage-gated calcium channels because these channels gate quickly (in the millisecond time range) and can produce large and rapid rises in $[Ca]_i$ that can be translated into a rapid secretory response. Cells that use ligand-gated calcium channels must secrete more slowly, because of delays caused by metabolic generation of the signals used to open the channels and/or the small magnitude and slow time course of the $[Ca]_i$ changes produced by the channels.

The mode of Ca^{2+} delivery may also depend upon *spatial* considerations. Studies on neurons suggest that the sites of neurotransmitter secretion must be located very near the voltage-gated calcium channels (e.g., Augustine et al., 1989). This results from the conclusion that the retarded diffusion of Ca^{2+} in cytoplasm will cause the large and rapid Ca^{2+} fluxes occurring in the immediate vicinity of these channels to be greatly attenuated and slowed over distances of only hundreds of nanometers (Simon and Llinas, 1985; Zucker and Fogelson, 1986; Smith and Augustine, 1988). This is likely to be true for other cells that use voltage-gated calcium channels to trigger rapid secretion. Cells that do not use voltage-gated calcium channels as secretory triggers can use Ca^{2+} sources that are spatially distant from the sites of secretion, although the $[Ca]_i$ changes at the sites of secretion will necessarily be relatively small and slow.

In summary, the selection of cellular Ca^{2+} sources appears to be determined by several factors: (1) the nature of the physiological signal that initiates secretion; (2) the kinetics of the secretory response; (3) the position of secretory sites; and (4) the magnitude and time course of the $[Ca]_i$ changes desired. Consideration of all of these factors suggests that secretory triggering by Ca^{2+} can be divided into two general categories (Table 1). The "fast" mode of triggering uses voltage-gated channels to yield rapid secretion, while the "slow" mode of triggering relies upon ligand-gated calcium channels (and/or other second messengers) to produce slower secretion. The differences in kinetics in these two modes can be appreciated by comparing the brief (less than 1 ms) latency between stimulation and secretion in neurons (e.g., Figure 2B) to the much longer (many seconds) secretory latency of mast cells (e.g., Figure 6). It is the "fast" mode of secretion that is described by the calcium hypothesis of Katz, Douglas and others who studied neurons and endocrine cells. While each of two modes of triggering may be more obvious than the other in a given type

TABLE 1
Two Modes of Secretory Responses

	Fast mode	Slow mode
Kinetics of secretion	Rapid (ms-s)	Slow (s or longer)
Types of cells	Neurons, endocrine cells	Eggs, mast cells, endocrine cells
Initiating signal	Action potentials	Slowly acting extracellular ligands
Sources of Ca^{2+}	Voltage-gated calcium channels in plasma membrane	Ligand-gated calcium channels in plasma membrane or organelles
Location of secretory sites	Very close (<1 μm) to calcium channels	Not necessarily close to channels
Nature of $[Ca]_i$ transient	Large ($1-1000$ μM) and rapidly rising (μs-s)	Smaller and slower
Role of $[Ca]_i$ transient in secretion	Primary signal for triggering secretion	May initiate or modulate, but other messengers may also initiate

of secretory cell (Table 1), it is clear from the example of endocrine cells that both triggering mechanisms can co-exist within single cells. We therefore prefer to distinguish these modes of secretion by kinetic criteria, rather than by whether they take place in excitable or nonexcitable cells.

C. Diversity of Voltage-Gated Calcium Channels

Calcium channels are a diverse group of voltage-gated channels (Hagiwara and Byerly, 1981; Tsien et al., 1988; Bean, 1989). The cells that rely on voltage-gated calcium channels to act as secretory triggers apparently have taken full advantage of this wealth of calcium channels: every type of calcium channel that has been identified appears to be associated with secretory activity in some sort of cell. One intriguing exception is that no neuron has yet been found to use the fast-inactivating, low-threshold (LVA or T-type) calcium channel for triggering neurotransmitter secretion (Smith and Augustine, 1988). However, such channels are probably important in endocrine cell secretion (e.g., DeRiemer and Sakmann, 1986; Matteson and Armstrong, 1986; Satin and Cook, 1988; Durroux et al., 1989). Much remains to be understood about the logic of implementing particular types of calcium channels as triggers of secretion in specific types of cells.

ACKNOWLEDGMENTS

Much of this chapter was written in the laboratory of Professor E. Neher; we thank him for his hospitality and his comments on the chapter. We also thank C. Augustine, M. Cahalan and D. McCulloh for helpful comments, S. Silva and B. van Loo for assistance in organizing references, M. Tom for helping with typing and several people for allowing us to use figures from their publications. Financial support was provided by NIH grant NS-21624 and fellowships from the Max Planck Institute and Alexander von Humbolt-Stiftung to GJA and NIH fellowship NS-08392 to SDH.

Chapter 8

Calcium Channels in Sensory Transduction: A Case Study in Photoreceptors

Juan I. Korenbrot and Andres V. Maricq

TABLE OF CONTENTS

I.	Introduction	108
II.	Physiology of Transduction in Vertebrate Photoreceptors	109
III.	Ionic Mechanisms of the Electrical Response to Light	111
IV.	Gain and Kinetics of Phototransduction	111
V.	Molecular Mechanisms of Phototransduction	114
VI.	Control of the Gain and Kinetics of Phototransduction	115
VII.	Regulation of Cytoplasmic Free Ca^{2+} Concentration in the Rod Outer Segment	116
VIII.	Properties of Ca^{2+} Permeable Ion Channels of the Rod Outer Segment Membrane	117
IX.	Signal Processing in Sensory Receptor Cells	120
	A. Cone Photoreceptors	120
	B. Hair Cells	121
X.	Conclusion and Perspective	122
Acknowledgments		123

I. INTRODUCTION

Organisms gather information about the features of their environment, both internal and external, through sensory receptor cells. These cells are classified according to the type of stimulus they best respond to as exteroreceptors, interoreceptors or proprioreceptors. Exteroreceptors collect information about the external environment (e.g., light, sound, smell, taste, pressure, temperature), interoreceptors about the internal environment (pressure, oxygen tension, pH) and proprioreceptors assemble data that allow the organism to perceive its position in space (vestibular receptors, muscle spindles, stretch receptors in tendons and joints). Each receptor cell type has developed, through evolution, unique anatomical and physiological qualities that enable it to respond with exquisite sensitivity and reliability to a single class of physical or chemical stimuli (Autrum et al., 1971). In spite of the variety of form and function all receptors share a common physiological task: they must transduce the attributes of a stimulus (intensity, duration, direction, position in space, etc.) into signals understandable to other information processing elements in the organism.

In this chapter we examine the role of calcium and calcium-permeable ion channels in the function of sensory cells that generate electrical signals in response to environmental stimulation. Transduction schemes in these cells have three sequential steps in common. First, the stimulus is bound, absorbed or trapped by a specific receptor element of the cell. Thus, for example, olfactory and gustatory receptor cells reversibly bind chemical stimuli to molecular receptors that are integral constituents of the plasma membrane (review in *Lancet,* 1986), photoreceptors absorb photons in membrane-bound visual pigments (review in Birge, 1981) and, in hair cells, ciliary bundles are directly displaced by mechanical stimuli (review in Roberts et al., 1988). Second, the molecular elements that capture the stimuli generate a second signal that couples their active state to the opening or closing of ion channels. While it is possible that the molecular receptors could function as ion channels, no examples of such behavior have been discovered. The coupling event between molecular receptors and ion channels may be physical, for example, in hair cells where the mechanical displacement of the ciliary bundle is believed to gate stretch-activated channels (Howard and Hudspeth, 1988), or may be chemical. In chemical coupling, activation of the molecular receptor causes changes in the concentration of cytoplasmic second messengers. For example, in olfactory receptors stimulation causes changes in cytoplasmic cAMP concentration *(Lancet,* 1986; Nakamura and Gold, 1987), in muscle stretch receptors stimulation may also change cAMP levels (Sokabe et al., 1988) as may stimulation of taste cells (Tonosaki and Funakoshi, 1988). In a third step, the cytoplasmic second messenger or the physical coupling event gate ion channels in the membrane of the receptor cells. This change in channel activity causes changes in the membrane electrical conductance and, therefore, generates membrane receptor potentials.

Because calcium ions are ubiquitous intracellular second messengers, it is an obvious possibility that they may play a role in chemically coupled sensory transduction. To investigate whether Ca^{2+} does indeed play a role in transduction it is common to measure stimuli-dependent electrical responses from sensory cells under conditions presumed to interfere with Ca^{2+} homeostasis, for instance, removal of external Ca^{2+} or addition of known blockers of calcium conductance. Also, changes in cytoplasmic Ca^{2+} concentration in response to stimulation are often sought. Positive results in such experiments suggest that Ca^{2+} may play a role in transduction in taste cells (Akabas et al., 1988), olfactory cells (Winegar et al., 1988), hair cells (Corey and Hudspeth, 1979; Ohmori, 1985; Ohmori, 1988) and muscle spindles (Sokabe et al., 1988). Many of these experiments have been possible only through the application of recently developed techniques, such as voltage clamping with patch-electrodes and the use of fluorometric calcium-sensitive dyes. For many sensory cells, therefore, investigation has just begun and molecular insights into the mechanism of action of Ca^{2+} or the mechanisms that regulate its concentration are generally lacking. In particular, calcium dependence of the transduction signal may not be sufficient proof of a role for Ca^{2+} in transduction, since such effects may arise from actions of Ca^{2+} on ion channels or on the coupling elements of the transduction process unrelated to the course of events that underlie transduction.

Photoreceptors have long been studied as examples of sensory cells in which Ca^{2+} ions play a role in transduction, starting with the original hypothesis of Hagins and Yoshikami (Hagins, 1972). While that hypothesis proved to be incorrect, it motivated important experimental work that has led to a profound understanding of the role of Ca^{2+} in photoreceptor transduction, unparalleled in other sensory cells. The story remains incomplete, but we review here the state of our understanding both as a detailed case study of the role of Ca^{2+} and Ca^{2+}-permeable channels in sensory transduction and as an example of the intricate interactions that intracellular messengers can have with each other.

In addition to a direct role in the process of transduction, calcium channels are critically important in two other aspects of sensory receptor cell function: synaptic activity and information processing (see Chapters 7 and 9, Section II). Sensory receptors not only act as transducers but, in some instances, are also the first stage at which the information conveyed by the stimuli is processed. In two sensory cells, cone photoreceptors and hair cells, the role of calcium channels in information processing is particularly well described. Thus, we also review these data with the expectation that they are but examples of more generalized features of sensory receptor cell physiology.

II. PHYSIOLOGY OF TRANSDUCTION IN VERTEBRATE PHOTORECEPTORS

In total darkness, rods can signal distinctly the absorption of single photons and the amplitude of their signal is saturated when only a few hundred photons are absorbed (Baylor et al., 1979a; Baylor et al., 1979b; review in Baylor, 1987). Equally impressive, however, is the fact that in the presence of continuous light, both rods (Fain, 1976) and cones (Baylor et al., 1974; Norman and Werblin, 1974) adjust the amplitude and waveform of their photoresponse, a process known as light-adaptation, and thus extend the range of light intensities to which they respond from 1 to about 10^7 absorbed photons. Photoreceptors respond to light with changes in membrane ionic currents (Hagins et al., 1970) and membrane potential (Tomita, 1970). The molecular mechanisms that couple the absorption of photons by the visual pigment of the sensory cells to the modulation of membrane currents are now understood in some detail and have been recently reviewed (Stryer, 1986; Pugh and Cobbs, 1986; Liebman et al., 1987; Yau and Baylor, 1989). It is now clear that visual pigment excitation is coupled to ion channel modulation by two intracellular second messengers: cyclic 3'-5' guanosine monophosphate (cGMP) and Ca^{2+} ions. Illumination causes a decrease in the concentration of both cGMP (Cote et al., 1984; Cote et al., 1986; Blazynski and Cohen, 1986; Cohen and Blazynski, 1989) and Ca^{2+} (McNaughton et al., 1986; Ratto et al., 1988), but whereas the cGMP concentration directly controls the activity of ion channels (Fesenko et al., 1985; Haynes and Yau, 1985), Ca^{2+} ions appear to act by modulating the cGMP concentration changes (Korenbrot and Miller, 1986; Torre et al., 1986).

The specialization of photoreceptors for the capture of light and its transduction into neural signals is first apparent in the highly polarized structure of the cells (Figure 1). Photon capture and transduction occurs at one end of the cell, the outer segment, while neural interaction with other retinal cells occurs at the opposite end of the cell, the synaptic pedicle (Cohen, 1972). The synaptic pedicle is at the end of a section of the cell termed the inner segment. The dimension of the outer segment varies among species and probably reflects evolutionary pressures in response to ambient light levels. For example, in human rods the

Figure 1. Scanning electron micrograph of a retina isolated from the tiger salamander eye. Shown are the rod and cone photoreceptors, recognizable by the shape of the outer segments. The outer segments are attached through a narrow cilium to the inner segment. The synaptic end of the photoreceptors is not resolved in the micrograph. The diameter of the rod outer segment is 12 μm. Photograph courtesy of S. Mittman and D. R. Copenhagen, Department of Ophthalmology, University of California at San Francisco.

outer segment is 20 μm long, but can be 200 μm long in some deep sea fish. Cone outer segments can be as short as 5 μm, in lizards for example, but as long as 30 μm in some fish that exist in murky waters. Regardless of their absolute dimensions, outer segments consist of an orderly array of membranes stacked upon each other and lying normal to the long axis of the outer segment (Olive, 1980). In the rods, the membranes form enclosed discs separated from the plasma membrane but mechanically attached to it (Roof and Heuser, 1982), whereas in cones the membranes are a continuous fold of the plasma membrane (Fetter and Corless, 1987). In both cells the outer segment membranes contain the visual pigments that capture photons and the enzymes involved in the phototransduction process (Daemen, 1973; Roof et al., 1982).

The transfer of the light-induced electrical signal from the outer segment to the synaptic pedicle is electrotonic and the characteristics of this transfer reflect the fact that the electrical properties of the outer and inner segment membranes are not the same. Since electrical properties are a manifestation of the activity of ion channels, it is not surprising that different ion channels are found in the outer and inner segment membranes. In rods, the outer segment plasma membrane contains only one class of ion channels: a cGMP-regulated cationic channel (Baylor and Lamb, 1982; Yau and Nakatani, 1985b; Baylor and Nunn, 1986; Fesenko et al., 1986). The inner segment membrane is not electrically passive, it sustains five different ionic currents: a K^+ selective delayed rectifier, a Na^+ and K^+ selective anomalous rectifier, a voltage-dependent calcium current, a Ca^{2+}-activated chloride channel and a Ca^{2+}-activated potassium current (Bader et al., 1982; Bader and Bertrand, 1984; Corey et al., 1984; Hestrin, 1986). In cones, the outer segment also appears to contain only cGMP-dependent cationic channels, but the evidence is not as strong as that in rods (Haynes and Yau, 1985). The inner segment membrane in cones contains much the same currents identified in the rods, although the Ca^{2+}-dependent K^+ current is often absent (Maricq and Korenbrot, 1988).

III. IONIC MECHANISMS OF THE ELECTRICAL RESPONSE TO LIGHT

In the dark, the membrane potential of rods and cones is between -35 and -45 mV (Tomita, 1970; Baylor and Fuortes, 1970). This potential is the same in the inner and outer segments. In the dark, a continuous current flows in the extracellular space, along the cell's length from the inner toward the outer segment (Penn and Hagins, 1969; Hagins et al., 1970). The extracellular current is sustained by two different transmembrane currents: an inward current in the outer segment and an outward current at the inner segment (Hagins et al., 1970; Hagins, 1972; Korenbrot and Cone, 1972; Baylor et al., 1979a; Bader et al., 1982). The two currents have approximately equal amplitude and, consequently, at the dark membrane potential, the net membrane current is near zero (Bader et al., 1979; MacLeish et al., 1984; Baylor and Nunn, 1986). The dark current is maintained by the activity of ATP-dependent and ouabain-sensitive Na^+ pumps located exclusively in the inner segment (Penn and Hagins, 1969; Stirling and Lee, 1980). The magnitude of the dark current varies among species from a value of about 12 pA in the monkey rod (Nunn and Baylor, 1982) to 75 pA in the tiger salamander rod (Cobbs and Pugh, 1985; Nunn and Baylor, 1986; Hestrin and Korenbrot, 1987).

Illumination of rods and cones reduces the dark current amplitude (Hagins et al., 1970; Baylor et al., 1979a) and hyperpolarizes the membrane potential (Tomita, 1970; Baylor and Fuortes, 1970). It is conventional to describe the dark-current suppression as a photocurrent identical in amplitude and time course to the suppressed current, but opposite in polarity. In both rods and cones, dark-current suppression is the result of a decrease in the conductance of the outer segment membrane (Hagins, 1972; Korenbrot and Cone, 1972; Brown and Pinto, 1974; Capovilla et al., 1981; Woodruff et al., 1982; Hagins and Yoshikami, 1985), generated by the closure of a single class of light-sensitive ion channels (Fesenko et al., 1985; Haynes and Yau, 1985).

IV. GAIN AND KINETICS OF PHOTOTRANSDUCTION

The amplitude and time course of the photocurrent is a function of the stimulus energy. Rods and cones respond to different wavelengths of stimulation and the spectral response characteristics of the cells depend on the absorption spectra of the visual pigment they contain (Knowles and Dartnall, 1977). Any photoreceptor contains only one class of visual pigments. In response to brief flashes of light (a few milliseconds in duration) of the appropriate wavelength, the photocurrent in rods and cones (Figure 2) increases to a peak and then returns to its starting value. The time course of the response generated by dim flashes is well described by the function (Penn and Hagins, 1972; Baylor and Hodgkin, 1973; Baylor et al., 1979a):

$$I(t) = AE\left(-\frac{t}{\tau}\right)\left(1 - \exp\left(-\frac{t}{\tau}\right)\right)^{n-1} \tag{1}$$

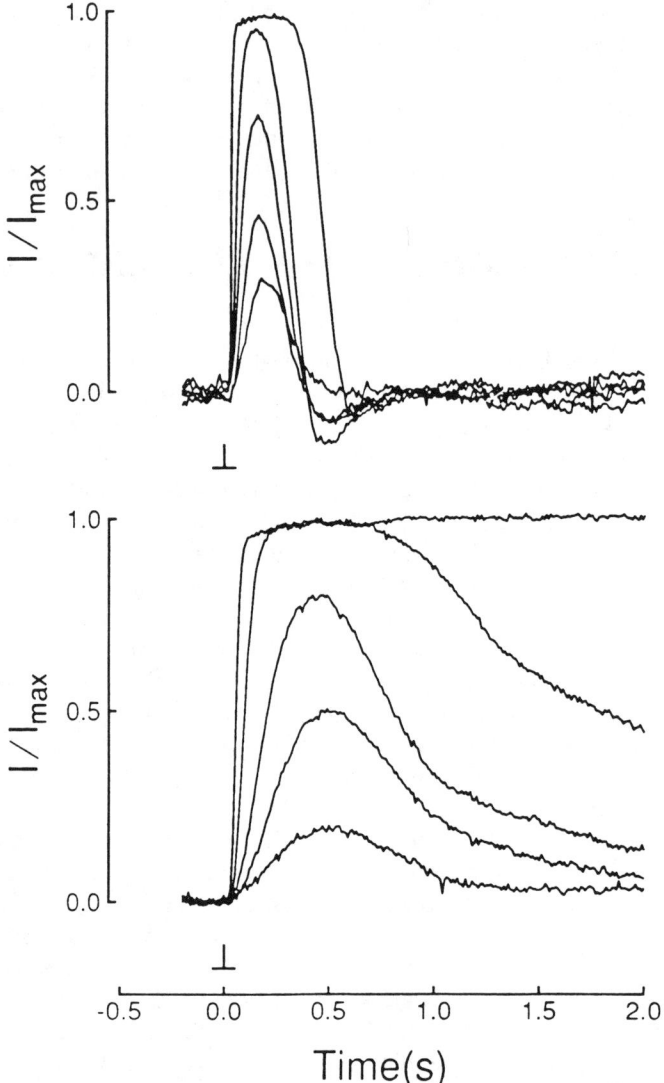

Figure 2. Photocurrents recorded in single cones (upper panel) and rods (lower panel) isolated from the tiger salamander retina. Currents were measured at room temperature with tight-seal electrodes in the whole-cell configuration. At time zero, cells were illuminated with 10 ms duration flashes of either 500 nm (rods) or 620 nm (cones) light. The responses in cones were generated in response to flashes that delivered 118.5, 354.3, 758.1, 1659, and 17,770 photons/μm^2. In rods the responses were generated by flashes that delivered 3.6, 7.9, 27.6, 180.8, and 846.4 photons/μm^2. The current amplitude was normalized by its value at saturation. In these cells, saturated photocurrent in cones was 35 pA and 71 pA in rods. The response of cones is both faster in time course and less sensitive to light than that of rods.

where I is the current, A is a proportionality constant, E is the energy of the flash and n and τ are adjustable parameters. This expression describes the kinetics of the impulse response of a linear sequence of n first-order chemical reactions each of time constant τ. For rods and cones in all species studied, n has been found to be in the narrow range of 4 to 6. On the other hand, the speed of the response, reflected in the value of τ is temperature dependent and varies over a large range among species: in rat rods at 36°C τ is about 30 to 70 ms

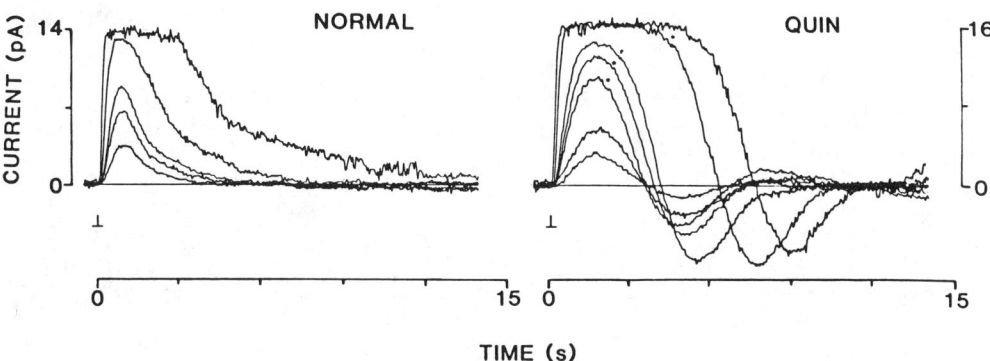

Figure 3. Photocurrents measured in single rods isolated from the toad retina. The panel labeled "normal" illustrates currents recorded under control conditions, while that labeled "Quin" illustrates currents recorded from rods loaded with the Ca^{2+} buffering agent, Quin2. Currents were measured at room temperature with suction electrodes. At time zero, cells were illuminated with 22 ms duration flashes of 500 nm light. The currents in normal cells were generated in responses to flashes that excited 15, 31, 63, 267, and 1630 rhodopsin molecules. In Quin2 loaded cells, the flashes excited 3, 8, 15, 31, 63, 267, and 690 rhodopsin molecules. The photoresponses in the presence of buffer, thus, are slower, oscillate at their offset, and are more sensitive to light than those in its absence. For example, the responses in Quin2-loaded cells identified by an asterisk, were generated by the four dimmest flashes used to excite the normal rods. (Data taken from Korenbrot and Miller, 1989.)

(Penn and Hagins, 1972), but it is about 530 ms in tiger salamander rods at room temperature (Baylor and Nunn, 1986; Hestrin and Korenbrot, 1987). Equation 1 does not describe the response generated by bright flashes and although the photocurrent amplitude saturates at low light levels, its initial rate of rise increases with light intensity and saturates only when over 10^6 photons are absorbed per cell both in rods (Penn and Hagins, 1972; Cobbs and Pugh, 1987) and in cones (Hestrin and Korenbrot, 1989). Although the molecular mechanism of transduction in cones is not fundamentally different than that in rods, in a given species and at a given temperature, the photocurrent in cones is invariably faster than that in rods (Figure 2).

The photoresponse in rods is more sensitive to light than that in cones. In complete darkness the amplitude of the photocurrent at its peak in both receptor types increases with the intensity of the stimulus up to a saturating value, when the dark current is entirely suppressed. Intensities above those that saturate the photocurrent amplitude prolong the time that the current remains saturated. The dependence of the photocurrent peak amplitude on light intensity is well described by the function (Baylor and Fuortes, 1970; Penn and Hagins, 1972; Baylor et al., 1979a):

$$I(E) = I_{max} \frac{E}{E + S} \qquad (2)$$

where I is the peak response amplitude, I_{max} is the saturated amplitude, E is the energy of the stimulus and S is the intensity at which the photocurrent amplitude is half its saturated value. The difference in sensitivity between rods and cones is apparent in the value of S. In the retina of tiger salamanders, for example, S in rods has an average value of 9 photons/μm^2, but in cones it is 299 photons/μm^2 (Hestrin and Korenbrot, 1989).

Photoreceptors respond to dim light flashes presented in the dark and also respond when the flashes are presented against a background of light. The response in the presence of a background, however, is less sensitive than that in darkness. The loss of sensitivity is referred to as "light-adaptation" and is evidenced by the fact that the response to a flash of constant intensity is smaller in the presence of a light background than in its absence (see Figure 4) (Fain, 1976; Baylor et al., 1979a; Lamb et al., 1981). Of course, when photoreceptors are

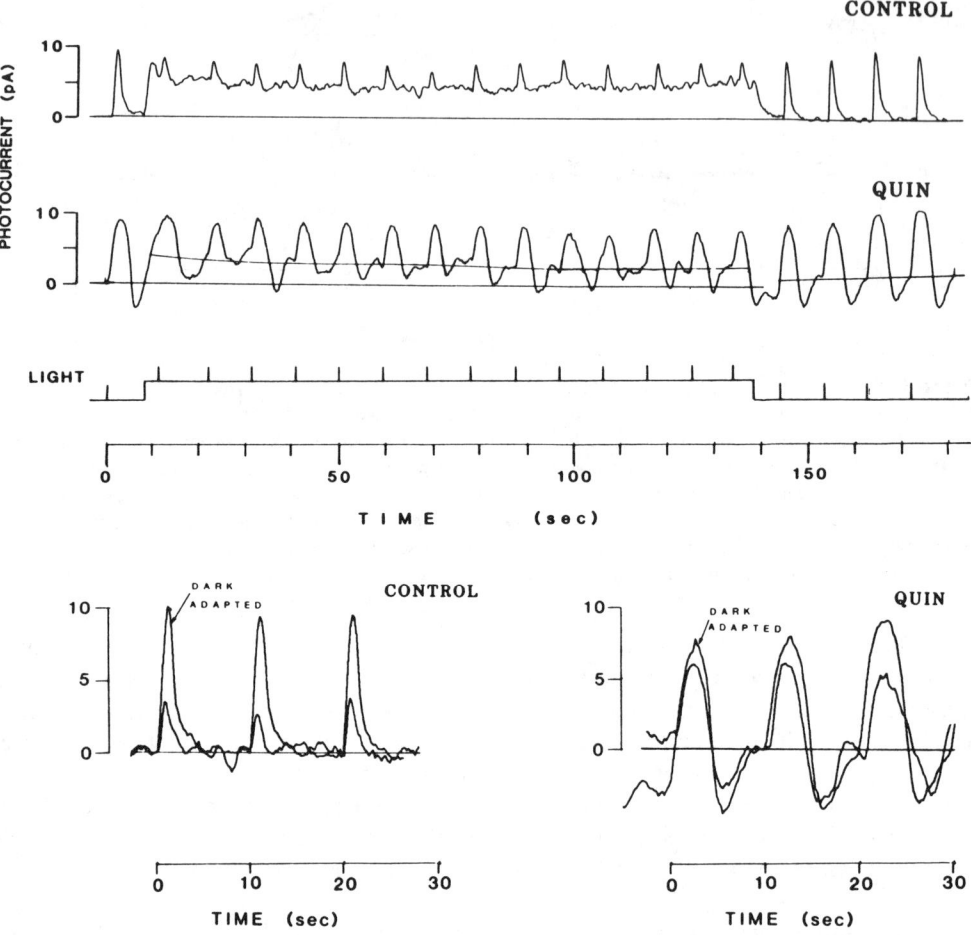

Figure 4. Effects of cytoplasmic Ca^{2+} buffers on light-adaptation in rods. Photocurrents measured in single rods isolated from the toad retina. The panels labeled "control" illustrate currents recorded under normal conditions, while those labeled "Quin" illustrate currents recorded from a rod loaded with the Ca^{2+} buffering agent, Quin2. Currents were measured at room temperature with suction electrodes. Flashes of constant intensity were repeated at 10 s intervals either in the dark of superimposed on a continuous background of light, as indicated by the light monitor. The lower panels illustrate and compare the flash photoresponses recorded either in the dark (dark-adapted) or superimposed on a background of light. In normal cells, the response to the flash was smaller in amplitude and faster in time course when superimposed on a background than when presented in the dark. These changes are the hallmarks of light-adaptation. In Quin2-loaded cells, the same background was much less effective in light-adapting the cell since the responses to the constant flash superimposed on the light-background were only slightly smaller and slightly faster than those recorded in the dark.

returned to darkness they recover their original sensitivity through a process of "dark-adaptation". Light-adaptation is itself a very sensitive process. In rods, absorption of 1 to 10 photons per second is sufficient to decrease the light sensitivity by half (Bastian and Fain, 1979). With light-adaptation not only do photoresponses become less sensitive, but the time course of the photocurrent is accelerated when compared to that of dark-adapted cells. In a light-adapted cell the time course of the photocurrent elicited by a dim flash is still described by Equation 1, but the speed of the response is faster.

V. MOLECULAR MECHANISMS OF PHOTOTRANSDUCTION

The mechanisms of phototransduction are more completely understood in rods than in cones, and the information reviewed below is that available for rods. Work reviewed elsewhere (Pugh and Cobbs, 1986; Pugh, 1987) has demonstrated that the kinetics of the photocurrent reflects the time course of closure of cGMP-gated channels in the plasma membrane of the outer segment. These channels close when the rod is illuminated because light causes a decrease in the cytoplasmic concentration of cGMP (Cote et al., 1984; Cote et al., 1986; Blazynski and Cohen 1986; Cohen and Blazynski, 1989). The concentration of cGMP is controlled by the dynamic balance of the activity of the enzyme that catalyzes its synthesis (guanylate cyclase) and that which catalyzes its destruction (cGMP-phosphodiesterase, PDE; Goldberg et al., 1983). In the dark, the balance of the two enzymatic activities maintains a steady-state total cGMP concentration of approximately 50 to 70 μM (DeAzeredo et al., 1978; Govardovski and Berman, 1981; Killbride, 1981; Cote et al., 1984; Blazynski and Cohen, 1986), of which only about 1 to 10% appears to be free, as judged by the magnitude of the dark current (Yau and Nakatani, 1985b). Absorption of a photon by the visual pigment, rhodopsin, results, within a few milliseconds, in activation of a GTP-binding protein, transducin (reviewed in Stryer, 1986; Hurley, 1987). Activation involves exchange of bound GDP for GTP. The "armed" transducin, in turn, activates the phosphodiesterase and thus initiates a decrease in the cGMP concentration (Liebman et al., 1987). For dim lights, a single photoexcited visual pigment molecule can activate as many as 500 to 2000 PDE molecules and catalyze the destruction of as many as 10^5 cGMP molecules/s (Liebman and Pugh, 1979; Liebman and Pugh, 1982). PDE activation is terminated by several reactions, each responsible for the deactivation of a step in the enzyme cascade. Thus, the visual pigment is inactivated by phosphorylation catalyzed by rhodopsin kinase, Transducin is disarmed by the hydrolysis of bound GTP into GDP by GTPase activity intrinsic to the protein molecule, and PDE activity is capped by binding of a specific inhibitory subunit (reviewed by Stryer, 1986; Liebman et al., 1987; Hurley, 1987).

Following its initial decrease, cytoplasmic cGMP must return to its starting value. The recovery of cGMP concentration demands not only the deactivation of PDE, but also the transient activation of guanylate cyclase (Goldberg et al., 1986). The mechanisms that generate this expected guanylate cyclase activation are unknown, but indirect arguments, presented below, suggest that light-dependent lowering of cytoplasmic Ca^{2+} concentration is involved in this process.

VI. CONTROL OF THE GAIN AND KINETICS OF PHOTOTRANSDUCTION

While the molecular mechanisms underlying phototransduction can be described in general terms, the processes that achieve the remarkable control of the kinetics and the gain of the various enzymatic reactions of the cGMP cascade are far from clear. Electrophysiological protocols have demonstrated that in both rods (Cobbs and Pugh, 1987) and cones (Hestrin and Korenbrot, 1989) the initial rate of rise of the photocurrent is probably limited by the rate of activation of the PDE and the kinetics of cGMP-gated channel closure. However, PDE appears to remain active for a period of tens of seconds after the membrane current or the membrane potential have returned to their starting, dark value (Kawamura and Murakami, 1986; Hodgkin and Nunn, 1988). That is, the time course of PDE activation appears to limit the activation phase of the photocurrent, but is unrelated to its deactivation phase. Indirect electrophysiological data also suggest that guanylate cyclase is activated at some point during the development of the photocurrent and that it, also, remains active even after the current has recovered its starting value (Hodgkin and Nunn, 1988). Thus, there is not a simple relationship between the time course of activation and deactivation of PDE or guanylate cyclase and the time course of the photocurrent.

Light-dependent changes in free cytoplasmic Ca^{2+} in the rod outer segment regulate the gain and kinetics of the photoresponse. This is made evident by the effects of cytoplasmic

Ca^{2+} buffering agents on the photocurrent. Ca^{2+} buffers, such as BAPTA, EGTA, and Quin2 all have qualitatively similar effects on the rod photoresponse when loaded either through patch electrodes (Lamb et al., 1986) or by cytoplasmic hydrolysis of hydrophobic esters of these molecules (Korenbrot et al., 1986; Korenbrot and Miller, 1986). The buffers alter both the photosensitivity and the time course of the response (Figure 3) (Torre et al., 1986; Korenbrot and Miller, 1986). In toad rods, for example, the light intensity necessary to half-saturate the photocurrent amplitude bleaches about 30 rhodopsin molecules in a normal rod, but only about 15 in one loaded with Quin2 (Korenbrot and Miller, 1986). The buffers cause large oscillations to occur at the end of the photocurrent. Also, the time to reach peak is about twice as long in the presence of buffer than in its absence and the time integral of the photocurrent may be as much as six- to eight-fold longer (Lamb et al., 1986; Korenbrot and Miller, 1986).

The control of the gain and kinetics of phototransduction, while exacting, is not stereotyped. As we discussed above, photoreceptors, light and dark, adapt by changing the time course and sensitivity of their response. If cytoplasmic Ca^{2+} is important in the modulation of the gain and speed of phototransduction, then light-adaptation, a functional expression of this modulation, should be altered by intracellular Ca^{2+} buffers. Indeed, both BAPTA and Quin2, when loaded into the rod cytoplasm, reduce the ability of the cell to light-adapt (Torre et al., 1986; Korenbrot and Miller, 1986): a given background of light is much less effective in causing light-adaptation in the presence of Ca^{2+} buffers than in their absence (Figure 4). A role for Ca^{2+} in light-adaptation has been directly demonstrated by the observation that photoreceptors, both rods and cones, fail to light-adapt under experimental conditions that are presumed to block light-dependent changes in the cytoplasmic Ca^{2+} concentration of the outer segment (Nakatani and Yau, 1988b; Matthews et al., 1988).

It is important to emphasize that interfering with light-dependent changes in cytoplasmic Ca^{2+} prevents light-adaptation, but also alters the time course of the photoresponse in dark-adapted cells (Torre et al., 1986; Korenbrot and Miller, 1986; Nakatani and Yau, 1988b; Matthews et al., 1988). That is, Ca^{2+} ions act to modulate the gain and speed of phototransduction under all conditions of background illumination. It is, therefore, inappropriate to think of Ca^{2+} ions simply as mediators of light-adaptation. Indeed, phototransduction can occur in the virtual absence of changes in cytoplasmic Ca^{2+} concentration but, again, the photocurrent under these conditions is abnormal in kinetics and photosensitivity (Nicol et al., 1987).

Since cytoplasmic Ca^{2+} alters the photocurrent, and the photocurrent reflects changes in cytoplasmic cGMP levels, it is not surprising that Ca^{2+} has been found to control the concentration of total cGMP in rods: a rise in Ca^{2+} lowers cGMP and a fall in Ca^{2+} increases the nucleotide concentration (Cohen et al., 1978; Killbride, 1980; Polans et al., 1981; Woodruff and Fain, 1982). It is obvious that such control could arise through an effect of Ca^{2+} on the activities of PDE, guanylate cyclase or both. It must be understood that such control may occur by direct interactions of Ca^{2+} with the enzymes or indirectly through the activity of regulatory elements that interact with the enzymes. Recent experiments have shown that guanylate cyclase activity in rod outer segment membranes is dependent on Ca^{2+} at concentrations in the range of those found in the cytoplasm (Lolley and Racz, 1982; Pepe et al., 1986; Pepe et al., 1986; Koch and Stryer, 1988). PDE activity, on the other hand, responds to changes in Ca^{2+} over a concentration range orders of magnitude larger than that of the cytoplasm (Robinson et al., 1980; Miller and Litman, 1989).

VII. REGULATION OF CYTOPLASMIC FREE Ca^{2+} CONCENTRATION IN THE ROD OUTER SEGMENT

The cytoplasmic free Ca^{2+} concentration in the outer segment of dark-adapted rods is around 270 nM (Ratto et al., 1988; Korenbrot and Miller, 1989), which is only about 0.005% of the total Ca^{2+} content of the organelle (Hagins and Yoshikami, 1975; Fain and Schroder, 1985). Light causes a decrease in free Ca^{2+} concentration, although the precise magnitude, time course, and photosensitivity of the change remain to be measured (McNaughton et al., 1986). In response to very bright illumination, this decrease follows an exponential time course (Ratto et al., 1988). The mechanisms that control the cytoplasmic Ca^{2+} concentration in the dark and its changes upon illumination are now understood in some detail (Yau and Nakatani, 1985a; Gold, 1986; Miller and Korenbrot, 1987; Nakatani and Yau, 1988b). In the dark, the Ca^{2+} concentration is maintained at a steady-state value by the dynamic balance between Ca^{2+} flux into and out of the outer segment. The influx occurs through the cGMP-dependent ion channels, the only channels in the plasma membrane. These channels, thus, are a class of poorly selective channels that allow both Na^+ and Ca^{2+} to enter the cell. Ca^{2+} efflux occurs through a $Na^+/K^+/Ca^{2+}$ carrier that transports 4 Na^+ ions into the cell in exchange for 1 Ca^{2+} and 1 K^+ ion (Cervetto et al., 1989). The coupling ratio of this exchanger, however, may not be constant since the rate of total Ca^{2+} efflux, measured directly with ion selective electrodes, is slower than that predicted from the electrogenic current generated by the Na/Ca/K exchanger transport (Gold, 1986; Miller and Korenbrot, 1987). This suggests, again, that the electrogenicity of the exchanger may not be constant or that other, nonelectrogenic, mechanisms of Ca efflux may operate in the rod.

Upon illumination, the light-sensitive channels close and the Ca^{2+} influx, therefore, decreases with a time course identical to that of the photocurrent and to an extent proportional to the photocurrent amplitude. In the presence of a continued Ca^{2+} efflux, the cessation of influx will reduce cytoplasmic Ca^{2+} concentration (Yau and Nakatani, 1985a). As the cytoplasmic Ca^{2+} is reduced below its level in the dark, Ca^{2+} efflux will then also decrease since active Ca^{2+} transport is proportional to cytoplasmic Ca^{2+} concentration. The quantitative characteristics of this model have been detailed in the toad rod, where it has been validated by its ability to predict light-dependent changes in extracellular Ca^{2+} concentration that accurately match experimental data (Miller and Korenbrot, 1987). Figure 5 illustrates a predicted time course for cytoplasmic Ca^{2+} concentration changes in the toad outer segment calculated from the model. Experimental data to test these predictions are not yet available. A potential shortcoming in the model is that it only considers the contribution of Ca^{2+} transport in the outer segment plasma membrane to the regulation of cytoplasmic Ca^{2+} concentration. Fluxes into or out of the discs might modify the profile of the concentration change (Fain and Schroder, 1987). However, rod outer segment disc membranes do not appear to have a light- or a cGMP-dependent calcium conductance (Noll et al., 1979; Cook et al., 1989).

VIII. PROPERTIES OF Ca^{2+} PERMEABLE ION CHANNELS OF THE ROD OUTER SEGMENT MEMBRANE

The light-sensitive channels of the rod outer segment membrane are, then, Ca^{2+} permeable channels. These channels, however, differ in most of their biophysical properties from the voltage-dependent and highly selective Ca channels discussed in other chapters of this volume. In contrast with conventional voltage-gated calcium channels, those in rods are gated by cGMP, are only weakly voltage-gated, are modulated by external and cytoplasmic Ca^{2+}, and are selective for other cations in addition to Ca^{2+}. Some of the relevant biophysical and physiological properties of the outer segment channels follow.

Ion selectivity—The ion selectivity of the light-sensitive channels in rods is one of its more intriguing features. As we discussed above, the principal carrier of the dark inward current in the outer segment is Na^+ (Hagins and Yoshikami, 1975; Yau and Nakatani, 1984; Hodgkin et al., 1985) but, along with the monovalent cation, Ca^{2+} ions also permeate the

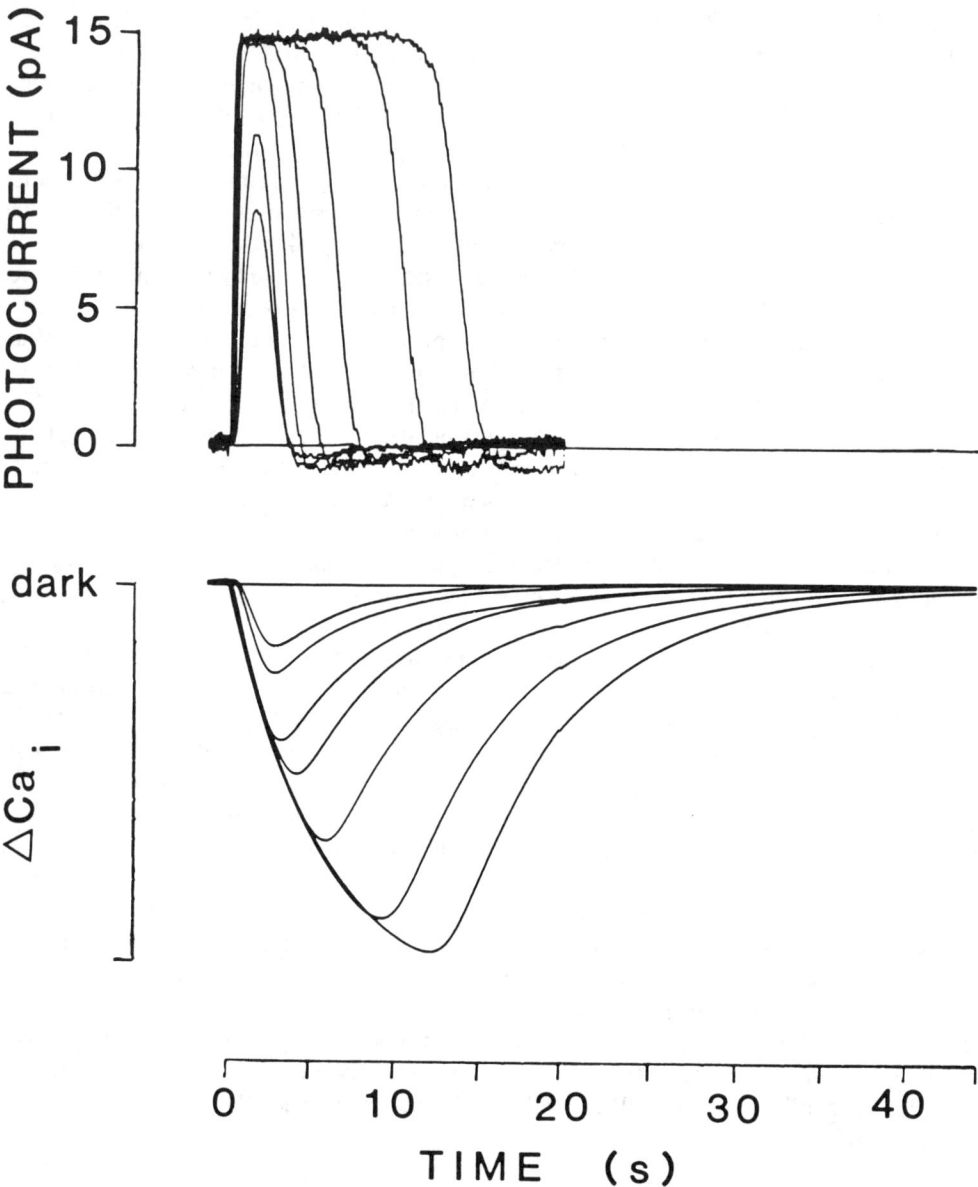

Figure 5. Model calculations of the time course of light-dependent decrease in cytoplasmic-free Ca^{2+} in the outer segment of toad rods. The top panel illustrates photocurrent measured in a rod in response to flashes that excited 33, 66, 275, 8.3×10^3, 1.8×10^4 and 7×10^5 rhodopsin molecules. The bottom panel illustrates the time course and extent of the decrease in cytoplasmic Ca^{2+} calculated for each photocurrent. Calculations are made using the model discussed in the text. (Data taken from Miller and Korenbrot, 1987.)

light-sensitive channels, as apparently do Mg^{2+} ions (Hodgkin et al., 1985; Nakatani and Yau, 1988b). In toad rods, for example, under normal conditions the fraction of the current carried by Na^+, Ca^{2+} and Mg^{2+} is around 0.7, 0.15, and 0.05, respectively, which can be calculated to arise from permeability ratios of P_{Ca}/P_{Na} of about 12.5 and P_{Mg}/P_{Na} of 2.5 (Nakatani and Yau, 1988b). The apparent permeability ratios for monovalent cations relative to Na^+ are: Li^+, K^+, Rb^+, Cs^+ 1.1, 0.7, 0.5, and 0.3, respectively (Hodgkin et al., 1984; Yau and Baylor, 1989).

The selectivity of the channels for divalent ions changes, apparently as a function of the cytoplasmic concentration of cGMP. Under normal conditions, the selectivity sequence

8. Calcium Channels in Sensory Transduction

is $Ca^{2+} > Sr^{2+} > Ba^{2+} > Mg^{2+} > Na^+$ (Menini et al., 1988). If the rods are treated pharmacologically with a protocol that increases cytoplasmic cGMP, the permeability of all divalent cations increase and some ions that are not permeable under normal conditions, like Mn^{2+}, can then flow through the light-sensitive channels (Cervetto et al., 1988).

Cyclic GMP- and voltage-dependence—Experiments on excised patches of plasma membrane from both rod (Fesenko et al., 1985; Zimmerman and Baylor, 1986; Haynes et al., 1986; Matthews, 1987), and cone (Haynes and Yau, 1985) outer segments have demonstrated that the light-sensitive channels are activated by cGMP and are not gated by Ca^{2+} concentration. That is, for channels to open the presence of cGMP is mandatory and Ca^{2+} does not affect the probability of channels to open. However, the conductance through the open channels is controlled by divalent ions. Mg^{2+} and Ca^{2+}, without distinction between them, can reduce the single channel conductance when present on the inside surface of the membrane (Lamb and Matthews, 1988; Nakatani and Yau, 1988b). The membrane conductance through these channels increases with cGMP concentration according to the sigmoidal function:

$$G(C) = G_{max} \cdot \frac{C^s}{C^s + K_{1/2}^s} \quad (3)$$

where G is the conductance, c is the cGMP concentration, $K_{1/2}$ is the concentration at half maximum conductance and S is a coefficient that reveals a cooperative dependence on cGMP concentration. The value of $K_{1/2}$ ranges between 10 and 30 μM in rods and is about 70 μM in a cone. The S coefficient is between 2 and 3 (Fesenko et al., 1985; Haynes and Yau, 1985; Zimmerman et al., 1985; Nakatani and Yau, 1985b; Stern et al., 1986; Matthews, 1986).

The channels are also gated by voltage (Nicol et al., 1984; Baylor and Nunn, 1986; Hestrin and Korenbrot, 1987). The current-voltage relationship of the photocurrent in rods has a reversal potential near 0 mV. For hyperpolarizing voltages, the current is nearly voltage independent. This implies that when the membrane voltage hyperpolarizes in response to light, the photocurrent will not be changed. On the other hand, the photocurrent exhibits strong rectification at depolarizing voltages. The outward rectification of the current has been explained as arising not from a direct effect of voltage on channel gating, rather by a voltage-dependent block of the current by divalent cations on the cytoplasmic surface of the channel (Yau and Baylor, 1989). The cooperative action of cGMP in opening the channel and the voltage dependence of the channel have been synthesized in a kinetic model that presumes the channel opening involves three diffusion-controlled cGMP binding steps followed by a voltage-dependent closed to open transition (Karpen et al., 1988).

Channel kinetics—Measurements of current noise in rod outer segment membranes in the dark first demonstrated that light-sensitive channels, under normal conditions, have an average lifetime in their open state of about 1 ms and a unitary conductance of about 0.05 pS (unitary current of about 2 fA) (Bodoia and Detwiler, 1985; Gray and Attwell, 1985). Similar conclusions are reached in the analysis of the noise activated by cGMP in excised membrane patches (Matthews, 1986).

The single channel activity of the light-sensitive channels has been measured in excised membrane patches (Zimmerman and Baylor, 1986; Yau et al., 1986; Matthews, 1987). Because divalent ions decrease the value of the open channel conductance, the amplitude of the single channel conductance is dependent on the concentration of divalent ions. Under physiological conditions, the single channel conductance is about 0.05 to 0.1 pS.

Openings of individual channels in the presence of low cGMP concentrations are infrequent and consist of brief bursts, several milliseconds in duration, of rapid flickering between open and closed states (Matthews and Watanabe, 1988). As the cGMP concentration increases, individual bursts remain of the same duration, but become more frequent until, at saturating cGMP concentrations, the channel is continuously and rapidly flickering between open and closed states. The lifetime of the open state during the burst of activity is about 0.2 ms at all concentrations of cGMP (Matthews and Watanabe, 1988). The opening rate of the channels, in response to nearly instantaneous changes in cGMP concentration, has been measured both in excised patches, when the cGMP is generated by laser photoexcitation of a "caged" cGMP compound (Karpen et al., 1988), and in intact rods, where cGMP is generated by brief, very bright flashes (Cobbs and Pugh, 1987). The channels respond to changes in cGMP concentration with a time constant of around 1 ms.

Molecular identity of channel protein—Biochemical purification has yielded a protein identified as the light-sensitive channel (Cook et al., 1986). Immunocytochemical studies have localized the protein in the outer segment plasma membrane (Cook et al., 1989). The protein is 68 kDa in molecular weight and, when incorporated into membrane vesicles, it acts as a cGMP-gated calcium conductance with a dependence on cGMP concentration similar to that of the channel in intact rods (Cook et al., 1987). In planar lipid bilayers, incorporation of this protein creates single channels that are also gated by cGMP (Hanke et al., 1988). The kinetics of the reconstituted channel, however, differ from those in the intact membrane in that the open channel lifetime is reported to be dependent on cGMP concentration. This is not the case in intact cells (Matthews and Watanabe, 1988).

Spatial distribution and membrane density—The density of the dark-current is uniform along the entire length of the rod outer segment (Hagins et al., 1970; Baylor et al., 1979). This implies that the surface density of open channels in the dark is constant throughout the outer segment. As discussed above, light-sensitive channels are the only channels present in the outer segment membrane (Baylor and Lamb, 1982; Hestrin and Korenbrot, 1987). In a tiger salamander rod, the dark-current is about 75 pA in amplitude (Hestrin and Korenbrot, 1987). Dividing this current by the single channel current, around 2 fA (Detwiler et al., 1982; Baylor et al., 1980), indicates that there are about 3×10^4 channels open in the dark at any given time. However, the number of channels open in the dark is but a small fraction, around 5%, of the total number of channels present in the membrane. This is evidenced by the fact that the average dark current is about 5% of the maximum dark-current measured in outer segments in the presence of saturating levels of cytoplasmic cGMP (Yau and Nakatani, 1985b; Hestrin and Korenbrot, 1987). The number of channels in a salamander rod outer segment, then, is about 6×10^5 from which a density of about 600 channels/μm^2 can be calculated.

IX. SIGNAL PROCESSING IN SENSORY RECEPTOR CELLS

In addition to its role in transduction and in synaptic function, Ca^{2+} may also participate in the processing of receptor signals. We next consider briefly in two sensory receptor cells, cone photoreceptors and hair cells, the way in which calcium channels are involved in information processing.

A. Cone Photoreceptors

Rods and cones hyperpolarize in response to light they absorb. In an intact retina, however, cones can also depolarize in response to illumination. Thus, a spot of light that illuminates only a single cone hyperpolarizes that cell. When the diameter of the illuminating spot is increased so that hundreds of neighboring photoreceptors are also stimulated, then cones, but not rods, exhibit a depolarizing component in their photoresponse (Baylor et al., 1973; O'Bryan, 1973; Fuortes et al., 1973). The depolarizing component can be isolated by stimulating with an annulus of light. Under these conditions the cones under the light annulus hyperpolarize, but those in the dark center of the annulus depolarize. If the stimulus is sufficiently bright, the depolarization may lead to action potentials in the cones. (O'Bryan, 1973; Piccolino and Gerschenfeld, 1978; Piccolino and Gerschenfeld, 1980; Gerschenfeld et al., 1980; Lasansky, 1981).

The interaction of distant cones with each other is mediated by horizontal cells which form both afferent and efferent synapses with the receptors. The synaptic interaction between cones and horizontal cells explains the depolarizing effect of light annuli on cones in the center of the annulus. In the dark, horizontal cells are depolarized and continuously release the neurotransmitter GABA (Ayoub and Lam, 1984) (review in Massey and Redburn, 1987). Cones have $GABA_A$ receptors in their synaptic pedicle and are hyperpolarized by the steady release of transmitter from the horizontal cell (Murakami et al., 1982; Kaneko and Tachibana, 1986). Upon illumination, horizontal cells connected to cones under the light annulus are hyperpolarized and, therefore, the release of GABA at their synapses is reduced (Murakami et al., 1982). The cones in the dark center of the annulus depolarize in response to the cessation of GABA release by the horizontal cell. Action potentials are generated if the depolarization is sufficiently large. Dissociated, single cones can also generate action potentials when depolarized (Maricq and Korenbrot, 1988).

The action potentials in single cones are generated by the activity of voltage-dependent calcium currents. This was first inferred in studies of intact retinas by the observation that action potentials occur only in the presence of extracellular Ca^{2+} (Piccolino and Gerschenfeld, 1978; Piccolino and Gerschenfeld, 1980; Gerschenfeld et al., 1980; Lasansky, 1981). Voltage clamp studies of dissociated single cones have affirmed that voltage-dependent calcium currents indeed underlie the depolarizing phase of the action potential and have resolved the following ionic mechanism of the action potential (Maricq and Korenbrot, 1988). The light-dependent depolarization of cones that follows cessation of GABA release by the horizontal cells, if sufficiently large, activates a voltage-dependent calcium conductance. This conductance is activated at around -40 mV and reaches half maximum activation at around -10 mV. Thus the current lies within the voltage operating range of the cones. The activated conductance allows Ca^{2+} to enter the cell and the elevated intracellular Ca^{2+}, in turn, activates a calcium-dependent chloride-conductance, producing an outward current that repolarizes the cell. The membrane repolarization terminates the calcium current. If membrane depolarization is sustained, cones can generate trains of action potentials because the calcium channels closed at the end of one action potential will be reopened by the sustained repolarization and thereby support another action potential (Maricq and Korenbrot, 1988).

The voltage-dependent calcium conductance of the cone inner segment has been studied in cells isolated from the retina of a lizard (Maricq and Korenbrot, 1988). The conductance is activated at voltages more positive than -40 mV, it shows little inactivation, it is permeable to Ba^{2+} as well as to Ca^{2+}, and it is blocked by cobalt. It is also blocked by the dihydropiridines, such as nitrendipine, and activated by others, such as Bay K 8644. In all respects, this conductance appears to arise from the activity of voltage-gated Ca^{2+}-selective channels with high activation voltage (HAV), similar to those reviewed in Chapter 3, Section I. It must be noted, however, that the calcium current in the cones exhibits a voltage-dependence that is matched to the voltage range over which the cone operates and that this range is not identical to that of the HAV channels in other cells.

B. Hair Cells

Hair cells are the sensory receptor cells of the vertebrate auditory, vestibular and lateral-line systems. These cells convert mechanical stimuli into electrical events by a process initiated by the displacement of the hair bundle located at the cell's apex. The hair bundle is made of tens to hundreds of stereocilia connected to each other and a single kinocilium located at the edge of the bundle. Individual hair cells respond best to a particular frequency of mechanical stimulation, referred to as the characteristic frequency (reviewed by Roberts et al., 1988; Howard et al., 1988).

In some species, frequency tuning of individual hair cells is explained by the features of the coupling between the stimulus and the displacement of the hair bundles in the cells: hair cells respond best to a given frequency because of the mechanical properties such as length, mass or stiffness of the bundles themselves (review in Roberts et al., 1988). In other species, tuning arises from the electrical behavior of the hair cell's membrane and calcium channels play a critical role in this function.

Electrical tuning of individual hair cell was first recognized by the observation that injection of current into single cells produces a sinusoidal oscillation in the membrane voltage with a frequency near that to which the cell responds best (Crawford and Fettiplace, 1981; Ashmore, 1983; Lewis and Hudspeth, 1983). Electrical resonance has been explained as arising from the kinetic interaction of two ionic currents: a voltage-dependent calcium current and a Ca^{2+}- and voltage-dependent potassium current (Art and Fettiplace, 1987; Ashmore and Attwell, 1985; Hudspeth and Lewis, 1988b). Depolarization of the hair cell produced by displacement of the hair bundle opens voltage-dependent calcium channels that allow Ca^{2+} to enter the cell and raise cytoplasmic Ca^{2+} concentration. The elevated intracellular Ca^{2+} and the depolarized membrane potential cooperate to open the Ca^{2+}- and voltage-dependent potassium channels. This produces an outward K^+ current that repolarizes the membrane potential. After a delay, calcium channels are activated again and the next cycle of oscillations starts. The kinetics of activation of the calcium currents and the Ca^{2+}- and voltage-dependent potassium currents vary among hair cells isolated from different mechano-transducing organs and even from the same organ (Roberts et al., 1986).

The voltage-gated calcium conductance in electrically resonant hair cells activate at membrane potentials more positive than -60 to -50 mV and they do not inactivate (Ohmori, 1984; Art and Fettiplace, 1987; Hudspeth and Lewis, 1988a). The time course of activation of the current is rapid (Ohmori, 1984; Art and Fettiplace, 1987) and can be described by a third order kinetic scheme (Hudspeth and Lewis, 1988a). The conductance is blocked by cobalt and is sensitive to dihydropiridines. In all respects, this conductance appears to arise from the activity of voltage-gated Ca^{2+}-selective channels of high activation voltage (HAV) of characteristics similar to those discussed in Chapter 3, Section I. Again, as in the case of the cones, the voltage dependence of the current is not identical to that of other cells but appears to be tuned to the membrane voltage range over which the hair cells operate.

X. CONCLUSION AND PERSPECTIVE

Understanding the role of Ca^{2+} in the physiology of sensory receptor cells addresses the following questions: What does Ca^{2+} do? How does it do it? and How does the cell control Ca^{2+}? The accumulated wisdom of research on the physiology of vertebrate photoreceptors lets us describe with some confidence the role of Ca^{2+} in transduction and information processing in these cells. The data reviewed here point out that while much is known, much also remains to be understood: we know that the gain and kinetics of the photo response are regulated by Ca^{2+}, but we do not know the mechanism of this regulation. We know that illumination causes a decrease in cytoplasmic free Ca^{2+} concentration, but we do not know the time course and extent of this change. We know Ca^{2+}-permeable channels are important in controlling the cytoplasmic Ca^{2+} concentration, but we do not know the role of intracellular fluxes in this process. These queries will be answered by future work. Nonetheless, the present information could serve as a rich guide in the analysis of the role of Ca^{2+} and Ca^{2+}-permeable ion channels in the transduction function of other sensory cells.

Consideration of the detailed biophysical features of the voltage-sensitive calcium channels of sensory cells discussed in this chapter and their comparison with similar channels in other cells (Chapter 3, Section I) reveals an important insight. While these channels appear to be a single class of transport proteins all described by rather similar kinetics, ion-selectivity and pharmacological properties, the voltage dependence of the channels is not identical from cell to cell. While in most cells, the voltage gating of the channels is well described by a Boltzmann function, the position of this function on the voltage axis changes from cell to cell. That is, the voltage threshold for channel activation and the voltage at which the channels are half-activated varies among cell types. Remarkably, however, the value of the threshold voltage is always well tuned to the range of membrane voltages that the cell is likely to experience in the course of its normal function. That such tuning should exist is hardly surprising, what is remarkable is that cells can specify the match between operating voltage and voltage-sensitivity of a channel that otherwise appears so exceptionally invariant.

ACKNOWLEDGMENTS

We thank Drs. C. Korenbrot, J. Miller, and A. Picones for their valuable comments on this chapter.

Chapter 9

Calcium Channels and Patterns of Electrical Activity in Neurons

Amy MacDermott and Richard Miles

TABLE OF CONTENTS

I. Introduction 126

II. Types of Calcium Channels 126
    A. Voltage-Activated Calcium Currents 126
        1. Transmitter-Regulated Calcium Currents 127
        2. Transmitter Modulation of Voltage-Gated Calcium Channels 128
    B. Ca^{2+}-Activated Currents 129

III. Localization of Calcium Channels 129
    A. Active Dendritic Responses 130
    B. Calcium-Mediated Events in Purkinje Cell Dendrites 130
    C. Dendritic Calcium Currents 130
    D. Visualization of Sites of Ca^{2+} Influx 130
    E. Synaptic Activation of Calcium Currents in Purkinje Cell Dendrites 131
    F. Ca^{2+} Influx and Synaptic Plasticity in Purkinje Cells 131
    G. Dendritic Calcium Channels in Other Neurons 131
    H. Micro-Organization of Calcium Channels 132

IV. Interactions Between Voltage-Gated Calcium Channels and Transmitter-Gated Channels 132

V. Calcium Channels and Burst Firing Patterns 133
    A. Calcium Currents of Burst Firing Invertebrate Neurons 133
    B. Role of a Ca^{2+}-Dependent Cation Current in Burst Firing 134

C.	Rhythmic Activity in Mammalian Neurons	134
D.	Is the Low-Threshold Current Responsible for Burst Firing in Mammalian Cells?	135
E.	The Low-Threshold Current as a Voltage-Dependent Switch	135

VI. Summary 135

I. INTRODUCTION

Ca^{2+} ions are essential for signaling within and between neurons. Intracellular Ca^{2+} acts as an essential trigger for transmitter release from presynaptic terminals, an interneuronal signal. As an intracellular signal, Ca^{2+} enters neurons through voltage- and transmitter-gated channels and carries a charge that depolarizes neuronal membrane. In addition, the Ca^{2+} influx influences the activity of other ion channels. In this chapter we examine how signal-gated calcium channels and Ca^{2+}-dependent channels contribute to the generation of specific neuronal firing patterns.

We will consider three facets of the complex role of Ca^{2+} ions in the control of neuronal activity. First is the cellular distribution of sites for Ca^{2+} influx. We examine the case that dendritic membrane forms an important locus of Ca^{2+} entry and the consequences of this localization for neuronal integration. Second, we will consider how interactions between transmitter-gated channels and voltage-dependent channels depend on membrane potential and intracellular Ca^{2+} concentration ($[Ca^{2+}]_i$). Finally, calcium currents have often been associated with slow rhythmic firing patterns. This occurs with calcium channels rather than sodium channels since calcium channel kinetics tend to be slower than those of sodium channels. We therefore explore how Ca^{2+} and Ca^{2+}-dependent channels are involved in the generation of rhythmic neuronal firing.

II. TYPES OF CALCIUM CHANNELS

Several types of calcium channels, differing in physiological and pharmacological characteristics, have been identified. Three groups of calcium channels may be clearly distinguished at present; low-threshold, voltage-gated channels, high-threshold, voltage-gated channels, and transmitter-gated channels that are permeable to Ca^{2+} (see Chapter 3, Section I). Neurotransmitters may also affect Ca^{2+} influx by changing membrane potential. Other neurotransmitters can modulate calcium channel activity through direct G-protein action on channels or indirect G-protein action via intracellular messengers. We shall consider briefly each of these routes for Ca^{2+} entry.

A. Voltage-Activated Calcium Currents

Regenerative membrane depolarizations observed in crustacean muscle in zero Na^+ solutions first suggested that voltage-gated calcium conductances exist (Fatt and Katz, 1953, for review; Hagiwara and Byerly, 1981). Recordings from starfish eggs provided the first evidence that in one cell there might be two calcium conductances with distinctly different voltage dependencies for activation and inactivation (Hagiwara et al., 1975). Two distinct Ca^{2+}-dependent potentials with low and high thresholds recorded from inferior olivary cells provided the first evidence for two types of calcium channel activity in mammalian neurons (Llinas and Yarom, 1981a,b).

Voltage clamp and single channel measurements have since shown that at least two types of voltage-gated calcium channel exist in many cells (Nilius et al., 1985; Bean, 1985) including mammalian and avian sensory neurons (Carbone and Lux, 1984; Bossu et al., 1985; Fedulova et al., 1985; Nowycky et al., 1985; see Chapter 3, Section I). These channels have different voltage dependencies, kinetics and conductances. They have different dependencies on intracellular metabolism and may also have different spatial distributions in neuronal membranes.

The high-threshold calcium current is also termed "L" current (Nowycky et al., 1985; Nilius et al., 1985) or high voltage-activated current (Carbone and Lux, 1987a,b). It generally begins to activate around -20 mV and current amplitudes are maximal close to $+20$ mV. This current seems to be regulated by intracellular metabolism. When the high energy phosphates inside a cell are maintained at a high level and the intracellular Ca^{2+} is buffered to a low level, this channel shows little inactivation (Chad and Eckert, 1986). The channel mediating high-threshold currents has a conductance around 20 to 25 pS, is more permeable to Ba^{2+} than Ca^{2+}, and is a binding site for dihydropyridines (see Chapter 13, Section III).

There have been several reports of another high-threshold channel, sometimes termed "N" channel (Nowycky et al., 1985). This channel appears to have a stronger, voltage-dependent inactivation and to be insensitive to dihydropyridines (review in Miller, 1987). Single channel conductance measured with Ba^{2+} as the permeant ion is 11 to 13 pS. From holding potentials near -90 mV, this current activates near -20 mV, peaks rapidly, then decays over hundreds of milliseconds. It is 50% inactivated at -60 mV (Nowycky et al., 1985).

Low-threshold currents have been characterized in a variety of cell types following their discovery in the eggs of starfish (Hagiwara et al., 1975). Detailed studies in sensory neurons of rat and chick show that this current activates around -60 mV, and that the corresponding single channels have conductances between 8 to 10 pS (Bossu et al., 1985; Nowycky et al., 1985; Carbone and Lux, 1984; Fox et al., 1987; Kostyuk et al., 1987). This current inactivates rapidly in a voltage-dependent fashion with a half maximal steady-state inactivation around -50 to -40 mV. It seems not to be regulated by Ca^{2+} or phosphorylation-dependent processes. Low-threshold currents have also been identified in embryonic hippocampal neurons in culture (Yaari et al., 1987; Meyers and Barker, 1989), and acutely dissociated trigeminal and dorsal horn neurons (Huang, 1989), but not acutely dissociated adult hippocampal CA1 neurons (Kay and Wong, 1987). Indeed, reports on calcium currents both in cultured and acutely dissociated hippocampal pyramidal cells suggest the presence of N, L, and T channels (Lipscombe, Madison and Bley, 1986; Meyers and Barker, 1989), high- and low-threshold currents (Brown and Griffith, 1983; Yaari, Harmon and Lux, 1988), or a single, high-threshold inactivating calcium current (Johnston et al., 1980; Kay and Wong, 1987; Alger and Doerner, 1988). One explanation for the differences may be that the T current is expressed transiently in the development of some cells (Fedulova et al., 1986; Haydon and Man Son Hing, 1988; Navarrete and Walton, 1989; Thompson and Wong, 1989).

1. Transmitter-Regulated Calcium Currents

Neurotransmitters can initiate Ca^{2+} entry into neurons in several ways. The most direct way is when the receptor/channel complex itself is permeable to Ca^{2+}. Three receptor-gated channels are permeable to the divalent cation, Ca^{2+}, as well as monovalent cations. They are the N-methyl-D-aspartate (NMDA) channel in central neurons (MacDermott et al., 1986;

Ascher and Nowak, 1986; and Jahr and Stevens, 1987; Mayer and Westbrook, 1987), the ATP channel in smooth muscle (Benham and Tsien, 1987), and the nicotinic channel at the neuromuscular junction (Miledi et al., 1980). The relative permeability to Ca^{2+} of the nicotinic channel is much lower than that of the NMDA and ATP channels.

At this time, the NMDA channel is the only well-established, agonist-gated channel permeable to Ca^{2+} in vertebrate central neurons. It is one of three channels activated by the excitatory amino acid and neurotransmitter, glutamate. One unusual feature of the NMDA channel, in contrast to the two non-NMDA channels, is that current flow in the presence of ligand is strongly voltage dependent. Thus the current-voltage relationship shows a region of negative slope conductance (MacDonald et al., 1982) resulting from a voltage-dependent channel block by Mg^{2+} ions (Nowak et al., 1984; Mayer et al., 1984). The more negative the membrane potential, the more effective the block.

There are two important physiological consequences of NMDA receptor activation in the presence of extracellular Mg^{2+}. One is use-dependent potentiation of NMDA-mediated depolarization (e.g., Collingridge et al., 1988). When membrane potential is in the region of negative slope conductance, as it would be near resting potential, a small depolarization will relieve some of the Mg^{2+} block and allow more inward current to flow. This can cause a highly nonlinear response to NMDA receptor activation and even suggests the possibility that an NMDA potential can become regenerative. That is, in the sustained presence of agonist, or with high frequency activation, the voltage dependence of NMDA currents is similar to the voltage dependence of voltage-gated sodium or calcium channels and may cause some form of regenerative activation. Furthermore, depolarization caused by excitatory transmitters or activation of other voltage-gated inward currents can enhance the use-dependent potentiation of NMDA channels.

The other consequence of voltage-dependent Mg^{2+} block of the NMDA receptor/channel is that Ca^{2+} entry is agonist gated and potential dependent (Mayer et al., 1987). This route is regulated in a different way than that of voltage-gated calcium channels. Ca^{2+} flow through the NMDA channel requires both activation by transmitter and sufficient depolarization to permit ion flow. The voltage dependence of the block produces a highly nonlinear relationship between synaptic activity and the elevation of $[Ca^{2+}]_i$. NMDA channels probably provide more focal elevations of $[Ca^{2+}]_i$ than voltage-gated channels since only channels that have bound agonist can open.

2. Transmitter Modulation of Voltage-Gated Calcium Channels

Neurotransmitters may also affect membrane Ca^{2+} fluxes indirectly by changing membrane potential. Activation of excitatory amino acid receptors, for example, can depolarize a neuron into a range of membrane potential where calcium channels open. In this way calcium currents may tend to amplify strong excitatory synaptic actions. Conversely, synaptic inhibition may close calcium channels by hyperpolarizing the neuronal membrane. The precise nature of these interactions will depend on the spatial relations between calcium channels and excitatory or inhibitory synapses. Different voltage dependencies of the low- and high-threshold calcium channels may complicate effects of transmitter-mediated shifts in membrane potential on Ca^{2+} influx. So, while membrane hyperpolarization tends to close high-threshold calcium channels, it also removes inactivation of the low-threshold calcium channels.

Activation of some transmitter receptors modulates voltage-gated calcium currents in a voltage-independent fashion (Dunlap and Fischbach, 1981). Isoprenaline, neuropeptide Y, and bradykinin receptors are coupled through G-proteins to voltage-gated calcium channels. Direct actions of the alpha subunit of the G-protein on the channel may underlie the coupling in some instances. For example, G_o can mediate the effects of isoprenaline on calcium channels in hippocampal neurons (Gray and Johnston, 1987) and neuropeptide Y on sensory neurons (Ewald et al., 1989) while G_{i2} and G_o both appear to mediate responses to bradykinin (Ewald et al., 1989). G-proteins may, in other cases, activate second messenger systems that act indirectly on voltage-gated calcium channels (see Chapter 14, Section III for a complete consideration of these phenomena).

B. Ca^{2+}-Activated Currents

Ca^{2+} acts as an intracellular second messenger as well as carrying depolarizing current. Intracellular Ca^{2+} affects the opening probability of several membrane channels that are permeable to ions other than Ca^{2+}. Ca^{2+}-dependent potassium channels are the most ubiquitous and best studied of this type of channel (Meech and Standen, 1975). The interaction between calcium channels and potassium channels, which depends on $[Ca^{2+}]_i$ and voltage (Barrett et al., 1982), can provide a negative feedback control of membrane potential. When calcium channels open, the neuronal membrane depolarizes and $[Ca^{2+}]_i$ rises. This, in turn, activates Ca^{2+}-dependent potassium channels that hyperpolarize neuronal membrane potential into a range where calcium channels close.

There are at least three pharmacologically and physiologically distinguishable forms of Ca^{2+}-dependent potassium conductances (Castle et al., 1989). These can have different Ca^{2+} affinities and perhaps different distributions and thus different dependencies on $[Ca^{2+}]_i$ and sensitivity to calcium channel activity. Other channels activated by elevation of $[Ca^{2+}]_i$ include the calcium-dependent Cl^- channels (Owen et al., 1984) and calcium-dependent nonselective cation channels (Yellen, 1982). The nonselective cation channels may generate depolarizing after potentials that contribute to pacemaker or bursting behavior in some neurons (Partridge and Swandulla, 1988).

III. LOCALIZATION OF CALCIUM CHANNELS

Membrane proteins that mediate different functions are not uniformly distributed on the surface of a neuron. Postsynaptic receptors are clustered facing presynaptic transmitter release sites (Triller et al., 1985) and the density of voltage-gated sodium channels is highest close to axonal sites of action potential initiation (Angelides et al., 1988). Presynaptic terminals are one site of high calcium channel density (Pumplin et al., 1981; Smith and Augustine, 1988). The density of calcium channels also seems to be higher on dendritic than on somatic membrane for many neurons.

The clustering of voltage-gated calcium channels in dendrites may provide a substrate for interactions between excitatory synaptic events and long-lasting Ca^{2+}-mediated depolarizations. The location of these interactions in dendritic processes, where the surface-to-volume ratio is much higher than in the soma, has two important consequences. First, since local input resistance is higher in dendrites than in the soma, small currents have a correspondingly greater effect on membrane potential. This may facilitate interactions between unitary synaptic currents and voltage-gated calcium channels. Second, the opening of a few channels may cause large local elevations in $[Ca^{2+}]_i$ due to the higher surface-to-volume ratio. Thus Ca^{2+}-dependent modifications of membrane or cytoplasmic proteins may occur within a local dendritic region without affecting the nucleus or soma.

We shall examine in detail the evidence for a dendritic location of calcium currents in one neuron—the cerebellar Purkinje cell. It seems clear that Purkinje cell dendrites do not function as a passive conduit for synaptic events but rather are regions where calcium currents play an active role in the integration of synaptic events. Dendritic calcium channels participate in the generation of rhythmic burst firing patterns that form the electrical signature of the Purkinje cell. Purkinje cell responses to dendritic synaptic inputs are shaped by the proximity of high calcium channel densities and terminals of specific afferent pathways. Finally, changes in intradendritic Ca^{2+} concentrations underlie short-term interactions between afferent synapses and can cause persistent changes in synaptic efficacy.

Three technical advances have driven this evolution in thinking about the role of Ca^{2+} in dendritic integration. Semi-intact slice preparations facilitate direct recordings of dendritic potentials with sharp electrodes. Low resistance suction electrode techniques have allowed precise measurements of membrane currents of cultured or acutely isolated neurons, and most recently in the brain slice. Finally, sites of Ca^{2+} influx have been visualized directly with molecular probes that emit Ca^{2+}-sensitive light signals.

A. Active Dendritic Responses

Intradendritic recordings were first made from Purkinje cells of alligator cerebellum (Llinas and Nicholson, 1971). Active depolarizing potentials, slower than somatic action potentials, occur spontaneously and are triggered by EPSPs. Membrane hyperpolarization fractionates these responses into smaller, all-or-none components. It has been suggested that multiple sites on Purkinje cell dendrites can generate active responses and that these sites are separated by regions of inexcitable membrane.

B. Calcium-Mediated Events in Purkinje Cell Dendrites

The role of calcium currents in Purkinje cell dendritic function was explored further in cerebellar slices (Llinas and Sugimori, 1980a,b; Crepel et al., 1981; Hounsgaard and Midtgaard, 1988). Suppression of sodium currents by TTX allowed two forms of Ca^{2+}-dependent activity to be distinguished. Small depolarizations evoked prolonged, low amplitude, Ca^{2+}-dependent plateau potentials. At higher threshold, larger, faster, fractionated action potentials appeared. The plateau potentials appeared to dominate in fine, distal dendrites, while larger calcium spikes were most prominent in proximal dendrites. Excitatory synapses formed by parallel and climbing fibers could trigger these dendritic responses. Furthermore rhythmic, TTX-resistant burst firing seemed to originate dendritically.

C. Dendritic Calcium Currents

Voltage clamp recordings with low resistance suction electrodes have been crucial in separating somatic calcium currents with different voltage dependencies and kinetic properties (Carbone and Lux, 1987a,b; Fox et al., 1987a,b). However, in neurons with extensive dendrites, adequate voltage control throughout the cell is difficult to achieve. Thus, when voltage control of regenerative Ca^{2+}-dependent events could not be maintained, the underlying currents have been assumed to originate in the dendrites (Schwindt and Crill, 1981; Brown and Griffith, 1983; MacLachlan and Hirst, 1985; Yaari et al., 1987).

Voltage-clamp studies have provided some information on the distribution of calcium currents in Purkinje cells. In the soma of cultured cells, calcium currents appear to be small (Bossu et al., 1989; Hirano and Hagiwara, 1989) or completely absent (Hockberger et al., 1989). Both low- and high-threshold calcium currents have been detected as single channel activity in somatic membrane patches (Hirano and Hagiwara, 1989). However, recordings from isolated somatic and dendritic membranes suggest that the density of high-threshold channels is significantly higher on dendrites (Bossu et al., 1989).

D. Visualization of Sites of Ca^{2+} Influx

It is now possible to see Ca^{2+} distribution inside a cell using fluorescent probes sensitive to Ca^{2+} concentration (Lipscombe et al., 1988; Ross and Werman, 1988; Tank et al., 1988; Hockberger et al., 1989; Regehr et al., 1989). Visually detected sites of Ca^{2+} influx agree strikingly with earlier studies on the location of calcium channels in Purkinje cell dendrites. Measurements with arsenazo III and Fura-2 suggest that Ca^{2+} influx into Purkinje cells is predominantly dendritic (Ross and Werman, 1988; Tank et al., 1988; Hockberger et al., 1989). Ca^{2+} entry coincides with slow membrane depolarizations, previously attributed to calcium currents. In small, distal dendrites, Ca^{2+} concentration oscillates between 60 nM and 3000 nM during spontaneous rhythmic Purkinje cell firing, while somatic Ca^{2+} remains below 100 nM. Activation of climbing fiber synapses produces large $[Ca^{2+}]_i$ transients. Variations of $[Ca^{2+}]_i$ within distances of about 40 μm along a dendrite that might correspond to the proposed clustering of calcium channels have been seen. Alternatively, spatial fluctuations in Ca^{2+} buffering, extrusion, or perhaps clustering of potassium channels might be responsible.

E. Synaptic Activation of Calcium Currents in Purkinje Cell Dendrites

Purkinje cell dendrites receive excitatory synapses made by parallel fibers from cerebellar granule cells and climbing fibers from medullary olivary neurons (Eccles et al., 1966a,b). Parallel fiber synapses terminate on fine distal dendritic processes, while climbing fiber synapses involve many terminals distributed along more proximal dendrites. The spatial distributions of these synaptic inputs are reflected in the active responses that they trigger. Climbing fiber EPSPs initiate calcium action potentials associated with Ca^{2+}-dependent plateau potentials (Llinas and Sugimori, 1980; Crepel et al., 1981; Campbell et al., 1983; Hounsgaard and Midtgaard, 1988). In contrast, parallel fiber EPSPs can have a local effect on calcium potentials in the tertiary dendrites without necessarily influencing the climbing fiber activity coming in on the proximal dendrite (Tank et al., 1988).

F. Ca^{2+} Influx and Synaptic Plasticity in Purkinje Cells

Calcium currents might underlie both short-term and long-lasting interactions between the two synaptic inputs. When climbing fibers are activated repeatedly, the resulting Ca^{2+} influx activates Ca^{2+}-dependent potassium currents to produce a membrane hyperpolarization lasting for several minutes (Hounsgaard and Midtgaard, 1989). This depresses subsequent parallel fiber EPSPs. Conversely, synaptic stimuli can regulate the duration of active dendritic responses. Parallel fiber stimulation also activates inhibitory stellate cells. When suitably timed, the resulting IPSPs curtail Ca^{2+}-dependent plateau potentials initiated by climbing fiber EPSPs (Campbell et al., 1983).

A heterosynaptic form of long-term synaptic plasticity seems to depend on Ca^{2+} influx subsequent to synaptic activation of Purkinje cells. When climbing fibers and parallel fibers are co-activated, parallel fiber EPSPs may be depressed for several hours—providing the climbing fiber stimuli activate calcium spikes (Ito et al., 1982; Kano and Kato, 1987; Sakurai, 1987). This heterosynaptic depression seems to depend on a reduced sensitivity of postsynaptic quisqualate receptors, one of the two non-NMDA receptors, that may result from the climbing fiber-induced elevation of $[Ca^{2+}]_i$ (Kano and Kato, 1987).

G. Dendritic Calcium Channels in Other Neurons

The layered anatomical arrangement of hippocampal pyramidal cells has also facilitated dendritic recordings (Wong et al., 1979; Wong and Prince, 1979; Schwartzkroin and Prince, 1980). Furthermore, it has been possible to surgically isolate dendrites of CA1 pyramidal cells and show unambiguously that calcium channels are located on the dendrites (Benardo et al., 1982; Masukawa and Prince, 1982). Even more recently, optical recording of dendritic calcium signals has been accomplished (Regher et al., 1989).

The question of differential somato-dendritic calcium channel distribution has also been explored in cultured hippocampal cells (Yaari et al., 1987). The low-threshold calcium current appeared to be predominantly somatic since it could be well clamped. High-threshold currents were not easily controlled and so were assigned to dendritic sites. However, only high-threshold currents were detected in acutely dissociated, adult CA1 pyramidal cells that possess little dendritic membrane (Kay and Wong, 1987). In this case, there may be differences in the complement of calcium channels in cultured (or young) and mature hippocampal neurons (Doerner and Alger, 1988; Thompson and Wong, 1989). However, clamp performance in extended neurons may need to be assessed.

In several other vertebrate neurons, current clamp records suggest that low-threshold calcium currents are predominantly somatic or proximal while high-threshold currents are preferentially expressed distally on dendrites. These include inferior olivary neurons (Llinas and Yarom, 1981), dorsal horn neurons from spinal cord (Murase and Randic, 1983), thalamic neurons (Jahnsen and Llinas, 1984), and neocortical pyramidal cells (Stafstrom et al., 1985). More detailed work may reveal whether this generalization holds true.

H. Micro-Organization of Calcium Channels

Studies of the micro-anatomy of calcium channel distribution may also be revealing. Calcium spikes in both cerebellar Purkinje cells and CA3 hippocampal pyramidal cells may possess multiple components (Spencer and Kandel, 1961; Llinas and Nicholson, 1971). These could reflect spatially distinct clusters of calcium channels. The presence of calcium channel "hotspots" and their spatial relations with excitatory and inhibitory synapses and other voltage-gated channels could critically influence dendritic integration (McLachlan and Bennett, 1986).

Although few details of a micro-organization for calcium channels are presently available, several techniques promise further progress (1) recording from membrane patches at different locations on the same cell (Bossu et al., 1989), (2) whole cell current recording coupled with selective perfusion of different regions of a neuron (Smith et al., 1985; Huguenard et al., 1989), (3) visualization of sites of Ca^{2+} influx with Ca^{2+} indicator dyes (Lipscombe et al., 1988; Ross and Werman, 1988; Tank et al., 1988; Regehr et al., 1989), and (4) calcium channel visualization with labeled molecules, such as fluorescent ω-conotoxin, that bind to calcium channel proteins (Jones et al., 1989).

IV. INTERACTIONS BETWEEN VOLTAGE-GATED CALCIUM CHANNELS AND TRANSMITTER-GATED CHANNELS

Evidence for direct participation of NMDA receptors in fast excitatory synaptic transmission lagged behind the initial identification of NMDA receptors as a subtype of excitatory amino acid receptor. Then a role for NMDA and both NMDA and non-NMDA excitatory amino acid receptors in monosynaptic transmission was demonstrated at the motoneuron of the *Xenopus* embryo spinal cord (Dale and Roberts, 1985). Synapses mediated by NMDA receptors alone have been identified in the cat neocortex (Thomson, 1986). There is now evidence for involvement of both types of excitatory amino acid receptors in monosynaptic transmission between spinal cord and hippocampal cells in culture (Forsythe and Westbrook, 1988), as well as neocortical neurons (Jones and Baughman, 1988).

Layer V cells in the cat neocortex have proven to be a good system to examine interactions between voltage-gated calcium channels and the NMDA receptor. The data regarding the localization of the voltage-gated channels are indirect but suggest a heterogeneity of calcium channel types and clustering (Stafstrom et al., 1985). A low-threshold calcium current is apparent with small voltage steps in these cells, suggesting that the channels are electrically near the soma. In addition, higher threshold Ca^{2+} transients that fire repetitively with depolarization appear with larger voltage steps. Since these events cannot be voltage clamped and appear in an all-or-none fashion with small differences in activation voltage, they are assumed to occur at electrically remote areas of the dendrites. However, whether the higher threshold of these events is due to the electrical distance from the current source during activation or due to inherent channel properties or both was not possible to determine.

Both NMDA and non-NMDA receptors must be present on the dendrites of these cells since EPSPs evoked by stimulation of layer II/III cells or white matter are mediated by both NMDA and non-NMDA receptors (Jones and Baughman, 1988). In 1 mM Mg^{2+}, the slower synaptic component mediated by NMDA receptors is highly voltage dependent. NMDA elicits a membrane depolarization in layer V cells *in vitro* which is accompanied by an increase in input resistance when tested with hyperpolarizing pulses (Flatman et al., 1986) due to the voltage-dependent, Mg^{2+} block of the channel. The agonist-induced depolarization can generate rhythmic depolarizations that in turn drive bursts of TTX-sensitive action potentials.

The mechanism driving the rhythmic depolarizations has not been determined. One possibility is that the depolarization, by activation of NMDA receptors, causes the activation of voltage-gated calcium channels that provide the driving force for the large, slow, rhythmic activity. The calcium current flow would depolarize the neurons, initiating the burst of TTX-sensitive action potentials. The elevated [Ca^{2+}]$_i$ would then activate Ca^{2+}-dependent potassium channels, causing the repolarizing phase of the oscillation. Alternatively, as suggested for a similar phenomenon observed in lamprey motoneurons, the voltage dependence of the NMDA channel itself may be sufficient to be the driving force behind the rhythmic depolarization, as well as the route of entry for the Ca^{2+} that triggers the repolarization (Brodin and Grillner, 1986). In layer V neurons, it is clear that both voltage-gated calcium channels and NMDA channels co-exit on the dendrites. Thus, even though the relative contributions of each channel to the oscillatory activity are not known, their interactions would be complex due to the different voltage-dependencies, kinetics, and localization of each channel type.

V. CALCIUM CHANNELS AND BURST FIRING PATTERNS

The firing pattern of a neuron reflects the voltage- and time-dependent activity of all of its membrane channels. The firing patterns of a few cells have been reconstructed from known properties of their membrane currents (Hodgkin and Huxley, 1951; Connor and Stevens, 1971; Barrett et al., 1980; Partridge, 1982). In other cells, specific currents have been assigned particular roles in the control of neuronal firing. Since calcium currents have slow kinetics, they have often been linked with the generation of rhythmic bursts of action potentials. In burst firing cells such as the R15 cell of Aplysia and hippocampal pyramidal cells, a number of action potentials occur at high frequency and are followed by a prolonged silent period before the next burst occurs. We shall examine the role of calcium currents in burst firing of invertebrate and vertebrate neurons.

A. Calcium Currents of Burst Firing Invertebrate Neurons

Several requirements for cyclic burst firing activity have emerged from work on cells from Aplysia, Helix, and Tritonia (Wilson and Wachtel, 1974; Eckert and Lux, 1976; Thompson and Smith, 1976; Johnston, 1976; Gorman et al., 1982; Kramer and Zucker, 1985a,b; Swandulla and Lux, 1985; Smith and Thompson, 1987). They are (1) a persistent, low-threshold inward current that provides a slow depolarization towards action potential threshold, (2) a way to generate depolarizing after-potentials and so sustain firing within the burst, and (3) a mechanism that hyperpolarizes the cell, terminating burst-firing.

Calcium currents seemed at first to fit some of these requirements. First, the kinetics and voltage dependence of the high-threshold channels seemed appropriate to generate the slow depolarization (Eckert and Lux, 1976; Kostyuk and Krishtal, 1977; Adams and Gage, 1979). Second, $[Ca^{2+}]_i$ was found to regulate a voltage-dependent potassium current (Meech and Standen, 1975; Thompson, 1977). Membrane potential could then oscillate. Activation of calcium channels would depolarize the cell and elevate $[Ca^{2+}]_i$ that would, in turn, cause a hyperpolarization as Ca^{2+}-dependent potassium channels opened.

This simple scheme led to the following predictions. Membrane oscillations should persist when sodium currents are suppressed but should be blocked by calcium channel blockers. Ca^{2+} influx should precede action potential discharge. Calcium currents should activate at more negative potentials than sodium currents and should show little inactivation in this potential range. The kinetics of Ca^{2+}-activated potassium currents should account for the after-hyperpolarization that follows the burst. Finally, suppression of Ca^{2+}-dependent currents should abolish rhythmic burst firing.

B. Role of a Ca^{2+}-Dependent Cation Current in Burst Firing

Burst firing invertebrate neurons do not meet all of these conditions. Instead, a crucial difference between burst firing neurons and repetitively firing cells may be that action potentials in burst firing cells activate a slow inward current whereas in repetitively firing cells, an outward current is generated. This current results in depolarizing after-potentials that sustain firing until counteracting potassium currents grow large enough to terminate the burst. It is a nonspecific cation current that is activated by intracellular Ca^{2+} (Lewis, 1984; Adams and Levitan, 1985; Kramer and Zucker, 1985a; Swandulla and Lux, 1985; Smith and Thompson, 1987; Partridge and Swandulla, 1988). Elevation of $[Ca^{2+}]_i$ entering during action potentials may directly cause calcium current inactivation (Eckert and Tillotson, 1981), possibly via a calcium-dependent dephosphorylation process (Chad and Eckert, 1986). The inactivation of calcium channels is removed slowly and may partly account for the postburst hyperpolarization in R15 cells of Aplysia (Kramer and Zucker, 1985b; Adams and Levitan, 1985).

C. Rhythmic Activity in Mammalian Neurons

The role of calcium currents in intrinsic oscillations of mammalian neurons can be examined in a similar framework. In cells that do not discharge bursts, such as motorneurons and dorsal horn neurons of the spinal cord, calcium currents exist but are normally dominated by strong potassium currents (Barrett and Barrett, 1976; Schwindt and Crill, 1980; Murase and Randic, 1983). In burst firing neurons such as cerebellar Purkinje cells (Llinas and Sugimoro, 1980) and CA3 pyramidal cells (Brown and Griffith, 1983), slow action potentials mediated by high-threshold calcium currents are initiated by strong depolarizations. High-threshold currents might then provide a means to sustain firing within a burst. However, only the low voltage-activated calcium current is activated at voltages that could provide a subthreshold depolarizing drive to initiate rhythmic burst firing. Although it inactivates quickly with large voltage steps, the low-threshold current decays slowly with smaller depolarizations (Carbone and Lux, 1987; Fox, Nowycky, and Tsien, 1987). Thus, it could provide a sustained depolarizing drive within a subthreshold potential range.

The contribution of calcium currents to the difference in firing patterns of hippocampal CA1 and CA3 pyramidal cells has not yet been addressed. CA1 pyramidal cells normally fire repetitively (Madison and Nicoll, 1986), while CA3 neurons discharge rhythmic bursts of action potentials (Wong and Prince, 1981). As in the case of invertebrate neurons, a depolarizing afterpotential follows action potentials in bursting CA3 cells while the afterpotential is hyperpolarizing in the repetitively firing CA1 cells. It remains unclear whether a calcium current mediates the depolarizing after potential (Wong and Prince, 1981; Storm, 1987).

D. Is the Low-Threshold Current Responsible for Burst Firing in Mammalian Cells?

In Purkinje cells, rhythmic firing persists in the presence of TTX and is blocked by calcium channel blockers (Llinas and Sugimori, 1980). Furthermore, fluctuations in $[Ca^{2+}]_i$ are correlated with the cyclic changes in membrane potential (Tank, Sugimori, Connor, and Llinas, 1988). However, while the calcium channel responsible for the slow depolarization preceding Purkinje cell bursts is activated at low voltages, it seems to differ from the low-threshold current of dorsal root ganglion cells (Carbone and Lux, 1987; Fox, Nowycky, and Tsien, 1987) in that it does not inactivate (Llinas et al., 1989).

Inferior olivary cells possess low-threshold, Ca^{2+}-dependent responses (Llinas and Yarom, 1981) that seem to conform to a T-type conductance. In the presence of pharmacological agents such as harmaline and serotonin that enhance the low-threshold current, olivary cells may show continuous, subthreshold, TTX-resistant membrane oscillations of amplitude up to 10 mV (Llinas and Yarom, 1986). However, the oscillation appears to be a network phenomenon. It occurs simultaneously in many neighboring cells and depends in part on electrotonic coupling between olivary cells. Electrotonically transmitted depolarizations reinforce the low threshold, Ca^{2+}-dependent responses of individual cells to generate the population rhythm.

E. The Low-Threshold Current as a Voltage-Dependent Switch

Thalamic cells illustrate one consequence of the voltage-dependent inactivation of low-threshold calcium responses (Jahnsen and Llinas, 1984). At resting membrane potential, where the low-threshold current is inactivated, thalamic neurons fire repetitively. However, when thalamic cells are hyperpolarized, T-current inactivation is removed and current injections evoke bursts of action potentials. Neurotransmitters that hyperpolarize thalamic cells can act as a switch between these two firing modes. Thus acetylcholine acts on muscarinic receptors to hyperpolarize thalamic cells, de-inactivating the low-threshold calcium conductance, and consequently transforming afferent responses into bursts (McCormick and Prince, 1986).

VI. SUMMARY

Calcium channels and Ca^{2+}-dependent channels regulate many of the complex electrical activities found in neurons. The dense localization of calcium channels on dendrites, intermixed with many of the synaptic inputs onto mammalian neurons, provides a physical substrate for neuronal integration. It is on the dendrites that different patterns of responses to synaptic inputs are generated due to either different types or densities of calcium channels. As methods of measurement of dendritic calcium events become increasingly sensitive and precise, it should be possible to achieve a more detailed understanding of the mechanisms associated with Ca^{2+}-dependent excitability in neurons.

Chapter 10

Calcium Channels in the Regulation of Cell Development and Cellular Interactions

Janet Holliday and Nicholas Spitzer

TABLE OF CONTENTS

I. Introduction 138

II. Calcium Conductances Appear Early in the Differentiation of Neurons and Myocytes and Ca^{2+}-Dependent Action Potentials are Often Developmentally Transient 138
    A. The Transient Nature of the Ca^{2+}-Dependent Action Potential in Development 138
    B. Amphibian Spinal Cord Neurons 139
    C. Myocytes 142

III. The Effects of Calcium Influx on Differentiation of Myocytes 143

IV. Ca^{2+} Influx and Neuronal Survival 145

V. Ca^{2+} Influx Effects on Neurite Outgrowth 147

VI. Ca^{2+} Effects on Neurotransmitter Metabolism and Neuronal Receptors 148
    A. Neurotransmitter Phenotype 148
    B. Neurotransmitter Sensitivity 150

VII. Possible Roles of Ca^{2+} Influx in Cell Interactions 150
    A. Gap Junctions 150
    B. Neuromuscular Interactions 152

VIII.	Evidence for the Role of Calcium in Gene Regulation	152
IX.	Summary	154

I. INTRODUCTION

Calcium influx through voltage-dependent channels may have effects on a cell in addition to membrane depolarization. Transient increases in intracellular Ca^{2+} concentrations may activate intracellular messenger systems and initiate a variety of cellular processes. Several excellent reviews that discuss the role of Ca^{2+} in the activation of second messenger systems are currently available (see Rasmussen and Barrett, 1984; Rasmussen, 1986a,b). Elevation of cytosolic Ca^{2+} can exert broad cellular effects via many routes (see Role and Schwartz, 1989 and Chapter 9 for a discussion of the effects of Ca^{2+} on enzyme activities). Two main pathways are especially well studied—the interaction of Ca^{2+} with calcium binding proteins, especially calmodulin, and the activation of protein kinase C. We will not discuss these pathways extensively, but it should be kept in mind that Ca^{2+} influx probably exerts many of its effects through these systems.

Our purpose in this review is to discuss various aspects of cellular differentiation that have been demonstrated to be affected by Ca^{2+} influx. The discussion will be limited to cells that are excitable (neurons and muscle), where Ca^{2+} influx is transient and is likely to occur through voltage-dependent calcium channels. Emphasis will be placed on systems in which Ca^{2+} influx changes with differentiation since this feature suggests that the influx is developmentally important. We will discuss alterations occurring at the cellular level and include Ca^{2+} effects of the interaction between pairs of cells. The second messenger systems involved in producing the effects initiated by Ca^{2+} influx will be noted briefly when such information is available. However, a more extensive discussion of second messenger mechanisms will be presented in the last section in which we will discuss Ca^{2+} regulation of gene expression.

II. CALCIUM CONDUCTANCES APPEAR EARLY IN THE DIFFERENTIATION OF NEURONS AND MYOCYTES AND Ca^{2+}-DEPENDENT ACTION POTENTIALS ARE OFTEN DEVELOPMENTALLY TRANSIENT

A. The Transient Nature of the Ca^{2+}-Dependent Action Potential in Development

Action potentials first observed in developing cells are often long in duration and depend to a substantial extent upon Ca^{2+} influx. The ionic dependence of the action potential in such cells frequently shifts to Na^+ during development, producing the short duration action potential observed in mature cells. This shift in ionic dependence has been demonstrated to occur in cells both *in vivo* and *in vitro* (reviewed by Spitzer, 1985). The transient nature of the Ca^{2+}-dependent action potential suggests that Ca^{2+} influx may have a specific role in development. The developmentally regulated mechanisms that produce the shift have been investigated in several systems, notably neurons of the amphibian spinal cord and avian and mammalian myocytes.

B. Amphibian Spinal Cord Neurons

In vivo studies of the Rohon-Beard neurons in *Xenopus* spinal cord have described the onset and maturation of electrical excitability (Baccaglini and Spitzer, 1977). Rohon-Beard cells are large sensory neurons located in the dorsal spinal cord, that become postmitotic during the midgastrula stage of development (Lamborghini, 1980). The cells are inexcitable until the neural fold stage when long duration (~100 ms) action potentials can first be evoked by the application of depolarizing current pulses (Baccaglini and Spitzer, 1977). The action potential is dependent upon Ca^{2+} since it is blocked by Co^{2+} or Mn^{2+} and is largely unaffected by tetrodotoxin or Na^+ removal. Neurite outgrowth begins shortly thereafter (Taylor and Roberts, 1983). The Ca^{2+}-dependent plateau of the action potential gradually disappears. As the embryo develops into a free swimming tadpole, the action potential has two components—a Na^+-dependent initial spike followed by a Ca^{2+}-dependent plateau. From the larval stage, achieved at one day after neural tube closure, the action potential is primarily Na^+-dependent and is of brief duration (1 ms).

Rohon-Beard neurons can be cultured from neural plate stage embryos. These cultures contain a variety of cell types, including myocytes, pigment cells, fibroblasts and notochord cells in addition to neurons. The neuronal population consists primarily of early birthdate neurons such as Rohon-Beard neurons and motor neurons (Lamborghini, 1980), and some interneurons that originate somewhat later (Lamborghini and Iles, 1985). Cultured neurons display the same pattern of action potential maturation in culture as they do *in vivo* (Spitzer and Lamborghini, 1976). At 6 h in culture (corresponding to a late neural tube stage embryo), when neurons can be recognized by their prominent processes, Ca^{2+}-dependent action potentials can be evoked. By 16 h in culture, evoked action potentials have become much shorter and by 1 day are predominantly Na^+-dependent. This schedule of maturation also occurs in neurons grown in single cell culture in minimal medium demonstrating that neither direct cell interactions nor released factors are necessary for normal differentiation after a cell is placed in culture (Henderson and Spitzer, 1986).

The cultured neurons acquire other differentiated characteristics, such as neurotransmitter sensitivity, a high affinity gamma amino butyric acid (GABA) uptake system, GABA-like immunoreactivity and neurite outgrowth, on the same developmental timetable they would have followed *in vivo* (Spitzer and Lamborghini, 1976; Bixby and Spitzer, 1982 and 1984a; Lamborghini and Iles, 1985; Spitzer et al., 1988). The apparent equivalence between *in vivo* and *in vitro* conditions for the maturation of early birthdate neurons provides confidence in the biological relevance of experimental findings from cells in standard culture conditions (Figure 1). Use of cultured neurons enables a more detailed study of the mechanisms that underlie the shift in ionic dependence of the action potential through the use of whole cell voltage clamp techniques.

The ionic currents that generate the action potentials of amphibian neurons have been isolated and examined during differentiation of cultured cells (O'Dowd, 1983a; Barish, 1986; O'Dowd et al., 1988). Calcium, sodium and potassium currents are observed in all morphologically identified neurons at the earliest stages examined. The characteristics of the voltage-dependent calcium current are relatively constant over the period during which the ionic dependence of the action potential is changing. The sodium current density increases two-fold over this period. The voltage-dependent potassium current increases in density by three-fold and activates twice as rapidly by the end of the first day in culture. Ca^{2+}-dependent potassium currents also increase in density. The decrease in the duration of the action potential during development is caused primarily by the change in potassium current densities; changes in potassium channel kinetics exert a smaller effect (Barish, 1986; Spitzer, 1988; Harris et al., 1988). This increase in potassium current produces a more rapid rate of repolarization of the neuron that decreases the amount of Ca^{2+} influx occurring during an action potential. The loss of Ca^{2+} dependence of the action potential is not due to changes in calcium current,

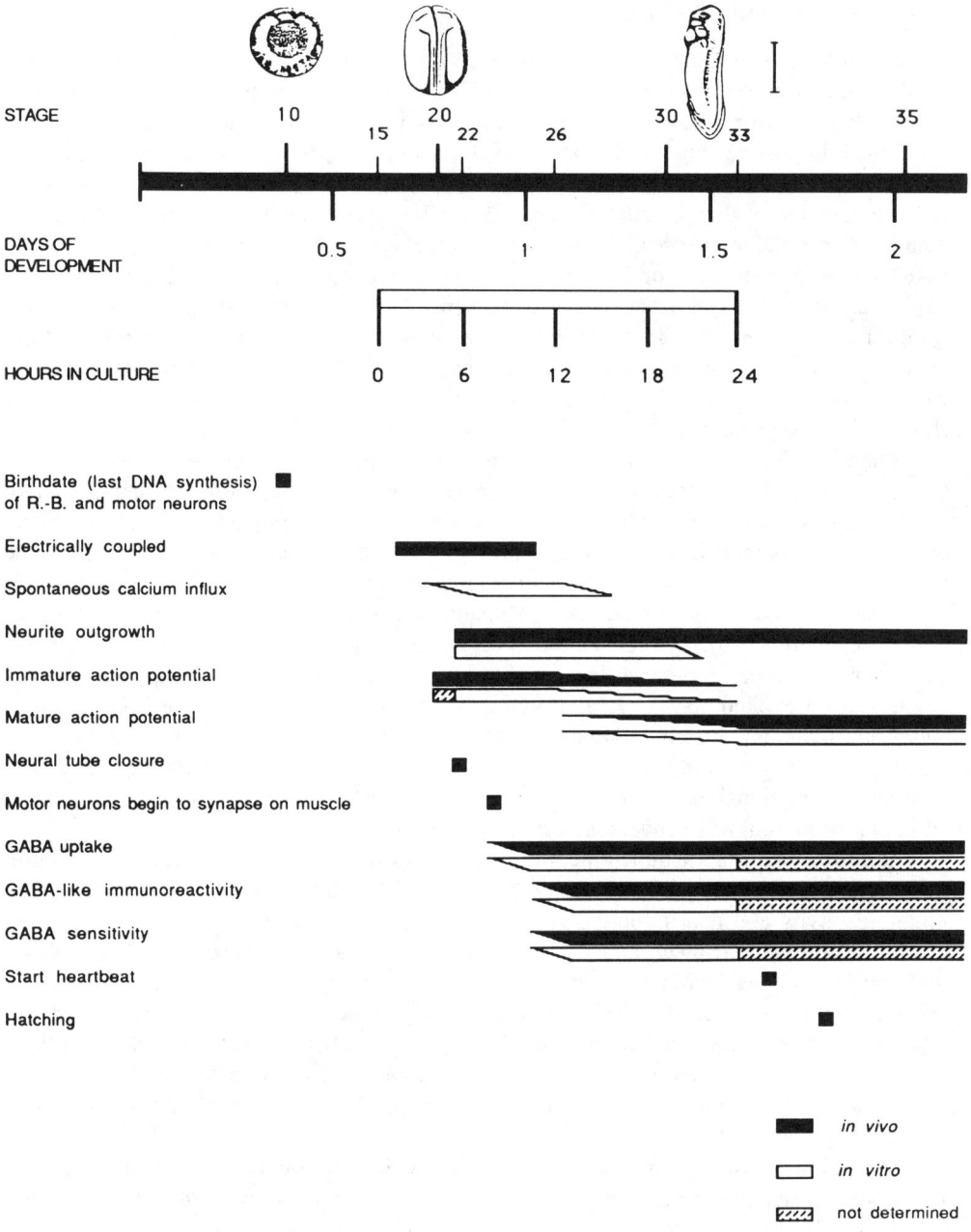

Figure 1. Developmental timetable for *Xenopus* neurons and myocytes: differentiation in culture parallels that is seen *in vivo*. The developmental acquisition of particular phenotypes of neurons and myocytes is charted for their development *in vivo* (solid bars) and *in vitro* (open bars). In all cases for which the comparison has been made a remarkable similarity is apparent. Scale bar is 1 mm. (Data drawn from Nieuwkoop and Faber, 1967; Spitzer and Lamborghini, 1976; Baccaglini and Spitzer, 1977; Lamborghini, 1980; Spitzer, 1982; Bixby and Spitzer, 1982, 1984a; Taylor and Roberts, 1983; Lamborghini and Iles, 1985; see review by Spitzer, 1985; Roberts et al. 1987; Spitzer et al., 1988.)

but to the more rapid repolarization of the neuron produced by the increased potassium current; the increase in sodium current allows the appearance of a brief, overshooting Na^+-dependent impulse (Figure 2).

The developmental shift in the ionic dependence of the action potential is dependent upon both protein and RNA synthesis (Blair, 1983; O'Dowd, 1983b). The action potential

10. *Calcium Channels in the Regulation of Cell Development*   141

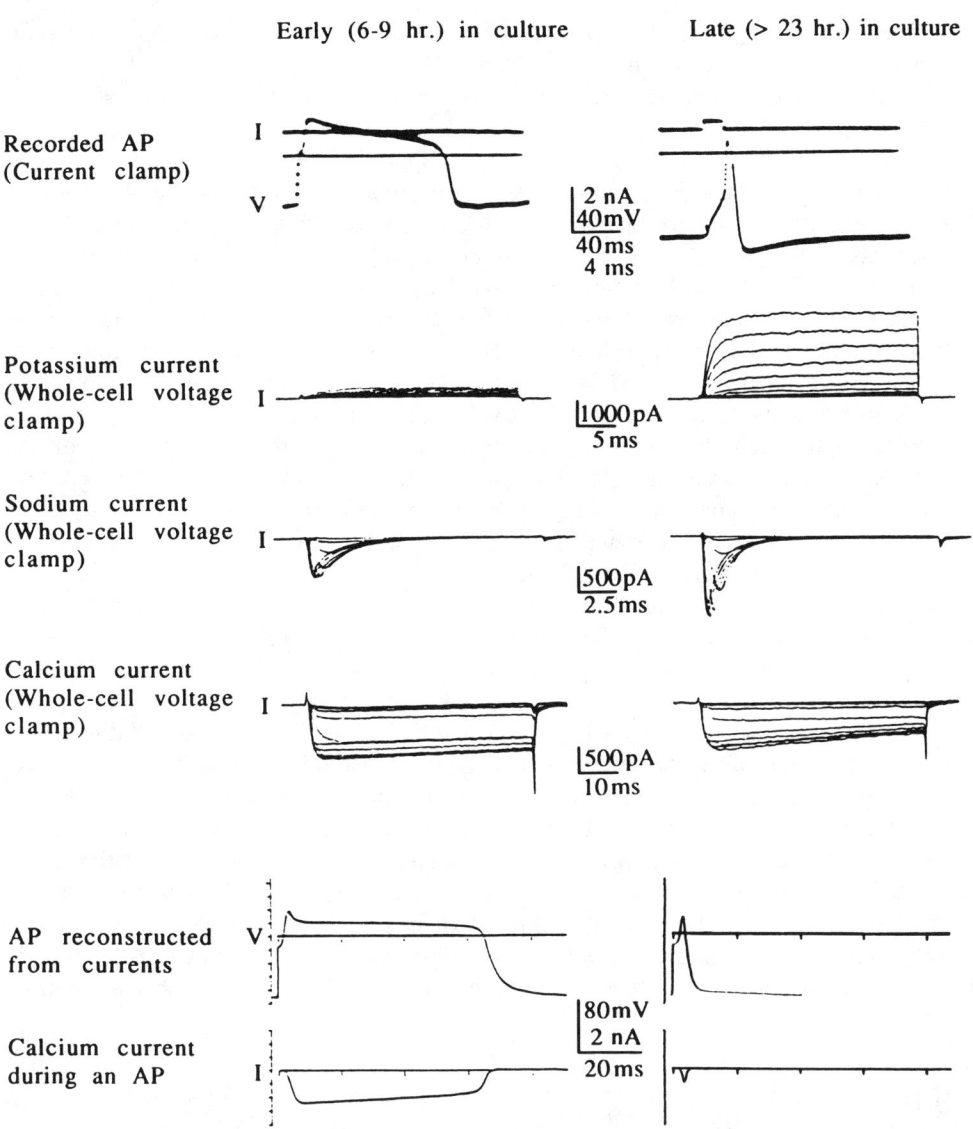

Figure 2. Maturation of the action potential of *Xenopus* spinal neurons is due to changes in ionic currents. The action potential is long in duration and principally Ca^{2+}-dependent when it is first recorded at the neural tube stage of development *in vivo* and at a comparable stage in culture. During the next day it matures to become brief and Na^+ dependent (top; Spitzer and Lamborghini, 1976). Whole cell recordings of voltage-dependent currents reveal an increase in the rate of rise and amplitude of delayed rectifier potassium current, an increase in the amplitude of sodium current, and an increase in the extent of inactivation of calcium current (middle; O'Dowd et al., 1988). A Ca^{2+}-activated potassium current also increases in amplitude during this period (not shown). Computer reconstruction of the action potentials from equations fit the whole cell currents, shows that these four currents are sufficient to reproduce the recorded potentials at both developmental stages (bottom, Spitzer, 1988). The action potentials at these two developmental stages involve rather different amplitudes and durations of calcium current, and indicate that substantially more Ca^{2+} enters the neuron during impulse production at the early stage of differentiation.

remains Ca^{2+}-dependent when either is blocked. Further, there is a defined period during which application of reversible inhibitors of messenger RNA synthesis permanently prevents the maturation of the action potential (Ribera and Spitzer, 1989). This finding indicates the existence of a critical period of transcription for its normal maturation.

While some developmental features may be influenced by Ca^{2+} influx at early times in culture, the maturation of the action potential is largely unaffected. Neurons grown in

Ca^{2+}-free medium to prevent Ca^{2+} influx appear to develop mature Na^+ dependent action potentials on the normal developmental schedule (Bixby and Spitzer, 1984b).

To suggest that Ca^{2+} influx plays a role in normal development, Ca^{2+} influx must occur without exogenous stimulation. Recent studies, using the Ca^{2+} indicator dye, Fura-2, suggest that influx occurs spontaneously in neurons developing in culture. This occurs during the time that neurons are capable of supporting Ca^{2+}-dependent action potentials (Holliday and Spitzer, 1988, 1990). Pharmacological agents known to inhibit voltage-dependent calcium channels also inhibit spontaneous elevations in intracellular Ca^{2+} ion concentration. The inhibition of the low-threshold T current (low-voltage activated, LVA, calcium current), by the application of Ni^{2+}, prevents spontaneous elevations of Ca^{2+} in cell bodies. The inhibition of N and L currents (high-voltage activated, HVA, calcium currents) with ω-conotoxin also eliminates spontaneous activity. These results suggest that the function of the T current may be to facilitate depolarization from a relatively hyperpolarized level. The depolarization effected through T current activation may then recruit the higher threshold N and L currents. Both types of currents appear to be necessary to produce detectable spontaneous Ca^{2+} elevations in the cell soma. The T current is expressed early in the differentiation of several cell types it precedes the appearance of N and L currents (Yaari et al., 1987; Gottmann et al., 1988; Lux, 1988; Barish, 1988; McCobb et al., 1989). The T current is transiently expressed in some neurons as it is in myocytes, suggesting that it may serve a developmental function to facilitate spontaneous Ca^{2+} influx. These observations suggest that Ca^{2+} influx plays a role in neuronal development and that influx occurs through voltage-dependent calcium channels.

Ca^{2+} levels appear to be elevated in the nuclear region relative to the surrounding cytosol in neurons growing under standard culture conditions (Holliday and Spitzer, 1990). This elevation is not observed in cells grown in the absence of extracellular Ca^{2+}. Nuclear elevation is observed in several cell types (Williams et al., 1985; Nicotera et al., 1989). Further, depolarization with high levels of K^+ causes the greatest Ca^{2+} elevation in this region (Plate 2).* This observation suggests that Ca^{2+} influx may act to regulate gene expression relatively directly rather than through other second messengers that must transmit a signal through the cytoplasm to the nucleus. While this is an interesting possibility, Ca^{2+} influx is likely to influence different developmental parameters through other mechanisms as well.

C. Myocytes

Developing myocytes have also been shown to display a shift in the ionic dependence of their action potential. Most chick skeletal muscle cells are inexcitable at embryonic day 13, but some cells are capable of sustaining a long duration action potential (23% of myocytes). By the time muscle cells have fused, few cells are inexcitable, 50% produce a long duration action potential, and 13% of cells generate a brief action potential. During this developmental period 9% of the myocytes have a mixed type of action potential—a spike followed by a plateau. By postnatal day 4, all of the excitable cells produce a short duration action potential. This spike is sensitive to tetrodotoxin while the long duration plateau is sensitive to Mn^{2+} or Co^{2+} ions, suggesting that the ionic dependence of the action potential in developing muscle has shifted from Ca^{2+} to Na^+ (Kano, 1975). A similar progression is observed in cultured myocytes (Kano et al., 1972; Kano and Shimada, 1973).

It has been postulated that there is a developmental increase in the density of sodium channels that accounts for the appearance of the fast spike in the action potential (Kano and Yamamoto, 1977). Studies of sodium channel ligand binding confirm a developmental

---

* Plate 2 appears after page 94.

increase in sodium channel number both in cultured chick myocytes and in rat muscle (Baumgold et al., 1983a,b; Sherman and Catterall, 1982). Studies on cultured *Xenopus* myocytes confirm the existence of a developmental increase in the sodium current density while no changes in channel kinetics were detected. However, this observation has less bearing on a shift in ionic dependence of their action potential since *Xenopus* myocytes support Na$^+$-dependent action potentials from the onset of excitability in contrast to chick and rat (Henderson and Spitzer, 1986; DeCino and Kidokoro, 1985). Whether there is a developmental increase in a repolarizing potassium current in chick muscle that accounts for the disappearance of the Ca^{2+} dependence of the action potential remains to be explored.

In rodent muscle cells, both low- and high-threshold calcium currents are expressed early in differentiation, but the low-threshold currents are only present transiently, while the amplitude of the high-threshold current increases with further development (Gonoi and Hasegawa, 1988; Beam and Knudson, 1988a,b). The disappearance of the T-type current is independent of innervation (Gonoi and Hasegawa, 1988).

The finding that the ionic dependence of the action potential in developing myocytes shifts from Ca^{2+} dependence to Na$^+$ dependence both *in vivo* and *in vitro* suggests a role for Ca^{2+} influx in the differentiation of myocytes. Further studies of myocytes have demonstrated that Ca^{2+} influx does have an influence on several developmental parameters. The shift of ionic dependence of the action potential during development is common to both neurons and myocytes indicating that transient Ca^{2+} influx may be a common regulator of differentiation in many excitable cells.

III. THE EFFECTS OF CALCIUM INFLUX ON DIFFERENTIATION OF MYOCYTES

Aspects of myocyte differentiation including postsynaptic specialization are influenced by Ca^{2+} influx. A specific role for calcium channels in the terminal differentiation of ascidian myoblasts has been suggested since calcium channels are expressed in myoblasts well before any contractile elements appear (Simoncini et al., 1989). At a later developmental stage, myoblast fusion can be prevented by the removal of extracellular Ca^{2+} or by blocking dihydropyridine-sensitive calcium channels, while fusion can be accelerated by the application of calcium ionophores or calcium channel activators (Shainberg et al., 1969; David et al., 1981; Rapuano et al., 1989). Increased Ca^{2+} influx, monitored by measuring Ca^{45} uptake, occurs during spontaneous myoblast fusion (David et al., 1981). In contrast, *Xenopus* myocytes, which do not produce early Ca^{2+}-dependent action potentials, do not fuse in culture. Three pathways that are able to initiate myoblast fusion are ACh receptor stimulation, depolarization, and prostaglandin stimulation. All three depend upon Ca^{2+} influx (Entwistle et al., 1988a,b). Other aspects of myocyte differentiation appear to be regulated by Ca^{2+} including muscle-specific gene expression in some species (see Shainberg et al., 1971; Endo and Nadal-Ginard, 1987).

Muscle gene expression during differentiation is altered in mice possessing the muscular dysgenesis mutation (Chaudhari and Beam, 1989). The myocytes are deficient in the slow, dihydropyridine-sensitive calcium current present in skeletal muscle transverse tubules while L-type calcium channels in heart muscle and in neurons are normal (Beam et al., 1986; Tanabe et al., 1988). These cells fail to express mRNAs that normally become elevated in myocytes by the time of birth. Further, mutant myocytes continue to produce transcripts through birth that are normally down regulated. This finding suggests that Ca^{2+} flux through a particular class of calcium channel promotes the normal differentiation of myocytes.

ACh receptor (AChR) synthesis appears to be decreased by Ca^{2+} influx in muscle cells. Since electrical activity is known to influence AChR number, and since changes in myoplasmic Ca^{2+} concentration occur in response to electrical activity, it has been hypothesized that the changes in Ca^{2+} levels cause the alterations of receptor number (Lomo and Rosenthal, 1972; Shainberg and Burstein, 1976; Betz and Changeux, 1979; Rubin, 1985). When intracellular Ca^{2+} concentration is increased in muscle, one would expect to observe a decrease in AChR number. This relationship may not be a simple one, as alterations of Ca^{2+} levels in specific subcellular compartments may be necessary for AChR regulation (see Birnbaum et al., 1980; McManaman et al., 1981).

There are many indications that Ca^{2+} may exert its regulatory effects at the transcriptional level (see Merlie et al., 1984; Goldman et al., 1985; Klarsfeld and Changeux, 1985; Fontaine et al., 1987; Evans et al., 1987; Shieh et al., 1987). Recent studies of cultured chick myotubes indicate that Ca^{2+} influx regulates not only the levels of mRNA coding for the alpha-subunit of the AChR but its precursor as well. Receptor levels are regulated largely by Ca^{2+} influx since blockade of the release of Ca^{2+} from intracellular stores with dantrolene produces little change in receptor levels and increases the effects of excitation blockade only slightly. Activity of protein kinase C is involved in receptor regulation since its inhibition increases both the number of receptors and the alpha subunit mRNA levels. The effects of excitation blockade and protein kinase C inhibition are not additive, suggesting that both effects are produced through a convergent regulatory pathway (Klarsfeld et al., 1989). It has been found that nuclear factors that bind to the regulatory sequences of the AChR alpha subunit gene vary with the activity of myocytes (Piette et al., 1989).

A relationship similar to that generally observed between Ca^{2+} levels and AChR number can be demonstrated for the regulation of sodium channels. Increased levels of intracellular Ca^{2+} decrease the number of TTX-sensitive channels while their number is increased by decreased electrical activity. The decrease in sodium channel number does not occur through changes in degradation rate, suggesting that the regulation of channel number occurs at some other step in the biosynthetic pathway. Recent evidence indicates that the number of sodium channel mRNA transcripts is affected by electrical activity (Cooperman et al., 1987). Other evidence suggests that the calcium effects may be mediated through the intracellular calcium receptor, calmodulin (Sherman et al., 1985).

While increased muscle activity decreases AChR and sodium channel levels, acetylcholinesterase (AChE) is regulated differently. Synaptic AChE increases with increased muscle activity (Rubin et al., 1980; Dennis, 1981). This increase appears to be mediated by increased muscle Ca^{2+} (Rubin, 1985).

In contrast to the effects seen in chick and rodent muscle, electrical activity exerts comparatively small effects on AChR number in *Xenopus* myocytes (Kidokoro and Gruener, 1982; Chow and Cohen, 1983) and no detectable effect on AChR subunit mRNA levels (Baldwin et al., 1988). The levels of AChE in *Xenopus* muscle cells remain relatively normal in the absence of neurons or neuronal stimulation (Moody-Corbett and Cohen, 1981; Cohen et al., 1984; Lappin and Rubin, 1985). The predominant effect of innervation in *Xenopus* muscle may be the localization of synaptic components rather than changes in levels of such components.

All of the aforementioned proteins are finally concentrated at the neuromuscular junction. Sodium channels appear to be concentrated in the synaptic region (Beam et al., 1985). Even in the absence of innervation, aggregates of AChRs and patches of AChE form on myocytes (Moody-Corbett and Cohen, 1981, 1982; Dennis, 1981; Weinberg et al., 1981; Kidokoro and Gruener, 1982). The aggregation of rat AChRs at neuromuscular junctions (or into patches in cultured myocytes) is sensitive to agents that are expected to alter intracellular Ca^{2+} levels (Bloch, 1979; Bloch and Steinbach, 1981; Bloch, 1983; Bursztajn et al., 1984). Growth of muscle in a low Ca^{2+} medium reduces the number of myofibers with AChR aggregates if the muscle is treated before adult junctional morphology has been established and AChRs are metabolically stabilized. Similarly, in *Xenopus*, the clustering of receptors either in patches or at sites of neuromuscular contact is prevented when development proceeds in the absence of extracellular Ca^{2+}, as determined with rhodamine-bungarotoxin localization of AChRs and by electrophysiological responses to bath and iontophoretically applied ACh (Henderson et al., 1984; but see also Davey and Cohen, 1986). These results indicate that the absence of Ca^{2+} will disrupt the clustering of AChRs in spite of neuronal innervation (Figure 3).

Positively charged polycation-coated latex beads are able to induce local AChR clustering in a Ca^{2+}-dependent manner in cultured *Xenopus* myocytes (Peng et al., 1981 and Peng, 1984). Either the absence of extracellular Ca^{2+} or the presence of calcium channel blocking agents (including D600 but not nifedipine) will prevent the formation of AChR clusters in response to this stimulus. In addition, calmodulin antagonists were found to prevent clustering. The polycation-coated latex beads stimulate a local increase in intracellular Ca^{2+} levels while uncoated beads, which do not induce clustering, produce no change in Ca^{2+} levels (Zhu and Peng, 1988). Clustering of AChRs appears to be a Ca^{2+}-dependent process that may involve the calcium binding protein, calmodulin. The polycation-coated latex beads are also able to induce the local accumulation of AChE, another characteristic of postsynaptic differentiation, although its dependence on intracellular or extracellular Ca^{2+} has yet to be investigated (Peng et al., 1988).

The requirement for Ca^{2+} influx during synaptic development may be partly supplied by activation of embryonic forms of AChRs in addition to the production of Ca^{2+}-dependent action potentials. The embryonic AChR has a longer mean channel open time and a larger conductance allowing greater Ca^{2+} entry than the adult type AChR (Sakmann and Brenner, 1978; Fishbach and Schuetze, 1980; Siegelbaum et al., 1980; Vicini and Schuetze, 1985; Mishina et al., 1986). The biochemical difference between the embryonic and adult form of the AChR appears to be substitution of the gamma subunit for the homologous epsilon subunit (Mishina et al., 1986). The embryonic form of the functional AChR itself is suppressed by electrical activity in the muscle cell (see Brehm and Henderson, 1988). Stimulation of the embryonic type of receptor is more likely to result in action potential production and muscle contraction (Jaramillo et al., 1988), which has been shown to promote the differentiation of cultured *Xenopus* myocytes (Kidokoro and Saito, 1988).

IV. Ca^{2+} INFLUX AND NEURONAL SURVIVAL

High levels of intracellular Ca^{2+} can be cytotoxic (see Mayer and Westbrook, 1987) while slightly elevated levels may exert physiological effects. The survival of cultured neurons is often enhanced by growth in depolarizing medium containing an elevated concentration of K^+ (Lasher and Zagon, 1972; Phillipson and Sandler, 1975; Scott, 1977; Chalazonitis and Fischbach, 1980; Nishi and Berg, 1981; Thangnipon, 1983). In several cases, the mechanism has been shown to rely upon Ca^{2+} influx through voltage-gated channels activated by depolarization with high K^+.

NGF promotes the survival of sympathetic neurons and its withdrawal causes neuronal death, yet depolarization and resultant Ca^{2+} influx can rescue NGF-deprived neurons (Koike et al., 1989). The depolarization allows Ca^{2+} influx through the L-type (dihydropyridine sensitive, noninactivating) channels and this influx is sufficient to promote neuronal survival.

Depolarization-promoted survival has been examined in cultured cerebellar granule cells (Gallo et al., 1987). Concentrations of potassium found to promote neuronal survival increased Ca^{45} uptake. The development of voltage-sensitive Ca^{2+} uptake paralleled the development of neuronal dependence on K^+ for survival. The blockade of Ca^{2+} influx with dihydropyridine agents in the presence of elevated K^+ inhibited neuronal survival while blockade of Na^+ influx had no effect, indicating that Ca^{2+} influx specifically enhances neuronal survival. N-methyl-D-aspartate (NMDA) receptor activation may provide another route of Ca^{2+} entry in these cells (Balazs et al., 1988, 1989). Opening of these channels

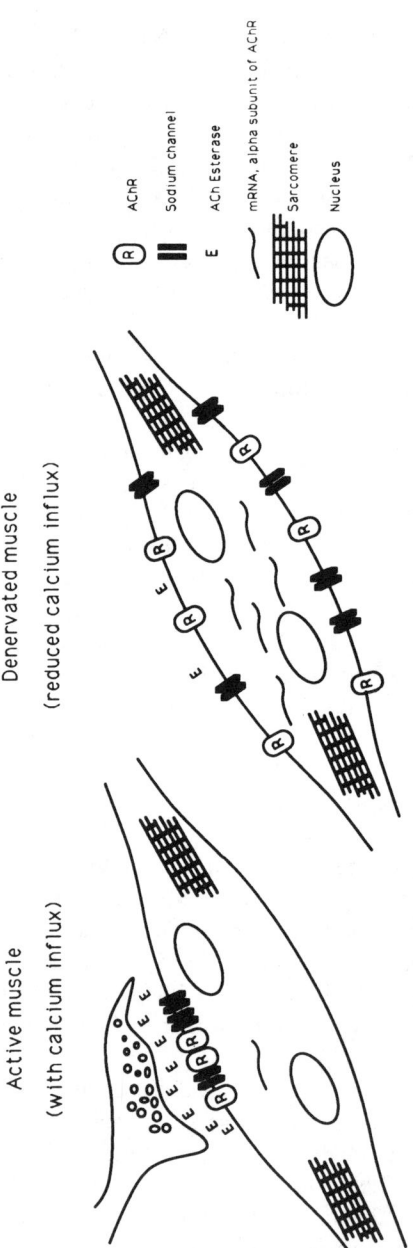

Figure 3. Denervation of muscle cells reduces Ca^{2+} influx and changes levels of membrane components and their localization. Innervated muscle undergoes periodic stimulation via ACh release from presynaptic nerve terminals that causes Ca^{2+} influx. Under these conditions, postsynaptic components are clustered beneath the nerve terminal. When Ca^{2+} influx is prevented (by denervation or other methods), postsynaptic components are no longer tightly aggregated at one site and levels of these components change. ACh esterase decreases while ACh receptor and sodium channel levels increase. It has been established that the levels of mRNA encoding the alpha subunit of the AChR are increased, indicating that some changes are transcriptionally controlled.

requires ligand activation of the receptors while the cell is depolarized and enables Ca^{2+} influx. Calmodulin inhibitors blocked the effect of depolarization suggesting that Ca^{2+} influx activates a calmodulin regulatory pathway to affect increased cell survival. The initial differentiation of granule cells in culture is unaffected by depolarizing conditions (Gallo et al., 1987). The number, morphological appearance of neurons, and the expression of the neuronal adhesion molecule, N-CAM, is equivalent under the two culture conditions over the first two days in culture.

Neuronal survival is enhanced by Ca^{2+} influx through dihydropyridine-sensitive calcium channels (L-type) in a variety of cell types. Chick parasympathetic (ciliary), sympathetic, and sensory (dorsal root ganglion) neuron survival is influenced by Ca^{2+} influx through L-type channels (Collins and Lile, 1989).

A similar role for Ca^{2+} influx can be postulated in the survival of cultured *Xenopus* spinal cord neurons. Growth of cells in the absence of Ca^{2+} accelerates the onset of neuronal death (Bixby and Spitzer, 1984b; Lamborghini and Iles, 1985). Such growth conditions prevent the spontaneous Ca^{2+} influx that occurs when cells are grown in the presence of Ca^{2+} (Holliday and Spitzer, 1988, 1990). Growth and morphological differentiation is accelerated in the absence of Ca^{2+} which may put a greater metabolic load on these cells causing them to deplete their energy stores more quickly, resulting in accelerated death. Growth of cultures in the absence of Ca^{2+}, preventing spontaneous Ca^{2+} influx, has been reported to increase the number of morphologically differentiated neurons (Bixby and Spitzer, 1984b). However, the apparent increase in differentiating *Xenopus* neurons in the absence of Ca^{2+} may be an indirect result of altered neuron-myocyte interaction under these conditions (Holliday and Spitzer, unpublished results).

V. Ca^{2+} INFLUX EFFECTS ON NEURITE OUTGROWTH

Ca^{2+} influx, whether through voltage-gated channels or through receptor-mediated mechanisms, affects neurite outgrowth. When extracellular Ca^{2+} is removed from the culture medium, *Xenopus* spinal cord neurons produce processes that are roughly three times as long as those found in cultures containing a standard concentration of Ca^{2+} (Bixby and Spitzer, 1984). This effect is independent of the presence of myocytes in the cultures (Holliday and Spitzer, unpublished observations). Measurement of resting levels of Ca^{2+} have been made using the calcium indicator, Fura-2. Cells grown in the presence or absence of Ca^{2+} do not display significant differences in Ca^{2+} concentration in the cell body, independent of the stage of development and whether or not the neurons can produce Ca^{2+}-dependent action potentials (Holliday and Spitzer, 1990).

Removal of extracellular Ca^{2+} is likely to have other effects in addition to prevention of influx. One concern is that vesicular release of a factor may be inhibited by altering Ca^{2+} influx thereby interfering with cell-cell communication. However, neurons grown in single cell cultures do not display any gross morphological differences from those grown in multicell cultures (Henderson and Spitzer, 1986). Some of the effects of Ca^{2+}-free medium may be mediated through adhesion molecules (such as *N*-cadherin, a calcium-sensitive adhesion molecule; Hatta et al., 1985; Tomaselli et al., 1988).

Experimental analysis demonstrates that inhibition of Ca^{2+} influx by hyperpolarization of the cultured cells, without altering the extracellular Ca^{2+} concentration, significantly increases neurite length. Further, removal of Ca^{2+} for the period of time that supports Ca^{2+}-dependent action potentials alters neurite length to the same extent as does Ca^{2+} removal for the entire culture period (Holliday and Spitzer, 1990). In parallel, depolarization reduces the percentage of cultured dorsal root ganglion neurons bearing processes and inhibits the extent of process outgrowth. The inhibition is prevented by blockade of L-type (dihydropyridine-sensitive) calcium channels (Robson and Burgoyne, 1989). These observations suggest that increased Ca^{2+} influx inhibits neurite outgrowth.

In contrast, studies of embryonic chick retinal neurons and a neuroblastoma cell line provide evidence that decreased Ca^{2+} influx inhibits neurite outgrowth (Anglister et al., 1982; Suarez-Isla et al., 1984). This apparent contradiction may be resolved by studies of Ca^{2+} concentration in processes and terminal growth cones. Direct studies of intracellular Ca^{2+} concentrations in growth cones of snail buccal ganglion cells indicate that there is probably an optimal Ca^{2+} concentration associated with process outgrowth, above or below which outgrowth is inhibited (Cohan et al., 1987; Mattson and Kater, 1987; Figure 4). Ca^{2+} concentration increases caused by influx generated either by direct electrical stimulation or by neurotransmitter stimulation will reversibly slow or stop the outgrowth of actively extending processes. With this view, both the resting level of intracellular Ca^{2+} and the magnitude of influx at the growth cone will determine whether outgrowth will be enhanced or inhibited by Ca^{2+} influx.

Neurotransmitter regulation of outgrowth appears to be dependent upon local depolarizations that activate voltage-sensitive calcium channels (Mattson and Kater, 1987; McCobb and Kater, 1988). Neurotransmitter regulation of outgrowth has also been observed in mammalian systems (Lipton et al., 1988). The specific mechanisms through which intracellular Ca^{2+} regulates outgrowth are beginning to be studied (Mattson et al., 1988). Ca^{2+} can affect the cytoskeleton and therefore alter process extension (Matsudaira and Janmey, 1988; Smith, 1988; Lankford and Letourneau, 1989). For example, high Ca^{2+} concentrations cause actin filaments to be severed by Ca^{2+}/actin binding proteins. Studies of Ca^{2+} effects on the cytoskeleton may elucidate mechanisms through which Ca^{2+} regulates neuronal process outgrowth (see also Weeds, 1982; Letourneau, 1984; Onuma and Hui, 1988).

VI. Ca^{2+} EFFECTS ON NEUROTRANSMITTER METABOLISM AND NEURONAL RECEPTORS

A. Neurotransmitter Phenotype

Studies of neurotransmitter phenotype in cultured rat sympathetic neurons have demonstrated the significance of Ca^{2+} influx in this aspect of differentiation. Depolarization was found to stabilize the adrenergic phenotype in cultured sympathetic neurons that might otherwise have become cholinergic in the presence of muscle conditioned medium (Walicke et al., 1977). The mechanism of this stabilization was shown to involve, in part, an influx of Ca^{2+} ions. Under depolarizing conditions, the removal of extracellular Ca^{2+} or the blockage of voltage-dependent calcium channels increases the expression of the cholinergic phenotype, thereby eliminating simple depolarization as a sufficient cause of adrenergic stabilization (Walicke and Patterson, 1981). When cells are grown in normal medium (without conditioned medium and without depolarizing levels of K^+) the inhibition of any spontaneous Ca^{2+} influx causes increased cholinergic differentiation.

Electrical activity in sympathetic neurons *in vivo* increases tyrosine hydroxylase (TH) activity, which may also be viewed as stabilizing the adrenergic phenotype since this enzyme is involved in its synthesis (Black et al., 1985). The increase appears to be due to an increase in the level of TH mRNA. In sympathetic neurons, depolarization with high K^+ *in vitro* is similar to electrical activity *in vivo*. While the Ca^{2+} dependence of the effects of depolarization *in vitro* has been shown, it is unknown whether the effects of electrical activity *in vivo* are Ca^{2+}-dependent. The effects of electrical activity on neurotransmitter synthesis are being studied in a number of laboratories (see Black et al., 1987).

Increases in TH activity are clearly due to increases in TH mRNA *in vitro*. Growth of cultured sympathetic cells in depolarizing medium depresses the development of choline

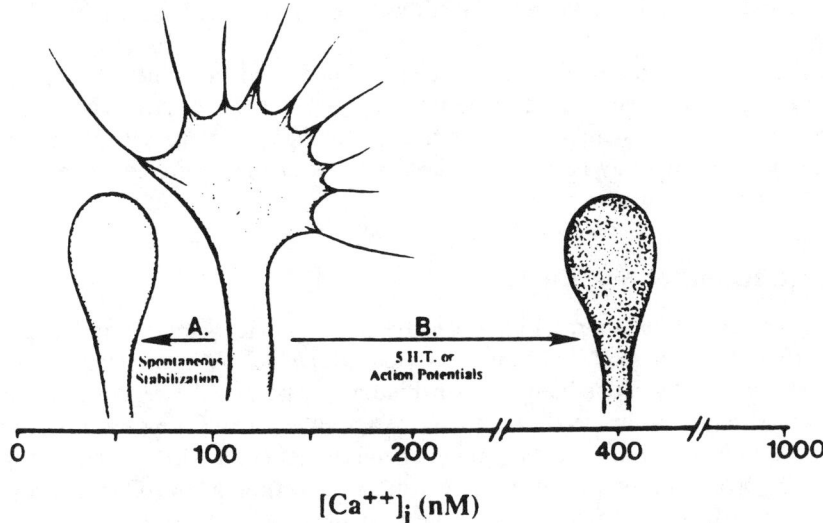

Figure 4. Ca^{2+} influx influences neuronal process outgrowth. The level of free intracellular Ca^{2+} appears to regulate growth cone motility and the rate of neurite outgrowth. Intracellular Ca^{2+} levels may regulate outgrowth by affecting the stability of the peripheral cytoskeleton. Extension of growth cones appears to be greatest at an optimal Ca^{2+} level. Higher or lower concentrations produced by agents such as action potentials or neurotransmitter (serotonin) stimulation result in graded changes in outgrowth until Ca^{2+} levels are reached that arrest extension completely (after Cohan et al., 1987).

acetyltransferase (the ACh synthetic enzyme) while it stimulates TH (Hefti et al., 1982; Raynaud et al., 1987a,b; Brice et al., 1989). The induction of TH can be blocked by the prevention of Ca^{2+} influx and by the inhibition of calmodulin activity (Hefti et al., 1982) while the depression of choline acetyltransferase activity by Ca^{2+} influx does not appear to be mediated by calmodulin (Vidal et al., 1989). The stimulation of TH is evident at the mRNA level since depolarization produces an increase in TH mRNA that is of the same order of magnitude as the increase in enzyme activity; this increase is mediated by Ca^{2+} influx through L-type calcium channels (Raynaud et al., 1987b; Vidal et al., 1989). In this system, Ca^{2+} influx regulates the choice of neurotransmitter phenotype by regulating message levels of the neurotransmitter synthetic enzyme.

In neurons cultured from chick ciliary ganglion as well as from the spinal cord of fetal mice, depolarization increases choline acetyltransferase activity in contrast to the relationship described above for sympathetic neurons (Nishi and Berg, 1981; Ishida and Deguchi, 1983). The increase in choline acetyltransferase activity is dependent upon Ca^{2+} influx suggesting that Ca^{2+}-dependent regulation of this enzyme is specific for the class of neurons.

Other systems are being described in which Ca^{2+} influx specifically appears to regulate neurotransmitter phenotype. For example, the presence of extracellular Ca^{2+} is required for the development of the GABAergic phenotype in neurons cultured from *Xenopus* neural plate in the presence or absence of myocytes (Spitzer et al., 1988 and unpublished observations). The dependence on Ca^{2+} influx for the expression of this phenotype is suggested by the coincidence of the temporal requirement for Ca^{2+} with the period in neuronal development that supports spontaneous Ca^{2+} influx and Ca^{2+}-dependent action potentials (Spitzer, C. de Baca and Holliday, unpublished observations). Since GABA uptake is unaffected by the absence of extracellular Ca^{2+} (Lamborghini and Iles, 1985), calcium regulation of the phenotype must occur at a different level. Further studies are necessary to determine the mechanisms by which Ca^{2+} regulates this neurotransmitter phenotype. Electrical activity in the visual cortex of the primate supports the expression of the GABA cell phenotype. Both GABA-like immunoreactivity and GAD (the GABA synthetic enzyme) immunoreactivity are reversibly reduced in the absence of electrical activity (Hendry and Jones, 1988).

The regulation of neurotransmitter phenotype may also be achieved by Ca^{2+} influx through a ligand-gated channel. Activation of the NMDA receptor stimulates the synthesis of phosphate-activated glutaminase in a transcription-dependent manner (Moran and Patel, 1989a,b). This enzyme may be used for the synthesis of the neurotransmitter, glutamate, in cerebellar granule neurons.

B. Neurotransmitter Sensitivity

Ca^{2+} can regulate neurotransmitter sensitivity as well as neurotransmitter phenotype. ACh sensitivity has been well studied in myocytes (see above). ACh receptors in chick retina cultures, as defined by alpha-bungarotoxin binding, appear to be regulated in a manner similar to that found in muscle (Betz, 1983). When other agents are used to identify ACh receptors in ciliary ganglion cultures, growth of cells under depolarizing conditions produces no change in surface receptor levels while causing a reduction in the ACh sensitivity (Smith et al., 1986; Halvorsen and Berg, 1987). The reduction in sensitivity is due to an alteration in receptor function rather than in receptor number (Margiotta et al., 1987).

Investigations of GABA sensitivity in cultured *Xenopus* spinal cord neurons demonstrate that cells grown in Ca^{2+}-containing medium develop GABA sensitivity on a developmental schedule paralleling that observed *in vivo* (Bixby and Spitzer, 1984a). In contrast, neurons grown in the absence of Ca^{2+} acquire GABA sensitivity normally, yet subsequent increases in the percentage of sensitive cells are blunted then reversed compared to cells grown in the presence of Ca^{2+} (Bixby and Spitzer, 1984b). Further, the concentration of GABA required to elicit a response is higher in many cells grown in the absence of Ca^{2+} than in cells grown in its presence or *in vivo*. The mechanism through which Ca^{2+} exerts its effects on GABA sensitivity is unclear. It may have an effect on receptor localization, analogous to its effect on ACh receptors on myocytes, since there are developmental changes in GABA receptor localization over development that could be altered (Harris and Spitzer, 1987). Further studies will be necessary to completely define the level at which Ca^{2+} alters neurotransmitter sensitivity in this system.

VII. POSSIBLE ROLES OF Ca^{2+} INFLUX IN CELL INTERACTIONS

A. Gap Junctions

Gap junctions allow direct communication between cells by coupling between them via channels through which small molecules can diffuse. The molecules comprising gap junction connexons are well studied (for a review see Hertzberg and Johnson, 1988). The regulation of gap junction function has also been studied intensively. The relative importance of Ca^{2+} and H^+ has been debated but the role of Ca^{2+} in determining the extent and strength of coupling is evident from the finding that calmodulin (or a calmodulin-like protein) is a necessary intermediary (Johnston and Ramon, 1981; Peracchia et al., 1981, 1983).

Embryonic cells are often electrically coupled at early stages and become uncoupled with further differentiation. The electrically coupled Rohon-Beard neurons of developing *Xenopus* embryos become permanently uncoupled during a period of development in which the ionic dependence of the action potential is changing (Spitzer, 1982). It was found that coupling is present from the earliest time studied (at the closure of the neural tube) and persists until the early tailbud stage of larval development (when the Na^+ component of the

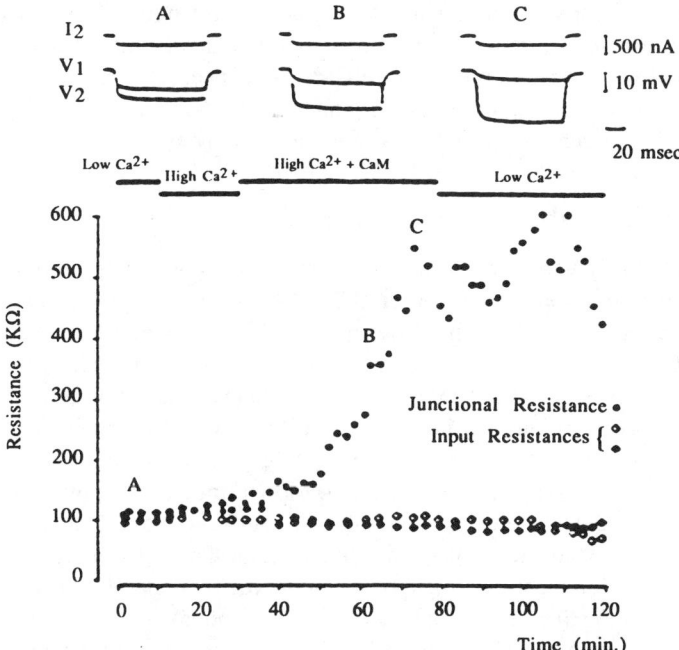

Figure 5. Ca^{2+} and calmodulin regulate gap junctions. Internal perfusion of crayfish axons that are in functional contact via gap junctions allows the controlled study of substances that regulate the resistance of the junction. A high Ca^{2+} perfusion solution alone does not alter the junctional resistance unless calmodulin (CaM) is added. This causes an increase in resistance that reaches a peak after about 40 min and persists despite washout with solution containing low Ca^{2+} and no calmodulin. The current and voltage traces corresponding to points A, B, and C in the graph are shown above (after Arellano et al., 1988).

action potential becomes prominent). The coupling is voltage dependent, and transient uncoupling may thus be observed during electrical stimulation, independent of the presence or absence of extracellular Ca^{2+}. Acidification was ineffective in uncoupling the cells but altered the voltage-dependent properties of the coupling. One might expect calcium-dependent uncoupling to occur with the production of calcium-dependent action potentials; yet even repetitive production of action potentials does not permanently uncouple the cells. This result suggests the involvement of a factor in addition to Ca^{2+} influx, such as a calmodulin-like protein, that is developmentally regulated.

Calmodulin has been shown to be involved in gap junction regulation, as indicated by the effect of calmodulin inhibitors in preventing electrical uncoupling elicited by elevation of intracellular Ca^{2+} (Peracchia et al., 1983; Peracchia, 1984; Verselis et al., 1986; Peracchia, 1987). Further reconstitution experiments demonstrate the necessity of both Ca^{2+} and calmodulin to regulate gap junction function (Girsch and Peracchia, 1985; Ramon et al., 1988; Arellano et al., 1988; Figure 5). Ca^{2+}-calmodulin probably interacts directly with the gap junction proteins, since calmodulin binds to gap junction proteins and the sequence of these proteins predicts a calmodulin binding site (Hertzberg and Gilula, 1981; Welsh et al., 1982; Gorin et al., 1984; Paul, 1986; Kumar and Gilula, 1986).

The role of gap junctions in development is not well defined but appears to be extremely important. Antibodies directed against the gap junction protein disrupt its function and can be used as an experimental probe. Intracellular injection of these antibodies, at the 8-cell stage of a *Xenopus* embryo, produces profound morphological disruption (Warner et al., 1984). Embryos become asymmetric, often fail to form eyes, and exhibit brain abnormalities. A similar type of perturbation during experimentally-induced head formation in hydra indicates that gap junctional communication is important for normal body patterning (Fraser et al., 1987). Refinements of these types of experiments will extend understanding of the role of gap junctions in development. Since Ca^{2+}, in concert with calmodulin-like proteins, appears to be an important regulator of gap junction function, its influx may have profound developmental consequences through this regulatory pathway.

B. Neuromuscular Interactions

Ca^{2+} is able to influence the ability of neurons to orient to target myocytes in cultures from *Xenopus* neural plate (Henderson et al., 1984). In the absence of extracellular Ca^{2+}, contact between neurites and neighboring myocytes is inhibited and contacts that result in terminations of processes upon myocytes are decreased. The decrease in terminations appears to be due to the absence of extracellular Ca^{2+} rather than to the prevention of influx. Growth of cultures in hyperpolarizing medium, designed to prevent the generation of spontaneous Ca^{2+}-dependent action potentials by stabilizing the resting potential below the threshold potential, does not affect the ability of neurons to form terminations upon contacted myocytes. However, growth under these conditions inhibits neuron-myocyte contact, suggesting that this is affected by Ca^{2+} influx rather than extracellular Ca^{2+} (Holliday and Spitzer, 1988, 1990). Consistent with these findings is the observation that neuron-myocyte contact is inhibited while termination is unaffected upon the removal of Ca^{2+} for the short period during which long duration, Ca^{2+}-dependent action potentials can be produced (Holliday and Spitzer, 1988, 1990). Myoblast conditioned medium has been found to attract neurite outgrowth (McCaig, 1986). It has yet to be determined whether the ability of neurites to respond and orient to myocytes is dependent upon Ca^{2+} influx as part of the neuronal response, or as part of the ability of myocytes to condition medium or both.

VIII. EVIDENCE FOR THE ROLE OF CALCIUM IN GENE REGULATION

Ca^{2+} influx may function to modify gene expression in developing systems. This is likely to be the case for expression of tyrosine hydroxylase in sympathetic neurons (discussed above). The study of gene expression in primary culture systems is facilitated by the availability of specific probes (such as those for tyrosine hydroxylase). Large, homogeneous populations of neurons are necessary to enable the use of techniques such as subtractive or differential hybridization to identify those messages that are regulated by Ca^{2+} influx. An important model is the PC12 system, a rat pheochromocytoma cell line that responds to nerve growth factor (NGF) in much the same way as do sympathetic neurons.

PC12 cells stop dividing, elaborate neurites, and display other differentiated neuron-like characteristics when they are grown in the presence of NGF, although the cells do not require this agent for survival (see Greene and Tischler, 1976). Studies of PC12 cells have elucidated the molecular mechanisms of NGF action. Some of the effects of NGF, such as neurite extension, can be mimicked by depolarization of the cells. These effects have been attributed to increased Ca^{2+} influx that occurs with depolarization (see Traynor, 1984; Schubert et al., 1978). However, NGF treatment does not cause large increases in Ca^{2+} influx (Schubert et al., 1978; Landreth et al., 1980) but can cause small intracellular Ca^{2+} elevations (Pandiella-Alonso et al., 1986).

NGF treatment of PC12 cells causes changes in gene expression, and some of these changes are also produced by Ca^{2+} influx (Kruijer et al., 1985; Morgan and Curran, 1986; Curran and Morgan, 1986; Milbrandt, 1986; Leonard et al., 1987; Milbrandt, 1987; Sukhatme et al., 1988; Morgan and Curran, 1988; Cho et al., 1989). NGF regulates two general classes of transcripts, encoded by early and late genes. Maximal induction of late genes occurs

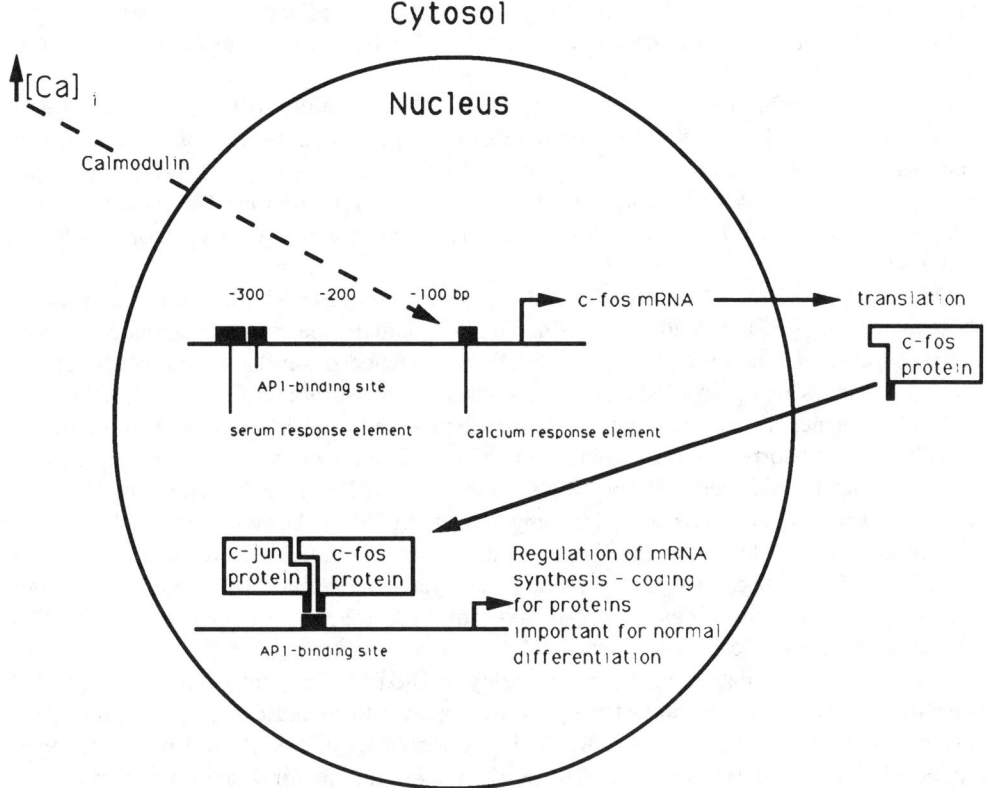

Figure 6. Ca^{2+} influx regulates the expression of c-fos and other genes. A calcium response element has been identified within 60 bp of the transcription start site of the transcription factor, c-fos (Sheng et al., 1988). Ca^{2+} influx activates c-fos transcription in a calmodulin-sensitive fashion with peak message levels reached by 1 h after stimulation. Once the c-fos protein is translated, it may then complex with the c-jun protein in the cell nucleus. This protein complex is capable of binding to the AP1 binding site, an upstream element involved in regulating the expression of many genes such as proenkephalin (Sonnenberg et al., 1989). Such gene activation may enable the expression of differentiated neuronal phenotypes.

about 1 day after NGF stimulation, while early genes are maximally induced by 1 h of treatment (Changelian et al., 1989). Late genes, in contrast to early genes, are not directly regulated by agents that may act through Ca^{2+} influx (Leonard et al., 1987; Changelian et al., 1989). However, two of the late genes regulated by NGF are very similar to those of Ca^{2+} binding proteins (Masiakowski and Shooter, 1988).

Early genes are likely to encode proteins that regulate transcription. One of the best studied of the rapidly induced messages is c-fos, which encodes a protein that can associate with the c-jun product to activate the specific transcription of other genes and is expressed early in normal development of the central nervous system (Rauscher et al., 1988; Curran and Franza, 1988; Caubet, 1989). C-fos is rapidly induced in PC12 cells by either NGF, phorbol esters, K^+ depolarization, or calcium ionophore and the induction involves calmodulin (Kruijer, et al., 1985; Milbrandt, 1986; Morgan and Curran, 1986; Curran and Morgan, 1986). While phorbol esters stimulate protein kinase C, K^+ depolarization and calcium ionophore treatment increase Ca^{2+} influx which could in turn stimulate calmodulin-dependent processes or protein kinase C (Figure 6).

Some of the elements required for Ca^{2+} regulation of c-fos expression have been investigated. The c-fos promotor sequence contains sequences similar to the serum response element (SRE) and the cAMP-regulatory element (CRE) (Treisman, 1986; Montminy et al., 1986). Deletion analysis demonstrates that the SRE region is necessary to enable a cell's response to growth factors and phorbol esters while a region containing the CRE is required for the calcium response in PC12 cells (Visvader et al., 1988; Sheng et al., 1988). The fos protein may inhibit its own expression, accounting for its transient appearance (Sassone-Corsi et al., 1988; Fisch et al., 1989).

Rapid, Ca^{2+} influx-dependent c-fos induction can be stimulated by nicotine treatment in differentiated PC12 cells, that have completed the transient expression of c-fos occurring at the onset of differentiation (Greenberg et al., 1986). This indicates that c-fos expression probably plays other roles in addition to those in early differentiation. The induction of c-fos by Ca^{2+} influx in other differentiated cells is being explored (see Connor, 1988 and Chapter 9).

Other early genes regulated by Ca^{2+} influx appear to encode other transcription factors. The sequence of NGFI-A predicts a "zinc finger" domain that is characteristic of some proteins capable of binding to DNA; the NGFI-B sequence is similar to that of the glucocorticoid receptors that also modulate gene transcription (Milbrandt, 1987 and Milbrandt, 1988). These genes are similar to c-fos in that their expression is stimulated by treatment with either NGF, phorbol ester or calcium ionophore. The upstream regulatory sequence of NGFI-A has been described and contains sequences related to cAMP responsive elements and serum responsive elements as well (Changelian et al., 1989). The entire isolated upstream sequence is sufficient to confer NGF inducibility on an unrelated reporter gene. The roles of the SRE and CRE regions of this promotor have yet to be defined with respect to various inducing agents, but similarities to the c-fos promotor are to be expected since the regulation of these genes appears to occur in parallel.

Specific Ca^{2+} regulatory elements are being studied in other cells. Glucose-regulated proteins are elevated in a fibroblast cell line in response to increases in intracellular Ca^{2+} produced by the calcium ionophore, A23187 (see Resendez et al., 1986; Lin et al., 1986; Wooden et al., 1988). Unlike c-fos and NGFI-A, stimulation for 2 to 3 h is required to cause increases in mRNA coding for these proteins and the induction is blocked by inhibitors of protein synthesis. The elevation of these transcripts by Ca^{2+} is inhibited by the application of calmodulin antagonists.

IX. SUMMARY

There is now good evidence that Ca^{2+} influx through voltage-dependent calcium channels plays an important role in the differentiation of nerve and muscle. Ca^{2+}-dependent action potentials have been observed during a developmentally transient period at early stages of differentiation. Significantly, recent evidence indicates that they can occur spontaneously, and that prevention of spontaneous activity disrupts normal development. A model for the regulation of differentiation by Ca^{2+} influx is proposed (Figure 7).

Ca^{2+} influx has been shown to affect neuronal survival, neurite outgrowth, neurotransmitter phenotype, and expression of neurotransmitter receptors. In developing myocytes AChR, AChE, and sodium channel syntheses are influenced by Ca^{2+} influx. Ca^{2+} can affect cell interactions in the form of gap junctions and cell-cell recognition and synapse formation. Information about the molecular basis of the action of elevations of intracellular Ca^{2+} is accumulating rapidly. The PC12 system has been particularly useful. Evidence points to the existence of nucleic acid sequences upstream of particular genes that serve as Ca^{2+} regulator elements, which may interact with Ca^{2+} in conjunction with Ca^{2+}-binding proteins such as calmodulin.

It will be of interest to identify the roles of different classes of calcium channels in the events of development. The effectiveness of dihydropyridines in specifically blocking L-type channels has implicated Ca^{2+} influx through these channels in cell survival, process outgrowth, and the choice of neurotransmitter phenotype. The availability of effective and specific agents to alter the function of other types of calcium channels will enable investi-

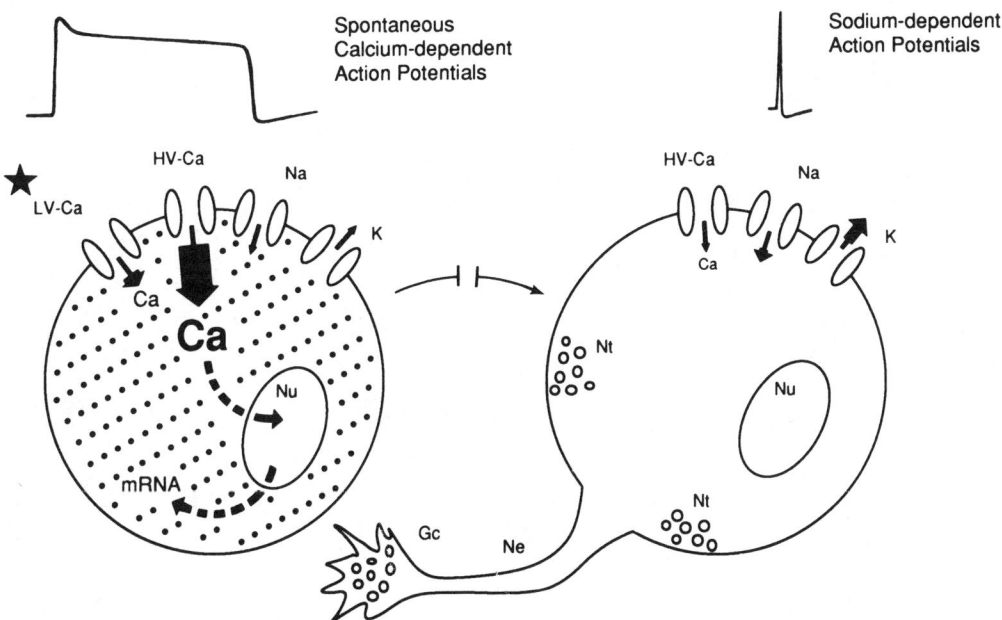

Figure 7. Ca^{2+} influx regulates neuronal differentiation in cultured *Xenopus* spinal neurons. A model for the regulation of differentiation by Ca^{2+} influx identifies spontaneous stimulation of low voltage threshold Ca^{2+} channels of young cells (*) as the trigger for long-duration Ca^{2+}-dependent action potentials leading to the transient elevation of cytoplasmic Ca^{2+} (left). Ca^{2+} stimulates specific transcription of messenger RNAs (such as those encoding neurotransmitter synthetic enzymes) and Ca^{2+} levels return to baseline. Low voltage-activated calcium channels later disappear and the effects of calcium modulation of neurite outgrowth and neurotransmitter synthesis are fully expressed (right). High voltage-activated calcium channels persist, but increases in density and changes in kinetics of potassium and sodium channels increase these currents and decrease Ca^{2+} influx, converting the action potential to a brief, sodium-dependent event. The model predicts a Ca^{2+}-dependent critical period for development of affected phenotypes. Ca, Ca^{2+} ions; Gc, growth cone; HV-Ca, high voltage-activated calcium channels; K, potassium channels; LV-Ca, low voltage-activated calcium channels; Na, sodium channels; Ne, neurite; Nt, neurotransmitters; Nu, nucleus.

gation of their roles in differentiation. Description of channel type expression patterns during development is providing initial clues to their roles. Modulation of calcium channel function provides another level for developmental regulation of Ca^{2+} influx that has yet to be fully investigated.

While the role of specific calcium channel types in the regulation of Ca^{2+} influx in development is being analyzed, the regulation of intracellular Ca^{2+} levels once influx has occurred is only beginning to be considered. Many Ca^{2+} binding proteins have been identified and their roles in buffering influx requires investigation. Expression of Ca^{2+} sequestration mechanisms may also be developmentally regulated. It will not be surprising if Ca^{2+} influx or release from intracellular stores is developmentally important for less traditionally excitable cells (e.g., lymphocytes, pancreatic β cells), and for cells in which calcium currents have yet to be identified. It will be useful to define the detailed molecular mechanisms through which elevation of intracellular Ca^{2+} affects transcriptional and translational processes that ultimately control the expression of differentiated states.

External stimuli are able to influence the differentiation of excitable cells by causing Ca^{2+} influx through voltage-dependent channels. Regulation of the temporal expression and spatial localization of ion channels of different types provides a mechanism to modulate Ca^{2+} influx during development so that cell depolarization will produce different effects dependent upon the developmental state of the cell. The availability of cloned calcium channel genes suggests that it will be possible to experimentally control their expression in the not too distant future.

Section III
Pharmacology of Calcium Channels

Chapter 11

The Effects of Chemical Agents on Calcium Channels: An Overview

Leon Hurwitz

The subject of Section III is the pharmacology of calcium channels. The next three chapters deal with the actions of a diverse group of chemical substances that modify the operation of calcium channels. Apart from the evident basic objective, namely to understand the mechanisms by which pharmacological agents modify ion channel activity, there are several reasons for the effort extended to study this particular group of chemical substances. They are based on observations that various selected members of the group: (1) are useful laboratory tools that can help to distinguish between different types of calcium channels; (2) are highly effective therapeutic drugs that can be used to treat a number of diseases, especially diseases of the cardiovascular system; or (3) are naturally occurring neurotransmitters, hormones, or second messengers, the actions of which must be explored in order to understand how the activity of a calcium channel is regulated in its natural environment.

Before one can embark upon a discussion of a drug-induced modification of some biological property or process, it is essential to give some consideration to the normal characteristics of that property or process. An important element in the normal activity of voltage-dependent calcium channels is the capacity of these channels to shift in probabilistic (stochastic) fashion from one functional state to another. Although it is somewhat of an oversimplification, one can say that calcium channels shift into and out of essentially three distinct functional states: (1) the deactivated state; (2) the activated state; and (3) the inactivated state (see Section I). The transformations among these functional states may be represented as follows:

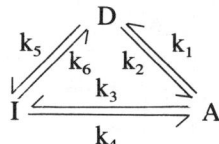

where D is the deactivated state; A is the activated state; I is the inactivated state; and k_1, k_2, k_3, k_4, k_5, and k_6 are rate constants that help determine the speed at which calcium channels are transformed from one functional state to another. When a calcium channel is in either the deactivated or inactivated state, its gates are closed and ion permeation cannot occur. When the channel is in the activated state, its gates are open and Ca^{2+} ions can rapidly traverse the cell membrane through the aqueous channel pore. In Sections I and II, as well as in subsequent chapters of this book, the term activation of calcium channels usually refers to the specific transformation $D \xrightarrow{k_1} A$. Similarly the term deactivation of calcium channels usually refers to the specific transformation, $A \xrightarrow{k_2} D$, and the term inactivation of calcium channels usually refers to the specific transformation, $A \xrightarrow{k_3} I$.

Clearly, the relative magnitudes of the rate constants depicted above and the manner in which they are modified by various physical and chemical stimuli determine, to a large degree, the level of activity that a channel will exhibit. Previous studies have shown that the rate constants, k_1 and k_2, which govern the equilibrium between the deactivated and activated states, have relatively large magnitudes. Equilibrium conditions between these two states are, therefore, established relatively quickly. By contrast, the rate constants, k_3, k_4, k_5, k_6, which govern the equilibria between the inactivated state and the other two states, have relatively small magnitudes. As a consequence, these equilibria are established relatively slowly. Moreover, when the cell membrane potential is at its resting level, i.e., -55 to -90 mV, depending on the cell type, and equilibrium conditions prevail, the equilibrium constants k_1/k_2, k_3/k_4, and k_5/k_6 are of such magnitudes that the vast majority of the calcium channels are transformed into the deactivated state. Thus, although all calcium channels are continuously undergoing stochastic shifts in state, when the distribution of functional states is measured at any instant in time, the mean percentage of calcium channels in the deactivated state will be large, whereas the mean percentage of calcium channels in the activated and inactivated states will be quite small. Consequently, the net calcium current that flows into the cell, under these circumstances, is small or imperceptible.

Since the gating mechanisms of a voltage-dependent calcium channel are sensitive to the electric field in which the channel is found, a large change in membrane potential from resting level to a more positive level has a profound influence on the activity of the channel. In effect, the change in membrane potential induces a substantial increase in the values of the equilibrium constants, k_1/k_2 and k_3/k_4, and a decrease in k_5/k_6. As a result, a redistribution of the three functional states ensues in the direction of the new equilibrium condition. Initially, because the rate constants k_1 and k_2 are large relative to the other constants, calcium channels undergo a rapid transformation from the deactivated to the activated state. During this initial period, the whole cell calcium current rapidly rises to a peak level. Following this initial period, the much slower shift of calcium channels to the inactivated state begins to become apparent and continues to progress until complete equilibrium is reached. At this point, a large percentage of the channels are in the inactivated state and very few are in the activated or deactivated state. The whole cell current under these new equilibrium conditions is, again, very small or imperceptible.

When considering the operation of a single channel, because of its stochastic behavior, one cannot predict when or how many times within a given time span the channel will shift from a closed (deactivated or inactivated) state to an open (activated) state and back again. Nor can one predict how long the channel will remain in any particular closed or open state before it undergoes another transformation. If, however, an individual channel is monitored many times (or a large number of channels are monitored simultaneously) and mean values for the experimental results are calculated, relationships between state transformations and the magnitudes of the rate constants become apparent. Assuming the data are obtained under conditions in which transformations of the channel to and from the inactivated state occur at a negligible frequency, these mean values will show that the average number of times that a channel will shift from a closed (deactivated) to an open (activated) position within a given time span is directly proportional to the value of $[k_1 k_2/(k_1 + k_2)]$. Moreover, the average length of time that a channel will remain open once it has shifted to the open state is equal to $[1/k_2]$. Thus, the mean fraction of the total time that a channel will be found in the open position is given by the expression $[k_1/(k_1 + k_2)]$, and is referred to as the probability of open time. The transitory increase in the probability of open time that occurs when the membrane is depolarized is a direct result of the changes that have been induced in the magnitudes of the rate constants. Similar links to rate constants also apply for the other functional states into which the channel shifts.

The presentation given above is meant to be a brief, general description of the dynamic behavior of calcium channels. It does not take into account the wide variations in functional activity that exist among different types of calcium channels. For example, there are some calcium channels in which the ultimate build up of channels in the inactivated state depends, not upon a change in membrane potential as indicated above, but upon an increase in the concentration of intracellular Ca^{2+} ions in the vicinity of the channel. In addition, there are some calcium channels in which the inactivation process is both voltage dependent and Ca^{2+} dependent. A number of other deviations from the above scheme could also be mentioned (see Section I).

Most pharmacological agents that modify the function of voltage-dependent calcium channels act as inhibitory agents. By interacting with the gating mechanism or with some other component of the permeation process (such as a binding site in the aqueous pore of the channel) they reduce the probability of open time and, as a result, lower the peak calcium current that would ordinarily be generated by a given change in membrane potential. Although fewer in number, pharmacological agents that increase the probability of open time and the peak calcium current have also been found. The former are referred to as calcium channel blockers and the latter as calcium channel activators.

It is convenient to group the chemical agents that affect calcium channel activity into three broad categories. The first of these consists of inorganic ions. Inorganic ions have been found to play several different roles in the operation of calcium channels. Some such as Ca^{2+}, Ba^{2+}, Sr^{2+}, and Na^+ ions act as substrates, i.e., substances that can be physically translocated by the action of the channel. These ions are the charged elements that carry electric current through the channel pore. Other ions such as Cd^{2+}, Co^{2+}, and La^{3+} have relatively high affinities for specific sites in various types of calcium channels, but cannot permeate the channel pore. Thus, they act as inhibitors of calcium channel activity. Still other inorganic cations and anions, including the H^+ ion, appear to influence, in some undefined manner, the stability of the membrane and the level of activity of the calcium channel. Chapter 12 provides a discussion of the effects of inorganic ions on calcium channel behavior.

A second category of chemical agents consists of a rather large number of organic substances that act in a highly selective manner on calcium channels. Examples of these are verapamil, nifedipine, diltiazem, and Bay K 8644. The pharmacological agents within this group are, for the most part, calcium channel blockers, although a few can increase the probability of open time and the peak calcium current. The agents within this group also differ with respect to: (1) their relative specificities for calcium channels vs. other ion channels and various types of pharmacological receptors; (2) their relative affinities for different types of calcium channels; and (3) their relative affinities for a calcium channel in each of its three functional states. These factors are important in determining the modifying action that any particular agent in the group will exert on a calcium channel. This subject is covered in Chapter 13.

The third category of chemical agents consists of naturally occurring compounds that influence calcium channel behavior. The neurotransmitters, hormones, and second messengers within this group include such agents as epinephrine, cyclic AMP, and G proteins. Some of these compounds act directly or indirectly to block calcium channel activity; while others exert an activating effect on the channel. Compounds within this group may also serve to recruit functional channels and, thereby, increase the density of operating channels within a given area of cell membrane. A discussion of the naturally occurring regulators of calcium channel activity is provided in Chapter 14.

Chapter 12

Effects of Inorganic Ions on Calcium Channels

N. Lakshminarayanaiah

TABLE OF CONTENTS

I.	Introduction	164
II.	Dependence of Structural Integrity of Calcium Channels on Inorganic and Organic Ions	164
III.	Inorganic Cations and Calcium Channels	165
	A. Cations Carrying Current Through Calcium Channels	166
	B. Cations that Act as Competitive Inhibitors of I_{Ca}	174
	C. Cations Exhibiting Different Actions in Various Types of Calcium Channels	175
	D. Effect of pH	178
IV.	Models of Calcium Channels	179
	A. Model I Contains Two Ion-Selecting Filtering Systems and Permits Permeation of Sodium Ions	180
	B. Model II Contains a Single Ion-Selecting Filter System and does not Permit the Permeation of Sodium Ions	183
	1. Mechanism of Saturation Kinetics	183
	2. Mechanisms of Inhibition of the Saturation Process	185
	3. Tabulation of Dissociation Constants for Various Divalent Cations that Interact with Calcium Channel Sites in Different Biological Systems	188
V.	Experimental Approach for Determining K_{Ca} and K_{Mg} in Barnacle Muscle	188

VI.	Experimental Determination of the Chemical Nature of Calcium Channel-Binding Site	189
	A. pK_a Derived from I_{Ca} vs. pH Relationship in which $(Ca^{2+})_o$ and $(Mg^{2+})_o$ are Kept Constant	190
	B. pK_a Derived from I_{Ca} vs. $(Ca^{2+})_o$ Relationships Obtained at pH 7.5 and pH 5.5	191
	C. Use of pK_a and Other Means to Deduce the Chemical Nature of the Calcium Channel Site that Regulates Channel Function	193
VII.	Summary	193

I. INTRODUCTION

Excitable membranes contain polar water-filled pores which are designated popularly as channels (Armstrong, 1975a; Hille, 1978). In general, these channels act as pathways for the selective flow of ions. Some allow K^+ to flow down its electrochemical gradient across the membrane. Similarly, others allow Na^+ to move down its electrochemical gradient and some others permit the flow of Ca^{2+} and other mono and divalent ions (see Chapter 2, Section I). Several agents have been used to study the characteristics of these various channels in many biological preparations. Inorganic and organic cations have played a major role in the study of the properties of the sodium and potassium channels (Hille, 1975; Armstrong, 1975b). In several early investigations, properties of the calcium channel, also called the slow channel, have been studied by using multifiber and single fiber muscle preparations (Reuter, 1973; Trautwein, 1973; Cranefield, 1975; Hagiwara, 1973, 1975). These studies have yielded considerable insight into the behavior of the calcium channel. In recent years isolated tiny patches of membranes from several biological preparations have been employed in voltage clamp (Patch Clamp Method: see Single Channel Recording, 1983) studies to follow the characteristics of the calcium channel at the molecular level. Interesting data related to single channel behavior have been derived. In many of the studies inorganic mono and divalent cations have been used to probe the calcium channel. These results are presented in the following discussion, placing emphasis on the behavior of the calcium channel in the barnacle muscle membrane when the channel is exposed to different ionic environments.

II. DEPENDENCE OF STRUCTURAL INTEGRITY OF CALCIUM CHANNELS ON INORGANIC AND ORGANIC IONS

With the introduction of intracellular perfusion techniques (Baker, Hodgkin and Shaw, 1961; Oikawa, Spyropoulos, Tasaki, and Teorell, 1961), biological cells can be investigated under a variety of intra and extracellular conditions. The compositions of the internal and external solutions used should be such that they allow the cells to remain viable for extended periods of time. Therefore, experiments have been performed using different intracellular solutions to determine their appropriate constituents so that resulting membrane potentials are maintained close to those predicted by the Nernst equation for potassium ions:

$$E_r \cong (RT/F)\ln \frac{(a_K)_o}{(a_K)_i} \qquad (1)$$

(where E_r is the resting membrane potential, $(a_K)_o$ and $(a_K)_i$ are the activities of K^+ in the extra and intracellular fluids, respectively. R, T and F are the gas constants, absolute temperature and Faraday constant, respectively). Maintenance of a stable resting membrane potential is a prerequisite for the generation of action potentials or of Ca-spikes in those excitable membranes in which calcium channels exist. Under these conditions, measurement of meaningful calcium currents (I_{Ca}s) becomes possible when the membrane is shifted to and held at different membrane potentials (voltage clamp).

Keynes, Rojas, Taylor and Vergara (1973), in their studies on the calcium and potassium systems in a giant barnacle muscle fiber, used potassium aspartate as the main constituent of the internal perfusing solution. Lakshminarayanaiah and Rojas (1973) carried out an extensive investigation of the effects of both inorganic and organic cations and anions on the resting membrane potential of internally perfused barnacle muscle fiber. According to expectation they found K^+ to be the preferred cation in the intracellular perfusion fluid, whereas TEA^+ and Cs^+ were less preferable, since their use led to low values for the resting membrane potential (TEA^+ gave -21 mV and Cs^+ gave -23.6 mV). However, these two ions have proved useful in the measurements of inward I_{Ca} in the barnacle muscle fiber membrane as either of them used in the internal perfusion solutions depressed the late outward I_K that interacted with the early inward I_{Ca} (Keynes et al., 1973; Beirao and Lakshminarayanaiah, 1979).

Of the several anions used in the internal perfusion fluid by Lakshminarayanaiah and Rojas (1973), acetate ion was the most effective in maintaining a steady resting potential (-56 mV). Isethionate (-55 mV), aspartate (-51 mV) and glutamate (-50 mV) were almost equally effective. With regard to the inorganic anions used in the study, F^- proved effective and the other anions decreased in the order $F^- > HPO_4^{2-} > SO_4^{2-} > FeCN_6^{4-} > ClO_3^- > NO_3^- > Cl^- > CNS^- > I^- > Br^-$ in their usefulness to serve as the constituents of the intracellular fluid in the barnacle muscle fiber. CN^- was very destructive in that the transparent muscle fiber became opaque and the resting membrane potential was completely abolished.

The stability of intracellularly dialyzed neurons isolated from snail, *Helix pomatia*, depended largely on the kind of anion used in the dialyzing solution (Kostyuk, Krishtal and Pidoplichko, 1975). With Cl^- in the dialyzing solution, the stability of the neuronal membrane could not be maintained. Leakage current increased and the ionic currents deteriorated. NO_3^- was also found to be less preferable, whereas F^- and $H_2PO_4^-$ maintained the membrane in a stable condition for a long period of time. In another study, Kostyuk et al. (1977) used F^- and $H_2PO_4^-$ in the dialyzing solution and found them to be satisfactory anions. On the other hand, Lee, Akaike and Brown (1978) found that F^- and $H_2PO_4^-$, each used in the internal perfusion solution of neurons located in subesophageal ganglion of *Helix aspersa*, were less favorable than either aspartate or Cl^-. The membrane currents measured by using the latter anions remained unchanged for more than 5 h and the leakage currents were small whereas the former anions generated unsatisfactory membrane currents. The preferability of the anion employed in the measurements of I_{Ca}s in voltage clamped neurons decreased in the order aspartate $\geq Cl^- \gg H_2PO_4^- > F^-$. This disagreement between the Russian and the American workers may be ascribed to differences in either the methods or the species used.

III. INORGANIC CATIONS AND CALCIUM CHANNELS

The effects of several inorganic cations on the calcium channels carrying I_{Ca} have been studied by a number of investigators using a variety of biological preparations. Kostyuk (1980, 1981) has reviewed some of the work related to calcium channels in excitable neuronal membranes. Reuter (1973) has given an account of the early work pertaining to the several effects of multivalent ions on the clacium channels in several biological preparations. Other reviews of the calcium channel have been presented by Hagiwara, (1973, 1975); Baker and Glitsch, (1975); Hagiwara and Byerly, (1981, 1983); and Tsien, (1983).

The different cations used in several studies may be roughly placed into four categories. The first category contains cations that carry current through calcium channels. The second category consists of cations that act as competitive inhibitors of I_{Ca}. The third category is composed of cations that exert different effects in various types of calcium channels. The fourth type consists of protons that exert, in a majority of cases, a depressing effect on the currents flowing through calcium channels. Some of the work dealing with these several aspects of the actions of cations on calcium channels are outlined below.

A. Cations Carrying Current Through Calcium Channels

Ca spikes can be generated in the membrane of the barnacle muscle fiber when the Ca concentration inside the muscle fiber is reduced ($<10^{-7}$ M). In the absence of Ca^{2+} in the external solution, action potentials are abolished. They can be restored by introducing Ba^{2+} or Sr^{2+} (Hagiwara and Naka, 1964). The overshoot of the spike potential has been observed to increase with increasing extracellular concentrations of Ba^{2+} or Sr^{2+}. When a concentration of 20 mM $(Ba^{2+})_o$ was introduced, prolongation of the spike potential occurred giving a plateau lasting several seconds. The effects of Sr^{2+} were less pronounced than those of Ba^{2+} but greater than those of Ca^{2+}. Under voltage clamp conditions Ba^{2+} carried more current than the other two ions and followed the sequence $Ba^{2+} > Sr^{2+} \approx Ca^{2+}$ (Hagiwara, Fukuda and Eaton, 1974). Cota and Stefani (1984) also followed the permeation of Ca^{2+} and Ba^{2+} through calcium channels in intact twitch muscle fibers of the frog by recording action potentials and membrane currents. Calcium action potentials were found to be 15 to 20 mV larger than barium action potentials. Under voltage clamp conditions, the maximal inward current increased from -39 mA/cm³ to -51 mA/cm³ (expressed as current per unit volume, a quantity independent of fiber diameter) and the I-V curve for the peak inward current shifted 15 to 20 mV to more negative potentials when 10 mM Ca^{2+} was replaced by 10 mM Ba^{2+}. Moreover, when the concentration of Ca^{2+} or Ba^{2+} was increased, both the threshold for the activation of the inward current (Th) and the membrane potential E*, at which the rate of rise of action potential ($\dot{V}_{max}$) was maximal, were shifted to more positive potentials along the voltage axis. The shifts in Th and E* were similar when either Ca^{2+} or Ba^{2+} were added to the bathing medium but were more pronounced or larger when Ca^{2+} rather than Ba^{2+} was present. $\dot{V}_{max}$ and conductance for Ca^{2+} or Ba^{2+} showed a tendency to saturate as the divalent cation concentration was increased. From these data, values for the dissociation constant K_M for Ca^{2+} and Ba^{2+} complexes that formed with the membrane sites have been derived and the values are given in Table 1.

In general, it has been observed in a number of preparations including artificial lipid bilayers that, as the divalent cation concentration is increased, the activation and inactivation curves for ion channels are shifted to more positive potentials along the voltage axis (Hille, Woodhull and Shapiro, 1975; Lakshminarayanaiah, 1979). These shifts arise as a result of a reduction in the external surface potential by the screening of negative membrane surface charges that are in a position to change the electric field near the channel. Also, further reduction of the surface potential is likely to occur as a result of binding of divalent cation to negative groups in the membrane. These effects have been quantified using the diffuse double layer theory including specific ion binding. For a brief account of the diffuse double layer theory developed by Gouy-Chapman and of its applications, see Lakshminarayanaiah (1984).

TABLE 1
Dissociation Constants of Inorganic Cation-Membrane Site Complex

Ion	Cell type	K_M (mM)	Ref.
Ba^{2+}	Mollusk neuron, *pomatia*	15.0	Kostyuk, (1980, 1981)
	Twitch muscle fibers of frog		
	From measurement of $\dot{V}_{max}$	12.5	Cota and Stefani, (1984)
	From measurement of conductance	8.0	
	Smooth muscle cells of guinea pig, *Taenia Caeci*	9.6	Ganitkevich et al., (1988)
	Egg cell membrane of starfish *Mediaster aequalis*		Hagiwara et al., (1975)
	Channel I	213	
	Channel II	73	
	Rat brain synaptosomes	1.56	Nachshen and Blaustein, (1979)
	Fast phase	4.2	Nachshen and Blaustein, (1982)
	Slow phase	7.9	
	Rat brain synaptosomes incorporated into lipid bilayer	2.7	Nelson, (1986)
	Mollusk Neuron, *Helix pomatia*	0.016	
	From measurement of permeation of monovalent cation		Kostyuk et al., (1983)
Ca^{2+}	Mollusk neuron, *Helix pomatia*	5.4	Kostyuk, (1980, 1981)
	Subesophageal ganglion, *Helix aspersa*		
	at 0 mV	5.4	Akaike et al., (1978)
	50 mV	10.4	
	75 mV	26	
	100 mV	82	
	Guinea pig ventricular cells		
	From measurement of I_{Ba}	10.0	Lansman et al., (1986)
	Japanese land snail, *Euhadra quaestia*	14.0	Kawa, (1979)
	Chick and rat dorsal root ganglion neurons:		
	low voltage and high voltage activated channels from whole cell recordings	3.3	Carbone and Lux, (1987a)
	low voltage activated channels from single channel recording	10.3	Carbone and Lux, (1987b)
	Rabbit sinoatrial node cells		
	Transient	0.95	Hagiwara et al., (1988)
	Long lasting	3.92	
	Barnacle muscle, *Balanus nubilus*	24	Hagiwara, (1973)
		<15	Hagiwara and Takahashi, (1967)
Ca^{2+}	Barnacle muscle, *Balanus nubilus*	5.2[a]	Beirao and Lakshminarayanaiah, (1979)
		3.7	Van Wagoner, (1985)
	Insect muscle fibers, *Carausius morosus*	7.6	Ashcroft and Stanfield, (1982)
	Twitch muscle fibers of frog		
	From measurement of $\dot{V}_{max}$	5.6	Cota and Stefani, (1984)
	From measurement of conductance	6.0	
	Smooth muscle cells of guinea pig, *Taenia caeci*	1.2	Ganitkevich et al., (1988)
	Ciliate *Stylonychia mytilus*	0.2	Ivens, (1986)
	Rat brain synaptosomes	0.15	Nachshen and Blaustein, (1979)
	Fast phase	0.21	Nachshen and Blaustein, (1980)
	Slow phase	0.18	
	Fast phase	0.32	Nachshen and Blaustein, (1982)
	Slow phase	0.43	
		0.3	Nachshen, (1984)

TABLE 1 (continued)
Dissociation Constants of Inorganic Cation-Membrane Site Complex

Ion	Cell type	K_M (mM)	Ref.
	Rat brain synaptosomes incorporated into lipid bilayer	1.2	Nelson, (1986)
	From measurements of permeation of monovalent cation		
	Helix pomatia	0.0002	Kostyuk et al., (1983)
	Skeletal muscle fiber at -20 mV	0.0007	Almers et al., (1984)
	Mouse neoplastic lymphocytes at -45 mV	0.002	Fukushima and Hagiwara, (1985)
Cd^{2+}	Helix pomatia	0.07	Kostyuk, (1980, 1981)
	Helix aspersa	0.013	Byerly et al., (1985)
		11.0	Akaike et al., (1978)
	Lymnaea neuron	0.003	Byerly et al., (1985)
	Cat dorsal root ganglion		
	High affinity site	0.016	Taylor, (1988)
	Low affinity site	0.106	
	Rat brain synaptosomes incorporated into lipid bilayer	0.0006	Nachshen, (1986)
	From measurement		Nelson, (1986)
	of I_{Ba}	4.1	
	of I_{Ca} and I_{Sr}	9.3	
	of I_{Mn}	21.6	
	calculated taking $K_{Ba} = 2.7$ mM	0.044	
	Skeletal muscle cell determined from movement of monovalent cation	0.7	Almers et al., (1984)
Co^{2+}	Helix pomatia	0.74	Kostyuk, (1980, 1981)
	Helix aspersa	20.9	Akaike et al., (1978)
		2.5	Byerly et al., (1985)
	Lymnaea neuron	0.9	Byerly et al., (1985)
	Barnacle muscle, Balanus nubilus	0.33[b]	Hagiwara, (1973)
	Starfish Mediaster aequalis		Hagiwara et al., (1975)
	Preparation (1) Channel I	7.0	
	Channel II	1.7	
	Preparation (2) Channel I	6.3	
	Channel II	1.7	
	Preparation (3) Channel I	4.7	
	Channel II	1.3	
	Rat brain synaptosomes	0.06	Nachshen, (1984)
Co^{2+}	Skeletal muscle cell	0.3	Almers et al., (1984)
	Determination from movement of monovalent cation		
Cu^{2+}	Rat brain synaptosomes	0.03	Nachshen, (1984)
Hg^{2+}	Rat brain synaptosomes	0.06	Nachshen, (1984)
La^{3+}	Helix aspersa	1.15	Akaike et al., (1978)
	Barnacle muscle, Balanus nubilus	0.027[b]	Hagiwara, (1973)
	Rat brain synaptosomes		Nachshen and Blaustein, (1980)
	Fast phase	0.0003	
	Slow phase	>0.1	
		0.0002	Nachshen, (1984)
	Rat brain synaptosomes incorporated into lipid bilayers		
	From measurement of I_{Sr}	0.15	Nelson et al., (1984)
	Taking $K_{Ba} = 2.7$ mM	0.0007	Nelson, (1986)
Mg^{2+}	Helix pomatia	18.2	Kostyuk, (1980, 1981)
	Barnacle muscle, Balanus nubilus	13.1[b]	Hagiwara, (1973)
		33.4	Van Wagoner, (1985)
	Ciliate Stylonychia Mytilus	0.7	Ivens, (1986)

TABLE 1 (continued)
Dissociation Constants of Inorganic Cation-Membrane Site Complex

Ion	Cell type	K_M (mM)	Ref.
	Rat brain synaptosomes	3.34	Nachshen and Blaustein, (1979)
		5.7	Nachshen, (1984)
	Determined from movement of monovalent cation		
	Helix pomatia	0.06	Kostyuk et al., (1983)
	Skeletal muscle cell	2.8	Almers et al., (1984)
	Mouse neoplastic lymphocytes	0.03	Fukushima and Hagiwara, (1985)
Mn^{2+}	*Helix pomatia*	0.36	Kostyuk, (1981)
	Barnacle muscle, *Balanus nubilus*	0.82[b]	Hagiwara, (1973)
Mn^{2+}	Rat brain synaptosomes		
	Fast phase	0.07	Nachshen and Blaustein, (1980)
	Slow phase	0.3	
		0.05	Nachshen, (1984)
	Rat brain synaptosomes incorporated into lipid bilayer	0.5	Nelson, (1986)
	Determined from movement of monovalent cation		
	Skeletal muscle cell	0.7	Almers et al., (1984)
Ni^{2+}	*Helix pomatia*	0.74	Kostyuk, (1980, 1981)
	Helix aspersa	0.15	Akaike et al., (1978)
		2.5	Byerly et al., (1985)
	Lymnaea neuron	0.6	Byerly et al., (1985)
	Insect muscle, *Carausisus morosus*	3.7	Ashcroft and Stanfield, (1982)
	Barnacle muscle, *Balanus nubilus*	1.23[b]	Hagiwara, (1973)
	Rat brain synaptosomes		
	Fast phase	0.037	Nachshen and Blaustein, (1980)
	Slow phase	0.027	
		0.05	Nachshen, (1984)
	Determined from movement of monovalent cation		
	Skeletal muscle cell	1.3	Almers et al., (1984)
Pb^{2+}	Rat brain synaptosomes	0.0004	Nachshen, (1984)
Sr^{2+}	*Helix pomatia*	10.0	Kostyuk, (1980, 1981)
	Smooth muscle cell of guinea pig, *Taenia caeci*	1.8	Ganitkevich et al., (1988)
	Rat brain synaptosomes	0.89	Nachshen and Blaustein, (1979)
	Fast phase	4.5	Nachshen and Blaustein, (1982)
	Slow phase	1.5	
		2.0	Nachshen, (1984)
Sr^{2+}	Rat brain synaptosomes incorporated into lipid bilayer	1.2	Nelson (1986)
	Determined from movement of monovalent cation		
	Helix pomatia	0.0035	Kostyuk et al., (1983)
Y^{3+}	Rat brain synaptosomes	0.0007	Nachshen, (1984)
Trivalent Lanthanides	Rat brain synaptosomes	(<1 μM)	Nachshen, (1984)
Zn^{2+}	Snail *Euhadra quaestia*	2.1	Kawa, (1979)
	Rat brain synaptosomes	0.04	Nachshen, (1984)

[a] Calculated using a value of 33.4 mM for K_{Mg}.
[b] Calculated using a value of 3.67 mM for K_{Ca}.

The Ca^{2+}- and Ba^{2+}-induced voltage shifts observed by Cota and Stefani (1984) can be attributed, as pointed out above, to the capacity of these divalent cations to screen and/or bind to the negative charges associated with the membrane surface and ion channels. Since Ca^{2+} has a greater affinity for membrane surface sites than does Ba^{2+} (see Table 1) its effects are more pronounced than those due to Ba^{2+}. Thus, Th for inward current shifted to more positive potentials when 10 mM Ba^{2+} was replaced by 10 mM Ca^{2+}. Cota and Stefani (1984) quantitatively described the shifts along the voltage axis by using the Gouy-Chapman theory calling for a surface charge density near the calcium channel of $0.2e/nm^2$ and a specific binding constant for the Ca^{2+}-membrane surface site complex of 45 M^{-1}. Also Ganitkevich, Shuba and Smirnov (1988) in their studies of I_{Ca} in smooth muscle cells of guinea pig, *Taeni caeci*, observed a shift in the I-V curve on the voltage axis when the concentration of the divalent cation was changed. The voltage shift again was fitted to the Gouy-Chapman theory assuming the surface charge density near the calcium channels to be $0.5e/nm^2$. The binding of Ca^{2+} to the calcium channel sites was stronger than those of Sr^{2+} and Ba^{2+} (see Table 1).

In voltage clamped hippocampal neurons (CA_1 and CA_3) of the guinea pig slow inward current was depressed by removal of Ca^{2+} but was increased by the addition of 1 mM Ba^{2+} to the external solution (Brown and Griffith, 1983).

Calcium channels in the *Helix aspersa* neuron also permit the passage of Sr^{2+} and Ba^{2+} as well as the passage of Ca^{2+} (Akaike, Lee and Brown, 1978). Equimolar substitution of Ba^{2+} for Ca^{2+} increased the transient current and made the null potential more positive. Sr^{2+} had similar effects and the order of their effects was $Ba^{2+} \approx Sr^{2+} > Ca^{2+}$. Similarly, each of the three alkaline earth cations readily permeated the calcium channels in snail neurons (Byerly and Hagiwara, 1982; Byerly, Chase and Stimers, 1985) and rat uterine smooth muscle (Jmari, Mironneau and Mironneau, 1987).

In saline solutions containing a single alkaline earth cation, the calcium channel current usually increased monotonically in a nonlinear fashion as the concentration of that ion was increased. Further increase in the external divalent cation concentration resulted in saturation of that current giving a plateau (see Section IV.B). In a mixture of two cations, for example Ca^{2+} and Ba^{2+}, in which the total divalent ion concentration is held constant, the measured membrane current will still vary as a monotonic function of the mole-fraction of either ion. This is the prediction made for the case of a cell-type whose calcium channel contains a single ion-binding site that controls ion permeation through the channel (Jmari et al., 1987). However, if the calcium channel contains two or more ion-binding sites, the channel current will not be a monotonic function of the mole-fraction of either ion. It has been experimentally observed in several biological preparations that, when the divalent cations Ba^{2+} and Ca^{2+} in the saline solution were present singly or as a mixture at a total concentration, for example of 10 mM, the peak current with 10 mM Ba^{2+} was larger than with 10 mM Ca^{2+}. When both ions were present in a (1:1) mixture, the current was smaller than when the less permeant ion, Ca^{2+}, was present as a single entity (Almers and McCleskey, 1984). Hess and Tsien (1984) obtained a minimum current in a saline solution that contained 1 to 3 mM Ca^{2+}. Jmari et al. (1987) observed a minimum current when 10 to 30% of $(Ba^{2+})_o$ was replaced by $(Ca^{2+})_o$. On the other hand, Byerly et al. (1985) obtained a minimum when there was 30% $(Ba^{2+})_o$ in the mixture of Ba^{2+} and Ca^{2+} in the saline solution. Thus the membrane current was smallest in a mixture of two cations. Partial exchange of Ba^{2+} for Ca^{2+} resulted in a decrease in membrane current that at some (Ba^{2+}/Ca^{2+}) ratio reached a minimum level. This effect has been described as an anomalous mole-fraction effect (see Section IV.A) which has been observed, in addition to other preparations, in heart muscle cells (Hess and Tsien, 1984), frog skeletal muscle fibers (Almers and McCleskey, 1984), snail neurons (Byerly et al., 1985) and rat uterine smooth muscle cells (Jmari et al., 1987). The effect is considered to be evidence for the existence of two or more binding sites occupied by permeant ions moving in "single file" through the calcium channel.

The bursting pacemaker neuron R-15 of *Aplysia* also permits a transmembrane flow of Ca^{2+}, Sr^{2+} or Ba^{2+} (Gorman, Hermann and Thomas, 1982). With a divalent cation concentration of 10 mM in the external saline, the peak inward currents at zero membrane potential were -210 nA (Ca^{2+}), -180 nA (Sr^{2+}) and -150 nA (Ba^{2+}). On this basis, the selectivity or the affinity of the ions for the calcium channel followed the order $Ca^{2+} > Sr^{2+} > Ba^{2+}$. However, the currents were measured under conditions in which intracellular accumulation of Ca^{2+} was prevented by EGTA which was injected into the neuron for 5 min at an intensity of 500 nA. Accumulation of intracellular Ca^{2+} has two effects. The first effect is called the calcium-dependent inactivation of the calcium channel (see Chapter 4, Section I) and the second effect is called the calcium-dependent activation of potassium channels. Under normal conditions, i.e., in the absence of EGTA, accumulation of intracellular Ca^{2+}, but not of Ba^{2+}, will inactivate calcium channels and thereby induce a reduction in I_{Ca}. Thus, in normal saline, membrane current is much larger in the presence of Ba^{2+} than in the presence of Ca^{2+} and can thereby reverse the effect of a greater affinity of the calcium channel for Ca^{2+} than for Ba^{2+}. An additional complication due to accumulation of Ca^{2+} in the cell is the increase in activity of $(Ca^{2+})_i$ activated potassium channels through which an outflow of K^+ occurs. This creates the possibility of an interaction between I_{Ca} and I_K which would lead to pacemaker oscillations (also see Keynes et al., 1973).

Calcium channels in the myenteric neuron of the guinea pig ileum maintain an action potential when Ba^{2+} was substituted for Ca^{2+} (Hirst, Johnson and van Helden, 1985). Voltage clamp experiments in 0.5 mM Ca^{2+} showed that the membrane current recorded in response to a 100 ms depolarizing step from a holding potential of -55 mV to a potential of -10 mV was inward and slowly became outward; whereas in 0.5 mM Ba^{2+}, the current remained inward and showed no outward current. These data suggest that in the case of Ca^{2+}, the outward current arose from the outflux of K^+ through the calcium-activated potassium channels and in the case of Ba^{2+} there was no such outward current because of the inability of Ba^{2+} to activate potassium channels. Furthermore, the magnitudes of the inward currents in the presence of Ca^{2+} or Ba^{2+}, each at a concentration of 0.5 mM, were similar indicating similar mobilities of the two ions in the calcium channels of the myenteric neurons of the guinea pig ileum. However, in the presence of Ba^{2+}, because membrane permeability to K^+ remained lower, the input resistance of the cells was greater, but the I-V relationship was similar to that of Ca^{2+}.

In chick and rat dorsal root ganglion neurons, Carbone and Lux (1987a) found that low voltage activated (lva) I_{Ca} was reduced from -0.18 to -0.12 nA (estimated from Figure 17 of their paper) at a membrane potential of -20 mV when 20 mM Ca^{2+} was replaced by 20 mM Ba^{2+}; whereas the amplitude of the high voltage activated I_{Ca} was increased from -0.9 to -1.7 nA (estimated from Figure 17 of their paper) at $+20$ mV. On the other hand, Sr^{2+} increased I_{Ca}s of both low and high voltage activated channels, respectively, from -0.12 to -0.15 nA at -20 mV and from -0.75 to -0.95 nA at $+20$ mV (values estimated from Figure 18 of their paper). In the case of cultured ganglion cells, patch clamp experiments showed that replacement of 20 mM Ca^{2+} by 20 mM Ba^{2+} produced little change in the amplitude of membrane current (-0.32 pA at -40 mV) generated in lva single calcium channels (Carbone and Lux, 1987b).

Two types of calcium channels, one activated at -75 to -70 mV and the other at -55 to -50 mV, were found to exist in new born rat dorsal root ganglion neurons (Fedulova, Kostyuk and Veselonsky, 1985). These neurons, when intracellularly dialyzed, allowed Ca^{2+}, Sr^{2+} or Ba^{2+} to pass through the channels. In 14.6 mM Ca^{2+} the amplitude of the current in the first type of channel was 0.2 to 0.9 nA whereas in the second type it was 1.7 nA. Replacement of Ca^{2+} by Sr^{2+} or Ba^{2+} increased the maximal current amplitude. The I-V curve for Ca^{2+} shifted by about 20 mV compared with those for Sr^{2+} and Ba^{2+}, indicating a stronger effect of Ca^{2+} on the surface charges of the neuronal membrane.

The immature egg cell membrane of starfish *Mediaster aequalis* was voltage clamped to study the currents of the calcium channel (Hagiwara, Ozawa and Sand, 1975). The cell membrane showed two different inward currents, one activated at -55 to -50 mV (channel I) and the other at -7 to -6 mV. Both channels allowed Ca^{2+}, Sr^{2+} or Ba^{2+} to serve as charge carriers. Channel II showed a greater tendency toward saturation of the current as the concentration of the cation was increased. The amplitude of the peak inward current showed maxima at -25 mV and $+15$ mV in Ca^{2+}, Sr^{2+} and Ba^{2+} solutions, although each of the three cations produced different maximum amplitudes.

Calcium currents in cultured dorsal root ganglion cells were investigated by Fox, Nowycky and Tsien (1987a,b) using the whole cell patch clamp technique. These investigators, by employing 3 to 10 mM $(Ca^{2+})_o$ or $(Ba^{2+})_o$, found three distinct types of I_{Ca}s. The three types of channels called L, T and N were distinguished on the basis of voltage-dependent kinetics and pharmacology. The L channel becomes active at relatively positive test potentials greater than -10 mV. When the holding potential is shifted from -60 mV to -40 mV, the T and L components are suppressed. The T channel can only be activated with weak pulses and at a holding potential greater than -60 mV the T channel is inactivated. Thus, the inactivation of the channel can be removed by setting the membrane potential between -60 and -95 mV. The N channel activates at relatively strong depolarizations with test pulses greater than -20 mV. The inactivation of this channel can be removed over a broad range of holding potentials (-40 to -110 mV). With 10 mM EGTA in the pipette solution, substitution of Ba^{2+} for Ca^{2+} does not alter the rates of activation or relaxation of L, T and N channels. T channels were observed to be equally permeable to Ca^{2+} and Ba^{2+}, while L- and N-type channels were found to be more permeable to Ba^{2+}. Raising $(Ba^{2+})_o$ to 110 mM largely increased the amplitude of the L current activated at a holding potential of -30 mV and there was little inactivation. Thus, the effects of several pharmacological agents on these currents have been studied by using a solution containing 110 mM isotonic $BaCl_2$ (Fox et al., 1987a,b). Similarly, Ba^{2+} has been used, (1) to study the inhibition of calcium channels by intracellular protons in ventricular myocytes of the guinea pig (Kaibara and Kameyama, 1988), (2) to follow the reversal of current flow in cat dorsal root ganglion (Taylor, 1988), (3) to study the single calcium channel properties in ventricular heart cells of the guinea pig (Lansman, Hess and Tsien, 1986) and (4) to follow the differences in barium and calcium action potentials and currents in heart cells of the guinea pig (Lee and Tsien, 1984).

Cavalie, Ochi, Palzer and Trautwein (1983) followed I_{Ca} and I_{Ba} through the calcium channels in adult guinea pig myocytes. Single channel current amplitudes recorded from patches of ventricular myocytes were larger when 50 mM Ba^{2+} was substituted for 50 mM Ca^{2+} in the pipette and increased more than twice when Ba^{2+} was increased to 90 mM. The open time of the channel was 1.52 ms at 30 mV and 3.25 ms at 50 mV when Ba^{2+} was the charge carrier. At corresponding voltages the open times for Ca^{2+} were 1.41 and 1.98 ms, respectively.

It is well known that channel gating is voltage dependent. The foregoing data indicate that both the magnitude of membrane depolarization and the type of divalent cation influence not only the probability of channel opening but also keep it open for longer periods of time. Also lipid bilayers, doped with a rat brain preparation, exhibited mean open times of 127, 385 and 454 ms when the channel carried Ba^{2+}, Sr^{2+} and Ca^{2+}, respectively (Nelson, French and Krueger, 1984). Single channel conductances were 5, 8.5 and 5 pS when the charge carriers were Ca^{2+}, Ba^{2+} and Sr^{2+}, respectively. These data suggest that the nature of the divalent cation carrying the current through the channel affected not only the single channel conductance but also the channel open time, with the mean open time being shortest for Ba^{2+}.

K^+-stimulated influxes of ^{45}Ca, ^{85}Sr and ^{133}Ba were measured in synaptosomes from rat brain (Nashshen and Blaustein, 1982). Two phases, fast and slow, of divalent ion entry into calcium channels were observed. Ion flow in fast channels lasted about 1 s and that in slow channels lasted more than 10 s. Influx through both channels saturated with increasing concentration of divalent ion whose relative permeation through the two types of channels were 6:3:2 (Ca^{2+}:Sr^{2+}:Ba^{2+}) for the fast channel and 6:3:1 for the slow channel. A rat brain membrane preparation when incorporated into planar lipid bilayers allowed each alkaline earth cation to pass through its calcium channels. The single channel conductances of the channels decreased in the order $g_{Ba} > g_{Ca} \approx g_{Sr}$ (Nelson, 1986).

Calcium channel currents were recorded in Cs^+-dialyzed, voltage clamped single smooth muscle cells of guinea pig, *Taenia caeci*, to determine the current carrying abilities of Ca^{2+}, Sr^{2+}, Ba^{2+} and Mg^{2+} (Ganitkevich, Shuba and Smirnov, 1988). Each of the three ions except Mg^{2+} carried an inward current through the calcium channels. Ba^{2+} carried the greatest current. Mg^{2+} depressed I_{Ba} in a competitive manner and to a greater extent than it did the I_{Ca}.

The patch electrode voltage clamp technique has been used to study the characteristics of the calcium channel in tissue cultured clonal cells (GH_3) isolated from rat anterior pituitary tumor (Hagiwara and Ohmori, 1982, 1983). In one cell submerged in 25 mM Ca^{2+}, Sr^{2+} or Ba^{2+}, the currents were -90, -155 and -190 nA (estimated from Figure 2 of Hagiwara and Ohmori, 1982). Averaged mean values of currents in five experiments conformed to the relative order of permeation of ions through the channel of 1 (Ca^{2+}), 1.6 (Sr^{2+}) and 2.7 (Ba^{2+}). The magnitude of I_{Ba} increased linearly with increases in $(Ba^{2+})_o$ up to 25 mM and, thereafter, showed a tendency to saturate. The single channel current was 0.2 pA at 25 mM and 0.7 pA at 100 mM. Similarly, the patch electrode voltage clamp technique has been applied to study the properties of the calcium channel in a hybridoma cell line (MA6-7B) formed by the fusion of S194 myeloma cells and splenic B lymphocytes from the mouse (Fukushima and Hagiwara, 1983, 1985). The relative magnitudes of the peak inward currents at 10 mM concentration, normalized to that for Ca^{2+} (-175 pA estimated from Figure 5 of Fukushima and Hagiwara, 1985 and equated to unity), were 1 (Ca^{2+}), 1.24 (Sr^{2+}), 0.99 (Ba^{2+}) and 0.07 (Mn^{2+}). Thus, Sr^{2+} carried the greatest current (Fukushima and Hagiwara, 1983). Both inward and outward currents flowed through the calcium channel and the reversal potentials measured at a concentration of 10 mM followed the order Ca^{2+}(38 mV) $>$ Ba^{2+}(26.5 mV) $\approx$ Sr^{2+}(26.2 mV) $>$ Mn^{2+}(24.4 mV). Similarly, Hess, Lansman and Tsien (1986) measured both reversal potentials and slope conductances in ventricular cells. The slope conductances followed the order Ba^{2+}(20 pS) $>$ Sr^{2+}(8 pS) $\approx$ Ca^{2+}(8 pS); whereas the reversal potentials followed the order Ca^{2+}(70.5 mV) $>$ Sr^2(66 mV) $>$ Ba^{2+}(59.2 mV).

The above data show that the behavior of reversal potentials and that of ion currents or conductances of several divalent cations are different following dissimilar sequences. The affinity or the selectivity of the ions for the channel sites is commonly evaluated by two methods. In the first method, the amplitudes of currents or conductances in a control and a test solution are compared. To be applicable, it is assumed that the number of open channels are the same in each solution and that there is no saturation or block of open channels by the control or test solution. This means that the independence principle of Hodgkin and Huxley (1952) is obeyed. The second method of calculating ionic selectivity or affinity of the membrane channel sites for ions uses the reversal potential. In this method, the result does not depend on the number of conducting channels. The change in reversal potential on changing the external saline from a control solution (e.g., Ca^{2+}) to a test solution (e.g., Ba^{2+}) gives a value for the permeability ratio of the two ions. In the case of monovalent cations, the constant field or Goldman-Hodgkin-Katz equation (see Lakshminarayanaiah, 1984) is commonly used to calculate the permeability ratio. In the case of divalent cations, the expression for the equilibrium potential, although more complex (see Lakshminarayanaiah, 1984b), yields a value for the permeability ratio. In essence, the results indicate that the cation with the highest affinity (i.e., highest permeability as revealed by the reversal potential) has the lowest mobility (i.e., lowest permeability as judged by channel conductance). This behavior can be illustrated by a parable that is narrated elsewhere (Spiegler and

Wyllie, 1956; see also Helfferich, 1962) to describe ion movements in an ion exchanger. The story is like that of the two politicians moving at an equal pace through a crowd of supporters to reach the podium. The more popular politician reaches the podium last, because his movement is hindered more than that of his opponent by the need to shake hands more frequently with his supporters. So, it is seen that the permeation of a divalent cation is controlled by the binding of the ion to a site in the calcium channel. Ca^{2+} has a higher affinity for the channel site than does Ba^{2+}, but Ca^{2+} carries less current in heart cells (Hess et al., 1986) and almost the same level of current in a hybridoma cell line, MA6-7B (Fukushima and Hagiwara, 1985).

Egg cell membrane of a tunicate, *Halocynthia roretzi* Drasche, exhibited I_{Ca}s that increased monotonically in a nonlinear fashion saturating above 50 mM $(Ca^{2+})_o$ (Okamoto, Takahashi and Yoshii, 1976). In 100 mM artificial sea water, the maximum peak current was 1 to 7 nA. Sr^{2+} or Ba^{2+} could be substituted for Ca^{2+} giving currents in nA of -7.2 (Ca^{2+}), -8.8 (Sr^{2+}) and -4.2 (Ba^{2+}) at a potential of 25 mV.

Current and voltage clamp techniques were used to study the action potential and the membrane current in calcium channels of mouse oocytes (Okamoto, Takahashi and Yamashita, 1977). Maximum peak inward currents in nA carried by Ca^{2+}, Sr^{2+} and Ba^{2+} were -6.71, -9.7 and -4.76, respectively, in 20 mM external divalent ion. They have also been shown to be -5.3, -6.25 and -5.6 in tunicate in 30 mM divalent ion and -11.43, -6.95 and -5.05 in sea urchin oocytes in 30 mM divalent ion. Similarly, ovarian oocytes isolated from adult mice developed action potentials in Ca^{2+}, Sr^{2+} or Ba^{2+} solutions, the overshoot of the spikes increasing by 28 mV for a ten-fold increase in concentration (Yoshida, 1983).

B. Cations that Act as Competitive Inhibitors of I_{Ca}

A host of ions act as inhibitors of I_{Ca}. It has been found that the valency and the concentration at which a particular cation is used determines the effectiveness of the ion in acting as a competitive inhibitor of I_{Ca}. At any given concentration the greater the valency of the cation the more effective it is as a competitive inhibitor, although exceptions to this general principle exist. When the concentration of an inhibiting cation is increased, the degree of inhibition of I_{Ca} increases.

In the barnacle muscle membrane, Hagiwara (1973, 1975) and Hagiwara and Takahashi (1967) found that the effectiveness of multivalent ions that competitively inhibit I_{Ca} decreased in the order La^{3+}, $UO_2^{2+} > Zn^{2+}$, Co^{2+}, $Fe^{2+} > Mn^{2+} > Ni^{2+} > Ca^{2+} > Mg^{2+}$, Sr^{2+}. Similarly, Ashcroft and Stanfield (1982) found that La^{3+} and Cd^{2+} at a concentration of 1 mM blocked the calcium current in the muscle fibers of insect, *Carausis morosus*. Co^{2+} (5 to 20 mM) and Ni^{2+} were found to be less effective than La^{3+} and Cd^{2+}. On the other hand, Ni^{2+} was found to be more effective than La^{3+} and Cd^{2+} as an inhibitor of I_{Ca} in molluscan neuron (*Helix aspersa*) in which the effectiveness of ion inhibitors decreased in the following order $Ni^{2+} \gg La^{3+} \gg Cd^{2+} > Co^{2+} \gg Mg^{2+}$ (Akaike et al., 1978). Moreover, these ions competitively blocked I_{Ca} in snail *Lymnaea* neurons (Bylerly and Hagiwara, 1982; Byerly et al., 1985), although a similar order of effectiveness was not observed (Byerly et al., 1985).

It was mentioned earlier that the calcium channels exhibiting monotonic variation of membrane channel current as a function of mole-fraction of either ion in a mixture of two divalent cations in the external saline solution characteristically contain a single ion-binding site. An anomalous mole-fraction effect is characteristic of a calcium channel that contains two ion-binding sites. Both of these sites may exist in the channel or one may exist inside and the other outside the channel. When blocking concentrations of an inhibiting ion such as Cd^{2+} are used in the measurement of I_{Ca} or I_{Ba}, both the sites are affected, although to different extents, leading finally to total block of the channel current. Taylor (1988) studied the blocking action of Cd^{2+} on I_{Ba} flowing through the calcium channels in cat dorsal root ganglion. Cd^{2+} acted on two sites. One was the high affinity site that was located within the membrane electric field and the complex it formed with Cd^{2+} had a dissociation constant around 16 μM at 0 mV. The cadmium block at this site was removed by hyperpolarization with a voltage dependence that corresponded to a divalent ion moving through about 75% of the membrane electric field. The other site that was blocked was a voltage-independent low affinity site and the complex it formed with Cd^{2+} had a dissociation constant around 106 μM.

In lipid bilayers incorporated with a rat brain membrane preparation, Mg^{2+} had no effect on the divalent cation movement through the calcium channels (Nelson, 1986). On the other hand, La^{3+}, Mn^{2+} and Cd^{2+} affected the current through the channel, Mn^{2+} being a less potent blocker than either La^{3+} or Cd^{2+}.

The several biological preparations in which multivalent ions have been used to inhibit or block I_{Ca} are: Co^{2+} or Mg^{2+} in the egg cell membrane of starfish (Hagiwara et al., 1975); H^+, La^{3+}, Co^{2+}, Mn^{2+}, Ca^{2+}, Sr^{2+}, Ba^{2+} or Mg^{2+} in the egg cell membrane of the tunicate (Okamoto et al., 1976); Cd^{2+} in the somatic membrane of mollusk neurons (Kostyuk and Krishtal, 1977); La^{3+}, Ni^{2+}, Mn^{2+} or Mg^{2+} in rat brain synaptosomes (Nachshen and Blaustein, 1980); La^{3+} in the pacemaker neuron R-15 of *Aplysia* (Gorman et al., 1982); Co^{2+}, Cd^{2+}, Mn^{2+} or La^{3+} in ovarian oocytes (Yoshida, 1983); Mg^{2+}, Co^{2+} or Cd^{2+} in twitch fibers of rat extensor digitorum longus and soleus muscles (Chiarandini and Stefani, 1983) and in mouse Swiss 3T3 Fibroblasts (Peres, Sturani and Zippel, 1988); Cd^{2+} or Mn^{2+} in hippocampal neurons of the guinea pig (Brown and Griffith, 1983); La^{3+} or Cd^{2+} in lipid bilayers containing a rat brain preparation (Nelson et al., 1984); Co^{2+}, Mn^{2+}, Cd^{2+} or Ni^{2+} in somatic membrane of newborn rat dorsal root ganglion neurons (Fedulova et al., 1985); Mn^{2+} or Co^{2+} in the myenteric neuron of the guinea pig ileum (Hirst et al., 1985); Cd^{2+}, Co^{2+} or Mn^{2+} in rat supraoptic neurosecretory neurons (Bourque and Renaud, 1985); Ca^{2+}, Cd^{2+}, Mg^{2+}, Co^{2+}, Mn^{2+} or La^{3+} in patch clamped guinea pig ventricular cells (Lansman et al., 1986); Ni^{2+} or Cd^{2+} in chick and rat sensory neurons (Carbone and Lux, 1987a); Mn^{2+}, Co^{2+} or Ni^{2+} in rat uterine smooth muscle (Jmari et al., 1987); and Ni^{2+}, Cd^{2+} or Co^{2+} in patch clamped sinoatrial node cells of the rabbit (Hagiwara, Irisawa and Kameyama, 1988).

In an extensive study Nachshen (1984) investigated the effects of 25 polyvalent metal cations on the K^+-stimulated calcium influx in rat brain synaptosomes. The influx of calcium was blocked by the ions with half inhibition constants (K) that fell into three distinct groups: (1) K $>$ 1 mM (Mg^{2+}, Sr^{2+} and Ba^{2+}); (2) K between 30 to 100 μM (Mn^{2+}, Co^{2+}, Ni^{2+}, Cu^{2+}, Zn^{2+}, and Hg^{2+}) and (3) K $<$ 1 μM (Cd^{2+}, Y^{3+}, La^{3+} and the other lanthanides and Pb^{2+}).

C. Cations Exhibiting Different Actions in Various Types of Calcium Channels

Crustacean muscle fibers are of great interest because of the role of Ca^{2+} in the generation of electrical activity and in the initiation of contraction. Muscle fibers from crab are able to generate long-lasting action potentials in the absence of external Na^+ (Fatt and Katz, 1953; Fatt and Ginsborg, 1958). While a variety of properties are displayed by several crustacean muscle fibers, the large muscle fibers of the barnacle are indifferent to the presence or absence of external Na^+. Artificial sea water commonly used in performing experiments with the barnacle muscle fibers usually contain about 430 mM of Na^+. Even then, no inward current is seen in the absence of Ca^{2+} in artificial sea water (Keynes et al., 1973; Murayama and Lakshminarayanaiah, 1977). When Ca^{2+} is restored, inward current appears both in presence and absence of Na^+ (Keynes et al., 1973). Barnacle muscle fibers display electrical properties that are explained on the basis of a model that contains a single ion-binding site in the calcium channel (see Section IV.B). There are other biological preparations, already referred to, that exhibit an anomalous mole-fraction effect and that also do not allow Na^+ to move through the calcium channel under normal conditions. Ca^{2+} occupies the active sites in the calcium channel and prevents the movement of Na^+ and other monovalent cations through the calcium channel. However, when the concentration of $(Ca^{2+})_o$ is reduced to a level below 10 μM or the membrane potential is shifted to a positive voltage, the calcium channels become permeable to monovalent cations.

In a large number of biological preparations, voltage and time-dependent ion channels co-exist with the calcium channels. To identify the current through the calcium channel the other channel currents must be minimized by using specific current suppressing chemical agents or by making ion substitutions. Even then, it is not easy to separate the current of the calcium channel from other currents. An additional difficulty arises because of the existence of different kinds of calcium channels probably existing side by side in the membrane of the same cell (Hagiwara and Byerly, 1981, 1983). It is possible that these different kinds of calcium channels may not display the same permeability to monovalent cations. Reuter and Scholz (1977) thought that at large positive membrane potentials, K^+ may carry the outward current through the calcium channel in heart muscle. In support of this, Lee and Tsien (1982) found that K^+ did carry outward current through the calcium channels in mammalian heart muscles. Similar conclusions have been reached in studies performed on preparations such as bovine chromaffin cells (Fenwick, Marty and Neher, 1982) and frog skeletal muscle fibers (Almers, McCleskey and Palade, 1984).

In recent years, inward currents carried by monovalent cations through the calcium channel have been recorded by using the voltage clamp technique. To record these currents, $(Ca^{2+})_o$ was reduced to submicromolar levels. Complete removal of the divalent cation from the external solution resulted in the calcium channels of mouse Swiss 3T3 Fibroblasts becoming permeable to monovalent cations (Peres et al., 1988). Russian workers separated the tetrodotoxin (TTX) resistant inward I_{Na} from the I_{Ca} in the somatic membrane of the mollusk neuron (Kostyuk and Krishtal, 1977). The relative permeabilities of monovalent cations through the calcium channel were observed to be 1 (Na^+), 0.8 (Li^+), 0.55 ($N_2H_5^+$), 0.21 (NH_3OH^+) (Kostyuk, Mironov and Shuba, 1983). The magnitude of I_{Na} decreased with increases in the divalent cation concentration in the external solution. Similarly, Na^+ permeation through the calcium channel has been shown to occur in rat uterine smooth muscle (Jmari et al., 1987), in fresh water ciliate *Stylonychia mytilus* (Ivens, 1986) and in low voltage activated calcium channels of chick and rat sensory neurons (Carbone and Lux, 1987b).

Using the frog semitendunosus muscle fiber, Almers et al. (1984) have carried out a detailed study of the membrane permeation of monovalent cations through calcium channels. Membrane currents were recorded in voltage clamped, EGTA-loaded muscle fiber conditions in which currents through ordinary sodium, potassium and chloride channels were prevented by drugs or by the absence of permeant ions (K^+, Cl^-). At 10 mM $(Ca^{2+})_o$, substitution of Na^+ for large impermeant organic cations (TMA^+ or TEA^+) gave no significant current. However, when $(Ca^{2+})_o$ was reduced to a level below 1 μM in the presence of Na^+, step depolarizations to negative potentials produced TTX resistant inward currents. Large inward currents were carried by Li^+, Na^+, Rb^+ and Cs^+ but not by TMA^+ and TEA^+. The permeability sequence was $Na^+ \approx Li^+ > Rb^+ > Cs^+ \gg TMA^+, TEA^+$. These currents could be blocked by divalent cations and the blocking potency followed the sequence $Ca^{2+} > Sr^{2+} \geq Co^{2+} > Mn^{2+} \approx Cd^{2+} > Ni^{2+} \approx Mg^{2+}$.

Single channel and whole cell recordings were used by Hess et al. (1986) to study ion permeation through calcium channels in isolated ventricular heart cells of the guinea pig. The permeability of monovalent cations was evaluated by measuring unitary current amplitude and by measuring reversal potential. Single channel conductances were 85 pS for Na^+ and 45 pS for Li^+. These measurements were made at a near zero membrane potential and with ion concentrations of 110 to 150 mM in the external solution. Values of the reversal potential for the several monovalent cations estimated from Figure 5 of Hess et al. (1986) followed the sequence Li$^+$ (-30 mV) > Na$^+$ (-42 mV) > K$^+$ (-55 mV) > Cs$^+$ (-60 mV). Again, an ion with a high permeability judged by its reversal potential had a low ion transfer rate. Similar types of studies carried out by Fukushima and Hagiwara (1985) using mouse neoplastic lymphocytes showed that the peak currents in one cell estimated from Figure 10 of their paper followed the sequence Na$^+$ (-35 pA) > K$^+$ (-25 pA) > Rb$^+$ (-5 pA) > Cs$^+$ (-3 pA). The relative permeabilities of the same ions calculated from the values of their reversal potential followed the order 1, Na$^+$ (-1.2 mV) > 0.8, K$^+$ (-6.7 mV) > 0.6, Rb$^+$ (-14.2 mV) > 0.35, Cs$^+$ (-27.0 mV) and thus contradicts the general rule noted above. Also, the same relative sequence was noted when the monovalent cations carried the outward current at large positive potentials.

Yoshida (1983) generated Ca spikes in the ovarian oocytes membrane of the mouse when the external solution was Na$^+$-free. Under Ca^{2+}-free conditions, he observed TTX resistant Na spikes which showed an overshoot of 39 mV for a ten-fold increase in (Na$^+$)$_o$. These spikes were blocked by calcium antagonists such as Co^{2+}, Cd^{2+}, Mn^{2+} or La^{3+}. In producing these spikes, Li$^+$ substituted for Na$^+$ while Rb$^+$ did not. Addition of Ca^{2+} to the bathing medium reduced both the magnitude of the overshoot and the maximum rate of rise of the Na spike in a competitive manner.

Mn^{2+}, in many biological preparations, acts as a competitive inhibitor of I_{Ca}. In the ovarian oocyte membrane of the mouse, it acted not only as an inhibitor of I_{Ca} but also as a charge carrier (Yoshida, 1983). During excitation, Mn spikes were detected in Na$^+$ and Ca^{2+}-free solutions and were blocked by calcium antagonists. Anderson (1979) generated Mn spikes in myoepithelial cells of a marine polycheate. The overshoot of the action potential increased 27 mV per ten-fold increase in (Mn^{2+})$_o$. The Mn spikes were blocked by Co^{2+} and La^{3+}. Ochi (1970) also observed that increases in (Mn^{2+})$_o$ in a Ca^{2+}-free solution increased the amplitude of the slow inward current during large depolarizations of the voltage clamped guinea pig myocardium. Moreover, he recorded propagated action potentials in the right ventricular papillary muscles of the guinea pig heart when it was exposed to Na$^+$ and Ca^{2+}- and Mg^{2+}-free solutions containing Mn^{2+} (Ochi, 1975, 1976). The overshoot of the Mn spike increased by 20 to 30 mV per ten-fold increase in (Mn^{2+})$_o$ over a range of 2 to 50 mM. The Mn spikes were depressed not only by La^{3+} but also by repeated electrical stimulation. In the neurons of the snail *Lymnaea stagnalis*, Byerly et al. (1985) were able to record small inward currents when (Ca^{2+})$_o$ was replaced by Mn^{2+}. Nelson (1986) found that Mn^{2+}, in the absence of other permeant ions, would pass through single channels in lipid bilayers that were doped with a rat brain membrane preparation. With Mn^{2+} as the charge carrier, channel conductance was 4 pS. However, when Mn^{2+} was added to a solution that contained another permeant divalent cation, the single channel current was reduced in a voltage-dependent manner.

Zn^{2+} acts as a competitive inhibitor of the calcium channel in many excitable cells, although in the Japanese land snail *Euhadra quaestia* Deshayes all or none action potentials were evoked in a saline solution containing 24 mM Zn^{2+} (Kawa, 1979). The overshoots were about 10 mV and the maximum rate of rise was 2.9 V/s. The action potentials were suppressed by Co^{2+}, La^{3+} and verapamil. Zn^{2+}, Mn^{2+}, Cd^{2+} and Be^{2+} were also observed to pass through the calcium channels of an insect muscle membrane (Fukuda and Kawa, 1977). Larval muscle fibers of a beetle, *Xylotrupes dichotomus*, produced Ca spikes that were maintained when the fibers were bathed in saline solutions containing Zn^{2+}, Mn^{2+}, Cd^{2+} or Be^{2+} instead of Ca^{2+}. The divalent ions Co^{2+}, Ni^{2+} and Mg^{2+} acted as competitive inhibitors.

Mg^{2+} or Na$^+$ added to a solution of 0.1 mM CaCl$_2$ prolonged Ca^{2+}-dependent action potentials in ciliate *Stylonychia mytilus* (Ivens, 1986). In a Ca^{2+}-free solution containing 2 mM MgCl$_2$, the cells generated repetitive spontaneous action potentials of small magnitude

(~17 mV). The I-V curve derived from voltage clamp experiments in a standard saline solution showed two maxima indicating the existence of two different voltage-dependent calcium channels. Mole-fraction experiments (alteration of Mg^{2+} and Ca^{2+} concentrations keeping the total divalent ion concentration constant) showed that Mg^{2+} and Ca^{2+} do not inhibit each others movement through channel I when the channel is activated by a 2 to 3 mV depolarization from a holding potential of -50 mV. Channel II, on the other hand, can only be activated at depolarizations larger than 10 mV. In nominally Ca^{2+}-free solution containing either 2 mM $MgCl_2$ or 20 Na^+, inward current passed only through channel I. This current was smaller and more prolonged at higher depolarizations than was the inward current in a standard solution containing 1 mM Ca^{2+} in which both channels I and II had been activated. The anterior part of the ciliate cell contains organelles called membranelles from which current I originates. This current and not current II disappeared when the cell released its membranelles. Additionally, these two types of calcium channels allow Ba^{2+} or Sr^{2+} to cross the cell membrane and are sensitive to the inhibitory actions of Co^{2+} and Cd^{2+}, channel II being more sensitive (Deitmer, 1984).

D. Effect of pH

Biological membranes are composed of phospholipids and proteins which are amphipathic molecules containing both polar and nonpolar chemical parts. The nonpolar parts are mostly hydrocarbon chains, while the polar parts contain negatively or positively charged chemical groups such as carboxylic, amino, etc. These ionogenic chemical groups existing in a biological membrane are usually oriented toward the aqueous phase bounding the cell membrane. These membrane charges associated with the chemical groups give rise to the formation of electrical double layers on both the inside and the outside of the membrane interfaces. A protein macromolecule in the membrane seems to be involved in channel formation. The open channel is filled with aqueous fluid. The ion traffic through the membrane occurs through these water-filled channels the walls of which carry a net negative charge.

Protons have played a unique role in establishing the nature of the charged chemical group present in the channel. They exert their effect primarily by binding to the chemical group (Spitzer, 1979) and thereby inhibiting one of the properties of the ion channel, namely ion conductivity or permeability. Thus, monitoring ion channel conductance or permeability as a function of pH has been used to derive an apparent value for the pK_a of the chemical group of the ion channel since half inhibition of that channel property is equated to pK_a.

Beirao and Lakshminarayanaiah (1979) measured the I_{Ca} in the voltage clamped barnacle muscle fiber as a function of $(H^+)_o$. In the pH range 8.8 to 6.5, I_{Ca} remained unaffected. When the pH was reduced below 6.5, I_{Ca} was depressed, the depression being about 2, 12 and 19% at pHs 6.1, 5.1 and 4.5, respectively (I_{Ca} = 100% at pH 7.5). The half inhibition of I_{Ca} occurred at a pH of 3.9 thus giving an apparent value of 3.9 for the pK_a of the chemical group of the calcium channel in the barnacle muscle membrane (Lakshminarayanaiah and Beirao, 1979). The apparent value, after correction for the presence of Ca^{2+} and Mg^{2+}, each at 100 mM, gave a value of 5.2 for the pK_a.

In guinea pig ventricular myocytes, binding and dissociation reactions of single protons and deuterium ions at a single site on the L-type calcium channel have been studied by Prod'hom, Pietrobon and Hess (1987). The experimental results were analyzed to determine protonation and deprotonation rates. For protons the rate constants for association and dissociation were 3.8×10^{11} $M^{-1}s^{-1}$ and 0.68×10^4 s^{-1}, respectively, giving a pK_a of 7.75. The values for deuterium were 4.09×10^{11} M^{-1} s^{-1} and 1.67×10^4 s^{-1}, yielding a pK_a of 7.40. Krafte and Kass (1988) employed L-type channels to study the effects of both alkaline and acidic pHs on the I_{Ca}. Alkaline pH enhanced I_{Ca} and acidic pH reduced it, the reduction being about 60 to 70% of that at pH 7.5. The observed shifts in I_{Ca} were considered to be consistent with the titration of negatively charged groups existing on the inside and/or outside of the membrane surface. These charged sites control not only gating but also ion permeation of the channels. The sites were estimated to have a pK_a of 5.8 and a density of $-1e/250$ Å^2. In addition, some consideration was given to the possibility that another mechanism by which H$^+$ may have acted on calcium channels was by physically blocking them.

The apparent pK_a of the chemical group of the calcium channel was found to be 5.8 in the neuronal membrane (Kostyuk, 1981). In other studies the apparent pK_a was found to be 6.3 in brain synaptosomes (Nachshen and Blaustein, 1979), 5.6 in immature egg cell membrane of starfish *Mediaster aequalis* (Hagiwara et al., 1975) and 6.6 in single ventricular myocytes of the guinea pig (Kaibara and Kameyama, 1988).

In response to a step increase in $(H^+)_o$, Konnerth, Lux and Morad (1987) noted a large transient inward current in chick dorsal root ganglion cells (also see Chapter 2, Section I). The cells clamped at a holding potential of -80 mV, responded rapidly by generating a large inward current when the external solution was switched from one buffered at pH 7.3 to another buffered at pH 6.7. The peak amplitude of the current ranged from 1 to 3 nA and it took 200 to 300 ms to reach the peak. The current decayed completely within 1 to 2 s. Also the current could be repeatedly activated provided the switch in solutions was executed so as to permit a rapid step increase in $(H^+)_o$. On the other hand, a step increase in $(H^+)_o$, i.e., switch of solution from pH 6.7 to 7.9, or a slow step increase in $(H^+)_o$ failed to induce any inward current. Between one step increase in $(H^+)_o$ and another step increase, a relaxing time of 2 to 3 s at pH 7.9 was required to reprime the system completely.

With $(Ca^{2+})_o$ at 1 mM, the dependence of current on pH was sigmoid with activation taking place at pH around 7 and reaching a maximum at pH 6 to 5.5. The data showed that the current induced by the proton was I_{Na} flowing through a transformed calcium channel. The increase in $(H^+)_o$ changed the calcium channel from one that was voltage gated to one that became proton gated. This transformation resulted in the generation of an ion channel that could preferentially transport Na$^+$ for about a second.

Byerly and Moody (1986) found, as expected, a decrease in I_{Ca} in internally perfused neurons of the snail *Lymnaea stagnalis* at low $(pH)_i$. Whether this decrease in I_{Ca} was due to a direct effect of $(pH)_i$ or due to some other secondary effects, they were unable to determine. A number of effects of $(pH)_i$ on the L-type calcium channel activity were noted in patch clamped ventricular myocytes of the guinea pig (Kaibara and Kameyama, 1988). They were: (1) protons reduced I_{Ba} of the calcium channels by 10 to 20% without any effect on the maximum slope conductance of the channel; (2) the number of channels in a patch remained unaltered when $(pH)_i$ was lowered; (3) lowering the $(pH)_i$ caused both the steady state activation and inactivation curves to shift toward more negative membrane potentials, the extent of the shifts was 10 to 15 mV, moreover, these reactions were reversible, and (4) the probability of channel opening was decreased when the $(pH)_i$ was lowered. Thus, the inhibition of I_{Ba} by a lowered $(pH)_i$ was attributed to the effect of protons on the gating mechanism of the channel. Similarly, lowering the external pH in a hybridoma cell line (mAb-7B) caused both activation and inactivation curves to be shifted toward less negative potentials (Iijima, Ciani and Hagiwara, 1986). This evidence suggests a pH-activated decrease of the external negative surface potential sensed by the gating mechanism.

IV. MODELS OF CALCIUM CHANNELS

There are essentially two models proposed to explain the foregoing results dealing with the functioning of calcium channels present in several biological preparations. The first model is considered here to exist in two forms and is applicable to a calcium channel through which monovalent cations can permeate. The second model is applicable to the calcium channel through which monovalent cations cannot carry a current. The latter type of calcium channel was discovered nearly three decades ago in crustacean muscle fibers (Hagiwara and Naka, 1964; Fatt and Ginsborg, 1985). The former type was found to exist in mollusk neurons (Kostyuk and Krishtal, 1977). Since then, both types of calcium channels have been found in many biological preparations. The principal features of the two models are outlined below.

A. Model I Contains Two Ion-Selecting Filtering Systems and Permits Permeation of Sodium Ions

There are two variations of this model—one form originally proposed by the Russian workers (Kostyuk et al., 1983) and the other form proposed by the American workers (Hess and Tsien, 1984; Almers and McCleskey, 1984). This classification into two variations of a model has been referred to in the literature as two distinct models (Fukushima and Hagiwara, 1985).

The model of the calcium channel in form one is postulated to contain two ion-selecting filters or sites, one located at the external surface of the membrane just outside the channel and the other inside the water-filled channel. Experimental results show that the calcium channel transforms into a monovalent cation transporting system only when calcium binding agents such as EDTA are used. The channel transformation is related to the removal of calcium from some highly specific binding site of the calcium channel. The experimentally determined dissociation constant of the calcium complex at this site as seen from the data of Table 1 is in the micromolar range. Kostyuk et al. (1983) found the dissociation constant to be unaffected by voltage. Consequently, they proposed that this high affinity, calcium-binding site was located outside the calcium channel and designated it as the external ion-selecting filter of the calcium channel. This is distinct from a low affinity site that is located inside the channel (the intrachannel ion selecting filter). The low affinity filter allows the passage of monovalent cations on the basis of their steric correspondence to the geometry of the filter and divalent cations on the basis of their relative binding capacity to the carboxyl group present in the filter. In general, when the binding capacity of the divalent cation is large, the probability of its passage through the filter decreases. Thus, Co^{2+}, which has a high affinity to the site (see Table 1), does not go through the filter. However, exceptions to this principle may exist. For example, Mg^{2+} which has a relatively low affinity to the site does not go through the selectivity filter in a number of biological preparations. In such cases, it appears as though some unknown factors, probably steric in nature, also seem to operate to prevent the passage of divalent cations through the filter.

According to the above model in form one, when external solutions containing mixtures of divalent ions (e.g., Ca^{2+} and Ba^{2+}), in which $(Ca^{2+})_o + (Ba^{2+})_o$ together are maintained at a constant concentration (e.g., 10 mM) are used to measure the channel current, the total current should be a monotonic function and should never be smaller than what it would be in either 10 mM $(Ca^{2+})_o$ or $(Ba^{2+})_o$ alone. However, the experimental results indicated that total current was not a monotonic function of the mole fraction (Hess and Tsien, 1984; Almers and McCleskey, 1984; Byerly et al., 1985). In ventricular cells, the total current went through a minimum when 70% of Ca^{2+} was replaced by Ba^{2+} (Hess and Tsien, 1984). Such anomalous mole-fraction effects have been accounted for in the other form of the model in which two high affinity sites are considered to exist within the channel. The calcium channel has been modeled by Hess and Tsien (1984) as a single file pore with two low energy valleys and three high energy barriers, the binding sites being situated one in each valley. The anomalous mole-fraction effect was accounted for by the fact that Ca^{2+} binds more strongly than Ba^{2+} and that Ca^{2+} and Ba^{2+} repel each other within the channel. When Ca^{2+} occupied the inner site, the Ba^{2+} that crossed the outer energy barrier of the channel was quickly repelled and could not enter or was forced to exit from the valley. Thus, replacement of $(Ca^{2+})_o$ by an equivalent amount of Ba^{2+} made the current smaller at first because Ba^{2+} was relatively impermeant. When $(Ca^{2+})_o$ was made low, Ba^{2+} started to dominate and began to contribute to the current.

The channel is permeable to monovalent cations when both sites are unoccupied by divalent cations. If one of the sites is occupied by a Ca^{2+}, the channel may become impermeable to both monovalent and divalent cations including Ca^{2+} (Fukushima and Hagiwara, 1985). The latter effect depends on the electrochemical character of the bond holding the Ca^{2+} to the channel site. Although the probability of this bond being completely in the covalent form is small, it will possess some covalent and some electrovalent characteristics. The exact fraction of the bonding that is covalent or electrovalent is determined by the level of $(Ca^{2+})_o$ in the bathing medium. When pCa is higher than 8, the fraction of the bond possessing a covalent character will be high. Under these conditions, the channel will be permeable to monovalent cations such as Na^+. As $(Ca^{2+})_o$ is increased, the chemical bond tends to acquire a relatively greater electrovalent character. When pCa is 2 to 3, a large fraction of the bond will be electrovalent. Under these conditions, Ca^{2+} at the site in the inner valley will be repelled by Ca^{2+} in the outer valley. The force of repulsion will overcome the bond energy to eject the ion from the site into the cytoplasm giving rise to I_{Ca}. Between the two extremes, there exists other states in which bound Ca^{2+} in the inner valley would not have sufficient net positive charge (i.e., the bond energy is relatively high) to develop enough repulsive force with the Ca^{2+} in the outer valley to overcome the bond energy to produce a current due to Ca^{2+}. Thus, Ca^{2+} becomes an impermeant ion, although there is a bound Ca^{2+} in the inner valley. Since the transition from the covalent character of the chemical bond to the electrovalent character is progressive with increases in $(Ca^{2+})_o$, one can expect a pCa range in which the calcium channel is permeable to both Na^+ and Ca^{2+}. Thus, in the pCa range 8 to 3 when the channel is permeable to Na^+ or to both Na^+ and Ca^{2+}, the assumption has been that the stoichiometry between Ca^{2+} and the channel site is one-to-one.

The foregoing considerations, although implied, are not considered explicitly in the literature. They are based on the experimental results obtained by Almers et al. (1984) using frog skeletal muscle fibers and by Fukushima and Hagiwara (1985) employing mouse neoplastic lymphocytes.

Almers et al. (1984) explored the effects of $(Ca^{2+})_o$ on the inward currents over a 10^8-fold range in concentration from 10^{-10} to 10^{-2} M. At $(Ca^{2+})_o < 10^{-8}$ M, the peak inward current carried by monovalent cations, for example Na^+, is high. As $(Ca^{2+})_o$ is increased the total current becomes smaller, reaches a minimum and then increases as Ca^{2+} becomes the current carrier. When the concentration of $(Ca^{2+})_o$ is less than 10^{-5} M, the inward current is carried entirely by Na^+. No significant contribution comes from Ca^{2+}, since the inward current decreases when $(Ca^{2+})_o$ is increased. At pCas between 2 and 3, the inward current is carried entirely by Ca^{2+}. At pCas between 3 and 5, both Na^+ and Ca^{2+} may contribute to the inward current. Similar results were obtained by Fukushima and Hagiwara (1985) who found a co-existence of Ca^{2+} and Na^+ currents under certain conditions. In a standard bathing medium containing 150 mM Na^+, 5 mM K^+ and 2.5 mM Ca^{2+}, they observed a peak inward I_{Ca} that was relatively high (about 100 pA). When the solution was switched to one containing 150 mM Na^+, 5 mM K^+ and 0.2 mM Ca^{2+} plus 1.5 mM Mg^{2+}, a lesser inward current (about 48 pA) was observed. When all Na^+ and K^+ in the buffered solution containing the Ca^{2+} and Mg^{2+} was replaced by 155 mM TMA^+, a significant inward current, about 20 pA, was seen. This inward current was a calcium current since TMA^+, an established impermeable ion, cannot generate inward current. Consequently, the lesser inward current, about 48 pA, must have been due to both Na^+ and Ca^{2+}.

The experimental points of peak inward currents obtained as a function of pCa in the range 8 to 3 by Almers et al. (1984) corresponded more closely to a curve they generated on the basis of one-to-one stoichiometry than to a curve generated on the basis of two-to-one stoichiometry of Ca^{2+} and the channel site.

Kostyuk et al. (1983), Almers et al. (1984), and Fukushima and Hagiwara (1985) measured the decreases in monovalent cation currents as a function of $(M^{2+})_o$ at low concentrations of divalent cations by using buffering agents such as EDTA, HEDTA, DPTA, etc. Decreases in monovalent cation current I_{M^+} with increases in divalent cation concentration $(M^{2+})_o$ can be approximated by the Langmuir's isotherm (see Equation 30) which for the present case can be written as

$$I_{M^+} = \frac{I_{M^+(max)} K_{M^{2+}}}{K_{M^{2+}} + (M^{2+})_o}$$

where $I_{M^+(max)} = I_{M^+}$ when $(M^{2+})_o = 0$. Rearrangement of this equation gives

$$\frac{I_{M^+(max)}}{I_{M^+}} = 1 + \frac{(M^{2+})_o}{K_{M^{2+}}}$$

Thus, a plot of $[I_{M^+(max)}/I_{M^+}]$ against $(M^{2+})_o$ gives a straight line whose slope is equal to $(1/K_{M^{2+}})$. This method was used by Kostyuk et al. (1983) to derive values for $K_{M^{2+}}$. On the other hand, Almers et al. (1984) and Fukushima and Hagiwara (1985) established the isotherms [plots of $(I_{M^+(max)}/I_{M^+})$ vs. $(M^{2+})_o$] for several divalent cations and determined the half blocking concentrations of $(M^{2+})_o$ for those divalent cations. According to the equation given above, the half blocking concentrations of $(M^{2+})_o$ are equal to $K_{M^{2+}}$. The values so derived for several divalent cations, given in Table 1, show that they are all in the micromolar range. As opposed to this, the values of dissociation constants for M^{2+} derived for several divalent cations by measurement of $I_{M^{2+}}$ as a function of $(M^{2+})_o$, also given in Table 1, show that the values of $K_{M^{2+}}$ are in the millimolar range. Most of these values are derived for biological systems (e.g., barnacle muscle fiber) that are known to contain a single site in the calcium channel. However, investigations of Kostyuk et al. (1983) with the *Helix pomatia* neuron show that this preparation has two site channels, one existing inside the channels and the other postulated to exist on the outside of the channels. The experimental data derived for this preparation by Kostyuk et al. (1983) conformed to Equation 10 given later. In the case of those preparations that are known to contain calcium channels with two sites both inside the channels (e.g., frog skeletal muscle fiber, ventricular heart cells of the guinea pig, mouse neoplastic lymphocytes, etc.), experimental data related to variation of $I_{M^{2+}}$ as a function of $(M^{2+})_o$ are sparse and the scant data that exist have never been analyzed from the standpoint of Equation 10 given later. There is, therefore, insufficient evidence to assert that the divalent cation flow through the two site calcium channels also follows the Langmuir adsorption isotherm.

Almers et al. (1984) obtained three I-V curves at 1, 3 and 10 mM $(Ca^{2+})_o$ and stated that increasing $(Ca^{2+})_o$ from 10 to 100 mM increased the peak current by a factor of only 3.5. A plot of these data [I_{Ca} vs. $(Ca^{2+})_o$] showed that I_{Ca} varied nonlinearly with $(Ca^{2+})_o$ and displayed a slight tendency to saturate. But a double reciprocal plot gave points that were scattered showing no trend towards a straight line. As these data are too few and far between in the concentration range 10 to 100 mM, it is difficult to draw any conclusion about the applicability of the Langmuir adsorption isotherm. Fukushima and Hagiwara (1985) have also presented I-V curves in Figure 4 of their paper. From this figure, the following peak values for I_{Ca} in pA can be estimated: 100 at 2.5 mM; 200 at 10 mM; 260 at 25 mM and 310 at 50 mM. These data show that I_{Ca} exhibits a tendency to saturate at higher $(Ca^{2+})_o$. Ignoring the effects of surface potential on this data, a plot of $(1/I_{Ca})$ vs. $1/(Ca^{2+})_o$ defines a straight line and values of 324 pA and 5.6 mM have been derived for $I_{Ca(max)}$ and K_{Ca}, respectively, by regression analysis. Although no correction for the effect of surface potential on I_{Ca} has been made (correction affects only the values of $I_{Ca(max)}$ and K_{Ca} but not the applicability of Equation 10), this exercise clearly demonstrates the applicability of the Langmuir adsorption isotherm to data derived for a biological preparation that is known to have two site calcium channels.

From the foregoing description, it is apparent that occupancy of both channel sites by Ca^{2+} leads to repulsion which promotes current flow. Ca^{2+} on the outer site ejects by electrostatic repulsion the Ca^{2+} on the inner site allowing it to enter the cell. Perpetuation of the current thus entails a number of dissociations and combinations akin to that postulated by Grotthus for the conduction of ions in solution (Glasstone, 1940).

This single file two site form of the model has been used to explain the results obtained with heart cells (Hess and Tsien, 1984), mouse neoplastic lymphocytes (Fukushima and Hagiwara, 1985) and frog skeletal muscle fibers (Almers and McCleskey, 1984). Although Byerly et al. (1985) observed anomalous mole-fraction effects in *Lymnaea* neurons, they were unable to support the model because there was a lack of data related to monovalent cation permeation of the calcium channel.

B. Model II Contains a Single Ion-Selecting Filter System and does not Permit the Permeation of Sodium Ions

It has been demonstrated experimentally in several biological preparations (Hagiwara and Takahashi, 1967; Hagiwara and Byerly, 1981) that the occupation of a site, S, by Ca^{2+} in the calcium channel is an important step in the transport of Ca^{2+} through the channel. When the chemical bond holding Ca^{2+} to the site is more electrovalent, the ion passes through the selectivity filter of the calcium channel to produce a current. However, as discussed above, when the chemical binding of an ion to the site is relatively high, e.g., the binding of Cd^{2+}, Co^{2+}, etc. that normally acts as a competitive inhibitor, and when other physical factors such as a mismatched geometry between the filter and the ion interfere with divalent cation passage through the filter, cation flow through the filter may not take place. With regard to permeable divalent cations such as Ca^{2+}, Sr^{2+}, etc. the dissociation constants of channel site(s)-M^{2+} complexes are, as mentioned above, in the millimolar range in various preparations (see Table 1) and, as a result, the bond holding M^{2+} to the site(s) is, for the most part, electrovalent.

It has been found experimentally that as the concentration of a permeable divalent cation such as Ca^{2+} is increased, I_{Ca} through the channel increases, but the increase becomes progressively less with each equal increment increase in $(Ca^{2+})_o$. At high $(Ca^{2+})_o$, the channel site(s) become highly saturated with Ca^{2+} since the concentration of these sites is very small compared to that of $(Ca^{2+})_o$. As a result, almost all the channel sites are in the form of a complex, CaS. Further increases in $(Ca^{2+})_o$ produce extremely little further increases in CaS. Consequently, the rate of dissociation of CaS, a factor that helps determine the magnitude of I_{Ca}, becomes independent of $(Ca^{2+})_o$.

1. Mechanism of Saturation Kinetics

The model to explain saturation kinetics corresponds to a single binding site situated in the well of the channel bounded by two energy barriers, one on either side. Accordingly, the approximate dynamic equilibrium can be described by the equation:

$$(Ca^{2+})_o + S \underset{k_{-1}}{\overset{k_1}{\rightleftharpoons}} CaS \underset{k_{-2}}{\overset{k_2}{\rightleftharpoons}} S + (Ca^{2+})_i \quad (2)$$
$$\text{Extracellular} \hspace{4cm} \text{Intracellular}$$

where k's are the rate constants; S is the level of free unbound calcium channel binding sites; CaS is the level of calcium ion-binding site complexes; $(Ca^{2+})_o$ is the extracellular calcium ion concentration and $(Ca^{2+})_i$ is the free intracellular calcium ion concentration.

The steady-state rate equation for (CaS) is given by

$$\frac{d(CaS)}{dt} = k_1(Ca^{2+})_o(S) + k_{-2}(S)(Ca^{2+})_i - (CaS)(k_{-1} + k_2) = 0 \qquad (3)$$

If the total number of sites is $(S)_t$, then

$$(S)_t = (S) + (CaS) \qquad (4)$$

Substituting for (S) from Equation 4 in Equation 3 gives on rearrangement

$$(CaS) = \frac{k_1(Ca^{2+})_o(S)_t + k_{-2}(Ca^{2+})_i(S)_t}{k_{-1} + k_2 + k_1(Ca^{2+})_o + k_{-2}(Ca^{2+})_i} \qquad (5)$$

The rate of transfer of metal ions across the channel is given by $v = \dfrac{d(Ca^{2+})_i}{dt}$ where v is the rate of transfer of charged Ca^{2+} into the cell. This is equivalent to current I_{Ca} which, according to Equations 2 and 4, is given by $\dfrac{d(Ca^{2+})_i}{dt} \times F \left[\dfrac{g.\ equiv}{sec} \times \dfrac{coulombs}{g.\ equiv} \right]$.

$$(I_{Ca}/F) = (CaS)[k_2 + k_{-2}(Ca^{2+})_i] - k_{-2}(Ca^{2+})_i(S)_t \qquad (6)$$

Substituting for (CaS) from Equation 5 gives on simplification

$$(I_{Ca}/F) = \frac{k_1 k_2 (Ca^{2+})_o (S)_t - k_{-1} k_{-2} (Ca^{2+})_i (S)_t}{k_{-1} + k_2 + k_1(Ca^{2+})_o + k_{-2}(Ca^{2+})_i} \qquad (7)$$

Since the internal concentration, $(Ca^{2+})_i$, is extremely small Equation 7 simplifies to

$$(I_{Ca}/F) = \frac{k_1 k_2 (Ca^{2+})_o (S)_t}{k_{-1} + k_2 + k_1(Ca^{2+})_o} \qquad (8)$$

The current I_{Ca} becomes $I_{Ca(max)}$ when all the sites are occupied by Ca^{2+} (saturation current). Thus

$$(I_{Ca(max)}/F) = k_2(S)_t \qquad (9)$$

Equation 8 can be written in the familiar form

$$I_{Ca}(V) = \frac{I_{Ca(max)}(V)(Ca^{2+})_o}{K_{Ca}(V) + (Ca^{2+})_o} \qquad (10)$$

where the voltage-dependent parameters are indicated by (V) and $K_{Ca} = \dfrac{k_{-1} + k_2}{k_1}$; K_{Ca} is called here the dissociation constant. As the mathematics of enzyme kinetics is applied to the steady-state analysis of reaction 2, K_{Ca} should be called the Michaelis-Menten constant. In the special case when k_2 is much smaller than k_{-1}, i.e., when the breakdown of CaS to $(Ca^{2+})_i$ and S is sufficiently slow to allow a true equilibrium of CaS with $(Ca^{2+})_o$ and S to

12. Effects of Inorganic Ions on Calcium Channels

be attained, then $(k_{-1} + k_2)/k_1$ approaches (k_{-1}/k_1) and now $K_{Ca} = K_s$ where K_s is the dissociation constant. K_{Ca} is always greater than K_s. It is unfortunate that K_{Ca} and K_s are used interchangeably in the literature. Some investigators have called K_{Ca} the Michaelis constant and others have called it simply the dissociation constant. In all the present considerations, K_{Ca} is called the dissociation constant.

Most of the alkaline earth ions (e.g., Ca^{2+}, Sr^{2+}, Ba^{2+}) that carry current through the calcium channel exhibit saturation phenomena in the manner described above.

Calcium channel processing of different divalent cations, when each is present singly or in the presence of another multivalent cation, is governed by several parameters. The important ones are the affinity of the ion for the channel site(s) and the channel conductance. The latter is a determinant of ion size, its mobility and the structural geometry of the channel itself. Ca^{2+} has a dissociation constant (see Table 1) and mobility in aqueous solution (see Robinson and Stokes, 1959) which are smaller than those of Ba^{2+} and in addition, has a larger hydrated ion size (see Lakshminarayanaiah, 1969). Consequently, I_{Ba} is larger than I_{Ca}. But in the case of an ion such as Mg^{2+}, although its dissociation constant and hydrated ion size are larger than those of Ca^{2+} and its mobility a little less, Mg^{2+} is not transported through the calcium channel in many biological preparations. Its presence, however, is very essential. In the barnacle muscle fiber, it seems to promote the structural stability of the muscle membrane (see later).

2. Mechanisms of Inhibition of the Saturation Process

Although ions such as H^+, Mg^{2+}, Co^{2+}, etc. do not permeate the calcium channel, they do bind to the channel site(s) in several biological preparations (see Table 1). Three situations must be distinguished. They are competitive, noncompetitive and anticompetitive inhibitions. Following Laidler and Bunting (1973), a brief account of classical inhibition mechanisms is presented.

If Q is the inhibiting ion (charge sign dropped), it acts on the site, S, to form the complex QS. In a simple competitive inhibition, existence of a complex CaSQ in which both Ca and Q are simultaneously attached to the site is ruled out. In the simple noncompetitive inhibition, formation of such a complex is permitted.

The classical scheme of reactions involved in competitive inhibition can be represented as follows:

$$(Ca)_o + S \underset{k_{-1}}{\overset{k_1}{\rightleftharpoons}} CaS \overset{k_2}{\rightarrow} S + Ca \quad \text{with} \quad Q + S \underset{k_{-3}}{\overset{k_3}{\rightleftharpoons}} QS \tag{11}$$

(superscripts on extracellular Ca and super and subscripts on intracellular Ca have been dropped). In Scheme 11, the site can bind either Ca or Q but not both. Steady-state equations for CaS and QS are

$$k_1(Ca)_o(S) - (k_{-1} + k_2)(CaS) = 0 \tag{12}$$

$$k_3(Q)(S) - k_{-3}(QS) = 0 \tag{13}$$

But

$$(S)_t = (S) + (CaS) + (QS) \tag{14}$$

Substituting for (S) and (QS) in terms of (CaS) from Equation 12 and 13 into Equation 14 gives

$$(S)_t = (CaS)\left[\frac{k_{-1} + k_2}{k_1(Ca)_o} + 1 + \frac{(k_{-1} + k_2)k_3(Q)}{k_{-3}k_1(Ca)_o}\right] \tag{15}$$

The current I_{Ca} [$= k_2$ (CaS) F] therefore is given by

$$I_{Ca} = \frac{k_2(S)_t(Ca)_oF}{\frac{k_{-1} + k_2}{k_1} + (Ca)_o + \frac{(k_{-1} + k_2)k_3}{k_{-3}k_1}(Q)} \tag{16}$$

This may be written as (see Equation 9)

$$I_{Ca} = \frac{I_{Ca(max)}(Ca)_o}{(Ca)_o + K_{Ca}\left[1 + \frac{(Q)}{K_q}\right]} \tag{17}$$

where $K_{Ca} = (k_{-1} + k_2)/k_1$ and $K_q = k_{-3}/k_3$, the dissociation constant of the complex QS. When K'_{Ca} the apparent dissociation constant,

$$K'_{Ca} = K_{Ca}\left[1 + \frac{(Q)}{K_q}\right] \tag{18}$$

is substituted, Equation 17 becomes equivalent to the hyperbolic Equation 10. These two equations written in reciprocal forms, $(1/I_{Ca})$ vs. $(1/(Ca)_o)$, become linear equations. A plot of $(1/I_{Ca})$ against $1/(Ca)_o$, a commonly employed procedure proposed long ago by Lineweaver and Burk (1934), will give a straight line with a slope equal to $K'_{Ca}/I_{Ca(max)}$ and the intercept on the $1/I_{Ca}$-axis equal to $1/I_{Ca(max)}$. Thus values for K'_{Ca} and $I_{Ca(max)}$ can be derived. In some of the experimental work described in Section III, values determined for K'_{Ca} have been equated to K_{Ca} when the influence of Q on K_{Ca} was found to be negligible. Where this is not so, as found in the barnacle muscle fiber, determination of K_{Ca} becomes a problem which is addressed later in Section V.

The classical scheme for simple noncompetitive inhibition can be represented by

$$\begin{array}{c} QQ\ k_2\\ (Ca)_o +\ +\ \rightleftharpoons\ +\ \rightarrow S + Ca\\ S\ \ k_{-1}\ \ CaS\\ k_3 \Updownarrow k_{-3}\ \ k_4 \Updownarrow k_{-4}\\ QS\phantom{\ \ k_{-1}\ }CaSQ \end{array} \tag{19}$$

The steady state equations for CaS, QS and CaSQ are

$$(Ca)_o(S) - K_{Ca}(CaS) = 0 \tag{20}$$

$$(Q)(S) - K_q(QS) = 0 \tag{21}$$

$$(Q)(CaS) - K_{Ca-q}(CaSQ) = 0 \tag{22}$$

where the dissociation constant $K_{Ca-q} = k_{-4}/k_4$. The total site concentration is now given by

12. Effects of Inorganic Ions on Calcium Channels

$$(S)_t = (S) + (CaS) + (QS) + (CaSQ) \tag{23}$$

Substituting from steady-state Equations 20 to 22 gives on rearrangement

$$(S)_t = \frac{(CaS)}{(Ca)_o}\left[K_{Ca}\left(1 + \frac{(Q)}{K_q}\right) + (Ca)_o\left(1 + \frac{(Q)}{K_{Ca-q}}\right)\right] \tag{24}$$

The current I_{Ca} therefore is given by

$$I_{Ca} = \frac{I_{Ca(max)}(Ca)_o}{K_{Ca}\left[1 + \frac{(Q)}{K_q}\right] + (Ca)_o\left[1 + \frac{(Q)}{K_{Ca-q}}\right]} \tag{25}$$

This is an important relation since all three types of inhibitions mentioned above follow from it.

When $K_{Ca-q} = K_q$, i.e., the bindings of CaS and S to the inhibitor are equal, Equation 25 becomes

$$I_{Ca} = \frac{I_{Ca(max)}(Ca)_o}{[K_{Ca} + (Ca)_o]\left[1 + \frac{(Q)}{K_q}\right]} \tag{26}$$

Thus this equation describes the mechanism for pure noncompetitive inhibition.

When $K_{Ca-q} \gg K_q$, Equation 25 reduces to Equation 17 applicable to pure competitive inhibition. However, when $K_q \gg K_{Ca-q}$, Equation 25 reduces to

$$I_{Ca} = \frac{I_{Ca(max)}(Ca)_o}{K_{Ca} + (Ca)_o\left[1 + \frac{(Q)}{K_{Ca-q}}\right]} \tag{27}$$

This equation describes the scheme for pure anticompetitive inhibition.

Equations 26 and 27 written in reciprocal forms become, respectively

$$\frac{1}{I_{Ca}} = \frac{1}{I_{Ca(max)}}\left[1 + \frac{(Q)}{K_q}\right] + \frac{1}{(Ca)_o}\cdot\frac{K_{Ca}}{I_{Ca(max)}}\left[1 + \frac{(Q)}{K_q}\right] \tag{28}$$

$$\frac{1}{I_{Ca}} = \frac{1}{I_{Ca(max)}}\left[1 + \frac{(Q)}{K_{Ca-q}}\right] + \frac{1}{(Ca)_o}\cdot\frac{K_{Ca}}{I_{Ca(max)}} \tag{29}$$

Plots of $(1/I_{Ca})$ against $1/(Ca)_o$ give straight lines in both cases. In the case of noncompetitive inhibition, addition of Q to the bathing solution containing Ca^{2+} increases both the slope and the intercept of the line by a factor of $[1 + (Q)/K_q]$. Whereas in anticompetitive inhibition, the slopes remain the same and the intercepts increase by a factor of $[1 + (Q)/K_{Ca-q}]$. According to Equation 17, double reciprocal plots for competitive inhibition give straight lines whose intercepts are the same but the slopes are different, increasing by a factor of $[1 + (Q)/K_q]$ for the presence of Q in the external bathing medium.

In studies involving voltage clamp measurements in biological systems, very few cases of noncompetitive and anticompetitive interactions have been described. However, in Section VI.B, an anticompetitive interaction that involves the inhibiting effect of protons on I_{Ca} in the barnacle muscle fiber is described.

3. Tabulation of Dissociation Constants for Various Divalent Cations that Interact with Calcium Channel Sites in Different Biological Systems

Most of the ion site interactions are described by Equation 17 (competitive inhibition) and thus the values for K'_{Ca}, as has been pointed out, may be easily derived. To obtain values for K_M according to Equation 18, values for K_q are required and these are not available in a large number of cases in the literature. Where values for K_q are available, values for K_M have been calculated. Consequently, the entries under column K_M in Table 1 include apparent values for the dissociation constant for a number of divalent ion-channel site complexes.

A direct way to obtain a value for K_M is to use Equation 10. This is possible when measurements of I_M are made as a function of $(M^{2+})_o$ in saline solutions containing only M^{2+} (see Ivens, 1986). Many biological systems under physiological conditions thrive in an environment that contains both Ca^{2+} and Mg^{2+}. When measurements are made under artificial conditions using only $(Ca^{2+})_o$ in the bathing solution, the system may not respond correctly (see Section V). So, one should be careful since fortuitous results may follow because of a lack of Mg^{2+} in the saline solutions. Consequently, investigators have always measured I_{Ca} in presence of both Ca^{2+} and Mg^{2+}. To evaluate K_{Ca} directly by employing Equation 10, they must first show that Mg^{2+} plays no role in controlling the magnitude of I_{Ca}. This may be done by conducting experiments designed appropriately to identify the role of Mg^{2+} as was shown by Akaike et al. (1978). With regard to M^{2+} or other multivalent cations that act as competitive inhibitors, values for K_M cannot be determined by employing Equation 10 since those cations produce little currents of their own. Determination of K_M for such cations is only possible with the use of Equation 18.

The significance of the values of the dissociation constants in generating and controlling several phenomena related to ion-site interactions and ion transport across cell membranes has already been mentioned in Sections III and IV.

V. EXPERIMENTAL APPROACH FOR DETERMINING K_{Ca} AND K_{Mg} IN BARNACLE MUSCLE

Although the technical difficulties pertaining to measurement of I_{Ca} in the barnacle muscle membrane have been overcome to a large extent (Keynes et al., 1973; Hagiwara et al., 1974; Murayama and Lakshminarayanaiah, 1977; Beirao and Lakshminarayanaiah, 1979; Lakshminarayanaiah, 1981), there is still a problem stemming from the complex structure of the muscle fiber which has an intricate network of clefts and tubules extending into the interior of the fiber (Hoyle, McNeill and Selverston, 1973). The extent of these invaginations differs from fiber to fiber and this leads to a situation in which it becomes difficult to obtain values for I_{Ca} that are not highly variable when different muscle fibers are used. Making all the I_{Ca} measurements on a single fiber is very frustrating because of the rapid deterioration of the muscle fiber. Consequently, one must resort to some normalizing procedure to overcome this difficulty. Without such a procedure, quantitative interpretation of results becomes unreliable.

The procedure used in our laboratory (Beirao and Lakshminarayanaiah, 1979; Van Wagoner, 1985) was to measure I_{Ca} at zero membrane potential in some given external saline solution. This procedure was followed because I_{Ca} was observed to be maximum at that membrane potential, provided the bathing solution had a high concentration of divalent ions particularly Ca^{2+} and Mg^{2+}.

At any given $(Mg^{2+})_o$, I_{Ca} was measured as a function of $(Ca^{2+})_o$ in the range (5 to 100 mM) using a single muscle fiber. The several I_{Ca}s measured were normalized by taking I_{Ca} measured in 20 mM $(Ca^{2+})_o$ as 100%. The measurements were repeated with at least

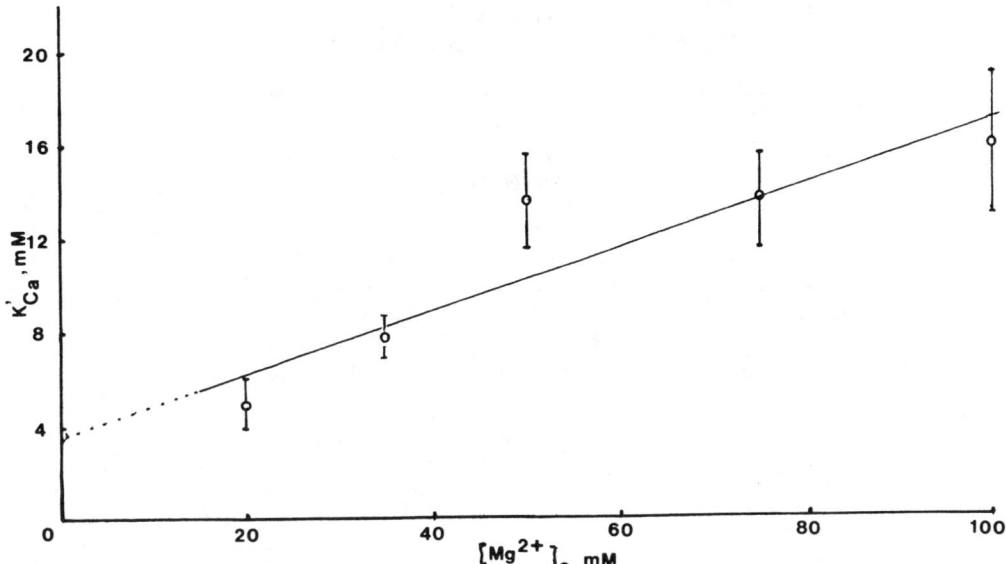

Figure 1. Values of K'_{Ca}, the apparent dissociation constant of membrane site-Ca^{2+} complex, plotted as a function of $(Mg^{2+})_o$. The bars indicate ±SE, one standard error of the mean. Regression analysis of the data gave an intercept on the ordinate, according to Equation 18, i.e., $(Mg^{2+})_o = 0$, a value of $K'_{Ca} = K_{Ca} = 3.67 ± 2.2$ mM. The coefficient of correlation (r) of this data is 0.915.

six muscle fibers. The values of I_{Ca} obtained for each $(Ca^{2+})_o$ were averaged, thereby obtaining a set of averaged I_{Ca}s as a function of $(Ca^{2+})_o$. These results showed the expected hyperbolic behavior (Van Wagoner, 1985).

The initial objective of the voltage clamp experiments was to obtain the value for K_{Ca} in the most straight forward manner possible. Therefore measurement of I_{Ca} as a function of $(Ca^{2+})_o$ were made in the complete absence of extracellular Mg^{2+}. Double reciprocal plots of such data yield, according to Equation 10, a straight line the slope and intercept of which will disclose the value for K_{Ca}. A series of voltage clamp experiments carried out in this manner gave very unsatisfactory results primarily for three reasons: (1) leakage currents, particularly when $(Ca^{2+})_o$ was low (5 or 10 mM), were very large, even larger than the current recorded in response to the depolarizing pulse; (2) muscle fibers showed a tendency to depolarize rapidly and (3) the measured currents were almost constant when $(Ca^{2+})_o$ was varied in the range (20 to 100 mM). I_{Ca} showed no tendency to saturate as $(Ca^{2+})_o$ was increased. This may be due to the effect of a surface potential arising from the lack of a sufficient concentration of divalent cations. The presence of a surface potential will tend to shift the activation potential leading to a shift in the membrane potential, from zero to some other potential, at which maximal current is obtained.

As a result of the foregoing failure, measurements of I_{Ca} as a function of $(Ca^{2+})_o$ were subsequently made in the presence of a constant concentration of $(Mg^{2+})_o$. These measurements of I_{Ca} were repeated at several constant concentrations of $(Mg^{2+})_o$ (e.g., 20, 35, 50, 75 and 100 mM). The values of K'_{Ca} obtained as a function of $(Mg^{2+})_o$ are shown in Figure 1 with their standard errors. Extrapolation to zero $(Mg^{2+})_o$ as per Equation 18 gave a value of 3.7 ± 2.2 mM for K_{Ca}. Using this value in Equation 18, values for K_{Mg} may be calculated for each $(Mg^{2+})_o$. The averaged value for K_{Mg} was 33.4 ± 7.0 mM.

VI. EXPERIMENTAL DETERMINATION OF THE CHEMICAL NATURE OF CALCIUM CHANNEL-BINDING SITE

The data of Table 1 show that, if cations are to bind to the site associated with the

TABLE 2
pK$_a$s of Titratable Groups in Ribonuclease

Group	Residue	pK$_a$ (expected)	(Found)
α-COOH	Carboxy-terminal	2.1—2.4	
β-COOH	Aspartate	3.7—4.0	4.7
γ-COOH	Glutamate	4.2—4.5	
-imidazolinium	Histidine	6.7—7.1	6.5
α-NH$_3^+$	Amino-terminal	7.6—8.0	7.8
-SH	Cysteine	8.8—9.1	—
ε-NH$_3^+$	Lysine	9.3—9.5	10.2
-phenolic	Tyrosine	9.7—10.1	9.95 > 12
$-NH-C\begin{smallmatrix}+NH_2\\ \\NH_2\end{smallmatrix}$	Arginine	>12	>12

calcium channel in the membrane, the site should be a negatively charged ionized or ionizable chemical group that depends on the pH in the region close to the group. Most of the chemical structures constituting the biological membrane are phospholipids and proteins that are amphipathic molecules. According to the fluid mosaic model of the biological membrane, lipid is the matrix in which integral proteins (globular) are incorporated randomly in the plane of the membrane (Singer and Nicolson, 1972). Some of the integral proteins distributed in the plane of the membrane probably act as channels. Other integral proteins may form aggregates that are partially embedded in and partly projecting from the membrane. Accordingly, the amino acid and/or other side chains that are oriented toward the water-filled channel and the lipoprotein side chains that are oriented toward the extracellular phase control the channel activity. The ionizable chemical groups of the several side chains commonly encountered are amino, carboxyl, imidazole, sulfhydryl, disulfide, phenolic, guanidino, indole and thioether. The pK$_a$s of some of these groups present in ribonuclease are shown in Table 2. Hence, it is possible to identify roughly the chemical group of the calcium channel from the value of its pK$_a$.

A. pK$_a$ Derived from I$_{Ca}$ vs. pH Relationship in which (Ca^{2+})$_o$ and (Mg^{2+})$_o$ are Kept Constant

In Section III.D, the work of Lakshminarayanaiah and Beirao (1979) was briefly mentioned. Using the method of least squares, they substituted their I$_{Ca}$ vs. pH data obtained in presence of (Ca^{2+})$_o$ and (Mg^{2+})$_o$, each 100 mM concentration, into the equation:

$$I_{Ca} = \frac{I_{Ca(max)}}{1 + \frac{(H^+)}{K_a'}} = I_{Ca(max)}\left[1 + 10^{pK_a' - pH}\right]^{-1} \tag{30}$$

By using 100% for I$_{Ca(max)}$, the pK$_a'$ was found to be 3.9. Recent calculations ignoring the experimental points beyond pH 7.2, because of the fact that other chemical groups which ionize at a higher pH might come into play, show that the experimental points at several pHs below 7.2 can be substituted into Equation 30 and when I$_{Ca(max)}$ is taken as 100%, the pK$_a'$ becomes 3.7. The value of pK$_a'$ (= 3.7) must be corrected for the presence of (Ca^{2+})$_o$ and (Mg^{2+})$_o$, each at 100 mM concentration. For this purpose, an extended form of Equation 18:

12. Effects of Inorganic Ions on Calcium Channels

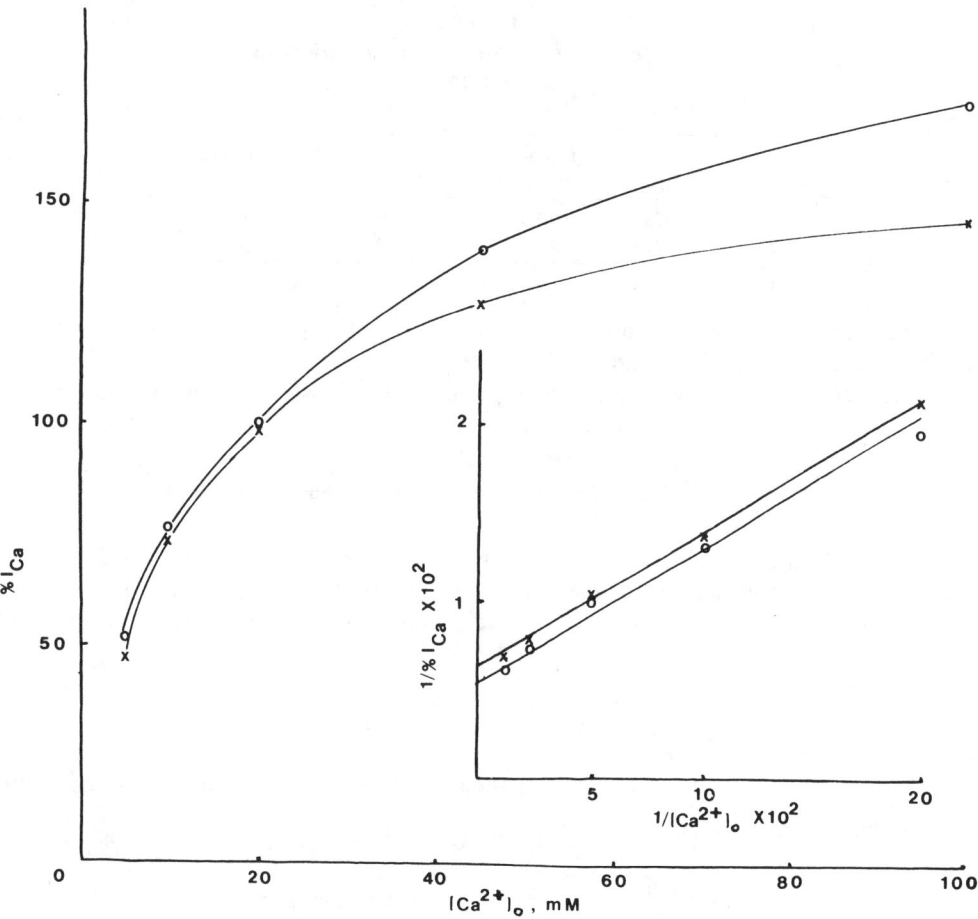

Figure 2. Effect of pH on normalized calcium currents plotted as a function of $(Ca^{2+})_o$. Four muscle fibers were used in artificial sea water containing $(Mg^{2+})_o = 50$ mM and varying concentration of $(Ca^{2+})_o$. The values of I_{Ca} for each fiber were normalized by taking I_{Ca} at 20 mM $(Ca^{2+})_o$, pH 7.5 as 100%. Each point is an average of measurements made on four muscle fibers used in salines pH 7.5 (o) and pH 5.5 (x). The inset is a double reciprocal plot of the same data from which it is seen that depression of I_{Ca} occurring as the pH is lowered is not competitive but rather anticompetitive. See Equations 33 and 34.

$$K_a' = K_a\left[1 + \frac{(Ca^{2+})_o}{K_{Ca}} + \frac{(Mg^{2+})_o}{K_{Mg}}\right] \tag{31}$$

may be used. Substituting the recent values of K_{Ca} (= 3.7 mM) and K_{Mg} (= 33.4 mM) in Equation 31 gives a value of 6.4×10^{-6} M for K_a and pK_a is 5.19.

B. pK_a Derived from I_{Ca} vs. $(Ca^{2+})_o$ Relationships Obtained at pH 7.5 and pH 5.5

Measurement of I_{Ca} as a function of $(Ca^{2+})_o$ at constant pHs e.g., 7.5 and 5.5, may also be used to derive a value for the pK_a. The results of this additional work should confirm the above value and thus prove the consistency of the experimental procedures.

The values of I_{Ca} measured as a function of $(Ca^{2+})_o$ using a constant $(Mg^{2+})_o$ of 50 mM, are shown in Figure 2. The values of I_{Ca} are the averages of measurements made on four different muscle fibers. The values of I_{Ca} are normalized by taking I_{Ca} measured in 20 mM $(Ca^{2+})_o$, pH 7.5 as 100%. The double reciprocal plots of these data are shown in the inset. Regression analysis gave the values shown in Table 3. These data, for the first time,

TABLE 3
Values for the Parameters of Equations 33 and 34

pH	Slope	$(1/I_{Ca})$-axis intercept	$I_{Ca(max)}$	Correlation coefficient
7.5	0.078	0.564	177%	0.998
5.5	0.076	0.631	158%	0.999

From Tanford and Hauenstein, 1956.

present an example of an anticompetitive inhibition (same slope and different intercepts on the $(1/I_{Ca})$ axis; see inset in Figure 2).

When the pH is 7.5, the data can be represented by Equation 17 rewritten as

$$I_{Ca} = \frac{I_{Ca(max)}(Ca^{2+})_o}{(Ca^{2+})_o + K'_{Ca}[1 + (H^+)/K'_a]} \quad (32)$$

where K'_{Ca} is the apparent dissociation constant for Ca^{2+}-site complex and K'_a is the apparent acid dissociation constant for H^+-site complex. For pH = 7.5 [i.e., $(H^+) = 10^{-7.5}$], Equation 32 simplifies to the usual hyperbolic equation which in reciprocal form becomes

$$\frac{1}{I_{Ca}} = \frac{K'_{Ca}}{I_{Ca(max)}} \cdot \frac{1}{(Ca^{2+})_o} + \frac{1}{I_{Ca(max)}} \quad (33)$$

Equation 27 when applied to pH 5.5 data for an anticompetitive inhibition can be written in reciprocal form as

$$\frac{1}{I_{Ca}} = \frac{K'_{Ca}}{I_{Ca(max)}} \cdot \frac{1}{(Ca^{2+})_o} + (1 + 10^{pK'_{a'} - pH}) \frac{1}{I_{Ca(max)}} \quad (34)$$

where $pK'_{a'}$ is the apparent dissociation constant of the complex H^+-CaS. The condition for the generation of anticompetitiveness as pointed out previously (i.e., $K_a \gg K_{Ca-q}$) is that $K'_a \gg K'_{a'}$. This, in essence, means that the relative affinity of H^+ for the free channel site(s) is so small as to be insignificant, whereas the affinity of H^+ for the channel site-Ca complex (CaS) is high enough so that the structure (CaSH) controls the subsequent reaction (i.e., the magnitude of I_{Ca}). Under these conditions, it is obvious that what is derived as a value for the apparent dissociation constant of the chemical group of the calcium channel is a complex parameter.

The values given in Table 3 after substitution into Equations 33 and 34 and after simplification gave $10^{pK'_a - pH} = 0.148$ (note that the distinction between K'_a and $K'_{a'}$ is dropped). Thus $pK'_a = 4.67$ or $K'_a = 2.14 \times 10^{-5}$ M. Applying the correction for the presence of 50 mM $(Mg^{2+})_o$ in the saline solution gives

$$2.14 \times 10^{-5} = K_a\left(1 + \frac{50}{33.4}\right)$$

Thus $K_a = 8.57 \times 10^{-6}$ or $pK_a = 5.24$. This value is in agreement with the value of 5.19 derived by the titration technique in Section VI.A. In view of this, it is suggested that in many biological systems where the titration technique (i.e., depression of I_{Ca} by H^+) is used (see Section III.D) to obtain an apparent dissociation constant for the chemical group existing in the calcium channel, the mechanism operating is not a simple neutralization reaction but the complex mechanism mentioned above.

C. Use of pK_a and Other Means to Deduce the Chemical Nature of the Calcium Channel Site that Regulates Channel Function

From the values of pK_a given in Table 2 for the several end groups of ribonuclease, it could be inferred that the chemical group associated with the calcium channel in the barnacle muscle membrane is either a carboxyl or a histidyl imidazole group, since the pK_a is 5.2. In order to identify the chemical group, the approach used was to selectively modify its structure by allowing it to react with those reagents that preferentially act on the biophysically active site. There are several chemical reagents that are known to specifically modify the end groups in proteins (Means and Feeney, 1971). Water soluble carbodiimide may be used in the presence of an amine to modify the carboxyl group that may be present in the calcium channel. Similarly, trimethyloxonium tetrafluoroborate (TMO) may be used (Spalding, 1980). Agents such as ethoxyformic acid, diazonium reagents may be used to modify the imidazole group of histidine (Means and Feeney, 1971). In addition, diethylpyrocarbonate has been used by Shrager (1974).

Some preliminary work has previously been done to modify the group of the calcium channel in the barnacle muscle fiber (Van Wagoner, 1985). Spalding (1980) used TMO to modify the sodium channel in single skeletal muscle fibers. This reagent reduced I_{Na} by two thirds, significantly lowered the pK_a of the site in the sodium channel and rendered a significant fraction of I_{Na} resistant to block by 90 μM TTX. The effect of TMO was attributed to esterification of a carboxyl group essential for the binding of the TTX toxin. The reaction is

$$\text{(P)} - C\begin{matrix}O\\ \diagdown\\ O^-\end{matrix} + R - \overset{R}{\underset{R}{O^+}}BF_4^- \rightarrow \text{(P)} - C\begin{matrix}O\\ \diagdown\\ O-R\end{matrix} + \overset{R}{\underset{R}{O}} + BF_4^-$$

$R = CH_3$, P = membrane protein and an additional reaction product is H^+.

The procedure used by Van Wagoner (1985) was to treat the barnacle muscle fiber with TMO in artificial sea water containing 80 mM of morpholine propane sulfonic acid buffer. This treatment had no effect on the action potential the physiological response monitored. The chemical nature of the active binding sites located in and around the calcium channel of the barnacle muscle fiber is different from that of the site of the sodium channel present in frog skeletal muscle fiber investigated by Spalding (1980). It has been suggested, therefore, that a carboxyl group is not involved in the functioning of the calcium channel. This must be confirmed by further work using other reagents specific for the carboxyl group (e.g., carbodiimide with an amine). Moreover, agents such as diethylpyrocarbonate that specifically modify the imidazole group of histidine must be used in further experiments to prove the existence of an active imidazole moiety in the calcium channel of the barnacle muscle fiber.

VII. SUMMARY

Inorganic and organic anions play a significant role in controlling the viability of the cell membrane and hence the structural integrity of the calcium channel. The resting membrane potential across the barnacle muscle membrane, using several inorganic and organic anions in the internal perfusion solution, has been measured to test the stability of the cell membrane. (See Chapter 13, Section III for further discussion of the effects of organic ions on calcium channels.) The effectiveness of various inorganic anions to maintain a stable resting potential followed the sequence $F^- > HPO_4^{2-} > SO_4^{2-} > Fe(CN)_6^{4-} > ClO_3^- > NO_3^- > Cl^- > CNS^- > I^- > Br^- \geq CN^-$.

In neurons of the snail, *Helix pomatia*, the preferred anion in the intracellular dialyzate was F^- or $H_2PO_4^-$; whereas these anions proved less favorable in neurons of *Helix aspersa*. Consequently, aspartate anion was used in the internal perfusion solution.

The extensive studies related to the effects of inorganic cations on the calcium channel existing in a variety of biological cells are reviewed. The several inorganic cations are grouped into four categories. Ca^{2+}, Sr^{2+} and Ba^{2+} that carry current through the calcium channel form category one. The second category is composed of cations that act as competitive inhibitors of I_{Ca}. The important ions in this group are La^{3+}, Co^{2+}, Ni^{2+} and Cd^{2+}. The third category consists of ions such as Mn^{2+}, Zn^{2+} and Mg^{2+} that act as competitive inhibitors in some biological cells but carry current in some other cells and produce their own action potentials. In addition, a discussion of the actions of monovalent ions such as Na^+, K^+, etc. that carry current through the calcium channel under conditions when the concentration of Ca^{2+} is low (in the millimolar range) is included. The fourth category consists of protons that generally inhibit action potentials and currents in *Helix* neurons, amphibian neurons of *Xenopus laveis* developing in culture and barnacle muscle cells and inhibit calcium fluxes in brain synaptosomes.

The two models proposed to explain the experimental results are discussed. The first model proposed appears to exist in two variations and is applicable to calcium channels through which monovalent cations penetrate. The second model is applicable to calcium channels through which monovalent cations cannot carry the current. The kinetic characteristics of the models, *viz.* saturation and several types of inhibition, are discussed in quantitative terms. The several values derived for the important kinetic parameter, the dissociation constant, in a variety of biological systems are tabulated.

Experimental and theoretical procedures directed to obtaining meaningful values for the dissociation constants, K_{Ca} and K_{Mg}, for the site complex in the calcium channel of the barnacle muscle membrane are presented.

Procedures for the determination of the pK_a of the site in the calcium channel by two methods are outlined. In the titration method in which I_{Ca} was measured as a function of pH, a value of 5.19 was derived for the pK_a. In the second method in which I_{Ca} was measured as a function of $(Ca^{2+})_o$ at two pHs, using a constant $(Mg^{2+})_o$, a value of 5.24 was obtained for the pK_a.

The value of pK_a thus derived for the site in the calcium channel of the barnacle muscle fiber indicate that the end group is either a carboxyl or a histidyl imidazole.

The finding that physiological responses of the barnacle muscle fiber were not modified when treated with a heavily buffered TMO solution, a reagent that esterifies specifically a carboxyl group, indicates that the active site may not contain a carboxyl group. Further work with other reagents that act on carboxyl and imidazole end groups needs to be done to establish the existence of the imidazole moiety of the histidine in the calcium channel of the barnacle muscle membrane.

Chapter 13

Drugs Acting on Calcium Channels

Ronald A. Janis and David J. Triggle

TABLE OF CONTENTS

I.	Introduction		197
	A.	Classes, Properties and Functions of Calcium Channels	197
		1. Subtypes of Channels	198
		2. Function	199
		3. Other Calcium Permeable Channels	199
	B.	Drugs Acting on Calcium Channels	200
		1. L-Channel Drugs and Their Therapeutic Uses	200
		2. Uses	200
		3. Allosteric Sites on L Channels	201
		4. Drugs Acting on N and T Channels	202
II.	Structure-Activity Relationships		206
	A.	Some Complexities	206
	B.	Drugs Active at L Channels	210
	C.	Drugs Active at N, T and Other Channels	215
	D.	Multiple Channel Classes and Multiple Channel States	216
III.	Sites of Action		217
	A.	Ligand-Binding Data	217
		1. Ligand-Binding and Physiological Data Correlate	217
		2. Location and Density of Binding Sites	218
		3. Voltage Sensor	218
		4. Neurons	220
	B.	Voltage Dependence of Binding	221

	C.	Several Allosterically Coupled Binding Sites are Associated with L Channels 221
	D.	Biophysical and Physiological Data 223
		1. Major Sites of Action 223
		2. Location of Binding Sites 223
		3. Smooth and Cardiac Muscle and Endocrine Cells 224
		4. Neuronal Cells 224
		5. Skeletal Muscle 224
		6. Endothelial Cells 226
		7. Blood and Exocrine Cells 226
		8. Other Cell Types 226
IV.	Modes and Selectivity of Action 227
	A.	Voltage Dependence of L Channel Block Results in Selectivity of Action 227
	B.	Frequency-Dependent Block 228
	C.	Other Mechanisms for Selectivity of Action 228
	D.	Mechanism of Action of BAY K 8644 and Related Compounds 229
	E.	Drugs Acting on T Channels 230
V.	Effects of Calcium Channel Ligands in Animal Models 232
	A.	Cardiovascular and Renal Systems 232
		1. Hypertension, Renal Effects and Myocardial Hypertrophy 232
		2. Angina and Effects on the Ischemic Myocardium 233
		3. Reperfusion 233
		4. Antiatherogenic Effects 234
	B.	Effects of Calcium Channel Ligands on Neurons and the Brain 234
		1. Lack of Behavioral Effects of Calcium Channel Blockers Used to Treat Cardiovascular Diseases 235
		2. Mechanisms for Selectivity for 1,4-Dihydropyridines 235
		3. Nimodipine can have Direct and Potent Effects on Neuronal L Channels 236
	C.	Treatment of CNS Disorders 237
		1. Subarachnoid Hemorrhage 237
		2. Seizures 238
		3. Learning and Improvement of Age-Related Behavioral Deficits 238
		4. Withdrawal Syndromes and Toxicity of Addictive Drugs 238
		5. Other Effects 239
VI.	Endogenous Substances Acting on Calcium Channel Ligand Receptors 239
	A.	Fractions that Inhibit Dihydropyridine Binding and Calcium Influx 239
	B.	Endogenous Substances that Inhibit Dihydropyridine Binding 239
	C.	Fractions and Substances that Effect Calcium Current 240
VII.	Chronic Regulation of Calcium Channels 241
	A.	L Channels as Pharmacological Receptors 241
	B.	Homologous Regulation 241
	C.	Heterologous Regulation 242
	D.	Alteration in Disease 243

13. Drugs Acting on Calcium Channels   **197**

VIII.	Actions at Other Sites..244	
	A. Effects on Other Receptors and Channels 245	
	B. Nucleoside Transporter..245	
	C. Effects on Platelets..245	
	D. Effects on Drug Metabolism..246	
	E. Other Effects of Calcium Channel Antagonists........................246	
	F. Effects on Drug-Resistant Cells 247	
	G. Other Effects of BAY K 8644...248	
IX.	Conclusions and Summary..248	
	Acknowledgments..249	

I. INTRODUCTION

The focus of this chapter is the preclinical pharmacology of organic drugs that bind directly to voltage-operated calcium channels. Other chapters in this volume cover the physiology of calcium channels, the clinical use of calcium channel blockers, and some aspects of the effects of these drugs in animal models.

A. Classes, Properties and Functions of Calcium Channels

The type, density, location, and function of calcium channels varies between different types of cells (Tsien et al., 1987, 1988; Kostyuk, 1989; Bean, 1989; Hess, 1990; Fox et al., Section III, Chapter 14). The physiological functions of calcium channels include excitation-contraction coupling, control of secretion, excitability, conduction, bursting and pacemaker activity. The effectiveness of calcium channel blockers in a large number of diseases likely reflects the pathological effects of excessive calcium influx in these states.

There are probably at least three classes of voltage-sensitive calcium channels that are high threshold, i.e., that require a large amount of depolarization before they are activated. These are the 1,4-dihydropyridine-sensitive L class, the N class, which may be either ω-conotoxin GVIA (ω-CgTX) sensitive or insensitive, and other types of calcium channels, including the P class, insensitive to both of these types of drugs. Those channels that are activated with small amounts of depolarization and inactivate rapidly, particularly with a large depolarization, are called low threshold, or T (for transient) channels. Sah et al. (1989) determined the amount of block of calcium current in several types of neurons (dorsal root ganglia, sympathetic ganglia, spinal cord, cerebral cortex, and hippocampal CA1) by ω-CgTX, 1,4-dihydropyridine, or both, and found similar results for each type. Block by ω-CgTX was 50 to 75%, by nimodipine, 5 to 30 and 20 to 30% remained unblocked by a combination of the two drugs at 3 and 10 μM, respectively. These ranges reflect the relative contributions of the L, N and what we will refer to as P channels to the total calcium current observed.

L channels—L channels play a key role in excitation-contraction coupling in smooth and cardiac muscle (see Section II, Chapter 6), and are probably also involved in modulating membrane excitability, and the release of some hormones and neurotransmitters. In many cell types the L channels tend to be long lasting, whereas T channels are transient, but there are exceptions in each case. For example, cardiac L channels show calcium-dependent inactivation and may inactivate quite rapidly under physiological conditions (with a half-time of 20 to 100 ms), although when Ba^{2+} rather than Ca^{2+} is used experimentally as the charge carrier, the rate of inactivation may be very slow (Bean, 1989).

N channels—The N channels have been found mainly or only in neuronal cells and are believed to play a major role in neurotransmitter release (Hirning et al., 1988; Tsien et al., 1988; Yoshikami et al., 1989). Their time constants for inactivation vary from 25 to several hundred milliseconds, depending on the type of neuron studied (for review, see Hess 1990). These channels are 1,4-dihydropyridine-insensitive, but some subtypes of them are directly and potently inhibited by ω-CgTX, and less potently by barbituates, neomycin and related drugs. Some N channels, like L channels, are long lasting, ω-CgTX-insensitive, and have a larger intrinsic conductance than that most frequently reported (Plummer et al., 1989; Hess, 1990). These recent results emphasize the importance of using activators, including Bay K 8644, to identify L channels in presence of N channels.

P channels—High threshold channels that are insensitive to both 1,4-dihydropyridines and ω-conotoxin, but are sensitive to certain spider toxins were called P channels (Llinas et al., 1989a; Bindokas and Adams, 1989; Salzberg et al., 1990; Cherksey et al., 1990). This class of channel has been termed "P" because it was first seen in Purkinje cells (Llinas et al., 1989a,b). This channel type, which may or may not be different from the conotoxin-insensitive N channels observed by Plummer et al. (1989), was not found in thalamic or inferior olivary cells. The lack of both 1,4-dihydropyridine and ω-CgTX binding sites in certain parts of the brain also suggests the presence of another type of calcium channel, such as P channels or an N channel subtype, that mediates neurotransmitter release (Maeda et al., 1989). Similarly, a voltage-sensitive calcium channel insensitive to these agents was detected when rat brain RNA was expressed in frog oocytes (Lester et al., 1989).

T channels—The T class of channels may be inactivated at the resting membrane potential at which many cells exist, and therefore may be activated following after-hyperpolarization subsequent to an initial depolarization (Llinas, 1988; Tsien et al., 1988). This channel type is most abundant in sinoatrial and Purkinje cells, the pacemaker cells of the atria and ventricles, respectively (Tseng and Boyden, 1989). T channels may play a key role in aldosterone secretion (Cohen et al., 1988; Barrett et al., 1989). These channels are also likely to be important in information processing by eliciting the burst firing of neurons (White et al., 1989). They may have some developmental role because they appear to be the dominant channel type in embryonic and immature cells (e.g., McCobb et al., 1989). T channel expression can be inhibited by transforming genes without modifying L channels, indicating their separate genetic control (Chen et al., 1988). Contrary to most reports of rapid inactivation of these channels, some channels that are low threshold and have small intrinsic conductances may not inactivate rapidly, particularly when exposed to a small amount of depolarization (Cohen et al., 1988). T channels, like sodium channels, are inhibited at somewhat higher concentrations, by many, if not all, of the same drugs that inhibit L channels. These results likely reflect the significant homology between these types of channels.

1. Subtypes of Channels

Subtypes of the various classes of voltage-sensitive channels exist (Pelzer et al., 1989; Hess, 1990). There are major differences in the L type channels between skeletal muscle and other types of excitable cells, and differences in the sensitivity of L, N, and T channels from different types of cells to various inhibitors. Of possible therapeutic significance is the greater sensitivity to 1,4-dihydropyridines of L channels in vascular smooth muscle relative to other cell types (Bean et al., 1986; McCarthy and Cohen, 1989; Yatani et al., 1987), and the decreased sensitivity of at least some neuronal cells relative to other cell types. Nucleic acid hybridization studies provide support for the existence of subtypes of L channels (Ellis et al., 1988; Koch et al., 1989; Mikami et al., 1989; Perez-Reyes et al., 1989; Slish et al., 1989). The presence of normal L channels in other tissues in dysgenic mice that lack dihydropyridine receptors in skeletal muscle is consistent with separate genes encoding these channel subtypes (Tanabe et al., 1988). L channels in neurons and GH_3 cells exhibit a slower rate of activation than those in smooth and cardiac cells. N channels may be subdivided based on the degree and rate of their inactivation, and their sensitivity to ω-CgTX (Plummer et al., 1989). Species differ in their proportions of subtypes of N channels in corresponding neurons (Yoshikami et al., 1989). There is some controversy about the sensitivity of L channels to ω-CgTX (Suzuki and Yoshioka, 1987; Aosaki and Kasai, 1989; Plummer et al., 1989; for a critical review, see Hess, 1990).

2. Function

The location on the cell, the voltage range over which a given channel type opens, and the duration of that opening defines, in part, the possible functions of a channel. T channels can be expected to play a role in pacemaker activity of some cells because these channels may open with small amounts of depolarization in very well-polarized cells, but can also be made to recover from inactivation and to re-open during hyperpolarization of a previously depolarized cell. T channels may also play a role in bursting and oscillatory activity (Llinas, 1988). P, N and L channels may mediate neurotransmitter release in response to the large depolarization associated with an action potential (Suzkiw et al., 1989). In those cell types in which L channels exhibit a longer lasting opening, they likely play a role in modulating the membrane excitability by modifying the activation and inactivation of voltage-dependent potassium channels. N and P channels may be the major presynaptic calcium channels involved in neurotransmitter release from neurons, but L channels probably also play an important role.

3. Other Calcium Permeable Channels

This chapter will review the pharmacology of drugs that act directly on the best studied, voltage-regulated calcium channels, the L, N and T classes. In addition to these three types of channels, and the P class for which there is little information, there are several other types of channels that are also permeable to calcium, but are not sensitive to the selective calcium channel blockers. These include the channels that are gated primarily by ligands (for references, see Bülbring and Tomita, 1987; Hosey and Lazdunski, 1988; van Breemen and Saida, 1989; Langs and Triggle, 1990), those channels gated by stretch (Laher and Bevan, 1989), as well as those for which there are no known physiological regulatory mechanisms (Rosenberg et al., 1988). A calcium-permeable, ATP-operated channel is present in smooth muscle (Benham and Tsien, 1987; Benham, 1989). Some of the data that supported the presence of adrenergic receptor-operated calcium channel in smooth muscle can now be interpreted as a shift in the voltage dependence of activation of the L channels (Nelson et al., 1988).

Although a channel may be classified according to its probable major regulatory mechanism, this does not mean that this channel is insensitive to other stimuli. For example, voltage-sensitive calcium channels in skeletal muscle may also be regulated by inositol 1,4,5-triphosphate (IP_3) (Vilven and Coronado, 1988). It can be expected that multiple mechanisms of calcium channel regulation will be the norm rather than the exception. The same second messengers that are known to regulate a variety of metabolic processes, including cyclic nucleotides, Ca^{2+}, inositol phosphates, 1,2-diacylglycerol and other lipids, also modify calcium channel function (Fox et al., this volume; Hess, 1990).

B. Drugs Acting on Calcium Channels

1. L-Channel Drugs and Their Therapeutic Uses

A key criterion for the classification of voltage-dependent calcium channels is pharmacologic specificity. In several respects these channels may be viewed analogously to conventional pharmacologic receptors; selective blockers and activators play an important role in calcium channel characterization. Subsequent to the pioneering work of Fleckenstein (Fleckenstein, 1983) in advancing the concept of calcium antagonism the major developments in calcium channel ligands have been for the L class of channels, but agents, including toxins, are becoming available for the other channel classes.

Calcium channel ligands are agents that bind directly to voltage-sensitive calcium channels and thereby decrease or increase the movement of Ca^{2+} through them. For the major drugs that are used clinically, nifedipine, diltiazem and verapamil, this binding results in decreased vasoconstriction, decreased cardiac arrhythmias, or decreased heart rate. These effects reflect two of the major roles of L channels in the cardiovascular systems, excitation-contraction coupling and electrical conduction.

2. Uses

The calcium channel blockers, including the clinically available verapamil, nifedipine, and diltiazem (Figure 1), are now established agents employed in a number of disorders including angina in its several forms, hypertension, peripheral vascular disorders and some types of cardiac arrhythmias (Table 1).

However, there are many other applications for which these agents have been used or suggested from achalasia through migraine and premature labor to urinary incontinence and vertigo (Table 2). Nimodipine represents the first drug approved for the treatment or prevention of neurological deficits associated with subarachnoid hemorrhage, and nicardipine is the first charged dihydropyridine approved for use in angina and hypertension in the U.S. (Figure 1). Several reviews document these clinical and potential clinical applications (Janis et al., 1987; Opie, 1988a,b; Nayler, 1988; Triggle, 1990a). That other chemical structures, including perhexiline, prenylamine, and terodiline (Figure 1) also possess calcium channel blocking properties was recognized early and Fleckenstein introduced the first classification of the calcium blockers for L channels:

"Group A compounds including verapamil and D600, nifedipine and diltiazem were the more specific, inhibiting calcium current with little effect on sodium current."

"Group B compounds including prenylamine, terodiline, perhexiline and fendiline, caroverine were less selective and had significant effects on sodium currents before calcium current block was complete." For various types of calcium modulators, detailed and more extensive classifications based on sites of action and structure have been presented elsewhere (see Tables 9 and 10 of Janis et al., 1987).

Fleckenstein, who carried out the first extensive studies on the L-channel inhibitors, has reviewed the historical development of these drugs in a major monograph (Fleckenstein, 1983). Other reviews offering historical perspective include Meyer et al. (1985), Godfraind et al. (1986), Janis et al. (1987), Glossmann and Striessnig (1988), Bossert and Vater (1989), and Triggle (1989a). The initial key observation was that prenylamine and verapamil mimicked the effect of calcium withdrawal on the heart and blocked contraction but produced little change in the action potential. The inhibitory effect of these drugs was found to be reversed by elevated calcium as well as drugs such as isoproterenol and ouabain which increase intracellular Ca^{2+}. Fleckenstein proposed that drugs inhibiting excitation-contraction coupling be called calcium antagonists.

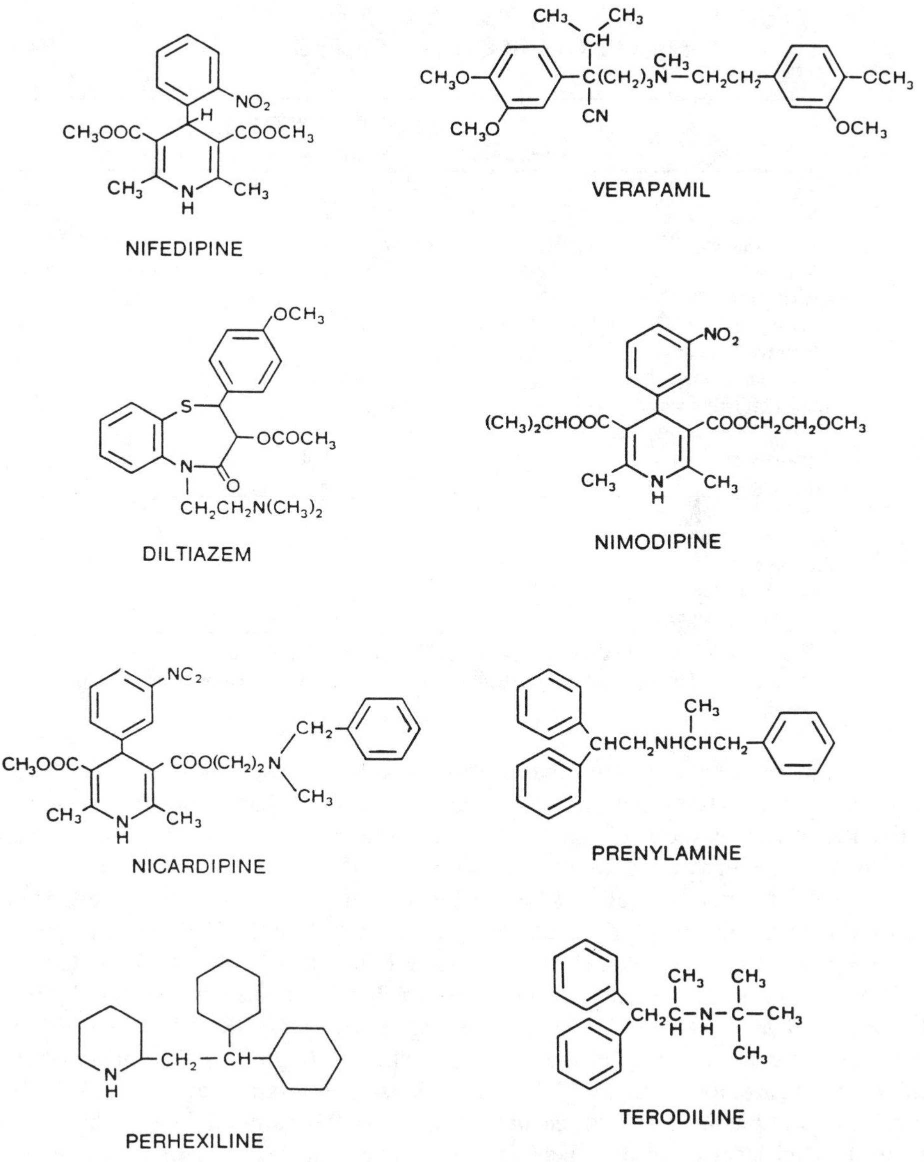

13. Drugs Acting on Calcium Channels

Figure 1. Structural formulae of calcium channel blockers. Nifedipine, nicardipine, nimodipine, diltiazem, and verapamil are approved for use in the U.S. Nimodipine has recently been approved for use in the U.S. for the improvement of neurological deficits due to spasm following subarachnoid hemorrhage, and, in Germany, for dementia.

3. Allosteric Sites on L Channels

The pharmacologic, therapeutic and chemical heterogeneities of these agents is inconsistent with their acting at a single site, and biochemical and radioligand binding approaches have shown that the calcium channel possesses a number of discrete binding sites with which structurally distinct chemical classes interact to modulate channel function. These sites are linked allosterically, one to the other, and to the permeation and gating machinery of the channel (Figure 2).

The consequences of binding to these separate sites result in different pharmacological profiles for these main classes of calcium channel blockers. Specifically, the 1,4-dihydro-

TABLE 1
Therapeutic Uses of Calcium Channel Blockers

	Blocking agents		
Use	Verapamil (I)[a]	Nifedipine (II)	Diltiazem (III)
Angina:			
Exertional	+++[b]	+++	+++
Prinzmetal's	+++	+++	+++
Variant	+++	+++	+++
Arrhythmias:			
Paroxysmal Supraventricular Tachyarrhythmias	+++	−	++
Atrial fibrillation and flutter	++	−	++
Hypertension	++	+++	+
Hypertrophic cardiomyopathy	+	−	−
Raynaud's phenomenon	++	++	++
Cardioplegia	+	+	+
Cerebral vasospasm (post hemorrhage)	−	+	−

[a] Classes I, II, and III as defined by World Health Organization.
[b] Number of plus signs indicates extent of use: +++, being very common; −, not used.

pyridines are more vascular selective than are the non-dihydropyridines. The marked diversity in their chemical structures and their potencies on L channels suggests differences in their effectiveness at various sites and this is observed. Table 3 gives the potencies of some drugs and toxins acting at each class of calcium channel.

It is likely that other major structural categories of ligands exist. The drugs in these categories include trans-diclofurime (Spedding et al., 1987), MD1 12,330A (Rampe et al., 1987a; Palfreyman et al., 1989), tetrandrine (King et al., 1988), McN 6186-11 (Rampe et al., 1989b), pimozide, fluspirilene, and other diphenylbutylpiperidines (Galizzi et al., 1986b), and HOE 166 (Qar et al., 1987, 1988; Striessnig et al., 1988) SR 33557 (Schmid et al., 1989) which interact primarily at the L class of channel (Figure 3).The diphenylbutylpiperidines, the benzothiazinone, HOE 166, as well as several other new drugs likely define new categories of binding sites at L channels (Figure 4). That a multiplicity of binding sites exist on the L channel is, in retrospect, not surprising. The sequence of the α_1-subunit of this protein is highly homologous to that of the sodium channel which possesses at least five distinct categories of binding sites (Tanabe et al., 1987; Strichartz et al., 1987).

In addition to the major classes of agents indicated, calcium blocker properties have been attributed to a very large number of other structures including α- and β-adrenoceptor antagonists, barbiturates, benzodiazepines, antidepressants and local anesthetics (Figure 5; for a more complete listing see Janis et al., 1987, Tables 10 and 26). The status of many of these agents as calcium blockers is dubious (Janis et al., 1987); they were characterized frequently on the basis of indirect evidence and certainly before the recent classification of calcium channels.

4. Drugs Acting on N and T Channels

In contrast to the rich pharmacology available for the L category of voltage-dependent calcium channel, there are comparatively few selective agents available for the T and N channels. The best studied agents for N channels are the ω-conotoxins, from the Conus genus of mollusks, including ω-CgTX GVIA (Table 4), a 27 amino acid peptide, which

TABLE 2
Potential Uses for Calcium Channel Blocking Agents

Cardiovascular
  Aortic and mitral valvular insufficiency
  Atherosclerosis
  Cardiac arrest, prevention of neurological damage
  Cardiac surgery, cardioplegia
  Cerebral ischemia and vasospasm
  Cerebral surgery
  Cor pulmonale
  Heart failure, congestive
  Headache, cluster and migraine
  Hypertension, associated with pregnancy and pulmonary hypertension
  Intermittent claudication
  Percutaneous transluminal angioplasty
  Peripheral vascular disease
  Provocative test for sick sinus syndrome (verapamil)
  Renal failure, inflammatory response to injuries
  Stroke
  Sudden hearing loss
  Ventricular tachycardia, exercise-induced
  Vestibular nystagmus (vertebrobasilar insufficiency)
Nonvascular smooth muscle spasm
  Achalasia, esophageal motor disorders
  Asthma
  Diarrhea due to carcinoid syndrome
  Dysmenorrhea (primary), myometrial hyperactivity
  Premature labor
  Urinary incontinence; ureteral spasms
Other
  Age-associated memory loss; Alzheimer's disease
  Aldosteronism, primary
  Allergic reactions
  Anticalcinotic, retardation of age-dependent calcinosis
  Cancer: enhancement of anticancer drug effect or decreased drug resistance
  Cirrhosis of the liver
  Cocaine intoxication
  Epilepsy
  External muscle pain syndrome (benign myalgia)
  Glaucoma
  Immunosuppressive therapy, adjunct
  Manic and schizomanic syndrome
  Panic disorders
  Psychosis, phencyclidine-induced
  Schistosomal infection
  Schizophrenia, chronic (as adjunct therapy)
  Seafood (ciquatera) poisoning
  Shock, *E. coli*-induced, endotoxic
  Sickle cell anemia
  Spinal cord injury
  Tinnitus
  Tourette's disorder (tics)
  Ulcers
  Urinary incontinence; bladder instability
  Vertigo
  Withdrawal symptoms (e.g., ethanol, morphine, phencyclidine)

From Tables 15 and 16 of Janis et al., 1987; Triggle, 1990. With permission.

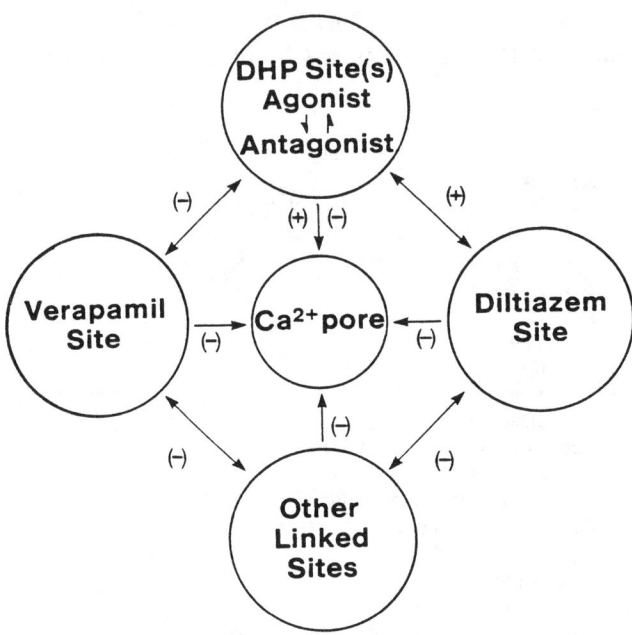

Figure 2. Arrangement of binding sites on the L class of voltage-dependent calcium channel for the three primary ligand classes, and a fourth site representing all other linked sites. The sites are represented as a set linked allosterically one to the other and to the gating and permeation machinery of the calcium channel. The positive and negative signs indicate positive and negative heterotropic interactions, respectively.

inhibits neuronal N channels (Olivera et al., 1985; Kasai et al., 1987; McCleskey et al., 1987; Oyama et al., 1987; Gray et al., 1988; Yoshikami et al., 1989). This toxin may also inhibit L channels of neurons and GH_3 cells (Suzuki and Yoshioka, 1987; but see Hess, 1990). ω-CgTX binds to high affinity sites in N channels in synaptic membranes from rat brain that are not observed in membranes from smooth or cardiac muscle (Feigenbaum et al., 1988; Yoshikami et al., 1989).

N channels are regulated by several different endogenous ligands, including dynorphin A (Gross and MacDonald, 1987; 1989); noradrenaline (Lipscombe et al., 1989), and $GABA_B$ agonists (Dolphin and Scott, 1987). Alpha-adrenoceptor-induced autoinhibition of norepinephrine release from frog sympathetic ganglia is mediated by G protein-coupled changes in rapid gating kinetics of presynaptic N channels (Lipscombe et al., 1989). At anesthetic concentrations, which are larger than those needed to produce antiepileptic effects, barbiturates block L and N, but not T channels (MacDonald and Meldrum, 1989).

Inhibitors of T channels include the synthetic agents amiloride (Tang et al., 1988), phenytoin (Twombly et al., 1988), tetramethrin (Yoshii et al., 1985) (Figure 6) and octanol (Llinas and Yarom, 1986). None of these agents demonstrate specificity (see Hess, 1990). However, very large, but nevertheless, therapeutic concentrations of the anticonvulsants ethosuximide and dimethadione are reported to selectively inhibit T current (Coulter et al., 1988; Gross et al., 1989). In contrast, barbiturates, diazepam, and phenytoin block presynaptic calcium current and neurotransmitter release only at supratherapeutic concentrations, which are higher than those needed to block sodium channels (MacDonald and Meldrum, 1989; but see Twombly et al., 1988).

Many early studies on smooth, cardiac and neuronal cells indicate a very poor sensitivity of the T channels to dihydropyridine-type calcium channel blockers (for review see Bean,

TABLE 3
Potency of Calcium Channel Ligands at Various Types of Voltage-Sensitive Calcium Channels

Cell type	Type of calcium channel	Ligand	Ed$_{50}$[a] ($\times 10^{-9}$ M)	Ref.
Vascular Smooth Muscle				
Artery	L	Nisoldipine	0.07	Nelson and Worley, 1989
	L	Nitrendipine	0.5	Bean et al., 1988
Aorta	L	BAY K 8644	2.4	Papaioannou et al., 1989
Aorta	T	Flunarizine	19	Kuga et al., 1989
	T	Nicardipine	48	Kuga et al., 1989
Vein	L	(+)-Isradipine	0.02	Loirand et al., 1989
	T	(+)-Isradipine	45	Loirand et al., 1989
Brain, synaptosomes	N	ω-CgTX	Ed$_{30}$ = 0.1	Reynolds et al., 1986b
	N	Neomycin	5000	Wagner et al., 1987
Neurohypophysis	N	ω-CgTX	1	Dayanithi et al., 1988
Hypothalamus	N	ω-CgTX	≈3	Takemura et al., 1989
Cardiac				
Ventricle	L	Nitrendipine	0.36	Bean, 1984
Atrial	L	Felodipine	0.67	Cohen et al., 1989
	T	Felodipine	13	Cohen et al., 1989
	T	Cinnarizine	900	Van Skiver et al., 1989
Endocrine, GH$_3$	L	Nimodipine	0.5	Cohen and McCarthy, 1987
	T	Nimodipine	600	Cohen and McCarthy, 1987
Dorsal root ganglion	L	Nimodipine	1.3	McCarthy, 1990
	T	ω-CgTX	(Weak transient block) 1 × 10^{-6} M	McCleskey et al., 1987a,b
Neurohypophysis (Xenopus)	"P"	FTX polyamine	<100 μM	Salzberg et al., 1990
Sensory neurons, cultured	T	Dimethadione	5 × 10^6	Gross et al., 1989
Sympathetic postganglionic neurons	N	ω-CgTX	3.8	Clasbrummel et al., 1989
Skeletal muscle, cultured myoballs	L	(+)-Isradipine	0.1	Cognard et al., 1986
	L	Fluspirilene	0.15	Galizzi et al., 1986b
	L	HOE 166	0.25	Qar et al., 1988
Thalamic	T	Ethosuximide	>200 μM	Coulter et al., 1989.

[a] For inhibition of calcium channel current or calcium influx, except for BAY K 8644, the ED$_{50}$ is for stimulation of ^{45}Ca influx, and ω-CgTX and neomycin, where it is for Ca^{2+} influx or neurotransmitter release.

1989). However, recent studies (e.g., McCarthy and Cohen, 1989; Van Skiver et al., 1989) have demonstrated a fairly high affinity block of T channels with selected 1,4-dihydropyridines in some cell types (Table 3). Stimulation of T channels by Bay K 8644 and related drugs has not been reported. Thus, in any given cell type, those high threshold calcium channels stimulated by Bay K 8644, and with high sensitivity to calcium channel antagonists, may be defined as L channels.

The current and potential therapeutic uses of the calcium channel blockers are outlined in Tables 1 and 2. Most of the therapeutic effects for those uses listed are believed to be mediated through the L class of channels, and the uses indicated reflect some aspects of the functions of these channels. However, a contribution from other sites of action, such as T channels, to the cardiovascular and central effects of these drugs has not been excluded.

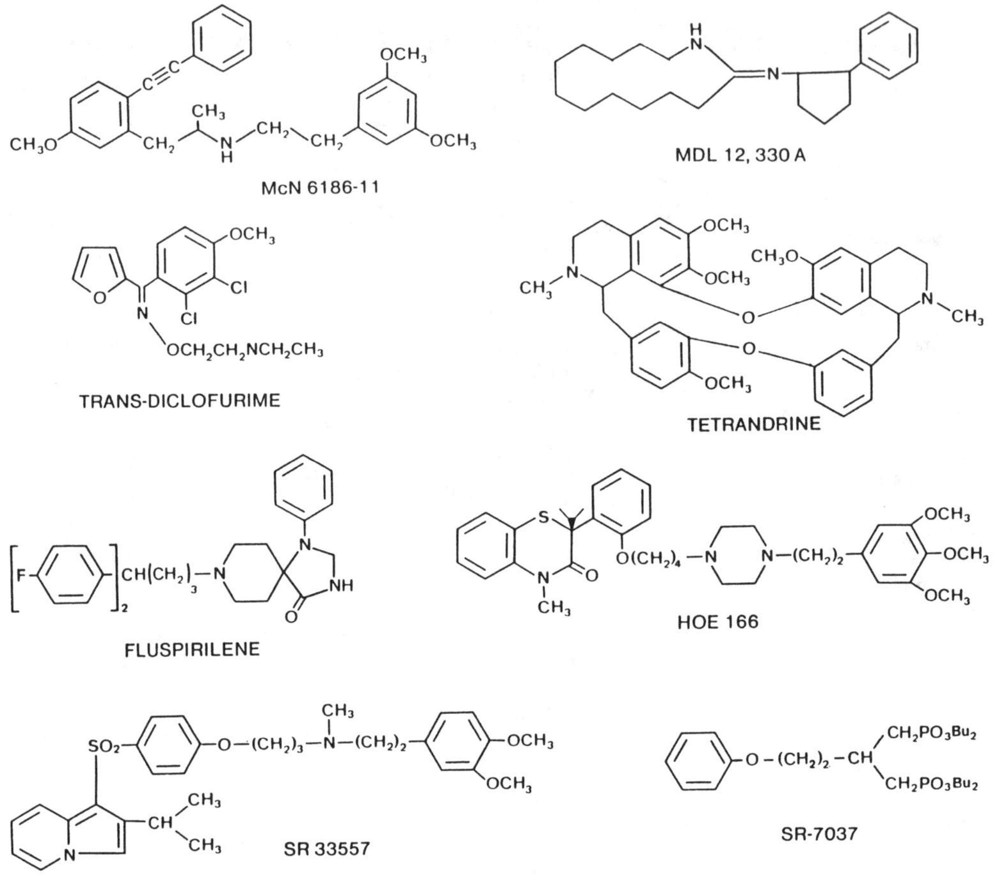

Figure 3. Structural formulae of other calcium channel blockers active at the L class of channel. As represented in Figure 4, many of these drugs may bind to sites different from those of the clinically used drugs.

II. STRUCTURE-ACTIVITY RELATIONSHIPS

A. Some Complexities

Drug interactions with ion channels present particular challenges to the elucidation of structure-activity relationships. First, channels exist in a series of states or families of states according to the membrane potential or biochemical modification. Each of these states may present a different binding site conformation or different access to drugs (Hondeghem and Katzung, 1984; Figure 7). Accordingly, drugs may selectively interact with one or other state of the channel and, conversely, the position of the equilibrium among these states will be an important determinant of the apparent affinity of a drug.

Second, the potency of these drugs will be determined not only by the state of the channel, and the interaction of the drug with that state, but also by the concentration of the drug in the bilayer at the level of the receptor. Most calcium channel blockers have large partition coefficients in membranes, such that the effective drug concentration in the membrane phase, which is the relevant concentration for determining the effects of drug structure on receptor binding, may be hundreds of thousand of times greater than that in the aqueous phase (e.g., Mason et al., 1989a, 1990a). Important to the determination of structure-activity relationships are marked differences in the partition coefficients of different drugs in a given series. For example, amlodipine, which is partially ionized at physiological pH, exhibits a

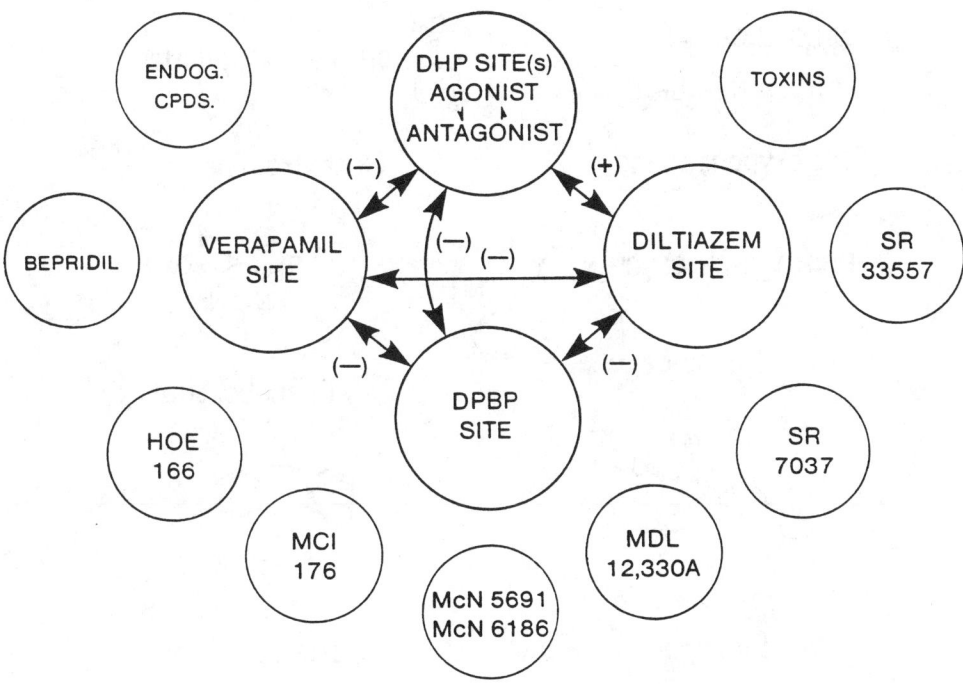

Figure 4. Diagrammatic representation of the multiple allosteric sites associated with the 1,4-dihydropyridine binding site, as well as other drug binding sites for organic ligands that may also be allosterically associated with these binding sites. (Modified from Janis et al., 1987; Rampe and Triggle, 1990.) Tetrandrine and lactamides may bind to the diltiazem site. DPBP is the diphenylbutylpiperidine binding site for fluspirilene, pimozide and related drugs. The multiple cation binding sites that are also associated with L channels were discussed in Janis et al., (1987), as were toxins that may act on L channels. See Section VII for a discussion of endogenous substances that modify 1,4-dihydropyridine binding.

partition coefficient of 19,000, which likely reflects not only hydrophobic but also ionic interactions with the bilayer (Mason et al., 1989b, 1990a). In contrast, the partition coefficients of nimodipine and nisoldipine are 6300 and 10,000 respectively (Herbette et al., 1989). The apparent affinity of these drugs for the calcium channel, as calculated from cytosolic concentrations, may be thousands of times higher than the value calculated from the average concentration in the bilayer. For example, when the cytosolic concentration of nimodipine is at its measured K_D value, 0.4 nM, the average concentration of drug in the bilayer will be near 1 μM, whereas that for amlodipine will be 10-fold higher. It remains to be determined whether those 1,4-dihydropyridines existing at relatively higher concentrations in the bilayer have a greater incidence of side effects. In atrially paced autonomically blocked dogs, amlodipine causes some prolongation in AV conduction times and, at equivalent levels of coronary vasodilation, its cardiovascular effects more closely resemble those of diltiazem or verapamil than those of other dihydropyridines (Dunlap et al., 1989).

Another important point is that the partition coefficients into anisotropic bilayers are much larger than those determined by an octanol/buffer distribution. Correction for differences in partition coefficients into the bilayer is necessary to determine the relative effect of a given change in structure on drug receptor binding. Carvalho et al. (1989) obtained low values for the partition coefficients of nitrendipine and desmethoxyverapamil into synaptic plasma membranes (2 to 3 × 10^2), but a very large value for flunarizine (23,000). These values may be lower than those of Herbette, Mason and co-workers because of the larger drug to membrane ratio used by Carvalho et al. (1989). The values obtained for liposomes were one half to one quarter of these values, indicating that drug binding to membrane bound proteins enhances uptake.

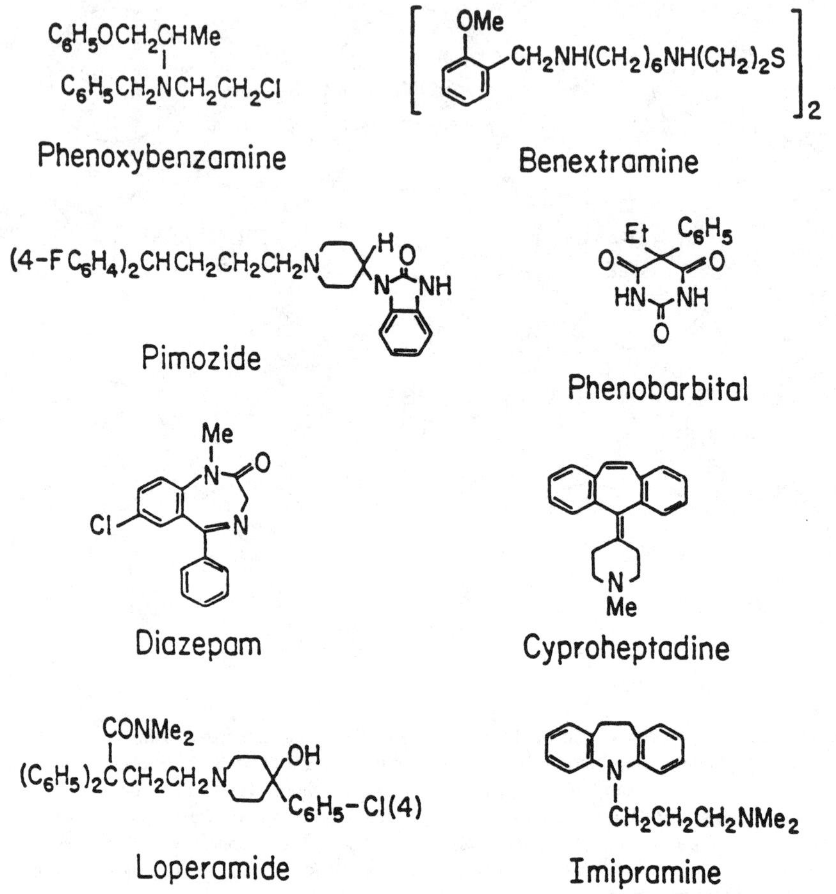

Figure 5. Structural formulae of miscellaneous agents with calcium channel blocking properties secondary to other activities.

TABLE 4
Sequences of Conus Toxins

GVIA (ω-CgTX):

```
 [1]           [2]              [3] [1]     [2]              [3]
C K S P G S S C S P T S Y N C C R o S C N P Y T K R C Y
      5           10                    20
```

GVIIA:
```
C K S P G T P C S R G M R D C C T o S C L L Y S N K C R R Y
```

MVIIA:
```
C K G K G A K C S R L M Y D C C T G S C R o o S G K C
```

Bracketed numbers indicate the positions of the disulfide bridges. o—gap for alignment.

From Gray et al., (1988). With permission.

Evidence that drug-lipid interactions, independent of lipophilicity, may be important determinants for the affinity of 1,4-dihydropyridine binding also comes from electrophysiological studies of Valdivia and Coronado (1988). Increasing the ratio of charged to uncharged phospholipids in the bilayer resulted a 5- to 10-fold increase in the apparent affinity

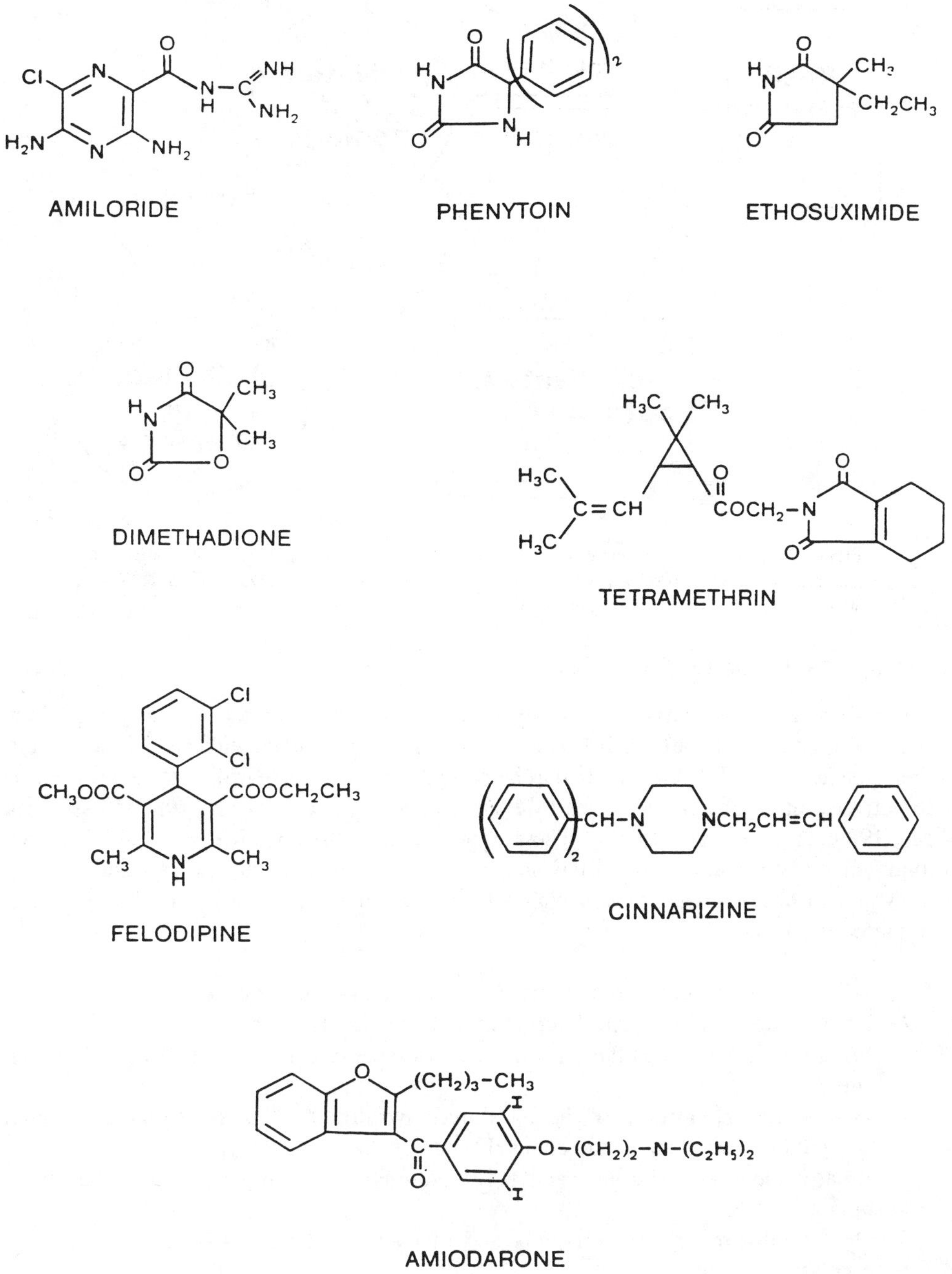

Figure 6. Structural formulae of some compounds active at the T class of calcium channel.

of binding of the positively charged 1,4-dihydropyridine, 207-180, but had no effect on the affinities of uncharged drugs.

The determination of structure-activity relationships is, therefore, dependent upon the experimental conditions that determine the state of the channel, the lipid composition of the bilayer, and the drug concentration of the bilayer. Thus, comparisons made between different tissue or cell types may be based on apparent rather than on real differences in drug-binding site interactions.

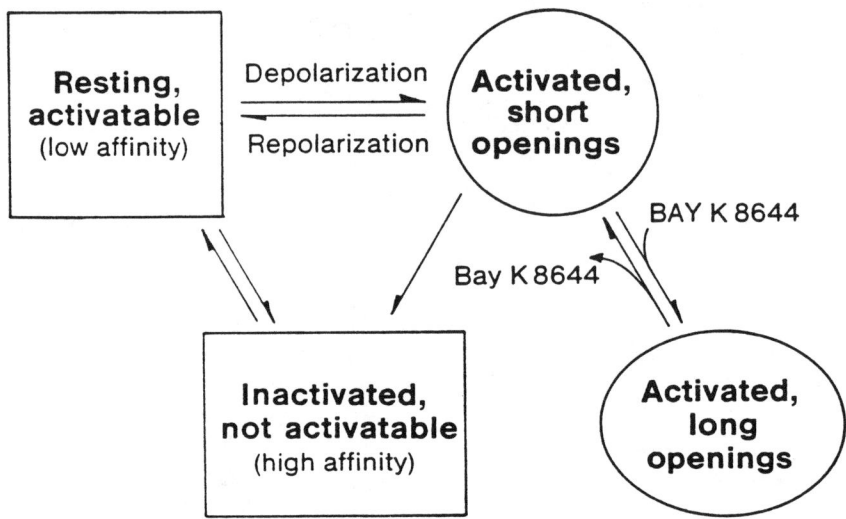

Figure 7. Schematic representation of channel equilibrium among resting, activated and inactivated states. Also shown is the promotion of long channel openings by BAY K 8644, an effect which is specific for L channels.

B. Drugs Active at L Channels

The most extensive structure-activity relationships exist for the 1,4-dihydropyridine analogs of nifedipine. In part this is because of their relative ease of synthesis, in part because of their potency and selectivity and in part because the 1,4-dihydropyridine nucleus embraces both activator and blocker properties (Meyer et al., 1985; Janis et al., 1987; Bossert and Vater, 1989; Triggle et al., 1989a). These agents include therapeutically valuable first and second generation cardiovascular drugs and, in addition, serve as valuable molecular probes with which to characterize this ion channel. A number of questions may thus be posed (Triggle et al., 1989a):

1. What are the structural requirements for activation and blockade?
2. Do activators and blockers interact at discrete sites?
3. What relationship does the 1,4-dihydropyridine site exhibit to other ligand binding sites?
4. What is the relationship of the 1,4-dihydropyridine binding site to the permeation and gating machinery of the channel?
5. What is the relationship between 1,4-dihydropyridine binding sites and pharmacologic activity?
6. Do 1,4-dihydropyridine activators and blockers mimic the actions of endogenous regulatory species?

The general structural requirements for activity of 1,4-dihydropyridine activity are summarized in Figure 8. These requirements have been reviewed in detail previously (Janis et al., 1987; Triggle et al., 1989a,b) and only critical points will be emphasized.

The 1,4-dihydropyridines may favor a membrane approach to a binding site rather than a conventional aqueous diffusion pathway (Rhodes et al., 1985; Chester et al., 1987; Figure 9). The implications for drug design are quite important: the diffusion and partitioning pathways may impose specific molecular requirements for both ligand binding and ligand access at the channel. Such considerations may underlie recent developments with charged 1,4-dihydropyridine analogs of nifedipine, including amlodipine (Figure 10), which have

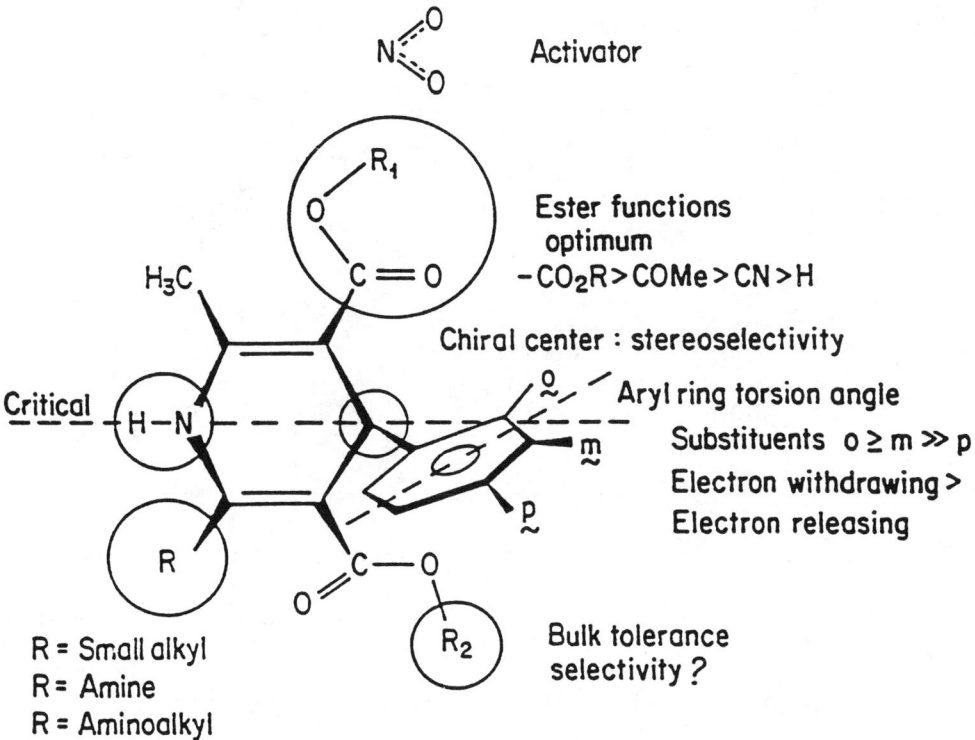

Figure 8. Structural requirements in the 1,4-dihydropyridines for activator and blocker activities.

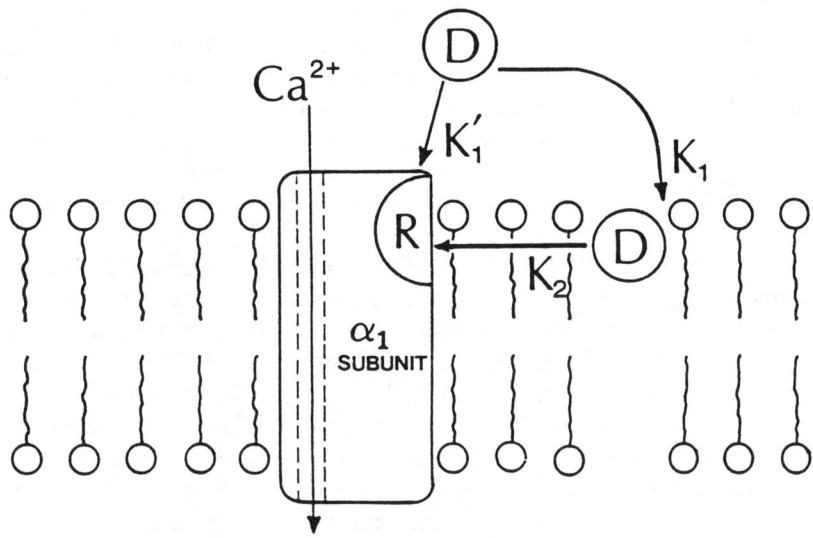

Figure 9. Access pathways for a drug (D) interaction with its receptor site on the α_1-subunit of the L-type calcium channel. Depicted is a pathway (k_1') where the ligand docks with the binding site directly from the aqueous environment, and a pathway involving initial partitioning into the membrane (k_1) followed by two-dimensional diffusion in the membrane (k_2). This two phase path is likely used by hydrophobic calcium channel blockers to reach their binding site in the bilayer. The concentration of 1,4-dihydropyridine at the location shown in the bilayer is much greater than that in the aqueous phase, and therefore the affinities calculated from the latter are likely to be higher than the true values.

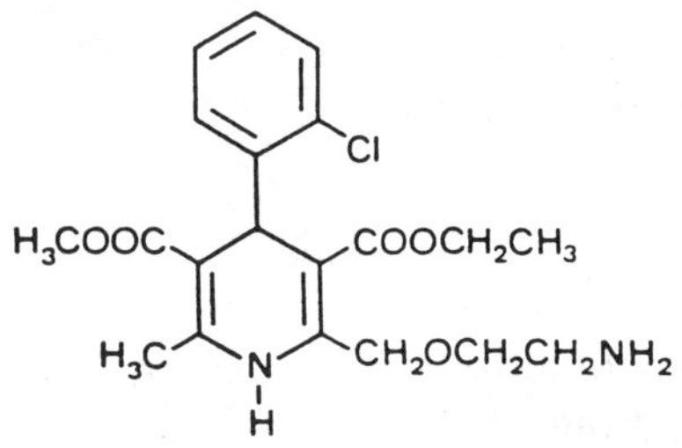

Figure 10. Structural formulae of neutral (isradipine, nitrendipine, nisoldipine) and charged (amlodipine) L channel antagonists. Other neutral antagonists include nifedipine and nimodipine, and charged, nicardipine (Figure 1).

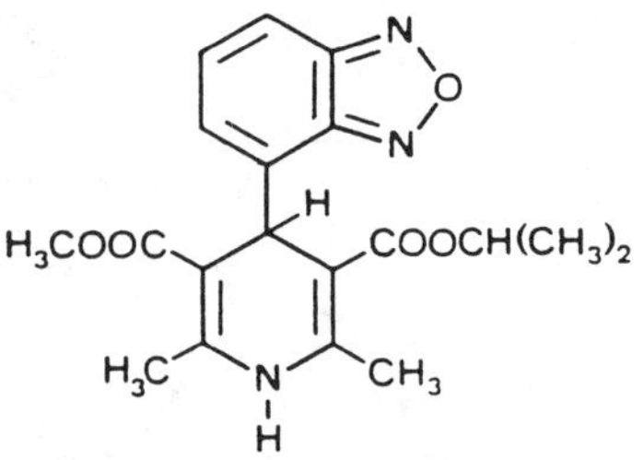

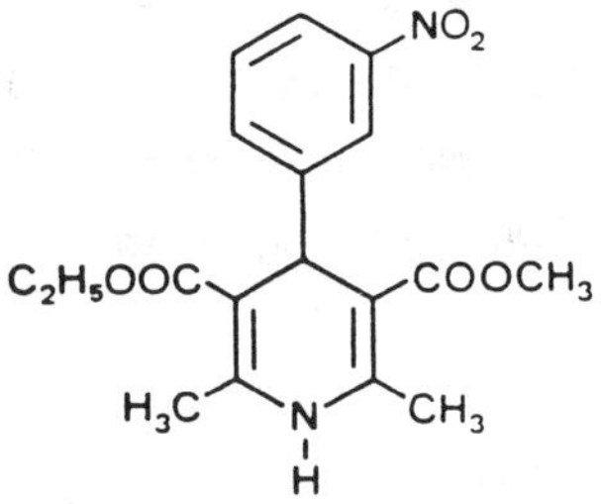

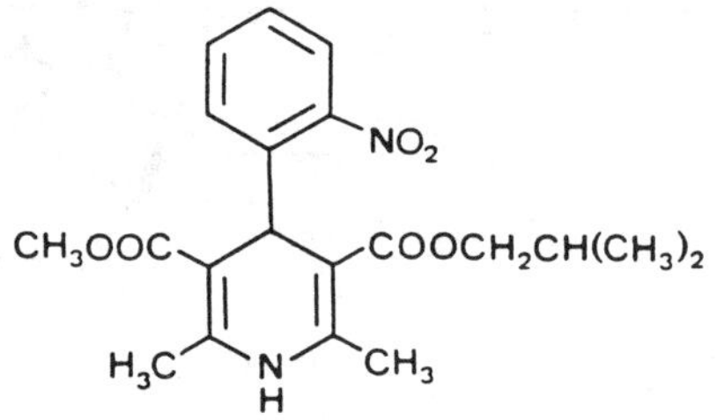

very slow onset and offset kinetics as the charged species and which may enjoy both a different orientation at the binding site and a different access pathway (Kwon et al., 1990; Figure 9). Mason et al. (1989b) have proposed that the positive charge on amlodipine, but not nicardipine (Figure 1), may interact with the charged anionic oxygen of the phosphate headgroup and that this interaction may account for its slower onset of action and its longer nonspecific association with membranes. The slower association rate might be due to a decreased rate of lateral diffusion in the bilayer because of this ionic interaction (Mason et al., 1989b).

The structural similarity of 1,4-dihydropyridine activators and blockers and their competitive interactions generally seen in pharmacologic and radioligand binding experiments are consistent with a common site of action at which opposing pharmacologic effects are mediated (Janis et al., 1987; Triggle et al., 1989a; Wei et al., 1989). It is possible, however, that discrete binding sites exist for activator and blocker species (Kokubun et al., 1986; for review, see Triggle and Rampe, 1990). Few data are available with which to compare structural requirements for activation and blockade (Figure 8). For a small series of activator 1,4-dihydropyridines the structural requirements for channel activation are qualitatively similar to those for calcium channel blockade and substituent effects in the 4-phenyl ring lie in the sequence o > m > p (Kwon et al., 1989). However, the activity enhancing effects of o- and m-substitution are much less marked in the activator than in the blocker series. Both activator and blocker series share the generally close agreement between binding and pharmacologic affinities in smooth muscle (Figure 11); however, discrepancies seen in cardiac muscle are less marked for the activators (Figure 12). Apparent differentiation between activator and blocker 1,4-dihydropyridines is provided by the opposing stereoselectivity exhibited by several enantiomeric pairs of 5-nitro-1,4-dihydropyridines where the S and R enantiomers are activator and blocker, respectively (reviewed in Janis et al., 1987; Triggle

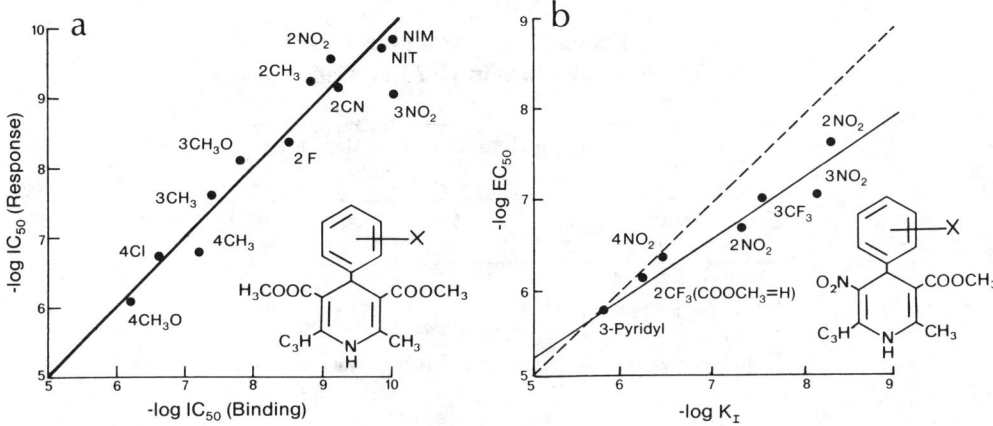

Figure 11. Comparisons of binding and pharmacologic activities of 1,4-dihydropyridine blockers and activators in guinea pig ileal longitudinal smooth muscle. (a) Comparison of binding (competition for specific [^{3}H]nitrendipine binding) and pharmacologic (inhibition of tonic component of K$^+$-depolarization-induced tension) activities for blockers. (b) Comparison of binding (competition for specific [^{3}H]PN 200 110 binding) and pharmacologic (tension generation) activities for activators. NIM = nimodipine, NIT = nitrendipine. (Data from Bolger et al., 1983 and Kwon et al., 1989. With permission.)

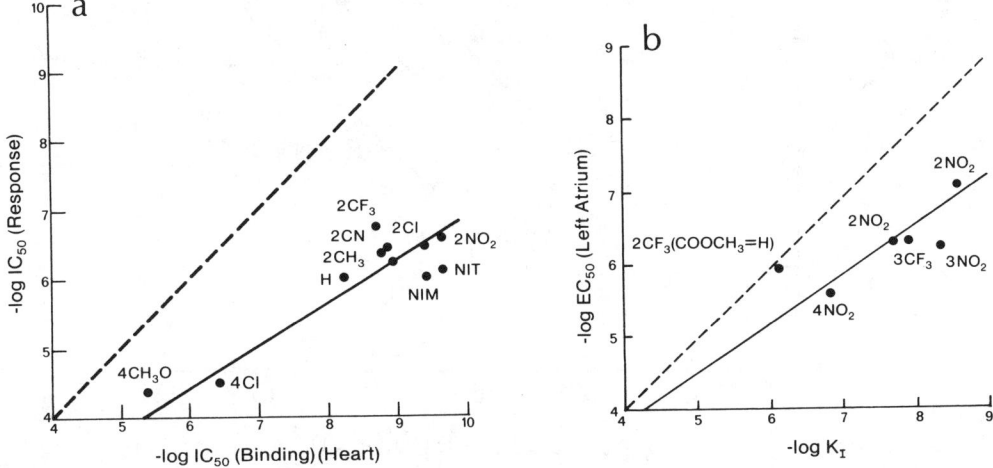

Figure 12. Comparison of binding and pharmacologic activities of 1,4-dihydropyridine blockers and activators in cardiac preparations. (a) Comparison of binding (competition for specific [^{3}H]nitrendipine binding in rat heart) and pharmacologic (paced cat papillary muscle) activities for blockers. (b) Comparison of binding (competition for specific [^{3}H]PN 200 110 binding in rat heart) and activator properties in rat left atrium. NIM = nimodipine, NIT = nitrendipine. (Data from Janis et al., 1984b and Kwon et al., 1989. With permission.)

et al., 1989a). This differentiation may, however, be a reflection of the voltage-dependent interactions of the 1,4-dihydropyridines. Confirmatory of electrophysiological observations (Sanguinetti and Kass, 1984; Bean, 1984), radioligand binding indicates that 1,4-dihydropyridine blocker binding is voltage dependent, being strongly enhanced by depolarization (Table 5). However, activator binding is relatively voltage independent (Wei et al., 1989; Kass, 1987; Figure 13). Thus, S enantiomers may behave as activators or blockers according to the level of membrane potential, whereas the R enantiomers will behave as blockers (Wei et al., 1986; Kass, 1987; Kamp et al., 1989).

Few quantitative structure activity studies are available for the 1,4-dihydropyridines.

TABLE 5
Voltage-Dependent Interactions of 1,4-Dihydropyridines in Cardiac Cells

	Polarized		Depolarized	
	$K_d{}^a$	$B_{max}{}^b$	$K_d{}^a$	$B_{max}{}^b$
[^{3}H]PN 200 110	3.57	50.1	0.06	47.2
[^{3}H]BAY K 8644	5.15	63.1	5.56	62.3

^a $\times 10^{-9}\ M$.
^b fmoles mg/protein.

From Wei et al., (1989); Ferrante et al., (1989). With permission.

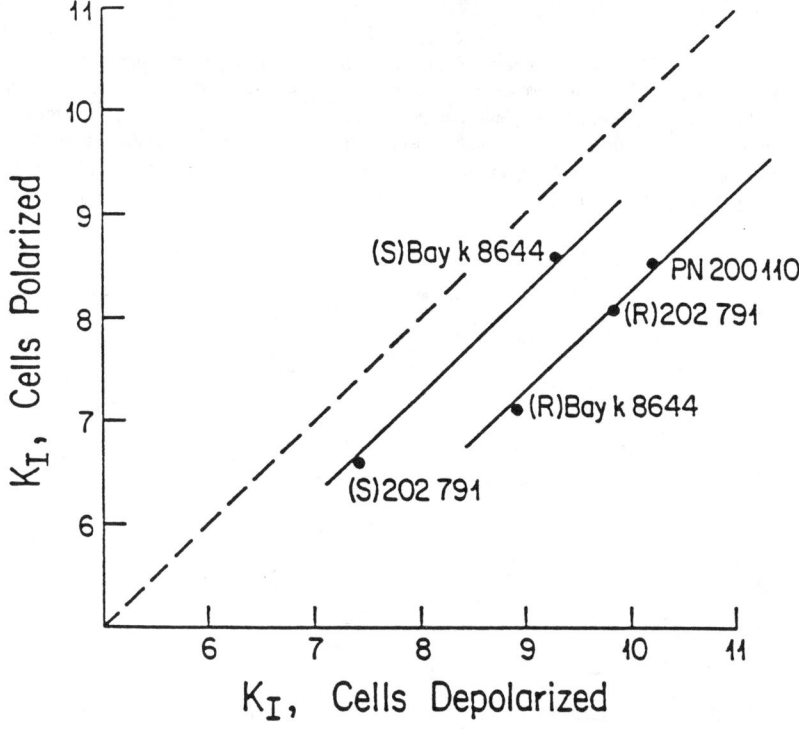

Figure 13. Correlation of binding in polarized (5 mM KCl) and depolarized (50 mM KCl) neonatal rat ventricular myocytes for a series of activator and blocker 1,4-dihydropyridines. The dashed line represent 1:1 equivalency. (Data from Wei, et al., 1989. With permission.)

The most comprehensive, for a series of nonchiral phenyl ring of substituted analogs of nifedipine in intestinal smooth muscle yielded (Coburn et al., 1988):

$$\log 1/IC_{50} = 0.62\pi + 1.96\sigma_m - 0.44L_{meta} - 3.26B_{1para} - 1.51L_{meta'} + 14.23$$

$$n = 46;\ r = 0.90;\ s = 0.67;\ F = 33.93$$

The equation indicates that steric interactions at the meta' position are more unfavorable than at the meta position. This is consistent with X-ray studies which show, quite generally, the favored synperiplanar orientation of phenyl ring substituents (Janis et al., 1987; Triggle et al., 1989a). The activities of the 1,3-benzothiazocines, rigid analogs of the 1,4-dihydro-

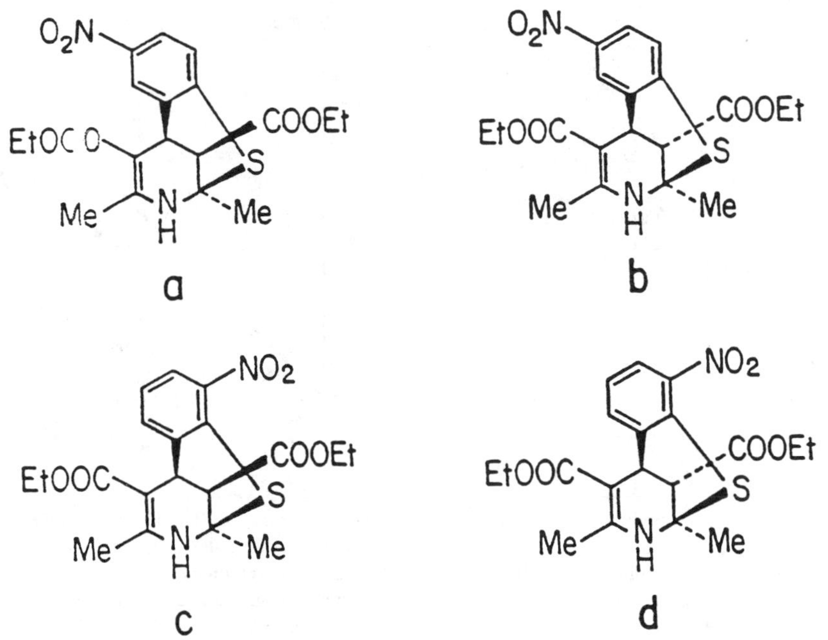

Figure 14. 1,3-Benzothiazocines as rigid 1,4-dihydropyridine analogs. See text for further explanation.

pyridines (Figure 14), show activity in the sequence a ≫ b = c > d consistent with the importance of the synperiplanar orientation of the 3-nitro group. Similarly, the 1,4-dihydropyridine receptor distinguishes between 2,3- and 2,5-disubstituted phenyl-1,4-dihydropyridines, the former being the more potent in both pharmacologic and radioligand binding assays.

For the structures and pharmacological activity of a variety of other calcium channel antagonists, see Janis et al. (1987); Scriabine (1987), Bossert and Vater (1989), Ohtsuka et al. (1989), and Rampe and Triggle (1990). A number of new compounds exhibit a combination of calcium channel inhibition and a second activity, such as adrenolytic or antiplatelet activity.

C. Drugs Active at N, T and Other Channels

Potent and selective synthetic drugs active at other than the L category of calcium channel exist. Structure-activity studies are correspondingly limited, but are becoming available for the ω-conotoxins (Gray et al., 1988; Yoshikami et al., 1989). The structure of the GVI, GVII and MVII toxins (Table 4) show significant differences aside from the invariant cysteines, glycine at position 5 and lysine at position 2. This may well be consistent with significant differences in biological activity observed in these toxins.

As previously mentioned, an apparent fourth category of calcium channel, designated P channels, has been characterized through its sensitivity to a low molecular weight toxin (FTX) from funnel-web spiders (Llinas et al., 1989a; Jackson and Parks, 1989). This channel, insensitive to the 1,4-dihydropyridines and the ω-conotoxins, may represent a distinct category of neuronal channel. In addition to the low molecular weight spider toxins, there are several peptides from spider toxins that inhibit transmission presynaptically. One, a 7.5-kDa peptide (ω-Aga-I) from the funnel web spider, *Agelenopsis aperta*, likely acts on calcium channels to potently inhibit (IC_{50} 7.5 nM) neuromuscular transmission in both vertebrates and invertebrates (Bindokas and Adams, 1989).

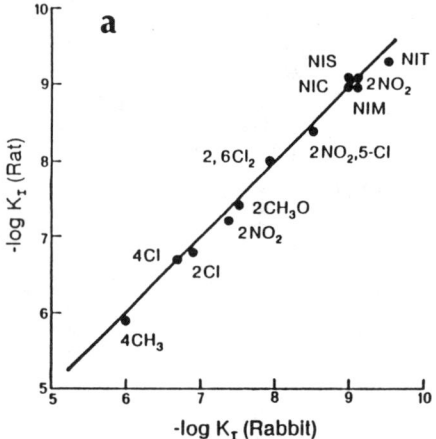 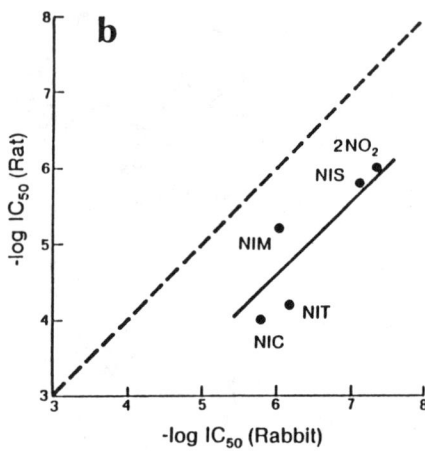

Figure 15. Correlations of binding and pharmacologic activities of a series of 1,4-dihydropyridines in rat and rabbit myocardium. The symbols indicate the position of the phenyl ring substituent in the nifedipine series and NIC = nicardipine, NIM = nimodipine, NIS = nisoldipine, NIT = nitrendipine. (a) Comparison of binding affinities (competition with specific [^{3}H]nitrendipine binding) in rat and rabbit cardiac membranes. (b) Comparison of pharmacologic activities in rat and rabbit myocardium. Dashed lined represents 1:1 equivalency. (Data from Boyd, et al., 1988. With permission.)

Maitotoxin has three actions: it activates L calcium channels, phospholipase C (resulting in the production of IP$_3$ and diacylglycerol, the release of intracellular Ca^{2+} and the activation of protein kinase C), and phospholipase C (Gusovsky et al., 1989; Choi et al., 1990). Other toxins have also been reported to act on L channels (see Janis et al., 1989).

D. Multiple Channel Classes and Multiple Channel States

The variable relationships seen between radioligand binding and pharmacologic potencies of 1,4-dihydropyridines and other calcium channel blockers have raised the issues of whether different subtypes or subclasses of the L channel exist (Janis et al., 1987; Triggle and Janis, 1987; Triggle et al., 1989a). By analogy to other receptor systems this is clearly possible (Triggle et al., 1989a). However, available data do not yet permit a clear distinction between variable affinity receptors, separate conformational states, and discrete subclasses. In a series of 1,4-dihydropyridine blockers, binding affinities were identical in rat and rabbit myocardial membranes, but pharmacologic potencies were some 10- to 100-fold lower in rat myocardium (Boyd et al., 1988; Figure 15). In a comparison of the binding and negative inotropic potencies of 1,4-dihydropyridines in cat papillary muscle the ratio of binding affinity to pharmacologic affinity ranged from approximately 100 to 3500 (Goll et al., 1986). The absence of a constant ratio in cardiac tissue contrasts with that observed in smooth muscle (Triggle et al., 1989b; Figure 16). Although these discrepancies between binding and pharmacologic affinities are interpretable in terms of state-dependent interactions, the differences between cardiac and smooth muscle suggest that channel or bilayer composition differences may exist in these two tissues which are recognized differentially by the 1,4-dihydropyridines.

Stereoselectivity may represent a further index of receptor discrimination and in a comparison of the activities of the enantiomers of verapamil, D600, and devapamil on coronary flow and the maximum systolic left ventricular pressure in the Langendorff heart it was observed that the (−)enantiomers were approximately equipotent in cardiac and vascular smooth muscle, but that the activities of the less potent (+)enantiomers differed significantly (Table 6). Additionally, stereoselectivity was higher in the cardiac than in the vascular preparations. The stereoselectivity index may be a useful probe, but it is not *a priori* possible to determine whether the index reflects state differences or channel subtypes.

13. Drugs Acting on Calcium Channels 217

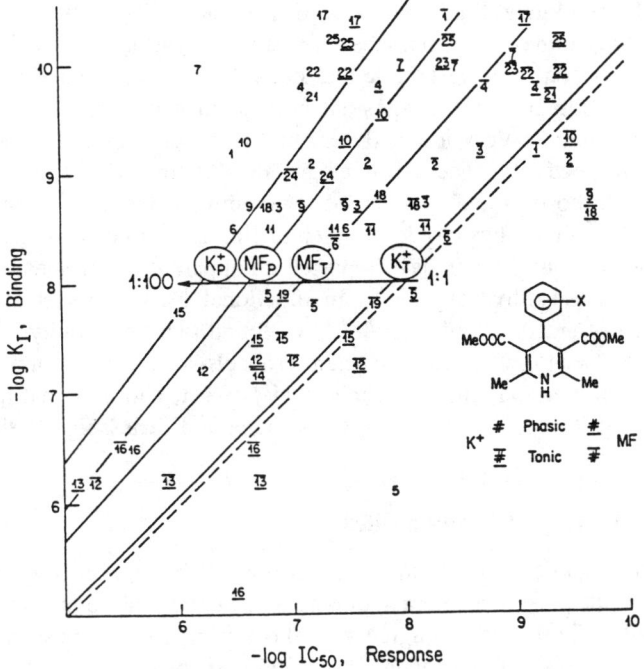

Figure 16. Correlations of binding and pharmacologic activities for a series of 1,4-dihydropyridine analogs of nifedipine bearing various ring substituents (1 to 25) as blockers of specific [³H]nitrendipine binding and as inhibitors of the phasic and tonic components of response of guinea pig ileal longitudinal smooth muscle to K^+ depolarization and muscarinic receptor activation by methylfurmethide (MF). (Data from Bolger et al., 1983. With permission.)

TABLE 6
Activities of Enantiomers of Verapamil, Gallopamil and Devapamil for Coronary Flow Increase and Maximum Systolic Pressure Decrease in the Langendorff-Perfused Heart

	Coronary flow		$EC_{50}MSLVP^a$	MSLVP	
	Ratio[+]/[−]	−logEC$_{50}$	EC$_{50}$ flow	−logEC$_{50}$	Ratio[+]/[−]
[+]Verapamil	2.45	7.31	43.6	5.67	31.6
[−]Verapamil		7.20	3.4	7.17	
[+]Gallopamil	3.3	7.12	15.8	5.92	56.2
[−]Gallopamil		7.64	0.93	7.67	
[+]Devapamil	19.9	7.21	4.5	6.56	61.7
[−]Devapamil		8.51	1.45	8.35	

[a] MSLVP = Maximum systolic left ventricular pressure.

From van Amsterdam and Zaagsma (1988). With permission.

III. SITES OF ACTION

A. Ligand-Binding Data

1. Ligand-Binding and Physiological Data Correlate

Cardiac and skeletal muscle contractile results notwithstanding, electrophysiological studies have confirmed ligand-binding data of the subnanomolar affinity of nitrendipine and related 1,4-dihydropyridines to membranes from smooth, cardiac and skeletal muscle (Figures 11 and 16; Janis et al., 1987; Triggle and Janis, 1989). An extensive listing of K_D and B_{max} values for approximately 150 studies has been published (Janis et al., 1987). Selected data are shown in Table 7. Very low and very high site densities are found in smooth and skeletal muscle, respectively. The affinities for the binding of the various radioligands generally agree with those estimates obtained for binding to the high affinity state of the L channel obtained by electrophysiologic or calcium flux measurements (Table 8).

A number of previous studies have reported an apparent lower sensitivity of neuronal calcium channels to 1,4-dihydropyridines in functional tests (Janis et al., 1987; Miller, 1987). However, Fry and McCarthy (1988) have found that the affinity of nimodipine for inhibiting neuronal (dorsal root ganglion) calcium channels is similar to that for [^{3}H]nimodipine binding to smooth and cardiac muscle (Janis et al., 1982), and neuronal membranes (Bellemann et al., 1983; Rampe et al., 1987a; Glossmann and Striessnig, 1988; Triggle et al., 1989a,b).

2. Location and Density of Binding Sites

Ligand-binding studies indicate that the location of the high affinity sites for calcium channel blockers is the plasma membrane of neural and endocrine cells, and smooth, skeletal and cardiac muscle. These findings agree with those from current measurements on single cells, as well as those from studies in which purified vesicles incorporated into bilayers were used (reviewed in Hosey and Lazdunski, 1988). High affinity binding exclusively to the plasma membrane channel is seen to occur, except in skeletal and perhaps cardiac muscle, where high affinity binding to the voltage sensor is also observed.

The density of high affinity binding sites is generally similar to the number L channels estimated by electrophysiology (reviewed in Janis et al., 1987). In chromaffin cells, the density of L channels estimated by ligand binding was $1.5/\mu m^2$ surface area, which is about one tenth that found by patch-clamp techniques (Castillo et al., 1989). However, the latter methodology is thought to overestimate channel density. This density is similar to that obtained for cardiac sarcolemma ($1/\mu m^2$). The methods and assumptions used for calculation were different, and the B_{max} values for the cardiac membranes, 1 pmol/mg protein (Colvin et al., 1985) are about 10 times that of the bovine adrenomedullary cells (Castillo et al., 1989). The density of channels in mesenteric arteries estimated to be 0.2 to $4/\mu m^2$ (Worley et al., 1986), and in cardiac cells, 0.1 to $0.5/\mu m^2$ (Reuter et al., 1983). In skeletal muscle there are $\simeq 230$ sites/μm^2, which is 35 to 50 times more than the number of L channels. The additional binding sites are believed to be on the voltage sensor (Schwartz et al., 1985).

3. Voltage Sensor

There is a similar (Lamb and Walsh, 1987; Brum et al., 1988; Erdmann and Lüttgau, 1989) or identical (Tanabe et al., 1987) high affinity binding site on the force-controlling voltage sensor on skeletal muscle t. tubular membranes. The affinity of (−)devapamil binding to this site is the same (1.7 nM) as that found in radioligand binding experiments using isolated skeletal muscle membranes. Devapamil likely stabilizes the inactivated state of the voltage sensor and the calcium channel in a similar way to gallopamil (Erdman and Lüttgau, 1989).

Reconstitution experiments demonstrated that the purified dihydropyridine binding site is a functional calcium channel (Flockerzi et al., 1986; Smith et al., 1987; Talvenheimo et al., 1987; Horne et al., 1988; Hymel et al., 1988; Nunoki et al., 1989). Direct evidence indicating that the 1,4-dihydropyridine binding sites are on both the calcium channel and the voltage sensor of skeletal muscle as recently been obtained in studies on dysgenic mice which lack the α_1-subunit of the 1,4-dihydropyridine receptor (Knudson et al., 1989).

TABLE 7
Comparison of Affinities and Site Densities for Various Radioligands Binding to L and N (ω-CgTX Binding) Channels in Membranes From Different Cell Types

Preparation and radioligand	K_D ($\times 10^{-9}$ M)	B_{max} (pmol/mg)	Ref.
Adrenal medulla			
Isradipine	0.04	0.12	Castillo et al., 1989
	0.1	0.14	Murphy et al., 1990
Nitrendipine	0.49	0.13	Ballesta et al., 1989
ω-CgTX	0.007[a]	0.24	Ballesta et al., 1989
(lower affinity sites)	0.36	1.2	Ballesta et al., 1989
Aorta sarcolemma			
(partially purified)			
Nitrendipine	0.12	0.16	Luchowski et al., 1984
Coronary artery			
Isradipine	0.04	0.31	Matlib, 1989
Brain: (synaptosomes)			
Isradipine	0.03	0.47	Rampe et al., 1987
Nimodipine	0.22	0.52	Rampe et al., 1987
Nitrendipine	0.12	0.48	Rampe et al., 1987
ω-CgTX	0.008	0.15	Takemura et al., 1989
(low affinity site)	0.69	4.2	Takemura et al., 1989
Cortex			
Nitrendipine	0.13	0.15	Wagner et at., 1988
ω-CgTX	0.07	8.3	
Hippocampus			
Nitrendipine	0.11	0.17	Wagner et al., 1988
ω-CgTX	0.05	6.4	
Midbrain			
Nitrendipine	—	<0.01	Wagner et al., 1988
ω-CgTX	0.11	3.5	
Brain stem			
Nitrendipine	—	<0.01	Wagner et al., 1988
ω-CgTX	0.04	0.55	
Olfactory bulb			
Nimodipine	0.17	0.56	Skattebol and Triggle, 1987
BAY K 8644	1.0	0.42	
Cardiac microsomes			
Nimodipine	0.24	0.40	Janis et al., 1982
Nitrendipine	0.25	0.39	Janis et al., 1982
BAY K 8644	2.4	0.55	Sarmiento et al., 1989
Cardiac sarcolemma			
Nitrendipine	0.36	1.55	Garcia et al., 1986b
Fluspirilene	0.6	≈1.5	King et al., 1989
Verapamil	38	1.65	Garcia et al., 1986
Diltiazem	80	1.95	Garcia et al., 1986
Skeletal muscle			
(transverse tubules)			
Fluspirilene	0.1	80 ± 15[b]	Qar et al., 1988
SR33557	0.08	78	Schmid et al., 1989
Nitrendipine	1.5	50 ± 15	Galizzi et al., 1986a
Gallopamil	1.5	70 ± 10	Galizzi et al., 1986a
Verapamil	10	50 ± 15	Galizzi et al., 1986a
Bepridil	20	75 ± 15	Galizzi et al., 1986a
Diltiazem	50	50 ± 15	Galizzi et al., 1986a

[a] Apparent K_D values are listed for ω-CgTX; equilibrium is not obtained (see Takemura et al., 1989).
[b] Mean ± S.E.

TABLE 8
Affinities ($\times 10^{-9}$ M) of Calcium Channel Blockers for L Channels of Various Cell Types

Values Estimated from Radioligand Binding (25°C), and, in Parentheses, Calcium Channel Current or Calcium Influx Inhibition

Cell Type	1,4-Dihydropyridines					Non-Dihydropyridines						
	Nt[a]	Nm	Ns	Is	Bay K 8644	Vp	D888	Dt	Fs	HOE 166	Beb	Flu
Nerve	0.1 (7)	0.2 (1.3)		0.03	3	94	0.5	50	20	12	1.5×10^4	500
Cardiac (Vent.)	0.2 (0.2)	0.27	0.07	0.03	2.4	4	0.2	80	0.6	4.6		360
Smooth Muscle	0.18² (0.02)	0.2 (0.02)	0.07 (0.07)	0.03 (0.09)	3.5 (2.4)	21	0.4	30	50	70	6×10^3	13
Skeletal Muscle	1.5	3.0		1.8 (0.9)		10 (10)	1.5	50 63	0.1 (0.1)	4 (0.2)	20	60
Endocrine	0.2 (2.5)	0.5 (0.5)		0.08 10	0.8 (4)					10 (10)		

[a] Abbreviations: Nt, nitrendipine; Nm, nimodipine; Ns, nisoldipine; Is, isradipine, or PN 200-110; Vp, verapamil; D888, (desmethoxyverapamil); Dt, diltiazem; Fs, fluspirilene; Beb, bepridil; Flu, flunarizine.

For binding data references, see Tables 7, 19 to 21, and Table 26 in Janis et al., 1987; for other references see Tables 7, 9, and 10, this chapter. With permission.

Injection of a plasmid expressing the 1,4-dihydropyridine receptor into cultured skeletal muscle cells from dysgenic mice was found to restore calcium channel current (Tanabe et al., 1988). In addition, excition-contraction coupling and charge movement were restored, reflecting the expression of the voltage sensor. These results suggest, but do not prove, that the α_1-subunit of the calcium channel complex serves both as a calcium channel and as a voltage sensor/excitation-contraction coupler in skeletal muscle.

4. Neurons

Calcium channels may be found on all parts of the surface membrane of neurons, but a given type is often highly localized in certain areas called "hot" spots (Tsien et al., 1988). The properties of the channels may change with the location on the cell because of changes in the intracellular environment, and their location on cells may change with cell isolation (see Section III, Chapter 14). ω-CgTX sensitive N channels are organized into multiple clusters on synapse-free soma of hippocampal CA1 neurons, but after innervation, are concentrated and immobilized at dendritic contact sites (Jones et al., 1989). Autoradiographic studies of brain slices have shown that the highest densities of high affinity binding sites are in those areas rich in synapses. The distribution of 1,4-dihydropyridine-antagonist sites is not significantly different from that for BAY K 8644 (Skattebol and Triggle, 1987). The localization seen by radioautography is consistent with that reported for binding to isolated fractions (reviewed in Janis et al., 1987).

ω-CgTX produces a potent, direct and irreversible inhibition of N (Reynolds et al., 1986b; McCleskey et al., 1987) and possibly L channels (see Plummer et al., 1989) in vertebrate neurons. The distribution of ω-CgTX and L channel-binding sites in the brain are clearly different (Takemura et al., 1989). ω-CgTX binding may reflect binding to both N and L channels, or to N channels only (Plummer et al., 1989). ω-CgTX binds only weakly and transiently to the T class of channels. The binding of this toxin is not inhibited by drugs acting specifically on L channels, but is inhibited by divalent cations (Be et al., 1987; Rivier et al., 1987). Aminoglycoside antibiotics such as neomycin are weak (IC_{50} 5 μM) inhibitors of toxin binding (Knauss et al., 1987; Atchinson et al., 1988), and their effect on calcium channels may cause neurotoxicity (Wagner et al., 1987). It remains to be determined whether modifications of the structures of these antibiotics will lead to clinically useful N channel inhibitors.

Prolonged exposure of CA1 hippocampal neurons to excitatory amino acids results in the complete loss of sensitivity of postsynaptic potentials to ω-CgTX (Krishtal et al., 1989). ω-CgTX binding is inhibited when the N channel is Ca^{2+}-free, in its Na^+-conducting state (Carbone and Lux, 1988).

B. Voltage Dependence of Binding

The affinity of tritiated 1,4-dihydropyridine binding has recently been found to be voltage dependent, agreeing with the results of voltage-clamp studies, and further supporting the view that tritiated ligand binding reflects drug association with the calcium channels (Kokubun et al., 1986). Similarly, binding studies, paralleling previous electrophysiological reports, have shown that the affinity (Schilling and Drewe, 1986; Kokubun et al., 1986; Wei et al., 1989) or apparent density (Kamp and Miller, 1987) of binding sites increases with increasing depolarization (Table 5, Figure 13).

In isolated membranes, calcium channels are likely in the inactivated state, as are channels in intact cardiac cells that are held at 0 mV under conditions where intracellular calcium accumulates and the channels inactivate. The possibility that L channels in isolated membranes can be frozen in the resting or open state has not been excluded.

The affinity of binding of tritiated nitrendipine or nimodipine to isolated (and therefore, depolarized) membranes from all cells studied, except for skeletal muscle, is 0.1 to 0.3 nM (Tables 3 and 8). Detailed studies on the thermodynamics of binding also indicate very similar recognition sites for 1,4-dihydropyridines on smooth and cardiac muscle and neuronal membranes (Rampe et al., 1987b). The affinities of the skeletal muscle sites for 1,4-dihydropyridines are about 10 times lower than those of the other excitable cells.

In general, there is an excellent agreement between the highest affinities obtained from intact tissue and cell studies and those obtained from ligand binding (Table 8, see Hamilton et al., 1987). There are many factors that may lower drug potency in intact tissue studies (see Section IV). If the highest affinity binding is to the open or inactivated state of the channel, then the extent of agreement between binding affinities and those determined in functional tests should reflect the degree to which the response being measured is dependent on the opening of L channels.

Ligand-binding studies using intact cells have provided evidence for two distinct 1,4-dihydropyridine-binding sites (Kokubun et al., 1986; see Kamp and Miller, 1987). The binding site for activators, e.g., 202 791 and Bay K 8644, may be different from that of the 1,4-dihydropyridine blockers (reviewed by Triggle and Rampe, 1990). Support for this hypothesis also comes from studies by Brown et al. (1989), who have argued that at least two binding sites exist for BAY K 8644, a higher affinity one responsible for producing an increase in the probability of channel opening, and a lower affinity site, mediating the longer opening of the channel.

C. Several Allosterically Coupled Binding Sites are Associated with L Channels

A major contribution of studies with [^{3}H]ligands has been the demonstration that there are several reciprocally coupled, allosteric sites associated with the 1,4-dihydropyridine receptor (Figures 2 and 4). It has been suggested that amlodipine may bind to both the dihydropyridine and the diltiazem sites (Matlib, 1989). Many new ligands have been reported that appear to act at sites on L channels distinct from those previously studied (Rampe and

Triggle, 1990). Not only are there distinct sites for verapamil, diltiazem, and 1,4-dihydropyridines, but there is also evidence for separate sites for bepridil, flunarizine (Hosey and Lazdunski, 1988; King et al., 1989), the benzothiazine, HOE 160 (Qar et al., 1988), KB 944, MDL 12330A (Garcia et al., 1987), cinnarizine and DPI 201-106 (Siegl et al., 1988), 1,3-diphosphonates (Rossier et al., 1989), and amiloride-like drugs (Suarez-Kurtz and Kaczorowski, 1988; Garcia et al., 1990). The very potent non-1,4-dihydropyridine, SR33557 (IC50, 20 pM), also binds to a site on the calcium channel that is separate from the 1,4-dihydropyridine and verapamil binding sites (Nokin et al., 1989). Various natural products that act on L channels have also been recently reviewed (Rampe and Triggle, 1990).

There is some controversy as to whether bepridil, verapamil and diltiazem bind to the same (Galizzi et al., 1986a) or to different sites (Garcia et al., 1986; Reynolds et al., 1986a; Balwierczak et al., 1987; Glossmann and Striessnig, 1988; Schoemaker and Langer, 1989). The complexities are seen in the temperature dependence of these interactions. For example, diltiazem and ($-$)devapamil may stimulate or inhibit 1,4-dihydropyridine binding depending on the temperature and the concentration of the non-1,4-dihydropyridines used (e.g., Reynolds et al., 1986a).

The ratio of the benzothiazepine, 1,4-dihydropyridine, diphenybutylpiperidine, and phenylalkylamine sites is 1:1:1:1 (Garcia et al., 1986; Galizzi et al., 1986a,b; Qar et al., 1988; King et al., 1989), consistent with their binding to the same protein (Table 7). It remains to be determined whether SC 33557 and fluspirilene bind to the same site or to a different site on the L channel. A diagrammatic representation of these sites is shown in Figure 4. King et al. (1989) suggest that there are at least seven separate sites for organic calcium channel antagonists on L channels.

The binding of calcium channel ligands is highly dependent on the presence of divalent cations (reviewed in Janis et al., 1987; Glossmann and Striessnig, 1988; Hosey and Lazdunski, 1988). Since even low concentrations of divalent cations can inhibit the binding of verapamil and diltiazem, a maximal potency of these drugs in electrophysiological experiments requiring elevated Ca^{2+} or Ba^{2+} concentrations cannot be determined. The binding site for fluspirilene is of special interest because Cd^{2+} increases both the affinity and B_{max} for drug binding. In contrast, Cd^{2+} inhibits the binding of the main classes of drugs binding to the L channel. Millimolar concentrations of Ca^{2+} and Ba^{2+} inhibit the binding of fluspirilene and of other non-1,4-dihydropyridines, but not of nitrendipine binding at 25 to 37°C (e.g., Schoemaker and Langer, 1989). Micromolar concentrations of Ca^{2+} or Mg^{2+} are required for binding of 1,4-dihydropyridines, but these same concentrations inhibit the binding of phenylalkylamines, diltiazem, or bepridil to skeletal muscle membranes in some studies (Galizzi et al., 1985).

The disparity in the K_D value of pimozide for its binding to the fluspirilene site and for its inhibition of 1,4-dihydropyridine binding reflects the allosteric nature of its inhibition of 1,4-dihydropyridine binding (King et al., 1989). Fluspirilene inhibits calcium channel current in GH_3 cells, consistent with its effects on skeletal muscle, but it increases calcium channel current in well-polarized atrial cells, indicating a qualitative difference in L channel pharmacology of these two cell types (Spires et al., 1988).

Belfosdil (SR-7037) represents another new class of L channel blocker, the 1,3-diphosphonates (Rossier et al., 1989). This class of compounds is known to include antiatherogenic agents (Henry, 1989). Although structurally unrelated to the commonly used calcium channels, belfosdil completely displaces isradipine (IC_{50} 19 nM, Hill slope 0.96), diltiazem and D888 binding from brain membranes. Detailed analysis of binding suggests that the diphosphonate may bind to a different site than the three prototypic drugs (Figure 4). Millimolar concentrations of divalent cations such as Ca^{2+} and Mg^{2+}, which inhibit D888 binding, stimulate the binding of belfosdil (Rossier et al., 1989).

Electrophysiological studies indicate that L channels are linked to GTP binding proteins

(Brown et al., 1988; Yatani et al., 1988, see Hess, 1990). In contrast, it has been difficult to directly demonstrate effects of GTP on 1,4-dihydropyridine binding to isolated membranes (Janis et al., 1987). Recently, it has been reported that GppNHp (5′-guanylylimidodiphosphate) caused a rightward shift of the BAY K 8644, but not nimodipine, binding curve to smooth muscle membranes (Higo et al., 1988).

The enhancement of 1,4-dihydropyridine binding produced by diltiazem may be reflected in the effect of these drugs on intact tissue (reviewed in Janis et al., 1987; Triggle and Janis, 1987). A possible recent example of this is the enhancement and inhibition by diltiazem of the antiseizure effect of nimodipine (Moron et al., 1989).

D. Biophysical and Physiological Data

1. Major Sites of Action

L channels of vascular smooth muscle resistance vessels and of cardiac cells are the major sites of action of the clinically used calcium channel antagonists. The inactivated state of the channel is thought to be the specific binding site. Electrophysiological studies demonstrate inhibition of these channels at the same concentrations at which these drugs bind to isolated organs and inhibit potassium-induced smooth muscle contraction (Bolger et al., 1983; reviewed in Janis et al., 1987; Triggle and Janis, 1989). Thus it is likely that the beneficial effects of these drugs in angina, hypertension and arrhythmias are mediated mainly or completely via the high affinity binding of these drugs to L channels of those vascular smooth muscle and cardiac cells that are involved in the maintenance of either elevated vasoconstriction or irregular cardiac electrical activity. For the treatment of angina and hypertension, those arteries and arterioles contracted because of the influx of extracellular Ca^{2+} into cells exhibiting a prolonged depolarization are the major cellular sites of action. For treatment of arrhythmias, cardiac cells exhibiting a high rate or degree of depolarization are the most susceptible.

In some studies of vascular smooth muscle, only L channels and a putative 1,4-dihydropyridine sensitive subconducting L type of channel were seen (Worley et al., 1985). L channels in an aortic cell line are activated by phorbol esters, but not cAMP, whereas T channels in these cells were insensitive to both agents (Fish et al., 1988).

The other likely sites of action of these drugs, particularly for neuronal disorders, are the L, and possibly T channels in neuronal cells. The beneficial effects of 1,4-dihydropyridines in epilepsy, age-associated learning deficits, and certain models of ischemia appear to be independent of effects on blood flow (Section V).

2. Location of Binding Sites

The exact location of the binding site on the α-subunit of L channels for calcium channel antagonists is not known, but since binding is voltage sensitive, the site may be tightly linked allosterically with either the voltage sensor or another part of the channel sensitive to channel state. In vascular smooth muscle, unlike cardiac and neuronal cells, gallopamil appears to bind to an intramembranal site that is more easily accessible from the external side (see Leblanc and Hume, 1989). Similarly, planar bilayer studies indicate that the highest affinity binding sites are buried in the lipid phase adjacent to the external end of the channel (Valdivia and Coronado, 1988). The threshold for (−)BAY K 8644 is 10 nM from the external side of the bilayer but 50 times larger from the internal side. Several blocking agents were also found to be 2 to 10 times more potent when added to the external side. Exceptions were verapamil, which showed no preferential block, and the quaternary ammonium derivative D890, which blocked only when added internally. More recent reconstituted bilayer studies suggest that there are two 1,4-dihydropyridine-binding sites, and that the one more rapidly accessible from the cell exterior is of apparent higher affinity (Valdivia and Coronado,

1990). Studies (Kass et al., 1988; Kass and Arena, 1989) are consistent with the hypothesis that the amlodipine-binding site is near the external face of the ventricular sarcolemma.

3. Smooth and Cardiac Muscle and Endocrine Cells

Electrophysiological studies on excitable cells, excluding skeletal muscle, have demonstrated a very high affinity (0.1 to 1 nM) block of L channels by nitrendipine, nisoldipine, nimodipine and isradipine (Tables 8 and 9). The affinities of binding calculated from biophysical measurements agree with those obtained from ligand-binding studies. The influx of Ca^{2+} through L channels is the mediator of calcium-induced Ca^{2+} release from the SR. This released calcium then activates the contractile proteins of cardiac muscle (see Best et al., Section II, Chapter 6).

Studies from cardiac cells, which first demonstrated the 1000-fold change in binding affinity of these drugs that occurred after marked depolarization, suggested that drug binding was likely to be to the inactivated state of the calcium channel (Bean, 1984; Sanguinetti and Kass, 1984). In contrast, studies with an anterior pituitary cell line, in which little inactivation of the channels occurred under the conditions used, showed that most binding was to the open state of the calcium channel (Cohen and McCarthy, 1988).

It is well established that both L and N channels can mediate hormone secretion (Godfraind et al., 1986; Janis et al., 1987). Adrenal chromaffin cells appear to have 1,4-dihydropyridine-sensitive L channels as well as a channel insensitive to both 1,4-dihydropyridines and ω-CgTX (Rosario et al., 1988). Of special interest is the recent report that submicromolar concentrations of nitrendipine, applied to depolarized cells, block T channels and aldosterone secretion in adrenal glomerulosa cells, indicating that this channel type can mediate secretion (Cohen et al., 1988; Barrett et al., 1989; also see Hescheler et al., 1988, for results in a related cell line). Calcium channel blockers used clinically have generally been free from detrimental effects on endocrine cells (Schoen et al., 1988).

4. Neuronal Cells

Autoradiographic and microdissection studies have demonstrated that 1,4-dihydropyridine-binding sites are localized in brain regions rich in synaptic contacts such as the hippocampus, striatum and cerebral cortex (reviewed in Janis et al., 1987). It has been suggested (Gray et al., 1988) that ω-CgTX may interact with both N and L channels; both GVI and MVII are proposed to interact with L channels on nerve soma, and also to interact, but in a weak reversible way with T channels. Plummer et al. (1989) reported inhibition by ω-CgTX of N but not L channels in dorsal root ganglion (DRG) cells. Both GVI and MVII conotoxins may interact with channels controlling the release of inhibitory transmitters in several species of vertebrates, including man and frog, but channels in nerve terminals controlling the release of excitatory transmitters may be differentially sensitive to these toxins according to species (Yoshikami et al., 1989). The binding affinity of ω-CgTX is in the subnanomolar range (Table 7), but micromolar concentrations are generally used to block calcium currents because of the inhibitory effect of cations on toxin binding. Considerable work is further needed to define these toxin effects. Many studies have shown that ω-CgTX can partially inhibit calcium entry and neurotransmitter release from nerves and synaptosomes (e.g., Reynolds et al., 1986b; Hirning et al., 1988; Woodward et al., 1988; Thate and Meyer, 1989). Potassium-stimulated calcium uptake in astrocytes is inhibited by 3 nM nimodipine (Hertz et al., 1989), suggesting the presence of L channels in these cells.

5. Skeletal Muscle

Although the highest densities of binding sites for 1,4-dihydropyridines are in skeletal muscle (Table 7) there are no clinically important side effects of these drugs obviously

TABLE 9
Affinities of Drugs Binding to Different States of the L-Class of Calcium Channels

Drug	K_i ($\times 10^{-9}$ M)[a]	Channel state[b]	Cell Type[c]	Ref.
Nimodipine	1.3	I	Nerve, DRG	McCarthy, 1989
	40	R		
	3	I	DRG cell line	Boland and Dingledine, 1990
	2000	R	(F-11)	
	0.5	O	Endocrine, $GH_4 C_1$	Cohen and McCarthy, 1986a, 1987
	7000	R		
	0.02	O	A7r5	
	1000	R		McCarthy and Cohen, 1989
Nisoldipine	0.07	I	Mesenteric artery	Nelson and Worley, 1989
	3.0	R		
	1	I	Cardiac	Sanguinetti and Kass, 1984a
	1340	R		
Nitrendipine	0.36	I	Ventricle	Bean, 1984
	700	R		
	0.5	I	Artery	Bean et al., 1986
	222	R		
	0.25	I	Vein	Yatani et al., 1987
	100	R		
	108	R	Vein	Hermsmeyer et al., 1989
	7	I	Cerebral granule cells	Carboni et al., 1985
	2.5	I	Endocrine, $GH_4 C_1$	Enyeart et al., 1987
Nifedipine	0.1	I	AtT-20/D16-16	Luini et al., 1986; Lewis et al., 1986
Nicardipine	160	R	Aorta	Kuga et al., 1989
Flunarizine	90	R		Kuga et al., 1989
(+)-Isradipine	0.03	I	Vein	Dacquet et al., 1989
	0.15	R		
(+)-Isradipine	0.1	I	Cultured skeletal muscle myoballs	Cognard et al., 1986
	13	R		
	0.02	I	Portal vein	Loirand et al., 1989
	0.15	R		
Fluspirilene	0.15	R & O	Skeletal muscle myoballs	Galizzi et al., 1986b
	10	I	Endocrine, GH_3	King et al., 1988
HOE 166	0.25	I	Skeletal muscle	Qar et al., 1988
Pimozide	100		$GH_4 C_1$	Enyeart et al., 1987
Amiodarone	360	I	Ventricle	Nishimura et al., 1989
	5800	R		
Calmidazolium	500			
(+)-(R)-BAY K 8644	4	O	GH_3	McCarthy and Cohen, 1986
	>1000	R		

[a] Most values less than 1 nM are estimated k_i values calculated from most experiments using a larger drug concentration.

[b] I = inactivated, R = resting, O = open. In some studies L- and T-type channels or currents were defined only by their activation voltage and their duration.

[c] DRG = dorsal root ganglion (sensory) neuronal cells; GH_3, $GH_4 C_1$ and AtT-20/D16-16 = pituitary cell lines; A7r5 = clonal rat thoracic aortic muscle.

attributable to effects on voluntary muscle. This is because inward calcium currents are not immediately necessary for activation of this muscle and the action potential is too short to allow drug binding to the inactivated state of the voltage sensor. For a review of calcium blocker effects on excitation-contraction coupling and calcium current in skeletal muscle (see Best et al., Section II, Chapter 6).

6. Endothelial Cells

It has been proposed that the protective effect of nilvadipine and other calcium channel blockers in 1-α-hydroxyvitamin D_3 aortic calcium deposition is due to an action on endothelial cells (Mutoh et al., 1988). Previous studies demonstrated that nisoldipine protected against the increase in microvascular permeability caused by ischemia-reperfusion (McDonagh and Roberts, 1986).

There are contradictory reports indicating both inhibitory effects or no effect of calcium channel blockers on the release of EDRF (see Whitmer et al., 1988; Vanhoutte, 1988). The latter report suggests that aortic endothelial cells have only nonspecific, low affinity (K_d >50 nM) binding sites for isradipine, and that calcium channel blockers do not inhibit the influx of calcium (Whitmer et al., 1988).

7. Blood and Exocrine Cells

Most evidence indicates that platelets, erythrocytes and exocrine cells studied lack L channels (see Janis et al., 1987; Ogawa and Ono, 1989; see Section VIII.C). In addition, there is evidence for calcium-independent secretion in platelets, neutrophils, mast cells, and adrenal medullary cells (Knight et al., 1989). Treatment of patients with low doses of nifedipine, diltiazem and verapamil inhibits granulocyte activation during hemodialysis, suggesting the presence of L channels on these cells (Haag-Weber et al., 1988).

The effects of calcium channel blockers on exocrine cells have been previously reviewed (Godfraind et al., 1986; Janis et al., 1987). Pancreatic enzyme secreting cells, in contrast to endocrine cells (e.g., β cells, adrenal glomerulosa, thyroid C, and juxtaglomerular cells), appear to lack L channels. Herling and Ljungstrom (1988) concluded that the doses of verapamil needed to produce nonspecific inhibition of gastric acid secretion in normal rats were larger than those needed to produce serious cardiovascular depression. Both verapamil and diltiazem decrease cold-restraint, stress-induced gastric ulcers in rats, but worsen ethanol-induced ulcers (Glavin, 1989). Nitrendipine's potent inhibition of stress-induced ulcers was attributed to inhibitory effects on both stomach contractions and gastric acid secretion. The partial calcium activator, CGP 28392, produced modest preventive effects against both types of lesions, but did not effect gastric acid secretion (Glavin, 1988). In a related study, pretreatment of rats with either verapamil or diltiazem protected against both ethanol- and indomethacin-induced gastric lesions (Ghanayem et al., 1987). Clinical studies indicate that these drugs do not have side effects related to inhibition of gastric motility or acid secretion.

8. Other Cell Types

Several nonexcitable cells such as fibroblasts, glial, myeloma, and osteoblasts contain both L- and T-like channels (see Bean, 1989). Human epidermal cells (keratinocytes) also contain L channels (Reverdin et al., 1989). Interleukin-8-induced lymphocyte migration is potently (IC50 0.1 nM) inhibited by the blocking $(-)$-isomer of 202 791 but not the activator, $(+)$-202 791, suggesting the presence of L channels (Bacon et al., 1989). There are no established clinical effects of calcium channels blockers on these cell types. Although erythrocytes, platelets and mast cells appear to lack L channels, calcium channel blockers have nevertheless been reported to exert effects on these cells (see Janis et al., 1987, and Section VIII).

Osteoblast-like cells contain calcium channels sensitive to verapamil and very high concentrations of dihydropyridines (Guggino et al., 1988, 1989; Wagner et al., 1988; Yamaguchi et al., 1989). Bay K 8644 stimulates the secretion of the bone matrix protein, osteocalcin, from these cells, which enhances bone matrix production. Injection of Bay K 8644 into intact fetal rats results in enhanced bone resorption, perhaps due to the stimulation of the release of factors from osteoblasts. Stimulation of the slowly inactivating calcium channel current in osteoblastic cells suggests the presence of some type of L channels

(Grygorczyk et al., 1989). Effects of clinically used calcium channel blockers on osteoporosis are not known.

At subnanomolar concentrations 1,25-$(OH)_2$-vitamin D_3, the active form of the steroid, stimulates L channel current in resting osteosarcoma cells, but at concentrations of 5 to 10 nM, inhibition predominates (Caffrey and Farach-Carson, 1989). This vitamin D_3 metabolite and BAY K 8644 stimulate, to a similar extent, calcium uptake across the basal lateral membrane of the duodenum (de Boland et al., 1990).

IV. MODES AND SELECTIVITY OF ACTION

A. Voltage Dependence of L Channel Block Results in Selectivity of Action

The clinically used organic calcium channel blockers do not act by occupying the ion pore and physically blocking the channel, but rather act allosterically to make the channels unavailable for opening. In many cell types this may be accomplished by preferentially binding to the inactivated state of the channel. The voltage- and time-dependence of inhibition of L channels are primary mechanisms determining their selectivity of action. Initial studies by Bean (1984) and Sanguinetti and Kass (1984) demonstrated that the affinity of nisoldipine for depolarized cardiac cells was 1000-fold greater than to well polarized cells. Similar observations have also been made for several other calcium channel blockers (Table 9). At therapeutic concentrations of 1,4-dihydropyridines (approximately nanomolar for nitrendipine or nimodipine) depolarization to 0 mV of several seconds will be required for sufficient drug binding to occur to produce maximal potency for channel inhibition (McCarthy and Cohen, 1989). Amlodipine represents an even more extreme case. This drug, whether in its charged or neutral form, produced no tonic block at negative (-80 mV) membrane potentials (Kass et al., 1988).

Armstrong and Kalman (1990) have attributed the voltage-dependence of binding of dihydropyridines to GH_3 cells to a voltage-dependent dephosphorylation of the L channel. In these cells, both run-down (loss of current with time), and inactivation were reversed by cAMP-dependent phosphorylation. They suggest that BAY K 8644 acts, in part, by inhibiting the dephosphorylation of L channels or related proteins, whereas the blockers bind to the dephosphorylated, or inactivated state, and reduce the probability of its reopening. It is important to test this hypothesis by measuring phosphorylation directly, and to extend the observations to other cell types.

Another mechanism that may contribute to the selectivity of action of 1,4-dihydropyridines for vascular smooth muscle is the greater sensitivity of these cells relative to others (Bean et al., 1986; Yatani et al., 1987; McCarthy and Cohen, 1989). In addition, the L channel resting state of smooth muscle cells may exhibit a much greater affinity for these drugs than that of other cell types (Hering et al., 1988; Table 9). However, gallopamil does not show a significant resting state block of long lasting channels in ear artery (Hering et al., 1989).

Drug binding to either the open or inactivated state of the channel could mediate inhibition. The time required for this binding to occur is sufficiently long that cells which are depolarized very briefly, such as neurons during a normal action potential, would not be inhibited by these drugs. Similarly, cardiac action potentials are normally too short to allow significant binding of 1,4-dihydropyridines at the usual clinical doses, whereas some inhibition of calcium influx into cardiac cells may be produced by calcium channel blockers of the non-1,4-dihydropyridine type. In ischemic tissue, cell depolarization may occur subsequent to alterations in ion flux, pH, or to the release of K^+ from surrounding cells.

Thus, calcium channel blockers bind preferentially to cells which are markedly depolarized for several seconds. This prolonged depolarization is more likely to occur abnormally in ischemic cells. Verapamil and diltiazem, but not nifedipine, were more effective negative inotropic agents in the ischemic heart (Smith and Briscoe, 1985). Bay K 8644 produces contraction of middle cerebral arteries in the absence of depolarizing concentrations of K^+, whereas such depolarization was required for peripheral arteries (Fernández-Alfanso et al., 1988). These results suggest that, at least under the *in vitro* conditions used, the cerebral vessels were more depolarized.

As discussed previously, some of the early studies of the voltage-dependence of 1,4-dihydropyridine actions suggested that binding was to the inactivated state of the channel. In contrast to cardiac cells where high affinity binding is to the inactivated state, cultured anterior pituitary cells ($GH_4 C_1$), A10 and A7r5 clonal cells exhibit a high affinity binding to the open state (Cohen and McCarthy, 1987; McCarthy and Cohen, 1989). If so, then nimodipine binding in these types of cells is independent of channel inactivation.

B. Frequency-Dependent Block

A second mechanism by which selectivity of action is achieved is the frequency dependence of block exhibited by some of the non-1,4-dihydropyridine calcium channel antagonists, including verapamil and other members of the phenylalkylamine drugs. Inhibition of calcium current is obtained in heart cells only at a sufficient rate of channel opening. The effectiveness of these drugs for the treatment of certain types of arrhythmias may be accounted for by this mechanism. As discussed previously, most of the 1,4-dihydropyridines are uncharged at normal pH, and access to their receptor may be by a lipophilic pathway through the membrane bilayer (Rhodes et al., 1985; Herbette et al., 1989). In contrast, for charged 1,4-dihydropyridines such as amlodipine, and non-1,4-dihydropyridines such as verapamil (pKa 8.7), access to the receptor may occur, in part, through the open channel. Block of L channels by the charged form of amlodipine and verapamil (existing at physiological pH) is enhanced by prolongation of pulse duration during application of pulse trains. It remains to be determined if this reflects binding to open channels or increased access through open channels to the inactivated state of the channel (Kass et al., 1988). In electrically-stimulated guinea-pig hearts, hypoxia increases the potency of verapamil approximately 10-fold but does not change that of nifedipine (Robertson and Lumley, 1989). The distribution of the drug between cellular compartments is another mechanism that likely contributes to the pharmacological profile of these drugs; distribution will change as intracellular pH falls in ischemic or hypoxic conditions.

C. Other Mechanisms for Selectivity of Action

A variety of factors account for the selective effect of the clinically used calcium channel blockers. First, only those cellular functions that depend on extracellular Ca^{2+} entry through L channels are expected to be potently blocked by calcium channel blockers. Therefore, any process, such as skeletal muscle contraction or secretion, dependent only on intracellular Ca^{2+} release, will not be blocked by a specific inhibitor of L channels. Similarly, neurotransmitter secretion that is triggered by Ca^{2+} influx through N or P channels is not blocked by these drugs. Second, as mentioned above, even if the process is dependent on L channels, if the channel is not in the correct state, either open or inactivated, long enough for drug binding to occur, then the calcium channel blockers will be expected to be ineffective. In the case of verapamil and other drugs that exhibit a large frequency dependence, channel block is dependent on an appropriate stimulation rate. The ability of calcium channel blockers to enter either the cell membrane or pass the blood brain barrier may also be important for their selectivity of action.

McCarthy and Cohen (1989) have proposed that the lipid composition may be of major importance in determining the apparent greater affinity of nimodipine block in clonal smooth muscle cell lines relative to that in clonal anterior pituitary cells ($GH_4 C_1$). In the latter cells, the calculated affinity of binding to the open state was 30 times larger than that for A7r5 cells. Since the off-rate was only twice as fast, the major factor resulting in the lower affinity was a 15 times slower on-rate, rather than a difference in dissociation rate. It was proposed that varying lipid compositions of the plasmalemmas produced this difference. The lipid composition could result in either a greater partition coefficient of nimodipine into the membrane, or a greater rate of diffusion. Mason et al. (1990b) found that the partition coefficient of nimodipine into dioleoyl phosphatidylcholine membranes increased five-fold as the cholesterol content was decreased from 35 to 0 mol% of total phospholipid.

Biochemical analysis of the isolated 1,4-dihydropyridine binding proteins from various tissues indicates differences for each of the cell types compared (e.g., Hosey et al., 1989; Nunoki et al., 1989). In addition to the highly potent inhibitors of L channels shown in Table 9, there are a variety of substances with other pharmacological activities that block L channels with low affinity (Table 10). In many cases, inhibition by these drugs does not show the voltage-dependence observed by the more potent and selective calcium blockers.

D. Mechanism of Action of BAY K 8644 and Related Compounds

Calcium channel activators such as BAY K 8644 and (+)-(S)-202 791 have been of key importance in defining the L class of channels (Bechem and Schramm, 1988; Bechem et al., 1988, 1989). These drugs produce long openings that may outlast the repolarizing voltage-clamp step, and increase the probability for opening of calcium channels. This unmistakable activator response is the best test for the presence of L channels, and the lack of this effect is of key importance in detecting ω-CgTX-insensitive N channels (Plummer et al., 1989). Blocking agents produce less definitive information because of the problems associated with incomplete block and their marked voltage dependence of binding. In contrast, the long openings produced by activators can be seen with nearly equal probability over a large voltage range, e.g., from -90 mV and -30 mV (Plummer et al., 1989). Bay K 8644 does not increase the intrinsic channel conductance nor does it recruit new channels (Fox et al., 1987).

Hess et al. (1984) suggested that BAY K8644 promotes the formation of a mode of gating of L channels that exhibits long openings, and β-adrenergic stimulation of cardiac cells appears to produce a similar effect (Yu et al., 1990). Sanguinetti et al. (1986) proposed that BAY K 8644 produces a 10-fold decrease in the rate of deactivation, k_{-1}

$$R \underset{k_{-1}}{\overset{k_1}{\rightleftharpoons}} O \rightleftharpoons I$$

This model is believed to account for the long openings, the increase in the rate of inactivation, as well as the equal shifts in activation and inactivation to more negative potentials. From this model it can be seen that partial activation of the channel is required before agonists can exert an effect. In the model proposed by Bechem and Schramm (1988), Bay K 8644 binds only to the open state of the channel, and the prolonged open state is attributed to the time of drug dissociation. In 10-fold concentrations than used for effects on L channels, Bay K 8644 also increases ion flux through sodium channels by increasing both the frequency and duration of openings (Yatani et al., 1988b). Lacerda and Brown (1989) observed non-modal gating of cardiac calcium channels in the presence of dihydropyridines.

GTP-binding proteins mediate the inhibitory effect of neurotransmitters by at least two mechanisms, one involving adenylate cyclase activation, and the other a direct inhibition of L and N channels. For example, norepinephrine, $GABA_B$, DADLE, inhibit L or N channels in certain cell types without the involvement of a second messenger (Holz et al., 1986;

TABLE 10
Low Affinity Inhibitors of the L Class of Calcium Channels[a]

Drug	IC_{50} or conc. used ($\times 10^{-6} M$)	Approx. % inhibition	Cell Type[b]	Ref.
Arachidonic acid	<1	≈50	Atrial	Cohen et al., 1990
Aconitine	7	50	GH_4C_1	Enyeart et al., 1987
Amiloride analog[c]	5	100	GH_3	Garcia et al., 1990
Amiodarone	2	48	Ventricle	Nishimura et al., 1986
2-Chloroadenosine	50	69	DRG	Dolphin and Scott, 1986
Calmidazolium	0.5	50	GH_4C_1	Enyeart et al., 1987
Haloperidol	3	50	GH_4C_1	Enyeart et al., 1987
Veratridine	1	50	GH_4C_1	Enyeart et al., 1987 1986a,b
3',4-Dicholorobenzamil	0.8	50	Atrial	Bielefeld et al., 1986
Chlorpromazine	20—50	50	NIE-115	Ogata et al., 1986; Ogata and Narahashi, 1987
OAG	5	58	DRG	Hockenberger et al., 1990
	10	100	GH_3	Wolfe and Brostrom, 1986
Pimozide	1	100	Ventricle Myocytes	Enyeart et al., 1986
Pinaverium	1.5	50	Jejunum	Beech et al., 1990
Trifluoperazine	10	100	GH_3	Wolfe and Brostrom, 1986
	50	44	Helix	Bernal et al., 1987
Diazepam	100	30	NIE-115	Watabe et al., 1986
	60	58	Cerebral	Rampe and Triggle, 1987
Ro5-4864	3	25	Ventricle	Holck and Osterrieder, 1985
Gallopamil	3	50	Smooth muscle	LeBlanc and Hume, 1989
	5	50	Skeletal muscle	Affolter and Coronado, 1986
D890	3 (Cytostolic side)	50	(In planar bilayer in the presence of 3×10^{-6} M BAY K 8644)	Affolter and Coronado, 1986
	75 (Extracellular side)	50		
Verapamil	10	50	Skeletal muscle	Walsh et al., 1986
	100	35	Hippocampal	Yaari et al., 1987
Diltiazem	63	50	Skeletal muscle	Almers and McCleskey, 1984
Phenytoin	100	60	Hippocampal	Yaari et al., 1987
Halothane	450	34	Cardiac	Eskinder et al., 1990
	900	63	Purkinje	

[a] In most studies L- and T-type channels or currents were defined only by their activation and their duration.
[b] DRG = dorsal root ganglion cells; NIE-115 = a neuroblastoma cell line; PC 12 = pheochromocytoma cell line; GABA = γ-aminobutyric acid; studies with synaptosomes measure K^+-induced ^{45}Ca uptake. [c]N^5-(ethyl, isopropyl) amiloride; data are for 30 min exposures.

Hescheler et al., 1987). Recently, Scott and Dolphin (1987; 1989; Dolphin and Scott, 1987, 1989) found that in the presence of GTPγS, 5 μM nifedipine, 10 μM gallopamil, or 30 μM diltiazem potentiated L channel current in DRG cells. They proposed that this was due to the stabilization of a resting state, presumably one which could be easily activated by these drugs (but see Hess, 1990). Several other studies have also shown stimulatory effects of phenylalkylamines on calcium channels (Frank, 1984; Katzka and Morad, 1989; Mcdonald et al., 1989).

E. Drugs Acting on T Channels

Voltage-dependent block of calcium channels by blocking agents is a property not only of L but also of T channels. McCarthy and Cohen (1986, 1987) first reported a relatively

TABLE 11
Inhibitors of the T-Class of Calcium Channels[a]

Drug	Conc. ($\times 10^{-6}$ M)	Percent inhibition	Cell Type	Ref.
Felodipine	0.013	50	Atrial	Cohen et al., 1989
	0.68	50	GH$_3$	
Nicardipine	0.048	50	Aorta	Kuga et al., 1989
	1.1	50	Hippocampal CA1	Takahashi et al., 1989
Nimodipine	0.2	50	A7r5	McCarthy and Cohen, 1989
	0.6	50	GH$_3$	
	0.3	50	DRG cell line (F11)	Boland and Dingledine, 1990
	7.0	50	Hypothalamic neuron	Akaike et al., 1989
				Kostyuk et al., 1989
Isradipine	0.045	50	Portal vein	Loirand et al., 1988
Tetramethrin	0.1	100	SA node	Hagiwara et al., 1988
Flunarizine	0.019	50	Aorta	Kuga et al., 1989
	0.7	50	Hypothalamic neuron	Akaike et al., 1989
	5	80	Cardiac-ventricle	Tytgat et al., 1989
(+) BAY K 8644	0.14	50	GH$_3$	McCarthy and Cohen, 1986
Nitrendipine	0.17	50	Adrenal glomerulosa	Cohen et al., 1988
Nisoldipine	1	50	Cardiac Purkinje	Tseng and Boyden, 1989
Cinnarizine	0.9	50	Atrial	Van Skiver et al., 1988
Octanol, nonanol	<1	+	Inferior-olivary	Llinás and Yarom, 1986
Amiodarone	3	50	Atrial	Van Skiver et al., 1988
Quinidine	3	50	Atrial	Van Skiver et al., 1988
Gallopamil	3.5	50	Hypothalamic neuron	Akaike et al., 1989
Nicardipine	3.5	50	Hypothalamic	Akaike et al., 1989
Nifedipine	7	50		Akaike et al., 1989
Methoxyverapamil	50	50		Akaike et al., 1989
Diltiazem	70	50		Akaike et al., 1989
Arachidonic acid	<5	50	Atrial	Cohen et al., 1990
Amiloride analog[b]	5	100	GH$_3$	Garcia et al., 1990
Amiloride	30	50	Neuroblastoma	Tang et al., 1988
OAG	40	60	GH$_3$	Manchetti and Brown, 1988
Clonazepam	100	50	NIE-115	Watabe et al., 1986
Phenytoin	100	35—80	NIE-115	Twombly and Narahashi, 1986
Ethosuximide	1000	10	Cultured sensory neurons	Gross et al., 1989
	500	35		
Dimethadione	5000	50	Thalamic	Coulter et al., 1989
Ethanol	>100	+	Inferior-olivary	Llinas and Yarom, 1986
Halothane	450	33	Cardiac	Eskinder et al., 1990
	900	56	Purkinje	

[a] Most drugs also inhibit L channels at lower concentrations than they inhibit T channels; putative T-channel selective antagonists, include amiloride, tetramethrin, octanol (but see Hess, 1990), ethosuximide, dimethadione and tetramethrin. Inhibition by OAG (1-oleoyl-2-acetyl-*sn*-glycerol) may be indirect via protein kinase C, or direct (Hockberger et al., 1989).

[b] N^5-(ethyl, isopropyl)amiloride, data for a 30 min exposure time.

potent block of T channels by 1,4-dihydropyridines (Table 11). The potency of felodipine for blocking the inactivated state of the L channels is not much greater than that for blocking T channels over part of the physiologically relevant voltage range (between -70 to -50 mV; Cohen et al., 1988). The most potent blockers of the T channel are felodipine and its congeners which are likely to bind to the inactivated state of T channels in atrial cells. Tissue selectivity was also found, felodipine had dissociation constants of 13 nM and 680 nM for atrial and GH$_3$ cells, respectively.

Hess (1990) has suggested that the inhibition of T channels by nitrendipine in adrenal cortical cells might actually be due to a contribution of L channels to the slowly deactivating tail current which was used to estimate T channel block by Cohen et al. (1988). If so, then this is a BAY K 8644-insensitive L channel current, and such currents have not yet been reported. In any case, there are a large number of reports of the block of T channels by this class of drugs (Table 11).

A variety of other drugs also block T channels (Figure 6, Table 11), and some of these, such as amiodarone and cinnarizine (Cohen et al., 1989), block atrial T channels without markedly inhibiting L channels under certain conditions. Amiodarone could exert at least part of its effect on supraventricular tachycardia by blocking T channels. More recently, the flunarizine-like drug, KB-2796, was reported to inhibit more potently T than L channels (N. Akaike, unpublished, quoted in Handa et al., 1990). Other drugs that have been reported to have some selectivity for T relative to L channels include amiloride, nonanol, ethosuximide and dimethadione (Table 11). It has been proposed that amiloride is a relatively selective drug for the inhibition of T channels (Tang et al., 1988). However, recent studies by Velly et al. (1988) demonstrate that amiloride blocks sodium channels, and Garcia et al. (1990) have demonstrated block of both L and T channels by amiloride and related drugs. Amiloride displaces [^{3}H]tetracaine binding with an IC_{50} value for 3 μM in heart, which is 10 times lower than the corresponding value for blocking T channels. Further studies are required to compare the relative potency of amiloride on sodium and T calcium channels. To date most of the reports of selectivity of action of T channel inhibitors have not been generally confirmed. It is necessary that putative selective T channel blockers be studied in several laboratories in different types of cells before it can be stated that generally useful selective drugs for this channel type are available. At this time there is little evidence that block of T channels contributes to the antianginal or antihypertensive effects of felodipine or other calcium channel blockers.

V. EFFECTS OF CALCIUM CHANNEL LIGANDS IN ANIMAL MODELS

A. Cardiovascular and Renal Systems

Various aspects of the effects of calcium channel blockers on the cardiovascular system of animals have been recently reviewed (Godfraind et al., 1986; Janis et al., 1987; Nayler, 1988). The clinical effects of these drugs, as well as some of the effects obtained in animal models, are covered in several chapters in this volume, and the individual drugs have been the subject of several recent reviews and symposia (Table 12). Recent reviews, symposia, and books covering the actions of calcium channel blockers in various animal models and disease states are listed in Table 13.

1. Hypertension, Renal Effects and Myocardial Hypertrophy

The major mechanism responsible for the antihypertensive effects of these drugs appears to be the lowering of total peripheral resistance by vasodilation in several major vascular beds (Janis et al., 1987; Nayler, 1988; Scriabine and Kazda, 1989). Other beneficial effects, such as a decrease in vascular lesions, are also seen in experimental animals (Kazda, 1988; Fleckenstein et al., 1989; 1990).

An important effect of calcium channel blockers in the treatment of hypertension is the prevention and regression of cardiac hypertrophy. Both animal (e.g., Garthoff et al., 1983; Kazda et al., 1987) and clinical (Frishman et al., 1989; Messerli et al., 1989) studies have shown regression of left ventricular hypertrophy.

Calcium channel blockers produce diuresis by at least two mechanisms, by decreasing

TABLE 12
Recent Reviews on Individual Drugs and Extensive General Reviews[a] on Calcium Channel Blockers

Individual Drugs	
Amlodipine	Schwartz and van Zwieten, 1989; Burkart, 1989, Burgess and Dodd, 1990
Bepridil	Flaim and Cummings, 1986; Hasegawa, 1988
CD-349	Tsuchida et al., 1990
Diltiazem	Chaffman and Brogden, 1985
Felodipine	Saltiel et al., 1988
Fendiline	Bayer and Mannhold, 1987
Flunarizine	Todd and Benfield, 1989
Isradipine	Sauter et al., 1989
Nicardipine	Sorkin and Clisold, 1987; Singh and Josephson, 1990
Nifedipine	Lichtlen, 1986
Nimodipine	Langley and Sorkin, 1989; Scriabine et al., 1989; Traber and Gispen, 1989
Nisoldipine	Knorr, 1987; Friedel and Sorkin, 1988; Lichtlen and Hugenholtz, 1988
Nitrendipine	Goa and Sorkin, 1987
Verapamil	McTavish and Sorkin, 1989

Books and Extensive General Reviews
Godfraind et al., 1986; Janis et al., 1987; Triggle and Venter, 1987; Morad et al., 1988; Nayler, 1988; Vanhoutte et al., 1988; Wray et al., 1989a

[a] See Table 12 in Janis et al., 1987 for list of reviews on other Ca^{2+} antagonists.

preglomerular more than postglomerular resistance, and by a direct effect on tubular Na^+ reabsorption (Kauker et al., 1987; Johns and Mantius, 1987; Zimmerman and Raich, 1988; Loutzenhiser, 1989; Loutzenhiser et al., 1989a,b). Further studies are needed to determine the importance of renal effects to the overall chronic antihypertensive effects of these drugs (Kazda et al., 1988). Independent of the diuresis, nisoldipine and related drugs have protective effects in animal models of acute renal failure (Garthoff et al., 1987).

2. Angina and Effects on the Ischemic Myocardium

Calcium channel blockers relieve the pain of vasospastic angina by coronary vasodilation, whereas reductions in afterload (decreased arterial pressure), inotropy and chronotropy may be important in the treatment of effort angina. A reduction of oxygen demand due to decreased afterload, and an increase in oxygen supply, due to coronary vasodilation, are of major importance for the 1,4-dihydropyridines. For non-1,4-dihydropyridines that have a direct effect on the myocardium, such as diltiazem and verapamil, slowing of heart rate and decreased contractility are of major importance. Diltiazem tends to have more effect on heart rate, and verapamil, more effect on contractility. Opie (1988b) has noted that many of the putative direct anti-ischemic effects of these drugs occur only at higher concentrations than those achieved clinically.

In general, calcium channel blockers have not been useful in the treatment of acute myocardial infarction, but they are beneficial in procedures such as PTCA (percutaneous transluminal coronary artery angioplasty) when ischemia is followed by reperfusion after therapeutic treatment (Hugtenburg, 1989). Studies in animal models have shown that calcium channel blockers improve blood flow to ischemic areas, and improve mechanical function when given before, but not necessarily when given after the period of ischemia. The mechanisms responsible for the beneficial effects of calcium channel antagonists in experimental ischemia depend on the choice of the model.

3. Reperfusion

There are a variety of potential mechanisms by which these drugs may prevent ''no

TABLE 13
Recent Books and Reviews on the Effects of Calcium Blockers in Animal Models and Human Disease

Cardiovascular	
Atherosclerosis	Nayler et al., 1988; Henry, 1988; Nayler, 1988; Weinstein, 1988; Jackson et al., 1989; Kjeldsen and Stender, 1989; Triggle, 1989; Fleckenstein et al., 1990 Akhtar et al., 1989
Arrhythmias	
Hypertension, and cardiac hypertrophy	Kaplan, 1989; Frishman et al., 1989; Scriabine and Kazda, 1989; Messerli et al., 1989; Fleckenstein et al., 1990
Ischemic, heart disease	Hugtenberg, 1989; O'Rouke, 1989
Renal effects	Giebisch et al., 1987; Zimmerman and Raich, 1988; Epstein and Loutzenheiser, 1990
Central Nervous System	
Age-associated memory and Alzheimer's disease	Bergener and Reisberg, 1989
Cerebral ischemia and spreading depression	Hossman, 1988; Siesjo and Bengtsson, 1989
Epilepsy	Binnie, 1989; Meyer, 1989; Macdonald and Meldrum, 1989
Migraine	Amery, 1989
Subarachnoid hemorrhage (associated neurological deficits)	Pickard et al., 1989
Other	
Obstructive airway diseases	Howard, 1989
Ulcers	Hertz and Cloarec, 1989

reflow phenomenon'', the failure of tissues to reperfuse following temporary ischemia. These include dilatation of constricted coronary vessels, which is the likely major mode of action for the 1,4-dihydropyridines. The prevention of myocardial contracture, edema, endothelial cell swelling, and obstruction of the microvessels by formed elements are also likely contributing factors. The dramatic prevention of ischemic damage by low concentrations (1 nM) of nisoldipine in isolated hearts subjected to ischemia and reperfusion is likely due to coronary vasodilation and is not associated with changes in ischemic contracture (Watts et al., 1987; 1990).

4. Antiatherogenic Effects

Many, but not all animal studies have shown at least some antiatherogenic effect of calcium channel antagonists. Mechanisms by which these drugs could exert beneficial effects in experimental atherosclerosis include effects on cell proliferation (Nara et al., 1987) and migration (Nomoto et al., 1989), effects on matrix synthesis (Heider et al., 1987), cholesterol ester hydrolysis (Etingen and Hajjar, 1990), endothelial-dependent relaxation (Kappagoda and Thomson, 1989), prevention of Ca^{2+} overload (Fleckenstein et al., 1989; 1990) and effects on platelets (see also Berini et al., 1989; Sections III and VIII). There is an increased uptake of Ca^{2+} by aorta in experimental atherosclerosis (Strickberger et al., 1988). Nisoldipine lowers serum lipids in a rabbit model of atherosclerosis (Kappagoda and Thompson, 1989). Enhanced cholesterol ester hydrolysis and decreased total cholesterol accumulation in human aortic tissue from patients taking nifedipine or diltiazem is associated with elevated cyclic AMP levels (Etingin and Hajjar, 1990).

B. Effects of Calcium Channel Ligands on Neurons and the Brain

Potential channel nervous system (CNS) uses for nimodipine, flunarizine, and related calcium channel antagonists are listed in Tables 2 and 13. Future drug development in this area will be aimed at the development of drugs that are specific for CNS disorders and are free from any detrimental cardiovascular effects (Scriabine et al., 1989).

1. Lack of Behavioral Effects of Calcium Channel Blockers Used to Treat Cardiovascular Diseases

The major calcium channel blockers used clinically are relatively free of CNS side effects (Bem et al., 1988). These results are consistent with a variety of *in vitro* experiments demonstrating little or no effect of low concentrations of many calcium channel blockers on neurotransmitter release (Miller, 1987; Barnes and Davies, 1988; Yu et al., 1988), and with many electrophysiological studies indicating a low sensitivity of neuronal calcium current to these drugs (Bean, 1989). Some of the reason for this general lack of effect are considered below and others were previously discussed (Section IV). Nifedipine, the clinically most used 1,4-dihydropyridine, exhibits a lower partition coefficient into the brain than does nimodipine (Scriabine et al., 1989); the latter drug is used to prevent or treat neurological deficits following subarachnoid hemorrhage (Pickard et al., 1989).

The lack of behavioral effects of 1,4-dihydropyridines would not be predicted from binding studies because the affinities of [^{3}H]1,4-dihydropyridines for isolated membranes from brain are the same or very similar to those of smooth and cardiac membranes, and the density of sites is higher than in blood vessels (Table 7). There are insufficient electrophysiological data to determine the similarities between high threshold channels in central and peripheral neurons. A few studies have demonstrated that high K^+-induced neurotransmitter release can be blocked by calcium channel blockers, but the relevance of these studies to normal physiological release has been questioned (Bean, 1989).

2. Mechanisms for Selectivity for 1,4-Dihydropyridines

The general lack of effect of calcium channel blockers on neurotransmitter release is due to a variety of factors. A major factor is that the duration of the nerve action potential is too short for drug binding to occur to the open or inactivated state. The affinity of the resting state of the L channel, which normally predominates in well-polarized cells is too low (Table 9) for binding to occur (Holz et al., 1988; McCarthy, 1989; Dingledine and Boland, 1990). Other factors include the small contribution of L channels to total calcium influx in many neurons (Plummer et al., 1989) and perhaps their possible location away from release sites (Hirning et al., 1988).

The apparent insensitivity of high threshold neuronal calcium channels to nifedipine and other dihydropyridines in earlier studies is likely also due to the presence of other classes of channels, such as N and P channels, that are insensitive to 1,4-dihydropyridines. In some types of cells, N channels appear to have a dominant role in secretion (Reynolds et al., 1986b; Obaid et al., 1989; Plummer et al., 1989). Sustained current elicited from depolarizing pulses is made up of other classes of channels (Plummer et al., 1989) making the estimations of affinities for 1,4-dihydropyridines difficult.

Calcium channel ligands have been valuable in defining the different classes of channels in neurons, particularly for distinguishing between L- and N-type channels, since both are high threshold, but the latter are not sensitive to calcium channel activators or blockers. As in the case of most other excitable cells, the T channels in neurons have generally been reported to have little or no sensitivity to 1,4-dihydropyridines (Fox et al., 1987, Bean, 1989). However, the most potent of the many drugs that inhibit T channels are dihydropyridines and flunarizine (Table 11).

In cells in which blockers are inactive, Bay K 8644 effects on neurotransmitter release may be due to a flood of calcium that spreads to active zones in neurons (Miller, 1987). Nimodipine's block of Bay K 8644-induced release of norepinephrine from cultured sensory and sympathetic neurons, but not K^+-induced release, is consistent with this hypothesis

(Perney et al., 1986). In cultured spinal cord and dorsal root ganglion neurons, Bay K 8644 increases calcium current but does not augment electrically-evoked release of neurotransmitter, indicating that L channels are probably not involved in this action potential-dependent transmitter release (Yu et al., 1988). Other recent studies have also concluded that the calcium channels required for secretion from nerve terminals are ω-CgTX sensitive, but relatively insensitive to 1,4-dihydropyridines (Obaid et al., 1989).

Another factor that might contribute to the insensitivity of neurotransmitter release to these drugs is the possible existence of endogenous substances that inhibit 1,4-dihydropyridine binding but do not exert effects on calcium channels. Hess and co-workers (1984) proposed the existence of a mode of gating called mode 1 which might be stabilized by substances to exert this type of action. Drugs which stabilize this mode would block 1,4-dihydropyridine effects without modifying channel function. For example, certain basic proteins and lipids, inhibit 1,4-dihydropyridine binding but not calcium channel current (Janis et al., 1988b; Spedding, 1989). If related proteins had access to the appropriate binding site *in situ*, then they could inhibit the effect of 1,4-dihydropyridines without themselves exerting any other effect. It might be expected that Bay K 8644 effects would also be blocked by such putative peptides, but clearly BAY K 8644 does exert central effects. These results suggest that either there are no such endogenous inhibitors of 1,4-dihydropyridine-type calcium channel blockers, or that 1,4-dihydropyridine activators can act at different sites than do the blockers.

Some of these studies could also be confounded by the fact that the same ligand can produce both inhibitory and stimulatory effects depending on the conditions used. For example, Scott and Dolphin (1989) found that internal perfusion of DRG cells with GTPγS converted 1,4-dihydropyridine blocker effects to activator effects. However, these results were not confirmed by Plummer et al. (1989).

Calcium channel blockers have direct behavioral effects in certain animal studies, for example, those involving seizures and drug withdrawal. One hypothesis that might explain the effects seen is that the latter conditions mimic either the effect of Bay K 8644, producing long openings of calcium channels, or that they mimic the effect of high K^+ concentrations, facilitating or generating an inactivated state of the channel (see Janis et al., 1987).

Hakim et al. (1989) have shown that [^{3}H]nimodipine accumulates in areas surrounding the experimental infarct. This likely represents the selective high affinity binding of the drug to ischemic, and therefore, depolarized neurons, consistent with the known voltage dependence of calcium channel blockers drug binding.

3. Nimodipine can have Direct and Potent Effects on Neuronal L Channels

In contrast to the lack of neuronal effects of calcium channel blockers in clinical use for cardiovascular disorders, calcium activators such as Bay K 8644 produce easily demonstrable behavioral effects that are associated with changes in neurotransmitter release (e.g., O'Neill and Bolger et al., 1989; Bourson et al., 1989). If the activators and blockers bind to the same site in neurons, then the large number of binding sites for the calcium channel blockers that are seen in those areas of the brain rich in synapses appear to be functionally coupled to calcium channels. Further support of this view comes from the behavioral and cellular effects of calcium channel blockers in a variety of animal models (see Tables 22 and 24 in Janis et al., 1987).

Inhibition of the effects of Bay K 8644 by other 1,4-dihydropyridines suggests that this activator may act at the same channels as the blockers. The general lack of observed pharmacologic or therapeutic effects of blockers is to be expected if these drugs bind with high affinity only to the open or inactivated state of the channel, since, as discussed above, this state would be expected to be of too short a duration during a neuronal action potential to permit drug binding. In some *in vitro* studies where long holding potentials at depolarizing levels were used to maximize inhibition, the amount of block obtained using nifedipine on nerve L channels was still less than that obtained on muscle (Fox et al., 1987; Rane et al., 1987; Bean, 1989).

Rane et al. (1987) found that the inhibitory effects of nifedipine on DRG cells were markedly time- and voltage-dependent. Subsequently, Fry and McCarthy (1988) demonstrated that calcium channel blockers can have very potent effects on neurons that are sufficiently depolarized. Nimodipine blocked L channels in freshly dissociated DRG neurons at nanomolar concentrations (Table 9). These latter results demonstrate for the first time that this drug can have direct neuronal effects at concentrations that match its affinity observed in radioligand binding studies. Comparative studies with nifedipine have not been reported; it is known that the volume distribution of nimodipine in the brain is higher than that of nifedipine (van den Kerchoff and Drewes, 1989), and its affinity for L channels is also higher (Janis et al., 1987). In contrast to the binding of blockers, that of Bay K 8644 is relatively independent of membrane potential (Section II; Ferrante et al., 1989), and only a small amount of depolarization is required for its effects on channel function (Rane et al., 1988).

Studies on peripheral nerves have shown that 1,4-dihydropyridines inhibit K^+-but not electrically induced neurotransmitter release (Holz et al., 1988). Similarly, Herdon and Nahorski (1989) have shown that isradipine inhibited K^+-induced (25 mM), but not spontaneous or electrically evoked dopamine release from rat striatal slices. These results presumably reflect the need for prolonged depolarization for isradipine binding to occur. Electrically induced release is not affected because the time of depolarization is too short to allow drug binding to the inactivated state, or this type of stimulation produces a greater increase in N channel opening which is insensitive to these drugs.

C. Treatment of CNS Disorders

The roles of excess calcium and excitatory amino acids in cell injury are discussed by Rosenberg (Section III, Chapter 16) and the antimanic effects of verapamil in affectively disordered patients, by Dose and Emrich (Section III, Chapter 17).

1. Subarachnoid Hemorrhage

Nimodipine is the only drug approved in the U.S. for the treatment of neurological deficits associated with vasospasm in patients who have had a recent subarachnoid hemorrhage. Its site of action in this disorder may be either vascular or nonvascular (Scriabine et al., 1989; Pickard et al., 1989). Nimodipine produces an increase in the perfusion of ischemic cerebral tissue without increasing intracranial pressure, producing edema, or producing significant effects on the peripheral circulation (Langley and Sorkin, 1989; Scriabine et al., 1989). The protective effects of both nicardipine (Hadani et al., 1988) and nimodipine (Uematsu et al., 1989) against ischemic brain damage have been ascribed to direct blockade of calcium influx. Both the increase in cytosolic calcium and the histologic damage following focal cerebral ischemia and reperfusion in cats was attenuated by nimodipine (Uematsu et al., 1989). A variety of studies suggest that the neuroprotective effect of calcium channel blockers is not dependent on detectable increases in blood flow (e.g., Meyer et al., 1986, 1988; Nuglisch et al., 1989; Tegtmeier et al., 1989). Conflicting results have been obtained using animal models of focal and global ischemia with regard to improved neurological outcome or the prevention of edema following treatment with various calcium antagonists (Fleischer et al., 1987; Germano et al., 1987; Alps et al., 1988; Nishikibe, 1988; Hadley et al., 1989; Sauter et al., 1989; Siesjö, 1989; for the results of a recent symposium, see Krieglstein, 1989). Recent studies indicate that there are potent vasodilatory effects of some of these drugs (ED_{50}, nimodipine, 1.6 nM and nifedipine, 8.6 nM) but not others (ED_{50}, verapamil, 0.11 μM, diltiazem, 1.5 μM) on intracerebral penetrating arteries that are not angiographically visible (Takayasu et al., 1988).

2. Seizures

Calcium channel blockers can prevent some types of seizures in animals and man (Meyer, 1988; Dolin et al., 1988; Bingmann and Speckmann, 1989). As previously discussed, effects on both T (Macdonald and Meldrum, 1989) and L channels (Janis et al., 1987; Binnie, 1989) are likely mechanisms, depending on the drug used. Ischemia may produce neuronal hyperexcitability which can lead to seizures. Reperfusion seizures following temporary focal ischemia are blocked by nimodipine (Meyer et al., 1988). Excess calcium influx through either L channels or NMDA-operated channels may be an initiating event for seizure activity (Meyer, 1989). Supporting this view are results demonstrating that nisoldipine is more potent than MK 801 in preventing NMDA-induced seizures in mice (Palmer et al., 1990). Nanomolar concentrations of nimodipine protect cultured fetal serotonergic neurons against NMDA- and MDMA (3,4-methlenedioxymethamphetamine)-induced toxicity, suggesting that closing the L channel allows the neuron to more easily regulate pathological increases in intracellular Ca^{2+}, independent of the mechanisms that mediated the increase (Azmita, 1989).

3. Learning and Improvement of Age-Related Behavioral Deficits

Chronic treatment with nimodipine can improve several aspects of the neuronal functioning of aging and brain-damaged rats, including sensorimotor control. It also increases learning of certain tasks, and prevents some aspects of age-associated degeneration, including the decrease in sciatic nerve density (Gispen et al., 1988; Schuurman and Traber, 1989; Traber and Gispen, 1989; Scriabine et al., 1989). Several double-blind studies have shown that nimodipine may be effective in the treatment of cognition deficits in elderly patients (Schmage et al., 1989). Of particular interest is the demonstration that nimodipine dramatically facilitates associative learning in aging rabbits (Deyo et al., 1989). The mechanism producing this effect remains to be established, but it is known that nimodipine can increase the activity of hippocampal pyramidal neurons in aging rabbits at a dose that facilitates learning (Thompson et al., 1990).

Landfield (1989) has hypothesized that there may be an abnormal increase in intracellular calcium in the aging brain. This may lead to both a greater feedback inactivation of transmitter-releasing processes, and an accumulation of damage due to calcium overload. Results of a voltage-clamp study of intact hippocampal neurons in aged rat (Reynolds and Carlen, 1989) agree with this hypothesis. Age-related increases in L channel current in rat hippocampal slices have also been observed (Campbell et al., 1989). It is possible that elevated levels of intracellular calcium may result in the observed apparent down-regulation of L channels in the hippocampus and forebrain of these rats (Landfield et al., 1989), as well as that observed in the senescence accelerated mouse, SAM-P/8 (Kitamura et al., 1989).

4. Withdrawal Syndromes and Toxicity of Addictive Drugs

Certain calcium channel blockers are effective in animal studies in the treatment of toxicity caused by psychostimulants (see Janis et al., 1987). For example, chronic morphine treatment increases the number of 1,4-dihydropyridine binding sites in rat cortex (Section VII), and naloxone enhances K^+-evoked norepinephrine release from cortical slices taken from these rats (Pellegrini-Giampietro et al., 1988). Nimodipine reverses this effect of naloxone, but does not effect the release of this neurotransmitter in slices from naive rats. Nitrendipine and flunarizine were effective in the treatment of cocaine intoxication of rats (Nahas et al., 1989).

5. Other Effects

Nimodipine increased vessel growth in the pial microvasculature of rats (Hutchins et al., 1989), and induced an increased vascularization and graft volume following injection into the dopamine-depleted striatum of rats of embryonic ventral mesencephalic cells (Finger, 1989). In addition, nimodipine promoted neuronal growth (Nyakas et al., 1989).

VI. ENDOGENOUS SUBSTANCES ACTING ON CALCIUM CHANNEL LIGAND RECEPTORS

Since voltage-dependent calcium channels may be regarded as pharmacologic receptors, it is reasonable to ask whether endogenous ligands exist that are genetically determined to act at the calcium channel blocker-binding sites. The arguments for the existence of 1,4-dihydropyridine-like regulators of L channels are similar to those made for the existence of an endogenous digitalis-like substance. In that case, it has been argued that the hypothesis for the existence of such substances rests on unproven assumptions. The existence of endogenous ligands for a relatively ubiquitous protein that acts both as its receptor and effector mechanism at the cell membrane, thereby directly affecting transmembrane ion flux would be unprecedented for a hormone or autocoid (Kelly and Smith, 1989). A recent report, however, suggests that an endogenous ligand for the digitalis receptor has been isolated (Hamlyn et al., 1989).

A. Fractions that Inhibit Dihydropyridine Binding and Calcium Influx

Preliminary reports on the isolation of endogenous ligands for calcium channel drug-binding sites have appeared but no substance has yet been definitively purified, or chemically identified (Janis et al., 1988a; Triggle, 1988). None of the fractions isolated to date have been reported to be competitive inhibitors at the 1,4-dihydropyridine or verapamil binding sites. An initial report suggesting that endothelin directly activates L channels was not confirmed (e.g., D'Orleans-Juste et al., 1989; Kasuya et al., 1989; Mitsuhashi et al., 1989). At this time it is not possible to state definitively that an endogenous ligand exists for a calcium channel drug-binding site. There is, however, evidence that GTP-binding proteins can activate L and N channels, and that vitamin D metabolites can modulate L channels.

All of the 1,4-dihydropyridine-displacing fractions that have been isolated differ in their characteristics from 1,4-dihydropyridines and verapamil in some key feature. For example, the low molecular weight, protease-sensitive substances isolated by Hanbauer and co-workers (Callewaert et al., 1989) and Ebersole et al. (1988) either do not modify or are poor inhibitors of 1,4-dihydropyridine binding to rat heart membranes, in contrast to the other drugs known to directly bind to brain L channels. The substance isolated by Hanbauer et al. inhibits 1,4-dihydropyridine binding noncompetitively, and inhibits both the L and T classes of channels. The substance isolated by Janis et al. (1988a), inhibits 1,4-dihydropyridine binding with a Hill slope greater than one. None of the known drugs (Figure 4) bind to sites allosterically coupled to the 1,4-dihydropyridine receptor have a steep Hill slope. Thus, if this substance is an endogenous regulator of calcium channels, it acts either on multiple binding sites or on an as yet unknown site.

B. Endogenous Substances that Inhibit Dihydropyridine Binding

A number of endogenous substances inhibit 1,4-dihydropyridine binding, but, with one possible exception, these do not include any of the known neurotransmitters or peptide hormones (Janis et al., 1987; 1988a,b). Basic peptides such as myelin basic protein, protamine, and histones inhibit 1,4-dihydropyridine binding in micromolar concentrations, but those that have been studied electrophysiologically were inactive on L channel current, even when the peptides were added into the recording pipette so that they had access to the inside of the cell (Janis et al., 1988b). There has been one report that substance P modifies the binding of D888 and nitrendipine to hippocampal membranes (Govani et al., 1988).

The existence of a large number of endogenous substances that inhibit 1,4-dihydropyridine binding, but are unlikely to be endogenous regulators, makes the search for true modulators of L channels particularly difficult (Janis et al., 1988b). In addition to the lipids and basic proteins mentioned above, divalent cations, including Cu^{2+}, Ni^{2+}, Cd^{2+}, Fe^{2+}, which are both endogenous, and/or are commonly found as contaminants of chromatography systems, inhibit 1,4-dihydropyridine binding. Some of these, such as Cu^{2+} and Zn^{2+} are also potent inhibitors of calcium channel current. Thus, it is essential that inhibition produced by all of these sources, including organic complexes of the above cations, is carefully considered before any claims are made for endogenous ligands.

The fact that many basic peptides, lipids, and divalent cations inhibit 1,4-dihydropyridine binding means that this binding assay can only be used for a relatively preliminary screen of isolated fractions. Furthermore, an accurate accounting of activity from crude fractions will not be possible until these substances are removed.

Long chain fatty acids and their acyl choline and carnitine derivatives inhibit 1,4-dihydropyridine binding at approximately 10^{-5} to 10^{-4} M (Janis et al., 1988b). They also inhibit L channels, but at least some of this effect is via protein kinase C activation (Linden and Routtenberg, 1989). Arachidonic and linoleic acid also block L channels in a manner that is similar to that of the clinical used drugs: (1) the rate of current decay during a test pulse is increased and this effect increases with the degree of channel activation; (2) there is a similar voltage-dependence of block, and an activator effect is sometimes seen when the fatty acid is first applied (Cohen et al., 1990). However, arachidonic and linoleic acid are approximately equally potent modulators of L and T channels, as well as sodium and potassium channels in a variety of cell types (Chen et al., 1990; Katz et al., 1990; Leibowitz et al., 1990). Spedding and Mir (1987) found that palmitoyl carnitine inhibited the binding of 1,4-dihydropyridines, diltiazem, and verapamil ($IC_{50} \simeq 0.1$ μM) and also produced smooth muscle contraction. However, at even lower concentrations than those used, palmitoyl carnitine is known to dissolve membranes. This makes it unlikely that this lipid is a physiological regulator of calcium channels; however, it has a likely pathological function.

Intact tissue assays will not be specific for assaying crude fractions for specific direct effects on calcium channels because there are so many possible sites at which an endogenous substance might act at high concentrations to produce a change in calcium flux or muscle contractions (Janis et al., 1987). For example, Wolf et al., (1989) found that a 10 μM micellar solution of lysophosphatidylcholine (mainly palmitic and stearic) produced a dose-dependent relaxation of rabbit aorta which was associated with increases in cGMP. The relaxation was not due to a solubilization of the cell membranes because repeated additions of this lipid produced an identical degree of relaxation, and normal relaxation to acetylcholine was still seen, demonstrating the integrity of the endothelium. Menon and co-workers (1988) suggest that a 10 μM concentration of lysophosphatidylcholine has antihypertensive and coronary vasodilator effects without causing arrhythmias or hemolysis. It would be interesting to determine whether this lipid produces inhibition of calcium channels since we have found that this and related lipids inhibit 1,4-dihydropyridine binding in the 10 to 100 μM range (Janis et al., 1988b, and unpublished studies).

C. Fractions and Substances that Effect Calcium Current

A putative 700 to 900 Da peptide fraction, HF ("hypertensive factor") isolated from rat erythrocytes produced increases in calcium channel current in frog ventricular myocytes

(Simmons et al., 1989). It shifts the current-voltage and inactivation relationships in the hyperpolarizing direction, similar to the effects of BAY K 8644. It was estimated by amino acid analysis that K^+- and norepinephrine-induced contractions of rat aorta were enhanced by 10 µM of HF. Further studies on the defined substance are required to determine its significance. Many endogenous compounds, including angiotensin II and thrombin, modulate the activity of calcium channels but most of these substances are thought to act indirectly on the channel protein (Janis et al., 1987).

Although they are not endogenous ligands for known drug-binding sites on L channels, several antibodies have effects on channel function. IgG antibodies against calcium channels cause the symptoms of Lambert-Eaton myasthenic syndrome (Vincent et al., 1989). The antibodies produce inhibition of presynaptic calcium channel effects at the neuromuscular junction and thereby reduce the evoked quantal release of acetylcholine. These antibodies inhibit L, but not T channels, in other cell types, e.g., certain neuronal and endocrine cell lines but not muscle cells. However, the channel type inhibited in Lambert-Eaton syndrome is probably not L, but only related to L channels, because the organic calcium channel blockers do not effect K^+-induced acetylcholine release in this disease (Wray et al., 1989b). Recent studies suggest that these antibodies, like other anti-receptor antibodies, cause down-regulation, in this case of ω-CgTX binding sites in a neuroblastoma cell line, IMR-32 (Sher et al., 1989).

Polyclonal, monospecific mouse antibodies to highly purified $α_1$-subunit of the skeletal muscle L channel modulate either inhibition or stimulation of parathyroid hormone (Fitzpatrick and Chin, 1988). Monoclonal antibodies to the the α-, γ- and β-subunits have been reported to modulate calcium channel current in reconstituted planar lipid bilayers, the first two causing inhibition, and the latter, stimulation (Campbell et al., 1988; Vilven et al., 1988).

VII. CHRONIC REGULATION OF CALCIUM CHANNELS

A. L Channels as Pharmacological Receptors

The regulation of the numbers and function of receptors for hormones and neurotransmitters during growth and development, during homologous and heterologous drug influence and in disease states is well known. The voltage-dependent calcium channel may, in a number of important respect, also be considered as a pharmacologic receptor. In particular, both activator and blocker ligands exist with defined binding sites described by discrete structure-activity relationships. It is likely that ion channels and drug receptors are regulated in similar fashion and that channel expression and function are subject to homologous and heterologous regulation. However, there may be significant differences in mechanisms of regulation since receptors for hormones and neurotransmitters have defined physiologic substrates, whereas the existence of endogenous ligands for calcium channels remains to be established (Section VI). In any event the tissue of channel regulation has been analyzed primarily for the L class of channel, though it will be surprising if the other channel classes do not show similar patterns of regulation. Potential consequences of therapeutic importance to drug regulation of calcium channels include the development of tolerance and of withdrawal phenomena (Raftery, 1984).

B. Homologous Regulation

Although there are few objective reports of withdrawal phenomena following cessation of therapy (Triggle and Janis, 1987), several experimental studies document both up- and down-regulation of ligand binding site numbers following chronic drug administration (Table

TABLE 14
Effects of Chronic Drug Treatment, Lesions, and Disease on [^{3}H]1,4-Dihydropyridine Binding

System	K_D	B_{max}	Ref.
Homologous Regulation			
28-day oral treatment of mice-nifedipine or verapamil in brain	nc	↓ 48%	Panza et al., 1985
20-day i.v. treatment of rats with nifedipine	nc (brain) (heart)	↓ 23% ↓ 49%	Gengo et al., 1988
5-day treatment of PC12 cells with nifedipine or (S)-BAY K 8644	nc	↑ 29% (nif) ↓ 24% (BAY K)	Skattebol et al., 1989
Heterologous Regulation			
Atropine (23-day) or DFP (14-day): rat heart and brain	nc	nc	Skattebol et al., 1989
Morphine-tolerant mice:brain	nc	↑ 60	Ramkumar and El-Fakahany, 1988
Thyroid hormone:rat heart, T4 excess,	nc	↓ 42%	Hawthorne et al., 1988
T4 deficient	nc	↑ 26%	
Alcohol tolerance:rat brain	nc	↑ 50%	Dolin et al., 1987
Lead treatment:rat brain	nc	↑ 48%	Rius et al., 1988
Lesions			
6-OH-Dopamine:rat heart	nc	↑ 31%	Skattebol and Triggle, 1986
Kainic acid:rat brain	nc	↓ 43%	Skattebol et al., 1988
Denervated:rat skeletal muscle	nc	↑ 200%	Schmid et al., 1984
Disease States			
Hypertension:SHR:heart	↑ 55%	↑ 43%	Chatelain et al., 1984
DOCA-NaCl rat:heart	nc	nc	Ishii et al., 1983
brain	nc	↑ 57%	
Cardiomyopathy:human heart	nc	↑ 162%	Finkel et al., 1988
Ischemia:rat brain	64%	↑ 62%	Magnoni et al., 1988
Muscular dysgenesis	nc	↓	Pincon-Raymond, et al., 1985
Malignant hypothermia	nc	↓ 48%	Ervasti et al., 1988
Parkinson's disease: human brain	nc	↓ 49%	Nishino et al., 1986

Note: nc = no change.

14). Not all studies have also measured calcium channel function. A more comprehensive review of channel regulation is available (Ferrante and Triggle, 1990).

Both up- and down-regulation of channels have been described following chronic administration of channel ligands *in vivo* and *in vitro* (Panza et al., 1985; Garthoff and Bellemann, 1987; Gengo et al., 1988; Skattebol et al., 1989). A study in PC12 cells revealed up- and down-regulation of both channel numbers and function following nifedipine or S-Bay K 8644, respectively (Skattebol et al., 1989), thus paralleling many studies of receptor regulation by chronic blocker or agonist administration, respectively. Differences between *in vivo* and *in vitro* studies may arise from the reflex cardiovascular effects activated *in vivo* (Gengo et al., 1988).

C. Heterologous Regulation

Few studies document co-regulation of calcium channels and associated modulatory receptors. Thus, chronic administration of atropine to rats, DFP (diisopropylflurophosphate), propranolol or isoproterenol produces the anticipated up- and down-regulation on the receptor systems, but is without effect on channel numbers (Gengo et al., 1988). However, treatment of cultured chick embryonic ventricular cells with isoproterenol for 4 h reduced β-receptors, isradipine binding sites, and BAY K 8644-stimulated calcium influx (Marsh, 1989). These effects were likely due to the elevation of cAMP and its protein kinase.

Several lesions including 6-hydroxydopamine and kainic acid simultaneously altered both receptor and channel site numbers (Table 14). These findings suggest roles for neuronal influences on channel-ligand binding sites other than, or additional to, tonic activity. Trophic influences or endogenous ligand activity may be involved. It is of interest that thyroid hyper- or hypoactivity is associated with reciprocal changes in cardiac β-adrenoceptor and calcium channel numbers (Hawthorn et al., 1988), and that cardiac tissue from humans treated chronically with calcium blockers contains β-adrenoceptor density (Hedberg et al., 1985) consistent with co-regulation.

Both morphine- and alcohol-tolerant animals show up-regulation of brain calcium channels (Table 14; Ohnishi et al., 1989), and in both instances calcium channel blockers reduce the withdrawal symptoms from the precipitating drug (Dolin et al., 1987; Ramkumar and El-Fakahany, 1988). Concurrent treatments with ethanol and nitrendipine prevented both the development of tolerance to the ataxic action of ethanol and the increased number of 1,4-dihydropyridine binding sites, suggesting that the change in the number of sites is involved in the adaptations that occur in response to chronic ethanol (Dolin and Little, 1989). As is the case for several other receptors that are altered in animals treated with ethanol, changes in the number of 1,4-dihydropyridine binding sites were not seen in the cortex from human alcoholics (Kril et al., 1989). Verapamil prevented the development of myocardial depression and preserved normal energy metabolism in hearts of hamsters drinking 50% ethanol (Garrett et al., 1987). Memory-enhancing drugs of the pyrrolidinone series, piracetam and oxiracetam, elevate the density of 1,4-dihydropyridine binding sites in rat brain cortex (Soldatov et al., 1990). In contrast, diltiazem, riodipine and verapamil produced only a transient decrease in 1,4-dihydropyridine receptors.

D. Alteration in Disease

Calcium channels appear to be altered in a number of disease states, although the relationship of these changes to the etiology of the disease has not, in every case, been established. Thus, some, but not all, studies indicate an increase in cardiac calcium channel numbers in spontaneously hypertensive rat heart (Ishii et al., 1983; Chatelain et al., 1984; Ferrante and Triggle, 1990). A decrease of binding sites has been reported for the brain stem of DOCA hypertensive rats (Lee et al., 1985), but increases have been reported in cerebral membranes (Huguet et al., 1987) and in certain areas of the nucleus tractus solitarius of the SHR (Krukoff and Scott, 1984). The densities of 1,4-dihydropyridine and ω-CgTX binding sites were unchanged in human frontal cortex from dementia patients (Colvin et al., 1989; 1990), and were unchanged in various brain areas in samples from Alzheimer's patients (Quirion and Nair, 1989). A 75% reduction in isradipine binding sites was found in autopsy striatal tissue from patients that had Huntington's disease, and a decrease was seen in samples from Parkinson's disease specimens in one study (Nishino et al., 1986; but see Greenberg et al., 1988; Quirion and Nair, 1989).

Calcium channel numbers in skeletal and cardiac muscle and brain are reported to be increased in the Syrian hamster model of human cardiomyopathy (Wagner et al., 1986; 1989), and increases have been documented for human cardiac tissue (Finkel et al., 1988; Wagner et al., 1989; Table 14). However, some studies have not shown increases in binding sites in the hamster model (Howelett et al., 1988; Bazan et al., 1987), and electron probe microanalysis of cardiomyopathic hamster hearts does not support the hypothesis of generalized Ca^{2+} overload (Bond et al., 1989). Depolarization of chick myotubes was reported to trigger the appearance of isradipine binding sites (Pauwels et al., 1987).

Although the 1,4-dihydropyridine binding sites in skeletal muscle may serve as charge sensors as well as calcium channels Flockerzi et al., 1986), they appear to be altered in at least two additional disease states. In embryonic muscular dysgenesis of mice a lethal gene impedes normal skeletal muscle development and this is accompanied by dramatic decreases in numbers of 1,4-dihydropyridine binding sites. Microinjection of an expression plasmid carrying complementary DNA encoding the 1,4-dihydropyridine receptor into dysgenic muscle cells restored both excitation-contraction coupling and 1,4-dihydropyridine sensitive current (Beam et al., 1988). In malignant hyperthermic pigs a significant decrease in the number of 1,4-dihydropyridine binding sites may be associated with the defective calcium release/sequestration processes of the sarcoplasmic reticulum (Ervasi et al., 1989).

Current studies on the regulation of calcium channels are in a preliminary stage. It is clear that voltage-dependent channels of at least the L class are regulated by a number of procedures including chronic drug administration, lesions and disease states (Ferrante and Triggle, 1990). It is also clear that channels may be down-regulated in both number and function by chronic depolarization (DeLorme et al, 1988). The signals for this down-regulation may be membrane potential and/or Ca^{2+} suggesting that the calcium channel, in common with other pharmacologic receptors, is a plastic species that is likely subject to constant regulation by the physiological state and demands of the system.

VIII. ACTIONS AT OTHER SITES

As emphasized above, the major therapeutic sites of action of the calcium channel blockers acting on the cardiovascular system are the L channels of vascular smooth muscle, and for the non-1,4-dihydropyridines, both vascular smooth muscle and cardiac cells. These channels are also likely to be the most important and perhaps the only site of action of most of the current used calcium channel blockers acting on the brain. The therapeutic importance of other sites of action of these drugs, such as those on renal tubular and endothelial cells, remains to be established. Nimodipine appears to have a relatively high affinity for the nucleoside transporter, and therefore some of its vasodilator and antiseizure activity might be due to this effect.

For non-1,4-dihydropyridines, particularly drugs related to flunarizine, cinnarizine, and bepridil, other sites of actions in or on neurons and smooth muscle are likely, but these remain to be defined. Flunarizine, which is approved for clinical use in some countries outside the U.S., is likely to have important blocking effects on serotonin receptors. The inability of BAY K 8644 to reverse the inhibitory effect of these drugs on smooth muscle, and its inability to reverse the effects of flunarizine, lidoflazine, and bepridil on cardiac inotropy, suggests that intracellular sites of action, in addition to the L channel, are involved in the effects of these non-1,4-dihydropyridines (Spedding, 1985; Boddeke et al., 1988).

A variety of sites of action have been identified for verapamil and diltiazem at the high concentrations that are often used in animal and *in vitro* studies. These include several neurotransmitter receptors (for an extensive listing of these other sites of action, see Tables 32 to 35 in Janis et al., 1987).

Although the 1,4-dihydropyridines have partition coefficients in bilayers of 5 to 10,000, their abundance in membranes relative to lipid is not high. For example, with a concentration equal to the K_D (0.4 nM) of nimodipine in the cytosol, there will be only one molecule of drug per million lipid molecules in the cardiac sarcolemmal membrane (Herbette et al., 1989). However, more lipophilic non-1,4-dihydropyridines, that are used at a hundred or thousand times higher concentration may achieve, on average, millimolar concentrations in the bilayer, even without considering the higher concentrations that may be achieved in specific loci or planes within the membrane. Given these high concentrations, and the well-known local anesthetic and anticalmodulin effects of these drugs, it is not surprising that a variety of effects of these drugs have been reported (Janis et al., 1987).

A. Effects on Other Receptors and Channels

The most potent effects of 1,4-dihydropyridines on other receptors are the reported stimulation of [^{3}H]baclofen binding by (+)-202 791 (ED_{50} 7.6 nM), and the inhibition of the binding of this $GABA_B$ ligand by nifedipine ($IC_{50} \simeq$ 17 nM). These may be allosteric effects since the changes in binding are reflected mainly in changes in B_{max} (Al-Dahan and Thalmann, 1989).

Those calcium channel antagonists that are neutral at physiological pH generally have much less effect at other sites than do those that may be partially charged drugs. Thus, verapamil and diltiazem, as well as nicardipine, have effects on other receptors in the 0.1 to 10 μM range. Nayler (1988) considers plasma levels of free verapamil to be in the 20 to 80 nM range, indicating that other neurotransmitter receptors will not significantly bind verapamil in concentrations that are used clinically. The mechanism by which high concentrations of verapamil enhance the formation of certain types of experimental lesions remain to be determined (Espluges et al., 1988). The protection against calcium-mediated cell death afforded by flunarizine in neuronal cultures was attributed to inhibition of sodium channels (Pauwels et al., 1989).

The sodium channel is a major site of action of 1,4-dihydropyridines at concentrations about 10 times higher than those needed to half maximally inhibit L channels. The binding site on sodium channels exhibits stereoselectivity, and is stimulated by BAY K 8644 (Yatani et al., 1988). Considering the homologies between sodium and calcium channels (Tanabe et al., 1987), it is not too surprising that 1,4-dihydropyridines may activate and inhibit sodium channels at somewhat higher concentration than they act on calcium channels. These results suggest that the marked homology in amino acid sequence reflects a similarity in the binding sites for these drugs.

B. Nucleoside Transporter

Other sites of action of 1,4-dihydropyridines include the nucleoside transporter (Hu et al., 1987), for which nimodipine has been reported to have an affinity as high as 20 nM (see Table 35 in Janis et al., 1987). It is not known if this represents a significant site of action of nimodipine, but it is interesting that nimodipine is unique among the 1,4-dihydropyridines studied in exhibiting a relatively high affinity for this site. This drug, unlike some calcium channel blockers, such as diltiazem, exhibits a particularly good antiseizure effect, and adenosine also has anti-ischemic activity. Plasma levels of free nimodipine are, however, less than the 10 nM range, suggesting that a clinical effect on the nucleoside transporter is unlikely. The most active isomer at this site in erythrocytes is the less active isomer on the L channel (see Glossmann et al., 1989).

Recent studies suggest that gallopamil is more effective than other calcium channel blockers in the treatment of antigen-induced bronchoconstrictor responses in allergic sheep (D'Brot et al., 1989). The mechanisms involved may be prevention of the release, or inhibition of the effect of histamine. Given the general lack of promising clinical results with calcium channel blockers in the treatment of allergic disorders, the effects observed with gallopamil may not be on the L channel.

C. Effects on Platelets

There is controversy about the presence of L channels in platelets and whether the calcium channel blockers exert any effects on platelets in its clinical use (Janis et al., 1987). In a recent study, no devapamil binding sites were detected on human platelets, and no evidence for antiaggregatory effects of these drugs in reasonable concentrations were detected in a variety of studies (Pannocchia et al., 1987).

Clinical studies indicate that ADP-induced platelet aggregation *ex vivo* was reduced in platelets from patients taking nitrendipine (e.g., Fritschka et al., 1987; Kribben et al., 1987). Extremely large concentrations of nitrendipine, nifedipine and diltiazem will inhibit human platelet function *ex vivo* (Blache et al., 1987; Moriyama et al., 1988). Some 1,4-dihydropyridines that exhibit antithrombotic activity are essentially devoid of cardiovascular effects (Sunkel et al., 1988). Consistent with the view that the effects observed on platelets are independent of L channels is the observation that Bay K 8644 ($K_i \approx 1.5\ \mu M$) inhibits platelet activation by competitive antagonism at the thromboxane A_2-prostaglandin receptor (Johnson et al., 1988).

Diltiazem added to human platelets *in vitro* in therapeutic concentrations (50 to 200 ng/ml), inhibited platelet aggregation caused by epinephrine (Mehta et al., 1986). In another study, 100 ng/ml diltiazem potentiated the effect of aspirin on aggregation caused by arachidonate plus platelet activating factor (Altaman et al., 1988). Similarly, nifedipine in oral doses may slightly reduce collagen-induced aggregation (e.g., Walley et al., 1989). The mechanisms involved in these inhibitory effects are unknown.

D. Effects on Drug Metabolism

Verapamil and diltiazem, but not nifedipine, inhibit the hepatic metabolism or renal clearance of several drugs (for review, see Hunt et al., 1989; Kirch et al., 1990). For example, diltiazem has been shown to elevate carbamazepine levels by 40 to 50% in some patients. Verapamil may increase digoxin levels by 60 to 80%, predominantly by decreasing renal clearance. Verapamil and diltiazem, but not 1,4-dihydropyridines, have depressant effects on the hepatic mixed function oxidase system by direct competition at cytochrome P450 (Rocci et al., 1989). This may be of clinical importance when these drugs are used in combination with those having a narrow therapeutic index, such as lidocaine, quinidine and warfarin, and in patients with decreased drug metabolism.

E. Other Effects of Calcium Channel Antagonists

Calcium channel blockers may exert effects that are dependent on relatively nonspecific interactions with the cell membrane. For example, 1 μM flunarizine prevented particle and membrane aggregation, and blebbing caused by elevated calcium concentrations acting on erythrocyte membranes (Thomas et al., 1988). The contribution of this type of effect to the pharmacology and toxicology of flunarizine remains to be established. Nifedipine treatment reduces the mean osmotic fragility of erythrocytes from uremic patients (Shasha et al., 1988). The mechanism for this effect may be related to lowered intracellular Ca^{2+} levels, but this effect is not mediated by L channels because erythrocytes, like platelets, lack this type of channel.

Another low affinity site of calcium channel blockers at micromolar concentrations in calmodulin and certain calmodulin-dependent enzymes (see Janis et al., 1987). The structure-activity relationships for 1,4-dihydropyridines inhibition of nitrendipine binding are clearly different from those for inhibition of calmodulin-dependent phosphodiesterase (van Inwegen et al., 1984; Schaeffer et al., 1988). The relative lack of stereoselectivity of the isomers of bepridil as vasodilators is consistent with inhibition at these nonspecific sites (Winslow et al., 1989). The very low affinity (0.6 μM) binding site for 1,4-dihydropyridines found on mitochondrial membranes (Zernig and Glossmann, 1988; Glossmann et al., 1989) is unlikely to bind any currently used 1,4-dihydropyridines at their therapeutic doses. Several reports of low affinity binding sites for 1,4-dihydropyridines have been published (Brush et al., 1987; Kunze et al., 1987). Low affinity sites for BAY K 8644 were removed with further purification of cardiac sarcolemma (Rampe et al., 1989a).

Flunarizine and cinnarizine cause extrapyramidal signs (Cappela et al., 1988; Lugaresi et al., 1988). Dopamine-induced antagonism has been attributed to decreased release of norepinephrine mediated by presynaptic dopaminergic receptors. The results suggest a possible mechanism (antagonism of dopamine at central synapses) for the tendency of flunarizine to cause Parkinsonism, and an explanation as to why 1,4-dihydropyridines are unlikely to have this side effect (Scriabine and Pan, 1988).

In the micromolar range, several calcium channel blockers inhibit *in vitro* production of superoxide anion by polymorphonuclear leukocytes (PMNs) by inhibiting the activation and activity of PMN NADPH oxidoreductase (Zimmerman et al., 1989). The concentrations needed (e.g., IC_{50} for lidoflazine is 7 μM) are likely to be at least 10 times higher than the free plasma levels obtained clinically. This effect is unlikely to contribute to the protective effect of these drugs in ischemia-reperfusion resuscitation. Since PMN activation is independent of Ca^{2+} fluxes, a site of action other than L channels is suggested (Zimmerman et al., 1989). The mechanism by which calcium channel blockers produce gingival overgrowth is not known; incubation of fibroblasts in the presence of verapamil reduced protein and collagen synthesis (Pernu et al., 1989).

F. Effects on Drug-Resistant Cells

There is a large and rapidly growing literature on the reversal by calcium channel blockers of multidrug resistance (Baeyens, 1988; Endicott and Ling, 1989). Several studies have demonstrated that calcium blockers bind to the 160 kDa P-glycoprotein responsible for outward drug transport when it is overexpressed, and thereby, inhibits the development of this form of multidrug resistance. The amount of this peptide increases dramatically in certain resistant cells which results in augmented efflux of anticancer drugs from the cell (Naito and Tsuruo, 1989).

This mechanism may also apply to antimalarial drugs, since verapamil has been shown to reverse chloroquine resistance (Martin et al., 1987). Structurally unrelated calcium channel blockers, such as tetrandrine, are also effective against resistant *Plasmodium falciparum* (Ye and Van Dyke, 1989).

The effects of calcium channel blockers on resistant cancer cells exhibiting P-glycoprotein overexpression are independent of L channels. Bay K 8644, like various calcium channel blockers and certain anticancer drugs, inhibit P-glycoprotein photolabeling by photoactive analogs of verapamil (for review, see Glossmann et al., 1989). A variety of other drugs that reverse multidrug resistance, many of which exhibit some calcium channel blocking activity, also inhibit the photoaffinity labeling of P-glycoprotein by a vinblastine analog (Akiyama et al., 1988).

There have been few clinical studies reporting beneficial effects of combination calcium channel blockers and anticancer drugs (Baeyens, 1988). Concentrations of verapamil greater than micromolar are generally needed to exert effects on anticancer drug accumulation, and these levels are not achieved in patients (Gruber et al., 1989) without impairment of cardiac function manifested as arrhythmia and hypotension (Dalton et al., 1989). The proposal that different classes of chemotherapeutic agents (i.e., vinca alkaloids, cisplatin, etc.) require different classes of calcium channel blockers for optimal enhancement of antitumor efficacy (Onoda et al., 1988; 1989) requires clinical testing. Nifedipine, but not nicardipine or nimodipine, enhances the antitumor effect of cisplatin in B16a-Pt tumor cells, providing further evidence that the structural requirements of this effect are unrelated to those for blocking L channels (Onoda et al., 1989). Because calcium antagonism is not required for the modulation of multidrug resistance, further drug development is needed to design agents that block the drug transporter without exerting effects on the cardiovascular or nervous systems (Zamora et al., 1988; Ford et al., 1989; Nogae et al., 1989).

G. Other Effects of BAY K 8644

One of the best tests for the presence of 1,4-dihydropyridine-sensitive L channels is the stimulation of calcium influx by Bay K 8644 (Janis et al., 1987). This activator is free from the requirement for a large degree of depolarization that is necessary for high affinity binding of blockers to L channels.

An interesting effect that appears to be unrelated to L channels is BAY K 8644s impairment of post-rest tension development in the heart. Bay K 8644 (and its activating isomer at 0.1 μM), like ryanodine, appears to promote the loss of Ca^{2+} from the SR during diastole (Bouchard et al., 1989; Hryshko et al., 1989). Neither nitrendipine, nifedipine, nor (+)-Bay K 8644 reversed this phenomenon, suggesting that it is unrelated to the activating activity of Bay K 8644 (Saha et al., 1989).

Bay K 8644 also increases the frequency of spontaneous and evoked neurotransmitter release from the neuromuscular junction in the absence of extracellular calcium and detectable depolarization (Atchinson, 1989; Pancrazio et al., 1989). These results indicate that Bay K 8644 can release intracellular calcium from the presynaptic nerve terminal.

IX. CONCLUSIONS AND SUMMARY

There are several classes of calcium channels but only one of these, the L class, exhibits an extremely high sensitivity to the clinically available calcium channel blockers. Selectivity of these drugs on the cardiovascular system is derived from several factors. The presence of particularly high affinity binding sites that are preferentially available at a time when the drugs will be most effective, i.e., during excessive vascular smooth muscle contraction or cell depolarization, or during cardiac arrhythmia, is of key importance.

Much is known about the beneficial effects of these drugs in preventing the cellular damage normally associated with calcium overload. However, many aspects of the mechanism of action of these drugs remain controversial, and there remains a need for development of drugs more effective in preventing cardiac and cerebral ischemia, infarction, and end-organ damage. Fleckenstein and co-workers (1990) have emphasized the potential effectiveness of calcium channel blockers in combination with ACE inhibitors for the prevention of calcium-overload-induced hypertension and arteriosclerosis.

As further importance is placed on the prevention of end-organ damage in the treatment of hypertension, new drugs may be discovered that produce regression of atherosclerosis, left ventricular hypertrophy and medial hypertrophy in resistance vessels, as well as preventing loss of arterial compliance. Studies on drugs that prevent organ damage independent of blood pressure lowering may provide clues to the development of new types of calcium channel blockers. Positive inotropic agents should be developed that will act on cardiac tissue without causing atrioventricular block or detrimental effects on vascular, neural, or endocrine cells.

Although the number of calcium channel blockers already studied is in the thousands (Bossert and Vater, 1989), there remains a need for the development of more selective drugs acting at both central and peripheral sites. For example, drugs that selectively dilate renal afferent arteries may be particularly useful antihypertensive agents. If the kidney is an important site of action of those calcium channel blockers used to treat hypertension, such renal selective drugs may lower blood pressure without causing side effects due to cerebral vasodilation or ankle edema. Additionally, there is a need to develop drugs that produce selective effects on central neurons, because of their selective binding to those sites, or their preferential accumulation in the brain, or both.

Many of the same drugs that inhibit the L class of channels may also be inhibitors of the T class of channels, albeit at somewhat higher concentrations. However, under hyperpolarizing conditions, certain calcium channel blockers are much less potent on L channels, and, at least in some types of cells, relatively more effective on T channels. It remains to be demonstrated whether additional therapeutic agents can be developed that act selectively on the T class of channels. Further studies are needed with analogs of the antiepileptics, ethosuximide and dimethadione, to determine if T channel block is predictive of efficacy.

T channel inhibitors may be useful not only in CNS disorders such as epilepsy, but also in the treatment of cardiac arrhythmia. Other channel types that have been recently described, but for which no known directly acting therapeutic agents have been developed include N and P channels. The former may be inhibited by ω-conotoxins, and the latter by certain spider toxins. The development of potent and selective drugs for ω-CgTX-insensitive N and the putative P channels is of major importance.

The core structure of various types of calcium channel blockers may be useful for the design of new drugs as adjunct therapy for the treatment multidrug resistant cancers and parasitic diseases. Similarly, there is a potential for new calmodulin and phosphodiesterase inhibitors based on these structures, although the site of action of these drugs may not be related to L channels.

Other poorly developed areas of calcium modulator pharmacology are: (1) drugs acting on G proteins, which regulate not only L and N channels, but also both IP_3-dependent and independent calcium release channels in smooth muscle, as well as other effectors, including adenyl cyclase and L channels (Robishaw and Foster, 1989; Ewald et al., 1989); (2) drugs blocking the synthesis and binding of IP_3. The possibility that some of the neuronal effects of drugs such as nimodipine are unrelated to effects on L channels also requires further investigation; (3) drugs acting by up- and down-regulation of calcium channels.

The accumulated evidence indicates that there is an enormous opportunity at this time for the development of new drugs acting on calcium channels. Structural and immunological studies demonstrate that the L channels are different in skeletal, smooth and cardiac muscle, and these types are different from the predominant type studied to date in brain. Increasing evidence for several types of calcium channels, and tissue specific subtypes, highlight the clear opportunity that exists for the design of more selective blockers for each channel and tissue type. It is expected that detailed information of the three-dimensional structure of the binding sites for calcium channel blockers in various tissues will lead to the development of more selective drugs than are currently available.

The isolation and characterization of new types of calcium channels will greatly expand the scope of drug development aimed at calcium channels. Current work on the genetic manipulation of ion channels will lead to new insights into their structure-function relationships, and that information will be valuable in studies on the pharmacological regulation of channel activity. Identification of the key residues involved in drug binding, as well as the mechanism by which drugs produce activator or blocking effects will be facilitated by studies utilizing the expression of mutant genes.

ACKNOWLEDGMENTS

We thank J. Chisholm, R. J. Fanelli, R. T. McCarthy, A. Scriabine, and J. A. Watts for their helpful suggestions on an earlier draft of this manuscript.

Chapter 14

Modulation of Calcium Channels by Neurotransmitters, Hormones and Second Messengers

Aaron P. Fox, Lane D. Hirning, David J. Mogul, Cristina R. Artalejo, Nicholas J. Penington, Reese S. Scroggs, and Richard J. Miller

TABLE OF CONTENTS

I. Introduction 252
    A. Mechanism of Calcium Channel Modulation 253
    B. Different Types of Voltage-Dependent Calcium Channels 253

II. Modulation of Calcium Channels 256
    A. Epinephrine Up-Modulates Cardiac L-Type Calcium Channel Activity 257
    B. What is the Site of Protein Kinase A Phosphorylation in Heart? 257
    C. Other Cardiac Responses to Protein Kinase A Activation 258
    D. Other Tissues Respond to β-Adrenergic Stimulation 258
    E. Protein Kinase C Also Modulates Calcium Channels 259
    F. Neurotransmitters and GTP-Binding Proteins 261

III. Summary 263

Acknowledgments 263

I. INTRODUCTION

Ca^{2+} ions entering cells via multiple types of voltage-dependent calcium channels and propelled by a steep electrochemical gradient, play a fundamental role both as charge carriers and as essential reactions in many cellular processes. For instance, Ca^{2+} ions are involved in synaptic transmission, muscle contraction, secretion, neurite extension, gene regulation, and regulation of Ca^{2+}-dependent ion channels (see various chapters in Section II). The few examples listed above illustrate the dual role that calcium channels play within cells. They are capable of responding to and influencing a cell's electrical properties; in many cells calcium channels in combination with sodium channels generate the cellular action potentials. Calcium channels also act as transducers carrying messages into cells. By opening or closing in response to the cell membrane potential, the calcium channels help regulate, on a millisecond time scale, the influx of Ca^{2+}, one of the best characterized second messengers, into cells. Intracellular Ca^{2+} activates other ion channels such as Ca^{2+}-activated potassium, chloride, or nonselective channels, leading to complex electrical behaviors. It is also possible that certain Ca^{2+}-dependent enzymes, such as Ca^{2+}/calmodulin kinases or protein kinase C, are governed directly or indirectly by calcium channels. Elevated intracellular Ca^{2+} levels may also play a role in various pathologic processes such as excitotoxicity produced by stroke and perhaps epilepsy (Chapter 22, Section IV). Indeed, the recent introduction of Ca^{2+} sensitive fluorescent dyes like Fura-2 and Quin-2 has shown that the intracellular Ca^{2+} signals may be surprisingly complex in nature exhibiting multiple oscillations and spatial heterogeneities (Hagiwara and Byerly, 1981, 1983; Pallotta et al., 1981; Yellen, 1982; Reuter, 1983, 1985; Hille, 1984; Nishizuka, 1984, 1986; Carafoli et al., 1985; Tsien, 1986; Augustine et al., 1987; Tsien et al., 1988; Berridge and Galione, 1988; Smith and Augustine, 1988; Bean, 1989a,b).

One of the most remarkable properties of calcium channels is the plasticity of these proteins, manifested as modulation by neurotransmitters, hormones and drugs that, in turn, affect the cells' electrical properties. In what is surely to be an intricate system (once all the parts are known), calcium channels regulate many aspects of cell physiology, and are themselves governed by a wide variety of external and internal stimuli. Modulation of calcium channels in the surface membrane of excitable cells will in turn regulate $[Ca^{2+}]_i$ (Dunlap et al., 1978, 1981; Reuter, 1983; 1985; Tsien, 1983; Canfield et al., 1984; Tsien et al., 1988; Bean, 1989; Miller and Fox, in press).

The main focus of this chapter will be on calcium channel modulation. We will review some of the large number of hormones, neurotransmitters, drugs, and second and third messenger systems that are known to alter Ca^{2+} entry into cells by increasing or decreasing calcium channel activity. We will consider the following questions: what is the biological significance of the modulatory response of calcium channels, and how does it fit in with the modulation of other channels? What type of voltage-dependent calcium channel is modulated? What is the underlying molecular mechanism of the effect? We will deal exclusively with voltage-dependent calcium channels and will not review receptor-operated calcium channels or ligand-gated channels that are Ca^{2+}-permeant. We will also omit any discussion of regulatory mechanisms in cells that maintain intracellular Ca^{2+} homeostasis. Normally the $[Ca^{2+}]_i$ in cells is low, approximately 10^{-7} M, and is maintained at these levels by a variety of buffering systems. These include Ca^{2+} pumps, exchange systems and Ca^{2+}-binding proteins (Carafoli, 1987; Volpe et al., 1988). These systems either expel Ca^{2+} from the cell or sequester the divalent ions in intracellular compartments such as mitochondria or other organelles including the recently described "calciosome" (Volpe et al., 1988; Pozzan et al., 1988). The $[Ca^{2+}]_i$ in these compartments and in the extracellular milieu may be very high ($>10^{-3}$ M).)

A. Mechanism of Calcium Channel Modulation

It is important to review the various factors that determine the overall calcium channel current (I) in a cell. This can be states as follows:

$$I = N \cdot P \cdot i$$

where N = the total number of channels; P = is the probability that the channels will be open and i = the unitary current (i.e., current through a single calcium channel).

Based on these criteria, we would expect single-channel modulation experiments to show either increased unitary current amplitudes, increased probability of single-channel openings with a fixed number of channels, an increase in the number of functional channels, or some combination of the three. In fact, while changes in probability and possibly the number of channels have both been reported, changes in unitary channel conductance have not yet been observed (Tsien, 1986).

Experimental tests of these hypotheses became possible with the advent of patch clamp methods for studying the unitary properties of calcium channels. In the investigations of calcium channels that have been reported two complementary variations of this technique have been used: (1) recordings from cell-attached patches to study the properties of individual channels and (2) whole cell recordings designed to assay the current from the entire pool of functional channels. More recently studies with excised patches have yielded important new insights (Fenwick et al., 1982; Kalman et al., 1988).

B. Different Types of Voltage-Dependent Calcium Channels

Before continuing with our description of calcium channel pharmacology, we need to explore the issue of calcium channel heterogeneity. It is important to consider which kinds of calcium channels are being modulated. Calcium channels have been identified in a number of cell types. Traditionally they have been associated with excitable cells such as the various types of muscle, neurons and endocrine cells. Calcium channels have now been shown to exist in a number of nonexcitable cell types including various types of glial cells (Barres et al., 1988), myeloma cells (Fukushima and Hagiwara, 1985), osteoblasts (Chesnoy-Marchais and Fritsch, 1988) and fibroblasts (Villereal and Jamieson, 1988; Chen et al., 1988). In the latter case, the observed modulation of the calcium channels by growth factors and oncogenes suggest that they might have important functions in growth control.

One important theme running through recent work on calcium channels is channel diversity. It is quite certain that different types of calcium channels exist (Llinas et al., 1980, 1981; Hagiwara and Byerly, 1981; Fox and Krasne, 1981, 1984; Hagiwara, 1983). At least three different types of vertebrate calcium channels have been described in chick dorsal root ganglion (DRG) neurons named T-, N-, and L-type calcium channels (see Chapter 13, Section III). There is a fourth type of calcium channel found in cerebellar Purkinje cells called P-type (Llinas et al., 1989). When rat brain RNA was expressed in frog oocytes a non-dihydropyridine, non ω-ConoToxin (ω-CgTx) sensitive channel was discovered (Lester et al., 1989). Is this newly discovered channel similar to the P-type calcium channel mentioned above? Or is it yet another type of calcium channel? Other types of calcium channels probably are still waiting to be found. Invertebrate calcium channels represent further calcium channel diversity. This should not be surprising as marine invertebrates have ion channels that operate in an ionic environment quite different from those faced by mammals. It is not yet known whether each class of channels mentioned above (T, N, L, P) represent a unique protein structure or whether each class, in fact, represents a family of closely related gene products. Briefly, the properties of the channels are as follows: (1) the lowest conductance type of calcium channel is known as the "T" type; (2) the best characterized type of calcium channel is known as the "L" type; (3) "N"-type channels are widely distributed in the nervous system and (4) The "P"-type calcium channel has not been extensively characterized.

The T type is characterized by a single channel conductance of about 5 to 10pS in 100 mM Ba^{2+}. Both Ca^{2+} and Ba^{2+} permeate the channel approximately equally well. These channels usually have a low threshold for activation, a negative inactivation range, exhibit rapid inactivation kinetics and are slowly deactivating (slow tail currents) (see Chapters 3 and 4, Section I for further discussion of channel deactivation, activation, and inactivation). An extensive pharmacology for T channels has not, as yet, been developed. Some compounds such as octanol (Llinas and Yarom, 1986), amiloride (Tang et al., 1988), and diphenylhydantoin (Yaari et al., 1987; Twombly, et al., 1988) will block T channels, but they are not very specific. T channels are not normally very sensitive to drugs or toxins that block other types of calcium channels such as the dihydropyridines (see Cohen et al., 1988; Akaike et al., 1989) or ω-conotoxin. However, they are potently blocked by Ni^{2+} and are less potently blocked by Cd^{2+} than are N or L channels. T channels are very widely distributed. They are also termed "low threshold", type I, slow deactivating or "fast" Ca^{2+} channels, by various authors.

The L type has the highest conductance (approximately 22 to 28pS in 100 mM Ba^{2+}) and is probably the most widely distributed. Ba^{2+} permeates L channels better than Ca^{2+}. These calcium channels have a high threshold for activation and are rapidly deactivating (fast tail currents). Unlike T currents, L currents tend to show less pronounced voltage-dependent inactivation (much slower and less complete), but can exhibit dramatic Ca^{2+}-dependent inactivation. Thus when Ba^{2+} is the current carrier and the cells are loaded with an efficient Ca^{2+}-chelator like EGTA or BAPTA, L currents can be "long lasting". Under conditions in which $[Ca^{2+}]_i$ is less well buffered L-type calcium channel currents are probably not long lasting. Ca^{2+} currents carried through T and L channels can be distinguished by employing the effects different holding potentials have on steady state inactivation. When a cell is held at fairly depolarized potentials (> -30 to -40 mV), T channels are generally inactivated, N channels are partially inactivated, but, in many cells, L channels are hardly inactivated at all. One major reason why L currents have been the most thoroughly investigated of all calcium channels is that an extensive L channel pharmacology exists (see also Chapter 13, Section III). Thus, there are many types of organic compounds such as the dihydropyridines that will either inhibit (channel blockers) or enhance (channel activators) L currents. Furthermore, at least one type of toxin (ω-conotoxin) is also a powerful blocker of L currents in some cell types. Note that there is some controversy regarding this point; McCleskey et al., (1987) suggest that L-type calcium channels are blocked by ω-CgTx, while Kasai et al., (1987) and Plummer et al., (1989) claim that L-type calcium channels show no ω-CgTx sensitivity. These compounds have played a key role in the isolation and characterization of L channels. Divalent cations, particularly Cd^{2+}, are also good L channel blockers. Indeed another potential way of distinguishing between L and T channels is by comparing the relative blocking effects of Cd^{2+} and Ni^{2+}, Cd^{2+} being a much more potent blocker of the former and Ni^{2+} a much more potent blocker of the latter. L-type calcium channels, but not yet T type or N type, have been cloned and functionally reconstituted. Calcium currents could be reconstituted by the dihydropyridine receptor α_1-subunit alone; (Tanabe et al., 1988; Mikami et al., 1989; Perez-Reyes et al., 1989). The L-type calcium channels have also been termed high threshold, type II, fast deactivating or DHP receptor calcium channels, by various authors.

Although N-type channels were first described in DRG neurons, they are in fact widely distributed in the nervous system. Some recent papers suggest that N-type channels are the predominant high threshold calcium channel in assorted neurons (Aosaki et al., 1988; Bean, 1989; Plummer et al., 1989). This channel has a conductance intermediate between T and L (approximately 13 to 17pS in 110 mM Ba^{2+}). N channels are considered high threshold channels that usually activate at relatively positive membrane potentials; voltage-dependent inactivation typically occurs over a very broad range of membrane potentials. These channels probably show Ca^{2+}-dependent inactivation. In chick DRG cells N-channel inactivation is quite rapid. Further, a large proportion of the current can be inactivated by holding cells at depolarized potentials (> -40 mV). However, N-current inactivation rates seem to be somewhat variable. For example, in superior cervical ganglion (Hirning et al., 1988) and myenteric plexus neurons (unpublished observations), N currents inactivate at slower rates than those initially observed in chick DRG cells. N channels are potently blocked by both Cd^{2+} and by ω-conotoxin but appear to be relatively resistant to the effects of dihydropyridines. Highly specific drugs acting at N channels have not yet been described although some compounds with relatively specific N channel blocking activities have been reported, e.g., Gd^{3+} ions (Docherty, 1988) and some aminoglycosides (Wagner et al., 1987). Some groups believe that ω-CgTx is specific for N-type channels (Kasai et al., 1987; Plummer et al., 1989). Until recently N channels had only been described in neurons but now Durroux et al., (1988) have also tentatively described their presence in adrenal glomerulosa cells. In addition, if ω-CgTx binding is considered to be a specific marker for N-type calcium channels, the reported binding of ω-CgTx to adrenal chromaffin cells would indicate the presence of N channels in these cells.

The P-type calcium channel has not been as extensively characterized as have the calcium channels mentioned above. This channel is blocked by the spider venom toxin FTX and has a single channel conductance similar to that of N-type calcium channels (14 pS; Llinas et al., 1989). No whole-cell reports have yet been published.

If there are more than four types of calcium channels then how useful is such a four-way categorization? The answer is clearly that other types of calcium channels certainly exist. Nevertheless, there are calcium channels in many cell types that possess greater similarities than differences and that fit quite well into these four categories. Thus, dihydropyridine-sensitive calcium channels exist in skeletal, cardiac and smooth muscle, in most types of neurons, in various types of excitatory endocrine cells, in fibroblasts and in type 2 astrocytes. T channels are also found in many of these cell types. N channels appear to be present in many types of neurons but possibly not in all (see Miller, 1987; Hosey and Lazdunski, 1988; Tsien et al., 1988; Bean, 1989; for many references on calcium channel distributions). However, even though many of these calcium channels fall comfortably into the four channel categories described, clear differences are also often apparent within each class. It is therefore likely that the four "classes" actually represent closely related families of channels rather than a single invariant molecular type. This would not be at all surprising considering what is currently known about the structure of voltage-sensitive sodium channels (Catterall, 1988), voltage-sensitive K^+ channels (Catterall, 1988), nicotinic receptors (Deneris et al., 1988) and GABA-A receptors (Wisden et al., 1988). For example, voltage-sensitive sodium channels are coded for by several genes that give rise to closely related but not identical proteins (Catterall, 1988). Differences in the properties of these voltage-sensitive sodium channels are apparent from their subtly different biophysical characteristics or their disparate sensitivities to toxins such as tetrodotoxin or μ-conotoxin (Moczydlowski et al., 1986; Auld et al., 1988). Families of similar channels can also be produced by alternate modes of gene splicing as in the case of the potassium channel known as IA (Schwarz et al., 1988). It is more than likely that similar processes are involved in producing families of closely related calcium channels. Furthermore, it is also possible that some members of these families may not even be normally used as calcium channels at all but as other voltage-sensitive cellular transducing elements that regulate functions such as stimulus/contraction coupling in skeletal muscle (see below) (Rios and Brum, 1987; Tanabe et al., 1988).

As far as the distribution of the various kinds of calcium channels is concerned, it is clear that different kinds of cells possess different complements of channels, sometimes in radically different amounts. Such a differential distribution also encourages the view that these channels truly represent different molecular entities with different functions. A further aspect of calcium channel organization to be considered is their distribution *within* a particular cell rather than between cells. Given the fact that many cells possess more than one type of calcium channel such micro-organization would not be surprising. Indeed in a cell as a neuron that may have an extremely complex morphology, different parts of the cell may be simultaneously engaged in quite separate activities. Some preliminary studies utilizing both patch clamp and Fura-2 based microfluorimetry/imaging techniques have been initiated to explore these possibilities. In bullfrog (Lipscombe et al., 1988 b,c) and rat (Thayer et al., 1987) superior cervical ganglion neurons N and L channels appear to be distributed over most portions of the cell including the soma, processes and growth cones of growing cells. On the other hand, Tank et al., (1988) have presented dramatic images suggesting that in Purkinje cells from guinea pig brain cerebellar slices most calcium channels occur in the dendritic arborization rather than the cell soma or axon. Llinas et al., (1980, 1981) also felt that calcium channels in Purkinje cells were segregated, because low threshold Ca spikes were recorded in the cell bodies of these neurons while high threshold Ca spikes were recorded in the dendrites. It should be pointed out that these conclusions differ somewhat from electrophysiological studies using acutely dissociated Purkinje neurons which indicated the presence of high threshold calcium channels in the cell soma as well (Regan, 1987). However, a recent study has indicated that the distribution of calcium channels in neurons may undergo a change *in vitro* depending on conditions (Haydon and Man-Son-Hing, 1988). Other electrophysiological studies have clearly demonstrated a micro-organization governing the distribution of calcium channels in neurons. Thus, during recording of calcium channels in on-cell patches from neurons which normally contain one or at most a few calcium channels, patches are occasionally observed that contain tens or even hundreds of channels (Fox et al., 1987; Thompson and Coombs, 1988). The physiological functions of such "hot spots" are not known. They may represent regions of the cell to which high concentrations of Ca^{2+} must be rapidly delivered under particular circumstances. These might include regions near the active zones for neurotransmitter release (Pumplin et al., 1981). In an interesting study, Thompson and Coombs (1988) mapped the distribution of calcium channels at several different points around the cell soma of giant invertebrate neurons. They showed that the degree and rate of inactivation of the calcium channel current differed at different points around the cell soma. Inactivation of calcium channels is thought to be at least partially dependent on the $[Ca^{2+}]_i$ (see below). It may be that the intracellular processes involved in Ca^{2+} buffering are also differentially distributed and that these are more efficient in some regions of the cell than others. This would allow the internal $[Ca^{2+}]_i$ to increase to a greater extent in the proximity of some of channel microdomains than others and different rates of calcium channel inactivation might therefore occur. Thus, the relative distribution of both calcium channels and intracellular Ca^{2+} buffering systems, particularly those close to the plasma membrane, may allow local shaping of cellular Ca^{2+} signals (Tillotson and Gorman, 1980).

II. MODULATION OF CALCIUM CHANNELS

The influx of Ca^{2+} into cells via calcium channels is a major mechanism for regulating $[Ca^{2+}]_i$ and therefore a major way of regulating many cell functions. Thus, it is not surprising that the calcium channels are themselves capable of regulation. Modulation of calcium channel activity can change the moment to moment influx of Ca^{2+} into the cell. There are now numerous examples of this type of phenomenon and regulation of calcium channels seems to be able to take many forms. The activity of calcium channels can be increased or decreased by a variety of neurotransmitters that operate through multiple second messenger and G-protein linked pathways. (Dunlap et al., 1978, 1981; Reuter, 1983, 1985; Tsien, 1983; Canfield et al., 1984; Dolphin et al., 1986, 1987, 1988; Tsien et al., 1988; Bean, 1989; Miller and Fox, in press).

A. Epinephrine Up-Modulates Cardiac L-Type Calcium Channel Activity

The cardiac L-type calcium channel response to epinephrine is perhaps the most widely studied case of hormonal modulation of ion channels known. Both the biological function and the molecular mechanisms have been studied in depth. The cardiac calcium current was the first recognized example of a voltage-gated channel being governed chemically (Reuter 1967, Vassort et al., 1969). Cyclic AMP (cAMP) was shown to be the intracellular second messenger carrying the message into the cell (Tsien et al., 1972; Tsien, 1973; Watanabe and Besch, 1974). The observation that an increased probability of L channel opening was still seen when β-agonists were added outside the patch pipette was one piece of evidence indicating that a diffusible second messenger was involved in the modulation of channel function (Trautwein and Pelzer, 1988; Tsien et al., 1988). Indeed it is widely believed that cAMP is responsible for most of the stimulation of L-channel activity in the heart. Thus cAMP and its analogs mimic the effects of β-adrenergic agonists on the heart (Kameyama et al., 1986, 1988). Furthermore, other agents that stimulate adenylate cyclase in cardiac tissue also stimulate the cardiac L current whereas agents, such as muscarinic agonists, that inhibit adenylate cyclase in ventricular cells decrease the calcium current (Reuter, 1983; Hescheler et al., 1986). Injection of activated cyclic AMP-dependent protein kinase (protein kinase A, [PKA]) into myocytes also mimics the effects of β-agonists (Kameyama et al., 1986, 1988, Hescheler et al., 1987b) whereas injection of a polypeptide that specifically blocks PKA inhibits the agonist induced activation of the calcium current (Kameyama et al., 1986). The effects of β-agonists could also be prevented or reversed by perfusion of myocytes with certain phosphatases (e.g., phosphoprotein phosphatase 1 or 2 A) (Kameyama et al., 1986; Hescheler et al., 1987b). These data certainly suggest that cAMP mediated phosphorylation can increase L-channel activity in the heart. The known steps in the response of the cardiac calcium channel to epinephrine are as follows. Epinephrine binding at surface β receptors causes the G-protein G_s to dissociate into the α- and βγ-subunits. The α-subunit of G_s stimulates the enzyme adenylate cyclase, thereby catalyzing the conversion of ATP into cAMP + PPi. Normally PKA is found in cells in the inactive form as C_2R_2, that is two active catalytic subunits of PKA are associated with two regulatory subunits thereby keeping the complex inactive. In the presence of cAMP, the catalytic subunit dissociates from the regulatory subunits according to the following reaction:

$$C_2R_2 + cAMP \rightarrow \text{Active PKA} + R_2\text{-cAMP}_4.$$

The active PKA in the presence of ATP is thought to directly phosphorylate the L-type calcium channel or some closely associated regulatory protein (see below). The phosphorylation step greatly increases the probability of channel opening (Trautwein and Pelzer, 1988; Tsien et al., 1986). It has recently become clear that β-adrenergic stimulation can induce very long-lived openings in the L-type calcium channels (Marban et al., 1989), similar to the effects that BAY K 8644 have on L-type calcium channels (Hess et al., 1984; Kokubun and Reuter 1984; Nowycky et al., 1985). In addition, it has been shown recently that some part of the response cascade to epinephrine is carried by the G-protein G_s (Yatani et al., 1988, 1989). The α-subunit of G_s seems capable of directly stimulating cardiac L-type channels directly. The up-modulation of the Ca current is extremely rapid; there is increased current flow almost instantly (Yatani et al., 1989).

B. What is the Site of Protein Kinase A Phosphorylation in Heart?

Are the L channels themselves the locus of phosphorylation? The answer is probably yes. The α_1-subunit of the purified skeletal muscle dihydropyridine receptor (Curtis and Catterall, 1985; Hosey et al., 1987; Nastainczyk et al., 1987) or the receptor localized in purified t. tubular preparations (Hosey et al., 1988) is a good substrate for PKA. Inspection of the primary sequence of the α1-polypeptide reveals that it contains at least seven consensus sequences for the cAMP-dependent phosphorylation of serine or threonine residues (Tanabe et al., 1987). Modeling studies suggest that these residues are all situated on the cytoplasmic side of the membrane (Catterall, 1988). The β-subunit of the dihydropyridine receptor is also a good substrate for cAMP-dependent phosphorylation, whereas the α2-subunit is not (Curtis and Catterall, 1985; Hosey et al., 1986; Nastainczyk et al., 1987). Functional evidence that the dihydropyridine receptor is actually the locus of cAMP-dependent phosphorylation comes from studies using purified dihydropyridine receptor preparations that clearly only contain the normal complement of subunits. PKA increased the probability of channel opening following reconstitution into black lipid membranes and also increase the rate of $^{45}Ca^{2+}$ uptake following reconstitution into phospholipid vesicles.

C. Other Cardiac Responses to Protein Kinase A Activation

Modulation of calcium channels does not occur in isolation. Positive responses are mediated by β1 receptors for the β-adrenergic agonists epinephrine and norepinephrine. Similar responses mediated by the activation of adenylate cyclase are produced by calcitonin-gene related peptide (CGRP; Ono et al., 1989). Inhibitory responses are mediated by Ach or adenosine binding through the appropriate receptors. Both the excitatory and inhibitory responses are mediated by G proteins, G_s for the positive responses and G_i for inhibitory. G_i works by inhibiting adenylate cyclase thereby lowering intracellular cAMP levels and thus regulating PKA. Other cellular processes are also regulated by intracellular cAMP. In the heart, enhancement of the potassium channel current keeps the action potential from becoming too prolonged by the calcium currents. Stimulation of Ca^{2+} pumps in the sarcoplasmic reticulum (SR) keeps the SR loaded with calcium but also helps terminate the intracellular calcium transient. Decreased responsiveness of the contractile machinery to the lowered free $[Ca^{2+}]_i$ favors rapid relaxation and allows the heart to fill with blood between beats, even when the heart is beating at an accelerated heart rate. There are, in these latter circumstances, increased supplies of ATP to meet the increased energy demands. In combination with the calcium channel currents these other responses to cAMP demonstrate a coordinated response to situations in which sympathetic stimulation is elevated (Tsien, 1986).

D. Other Tissues Respond to β-Adrenergic Stimulation

The heart is not the only tissue in which calcium currents are up-modulated in response to β-adrenergic stimulation. It has been shown that L currents in skeletal muscle are augmented by β-adrenergic/cAMP mediated stimulation as they are in cardiac muscle (Arreola et al., 1987). This is not surprising as the purified skeletal muscle α1-subunits are substrates for cAMP-dependent protein kinase. Interestingly, a second dihydropyridine insensitive, low threshold skeletal muscle calcium current can also be augmented by β-adrenergic agonists. It is tempting to describe this second current as a T current, but its slow rate of inactivation distinguishes it from most T currents. Norepinephrine also increases the calcium current in some smooth muscle cells (Benham and Tsien, 1987; Pacaud et al., 1987; Nelson et al., 1988). However, in one case this is clearly a T-type calcium channel current and in another it does not seem to be a β-adrenergic effect. Calcium currents in hippocampal granule cells are also increased by β-adrenergic agonists (Gray and Johnson, 1987). It is possible that reductions in cAMP levels suppress N-type calcium channels in sympathetic neurons (Lipscombe et al., 1989). cAMP effects may play a role in DRG neurons as well. CGRP greatly potentiated inward Ca currents in DRG neurons (Ryu et al., 1988). Cardiac cells have an analogous CGRP response that is mediated by PKA activation. It is possible that L-type calcium channels in rat DRG neurons can be up-modulated by PKA in the same way that they are in heart. On the other hand, a completely separate mechanism may be functioning. Interestingly, L-type calcium currents recorded in pancreatic cells were larger when 20 mM glucose was in the bath, than when O mM glucose was in the bath. It was felt that intracellular ATP levels in the cells were changed by the two concentrations of glucose (Smith et al., 1989). Perhaps lowering ATP levels (low glucose case) changed the phosphorylation state of the L-type calcium channels. In general, though, addition of cyclic AMP analogs to neurons or excitable endocrine cells does not increase the L current greatly, although there are certainly some exceptions to this rule, (e.g., Gray and Johnston, 1987; Levitan et al., 1987; Doerner and Alger, 1988; Lipscombe et al., 1988a) (some groups might argue that L-type Ca currents are not increased greatly by activating PKA because neurons in general do not contain many L-type calcium channels (Bean, 1989; Aosaki et al., 1988; Plummer et al., 1989). Nevertheless, cAMP-mediated phosphorylation may play an important role in the regulation of neuronal L channels. For example, it has been frequently observed that in whole-cell patch clamp recordings or excised patch paradigms the L currents gradually "rundown", suggesting that certain cytoplasmic factors are responsible for the maintenance of channel activity (Chad and Eckert, 1986). Furthermore, it has been frequently observed that inclusion of an ATP regenerating system in the internal perfusing medium considerably slowed or abolished such rundown (e.g., Forscher and Oxford, 1985). Exactly why this should be is not completely clear. However, in some cases cAMP appears to be necessary for the full expression of the retardation or abolition of rundown (Chad and Eckert, 1986; Doroshenko et al., 1986; Armstrong and Eckert, 1987). Indeed, one school of thought believes that cAMP-dependent phosphorylation is an absolute requirement for the maintenance of active L-channel function (Armstrong and Eckert, 1987; Kalman and Armstrong, 1988). Thus, in GH3 clonal pituitary cells exogenously added cAMP-dependent kinase maintained active L channels (Armstrong and Eckert, 1987; Kalman and Armstrong, 1988). On the other hand, perfusion of neurons with the Ca^{2+}-dependent phosphatase, calcineurin, abolished the activity of the calcium channels (Chad and Eckert, 1986). It is, therefore, conceivable that in GH3 cells and possibly other cases as well, cAMP and Ca^{2+} have opposing effects on the maintenance of L-channel activity (Kalman et al., 1988). The observation that calcineurin-like enzymes abolish L-channel activity is interesting in relation to the two proposed mechanisms of L-current inactivation (Eckert and Chad, 1984). In many instances, L currents exhibit only slow or limited voltage-dependent inactivation (Brown et al., 1981; Lee et al., 1985). Inactivation of L currents often appears to be "current" dependent, that is it depends on the influx of Ca^{2+} into the cell (in some preparations like myocardial cells there seems to be a voltage-dependent component as well, while in others, like GH3 L-type calcium channel inactivation seems to be exclusively current dependent). Consequently, the substitution of Ba^{2+} for Ca^{2+} as a current carrier is frequently found to slow L-current inactivation (but see Kasai et al. for an exception; they find the current-dependent inactivation of N-type calcium channels proceeds similarly in the presence of either Ba^{2+} or Ca^{2+}). In DRG cells, a rapid increase in the internal $[Ca^{2+}]_i$ induced by releasing Ca^{2+} from a caged complex using a laser was found to lead to rapid inactivation of the sustained L current (Morad et al., 1988). In contrast, it has also been observed that calmodulin antagonists enhanced neuronal calcium currents under some circumstances (Doroshenko et al., 1988). It may, therefore, be that Ca^{2+}-dependent inactivation is a manifestation of Ca^{2+}-activated phosphatase activity and the removal of phosphates that are important for the expression of normal L-channel function.

E. Protein Kinase C Also Modulates Calcium Channels

Although the role of the diacylglycerol/protein kinase C system has not been studied in as great a detail as the β-adrenergic pathway, the literature is replete with observations suggesting that calcium channels can be modulated by this pathway as well. Much of the data is very controversial and contradictory at this point. Indeed reports vary as to how good a substrate the α1-subunit of the skeletal muscle dihydropyridine receptor is for protein kinase C, although recent studies do indicate that it can be efficiently phosphorylated by this enzyme (Hosey and Lazdunski, 1988; Nastainczyk et al., 1987). Variable effects of protein kinase C activation on calcium channels in the heart have been reported. However, Lacerda et al. (1988) recently demonstrated that phorbol esters have a biphasic effect on the L current of ventricular myocytes, activating it at short incubation times and inhibiting it at later times. As long term exposure of cells to phorbol esters leads to the down regulation of protein kinase C, these authors suggested that the long term inhibitory effects may be due to such a phenomenon. This implies that the presence of protein kinase C is normally required for basal L-channel activity in these cells. In general, the role of protein kinase C in mediating the activation of cardiac L channels by neurotransmitters is unclear. However, the enzyme could mediate the activating effects of some agents such as thrombin that are known to generally activate phosphatidylinositol bisphosphate breakdown in a number of cells (Markwardt et al., 1988). Activation of protein kinase C has also been reported to have variable effects of calcium channels in smooth muscle. Most of the effects reported have suggested an augmentation of L-channel activity. Thus phorbol esters stimulated the L current in the A7r5 vascular smooth muscle cell line (Sperti and Colucci, 1987; Fish et al., 1988) and increased the L current in smooth muscle cells from the stomach of the toad *Bufo marinus*. In the latter case, there is the possibility that diacyglycerol normally mediates the muscarinic receptor induced augmentation of the L current in these same cells. Norepinephrine also increased the activity of dihydropyridine sensitive calcium channels in arterial smooth muscle cells (Nelson et al., 1988). This is an α-adrenergic effect that may also be mediated by diacyglycerol. These results are all consistent with observations that phorbol esters are often found to produce contraction of smooth muscle cells. Such effects may include an intracellular locus of action but also seem to be at least partially dependent on an increased dihydropyridine sensitive Ca^{2+} influx (Forder et al., 1985). In contrast to these reports, one other study found that phorbol esters decreased L currents in the A7r5 cell line (Galizzi et al., 1987). Importantly, the endogenous generation of diacylglycerol following stimulation of the cells with vasopressin had a similar effect. Why this discrepancy exists with the other reports discussed above is not clear.

A large number of effects of protein kinase C activation on neurons have also been observed. Again, the nature of these effects are extremely variable. However there are a large number of reports that phorbol esters/protein kinase C can inhibit calcium channels in neurons, in neuronal cell lines and in excitable endocrine cells (Rane and Dunlap, 1986; Harris et al., 1986; Werz and MacDonald, 1987a,b; Gross and Macdonald, 1988; Marchetti and Brown, 1988; Doerner et al., 1988). There is a report that some of the inhibitory effects of phorbol esters are due to direct block of calcium channels and not due to activation of protein kinase C (Hockberger et al., 1989). In contrast, other authors have reported protein kinase C mediated increases in calcium currents in sympathetic neurons (Lipscombe et al., 1988a; Malhotra et al., 1988), in hippocampal neurons (Madison et al., 1989), in oocytes (Leonard et al., 1987; Sigl and Baur, 1988) and some *Aplysia* neurons (Strong et al., 1987; Braha et al., 1988). The latter effect is particularly interesting as it was found to be due to the expression of covert calcium channels rather than to the modulation of previously existing channels. There is also little agreement as to the type of neuronal calcium channels that are modulated following the activation of protein kinase C. For example, Rane and Dunlap (1987) suggested that protein kinase C activation resulted in the inhibition of L currents in

DRG neurons (by decreasing probability of channel opening) whereas MacDonald et al. have claimed that such effects are all directed at N currents (Gross and MacDonald, 1988). Ewald et al. (1988b) found that removal of protein kinase C from DRG cells by down regulation with phorbol esters reduced somewhat the ability of neuropeptide Y (NPY) to inhibit the sustained (L type?) calcium current, but did not alter its ability to inhibit the transient (N type?) calcium current (see also Walker et al., 1988). In hippocampal neurons Doerner et al. (1988) observed that protein kinase C activation primarily led to the inhibition of the transient (N type?) portion of the calcium current. Recently, it was observed that protein kinase C stimulation increased the probability of N-type (and perhaps L-type) calcium channel opening in cell attached patches, but these increases were not seen in whole-cell current recordings when protein kinase C was activated (Doerner et al. 1988; Madison et al., 1989). It was suggested that some cytoplasmic factor required in the protein kinase C response cascade was being washed out of the cells (perhaps the protein kinase C itself). Similar results were obtained in Aplysia bag cell neurons where whole-cell currents could not be increased by the addition of phorbol esters. Nonetheless, when microelectrodes were used in the bag cells, large changes in Ca^{2+}-dependent action potential height could be detected after activating protein kinase C (De Riemer et al., 1985; Strong et al., 1987). In other experiments rat brain mRNA injected into oocytes expressed calcium channels that were not L- or N-type calcium channels; these channels were stimulated by phorbol esters (Leonard et al., 1987). It may well be that this variability is a true reflection of the great heterogeneity of neurons and the many important roles that calcium channels have in these cells. However it may also reflect the difficulty of accurately assigning different portions of the neuronal Ca^{2+}-current to different classes of calcium channels based on experimental paradigms employing changes in holding potential. This type of approach may not be a completely reliable way of classifying calcium channels in neurons where the inactivation rates of N currents may be quite variable, thus allowing them to be confused with L currents under same circumstances (see Bean, 1989 for discussion).

F. Neurotransmitters and GTP-Binding Proteins

The majority of effects of neurotransmitters on neuronal calcium channels are inhibitory, (see Tsien et al., 1988 for a comprehensive list of references) although some stimulatory effects have also been observed (e.g., Gray and Johnston, 1987; Kramer et al., 1988; Lipscombe et al., 1988a; Nedergaard et al., 1988; Ryu et al., 1988; Saleh et al., 1988). The inhibitory effects of neurotransmitters have been reported to be directed against T, L and N currents or various combinations of the three (Tsien et al., 1988). Are any of these effects mediated through the activation of protein kinase C? For that matter is it actually necessary to propose the participation of diffusable second messengers in mediating the action of neurotransmitters on calcium channels in neurons? This question requires us to examine the possible role of G proteins in these events. It is now quite clear that some potassium channels can be directly coupled to receptors via these proteins (Pfaffinger et al., 1985; Breitweisser et al., 1985; Yatani et al., 1987a). Can the same arrangement occur for calcium channels as well? Anderson and Dunlap (1988) reported that norepinephrine, added outside the patch pipette, still inhibited L channels in chick DRG cells in on-cell patches, suggesting that a diffusable second messenger (diacylglycerol?) was involved. Furthermore, as discussed above Ewald et al. (1988b) found that down-regulation of protein kinase C in rat DRG cells partially abolished the effects of NPY on the sustained Ca^{2+} current. Moreover, NPY has been shown to stimulate the production of diacylglycerol in DRG cells (Perney and Miller, 1989). Bradykinin, another activator of diacylglycerol synthesis, in these cells has also been shown to inhibit the DRG Ca^{2+} current (Ewald et al., 1989). However, in contrast to these observations, others have found that norepinephrine (Forscher et al., 1986) and gamma amino butyric acid (GABA) (Green and Cottrell, 1988) did not inhibit calcium channels in on-cell patches from DRG neurons when added outside the pipette, although norepinephrine was reported to be effective in inhibiting calcium currents in excised patches (Marchetti et al., 1986). Furthermore, in rat myenteric plexus neurons, where the effects of inhibitory neurotransmitters appear to be directed primarily against the N current, NPY did not inhibit N-channel activity in on-cell patches. However, NPY produced strong inhibition of N currents recorded from whole cells (Hirning, Fox and Miller, unpublished observations). Thus, it appears that in many cases diacylglycerol or another diffusable second messenger is not required for the inhibitory modulation of neuronal calcium channels. Certainly in the study of Ewald et al. (1988a,b) NPY still produced substantial inhibitory effects on the DRG calcium current when there was no protein kinase C left in the cell to mediate this inhibition.

Whether or not diffusable second messengers are involved, it has normally been found that the neurotransmitter induced inhibition of neuronal calcium currents is abolished by pretreatment of cells with pertussis toxin (Tsien et al., 1988; Ewald et al., 1989). This implies that such effects are mediated by a G protein of the G_o or G_i type. Furthermore, the inhibitory effects of neurotransmitters can be reinstated if the purified α-subunits of pertussis toxin substrates are perfused into toxin treated cells with a patch pipette (Ewald et al., 1989; Harris-Warwick et al., 1988; Hescheler et al., 1988). Exactly which pertussis toxin substrate normally mediates the effects of inhibitory neurotransmitters on neuronal calcium channels is not known. Different receptors may, in fact, use different G proteins. In rat DRG cells for example, the coupling between calcium channels and bradykinin or NPY receptors can be differentially reconstituted by different pertussis toxin substrates (Ewald et al., 1989). Further indications showing that G proteins mediate the inhibition of calcium channels in neurons comes from observations in which stable GTP analogs, e.g., GppNHp or GTPγS are perfused into cells. These substances, by replacing GTP components that bind in a reversible fashion, inhibit neuronal calcium currents or enhance the ability of neurotransmitters to do so in irreversible fashion (Dolphin and Scott, 1987; Dolphin et al., 1988; Lipscombe et al., 1989; Penington et al., in press). These effects are usually abolished by pertussis toxin treatment. Considered overall these various results indicate that some neuronal calcium channels may be directly coupled to neurotransmitter receptors via pertussis toxin sensitive G proteins. Activation of such receptors would lead directly to the inhibition of these calcium channels. Such a model is also consistent with a variety of reports that guanyl nucleotides can alter dihydropyridine binding (Higo et al., 1988) and also the manner in which dihydropyridines and other drugs interact with calcium channels in electrophysiological studies (Scott and Dolphin, 1987, 1988). Interestingly, the inhibition produced by neurotransmitters may be due to a shift in the voltage dependence with which channels are opened. In his studies Bean found that the inhibitory neurotransmitters profoundly altered the voltage dependence of activation of the calcium channels but the number of channels present and their functional activity were unchanged (Bean, 1989). Could this be the mechanism by which G_o or G_i inhibit neuronal calcium channel currents?

A more surprising example of a direct calcium channel/G-protein interaction comes from recent reports of L-channel function in cardiac and skeletal muscle (Yatani et al., 1987b, 1988). These studies have demonstrated that L channels incorporated into black lipid membranes from SR vesicles can be activated by GTPγS, apparently through a direct effect of G_s that does not require the production of cAMP. It is interesting to note that in some other cells types, receptor mediated *activation* of calcium channels can be blocked by pertussis toxin treatment indicating the participation of a G protein of the G_o or G_i type rather than G_s (Hescheler et al., 1988; Rosenthal et al., 1988).

There may be further mechanisms by which receptors can modulate calcium channels. For example, regulation by a cyclic GMP-dependent kinase has also been reported in one instance (Paupardin-Tritsch et al., 1986). Furthermore, this review did not cover exciting new developments concerning modulation of cardiac calcium channels by endogenous peptides as has been reported (Callewaert et al., 1989). Nor did we focus on channel gating changes (except in cardiac cells), accompanying neurotransmitter inhibition of calcium channels (see for instance Bean, 1989; Lipscombe et al., 1989). Such possibilities remain to be fully investigated.

III. SUMMARY

This chapter explores the literature on calcium channel modulation by neurotransmitters, hormones and second messengers. The chapter starts with a brief description of the importance of calcium channels and the vital roles they may play in normally functioning cells. The next section deals with the various mechanisms of ion channel modulation, i.e., possible molecular descriptions of modulation. This section is followed by one that describes four types of calcium channels (T, N, L and P type) that have been described in mammalian cells and some of their biophysical and pharmacological properties. This section includes speculations about other kinds of calcium channels that may yet be discovered, and compares calcium channels to voltage-dependent sodium and potassium channels. This section ends with a short review of calcium channel distribution, in various experimental preparations.

The rest of the chapter deals with calcium channel modulation. Because it is the most widely studied modulatory response, special attention is given to the epinephrine modulation of cardiac L-type calcium channels. Details including receptors (β-adrenergic receptors), GTP-binding proteins (G_s), enzymes (adenylate cyclase), intracellular messengers (cyclic AMP), and kinase activation (protein kinase A) are provided. Possible sites of action for protein kinase A phosphorylation are presented. The calcium channel modulatory response is placed in context with other cardiac responses to epinephrine, in order to better understand cardiac physiology and the role calcium channels and Ca^{2+} ions may play. Finally, β-adrenergic stimulation of other types of cells are examined.

The next section of the chapter provides a brief summary of protein kinase C modulation of various calcium channels. The literature on this topic is somewhat diffuse. There are reports of calcium channel inhibition and stimulation. There is also a report that protein kinase C activators directly block calcium channels independent of protein kinase C activation.

The chapter ends with two different topics. One is the inhibition of neuronal calcium currents induced by a variety of neurotransmitters including NPY, norepinephrine and GABA. The inhibition may be directly mediated by a GTP-binding protein acting directly on calcium channels. This inhibition can frequently be suppressed by incubating cells in pertussis toxin. The final section of the chapter deals with direct GTP-binding protein effects on calcium channels (both inhibitory and stimulatory).

ACKNOWLEDGMENTS

Supported by PHS grant NS26189 and grants from the McKnight, Klingenstein, Whitaker and Sloan Foundations and Miles Labs, to Aaron P. Fox and PHS grants DA02121 and MH40165 and grants from Marion Inc. and Miles Labs; to Richard J. Miller; Nicholas Penington is a Wellcome Trust fellow.

Section IV
Modulation of Calcium Channels in Clinical Medicine

Chapter
15

Modulation of Calcium Channels in Clinical Medicine: An Overview

John K. Leach

TABLE OF CONTENTS

I.	Introduction	268
II.	Cardiovascular Disease	268
	A. Ischemic Heart Disease	268
	B. Cardiomyopathy	268
	C. Hypertension	269
	D. Arrhythmias	269
III.	Peripheral Vascular Disorders	269
IV.	Neurologic and Psychiatric Disorders	270
V.	Other Potential Clinical Indications	270

I. INTRODUCTION

Experimental studies in the laboratories of Fleckenstein and Godfraind in the 1960s led to the development of a group of agents that selectively inhibit inward calcium current in cardiac and smooth muscle (Fleckenstein, 1985; Godfraind et al., 1986). These agents, the calcium channel blockers, depressed impulse formation in the sinoatrial (SA) node and slowed conduction through the atrioventricular (AV) node. They also reduced myocardial contractility, thus lessening ventricular work and myocardial oxygen consumption, and improved myocardial relaxation. Inhibition of calcium current in smooth muscle resulted in coronary, systemic, and pulmonary vasodilation. As can be appreciated from these effects, it is not surprising that calcium channel blockers found practical use in the management of patients with many types of cardiovascular disease. Furthermore, the potential therapeutic indications for these agents were subsequently expanded to include pulmonary, neurologic, psychiatric and endocrine disorders. The following chapters will address the use of calcium channel blockers in several of these areas.

II. CARDIOVASCULAR DISEASE

A. Ischemic Heart Disease

A major application of calcium channel blockers has been in the management of coronary artery disease, particularly angina pectoris. The pathogenesis of myocardial ischemia and angina has been the subject of intense investigation for over 100 years. In the late 1800s and early 1900s it was believed that ischemia was the result of both structural and functional changes of the coronary vessels. The concept of "coronary spasm", while not objectively documented, was accepted as the probable cause of rest angina. Subsequently, however, this theory was refuted, except for scattered reports in the literature, until the 1970s. Since then, there has been an increased interest in, and documentation of, coronary artery spasm as a cause of ischemia. There is now evidence that angina pectoris is frequently the result of both fixed obstruction, which prevents increased coronary flow to meet increased myocardial oxygen demands, and transient constriction or spasm of coronary arteries, causing reduced coronary flow and myocardial oxygen supply (Maseri and Chierchia, 1982). Chapter 16 addresses the beneficial effects of calcium channel blockers in improving the balance between myocardial oxygen supply and demand, and other possible antiischemic effects of these agents. In addition to the established use of these agents in the management of angina, their potential value in limiting infarct size and reducing ischemic injury has been the subject of much investigation, and will also be addressed in this chapter.

B. Cardiomyopathy

In 1980, cardiomyopathies were defined as heart muscle disease of unknown cause, and were classified as dilated, hypertrophic, and restrictive (Report of the WHO/ISFC task force on the definition and classification of cardiomyopathies, 1980). Thus, the term, cardiomyopathy, excludes myocardial dysfunction due to valvular disease, coronary artery disease, systemic or pulmonary hypertension, and congenital cardiac anomalies. The cardiomyopathies have varied manifestations, as would be expected from the different pathophysiologic classifications, but all may be associated with serious arrhythmias and progressive impairment of cardiac function. In Chapter 17, the basic abnormalities of these disorders are described, and the rationale for, and results obtained by the use of calcium channel blockers are discussed. This chapter also addresses the disturbed Ca^{2+} metabolism with resultant increased intracellular Ca^{2+} that is found in Syrian hamster cardiomyopathy, and the changes that calcium channel blockers bring about in this experimental model.

C. Hypertension

During the past several years there has been increased appreciation of the role of Ca^{2+} in tension development of vascular smooth muscle. There is evidence for increased levels of intracellular Ca^{2+}, altered Ca^{2+} storage, and altered Ca^{2+} transport in arterial hypertension. Studies comparing spontaneously hypertensive rats (SHR) with normotensive Wistar rats (WKY) indicate that SHR are more dependent on extracellular Ca^{2+} for aortic smooth muscle contraction than are WKY, and that SHR are more responsive to the relaxing effects of the calcium channel blocker nifedipine (Lederballe Pedersen, 1981). Of major clinical interest in this regard is the observation that the hypotensive effect of calcium channel blockers may not be as pronounced in normotensive or mildly hypertensive subjects as in those with severe hypertension. Thus, calcium channel blockers have become an important group of agents for antihypertensive therapy. Mechanisms of action, efficacy of, and indications for the use of these drugs will be described in Chapter 18. In addition, effects of these agents on cholesterol metabolism, renal function, and calcium balance will be discussed.

D. Arrhythmias

The effects of calcium channel blockers on SA and AV nodal function led to their use in the management of supraventricular arrhythmias. However, all calcium channel blocking drugs do not exhibit the same level of activity on all cardiac tissues, thus they are not equally effective as antiarrhythmic agents. Several classifications of calcium channel blockers have been proposed (Godfraind et al., 1986) based on clinical, pharmacologic, chemical and electrophysiologic effects. In 1985, Singh et al. proposed a classification based on the variable specificity of these agents for cardiac and peripheral activity, in order to provide a rational basis for clinical use. Type 1 agents, verapamil, tiapamil, gallopamil, and diltiazem prolong AV nodal conduction and refractoriness, which accounts for their antiarrhythmic effects, and are moderately potent vasodilators that are useful for the control of mild to moderate hypertension. The type 2 agents, nifedipine and other dihydropyridines are potent peripheral vasodilators with little, if any, direct electrophysiologic effects in the heart. They do, however, produce significant reflex increases in heart rate and contractility. Type 3 agents are the piperazine derivatives flunarizine and cinnarizine. They exhibit potent peripheral vasodilating effects, without calcium-blocking actions in the heart. The type 4 calcium channel blockers include lidoflazine, perihexilene, and bepridil and have complex pharmacologic actions. Lidoflazine causes relaxation of vascular smooth muscle, but does not block myocardial calcium channels, while perhexiline and bepridil have calcium channel blocking effects on both cardiac and smooth muscle. All three of these drugs, however, block the fast sodium channel and prolong repolarization in cardiac muscle, thus making them potentially useful in the treatment of both supraventricular and ventricular arrhythmias. In Chapter 19, the role of Ca^{2+} in the genesis of arrhythmias, and experiences with the use of calcium channel blockers in the management of supraventricular and ventricular arrhythmias will be thoroughly discussed.

III. PERIPHERAL VASCULAR DISORDERS

The effects of calcium channel blockers on smooth muscle and platelets have led to investigation of their use in peripheral vascular disorders. Blood platelets, which are necessary for the normal clotting of blood following injury, also have a major role in the pathogenesis of occlusive vascular disease (Ahn and Harrington, 1987). Since Ca^{2+} plays a major role in platelet activation, it would be expected that reduced aggregation of platelets and the vasodilation brought about by calcium channel blockers would make them useful in clinical conditions associated with vascular spasm and intravascular thrombosis. In the first part of Chapter 20, the role of Ca^{2+} in the platelet activation process and the effects of calcium channel blockers on platelet function are described in detail. The second part of this chapter includes a discussion of mechanisms of vasoconstriction, the vasodilator and antiatherogenic properties of calcium channel blocking drugs, and experiences with the use of these agents in peripheral vascular disorders.

IV. NEUROLOGIC AND PSYCHIATRIC DISORDERS

Calcium plays a major role in normal brain function and marked increases in intracellular Ca^{2+} levels can lead to irreversible neuronal damage. In Chapter 21 normal Ca^{2+} transport and metabolism in the central nervous system (CNS) and the specific effects of pathological accumulation of Ca^{2+} are described in detail. Experimental and clinical evidence indicating potential beneficial effects of calcium channel blockers in ischemic stroke and subarachnoid hemorrhage are presented.

For many years there has been much interest in the possible association of abnormal Ca^{2+} regulation with neuropsychiatric symptoms. In the final chapter, *Use of Calcium Channel Blockers in Psychiatric Disorders,* the evidence for such an association is presented, and the effects of calcium channel blocking drugs in patients with psychiatric and seizure disorders are discussed. As the authors point out, the data are conflicting, and further research will be expected to provide a better understanding of the effects of abnormal Ca^{2+} metabolism in these conditions.

V. OTHER POTENTIAL CLINICAL INDICATIONS

The calcium channel blocking drugs have already clearly become established as important therapeutic agents. In addition to their use in the areas cited, there is evidence that they may become useful in the management of some types of pulmonary, endocrine and hepatic disorders.

Bronchial asthma is characterized by airway smooth muscle constriction, release of histamine from mast cells, and mucus gland secretion, which are all Ca^{2+} mediated events. Although the data are controversial, studies in both experimental animals and humans suggest that calcium channel blockers may prevent or reduce the severity of allergic and exercise-induced bronchospasm. However, for reasons that are not entirely clear, much larger doses of these drugs are required than those used to relax vascular smooth muscle (Townley et al., 1988).

Primary dysmenorrhea is associated with increased uterine contractile activity which may compromise uterine blood flow and lead to ischemia and pain. It has been well established that calcium channel blockers can inhibit spontaneous contractions in nonpregnant and pregnant myometrium. Clinical trials with nifedipine and diltiazem in women with primary dysmenorrhea have yielded encouraging results, and investigation continues (Andersson, 1988).

Extensive studies have been carried out in rats *in vivo* and in the perfused liver to determine if calcium channel blockers can offer protection against chemical and hypoxic hepatic damage. Hepatotoxicity due to chemical injury can be prevented by verapamil, diltiazem and dihydropyridine-type calcium channel blockers, while only the dihydropyridine type appears to protect against hypoxic and ischemic damage. Also, there is preliminary evidence that these agents may offer some protection against alcohol-induced liver damage (Thurman et al., 1988).

Finally, preliminary studies in rats suggest that nitrendipine may offer protection against cocaine-induced cardiac arrhythmias, myocardial infarction, stroke and intracranial hemorrhage (Nahas et al., 1988).

It is thus readily apparent that the calcium channel blockers will become even more valuable as further indications for their use develop.

Chapter 16

Modulation of Calcium Channels in the Treatment of Ischemic Heart Disease

Lionel H. Opie

TABLE OF CONTENTS

I.	Introduction	274
II.	Mechanisms and Experimental Background	275
	A. Coronary Vasodilation and Increased Oxygen Supply	275
	B. Decreased Myocardial Oxygen Demand and Anti-Ischemic Effect	275
	C. Relief of Increased Diastolic Tension and a Favorable Redistribution of Blood to the Ischemic Zone	276
	D. A "Direct" Cellular Anti-Ischemic Effect?	277
III.	Chronic Stable Effort Angina	277
	A. Verapamil for Effort Angina	277
	B. Nifedipine for Effort Angina	278
	C. Diltiazem for Effort Angina	279
	D. Withdrawal of Calcium Blockers in Effort Angina	279
	E. Comparison of Calcium Blockers with Each Other in Effort Angina	280
	F. Calcium Blockers Compared with or Added to β-Adrenergic Blockade	280
	G. Efficacy of Calcium Blockers in Effort Angina: Summary	280
IV.	Angina Caused by Coronary Spasm: "Vasospastic Angina"	282
	A. Complexities in Vasospastic and Prinzmetal's Angina	282
	B. Calcium Blockers for Vasospastic Angina	282

		C.	Calcium Blockers for Vasospastic Angina: Placebo-Controlled and Comparative Studies 284
		D.	Vasospastic Angina: Summary 285

V. "Silent" Ischemic Episodes 285
    A. "Silent" Ischemia in Effort Angina 285

VI. "Mixed" Angina 286

VII. Angina at Rest 286
    A. Stable Angina at Rest 287
    B. Unstable Angina with Threatened Infarction (True Unstable Angina) 287
    C. Pathophysiology 287
    D. Unstable Angina: Summary 288

VIII. Acute Myocardial Infarction 288
    A. Ventricular Fibrillation: Experimental Data 288
    B. Infarct Size: Experimental and Clinical Data 289
    C. Nifedipine for AMI 289
    D. Verapamil for AMI 289
    E. Diltiazem for AMI—a Special Case? 290
    F. Calcium Blockers in Post-Infarction Follow-Up 290

IX. Reperfusion Injury 291

X. Failure of Calcium Blockers to Benefit Acute Myocardial Infarction and to Give Post-Infarction Protection? 291

XI. Percutaneous Transluminal Coronary Angioplasty (PTCA) 291

XII. Summary: Calcium Blockers in Anginal and Ischemic Syndromes 292

Acknowledgments 293

I. INTRODUCTION

Modulation of the activity of calcium channels in vascular smooth muscle and in the myocardium can be expected to play a major role in the therapy of ischemic heart disease. Myocardial ischemia is a state of inadequacy of the myocardial blood flow, almost always found in the presence of severe coronary artery disease, resulting in the two-fold tissue metabolic consequences of an inadequate supply of oxygen and an inadequate washout of metabolites. Myocardial ischemia results from an imbalance between the oxygen supply and demand ratio. Coronary vasodilation, a consequence of therapy with calcium channel blockers, is one mode of improving the blood supply to the ischemic myocardium, because calcium channel blockers ultimately interfere with the capacity of coronary arteries to constrict. Another way whereby calcium blockers could relieve myocardial ischemia is by decreasing the oxygen demand of the myocardium, as results when the load against which the heart works (the afterload) is reduced as the calcium channel blockers cause peripheral arterial vasodilation as a result of interference with calcium entry and, ultimately, the capacity of the arteries to constrict.

Thus the prime site of action of calcium blockers in ischemic heart disease is the vascular smooth muscle of the coronary arteries and the peripheral circulation. In addition, some calcium blockers have a direct negative inotropic effect on the myocardium as a result of inhibition of myocardial calcium channels and ultimately a decreased availability of calcium to the contractile proteins. This negative inotropic effect will decrease the myocardial oxygen demand, and thereby have an indirect anti-ischemic effect.

Today, calcium channel blockers are regarded as among the agents of first choice for the therapy of angina pectoris, a clinical syndrome associated with transient myocardial ischemia and resulting from the myocardial oxygen demand exceeding the supply. All of the three first generation calcium blockers, verapamil, nifedipine and diltiazem, have established antianginal potency. Of these three, verapamil was historically the first to be used against angina, followed by nifedipine and lastly diltiazem. The mechanisms of the antianginal effect of these three agents are very complex and fully reviewed by Opie (1988a), from which the following information is extracted and condensed. The second generation calcium blockers such as nicardipine, felodipine, isradipine and amlodipine (as well as many others) have been separately reviewed (Opie, 1988b) so that their use is not considered here. The antianginal mechanisms and effects of these newer dihydropyridines are likely to be similar to those of nifedipine, the "gold standard" for this group of compounds. (For a discussion of organic calcium channel blockers, see Section III, Chapter 13).

II. MECHANISMS AND EXPERIMENTAL BACKGROUND

The chief mechanisms proposed for the antianginal effects of calcium blockers vary among verapamil, nifedipine and diltiazem and include (1) coronary vasodilation, (2) decreased myocardial oxygen demand (decreased afterload, a decreased heart rate and a negative inotropic effect), (3) relief of an increased diastolic tension caused by ischemia with improved diastolic function and (4) a "direct" cellular anti-ischemic effect.

A. Coronary Vasodilation and Increased Oxygen Supply

Because the calcium blockers are coronary vasodilators, they are especially effective when coronary spasm is the cause of myocardial ischemia. Likewise, when spasm is added to organic stenosis to create "dynamic stenosis", then calcium blockers should also relieve the ischemia. Various degrees of coronary spasm (or failure of coronary relaxation) may also be invoked by exercise, so that coronary dilation may relieve exercise-induced angina. Experimentally, when exercise-induced coronary vasoconstriction is relieved by calcium blockers, blood flow increases to the subendocardial zones (Thaulow et al., 1988).

B. Decreased Myocardial Oxygen Demand and Anti-Ischemic Effect

Calcium blockers influence three of the determinants of the myocardial oxygen uptake. First, the afterload is reduced as peripheral vasodilation brings down the blood pressure. Second, some agents (especially diltiazem and in some studies verapamil) reduce the heart rate, whereas nifedipine may increase the heart rate. Third, there may be a direct negative

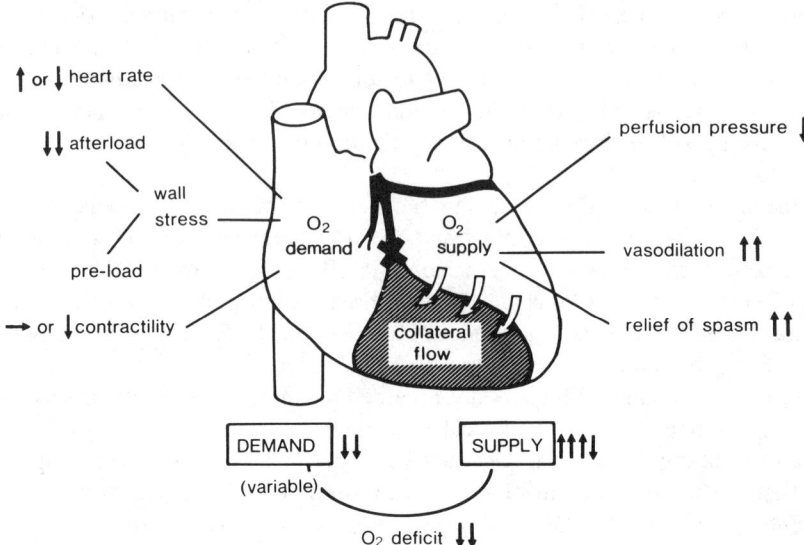

Figure 1. Possible beneficial effects of calcium channel blockers, as a group, on the ischemic myocardium. Upward arrows indicate increases, downward arrows decreases, and horizontal arrows no change. Note prominent role for vasodilatory effects: afterload reduction, coronary vasodilation and/or relief of vasospasm. Figure copyright L. H. Opie.

inotropic effect to reduce the oxygen demand, evident especially in the case of verapamil. In isolated preparations it is nifedipine that has the greatest negative inotropic effect of the three first-generation agents; however, this direct negative effect is usually more than offset by powerful afterload reduction.

When considering the possible protection by calcium blockers against regional ischemia, the effects of calcium blockers can usually be explained by the postulated favorable effects on the oxygen supply-demand equation (Figure 1). It is striking that the majority of studies showing myocardial protection relate to experiments in which calcium blockers were given before coronary occlusion and not after (De Leiris et al., 1984; Nayler, 1987). There are, however, many exceptions to this rule (Kloner and Braunwald, 1987) so that generalizations are not always possible. In all, it seems that these drugs have only a modest effect in limiting infarct size and may be most effective when the ischemic zone is small and the collateral flow adequate, as well as when there is early reperfusion (Kloner and Braunwald, 1987).

C. Relief of Increased Diastolic Tension and a Favorable Redistribution of Blood to the Ischemic Zone

During effort-induced angina or "demand" ischemia there is a decreased myocardial compliance with a rise of end-diastolic tension. The cellular basis of demand ischemia is thought to be an increased free cytosolic calcium (Lee et al., 1987), the origin of which may be internal redistribution of calcium as the glycolytic metabolic pathways fail (Opie, 1988a). Another intriguing possibility is that at least some of the calcium derives from influx through calcium channels. Accordingly, one hypothesis is that calcium blockers may improve diastolic function in angina with a subsequent relief of myocardial tension and removal of external pressure on the coronary arteries, with the ultimate improvement of blood supply to the ischemic zone. For example, the early increase in diastolic stiffness during angina pectoris (Apstein and Grossman, 1987) may be improved by verapamil so that left ventricular diastolic filling improves (Bonow et al., 1981). Through this mechanism diastolic function can improve while systolic function is unchanged.

D. A "Direct" Cellular Anti-Ischemic Effect?

Whether, in addition to benefits dependent on hemodynamic changes, the calcium blockers also "directly" protect against ischemia is not clear. For example, work on hypoxic isolated guinea pig papillary muscle (Nayler et al., 1976) shows that perfusion with verapamil has a direct protective effect. Isolated mitochondria from ischemic rabbit hearts pretreated by verapamil function better than those from ischemic hearts not pretreated by verapamil (Nayler et al., 1980). However, many such data can be explained by a negative inotropic effect of verapamil and, in the case of *in vivo* experiments, beneficial effects of pretreatment on the product of the heart rate and the systolic blood pressure, both determinants of the myocardial oxygen demand. Studies in which there is an additional protective effect, not explicable by changes in the myocardial oxygen supply-demand ratio, are listed by Opie (1988a). In each case the "protective" dose was well above the therapeutic free blood levels in humans.

Dose-response curves have been undertaken in only a few studies so that the efficacy of the calcium blockers at concentrations corresponding to the levels therapeutically effective in humans usually cannot be assessed. Watts et al. (1985) varied verapamil from $7.5 \times 10^{-8} M$ to $2 \times 10^{-6} M$ in experiments on the isolated ischemic rat hearts and found improved function upon reperfusion. Nifedipine $3 \times 10^{-8} M$ started to have some protective effect in the rat heart with subtotal global ischemia followed by reperfusion (De Jong et al., 1982), but a much better effect was found at $3 \times 10^{-7} M$ nifedipine. In a coronary ligated isolated working rat heart model of regional ischemia and developing infarction, $10^{-7} M$ nifedipine or diltiazem was required to reduce enzyme release (Hamm and Opie, 1983). In such experiments with regional ischemia, it is impossible to exclude an apparently minor, but possibly critical, redistribution of myocardial blood flow.

During the reperfusion period after severe myocardial ischemia, there may be a vigorous uptake of calcium by the myocardium with risk of cellular *calcium overload*. Only part of this excessive uptake of calcium is susceptible to calcium blockers (Nayler et al., 1980). The calcium blockers may act chiefly by limiting the extent of ischemic damage which in turn decreases the severity of reperfusion damage (Weishaar and Bing, 1980; Nayler, 1982).

In general, the concentrations of calcium blockers with "direct" anti-hypoxic or anti-ischemic effects are well above the therapeutic levels in man (about $10^{-8} M$). Thus the mechanisms for the antianginal effects of these agents should first be sought in favorable alterations in the myocardial oxygen supply-demand ratio, including coronary vasodilation, afterload reduction and improved diastolic function.

III. CHRONIC STABLE EFFORT ANGINA

It is conventionally held that the benefit of β-adrenergic blockade in this condition is predominantly mediated by a reduction of the oxygen demand, whereas calcium blockers may increase the oxygen supply. However, the complex mode of action of calcium blockers as described means that the contrasting modes of actions of these two types of agents are, in fact, an oversimplification. Furthermore, the calcium blockers differ among themselves. For example, nifedipine may be a more powerful vasodilator and afterload reducer than the other two first-generation agents, whereas a negative inotropic effect may be more prominent among the clinical actions of verapamil.

A. Verapamil for Effort Angina

Verapamil was used as an antianginal agent before it became popular in the control of supraventricular arrhythmias (Knoch et al., 1963; Tschirdewahn and Klepzig, 1963). In angina, the earliest double-blind control studies were done by Sandler et al. (1968) and Livesley et al. (1973) who compared verapamil with propranolol and found that 120 to 360 mg verapamil daily was the approximate equivalent of propranolol 300 mg daily. Further studies have shown that a similar dose of verapamil (360 mg) is also the approximate equivalent of metoprolol 200 mg 2× daily (Arnman and Ryden, 1982) or nifedipine 60 mg daily (Dawson et al., 1981). By now verapamil is very well documented for its benefit in effort angina (Andreason et al., 1975; Subramanian et al., 1980; Hecht et al., 1981; Subramanian et al., 1982; Subramanian, 1983; Weiner et al., 1987).

Verapamil compared with propranolol—In a review of seven trials on 117 patients, Opie (1988a) concludes that verapamil 360 mg daily is the approximate equivalent of propranolol 160 to 320 mg daily. Thus all the trials together have merely confirmed what the earlier studies show. When the dose of verapamil is even higher (480 mg daily), it is more effective than propranolol; the disadvantage of the high dose is an increasing incidence of side effects of verapamil.

Verapamil plus propranolol—Although there is the possibility of added negative inotropic, chronotropic and dromotropic effects, this combination is hemodynamically acceptable provided that very careful patient selection is made (Leon et al., 1981; Subramanian, 1983). Antianginal effects are more marked than with either agent given singly (Subramanian, 1983; Johnston et al., 1985; Winniford et al., 1985; Findlay et al., 1987). Details of the verapamil-β-blocker interaction are considered later.

Mechanisms of antianginal effects of verapamil—The hypotensive effect of verapamil means that afterload and double-product during exercise are reduced especially when the heart rate also falls (Subramanian, 1983). The prominent negative inotropic effect of verapamil should also decrease myocardial oxygen demand (Leon et al., 1981). The preferential effect of verapamil on the ischemic rather than on the nonischemic myocardium (Smith et al., 1976) enhances the benefits of its negative inotropic effect. Verapamil is a coronary vasodilator although only moderately powerful when compared with nitroglycerin (Chew et al., 1983). Verapamil can improve left ventricular diastolic filling (Bonow et al., 1981) probably by enhancing early relaxation.

B. Nifedipine for Effort Angina

Nifedipine likewise has been shown to be an effective antianginal agent. When given acutely, as a dose of 10 or 20 mg, the time that the patient can exercise till the onset of angina or shortness of breath or some other limiting aspect (the exercise time) improves 30 min after the dose (Atterhog et al., 1975) and the effect lasts for at least 3 and possibly up to 6 h (Ardissino et al., 1983; Chaitman et al., 1984).

Of the *chronic studies,* the more important are as follows. First, nifedipine 60 mg daily rendered 29% of patients angina free and prolonged exercise time by 39% (Subramanian, 1983) (a significant difference from placebo). In some studies a total daily dose of 30 mg nifedipine is as effective as higher doses (Lynch et al., 1980; Sherman and Liang, 1983). In a placebo-controlled crossover and parallel designed study (Mueller and Chahine, 1981), upward titration from nifedipine 30 mg daily to a mean dose of 51 mg daily was required to prolong exercise time, whereas the lower dose only reduced anginal attacks and nitroglycerin use without any effect on exercise time. However, this study (Mueller and Chahine, 1981) only used one-tailed p values, lessening the statistical impact of the data. The antianginal effect of nifedipine is considerably reduced in patients who smoke (Deanfield et al., 1984). This surprising interaction is completely different from the benefit of nifedipine in relieving coronary vasoconstriction induced by smoking (Winniford et al., 1987).

Nifedipine compared with or combined with β-blockade—In a comparison of nifedipine with β-blockade in five studies on 128 patients, Opie (1988a) concluded that β-blockade appeared to be the superior modality of therapy in three studies, whereas equivalent benefit between these two types of agents was found in the other two studies.

Provocation of angina by nifedipine—One reason why nifedipine may not always work is that careful dose titration is required to avoid precipitation of ischemia (Jariwalla et al., 1978; Subramanian, 1983; Deanfield et al., 1983). Presumably powerful vasodilation in the nonischemic zone "steals" blood from the ischemic zone.

Mechanisms for antianginal effects of nifedipine—As in the case of the other calcium blockers, the mechanisms whereby nifedipine benefits effort angina are very complex. Some of the factors to consider are: (1) the extent of afterload reduction, variable from patient to patient; (2) the extent of reflex tachycardia, more marked during the acute than chronic phases of nifedipine administration (Kiowski et al., 1986); (3) the contribution of coronary artery spasm and "dynamic stenosis" to the symptom of angina; (4) the risk of "coronary steal" with production of ischemic chest pain and (5) the coronary artery anatomy. The presence of adequate pre-existing myocardial collaterals tends to diminish the effect of nifedipine (Schulz et al., 1985), suggesting that dilation of potential collaterals to the ischemic zone is an important mechanism of action. Taking all these factors together, two of the most important mechanisms for relief of angina could be an improved myocardial oxygen supply, and lessened myocardial diastolic stiffness.

C. Diltiazem for Effort Angina

Diltiazem also is an effective agent for treating angina pectoris. Although the last of the three first-generation calcium blockers to appear in the U.S., diltiazem has established itself well. Its antianginal effect can be shown both acutely and chronically. When 120 mg diltiazem was given *acutely,* the time to peak exercise was increased by 29%, 3 h after the acute dose and during the submaximal period of exercise, the rate-pressure product fell, probably an indicator of a decreased oxygen demand (Wagniart et al., 1982). In another study, exercise time increased by 31% and improved work capacity was still present 8 h after 120 mg diltiazem (Chaitman et al., 1984). In patients with ischemic heart disease and with mild impairment of myocardial contraction, diltiazem given acutely (again 120 mg) improved the systolic function of the myocardium (Anderson et al., 1984). During *chronic therapy,* doses of 240 to 360 mg daily have been shown to benefit patients with effort angina (Hossack et al., 1982; Hung et al., 1983; Lindenberg et al., 1983; Go and Hollenberg, 1984; Strauss and Parisi, 1985; Findlay et al., 1987). In a dose-titration study (Lindenberg et al., 1983), anginal frequency was reduced more by a higher dose (360 mg daily) than by 240 mg daily; however, even 120 mg daily had an effect. The antianginal benefit of diltiazem 360 mg daily has been shown over 4 months (Khurmi and Raftery, 1987a) and up to 52 weeks (Khurmi et al., 1984).

Diltiazem compared with propranolol—In 5 studies on 108 patients, Opie (1988a) concluded that diltiazem 240 to 360 mg daily was the approximate equivalent of the same dose of propranolol or, in some studies, was better.

Mechanism of angina relief by diltiazem—Again a complex mode of action is invoked (Hung et al., 1983; Anderson et al., 1984; Weiner et al., 1986; Joyal et al., 1986), including enhanced mechanical performance of the ischemic myocardium, a reduced heart rate and blood pressure and improved diastolic function.

D. Withdrawal of Calcium Blockers in Effort Angina

Few general statements can be made because the rebound of anginal pain is more likely to occur in patients with unstable spasm-related pain. There are several studies suggesting that verapamil can be safely stopped (Frishman et al., 1982; Danish Study Group, 1986). When, however, subtle signs of ischemia are searched for after stopping verapamil, such as the number of asymptomatic episodes of ST deviation on the ambulatory monitoring system ("silent ischemia"), there is some evidence for a verapamil withdrawal effect (Lahiri et al., 1987). In the case of nifedipine, there appears to be no study in which patients with effort angina had their nifedipine suddenly stopped. In contrast, in patients who also received therapy with nitrates and propranolol, nifedipine could be abruptly withdrawn, seemingly without any symptomatic effects (Gottlieb et al., 1984). In the case of diltiazem, there are no reports of a withdrawal syndrome nor have any studies been designed around this point (for diltiazem withdrawal in vasospastic angina, see Schroeder et al., 1985).

E. Comparison of Calcium Blockers with Each Other in Effort Angina

Verapamil vs. nifedipine—There are surprisingly few good comparative studies (Table 1; Opie, 1988a). Verapamil 360 mg is either the equal of nifedipine 60 mg (daily doses) or somewhat better. Diltiazem 360 mg, on the other hand, was the approximate equivalent of verapamil 360 to 480 mg daily, so that these two agents are more or less equipotent. Diltiazem seems marginally better than nifedipine 30 mg daily (without any comparison with nifedipine 60 mg daily).

F. Calcium Blockers Compared with or Added to β-Adrenergic Blockade

Three separate studies suggest that *verapamil-β-blocker* therapy may lead to hemodynamic problems. The first study sequentially compared the effects of diltiazem (240 mg daily), verapamil (360 mg daily) or nifedipine (60 mg daily) when added to propranolol 160 mg daily (Johnston et al., 1985). All these agents equally reduced the number of anginal attacks and the degree of ST-segment depression but *propranolol-nifedipine* was statistically the best. Most side effects occurred with propranolol-verapamil (chiefly constipation), a lesser number with propranolol-nifedipine (chiefly dizziness and leg swelling) and least with propranolol-diltiazem (with the incidence of side effects being close to placebo). From the hemodynamic point of view propranolol-nifedipine was superior, whereas subjectively propranolol-diltiazem was better. In the second study, aimed at comparing verapamil with nifedipine, a population of patients with angina not fully responsive to propranolol (mean daily dose 229 mg) had either verapamil 360 mg daily or nifedipine 60 mg daily added in a double-blind, cross-over manner (Winniford et al., 1985). Propranolol plus verapamil was a better antianginal combination than propranolol plus nifedipine, yet was more risky from the hemodynamic point of view because symptomatic bradycardia occurred in 2/10 patients on propranolol-verapamil (Winniford et al., 1985). In the third study, combining verapamil 360 mg daily with a cardioselective β-blocker, atenolol 100 mg daily, led to 4/15 patients withdrawing from the trial (Findlay et al., 1987). Therefore, the overall evidence is that the combination verapamil-β-blocker can only be given if there is close supervision and patient selection. The incidence of undesirable interactions is decreased by either reducing the dose of verapamil to 240 mg daily (Winniford et al., 1985) or the propranolol dose to 160 mg (Johnston et al., 1985).

G. Efficacy of Calcium Blockers in Effort Angina: Summary

As reviewed by Opie (1988a), *the antianginal benefits of these three agents are similar in most studies. However, imperfect evidence suggests that nifedipine might be marginally less effective on its own than is verapamil or diltiazem; further detailed controlled trials are required. In comparison with β-blockade, nifedipine was worse than β-blockade in 3/5*

TABLE 1
Comparative Effects of the Three First Generation Calcium Blockers in Chronic Effort Angina

Author	Trial design	Patients (n)	Drug test period	Daily doses of calcium blockers			End-point	Result
				Verapamil	Nifedipine	Diltiazem		
Dawson et al., 1981	Pl, R, DB, Seq, WOut	16	4 weeks	360 mg	60 mg	—	Clin	V = N
Subramanian et al., 1982	Pl, R, DB, CO, no WOut	28	4 weeks	360 mg	60 mg	—	Clin	V > N testing, Holter
Weiner et al., 1984	DB, CO, then parallel	46	6 weeks–9 months	480 mg	—	360 mg	Clin	D = V
Schurtz et al., 1983	Pl, R, DB, CO, no WOut	20	15 days	—	30 mg	180 mg	Ex	D = N (ex) D > N (ECG)
Khurmi and Raftery 1987a	Initial Pl, then open-label parallel	45	16 weeks	360 mg	—	360 mg	Ex	D = V
Khurmi and Raftery 1987b	Separate trials each vs. placebo	146	2–4 weeks	360 mg	60 mg	360 mg	Clin, ex testing	D, V > N (ex) V > N, D (pain)

Note: Abbreviations: Ex = exercise; Pl = placebo; R = randomized; DB = double-blind; Seq = sequential; WOut = washout; Clin = clinical; D = diltiazem; N = nifedipine; V = verapamil; ECG = electrocardiogram; V = N means that the effects of verapamil were similar to those of nifedipine; V > N means that verapamil efects were judged better than those of nifedipine.

From Opie (1988a) by permission of *Cardiovascular Drugs and Therapy* and Kluwer Academic Publishers.

studies, whereas verapamil did better than β-blockade in 4/6 studies and diltiazem was equal to β-blockade in 4 studies. *In contrast, nifedipine is easier and safer to combine with β-blockade than the other two agents.* The approximate daily dose equivalents of these agents are verapamil 360 to 480 mg, nifedipine 60 to 80 mg, diltiazem 360 mg and propranolol 240 to 360 mg daily. In the case of nifedipine, it should be realized that achievement of the optimal dose requires dose titration, generally not done in these trials. The highest dose required may be more than usually used and may be up to 120 mg daily. Furthermore, with prolonged use of nifedipine the initial reflex sympathetic stimulation seems blunted, at least during the chronic therapy of hypertension (Kiowski et al., 1986; Landmark, 1985). Thus, possibly prolonged antianginal therapy with nifedipine might have given different results from those in the acute trials.

IV. ANGINA CAUSED BY CORONARY SPASM: "VASOSPASTIC ANGINA"

In the genesis of coronary spasm, a critical interaction is that between the endothelium, the platelets and the vasoconstrictors released from the platelets such as serotonin and thromboxane-A_2. When the endothelium is damaged, certain vasodilator stimuli are blunted (Rapaport et al., 1983) and vasoconstrictor stimuli predominate. The endothelial relaxing factor (Furchgott, 1983; Griffith et al., 1988) is of crucial importance. Vasodilators may accordingly be divided into those that require the endothelium for their activity and those that do not, such as the nitrates (Rapaport and Murad, 1983). In conditions where there is endothelial damage, as in atheroma, stimuli which are normally vasodilatory may become vasoconstrictive. For example, serotonin released from damaged platelets becomes a vasoconstrictor when the endothelium is removed. Maintenance of endothelial integrity may be critical in avoiding coronary artery spasm. At present no therapeutic agents are available to maintain or to restore endothelial integrity.

A. Complexities in Vasospastic and Prinzmetal's Angina

The diagnosis in a typical case of Prinzmetal's angina with electrocardiographic ST-segment elevation is usually not in doubt; calcium blockers rather than β-blockers should be the basic therapy. Although coronary spasm is a proven hypothesis to explain Prinzmetal's angina (Opie and Maseri, 1986), it should not be forgotten that Prinzmetal described the combination of coronary spasm and organic coronary artery disease (Prinzmetal et al., 1959) and that two of his original three patients went on to develop a classical myocardial infarction. Thus it is difficult to exclude underlying coronary artery disease which may cause associated ischemic syndromes that respond well to β-blockade, even if the spasm itself were to respond poorly. Another problem is that the more subtle features of lesser degrees of spasm (Maseri et al., 1978) merge into "silent" ischemia and so called "mixed" angina (see next section) in which calcium blockers and β-blockers may both be effective. It requires emphasis that the role of coronary spasm in silent ischemia and mixed angina is still unproven. Therefore the "edges" of the clinical picture of coronary spasm are now increasingly blurred.

B. Calcium Blockers for Vasospastic Angina

Open-label studies—All three major calcium blockers are very effective in relieving Prinzmetal's variant angina (Table 2). Nifedipine (40 to 160 mg/day) was strikingly effective in an open-label study. It eliminated painful episodes in nearly two thirds of patients while side effects required withdrawal in only 5% of patients (Antman et al., 1980). However,

TABLE 2
Comparative Effects of the Three First Generation Calcium Blockers in Vasospastic Angina

Author	Patients (n)	Trial design	Drug test period	Evidence for spasm	Daily doses of calcium blockers			End-point	Result
					Verapamil	Nifedipine	Diltiazem		
Waters et al., 1981	27	R, seq, no WOut, open-label	3 days	Repeated ergonovine	480 mg	80 mg	360 mg	Angina or ECG changes	N = D (81% benefit) V = 66% benefit
Prida et al., 1987	15	DB, CO, open-label for N + D phase	16 weeks	Angiospasm spontaneous or ergonovine	—	30-120 mg (mean 82)	90-360 mg (mean 257)	Clinical; ambulatory ECG	N = D; D fewer side effects; D (206 mg mean) + N (61 mg mean) best[a]
Pepine et al., 1983	45	Open-label	1 year	ST-elevation with pain or angiospasm	(Mean 419 mg n = 16)	(Mean 68 mg n = 16)	(Mean 240 mg n = 13)	Pain	D = N = 69% benefit V = 56% benefit

Note: Abbreviations as in Table 1. Angio = angiographic.

[a] D + N caused frequent side effects while helping those patients who could tolerate the combination.

From Opie (1988a) with permission of *Cardiovascular Drugs and Therapy* and Kluwer Academic Publishers.

coronary artery spasm had been diagnosed by rigorous angiographic criteria. Another convincing, yet unblinded, study is that of Kimura and Kishida (1981). In 286 Japanese patients with typical Prinzmetal's angina, over 86% of patients benefited from verapamil about 240 mg daily, 94% from nifedipine 40 mg daily, and 91% from diltiazem 160 mg daily. Today all these doses could be regarded as "low". Calcium blocker therapy is associated with long-term survival over 15 months of more than 90% of patients receiving calcium blocker therapy (Waters et al., 1983); nifedipine 80 mg daily was similar to diltiazem 360 mg daily or verapamil 480 mg daily. However, such long-term studies cannot be properly controlled for obvious reasons.

C. Calcium Blockers for Vasospastic Angina: Placebo-Controlled and Comparative Studies

In an acute and placebo-controlled study, nifedipine (10 to 20 mg every 4 h) decreased the incidence of electrocardiographic ST-segment deviations (Previtali et al., 1980). In a complex study (Winniford et al., 1982), verapamil and placebo were first compared in a long-term, double-blind randomized trial lasting 9 months, and then open-label nifedipine was followed for 2 months. A mean nifedipine dose of 71 mg daily was about equivalent to verapamil 450 mg daily, both being given in 3 to 4 daily doses, as judged by the reduction of chest pain, usage of nitroglycerin, and decreased ST-segment deviation. The patients were treated throughout with isosorbide dinitrate (average daily dose 104 mg/day). Side effects of verapamil and nifedipine were similar, but subjectively worse for nifedipine in that seven of the patients had dose-limiting adverse effects and in one patient therapy was stopped because of orthostatic hypotension.

Diltiazem vs. nifedipine in coronary vasospasm—In 15 patients with angiographically proven spasm (spontaneous or ergonovine-induced), the effects of nifedipine (30 to 120 mg, mean 82 mg daily) were about equal to those of diltiazem (90 to 360 mg, mean 257 mg daily); end-points were (1) the rate of clinical attacks and (2) nitroglycerin usage (Prida et al., 1987). More nifedipine-treated patients had adverse reactions than those treated by diltiazem ($p < 0.05$).

Ergonovine-induced angina—Ergonovine is a coronary vasoconstrictor agent which may precipitate attacks of vasospastic angina. In a study on 27 hospitalized patients with typical clinical Prinzmetal's angina (Waters et al., 1981), provocative testing with ergonovine was used to assess the efficacy of nifedipine (80 mg daily), diltiazem (360 mg daily) and verapamil (480 mg daily), given in randomized order. Different patients, all with vasospastic angina, responded differently to the three calcium blockers and it appeared difficult to select the best agent in advance.

Calcium blockers compared or combined with long-acting nitrates in vasospastic angina—Generally, nifedipine (82 mg mean daily dose) was preferred to isosorbide (66 mg mean daily dose) because of the increased subjective benefit and fewer uncomfortable side effects (Ginsburg et al., 1982). In a second study which used the number of anginal attacks and the use of nitroglycerin as end-points, nifedipine in a mean dose of 65 mg daily was approximately equivalent to isosorbide dinitrate in a mean dose of 75 mg daily (Hill et al., 1982). In vasospastic angina, combination therapy by calcium blocker-nitrate dramatically reduced the frequency of angina and ST-segment deviation, and either verapamil or nifedipine combined with a nitrate had similar potency (Hill et al., 1982; Winniford et al., 1984).

Calcium blockers compared or combined with β-blockade in vasospastic angina—Although β-blockers are widely regarded as inappropriate therapy for vasospastic angina, strict studies are few. Some indirect evidence is that transient ECG ST-segment shifts in patients with angina at rest are considerably more improved by verapamil 400 mg daily than by propranolol 200 mg daily (Parodi et al., 1986). However, there was no direct proof that spasm was causing the ST-segment changes.

Rebound after cessation of calcium blocker therapy in vasospastic angina—Nifedipine therapy should not be abruptly halted because the frequency and duration of attacks may increase (Lette et al., 1984). No data are available on rebound after cessation of verapamil therapy. However, because nifedipine tends to increase circulating catecholamines (Kiowski et al., 1986) more than verapamil (Muiesan et al., 1986), there may be a greater chance of rebound with nifedipine than with verapamil. In the case of diltiazem, abrupt withdrawal of daily doses of 120 to 240 mg did not precipitate attacks in patients with Prinzmetal's angina (Schroeder et al., 1985).

Exertional vasospasm—Sometimes vasospasm can be provoked by exertion especially in the early morning (Yasue et al., 1979a). This topic appears to be poorly studied with no double-blind studies. In four patients with exercise-induced coronary spasm (Yasue et al., 1979b) anginal attacks were not inhibited by propranolol but were inhibited by diltiazem (90 mg) or nifedipine (20 mg).

D. Vasospastic Angina: Summary

As concluded by Opie (1988a): *"Therapy with all the major three calcium blockers is highly effective in vasospastic angina manifesting as Prinzmetal's variant angina.* The various agents appear to be approximately equally effective. However, strict double-blind comparisons are not available. Nifedipine 60 to 80 mg daily is the rough equivalent of isosorbide dinitrate 60 to 80 mg daily. Nifedipine or verapamil may be combined with isosorbide dinitrate (diltiazem apparently not tested). Whereas β-blockade on its own may exaggerate the condition or fail to benefit, the combination diltiazem-propranolol is as effective as diltiazem alone, so that diltiazem annulled the harmful effects of β-blockade. Efficacy of calcium blocker agents in the case of vasospastic angina must not be extrapolated to angina at rest and unstable angina."

V. "SILENT" ISCHEMIC EPISODES

A. "Silent" Ischemia in Effort Angina

Continuous ST-monitoring techniques can detect transient silent ischemia in angina pectoris. Such silent ST changes are as much an expression of the anginal syndrome as is pain. In oversimplified terms, silent ischemia is "angina without anginal pain". Possibly mild or moderately severe episodes may be silent, whereas more severe or prolonged episodes may reach the threshold for pain (Stone, 1987). The threshold for angina varies among patients; a generalized defective perception of all painful stimuli may explain some cases of silent ischemia (Glazier et al., 1986). Such silent ischemic episodes in patients with effort angina respond both to β-blockers and calcium blockers (Stone, 1987). Heart rate seems to be an important determinant of the development of silent ECG changes (Cocco et al., 1979; Quyyumi et al., 1984) so that atenolol which decreased the heart rate had a better effect on ECG ischemia than did pindolol. Logically, response to β-blockade excludes a specifically spastic cause for these silent episodes, because β-blockade is not known to relieve coronary spasm, rather it is thought to exaggerate any spasm present. In the study of Lynch et al. (1980), propranolol was more effective than nifedipine in relieving symptoms of angina and in decreasing the incidence of ST deviations (some of which were silent).

ST deviations in patients with transient short duration chest pain at rest—Some patients with angina at rest have only short-lived attacks of chest pain, which are benefited by calcium blockers (Parodi et al., 1982; Parodi et al., 1986; Rizzon et al., 1986). In a very carefully controlled study interspersed with three placebo periods, verapamil 400 mg daily was considerably better than propranolol 300 mg daily (Parodi et al., 1982).

However, there is no proof that calcium blockers by reversing silent ischemia can alter symptoms or prognosis in patients with ischemic heart disease.

ST deviations in unstable (pre-infarct) angina—In patients with genuine unstable angina, added silent ischemic episodes indicate a poor prognosis (Gottlieb et al., 1986). Specific aggressive therapy may be indicated; this topic is now being clarified.

VI. "MIXED" ANGINA

Although the term "mixed" angina has been widely used, the concept is now becoming controversial and the present ideal is to incorporate this category into other better established descriptions such as unstable angina or vasospastic angina, depending on the prominent clinical presentation.* Mixed angina was defined by Maseri et al. (Maseri et al., 1985) as angina in which two quite different basic pathophysiological mechanisms were at work, namely both an excessive increase in myocardial oxygen demand (secondary angina) and a transient impairment of coronary blood flow supply (primary angina). Stone et al. (1983) used the term "mixed angina" to describe patients who had both classic exertional angina as well as clinically suspected coronary vasospasm defined as occasional episodes of rest angina or by a variable effort threshold for the angina; however, patients with ST-segment elevation during pain were excluded and vasospasm was not proven. Nifedipine therapy was thought to be most effective in patients with "pure" vasospasm and least effective in patients with classical exertional angina. For Andre-Fouet et al. (1983) mixed angina meant the onset of spontaneous angina at rest in patients previously known to have effort angina; such patients responded equally well to diltiazem or propranolol therapy. However, these studies were not undertaken on an intention-to-treat basis but rather mixed angina was retrospectively defined as a subgroup of patients. Therefore the conclusions are subject to statistical reserve. *Formal proper therapeutic trials in patients with documented and proven "mixed" angina have not yet been reported.* To prove that a component of mixed angina is caused by coronary vasospasm is no easy task because, strictly speaking, angiographic evidence is required (Maseri et al., 1978; Antman et al., 1980).

The concept of preferential calcium blocker therapy for mixed angina is at present purely conjectural and not supported by the data of Andre-Fouet et al. (1983). It is only when true coronary spasm is thought to be the cause of spontaneous angina in a patient also with effort angina, that it is logical to prefer calcium blockers to β-blockade as first-line therapy. In other patients with effort angina and ST deviations, either calcium blockers or β-blockade may be selected according to other criteria such as side effects and expected tolerance (Andre-Fouet et al., 1983).

VII. ANGINA AT REST

Definitions are critical without being standard. Thus frequently "angina at rest" is confused with "unstable angina" which merges into "pre-infarction angina" and "threatened myocardial infarction". Both angina at rest and unstable angina have spontaneous anginal pain at rest, not evoked by any known external factor. Yet it is particularly important to distinguish between angina at rest with short-lived episodes of chest pain, usually less

---

* This and the following three sections closely follow Opie (1988a) and are reproduced by permission of *Cardiovascular Drugs and Therapy* and Kluwer Academic Publishers, Boston.

than 15 min (Parodi et al., 1986), from the longer lasting and much more serious attacks of true pre-infarction unstable angina (see next section). For example, it seems as if the patients studied by Parodi et al. (1982, 1986) and Rizzon et al. (1986) had repetitive stable short-lived attacks of chest pain or ECG episodes, with stable angina at rest, placing them in a different clinical category from unstable angina or the "intermediate coronary syndrome". Thus, there are important differences between stable angina at rest (Parodi et al., 1986) and unstable angina at rest. Stone (1987) likewise distinguishes between (1) "simply the presence of angina at rest with reversible ST-segment deviation", which is possibly due to coronary vasospasm, and (2) "unstable angina with its heterogeneous pathophysiology". In any given patient such distinctions may be arbitrary and there is, in fact, a spectrum of conditions extending all the way from Prinzmetal's angina through angina at rest to unstable angina and myocardial infarction (Maseri et al., 1978).

A. Stable Angina at Rest

In angina at rest, there are several different patient populations which may be involved, varying from (1) those with short-lived attacks of repetitive chest pain, hypothetically caused by coronary spasm or another cause of intermittent coronary obstruction, and accompanied by frequent ST-segment deviations, to (2) the situation in unstable angina with threatened infarction where the pain is longer in duration and the situation unstable so that infarction is truly a risk. When considering the patients in the first category, the evidence for the benefit of verapamil is strongest but all calcium blockers are likely to work. When considering the patients in the second category (see next section), nifedipine is less effective and is contraindicated unless accompanied by β-adrenergic blockade. In patients with stable angina at rest, characterized by very short-lived episodes of chest pain and numerous transient ST-segment deviations on the ECG, calcium blockers are better than propranolol which may be ineffective (Parodi et al., 1979; Previtali et al., 1980; Parodi et al., 1982; Andre-Fouet et al., 1983; Moll et al., 1984; Parodi et al., 1986).

B. Unstable Angina with Threatened Infarction (True Unstable Angina)

In true unstable angina at rest, one of the following is required (Rahimtoola et al., 1983): (1) crescendo angina, being the presence of anginal pain with a recent increase in frequency, intensity and duration; (2) acute coronary insufficiency (= "intermediate coronary syndrome") with prolonged anginal pain poorly relieved by nitrates yet without ECG or enzyme evidence of acute myocardial infarction or (3) spontaneous angina 3 to 30 days after the onset of acute myocardial infarction. This definition includes those patients with longer episodes of chest pain lasting 15 to 20 min or more, which may go onto myocardial infarction as end-point (Gazes et al., 1973; Fischl et al., 1973).

C. Pathophysiology

The heterogeneous nature of the pathophysiology of unstable angina is now becoming apparent. A ruptured or fissured or ulcerated atherosclerotic plaque may lead to a nidus for platelet aggregation and partial thrombosis, the latter usually being reversible in unstable angina (Forrester et al., 1987). The plaque can act as a trigger for spasm-induced "dynamic stenosis", possibly by an interaction between platelets and the damaged endothelium, leading to release of vasoconstrictive substances including thromboxane A_2 (Stone, 1987). The four critical factors in the mechanism of unstable angina are the atherosclerotic plaque, the platelets, a partial thrombus and coronary vasospasm. These heterogeneous factors demand a complex approach to therapy, especially when it is considered that prolonged pain and left ventricular failure, as found in some patients with the "intermediate coronary syndrome", may lead to catecholamine release with consequences such as tachycardia and a metabolically based increased myocardial oxygen demand (Opie, 1975).

Nifedipine for true unstable angina at rest—Although nifedipine is excellent therapy for short-lived episodes of chest pain and for Prinzmetal's angina, recent evidence shows that when used as sole therapy in true unstable angina, it is not as good as propranolol (Muller et al., 1984) and probably detrimental (HINT Research Group, 1986). In patients with ischemic pain exceeding 45 min, classified as having threatened myocardial infarction (Muller et al., 1984), nifedipine monotherapy actually increased mortality with p <0.02. These *reservations about the use of nifedipine in true unstable angina at rest* are supported by isolated case reports of adverse responses to this drug in unstable angina (Yokoyama et al., 1982; Sia et al., 1985). In the HINT Research Group (1986), the addition of nifedipine was clearly beneficial for patients with true unstable angina already receiving β-blockade, yet the combination nifedipine-metoprolol was no better than metoprolol alone for patients not already receiving β-blockade. Thus, in the latter category, β-blockade rather than nifedipine was recommended as first-line therapy.

Diltiazem for true unstable angina at rest—In "true" unstable angina in which Prinzmetal's variant angina was excluded and in which the chest pain was prolonged so that admission to a Coronary Care Unit was required, both the calcium blocker diltiazem (360 mg daily) and propranolol (240 mg daily) were equally effective (Theroux et al., 1985).

Verapamil for true unstable angina at rest—In the study by Capucci et al. (1983), some patients had only short-lived attacks of pain and others were seemingly on the way to infarction. Verapamil 480 mg/day was compared with propranolol 240 mg/day; both agents were effective, although verapamil seemed better. This study lends support to the early use of calcium blockers of the verapamil-diltiazem group in true unstable angina. In a follow-up study, verapamil continued to produce benefit without altering the natural history so that there was still a high incidence of death from myocardial infarction (Scheidt et al., 1982).

D. Unstable Angina: Summary

In true unstable angina at rest, where infarction is threatened, there are arguments against the use of nifedipine so that it is usually contraindicated unless used together with β-blockade (Muller et al., 1984; HINT Research Group, 1986). It seems preferable to use diltiazem or verapamil if calcium blocker monotherapy is desired, although there are no strict comparisons with nifedipine. *Therefore the closer the patient is to threatened infarction, the stronger is the case for β-blockade. On the other hand, the closer the patient is to Prinzmetal's vasospastic angina, the stronger is the case for calcium blockers. These potential differences appear to be particularly important in the case of nifedipine.*

VIII. ACUTE MYOCARDIAL INFARCTION

For any agent to be fully effective in acute myocardial infarction requires demonstration of a reduction in the death rate, as in the case of β-blockade. Failing that observation, a reduction in ventricular fibrillation and/or infarct size is the next best.

A. Ventricular Fibrillation: Experimental Data

In patients with previous myocardial infarction, sudden death is reduced by β-adrenergic receptor antagonism, presumably as a result of a decreased incidence of ventricular fibrillation. In acute coronary occlusion, an elevation of tissue cyclic AMP in ischemic tissue could be linked to the onset of ventricular fibrillation (Lubbe et al., 1978; Opie et al., 1979;

Opie et al., 1980). Calcium is the proposed active "messenger" of cyclic AMP. A calcium-dependent transient inward current may underlie the development of ventricular automaticity in mildly ischemic or nonischemic tissue (Coetzee et al., 1987; Coetzee and Opie, 1987). In the isolated rat heart model with ligation and with added adrenaline stimulation, the inhibition of the fall in the ventricular fibrillation threshold by l-verapamil rather than by d-verapamil (Thandroyen et al., 1986) favors the view that the calcium channel blocker effects of verapamil prevented the added effects of catecholamine stimulation in the presence of coronary ligation. However, without catecholamine stimulation, both d- and l-verapamil isomers were equally effective, so that specific calcium antagonism by itself does not prevent ventricular fibrillation.

B. Infarct Size: Experimental and Clinical Data

Although all three calcium blockers can decrease enzyme release from the isolated rat heart with coronary artery ligation, the doses used were supratherapeutic (Hamm and Opie, 1983). The unexpected finding that diltiazem actually improved the capacity of the ischemic left ventricle to work has not yet been confirmed. There are important distinctions between decreased ischemic injury and a reduction of the ultimate infarct size. Kingma and Yellon (1988) found that verapamil could improve experimental infarct size in an embolism-occlusion dog model, in contrast to the studies of Reimer et al. (1985). The differences between the two studies (Kingma and Yellon, 1988) lead to the inevitable conclusion that the infarct reducing capacity of calcium blockers depends on the experimental circumstances chosen.

C. Nifedipine for AMI

Nifedipine is among the best studied of the calcium blocker agents in patients with acute myocardial infarction (AMI). Although nifedipine (10 mg sublingually) may improve a low cardiac output and reduce a high wedge pressure, as well as bring down the blood pressure (Gordon et al., 1984; Gordon et al., 1986), upon formal testing nifedipine has shown no ultimate clinical benefit in two large multicenter trials (Muller et al., 1984; Sirnes et al., 1984). Indeed in one trial patients randomized to nifedipine showed some excess mortality (Muller et al., 1984). In the giant nifedipine TRENT Study, nifedipine given 10 mg 4× daily for 28 days showed neither benefit nor harm (Wilcox et al., 1986); patients on prior β-blockade therapy had a reduced mortality, thereby reconfirming the benefits of β-blockade as opposed to those of nifedipine. There is no indication for routine use of nifedipine in AMI or threatened myocardial infarction (Gordon et al., 1986). The possible adverse effects of nifedipine in true unstable angina also suggest that *nifedipine in the absence of β-blockade is not the therapy of choice in threatened myocardial infarction* or actual AMI, unless the specific hemodynamic changes induced by nifedipine are desired.

D. Verapamil for AMI

Regarding verapamil used for patients with acute myocardial infarction, in two clinical studies (Wolf et al., 1977; Bussmann et al., 1983), verapamil has apparently reduced parameters of myocardial infarct size when given intravenously and acutely. However, in a double-blind study on 217 patients (Thuesen et al., 1983), verapamil 0.1 mg/kg intravenously (mean 4 h after onset of chest pain) followed by 120 mg 3× daily did not reduce cumulative enzyme release.

In a large double-blind study on 1436 patients with AMI (Danish Study Group, 1986), half were treated with verapamil starting with an intravenous dose followed by 120 mg 3× daily; there was no difference in the acute nor chronic mortality. More patients were withdrawn from the verapamil group than from the placebo group due to the development of second and third degree heart block. Also, heart failure was more frequent in the verapamil-treated group. The single benefit of verapamil treatment was a reduction in intermittent atrial fibrillation. Further retrospective and subgroup analysis (Danish Study Group, 1986), which is a procedure open to severe criticism, suggests decreased re-infarction and mortality in the verapamil group when the late results are considered (22 to 180 days). This concept would suggest that early adverse effects of verapamil balance the later beneficial effects. However, a formal trial would be required to prove this point. Verapamil also is not effective in preventing early post-infarction angina and re-infarction (Crea et al., 1985). This small trial, which only studied 17 patients over 10 days, was stopped because of lack of benefit of verapamil.

E. Diltiazem for AMI—A Special Case?

Enthusiasm for the use of diltiazem in AMI has been fanned by a study in which patients with non-Q-wave myocardial infarction (formerly called nontransmural or subendocardial infarction) appear to benefit from diltiazem 90 mg every 6 h and initiated 24 to 72 h after the onset of infarction and continued for up to 14 days. Diltiazem reduced the incidence of re-infarction and the frequency of refractory post-infarction angina without changing the low mortality which was only 3 to 4% (Gibson et al., 1986). The study suffers from three defects. First, using the standard two-tailed Student's t-test instead of the one-tailed test that the authors preferred, no statistical significance is shown. In other words, the authors did not allow for the possibility that diltiazem might have *worsened* the clinical situation (a real possibility when judged by the post-infarct study; discussed in the next section). Second, nearly two thirds of the patients received β-blockade so that the real comparison was between diltiazem plus β-blockade vs. placebo plus β-blockade. Third, the study was limited to very early re-infarction, within 14 days and no reduction in mortality was achieved. Thus, there can be no comparison with the β-blocker studies in which mortality was reduced over a period of months and years.

F. Calcium Blockers in Post-Infarction Follow-Up

Recently, Yusuf and Furberg (1987) have summarized their meta-analysis of data on acute short-term and long-term studies in patients treated with calcium blockers following myocardial infarction. The overall data, which included one study on verapamil, four on nifedipine, one on diltiazem and one on lidoflazine, all indicated about a 6% excess in mortality by treatment. Another study of considerable importance has been recently published. The Multicenter Diltiazem Post-Infarction Trial Research Group (1988) showed that diltiazem had no overall benefit. The mean daily dose was 180 mg which followed a starting dose of 240 mg commenced 13 to 15 days after the onset of acute myocardial infarction; the trial was continued for 12 to 52 months (mean 25 months). Over half of the patients also received β-blockers. Retrospective subgroup analysis (a procedure against which even the authors of the article warn) suggests that diltiazem benefited 80% of the patients who did not have pulmonary congestion and also had a high ejection fraction exceeding 40%, whereas diltiazem could have harmed the other 20% of patients to such an extent that there was no overall benefit for the group as a whole. The recent DAVIT II study is commented on in Section X. It requires emphasis that there has been no proper study comparing diltiazem or any other calcium blocker with β-blockade in the post-infarct period. Where there is some positive clinical indication for the use of calcium blockers rather than for β-blockade (e.g., pulmonary disease), a calcium blocker continues to be chosen providing that there is no left ventricular failure.

IX. REPERFUSION INJURY

Reperfusion damage is, at least in part, calcium mediated (and in part mediated by free radicals). Experimental reperfusion injury may respond in part to calcium blocker treatment. Reperfusion arrhythmias are also in part ameliorated by calcium blockers although not consistently (Opie and Coetzee, 1987). What is not clear is whether calcium blockers given at the time of reperfusion (but not before) prevent reperfusion injury or whether they act merely by lessening of the severity of ischemic injury.

X. FAILURE OF CALCIUM BLOCKERS TO BENEFIT ACUTE MYOCARDIAL INFARCTION AND TO GIVE POST-INFARCTION PROTECTION?

With some impressive experimental effects of calcium blockers against myocardial ischemia (for reviews, see De Leiris et al., 1984; Nayler, 1987) and infarct size (for review, see Kloner and Baunwald, 1987) and reperfusion injury (Nayler et al., 1987), as well as the reduction of ischemic ventricular arrhythmias in some experimental models (for review, see Coetzee et al., 1987) and a possible anti-atherogenic effect (Sievers et al., 1987), *it is highly disappointing that no specific benefit of calcium blocker therapy has been shown either during or after AMI* (Yusuf et al., 1985).* Some possible explanations are as follows.

First, the antiarrhythmic concentrations required usually exceed the ordinary therapeutic blood levels; Clusin et al. (1982) delayed the onset but did not prevent ventricular fibrillation in dogs with therapeutic blood diltiazem concentrations. Second, in the case of nifedipine (not verapamil and diltiazem), reflex tachycardia and excess hypotension may limit some of the anti-ischemic benefit. Third, not enough careful studies have been done with calcium blockers to fully exclude antiarrhythmic and anti-ischemic protection during the very early phase of acute myocardial infarction when the effect might be greatest (Hugenholtz et al., 1986). Finally, as shown in the recent DAVIT II trial, a beneficial effect can be found if patients with heart failure are excluded (Danish Study Group, 1990).

Mortality in ischemic heart disease is complex and ill-understood, and probably is a combination of sudden death due to ventricular fibrillation, myocardial failure secondary to ischemic failure, and ventricular rupture. Presumably ventricular fibrillation is due, at least in part, to excess β-adrenergic activity acting through a complex variety of mechanisms, including a lowered arterial plasma potassium, a general enhancement of the intracellular calcium ion movements through an elevation of myocardial cyclic AMP levels, and increased ischemic injury resulting from an increased oxygen demand. Calcium blockers cannot be expected to counter all these effects. Put differently, enhanced calcium ion entry is only one of several possible arrhythmogenic mechanisms of β-stimulation. Thus, in acute infarct management, it would seem more logical to decrease potentially harmful excessive β-stimulation by β-adrenergic blockade than to use calcium blockers. Similar arguments can be applied to explain why post-infarct trials of calcium blockers have thus far been so disappointing.

XI. PERCUTANEOUS TRANSLUMINAL CORONARY ANGIOPLASTY (PTCA)

During this procedure there is a period of deliberate transient total ischemia at the site of coronary stenosis while balloon dilation takes place, followed by reperfusion. Both

---

* Section taken from Opie (1988a) with permission.

transient ischemia and reperfusion may respond to prophylactic therapy with calcium blockers, which may be given locally at high concentrations. Therefore calcium blockers are potentially promising in this situation (Hugenholtz et al., 1986) and need careful controlled evaluation.

XII. SUMMARY: CALCIUM BLOCKERS IN ANGINAL AND ISCHEMIC SYNDROMES

Calcium blockers are effective antianginal agents, acting in a complex fashion and offering an important and significant new advance in the therapy of these conditions. Nifedipine chiefly relieves arterial spasm and afterload; an acute compensatory tachycardia may offset some of the antianginal benefit. Verapamil, besides reducing afterload, has a negative inotropic effect partially offset by the afterload reduction; in addition, verapamil seems to act specifically on the ischemic zone. Diltiazem reduces afterload and has a mild negative chronotropic effect.

The mechanisms of action in *effort angina* are complex and may include a reduction of an exercise-induced increase in coronary tone, as well as a decreased myocardial oxygen demand and improved ventricular relaxation (Murakami et al., 1985). In *vasospastic angina*, coronary dilation is the mechanism of the relief of pain by calcium blockers. In *true unstable angina at rest with threatened infarction*, the calcium blockers, as a group, do not benefit. Probably the best result is that diltiazem is similar in its benefits to propranolol. The emerging complexity of the etiology of unstable angina with an increasing emphasis on microthrombi and coronary obstruction rather than on spasm, may explain the failure of calcium blockers to benefit unstable angina. In *threatened myocardial infarction,* nifedipine may be harmful when given as monotherapy, presumably because of tachycardia and hypotension, whereas its combination with β-blockade is helpful.

For *post-infarction protection,* it is difficult to explain the failure of benefit from calcium blockers. The established success of the β-blocking agents suggests that a specific catecholamine-mediated effect operates to contribute to mortality in the post-infarct period.

Comparisons of the efficacy of calcium blockers with each other are difficult. Verapamil is the calcium blocker in longest use and licensed for more indications than the others and hence the "drug for all seasons". However, it frequently causes constipation, an adverse effect in patients with cardiovascular disease in whom straining at defecation can be especially harmful. It combines somewhat poorly with β-blockade (added negative inotropic, chronotropic and dromotropic effects), even though with care this combination has been safely and extensively used. *Nifedipine* is excellent for vasospastic angina and short-lived attacks of rest angina, while in effort angina careful dose titration is reported to avoid the occasional precipitation of ischemia. In unstable pre-infarction angina, nifedipine seems contraindicated in the absence of β-blockade. Nifedipine usually combines well with β-blockers unless there is poor myocardial function when added negative inotropic effects can be harmful. However, side effects of nifedipine are frequent. *Diltiazem* is well tested in effort angina, vasospastic angina and unstable angina; there is also suggestive evidence for benefit in non-Q-wave myocardial infarction but not for post-infarction protection. Diltiazem traditionally has few side effects yet with the higher doses now frequently used, side effects including pedal edema are becoming more common. Combination of diltiazem with β-blockade is theoretically easier than in the case of verapamil because of the lesser negative inotropic effect of diltiazem. Yet in practice, adverse interactions have occurred, so that again somewhat more care is needed than with the nifedipine-β-blocker combination.

In individual patients, the hemodynamic status and the anticipated side effects of the various calcium blockers might be important, as well as possible co-therapy with β-blockade.

For example, in a patient with a resting bradycardia or with borderline heart failure, nifedipine is likely to be chosen. In a patient with supraventricular or sinus tachycardia, diltiazem or verapamil is likely to be chosen. In a patient tending to constipation, diltiazem is probably better than verapamil. In a patient with severe effort angina, combined β-blockade-calcium blocker therapy will probably be chosen, working up the doses to the limit of subjective and hemodynamic tolerance. In threatened infarction, nifedipine should be avoided unless combined with β-blockade. For post-infarction protection, β-blockade should be used rather than calcium blockade.

ACKNOWLEDGMENTS

Kluwer Academic Publishers, publishers of *Cardiovascular Drugs and Therapy*, are thanked for permission to reproduce Tables 1 and 2 and for sections of the text from Opie (1988a).

Chapter

17

Modulation of Calcium Channels in the Management of Cardiomyopathies

Gaétan Jasmin, André Pasternac, Ghassan Bkaily, and Libuse Proschek

TABLE OF CONTENTS

I.	Introduction	296
II.	Basic Abnormalities of Myocardial Perfusion in Cardiomyopathies	297
III.	Calcium Channel Blockers in Hypertrophic Cardiomyopathy	297
IV.	Calcium Channel Blockers in Dilated Cardiomyopathies	298
V.	Restrictive Cardiomyopathy	299
VI.	The Cardiomyopathic Syrian Hamster	299
VII.	Disturbed Ca^{2+} Metabolism in Cardiomyopathic Hamsters	299
VIII.	Assessment of Cardiac Necrosis	301
IX.	Comparative Effects of Calcium Channel Blockers Upon Myopathic Hearts	301
X.	Competition Between Gallopamil and Adrenergic Agents	302
XI.	Electrophysiological Studies in Isolated Cardiocytes from Newborn Cardiomyopathic (CM) Hamsters	303

XII. Patch Clamp Studies of Macroscopic Inward Currents by Whole Cell
     Voltage Clamp Recordings.. 303

XIII. Summary 307

I. INTRODUCTION

In recent years, the use of calcium channel blockers has been advocated for the treatment of numerous cardiovascular disorders, i.e., angina pectoris, supraventricular tachycardia and hypertension (Braunwald, 1982 and 1987). The use of these drugs also appears warranted for the treatment of hypertrophic and in some instances of dilated cardiomyopathies (Chatterjee, 1987; Baughman, 1986; Colucci, 1987). In this paper, the indications for calcium channel blockers will be reviewed in light of their pharmacological properties and clinical trials in patients with hypertrophic, congestive or restrictive cardiomyopathy.

The dihydropyridine derivatives, nifedipine, nitrendipine, nimodipine and nisoldipine are efficient coronary dilators with little negative inotropic, chronotropic or dromotropic effects. In contrast, the phenylalkylamines verapamil and gallopamil (D600) and the benzothiazepine diltiazem exert an equipotent coronary vasodilation and a negative chronotropic effect with a less significant negative inotropic effect (Fleckenstein, 1977 and 1983; Taira, 1987). These differences relate to the selectivity of channel blockers toward cardiac and smooth muscle as much as the sensitivity of the modulated calcium channel binding sites to the various channel blockers (Triggle and Swamy, 1983; Schwartz and Triggle, 1984). Their sensitivity depends largely upon membrane potential, frequency of stimulation, intracellular and extracellular Ca^{2+}, protein phosphorylation together with stimulation of β-adrenergic receptors (Janis and Scriabine, 1983; Reuter, 1987; Yatami and Brown, 1989). It has also been claimed that some calcium channel blockers inhibit vasoconstriction by mere activation of post-synaptic adrenergic α_2 receptors (van Zweiten and van Meek, 1983; Cavero et al., 1983). On the other hand, verapamil and its derivatives show a strong state-dependent inhibitory effect in contrast to dihydropyridines which mode of action is apparently related to the ventricular calcium channel state, i.e., activated, inactivated or resting. Moreover, nifedipine would exert a frequency-dependent blocking effect upon the atrioventricular (AV) nodal conductance (Kohlhardt and Happ, 1981), and the effect of diltiazem would lie between these two extremes (Schwartz and Triggle, 1984). Thus the therapeutic success of calcium channel blockers in the treatment of cardiomyopathies, vascular diseases, hypertension, angina pectoris and similar disease entities derives basically from the selective inhibition of Ca^{2+} transmembrane movements into the muscle cells and the specific responses of calcium channel receptors of each respective target cell toward the calcium channel blocker. Differences in therapeutic doses of the individual drug vary therefore with respect to cardiac muscle, atrioventricular tissue or smooth muscle interactions.

II. BASIC ABNORMALITIES OF MYOCARDIAL PERFUSION IN CARDIOMYOPATHIES

Chest pain is a frequent symptom in patients with cardiomyopathy. More than 50% of patients with idiopathic hypertrophic and/or dilated cardiomyopathies complain of anginal symptoms in the absence of evident obliterative changes in epicardial arteries. Often, these patients' electrocardiograms suggest myocardial ischemia or even a pattern of myocardial infarction (Braunwald et al., 1964; Massumi et al., 1965). Myocardial ischemia may play an important role in the pathophysiology of hypertrophic, dilated and congestive cardiomyopathies as revealed by coronary flow studies. A reduced myocardial perfusion per gram of left ventricular mass or an increased coronary resistance at rest (Weiss et al., 1976; Pasternac et al., 1982) with insufficient coronary reserve during pacing stress (Pasternac et al., 1982; Cannon et al., 1985) are common features in hypertrophic, dilated and congestive cardiomyopathies. Using thallium-201 emission computed tomography O'Gara et al. (1987) have identified regional perfusion defects after maximal exercise in 57% of patients with hypertrophic cardiomyopathy. While investigating the coronary vein flow by a thermodilution technique Cannon et al. (1985) noticed that a metabolic stress elicited by cardiac pacing can provoke ischemic pain due to an inadequate coronary vasodilatation. The resulting increased left ventricular diastolic pressure can be alleviated by a treatment with verapamil (Bonow et al., 1983). Chronic hypoperfusion equally results in progressive impairment of ventricular function often in the subendocardial area including papillary muscle with production of focal necrosis and fibrosis (Roberts and Ferrans, 1975). The defect was found to be reversible except in those patients with depressed left ventricular function at rest.

In addition, episodes of acute transient ischemia may occasionally occur in these patients resulting in the "stunned myocardium" or myocardial infarction (Maron et al., 1979; Fine et al., 1989). The repetition of these events may result in alterations in wall dynamics, wall thickness and ventricular volume (Pasternac, 1989).

Since calcium channel blockers improve coronary flow (Jolly and Gross, 1980) and myocardial oxygen supply through their coronary vasodilating properties and reduce wall tension through their peripheral vasodilating properties, they may be of benefit in patients with cardiomyopathy and possibly prevent the ventricular deterioration related to progressive myocardial ischemia.

III. CALCIUM CHANNEL BLOCKERS IN HYPERTROPHIC CARDIOMYOPATHY

The functional abnormality in hypertrophic cardiomyopathy basically results from a hypertrophic, often hypercontractile left ventricle, with decreased diastolic compliance. There is now general agreement that the main effect of calcium channel blockers consists in improving diastolic function. They may be used alone or in combination with β-blockers, which act predominantly on systolic function and tend to decrease contractility. Over the past decade, there have been several studies on efficacy of verapamil (Kaltenbach et al., 1979; Rosing et al., 1979; Hanrath et al., 1980; Tencate et al., 1983; Bonow et al., 1983 and 1987), nifedipine (Lorell et al., 1982) and more recently diltiazem (Isawe et al., 1987) in this disorder. Some studies have compared the effects of verapamil and propranolol on diastolic properties (Hess et al., 1983) and more recently the effects of combined nifedipine and propranolol (Hopf and Kaltenbach, 1987).

In their review papers, Rosing et al. (1985) and Chatterjee (1987) have carefully appraised the therapeutic value of calcium channel blockers in the management of hypertrophic cardiomyopathy. Verapamil appears to be the calcium channel blocker of choice. It induces a small decrease in systolic blood pressure and decreases the basal left ventricular outflow gradient (Rosing et al., 1985). In addition, it exerts a significant improvement in left ventricular relaxation and filling dynamics (Hanrath et al., 1980; Tencate et al., 1983; Bonow et al., 1983; Bonow et al., 1987). Radionuclide studies have revealed that the main factor responsible for increasing performance relates to improvement in left ventricular diastolic function. Many patients exhibit a reduction in symptoms and a significant increase in exercise tolerance after long-term verapamil therapy (Bonow et al., 1985; Rosing et al., 1985). Verapamil therapy alone resulted in a more evident symptomatic improvement than did propranolol or the combination of nifedipine and propranolol (Hopf and Kaltenbach, 1987; Kober et al., 1987). However, either monotherapies or combined therapies needed to be discontinued in some patients because of such adverse effects as sinus arrest and bradycardia, AV blocks, hypotension, edema, arrhythmias, weakness (Epstein and Rosing, 1981; Hopf and Kaltenbach, 1987). Furthermore, the following contraindications deserve to be mentioned: (1) an abnormal pulmonary wedge pressure reflecting left atrial pressure often associated with obstruction to left ventricular outflow; (2) a history of paroxysmal nocturnal dyspnea or orthopnea and (3) a sick sinus syndrome or AV junctional disease without an implanted pacemaker. A study by Isawe et al. (1987) has shown that diltiazem also improves the left ventricular diastolic behavior in patients with hypertrophic cardiomyopathy during dynamic exercise of mild intensity.

Interestingly, it has been reported that calcium-antagonist receptors are increased in hypertrophic cardiomyopathy, suggesting that abnormal calcium fluxes may play a role in the pathophysiology of this disease (Wagner et al., 1989). Thus, prophylactic use of calcium channel blockers may contribute to a better control of Ca^{2+} homeostasis.

IV. CALCIUM CHANNEL BLOCKERS IN DILATED CARDIOMYOPATHIES

Nifedipine, as mentioned previously, is a potent peripheral vasodilator with moderate or little negative inotropic effect. Therefore, this calcium channel blocker could, in theory, be quite suitable in the management of dilated cardiomyopathy (Baughman, 1986). According to Klugmann's studies (Klugmann et al., 1980), sublingual nifedipine reduces afterload and lowers filling pressures in patients with chronic congestive heart failure and dilated cardiomyopathy. The cardiac index and ejection fraction tend to increase together with the left ventricular dp/dt. This improvement in contractile function is secondary to reflex baroreceptor-mediated adrenergic responses. Improved resting and exercise hemodynamics up to 1 h after administration of sublingual nifedipine have been reported in patients with mild congestive heart failure (Hanrath et al., 1982). Baughman (1986) has summarized his observations regarding nifedipine and verapamil studies as follows: "Nifedipine appears to be an acceptable vasodilator; its dose should probably not exceed 20 mg, as higher doses may result in a clinically apparent negative inotropic effect. The dose interval should be 6 to 8 h as tolerated by the patient. The negative inotropic effect of verapamil appears to be too potent for patients with dilated cardiomyopathy".

The beneficial effects of diltiazem in dilated cardiomyopathy have been discussed recently by Figulla et al. (1989). It seems that the therapeutic effects afforded by calcium channel blockers are often "nonspecific", i.e., independent of the pathogenesis of the dilated cardiomyopathy. These include: (1) a reduction in afterload, an effect that can improve hemodynamics, survival, and the clinical state (Walsh et al., 1984); (2) a reduction in the incidence of arrhythmias, by inhibiting trans-sarcolemmal Ca^{2+} influx (Fleckenstein, 1977 and 1983); and (3) an inhibition of the effect of catecholamines (Lipkin and Poole-Wilson, 1985).

Colucci (1987) has pointed out the advantages and some serious limitations of the use of calcium channel blockers in patients with congestive heart failure due to dilated cardiomyopathy. Besides the possible deterioration in left ventricular (LV) function as a result of a direct negative inotropic action, there may be a potentially adverse effect due to activation of the renin-angiotensin system. Such activation may aggravate fluid retention in those patients with congestive heart failure. Thus, the systematic use of calcium channel blockers in patients with dilated cardiomyopathy appears to be debatable. Nevertheless certain dihydropyridine derivatives such as nicardipine, nitrendipine and felodipine on one hand and diltiazem on the other can improve ventricular function in selected cases of cardiomyopathy. In patients with myocardial ischemia due to coronary disease, calcium blockers may be advantageously combined with nitrates or angiotensin converting enzyme inhibitors.

V. RESTRICTIVE CARDIOMYOPATHY

This term applies to patients with elevated filling pressures of the ventricles in the absence of significant dilatation, hypertrophy or impaired systolic function of the ventricles. The term may also apply to certain idiopathic forms and more often to amyloid heart disease, hemochromatosis or any type of cardiac infiltrative diseases. Patients showing severe restrictive physiology with relatively mild hypertrophy or dilatation may also be included. The use of calcium channel blockers has been suggested in the idiopathic form of restrictive cardiomyopathy (Benotti et al., 1980) but studies documenting hemodynamic improvement are lacking.

VI. THE CARDIOMYOPATHIC SYRIAN HAMSTER

The hamster hereditary cardiomyopathy provides unique possibilities of studying the pathology and clinical course of primary congestive cardiomyopathies. This autosomal recessive disorder is readily transmissible with 100% incidence in the offspring but the defective gene has not yet been identified. The cardiomyopathy develops in characteristic, well-defined and predictable stages: (1) a necrotic phase with multifocal cardiac lesions occurring between 30 and 120 days of age; (2) a healing phase with scar formation and progressive dilatation of the atrial and ventricular wall occurring between 120 and 200 days and finally, (3) a terminal phase with moderate to severe heart failure between 200 and 300 days. Our early findings on the prevention of the heart necrotic changes by verapamil have led us to believe that the hamster cardiomyopathy derives from defective transmembrane ion movements (Jasmin and Proschek, 1984).

VII. DISTURBED Ca^{2+} METABOLISM IN CARDIOMYOPATHIC HAMSTERS

The membrane molecular changes leading to Ca^{2+} over-accumulation in cell injury are not fully understood (Hearse, 1980). While it has been clearly demonstrated that a Ca^{2+} load leads to an energy depletion, the causative factors interfering with membrane function seem to be nonspecific (Berridge, 1984; Sperelakis, 1988). In cardiomyopathic hamsters, it all seems that the primary defect is a sudden drop in mitochondrial oxidative phosphorylation as evidenced by a change in respiratory rate concomitant with a lower production of creatine phosphate (Proschek and Jasmin, 1982; Wikmann-Coffelt, et al., 1986a). Interestingly, verapamil and isoproterenol which have an opposite effect on Ca^{2+} transmembrane

TABLE 1
Body Weight Gain and Heart Body Weight Ratio of Cardiomyopathic Hamsters After Treatment with Calcium Slow Channel Blockers[a]

Treatment (no animals)	Dose (mg/kg/d)	Body weight gain (g) (Initial 48 ± 2)	Heart weight[b] (mg) / Body weight (g)
Untreated (38)	Saline or vehicles	30 ± 1	3.17 ± 0.06
Gallopamil (28)	2.6 s.c., saline	29 ± 2	3.29 ± 0.07
Verapamil (35)	20 s.c., saline	28 ± 1	3.68 ± 0.06 $p < 0.01$
Diltiazem (14)	150 p.o., saline	27 ± 2	3.21 ± 0.08
Nifedipine (21)	20 p.o., 30% DMSO	28 ± 1	3.10 ± 0.12
Bepridil (17)	150 p.o., 2.5% ethanol	27 ± 1	3.11 ± 0.07
Prenylamine[c] (14)	100 p.o., aq.	15 ± 2 $p < 0.005$	3.26 ± 0.08

[a] Means ± SE.
[b] Ventricles only.
[c] Treatment with prenylamine 3 weeks, body weight gain, and heart body weight ratio of the corresponding untreated group: 24 ± 1, and 3.23 ± 0.06, respectively.

$p > 0.05$ vs. untreated group is considered as nonsignificant, otherwise as indicated.

movements, were highly efficient in maintaining energy reserves and in preventing, at the same time, development of cardiac necrotic changes (Jasmin and Bajusz, 1973; Jasmin and Solymoss, 1975; Wikman-Coffelt et al., 1986b; Sievers et al., 1986). We have assumed that the integrity of membrane ion channels depends upon cytosolic phosphorylating processes (Jasmin and Proschek, 1984). In addition, it has been shown previously that cardiomyopathic hamsters exhibit an abnormal thermogenic response to β-adrenergic stimulation and this may be ascribed to a hypothyroid state observed in these animals (Horwitz and Hanes, 1974; Jasmin et al., 1987a). Hence, verapamil and isoproterenol modulate calcium channel activity by the phosphorylation of membrane protein receptors via cAMP dependent protein kinase (Vaghy et al., 1988), by adenyl cyclase and cAMP cascade transductions (Sperelakis, 1988; Stoclet et al., 1988) or by a direct pathway via the activation of G_s protein linking β-adrenergic receptors and calcium channels (Yatami and Brown, 1989).

Therapeutic studies reported here include both *in vivo* and *in vitro* assays. *In vivo* experiments were carried out in cardiomyopathic hamsters, using basically two different types of calcium channel blockers: (1) verapamil and gallopamil (D600) which exert a more direct effect on cardiomyocytes and (2) nifedipine, diltiazem and bepridil which are potent peripheral and coronary vasodilators. *In vitro* studies using the patch clamp technique deal with the pharmacology and kinetics of ion channels.

Young 27- to 30-day-old male and female UM-X7.1 cardiomyopathic Syrian hamsters were used. Drugs were solubilized in physiologic saline or in appropriate vehicles for twice-daily systemic or oral treatment. Data concerning doses, mode of administration and total number of animals are given in Tables 1 and 2. All experiments lasted 28 days in order for

TABLE 2
Interaction of Adrenergic Agonists and Antagonists with Gallopamil in Cardiomyopathic Hamsters[1]

Group	Dose (mg/kg/d)	Heart weight (mg)[2] / Body weight (g)		Cardiac necrotic lesions			
				Incidence (%)	Severity (Grade 0-3)	Incidence (%)	Severity (Grade 0-3)
		Control	Gallopamil[3]	Control		Gallopamil[3]	
Vehicle	—	3.18 ± 0.03 (30)	3.44 ± 0.04 (24)	100	2.10 ± 0.13	10	0.18 ± 0.05
Isoproterenol	3 s.c.	4.75 ± 0.15 (8)	4.29 ± 0.09[a] (8)	25	0.32 ± 0.12	0	0
	10 s.c.	4.57 ± 0.10 28 (8) 29	3.97 ± 0.07[a] (8)	50	0.66 ± 0.24	0	0
Methoxamine	3 s.c.	3.32 ± 0.09 (8)	3.49 ± 0.10 (8)	100	2.38 ± 0.33	0	0
Propranolol	20 i.p.	3.30 ± 0.11 (10)	ND[4] (15)	70	1.45 ± 0.13		ND[4]
Prazosin	200 p.o.	3.24 ± 0.13 (6)	3.46 ± 0.27 (12)	100	2.50 ± 0.16	100	2.29 ± 0.19[b]

Note: [1] Means ± SE.
[2] Ventricles only.
[3] The dose of gallopamil was gradually increased from 1 mg/kg/d during the first week up to 2 mg/kg during the 3 ensuing weeks; the number of animals is shown in parentheses.
[4] ND, no data because these animals died prematurely.

Significant differences between gallopamil control and gallopamil combined treatment are indicated by [a], $p < 0.05$; [b], $p < 0.001$.

the animals to reach the critical age of 55 days when the necrotic changes become fully expressed in nontreated animals.

At autopsy, the hearts were examined by the naked eye for an overall quantification of necrotic foci. They were then removed and the ventricles freed from atria, rinsed, blotted and weighed. The septal and adjacent ventricular segments were fixed in formol-sublimate and processed for routine histology.

VIII. ASSESSMENT OF CARDIAC NECROSIS

Microassessment of cardiac necrosis was based on a previously described arbitrary scale of 0 to 3 (Jasmin and Proschek, 1987b). Using a double-blind procedure, the degree of damage was rated 1, 2 or 3 (with half marks if necessary) to designate slight, moderate or severe necrosis, respectively. The highest score on tissue sections corresponds to 50% damage of the entire ventricle. A 2.2 average score for 10 animals corresponds to the upper limit of severity of necrotic changes. Higher readings in 80% of animals indicate aggravation by some form of treatment. Below 1.4, we consider that a drug exerts a protective effect. Total Ca^{2+} concentration was determined in acid extracts of ventricle homogenates by atomic absorption spectrophotometry (Proschek and Jasmin, 1982) and the protein content by the method of Lowry (Lowry et al., 1951). The statistical significance was determined following the Student t-test.

IX. COMPARATIVE EFFECTS OF CALCIUM CHANNEL BLOCKERS UPON MYOPATHIC HEARTS

As shown in Table 1, the growth rate of myopathic hamsters was unaffected by all

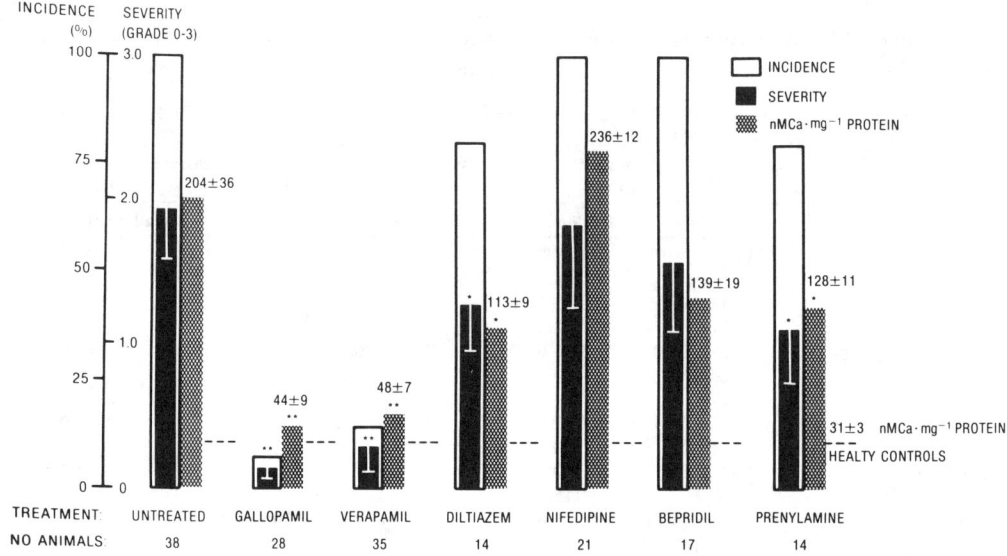

Figure 1. Effect of slow channel blockers on development of necrotic changes and calcium content in the heart of cardiomyopathic hamsters UM-7.1.

channel blockers at the doses utilized except for prenylamine. The verapamil treatment caused a slight but significant increase in ventricular/body weight ratio. However, it became evident that the different calcium channel blockers were not equally effective in preventing the cardiac necrotizing process (Figure 1). Cardioprotection was best achieved by gallopamil and verapamil. Diltiazem and to a lesser extent prenylamine reduced the incidence and severity of heart lesions whereas nifedipine and bepridil were totally ineffective. Interestingly, there was a close correlation between the ventricular Ca^{2+} content and the severity of the cardiomyopathy. We ought to mention however that none of these channel blockers interfered with the progression of the skeletal muscle lesions.

X. COMPETITION BETWEEN GALLOPAMIL AND ADRENERGIC AGENTS

The necrotizing process in cardiomyocytes is often associated with a disturbed catecholamine metabolism (Sole et al., 1975; Jasmin and Proschek, 1983). Allegedly, myocardial necrotization in CM hamsters relates to an imbalance between α- and β-adrenergic receptors (Karliner et al., 1981; Jones, et al., 1988). It remains to be seen however whether the cardioprotection afforded in these diseased animals is purely β-receptor mediated (Jasmin and Proschek, 1984).

In order to investigate the mechanism of cardioprotection elicited by inotropic agents, we undertook treating young myopathic hamsters with a combination of slow calcium channel blockers and adrenergic agonists or antagonists. The results in Table 2 show that the hypertrophy of myopathic hearts is not dose related in isoproterenol treated animals. Moreover, the isoproterenol preventive effects upon the necrotizing process were unchanged by the conjoint treatment with gallopamil. On the other hand, this calcium channel blocker proved to be efficient in antagonizing the adverse effects of methoxamine upon the cardiomyopathy.

Propranolol, as a nonspecific β-blocker, significantly reduced the severity of the cardiac necrotic changes in diseased hamsters. The combined treatment with gallopamil was, however, unsuccessful simply because of a lack of tolerance at any dose level. Prazosin, either given alone or in combination with gallopamil had no favorable effect on the development and the course of the cardiomyopathy. At the present time, we can only speculate on the mechanism of cardioprotection achieved by both, a β-adrenergic blocker and stimulator. At any rate, it demonstrates that the inotropic response in CM hamster become disturbed with progression of the disease until development of cardiac failure.

In summary, we have learned through these pharmacologic trials in CM hamsters that cardioprotection can be achieved either by interfering with excessive calcium influx (gallopamil) or through activation of cAMP phosphorylation processes (isoproterenol) in order to secure functional integrity of cell membranes. It is not unlikely that these same drugs, in normalizing sarcolemmal Na^+-Ca^{2+} exchange, also contribute in the maintenance of cardiac energy reserves (Makino et al., 1985; Wikman-Coffelt et al., 1986b). As the mechanisms of action of adrenergic drugs on cardiac cell membranes become better understood with regard to enzyme complexes involved in protein phosphorylation, it is anticipated that new strategies will develop in the management of human congestive cardiomyopathies.

XI. ELECTROPHYSIOLOGICAL STUDIES IN ISOLATED CARDIOCYTES FROM NEWBORN CARDIOMYOPATHIC (CM) HAMSTERS

Primary cultures of ventricular myogenic cells from newborn normal and CM hamsters were prepared as previously described (Bkaily et al., 1988). The hearts from 5 to 10 hamsters were quickly removed and washed; the dissected ventricles were separated from the atria, cut into small pieces and enzymatically digested in HMEM solution containing 0.1% trypsin. The cell digests were collected, pooled and centrifuged at 170 g/10 min. The pellet was resuspended in culture medium and centrifuged again to wash out the trypsin. The cells were then placed in a plastic dish for 30 min to allow fibroblasts to become attached. The myogenic cells that remained in suspension were transferred to a new dish and were kept in an incubator (95% air, 5% CO_2) at 37°C for 1 to 12 h and subsequently used for the voltage clamp recordings.

XII. PATCH CLAMP STUDIES OF MACROSCOPIC INWARD CURRENTS BY WHOLE CELL VOLTAGE CLAMP RECORDINGS

The fire-polished patch pipettes having resistances between 2 and 4 MΩ and the seal resistances ranging from 10 to 20 GΩ were connected to an axopatch amplifier (Axon Inst.) The series resistance compensation was adjusted near to the point of ringing. Current recordings were displayed, stored and analyzed by using an IBM-AT computer equipped with Axess software (Axon Inst.). The capacitive transient currents were not corrected and only experiments which showed almost no leak currents are presented. The extracellular solution contained (mM): 130 NaCl, 2.0 $CsCl_2$, 1.03 $MgCl_2$, 5.4 TEA, 5 HEPES and 5 glucose (pH 7.4); the intracellular pipette solution contained (mM): 20 NaCl, 2.0 $MgCl_2$, 120 $CsCl_2$, 5 HEPES, 5 EGTA and 5 glucose (pH 7.2). All experiments were carried out at room temperature.

As discussed above, the myocardium of CM hamsters show evidence of Ca^{2+} overload and it is believed that possible dysfunction(s) or modulation(s) of ionic channels play a substantial role in the pathogenesis of the disease. Relevant to this theory are the observations concerning Ca^{2+} transmembrane movements by drugs preventing or reducing the development of heart necrotic lesions (Jasmin and Proschek, 1984). Thus, in this study, we inves-

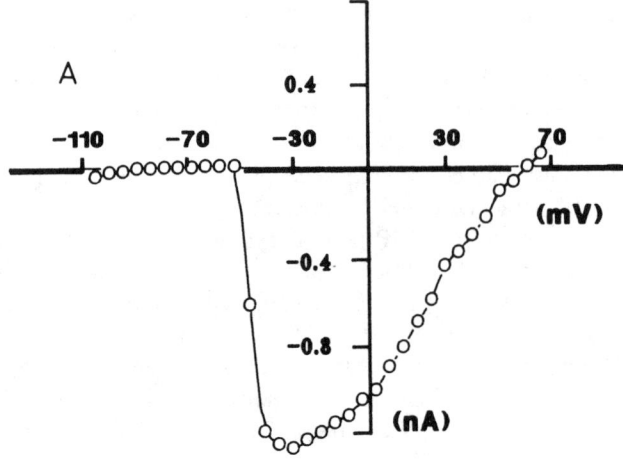

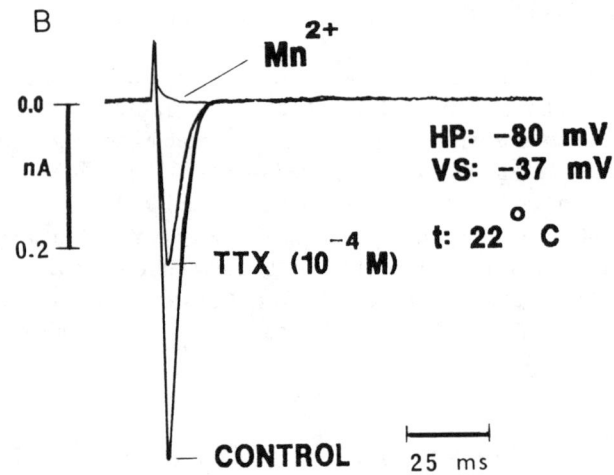

Figure 2. The current-voltage relationship of the TTX sensitive and TTX resistant I_{Na} in a single heart cell of newborn normal hamster. (A) The threshold potential of the whole fast I_{Na} in the control solution is -48.3 ± 2.6 mV (n = 3); the peak amplitude current is near -36.6 ± 1.5 mV (n = 3) and the reversal potential is $+51.6 \pm 3.5$ mV (n = 3). (B) The inward Na$^+$ current is decreased by 10^{-4} M TTX and the remaining TTX resistant I_{Na} is blocked by 2 mM Mn^{2+}.

tigated the ionic inward currents in isolated cardiocytes from newborn normal and CM hamsters.

In cardiocytes from normal hamsters, we found TTX-sensitive and TTX-resistant sodium inward currents (I_{Na}); the latter is completely inhibited by Mn^{2+} as shown in Figure 2B. On the contrary, the I_{Na} in CM hamsters was completely insensitive to TTX and only partially diminished after the addition of Mn^{2+} (Figure 3). Thus, in CM hearts there exists residual TTX and Mn^{2+}-insensitive I_{Na} having a similar threshold potential, peak amplitude and reversal potential as that found in the absence of Mn^{2+} and TTX as illustrated in Figure 4. This TTX and Mn^{2+} resistant I_{Na} (slow I_{Na}) depends on the holding (clamp) potential. As shown in Figure 5, the inactivation curve suggests that at a holding potential more negative

17. Modulation of Calcium Channels in the Management

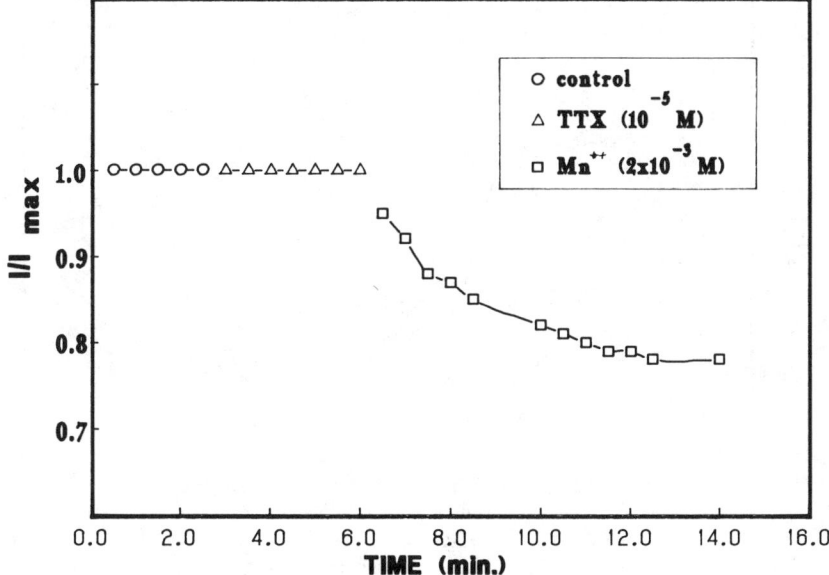

Figure 3. The TTX resistant and the TTX and Mn^{2+} insensitive I_{Na} in a single heart cell of newborn CM hamster. (○)—The relative inward Na^+ current recorded from holding potential of -80 mV; (△)—Addition of 10^{-5} M of TTX does not affect the peak amplitude of I_{Na}; (□)—Addition of 2 mM Mn^{2+} decreases the peak amplitude of the TTX resistant I_{Na} which remains constant afterwards.

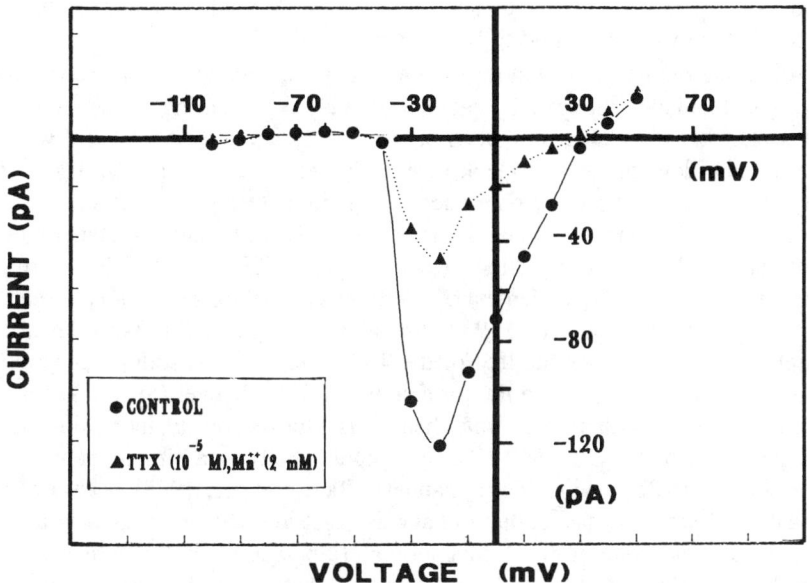

Figure 4. The current-voltage relationship of the TTX resistant and TTX and Mn^{2+} insensitive I_{Na} in a single heart cell of newborn CM hamster. (●)—In absence and in presence of 10^{-5} M TTX, the threshold potential is -45.0 ± 2.0 mV, the peak amplitude is reached at -26.0 ± 1.8 mV and the reversal potential at $+44.0 \pm 2.2$ mV; (▲)—Mn^{2+} has diminished the amplitude of the I/V curve of the TTX insensitive I_{Na} but the threshold and the reversal potential remain unchanged (holding potential -80 mV, n = 5).

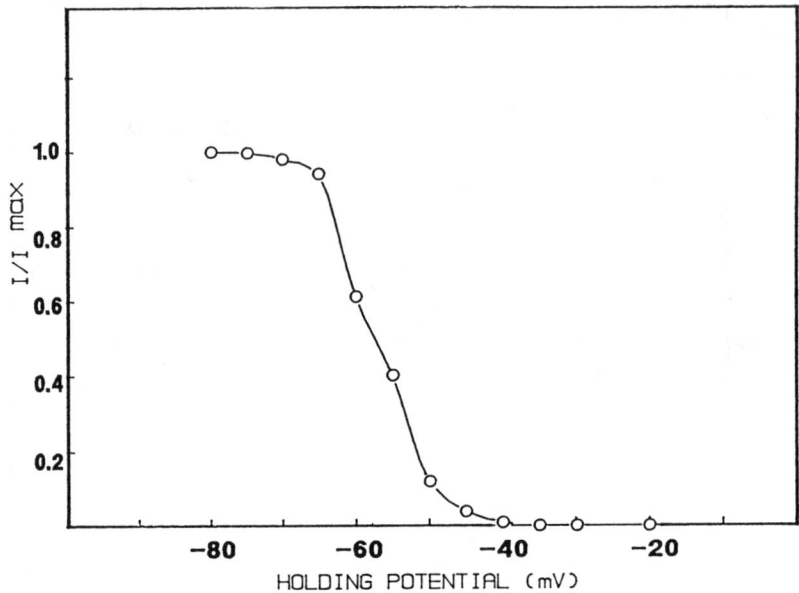

Figure 5. The inactivation curve of the TTX and Mn^{2+} insensitive I_{Na} in a single heart cell of newborn CM hamster. The TTX and Mn^{2+} insensitive channels are fully functional below -75 mV membrane potential and 50% are inactivated at -58 mV and no channels are available above -35 mV.

than -75 mV the Na^+ channels insensitive to TTX and Mn^{2+} might be fully functional whereas 50% of them become inactivated at -58 mV; at holding potentials more positive than -35 mV they were all found to be inactivated.

Regarding the ontogenesis of slow sodium channels, several ultrastructural, metabolic, pharmacological and electrophysiological changes may occur during the development of the cardiac muscle. Striking changes in electrical properties and the cAMP level were reported during embryonic development of chick hearts (Sperelakis et al., 1988). The 2- to 3-day-old embryonic hearts have slowly rising action potentials (10 to 30 V/s) that are dependent mainly on the extracellular $[Na^+]$ and are resistant to the fast sodium channel blocker TTX and an inorganic calcium channel blocker Mn^{2+}. This TTX and Mn^{2+} insensitive channel is considered to be an immature element of excitable cells (Shigenobu and Sperelakis, 1971; Bossu and Fletz, 1984; Anderson, 1987; Sperelakis et al., 1988; Bkaily et al., 1988a). Interestingly, its kinetics resemble that of the TTX-sensitive fast sodium channel while its pharmacology is slightly similar to that of a slow calcium channel (Sperelakis et al., 1988; Bkaily et al., 1990). Furthermore, this channel is blocked by an increasing intracellular concentration of cAMP and cGMP (Bkaily, unpublished) in a similar pattern to that of vascular smooth muscle slow calcium channels (Bkaily et al., 1988). These observations suggest that TTX-sensitive fast sodium channels and slow calcium channels are absent or less numerous in embryonic hearts (Shigenobu and Sperelakis, 1971; Bernard, 1976; Sperelakis and Bkaily, 1987; Bkaily et al., 1988) and newborn CM hamster hearts.

The calcium channel blockers, verapamil and gallopamil and the β agonist, isoproterenol that block the TTX and Mn^{2+}-insensitive sodium channel (Bkaily et al., 1991, Kojima and Sperelakis, 1983) were highly efficient in preventing the development of heart lesions (Jasmin and Proschek, 1984) whereas diltiazem and bepridil, which afforded less or no cardioprotection after *in vivo* administration (Jasmin and Proschek, 1987), were unable to block slow sodium channels as well (Kojima and Sperelakis, 1983). Thus, several compounds that are potent cardioprotectors are blockers of the TTX and Mn^{2+}-insensitive slow sodium channel in embryonic hearts. Should this type of sodium channel (among others) become active in the adult myocardium, it would contribute to the increase of intracellular [Na^+] which in turn may block the Na^+-Ca^{2+} exchanger. Once intracellular [Na^+] reaches a certain high level, the Na^+-Ca^{2+} exchanger may facilitate Ca^{2+} influx resulting in calcium overload.

Surprisingly, the skeletal muscle cells from Duchenne muscular dystrophy patients also possess a TTX and Mn^{2+} insensitive sodium channel with similar kinetics as that found in CM hamster hearts (Bkaily et al., 1990). Whether the dysfunction of calcium channels is the only factor contributing to the calcium overload remains to be seen. Many membrane events could contribute to Ca^{2+} loading in heart muscle, namely: (1) an increase in opening time of the calcium channels (Reuter, 1983; Hess, 1988); (2) an increase or a decrease in the rate of phosphorylation of calcium channel proteins (Tsien et al., 1986; Reuter, 1987); (3) a dysfunction of Na^+/K^+ and Ca^{2+}-ATPase pumps (Panagia et al., 1984; Makino et al., 1985); (4) an excessive release of Ca^{2+} from intracellular stores (Lee et al., 1985); (5) a defective Na^+/H^+ exchange system (Lazdunski et al., 1985) and (6) the presence of embryonic-like TTX and Mn^{2+}-insensitive slow sodium channels. At the present time, our results indicate that an embryonic-like slow sodium channel exists in newborn CM hamster hearts; consequently, it, in addition to other regulatory Ca^{2+} pathways, may contribute to an impaired Ca^{2+} homeostasis in myopathic cardiac cells.

In the hereditary hamster cardiomyopathy, at least three alternative explanations of the origin of the TTX and Mn^{2+}-insensitive slow sodium channel can be proposed. It is possible that this functional slow sodium channel: (1) derives from the same gene as that coding the TTX resistant and Mn^{2+}-sensitive sodium channel in the embryo; (2) may result from a minor structural modification of the otherwise normal TTX resistant sodium channel (or the TTX sensitive sodium channel), leading to a lower sensitivity to certain calcium blockers and (3) or simply derives from a gene that finds its expression in hereditary muscles diseases. Since this channel is functional only in embryonic hearts (Shigenobu and Sperelakis, 1971; Bkaily et al., 1988; Sperelakis et al., 1987), it is inferred that the gene that normally functions only in the embryonic phase continues to operate during the heart development. Hence, its presence in embryonic heart, in hereditary hamster cardiomyopathy and Duchenne muscle dystrophy would be consistent with the third hypothesis.

It is tempting to suggest that the TTX and Mn^{2+}-insensitive sodium channel or an immature element of excitable cells are common features of embryonic and of pathologic ill-differentiated cells. In order to clarify this hypothesis, more work is required at membrane and molecular biology levels concerning a TTX and Mn^{2+}-insensitive I_{Na} and of inward Ca^{2+} currents in hereditary cardiomyopathy.

XIII. SUMMARY

Different calcium channel blockers deriving from dihydropyridines, phenylalkylamines or from benzothiazepine have been compared for their efficacy in the treatment of human and hamster primary cardiomyopathy. Most of these drugs can afford some cardioprotection in so far as the impaired ventricular muscle function has not reached the stage of irreversibility. Thus, these agents should be used cautiously in the management of congestive heart failure. Therapeutic trials in hamster hereditary cardiomyopathy have revealed that phenylalkylamines on one hand and β-adrenoagonists and antagonists on the other were efficient in preventing the necrotizing process in myocardial cells. Electrophysiological studies in isolated cardiocytes from newborn cardiomyopathic hamsters indicate that there exists a residual TTX and Mn^{2+} insensitive slow inward sodium current which is absent in normal heart cells. This embryonic-like slow sodium channel which proved to be sensitive to isoproterenol and to a verapamil congener, D888, represents a basic pathogenic element in the hamster cardiomyopathy.

Chapter 18

Experiences with Calcium Channel Blockers in the Treatment of Hypertension

Michael A. Weber and William F. Graettinger

TABLE OF CONTENTS

I.	Introduction	310
II.	Types of Calcium Antagonists	310
	A. General Properties	311
	B. Mechanisms	311
	C. Cardiac Effects	312
	D. Vasoselectivity	312
III.	Left Ventricular Hypertrophy	313
IV.	Effects on Renal Function	314
V.	Metabolic Effects	315
VI.	Efficacy	316
VII.	Combinations with Other Drugs	318
VIII.	Importance of the Pre-Treatment Blood Pressure	320
IX.	Acute Administration	321
X.	Side Effects	324

| XI. | Beyond Antihypertensive Effects | 324 |
| XII. | Summary | 325 |

I. INTRODUCTION

Calcium channel blockers have taken on a major role in the treatment of essential hypertension. They are clearly efficacious, and appear to have side effect profiles that compare favorably with the more traditional classes of antihypertensive agents. Unwanted metabolic effects, impairment of exercise tolerance and such CNS symptoms as drowsiness and depression are only rarely encountered.

Calcium channel blockers interact effectively with other types of antihypertensive agents, including converting enzyme inhibitors, beta blockers, sympatholytic agents and even diuretics. Their natriuretic properties, however, enable calcium blockers to be effective in the absence of diuretic therapy. Beyond their use for the chronic management of hypertension, the rapid onset of action of these agents has made them highly effective for the immediate oral or intravenous treatment of hypertensive emergencies. There is also interest in the possible ability of the calcium channel blockers to prevent the formation of atheromatous lesions in peripheral vascular tissue, especially in view of the growing awareness of the accelerated process of vascular disease in susceptible hypertensive patients.

II. TYPES OF CALCIUM ANTAGONISTS

Ca^{2+} is an essential part of the activation of the intracellular contractile elements of vascular smooth muscle (see Chapter 6, Section II). Thus, drugs that work to prevent Ca^{2+} from reaching these elements will inhibit cellular contraction and thereby facilitate vasodilation. There are two principal ways in which calcium mechanisms can be interrupted: first, through blockade of the calcium channels of the cell membrane, thereby limiting entry of Ca^{2+} into cells; and, second, through blockade of intracellular calcium receptors such as calmodulin or troponin. Additionally, intracellular actions can include the facilitation of Ca^{2+} binding by intracellular organelles, preventing release of Ca^{2+} from its intracellular stores, promoting Ca^{2+} efflux from the cell, or by inhibitory effects at the contractile proteins themselves (Rahwan, 1983).

Both calcium channel blockers and the intracellular antagonists produce vasodilation and a consequent fall in blood pressure, although their other hemodynamic effects are quite different. The intracellular calcium receptor blocking agents include a number of drugs generally referred to as vasodilators that have long been in use as antihypertensive agents. Hydralazine, diazoxide, and nitroprusside are examples of these agents. A characteristic of their actions, especially during more chronic therapy, is that they reflexively activate the sympathetic nervous system and the renin-angiotensin system. In turn, this produces tachycardia and the vasoconstrictor effects of norepinephrine and angiotensin II. A further consequence, perhaps mediated through increased activity of aldosterone, is retention of sodium and water. Thus, such agents as hydralazine often lose their antihypertensive efficacy after the first few days of treatment unless sympatholytic agents and diuretics are added to the treatment regimen.

In contrast, the antihypertensive actions of calcium channel blockers are not usually associated with reflex sympathetic activation and sodium retention. These agents often are highly effective when used as monotherapy, and may actually produce a modest degree of natriuresis. The detailed characteristics and classification of the different calcium channel blockers are discussed in Chapters 12 and 13, Section III. At present, agents of this class available in the U.S. are verapamil, diltiazem, and the dihydropyridine agents nifedipine and nicardipine. Other drugs of the dihydropyridine group are expected to be approved soon for the treatment of hypertension.

A. General Properties

Beyond the question of antihypertensive efficacy, other characteristics of antihypertensive agents are now regarded as critical. For calcium channel blocking agents as a class, there still remains some concern that they might produce unwanted effects on myocardial contractility. Although this would not usually be of clinical importance in healthy younger patients, it is possible that individuals with some compromise of left ventricular systolic function could be adversely affected by these agents. There is also much interest in the metabolic and renal effects of these and other antihypertensive agents. The ultimate goal of treatment of hypertension is to prevent vascular complications, especially coronary artery disease. Some of these issues are discussed briefly below.

B. Mechanisms

Calcium channel blockers are a rather diverse group of agents that have specific effects on the voltage-dependent calcium channels that regulate calcium entry into vascular smooth muscle. This specificity is important, for it allows these drugs to produce their desired actions without affecting the total extracellular Ca^{2+} pool. They do not influence bone mechanisms or other aspects of overall calcium metabolism. Importantly, they do not alter skeletal muscle function despite the high density of calcium channels found in that tissue.

The three types of blockers now available each work at a different receptor domain of the calcium channel protein. In general, the dihydropyridines (such as nifedipine and nicardipine) have the greatest affinity for calcium channels of vascular smooth muscle, including the coronary circulation. The phenylalkylamine group (such as verapamil) probably has the greatest affinity for cardiac tissue, whereas the benzothiazepines (such as diltiazem) are intermediate between the other groups.

The possibility has been raised that calcium channel blockers might have additional effects on the vasoactive substances produced by the endothelium. It has been suggested that these agents might inhibit effects of endothelin, the powerful vasotonic substance that originates in endothelium (Godfraind et al., 1989). Moreover, diltiazem might stimulate release of the vasodilatory endothelium-derived relaxation factor.

It is reasonable to assume that a major part of the action of calcium channel blockers is to attenuate the pressor effects of the major vasoconstrictor hormones, chiefly norepinephrine and angiotensin II, upon the vascular smooth muscle. This action would have the effect of decreasing the tonic action of the sympathetic and renin systems on the circulation as well as reducing the responses to acute stimuli. It has been theorized that calcium channel blockers may be more effective against the effect of norepinephrine than those of angiotensin. Studies in patients treated chronically with the dihydropyridine, nitrendipine, have shown that the pressor dose-response curves to norepinephrine are shifted to the right by this calcium channel blocker, whereas the responses to angiotensin II are largely unchanged (Simon and

Snyder, 1984). Consistent with this observation is the finding that calcium channel blockers inhibit pressor effects mediated through both 1 and 2 receptors (Janssens and Verhaege, 1984). The effects of the blockers on alpha-2 receptors appear to be more complete than on alpha-1 receptors, but this may be simply a function of the relatively weaker pressor potency of the alpha-2 receptor (Cooke et al., 1985). Studies performed in an *in vitro* isolated vascular smooth muscle preparation obtained from the rabbit femoral artery similarly have shown that calcium channel blockers might be more effective in blocking contractile effects produced by sympathetic activity rather than those produced by angiotensin (Weber et al., 1989). Indeed, it has been suggested that the effects of angiotensin II might be mediated more directly by changes in intracellular Ca^{2+} (Weber et al., 1989). It should be emphasized, however, that these apparent differences between norepinephrine and angiotensin II mechanisms might vary according to the blocker being used or the tissue being tested.

It has been stated frequently that the available calcium channel blockers differ from each other in their relative effects on myocardial and peripheral vascular muscle. The dihydropyridines are thought to work primarily as vasodilators in the peripheral arterial circulation, and to have relatively less effect on the myocardium; verapamil, on the other hand, is considered to have more of an action on the myocardium than on the peripheral circulation; and diltiazem is considered to be intermediate between the other types of blockers. It is important to note, however, that verapamil appears to be just as effective as an antihypertensive agent as the other classes. Moreover, there is clear evidence that it has strong vasodilatory properties (Hulthen et al., 1982) and has been shown to be effective both during acute and chronic treatment of hypertension.

C. Cardiac Effects

Calcium channel blockers have clear effects on myocardial performance, coronary blood flow and cardiac conduction. Indeed, two of the principal indications for the use of these agents, other than for hypertension, is in the treatment of angina pectoris and certain types of supraventricular arrhythmias. These latter indications are discussed in detail in Chapter 16 and 19, Section IV. These cardiac effects of calcium channel blockers do not always produce therapeutic benefit. Effects on conduction, especially at the atrioventricular (AV) node, can occasionally cause electrocardiographic (ECG) changes and even complete heart block in susceptible patients. This problem can be troublesome when agents such as verapamil are used in combination with the beta blockers.

D. Vasoselectivity

Vasoselectivity is a measure of the preference of calcium channel blockers for affecting arterial smooth muscle as opposed to the myocardium. Because of concerns that actions on the myocardium can produce negative inotropic effects, with the associated danger of exaggerating any tendencies toward left ventricular systolic dysfunction, a high degree of vasoselectivity is considered a desirable property of calcium channel blockers used for the treatment of hypertension and angina pectoris.

Although extrapolation from animal data to clinical experience is often not accurate, studies that have used the ratio of IC_{50} for vascular to the IC_{50} for cardiac tissue as a measure of vasoselectivity are illuminating (Clarke et al., 1983). In a comparison of the four calcium channel blockers currently available in the U.S., the ratio for verapamil was 0.1, for diltiazem 3.8, for nifedipine 5.5, and for nicardipine 11.1 (Clarke et al., 1983). These results, of course, are not really surprising, for it has been well established that verapamil has a comparatively greater effect on the heart whereas the dihydropyridine agents have been shown to have their primary actions on arterial tissue. Clinical aspects of this property have been examined in patients with established coronary disease (Silke et al., 1985). Nicardipine was found to produce a greater increase in cardiac index than verapamil during both rest and exercise in these patients. Moreover, the nicardipine was less likely to increase pulmonary occlusion pressure (reflects left atrial pressure) during rest and exercise in patients with ischemic heart disease (Silke et al., 1983). Similarly, studies of left ventricular function in patients receiving nifedipine showed small but significant increases in both cardiac output and ejection fraction (Shen et al., 1985).

It must be emphasized, however, that the relatively favorable profile of dihydropyridine agents such as nicardipine does not justify their use in patients with known or probable left ventricular systolic dysfunction. Nevertheless, the potential of these drugs to enhance flow in the arterial circulation, including the coronary distribution, is interesting. Nicardipine has been compared with propranolol, a widely used beta blocker for the treatment of angina pectoris, in patients subjected to pacing-induced tachycardia (Rousseau et al., 1986). There were significant differences between the two drugs in both myocardial lactate extraction fraction and myocardial lactate uptake, indicating that the dihydropyridine agent more adequately sustained aerobic mechanisms during the periods of stress.

III. LEFT VENTRICULAR HYPERTROPHY

The presence of an increase in left ventricular muscle mass in hypertensive patients indicates a marked increase in the likelihood of cardiovascular complications. The availability of echocardiography has made it possible to discriminate relatively mild degrees of hypertrophy, and has confirmed that increased left ventricular wall thickness or mass might have a stronger impact on prognosis than even the hypertension itself. Studies in children and young patients have indicated that hypertrophy appears very early in the process of hypertension, and actually might precede it (Celentano et al., 1988). Left ventricular diastolic dysfunction, caused chiefly by a stiffening or lack of compliance in the ventricular walls—probably a forerunner of hypertrophy itself—also has been found in children with very minor elevations of blood pressure (Graettinger et al., 1987). Indeed, the presence of a family history of hypertension appears to be a strong predictor of the likelihood of left ventricular changes even in the absence of any increases in blood pressure. Because of the prognostic importance of left ventricular hypertrophy, which perhaps is also associated with arterial hypertrophy throughout the circulation, there has been growing interest in antihypertensive agents that might additionally cause regression of these vascular changes as part of their antihypertensive actions. Those drugs that appear most effective in producing regression are those that have the ability to block either the sympathetic nervous system or the renin-angiotensin system.

Experience with calcium channel blockers has been inconsistent. Preliminary reports with the dihydropyridine agent nifedipine have indicated that it can cause regression of hypertrophy and improve left ventricular diastolic function. On the other hand, other investigators using careful echocardiographic monitoring during a long-term study were unable to find significant decreases in left ventricular muscle mass during treatment with a dihydropyridine agent (Drayer et al., 1986). It is possible that the discrepancy between these studies might have reflected differing severities of the hypertension and the hypertrophy in the participating patients. In individuals with severe hypertension whose hypertrophy is due to the marked increases in left ventricular systolic wall stress, any drug that can effectively lower blood pressure should produce some reduction in left ventricular muscle mass. There have also been reports that verapamil and diltiazem can cause regression of hypertrophy. Moreover, there is recent evidence that diltiazem and other calcium channel blockers can improve diastolic function as well as reduce left ventricular mass (Weiss and Bent, 1987; Szlacheic et al., 1987).

TABLE 1
Diltiazem and Renal Function

	Normal renal function		Renal insufficiency	
	Control	Diltiazem treatment	Control	Diltiazem treatment
Glomerular filtration rate (ml/min/1.73 m^2)	92 ± 8	92 ± 5	60 ± 6	89 ± 6[a]
Renal blood flow (ml/min/1.73 m^2)	850 ± 74	861 ± 50	601 ± 76	837 ± 58[a]
Renal vascular resistance (units)	13.5 ± 1.7	10.5 ± 0.6	18.1 ± 2.8	10.3 ± 0.7[a]

Values are Mean ± SEM.

[a] $p < 0.025$.

IV. EFFECTS ON RENAL FUNCTION

It could be anticipated that calcium channel blockers would improve renal function, for their vasodilatory actions increase overall renal blood flow. Moreover, vasodilatory effects on the afferent renal arterioles should tend to increase glomerular filtration rate. Table 1 summarizes a study in which the calcium channel blocker diltiazem was administered to patients with normal renal function and others with impaired renal function (Bauer et al., 1985). In the patients with normal function, glomerular filtration rate, renal blood flow, and renal vascular resistance all remained constant during treatment. However, in the patients with renal insufficiency, there was a marked reduction in renal vascular resistance that was associated with significant increase in both renal blood flow and glomerular filtration rate. It has been shown that diltiazem produces dilatory effects in the afferent arterioles of the glomeruli (Loutzenhiser et al., 1986). These data are encouraging, but are somewhat difficult to interpret from a prognostic point of view. It is not yet entirely clear whether the improvement of blood flow and tubular function in the kidney is necessarily associated with protection against the long-term structural changes that can be associated with hypertension. Longitudinal studies are required to document renal changes during chronic antihypertensive therapy.

Calcium channel blockers also appear to produce a natriuretic effect. Balance studies, employing careful daily measurements of urinary sodium excretion, have shown that the initiation of therapy with dihydropyridine agents such as nitrendipine is associated with a clear increase in sodium excretion in patients previously brought into sodium balance on a constant sodium diet (Luft et al., 1985). The total amount of sodium depletion produced by this mechanism is less than that found with the use of a conventional diuretic agent, but it may be adequate to facilitate the other antihypertensive mechanisms of the calcium blockers. Moreover, it may explain partly the efficacy of calcium channel blockers when used in combination with other nondiuretic antihypertensive drugs. Interestingly, it has been reported that the blood pressure-lowering effects of calcium channel blockers, especially when administered acutely, appear to be enhanced in patients concurrently consuming a high sodium diet (Nicholson et al., 1987). Again, the mechanism for this action is not clear, but could possibly be explained by an inhibitory action of sodium upon the renin-angiotensin system. Presumably, through their ability to antagonize the mechanisms by which sodium retention raises blood pressure, calcium channel blockers are able to utilize the apparent blood pressure-lowering properties of sodium (through renin inhibition) while avoiding the consequences of the usual stimulatory effects of sodium on blood pressure.

Calcium channel blockers do not appear to cause long-term changes in the renin-aldosterone axis. The effects of nitrendipine on plasma renin activity and urinary aldosterone

TABLE 2
Effect of Nitrendipine on Renin and Aldosterone

	Placebo	Nitrendipine treatment		
		3 weeks	6 months	12 months
Plasma renin activity (ng/ml/h)	1.6 + 0.3	3.3 + 0.6[a]	2.7 + 0.6	2.1 + 0.4
Aldosterone excretion rate (μg/24 h)	5.3 + 0.6	6.9 + 0.8[a]	7.9 + 1.2[a]	5.4 + 0.9

[a] $p < 0.05$.

N = 18.

excretion rate in a group of 18 hypertensive patients are shown in Table 2 (Weber and Drayer, 1984). After three weeks of treatment, plasma renin activity was significantly increased when compared with baseline values, but then fell progressively during a 12-month treatment period to values that were not different from baseline. Similarly, aldosterone excretion rate was slightly but significantly increased for up to six months, but after the 12-month treatment period it had returned to its baseline values. Thus, unlike other classes of agents with direct vasodilatory properties, calcium channel blockers appear able to minimize or avoid stimulatory effects on the renin system. This may be due to their ability to prevent the full sympathetic response to their hypotensive effects, or alternatively it might reflect an ability of calcium channel blockers to inhibit the stimulatory effects of sympathetic factors on renin release.

V. METABOLIC EFFECTS

Because of their selectivity for the smooth muscle of the circulation, calcium channel blockers have only minimal direct metabolic effects. It is likely that even the effects on renal function described above are mediated primarily through actions on the vascular mechanisms of the kidney. There has recently been a growing interest in the importance of cholesterol as a cardiovascular risk factor, especially when increased levels occur concurrently with such other risk factors as hypertension. This is of practical relevance during the treatment of hypertension, for the various classes of antihypertensive agents have differing effects on cholesterol. For example, the blood cholesterol profile may sometimes be affected adversely by diuretics and beta blockers: the diuretics can increase concentrations of the LDL cholesterol fraction, whereas the beta blockers can reduce concentration of the protective HDL cholesterol fraction. Agents with sympathetic blocking activity, especially the peripheral alpha blockers, might actually decrease total cholesterol concentrations. The newer classes of antihypertensive agents, including the converting enzyme inhibitors and the calcium channel blockers, are generally regarded as being "lipid neutral".

Although there has been relatively little reported in this area, Table 3 shows the effects on cholesterol of diltiazem during a 30-week study (Pool et al., 1985). The calcium channel blocker did not change total cholesterol levels during treatment, but there was a significant increase in the protective HDL fraction. Consistent with this observation, the ratio of total cholesterol to HDL decreased significantly. Other investigators have reported lipid measurements during a comparison of the calcium channel blocker nicardipine with the beta blocker propranolol (Naukkarinen et al., 1987). There were significant differences between the two agents in their effects on HDL (decreased by propranolol, slightly increased by nicardipine), triglycerides (slightly decreased by nicardipine, increased by propranolol), and in the total cholesterol:HDL ratio (increased by propranolol, slightly decreased by nicardi-

TABLE 3
Changes in Lipid Concentrations

	Control	Diltiazem treatment
Total cholesterol (mg/dl)	227 ± 6	223 ± 7
High density lipoprotein (mg/dl)	52 ± 3	60 ± 3[b]
Low density lipoprotein (mg/dl)	150 ± 5	147 ± 7
Total: HDL ratio	4.7 ± 0.2	4.2 ± 0.2[a]

Note: Mean (+SE); measurements of cholesterol values in 31 patients with essential hypertension treated for approximately 30 weeks with diltiazem.

[a] $p < 0.05$.
[b] $p < 0.01$.

pine). It has yet to be established whether these small effects of calcium channel blockers on lipid measurements are of clinical importance.

As yet, there are no clear data on the effects of calcium channel blockers on insulin and glucose metabolism. It is becoming well accepted that hypertension intrinsically may be a condition of insulin resistance. As a group, hypertensive patients tend to have higher plasma insulin concentrations in response to glucose challenges than are found in normal controls. It has been reported recently that antihypertensive agents can have differing effects on insulin sensitivity. The effects of calcium channel blockers on these measurements are yet to be defined. Importantly, calcium channel blockers do not appear to cause metabolic effects related to changes in overall calcium balance. There have been no consistent findings of changes in bone or other tissues in which calcium is a major factor. There have been occasional reports of slightly elevated plasma alkaline phosphatase concentrations, but these are probably of hepatic origin and might reflect clinically unimportant changes in the tone of the biliary system.

VI. EFFICACY

It has been quite some time since the first calcium channel blockers were found to effectively decrease blood pressure and to be of clinical value in the management of hypertension (Heidland et al., 1962; Olivari et al., 1979). Despite the diversity of the types of calcium channel blockers, and the emergence of several new compounds in recent years, it seems likely that all of these agents exhibit similar efficacy. Indeed, comparisons within the same patients of verapamil, diltiazem, and nitrendipine have demonstrated that they produce virtually identical efficacy (Kiowski et al., 1985). Newer investigational calcium channel blockers also have been found to have similar blood pressure-lowering properties (Kiowski et al., 1985).

Clinicians have been interested in features that could predict the probable responsiveness of individual hypertensive patients to treatment with the differing classes of antihypertensive agents. In particular, the suggestion has been made that patient characteristics such as age or race might be helpful in choosing the antihypertensive agents most likely to be effective for a given patient. Moreover, there are some who claim that calcium channel blockers might be most effective in elderly patients. Investigators using a variety of these agents have demonstrated a correlation between patients' ages and the drug-induced decreases in blood pressure (Kiowski et al., 1985). When compared with beta blockers, for example, calcium channel agents appear to have a different profile. In patients between 20 and 40, the beta

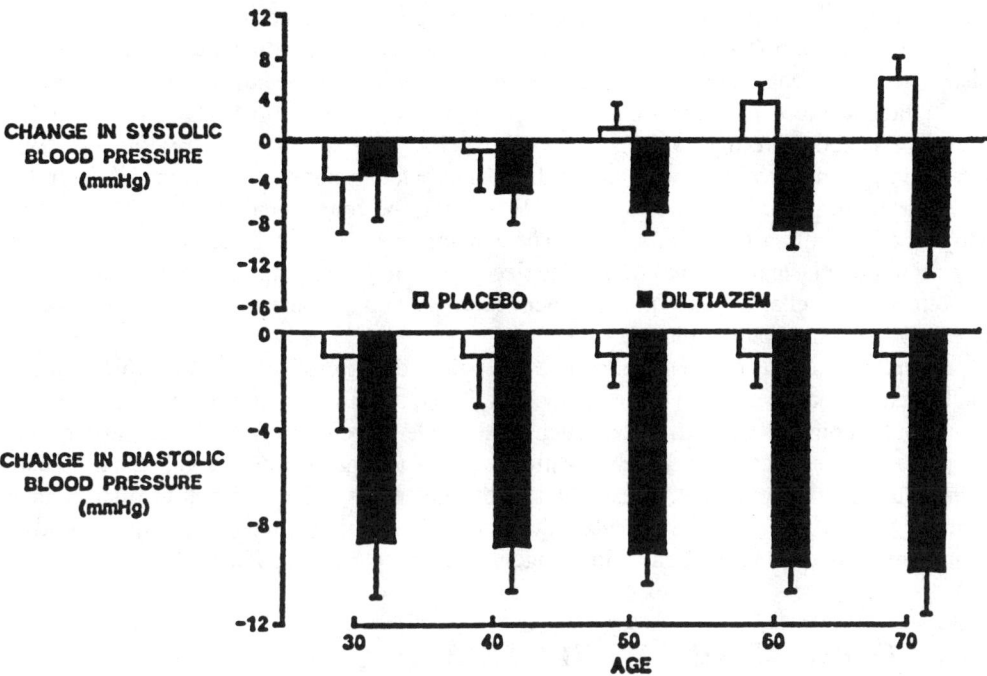

Figure 1. Effects of diltiazem or placebo on systolic and diastolic blood pressures in hypertensive patients subdivided by age.

blocking agents are more likely to produce meaningful reductions in blood pressure. Between 40 and 60 years of age there appears to be little difference between these two drug classes, but the calcium channel blockers appear to be superior in patients over 60 (Schwartz and Abermethy, 1987). Other investigators, however, have been unable to demonstrate such clear findings. In Figure 1 are shown the blood pressure effects of diltiazem when administered to patients of various age groups (Marion Laboratories). It is clear that this calcium channel blocker reduces diastolic blood pressure equally well across all age groups. There is a slight tendency for greater systolic blood pressure decrements in the older patients, but this is probably a reflection of the higher baseline levels in the older group. As discussed below, calcium channel blockers produce their greatest effects in those patients who are the most hypertensive prior to therapy. Thus, it is still not clear whether there is an age dependency of the antihypertensive effects of calcium channel blockers. From a practical standpoint, there would be no reason to avoid the use of calcium channel blockers in younger patients, nor any justification to be committed to their use only in the elderly.

Racial characteristics, especially when based on comparisons of white and black patients, have been thought to be of value in selecting antihypertensive drugs. The beta blockers might be more effective in white patients than black patients, whereas the opposite profile appears to be true of diuretic agents (Hollifield et al., 1978). The use of nifedipine in black African patients has been shown to be especially effective (Kiowski et al., 1985). In a further study performed in the U.S., the calcium channel blocker diltiazem was also found to be highly effective in black patients and to produce decreases in blood pressure that were not different from those produced by diuretic therapy (Moser et al., 1985).

Beyond these simple demographic characteristics, baseline plasma renin activity may be a useful predictor of response to treatment. Low renin patients appear to respond favorably to calcium channel blockers, whereas high renin patients exhibit poorer effects (Buhler et al., 1982). Previously it was also claimed that renin status can be used to predict responses to other classes of drugs. Beta blockers and converting enzyme inhibitors appeared to be most effective in high renin patients, whereas diuretics produced their best results in the low renin subgroup (Laragh, 1973). These findings are consistent with the demographic observations, for both elderly and black patients tend to have low renin values. These groups would then be expected to exhibit strong blood pressure responses to treatment with calcium channel blockers. The mechanisms by which calcium channel blockers are effective in the low renin patients have not been clarified. Plasma calcium concentrations tend to correlate with plasma renin activity and low Ca^{2+} levels may be found more commonly in elderly and black populations in whom there may be a dietary insufficiency of calcium. It is possible that decreased plasma concentrations of ionized Ca^{2+} may affect the membranes of vascular smooth muscle cells, actually making them more sensitive to the blood pressure-lowering effects of calcium channel blockers.

Despite these speculations, there are insufficient data to reliably predict which patients might best respond to calcium channel blockers. Thus, to some extent, their selection remains somewhat empirical. Considerations such as the side effects profiles of these agents, and the presence of concurrent medical conditions such as angina pectoris, might be additionally helpful in guiding treatment. It should also be noted, as discussed below, that calcium channel blockers work well in combination with other antihypertensive agents to produce control of blood pressure when a single agent appears to be inadequate.

VII. COMBINATIONS WITH OTHER DRUGS

Traditionally, most antihypertensive drug combinations have included a diuretic agent. A major reason has been the assumption that the blood pressure-lowering effects of many of the antihypertensive drug classes might be associated with sodium retention. Thus, quite apart from their own antihypertensive properties, diuretics compensate for the sodium retention that could be the cause of tolerance or poor long-term effects with other drugs. On general principles, it would be predicted that diuretics and calcium channel blockers would not comprise a very effective combination. First, as discussed earlier, calcium channel blockers tend to have natriuretic properties of their own, making it unlikely that sodium and volume retention would occur during their administration. Second, since calcium channel blockers and diuretics appear to work best in the same patients subgroups, including the elderly, the black, and the low renin patients, it could be anticipated that they would not provide complementary antihypertensive actions.

In reality, however, calcium channel blockers and diuretics form very effective combinations. In a carefully designed multifactorial study (Burris et al., 1990), differing doses of diltiazem and the diuretic hydrochlorothiazide were tested individually and in a variety of combinations. The combined treatment was clearly more effective than either drug alone, indicating that these two agents could provide powerful blood pressure-lowering effects when administered simultaneously in low doses. Figure 2 shows an experience with a dihydropyridine agent nitrendipine when used in combination with hydrochlorothiazide (Weber and Drayer, 1984). In this latter study, nitrendipine was initially administered as monotherapy but hydrochlorothiazide was then added in those patients whose blood pressure had not responded adequately to the nitrendipine alone. Addition of the hydrochlorothiazide produced a further significant reduction in blood pressure. Moreover, this two-drug combination provided effective antihypertensive therapy during a full year of observation.

Calcium channel blockers also work effectively when combined with beta blockers. Figure 3 describes nitrendipine-treated patients who were studied in exactly the same fashion in Figure 2. However, in this facet of the study, physicians could elect to add propranolol rather than the diuretic in those patients who had failed to respond to nitrendipine alone (Weber and Drayer, 1984). It is evident that the beta blocker produced further significant

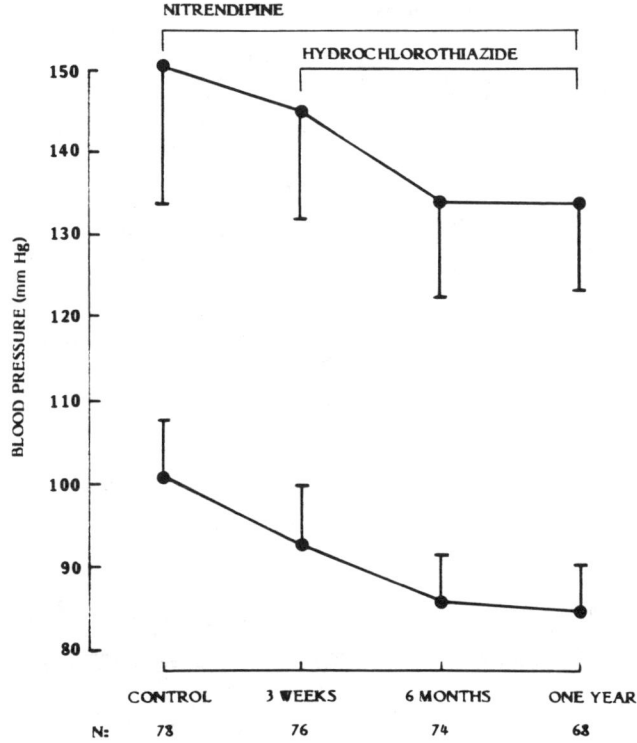

Figure 2. Effects of nitrendipine alone and in combination with hydrochlorothiazide on blood pressure in hypertensive patients treated for 1 year.

decrements in blood pressure, and the combination therapy continued to provide satisfactory efficacy during chronic therapy. It is likely that the efficacy of this type of combination will be found for calcium channel blockers as a class, for verapamil also is effective when used in combination with a beta blocker. There have been some safety concerns when these two classes of drugs are used together, especially when the calcium channel blocker employed is a verapamil-like agent. Each of these drug types can have inhibitory effects on AV conduction in the heart, and their combined use potentially could lead to such conduction disturbances as heart block. Although this is a rare outcome, it is considered prudent to use dihydropyridines when a combination with a beta blocker is selected for optimal control of blood pressure. There has been relatively little published concerning the use of calcium channel blockers with centrally acting alpha agonists or peripheral alpha blockers. Although a limited clinical experience indicates that these combinations are highly effective, further careful studies are required.

The use of converting enzyme inhibitors and calcium channel blockers in combination has also been reported as highly effective, especially in patients with severe hypertension (White et al., 1986). There appears to be a logic behind this combination, for calcium channel blockers can inhibit the alpha receptors that mediate sympathetic and other pressor mechanisms, whereas the angiotensin converting enzyme (ACE) inhibitors block the effects of the renin-angiotensin system. Moreover, the slight natriuretic properties of calcium channel blockers additionally enhance the efficacy of the ACE inhibitors. This type of combination is especially attractive because each of these drug classes is associated with favorable side effect profiles. Symptomatic complaints or reduced exercise tolerance are only rarely encountered when these two types of agents are given together.

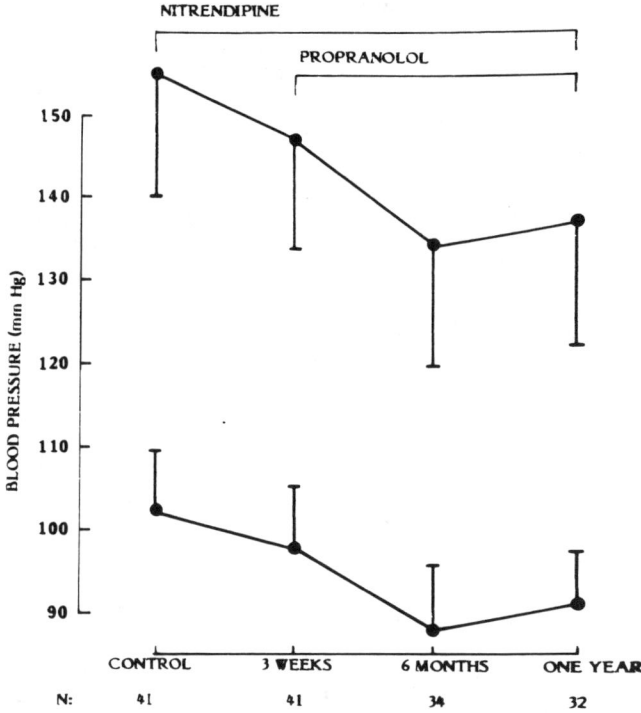

Figure 3. Effects of nitrendipine alone and in combination with propranolol on blood pressure in hypertensive patients treated for 1 year.

VIII. IMPORTANCE OF THE PRE-TREATMENT BLOOD PRESSURE

An early observation regarding calcium channel blockers was that they appeared to be most effective in those patients with the highest pre-treatment blood pressures. Indeed, when compared with other predictors of calcium channel blocker efficacy, including patient age and pre-treatment plasma renin activity, the baseline blood pressure correlated most strongly with the subsequent treatment-induced, blood pressure-lowering effects (Kiowski et al., 1985). This characteristic of these agents has two consequences. First, the calcium channel blockers can be recommended as agents likely to be of most benefit in patients with more severe forms of hypertension. Indeed, as discussed below, calcium channel blockers have been found to be very effective for the immediate treatment of patients in a hypertensive crisis. The second equally important feature is that calcium channel blockers are not likely to produce hypotensive responses in patients whose blood pressures are only slightly elevated. In view of the current interest in the phenomenon of the "J-shaped curve", which is based on the possibility that excessive blood pressure reductions in susceptible patients with prior ischemic heart disease can actually increase the risk of fatal cardiac events (Hansson, 1988), this second property of calcium blockers might be of therapeutic importance.

A study that focused on baseline blood pressures prior to calcium channel blocker treatment, is summarized in Figures 4, 5 and 6 (Weber et al., 1988). Figure 4 shows data derived from whole-day ambulatory blood pressure monitoring studies in which the effects of diltiazem were compared with pre-treatment baseline measurements. Unlike a group of placebo-treated patients, in whom there were no blood pressure changes, the diltiazem clearly produced a reduction in blood pressure throughout the full 24-h period. On more detailed examination of the pre-treatment 24-h blood pressure values, however, it was determined

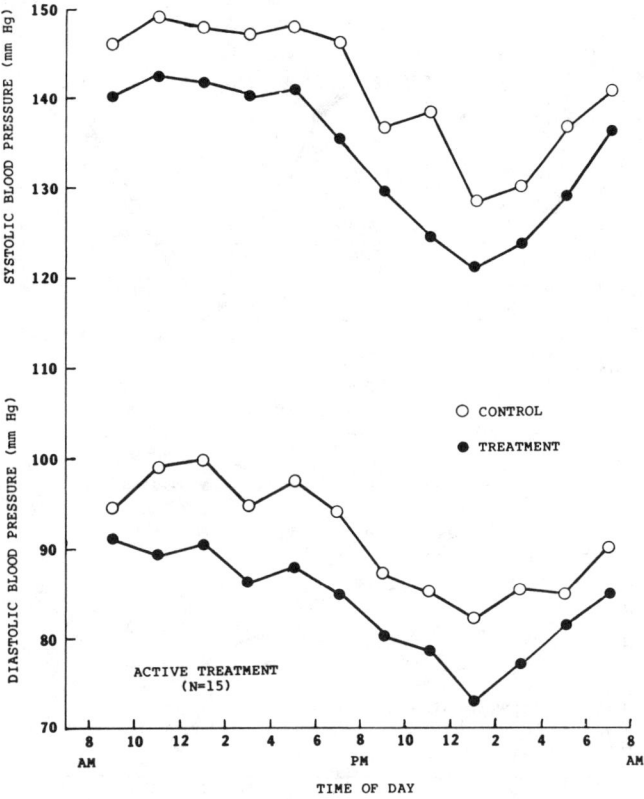

Figure 4. The effects of diltiazem on whole-day systolic and diastolic blood pressures in 15 hypertensive patients evaluated by ambulatory blood pressure monitoring.

that six of the 15 patients participating in the study actually appear to be "normotensive" when judged by this rigorous technique of blood pressure measurement. The effects of the diltiazem in this subgroup of nonconfirmed hypertensives are shown in Figure 5, where it is now evident that the drug had little, if any, blood pressure-lowering actions in these individuals with apparently normal baseline blood pressures. The effects of the diltiazem in the 9 patients with confirmed hypertension are summarized in Figure 6, where the antihypertensive responses were even more powerful than those portrayed for the group as a whole in Figure 4. Thus, this study showed that calcium channel blockers are especially effective in patients with clearly proven hypertension, whereas in individuals with normal or only marginally elevated blood pressures, these drugs can avoid the potentially hazardous consequences of hypotension. This latter property may add to the desirability of calcium channel blockers for the treatment of elderly patients in whom there is a higher likelihood of underlying coronary disease and the potential risks associated with excessive reductions in blood pressure.

IX. ACUTE ADMINISTRATION

Calcium channel blockers have been gaining in popularity for the rapid treatment of hypertensive emergencies or urgencies. This trend reflects both the efficacy of these agents when administered orally or parenterally, and also the potential safety factor associated with their ability to prevent excessive hypotensive responses. This later characteristic may be of

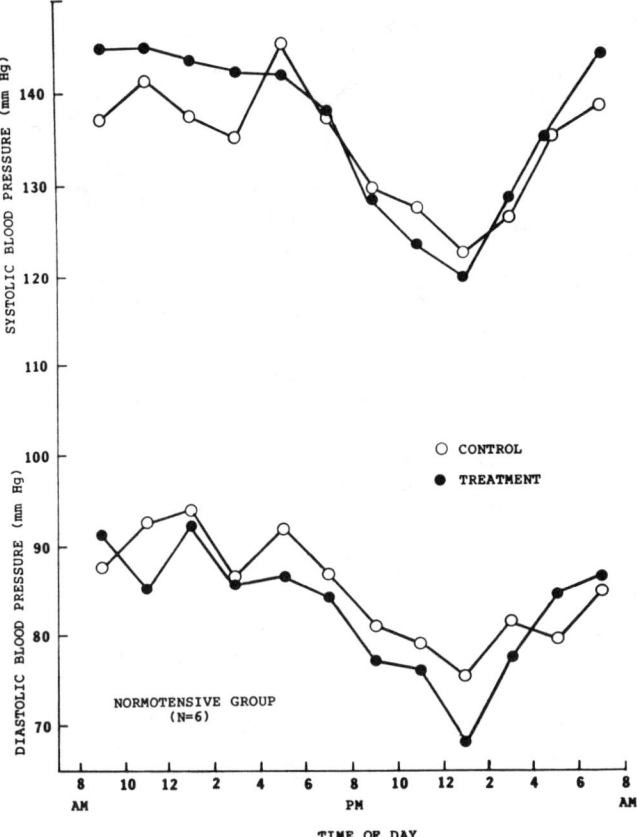

Figure 5. The effects of diltiazem on whole-day systolic and diastolic blood pressures in a subgroup of 6 hypertensive patients whose whole-day diastolic blood pressures averages were below 90 mmHg prior to the start of active treatment.

particular importance in patients with potential cerebrovascular complications. Earlier studies have shown that the cerebrovascular circulation has the capacity to autoregulate across a fairly wide spectrum of blood pressures. However, when pressure rises excessively there is a danger of direct damage to blood vessels; and when the blood pressure falls excessively and is below the lower limits of the autoregulatory range, there is a danger of vascular insufficiency and ischemia. In hypertensive patients, the range of autoregulation is generally shifted to a higher level; thus, there is a danger of precipitating symptoms or serious consequences if blood pressure is reduced rapidly to levels which, in a normotensive patient, might actually be in the normal range (Strandgaard et al., 1973).

Recent studies with an intravenous form of nicardipine have shown that the high efficacy of this agent is almost never associated with excessively low blood pressures or with symptoms suggestive of cerebrovascular insufficiency (Wallin et al., 1990). This excellent safety profile was maintained even though the drug rapidly reduced blood pressure to goal levels in over 90% of cases. For patients who require intravenous treatment of their severe hypertension, this approach clearly is highly beneficial. Moreover, nicardipine can be administered intravenously for periods of several days in hospitalized patients unable to take alternative oral forms of treatment. An important characteristic of this prolonged treatment is that its effects wear off within a short time after discontinuation, and the accumulation problems that might be anticipated following chronic intravenous therapy do not occur (DuPont Laboratories).

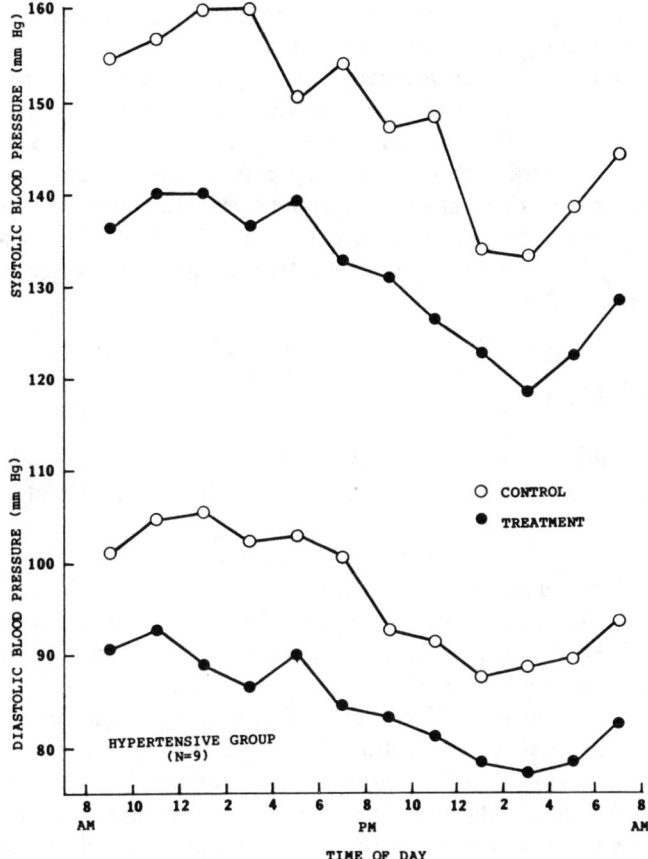

Figure 6. The effects of diltiazem in whole-day systolic and diastolic blood pressures in a subgroup of 9 hypertensive patients whose whole-day diastolic blood pressure averages were 90 mmHg or greater prior to the start of active treatment.

Although most experience with the administration of intravenous nicardipine has been by continuous infusion, recent studies have shown that small bolus doses also can be effective. The duration of the antihypertensive response to the boluses is probably dose dependent, and further research will establish protocols for this method of administration. This may be of value not only in the emergency room but also during surgery. Anesthesiologists have already accumulated experience with other types of intravenous antihypertensive agents administered as sequential boluses, and it seems likely that calcium channel blockers, used in this fashion, will also be of value (Cheung et al., 1990).

Most experience with calcium channel blockers for hypertensive emergencies has been with oral administration. Nifedipine, in its capsule form, has been widely used in emergency rooms and other acute settings. The efficacy rate of this agent is high, and relatively few serious symptoms or side effects have been reported. Although there have been claims that its action can be hastened by mechanically removing the nifedipine from its capsule and placing it directly on the buccal mucosa, more recent experience suggests that swallowing the capsule whole works with similar, and impressive, rapidity. Indeed, this form of oral therapy probably provides control of blood pressure that is as rapid as intravenous treatment, for there is no delay due to establishing intravenous access. Clearly, too, this form of therapy can be utilized in the physician's office or other nonacute environments. It must be stipulated, however, that hypertensive emergencies are serious conditions, and despite the apparent ease of controlling the blood pressure with a simple oral medication, it may still be necessary to hospitalize the patient for further close observation.

Verapamil, used orally, is also effective for hypertensive emergencies. Although not quite as rapid in onset as nifedipine, it provides useful antihypertensive effects within 30 min of administration. For those patients, probably a majority, who do not require very rapid blood pressure reduction, this approach appears to be satisfactory. One attraction of using oral agents for the treatment of antihypertensive emergencies is that the subsequent long-term phase of treatment can easily be continued with the same or similar oral agent. The difficulties often encountered in switching from intravenous therapy to oral treatment can thus be circumvented.

X. SIDE EFFECTS

As with all antihypertensive agents, it is difficult to determine whether symptomatic side effects are directly attributable to drug therapy or instead are simply reflective of the underlying condition. The complaints most commonly associated with calcium channel blockers are changes in bowel function, headache, dizziness, and peripheral edema. The principal bowel effect, constipation, is seen most commonly with drugs of the verapamil type. Headache and dizziness, however, appear to occur most frequently with the dihydropyridine agents. These effects may be, to some extent, idiosyncratic, for they do not appear to be dose related. On the other hand, peripheral edema emerges as the one meaningful complaint that is related to dose. The edema is interesting, for it does not appear to be associated with sodium and water retention. Indeed, as discussed earlier, calcium channel blockers may actually promote some natriuresis. It seems likely that the peripheral edema is a manifestation of calcium channel blocker dilatory action on the venous circulation.

There have been concerns about the effects of calcium channel blockers on the heart. ECG changes can occur as a result of the actions of these agents on conduction, and the symptomatic and clinical consequences of heart block sometimes occur. There has also been interest in the possibility that these agents might cause myocardial ischemia through a process of "coronary steal". The explanation for this phenomenon is that calcium channel blockers are most effective at dilating the healthy components of the coronary microcirculation, which thus enhance their filling at the expense of the diseased portions of the circulation. Whether this leads to true pathologic changes in the myocardium has never been fully established. Potential problems with left ventricular systolic dysfunction have been discussed earlier and will not be dealt with further.

Overall, the symptomatic side effects associated with the use of calcium channel blockers have been acceptable, especially when compared with the older classes of antihypertensive agents. Indeed, the palatability of calcium channel blockers has enhanced their use in clinical practice. Another positive attribute of these agents is that they do not appear to impair exercise tolerance (Klein et al., 1983). Although calcium channel blockers decrease blood pressure at rest, they do not prevent appropriate increases in systolic blood pressure, and decreases in diastolic blood pressure, that occur in response to vigorous exercise. Obviously, the ability to continue an active life style adds further to the acceptability of these agents to hypertensive patients.

XI. BEYOND ANTIHYPERTENSIVE EFFECTS

It is becoming more generally appreciated that the goal of treating hypertension is not simply to decrease blood pressure, but rather to decrease the frequency of strokes and coronary events. Although treatment with virtually any type of antihypertensive agents appears effective at decreasing the incidence of cerebrovascular disease, there has been disappointment in the poor results in preventing atheromatous-related conditions such as coronary disease. Thus, the treatment of hypertension, and perhaps the selection of blood pressure-lowering agents, should be influenced by the likely impact of the therapy upon atheromatous disease.

Calcium channel blockers may have some important properties in this context. The formation of atheromatous lesions, chiefly the classical plaques found in the walls in the arterial circulation, involves several processes. These include deposition of cholesterol, the proliferation and migration of cells, an increase in cell matrix and an increase in the incorporation of calcium. Potentially each of these processes involves mechanisms that can be inhibited by calcium channel blockers. Preliminary studies in animal models have shown promise. The dihydropyridine agents, for example, have been shown to attenuate the formation of atheromatous lesions in the aortas of rabbits fed cholesterol-rich diets (Henry and Bentley, 1981; Watanabe et al., 1987). Similar results have been shown in monkeys exposed to high cholesterol diets (Wesselinovitch et al., 1986). Verapamil (Blumlein et al., 1988) and diltiazem (Sugano et al., 1986) also have been shown to inhibit the formation of atherosclerosis. However, as yet there have been no reports showing that the calcium channel blockers can prevent atheromatous lesions in humans treated for hypertension.

XII. SUMMARY

Calcium channel blockers are now widely used in the treatment of essential hypertension. The various classes of calcium channel blockers, although having differing effects on the myocardium and on arterial smooth muscle, appear equally effective in decreasing blood pressure. The mechanisms by which they produce their vasodilatory effects in the arterial circulation are not completely defined, but it is clear that they can attenuate the vasoconstrictor effects of such hormones as norepinephrine and angiotensin II. Vasoselectivity may be an important property in agents used for the treatment of hypertension, for it implies that inhibitory effects on myocardial contractility and conduction may be minimized. It is prudent, however, not to use these drugs in patients with known left ventricular systolic dysfunction. On the other hand, some preliminary evidence indicates that calcium channel blockers actually can improve left ventricular diastolic function, and in some cases may produce regression of left ventricular hypertrophy. Actions of these drugs in the kidney produce the expected decrease in renal vascular resistance and the corresponding increase in renal blood flow. Glomerular filtration rate can be increased in patients with renal impairment. Interestingly, calcium channel blockers produce a modest natriuretic effect that might add to their blood pressure-lowering properties and enhance their efficacy when used in combination with other antihypertensive drugs. In fact, calcium channel blockers form effective combinations with several other drug classes, including beta blockers, angiotensin converting enzyme inhibitors, and even diuretics. The efficacy of the calcium blockers has been established in all subgroups of hypertensive patients, including both white and black patients and the young as well as the elderly. Interestingly, the blood pressure-lowering properties of calcium channel blockers are most marked in patients with more severe forms of hypertension, while being minimal in patients whose blood pressures are normal or only slightly elevated. This property might provide protection against unwanted hypotensive responses. There is growing interest in the use of calcium channel blockers, administered either orally or intravenously, in the treatment of hypertensive emergencies. Parenteral administration of calcium blockers also may be of value for blood pressure control during surgery and in the perioperative period. There are relatively few symptomatic side effects created with these agents, although headache, dizziness, alterations in bowel habit, and peripheral edema occasionally occur. Effects on cardiac conduction, especially at the AV node, can occur during therapy with these agents. Calcium channel blockers do not appear to produce unwanted metabolic changes, and preliminary studies of their effects on the lipid profile and on glucose and insulin metabolism indicate that they are neutral or even slightly beneficial. Because hypertensive patients have a heightened risk of premature cardiovascular events, there has been a growing focus on the potential ability of calcium channel blockers to modify the development of atheromatous lesions. Definitive studies of the effects of calcium channel blockers on the arterial circulation are awaited.

Chapter 19

Control of Cardiac Arrhythmias by Modulation of the Slow Myocardial Channel

Bramah N. Singh

TABLE OF CONTENTS

I.	Introduction	328
II.	The Role of the Calcium in the Genesis of Cardiac Arrhythmias	330
	A. Early Afterdepolarizations (EADs)	330
	B. Delayed Afterdepolarizations (DADs)	332
	C. Calcium Channel-Dependent Conduction, Myocardial Ischemia and Reentrant Arrhythmias	332
III.	Electropharmacologic Considerations	334
	A. Electrophysiologic Effects of Calcium Channel Blockers	334
IV.	Calcium Channel Blockers in Supraventricular Arrhythmias	336
	A. Electrophysiologic Mechanisms	336
	B. Acute Termination of PSVT	337
	C. Modes of Conversion of PSVT with Calcium Channel Blockers	339
	D. PSVT Conversion by Calcium Channel Blockers Relative to Plasma Drug Levels	341
	E. Acute Conversion of PSVT by Calcium Channel Blockers vs. Other Antiarrhythmic Agents	341
	F. Calcium Channel Blockers vs. Adenosine and ATP in the Acute Conversion of PSVT	342
	G. Response in Other Forms of Reentrant Ectopic Supraventricular Tachycardia	343

	H.	Calcium Channel Blockers and Multifocal Atrial Tachycardia (MAT) 343
	I.	Chronic Prophylaxis of PSVT with Calcium Channel Blockers..........344
	J.	Calcium Channel Blockers in Pre-Excitation Syndromes 345
	K.	Treatment of Other Atrial Tachyarrhythmias by Calcium Channel Blockers..346
		1. Atrial Fibrillation 347
		a. Maintenance of Sinus Rhythm After Conversion 348
		2. Atrial Flutter..348
V.	Ventricular Arrhythmias 349	
	A.	Premature Ventricular Contractions..349
	B.	Chronic Recurrent Ventricular Tachycardia.............................349
	C.	Ischemic Ventricular Arrhythmias 351
	D.	Exercise-Triggered Ventricular Tachycardia 352
	E.	Idiopathic Ventricular Tachycardia with Right Bundle Branch Block and Left Axis Deviation Morphology............................352
	F.	Calcium Channel Blockers and Torsades De Pointes 353
VI.	Calcium Channel Blockers and Sudden Death 353	
	A.	Choice Among Calcium Channel Blockers and Combination with Other Antiarrhythmic Agents..357
VII.	Side Effects, Contraindications and Precautions in the Use of Calcium Antagonists as Antiarrhythmic Drugs 359	
VIII.	Conclusions...361	
	Acknowledgments 361	

I. INTRODUCTION

Since the early 1970s evidence has gradually accumulated to establish the role of calcium channel blockers as potent agents for the control of certain supraventricular and ventricular arrhythmias (Singh et al., 1978; Ellrodt et al., 1980; Stone et al., 1980; Lazzard and Scherlag, 1980; Singh, 1982; Singh et al., 1982). The antiarrhythmic actions of this class of compounds (Class IV) have been elucidated relative to the action of verapamil. In 1972 Singh and Vaughan Williams found that the properties of this compound differed significantly from those of beta-adrenoceptor blocking drugs (Class II) as well as those of the local anesthetic type of antiarrhythmic (Class I) compounds (Hauswirth and Singh, 1979). Verapamil had little or no effect on the action potential duration or refractoriness; thus, it was devoid of Class III actions (e.g., amiodarone). Its antiarrhythmic action therefore was thought to represent a discrete mechanism for the control of cardiac arrhythmias (Singh and Williams, 1972; Hauswirth and Singh, 1979). Subsequently, it has been established that the control of arrhythmias by calcium channel blockade correlates reasonably well with the *in vivo* electrophysiological properties of individual compounds (Lazzara and Scherlag, 1980; Singh et al., 1982). An understanding of the electropharmacologic effects of calcium channel blockers followed in the wake of the knowledge that the inward depolarizing current in the heart muscle is separable into two discrete components. The first, a kinetically fast current ("fast response"), is carried by sodium ions and is activated at -65 to -75 mV; it is sensitive to changes in Na^+ concentration and is selectively inhibited by tetrodotoxin. The fast-channel activity is the basis for depolarization in most normally conducting tissues in the heart: atrial and ventricular muscle, His-Purkinje fibers and atrial internodal pathways, as well as anomalous tracts in the pre-excitation syndromes. The inhibition of the fast channel (Hauswirth and Singh, 1979) and the modification of its kinetics by selective agents (e.g., local anesthetic or Class I antiarrhythmic agents) produces an increase in the time-dependent refractoriness of cardiac muscle, a fundamental mechanism of action of many antiarrhythmic compounds. In such fast channel dependent fibers, the so called slow channel effect ("slow response") is overshadowed by the fast channel activity. The slow channel, the charge carrier for which is predominantly Ca^{2+}, is slowly activated at a threshold potential of -35 mV and is equally slowly inactivated.

Two types of calcium channels have now been identified. The first, the T-type channel, conducts a transient type of calcium current; the other, the L-type channel, has the property of conducting a long-lasting calcium current (see Section I). The latter type channel forms the basis for the slow-response action potentials and accounts for the upstroke velocity of the action potentials in the automatic cells in sinoatrial (SA) and atrioventricular (AV) nodal tissues (Hagiwara et al., 1988). The L-type channels are sensitive to the action of sympathomimetic amines whereas the T-type channels are not. Thus, the increases in heart rate induced by catecholamines are mediated by L-type channels, although, the T-type channels are involved in the second half of the slow diastolic depolarization in the nodal cells (Hagiwara et al., 1988).

For the purposes of discussion of the antiarrhythmic actions of calcium channel blocking drugs, it is important to emphasize that nickel and teramethrin block the T-type channels whereas the dihydropyridines, verapamil and diltiazem block the L-type channels. A great deal of progress has recently been made regarding the characterization of these channels by the use of single channel recordings using the patch-clamp techniques. Experiments in many different laboratories have clarified the functional roles of different types of calcium channels and their distribution within a given cell and within different parts of the heart. With further advances, it may be possible that many electrophysiologic phenomena in the normal heart may be understood more precisely in terms of the selectivity and the kinetic properties of myocardial ionic channels. Whether this information will have a direct application to the understanding and control of disorders of cardiac rhythm and conduction in disease states and their control remains conjectural.

An understanding of the properties of calcium channels in cardiac muscle appears nevertheless crucial to the delineation of the electrophysiologic and antiarrhythmic effects of calcium channel blockers. It is the selective inhibition of calcium channel activity with the consequent increase in the time-dependent refractoriness in the slow response-dependent fibers in the heart that characterizes the fundamental property of the so called calcium channel blockers. Myocardial fibers, which are dependent on calcium channel activity, under either physiological (SA and AV nodes) or pathological (ischemia, infarction) conditions, thus become the primary focus of action for calcium channel blockers. The role of slow response potentials in the origin of cardiac arrhythmias is therefore of paramount clinical importance.

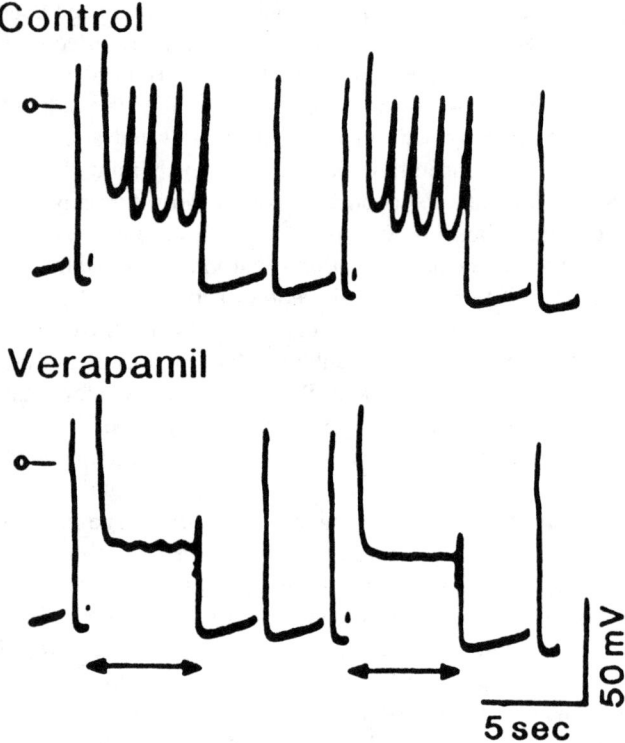

Figure 1. Recordings of automaticity induced by changing resting membrane potential from −85 to −50 mV (between arrowheads) in cardiac Purkinje fiber from the dog. Verapamil (3 μM) suppressed automaticity. (Modified with permission from Elharrar V, Zipes DP: Voltage modulation of automaticity in cardiac Purkinje fibers, in Zipes DP, Bailey JC, Elharrar V (eds): *Slow Inward Current and Cardiac Arrhythmias*. The Hague, Martinus Nijhoff Publishing, 1980; 357—373.)

II. THE ROLE OF THE CALCIUM CHANNEL IN THE GENESIS OF CARDIAC ARRHYTHMIAS

There has recently been increasing agreement that most arrhythmias result from perturbations in impulse generation (ectopic or automatic tachyarrhythmias), in impulse conduction (reentrant arrhythmias), or both (Hoffman and Rosen, 1981; Zipes et al., 1983). For the present, it is not entirely clear under what experimental and clinical conditions disturbances in impulse generation and in conduction that lead to various cardiac arrhythmias are mediated via calcium myocardial channel. Recent experimental studies have focused on the role of the calcium channel in the genesis of triggered automaticity (Cranefield, 1977; Rosen and Reder, 1981) in the form of early afterdepolarizations as well as delayed afterdepolarizations. The typical recordings illustrating the electrophysiologic phenomena representing the two types of triggered automaticity are reproduced in Figures 1 and 2. Their nature will be discussed briefly.

A. Early Afterdepolarizations (EADs)

Although EADs can be produced by a variety of pathophysiologic interventions such as hypoxia and acidosis and experimentally by blocking repolarizing currents (e.g., Cs^+ and aconitine), perhaps the most striking clinical examples are from the use of antiarrhythmic

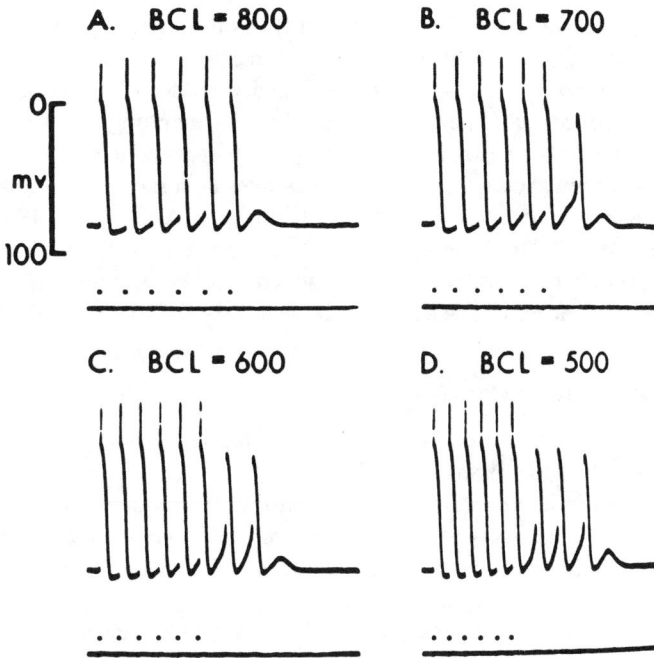

Figure 2. Graphic representation of appearance of delayed afterdepolarization (DADs) in isolated strip of canine Purkinje fibers. In each panel, the strip was given six pacing stimuli (dots) at specific basic cycle lengths (BCL). (A) DADs become progressively more prominent after each successive pacing stimulus (BCL = 800 ms) but do not reach threshold. (B) When BCL is reduced to 700 ms, DAD after last driven beat reaches threshold and triggers spontaneous action potential. (C) When BCL equals 600 ms, two triggered action potentials appear. (D) When BCL equals 500 ms, three triggered action potentials appear. (Reprinted from Ferrier GR, Saunders JH, Mendez C: Cellular mechanisms for the generation of ventricular arrhythmias by acetylstrophanthidin. *Circ. Res.* 1973; 32:600—609, with permission from the American Heart Association.)

agents that prolong cardiac repolarization (quinidine, procainamide, disopyramide, *n*-acetyl procainamide, sotalol, amiodarone and a growing list of newer Class III agents such as sematilide, E-4031, and UK-68,798). EADs, which generally occur in the setting of prolonged cardiac repolarization, tend to occur in salvoes, each EAD being initiated before repolarization is completed. The development of EADs is not related quantitatively to the degree of lengthening of repolarization. The development of EADs in the experimental setting is facilitated by the presence of hypokalemia and hypomagnesemia and by low frequencies of stimulation.

Monophasic action potential recordings in experimental animals and preliminary data in man using contact electrode catheters have indicated that EADs most likely mediate the clinical syndrome of polymorphic ventricular tachycardia known as *torsades de pointes* (Levine et al., 1985). However, the precise ionic mechanisms mediating the development of EADs and hence torsades de pointes is unknown (Brachmann et al., 1983; Takanaka and Singh, 1990). It is of interest that EADs, induced by Cs^+, develop at two levels of membrane potentials, one at 0 to -35 mV, the other at -60 to -70 mV (B. V. Damiano and M. R. Rosen). This suggests that at least two different ionic mechanisms might be involved in the genesis of EADs. Studies with the calcium channel agonist Bay K-8644 have recently provided evidence that EADs induced at a membrane potential level of 0 to -30 mV are most likely due to an inward calcium current that flows through the L-type calcium channels (January et al., 1988). Therefore, it is likely that at least certain forms of torsades de pointes are due to the augmentation of calcium current (January and Riddle, 1989). This is supported by the growing evidence that calcium channel blockers markedly reduce the tendency for the development of EADs in isolated Purkinje fiber preparations (Hiramasu et al., 1988). Recently, it has also been shown that torsades de pointes documented by monophasic catheter recordings may be reversed by calcium channel blockers (Aliot et al., 1985). However, it must be emphasized that EADs leading to the development of torsades in intact animals have also been produced by anthopleurin A which markedly lengthens repolarization by increasing the sodium current in ventricular myocardium (El-Sherif et al., 1988).

B. Delayed Afterpolarizations (DADs)

These develop toward the very end of myocardial repolarization (Hoffman and Dangman, 1987; January and Fozzard, 1988); there is substantive evidence that certain clinically occurring cardiac arrhythmias (e.g., those due to cardiac glycoside intoxication) are mediated by DADs (Figure 2). There is a wealth of experimental evidence indicating that this form of triggered automaticity may develop in the wake of interventions that elevate the intracellular level of Ca^{2+} above a certain critical level (January and Fozzard, 1988). This may result from a high extracellular concentration of Ca^{2+}, high stimulation frequencies, digitalis, or catecholamines (Ferrier, 1977). As indicated, DADs manifest as oscillatory, positive deflections from the resting membrane potential; when such deflections reach threshold, action potentials, occurring in isolation or in a repetitive fashion, are generated. DADs reach threshold and trigger a series of action potentials that occur more readily as the cycle length of stimulation is shortened. As indicated, the initiating factor in the genesis of DADs is the critical increase in the level of cytosolic Ca^{2+} load, which may strain the uptake capability of the sarcoplasmic reticulum (SR); in the event of overloading of the SR, some leakage of Ca^{2+} into the cytosol occurs leading to the production of aftercontractions and DADs. Kass et al. (1978) have shown that the secondary increase in the level of Ca^{2+} in the cytosol elicits a transient inward current (I_{ti}) of ions (mostly Na^+ and some K^+) across the membrane. The data suggest that the rise in the intracellular Ca^{2+} mediates an increase in the ionic conductance of nonselective "leak" currents through which the ionic current (I_{ti}) of Na^+ and K^+ can pass. In the studies by Kass et al. (1978) D600 completely blocked the I_{ti} and the associated aftercontractions during voltage clamp protocols in isolated Purkinje fibers intoxicated with strophanthidin. There is also evidence to suggest that DADs play an important role in myocardial reperfusion arrhythmias (Opie et al., 1988), again linking cytosolic Ca^{2+} to the development of this type of arrhythmia in experimental models of coronary artery occlusion and release (Opie et al., 1988).

The relationship between the concentration of intracellular Ca^{2+} and the development of certain cardiac arrhythmias is shown in Figure 3. The schema is based on numerous lines of experimental evidence but as yet a direct clinical extrapolation of the data is less than convincing. From the standpoint of the antiarrhythmic actions of calcium channel blockers, the development of reentry on the basis of slowed conduction as a result of slow response cells is of more direct relevance.

C. Calcium Channel-Dependent Conduction, Myocardial Ischemia and Reentrant Arrhythmias

There are areas in which the calcium channel activity might be of potential significance in mediating experimentally and clinically occurring cardiac arrhythmias. With the increasing use of calcium channel blocking drugs in patients with ischemic heart disease, the question

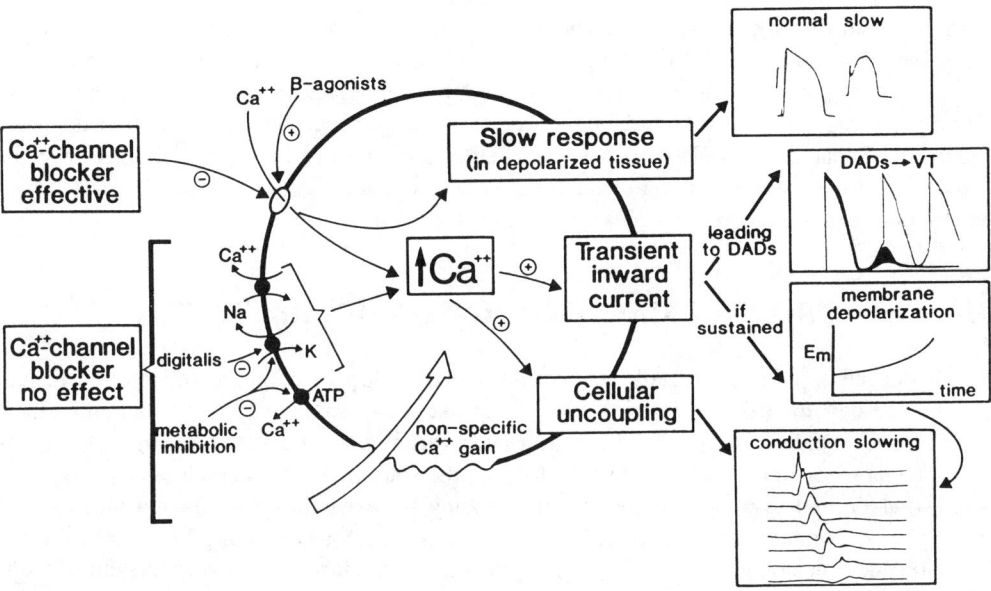

Figure 3. Cell Ca^{2+} and arrhythmias. Assuming that intracellular Ca^{2+} does increase during ischemia (see text) there are a number of potential mechanisms by which the calcium current could be arrhythmogenic: (1) Ca^{2+} influx might elicit slow responses; (2) an additional increase in cytosolic Ca^{2+} could activate oscillatory inward current flow and precipitate DADs; (3) if the Ca^{2+}-dependent activation of this inward current were sustained, then accelerated depolarization might increase injury current across the ischemic boundary and precipitate fibrillation; (4) through promoting depolarization and increasing the coupling resistance between cells, a raised cytosolic (Ca^{2+}) might also slow conduction and increase the likelihood of reentry. Ca^{2+} channel blockers should be effective against the effects of a gain of intracellular Ca^{2+} resulting from enhanced Ca^{2+} entry through the calcium channel. However, these agents should not affect the consequences of sodium pump or metabolic inhibition or nonspecific Ca^{2+} gain except by their antiischemic effect. (Reproduced with permission from Opie et al. and the *Annals of the New York Academy of Sciences*.)

has arisen whether slow response potentials might play a role in the genesis of reentrant ventricular arrhythmias. The development of focal slow responses due to ischemia may produce depressed conduction and unidirectional block. It has been shown that slow response potentials can be generated in isolated cardiac muscle using high extracellular K^+ concentration to partially depolarize the cell and thereby inactivate the fast sodium channels in the presence of catecholamines (Cranefield, 1977). In such *in vitro* models, the requirements for the occurrence of reentrant ventricular arrhythmias are fulfilled (Hoffman and Rosen, 1981; Zipes et al., 1983). Such slow response potentials in this setting, have been found to be very sensitive to the depressant effects of calcium channel blocking drugs such as verapamil (Cranefield et al., 1974; Wit and Cranefield, 1974). Thus, the question has arisen whether the K^+-depolarized and catecholamine-stimulated myocardial cell in the *in vitro* model is comparable to what is obtained in the context of myocardial ischemia *in vivo*. Regional concentrations of K^+ and catecholamines are certainly high in the early stages of myocardial ischemia following coronary occlusion (Lazzara et al., 1978), and it is known that pretreatment with verapamil before coronary artery ligation nearly eliminates the incidence of ventricular fibrillation in the dog (Kaumann and Aramendia, 1968). However, at present it is not known whether this is a primary antifibrillatory effect of calcium channel inhibition by verapamil or merely a secondary consequence of improvement in the intensity of myocardial ischemia by the agent's *anti-ischemic* effect. Overall, it appears unlikely that the calcium channel activity plays a dominant role in the genesis of arrhythmias in the later stages of myocardial infarction (Lazzara et al., 1978; El-Sherif and Lazzara, 1979). At this stage, local increases in K^+ and catecholamines accompanying ischemia or fibrosis are not significant and it is known that ventricular arrhythmias in this chronic setting are not responsive to calcium channel blocking drugs (see below). Nevertheless, it is of interest that slow response potentials have been demonstrated in diseased atria biopsied during open heart surgery (Hordof et al., 1976) and in ventricular tissue removed during ventricular aneurysmectomy (Spear et al., 1979). However, whether the presence of such potentials provides the basis for reentrant excitation or for membrane oscillatory activity for automatic atrial or ventricular arrhythmias for the present remains speculative.

III. ELECTROPHARMACOLOGIC CONSIDERATIONS

From the experimental findings discussed above, the role of intracellular Ca^{2+} in mediating the development of certain types of cardiac arrhythmias is reasonably compelling. In contrast, as judged by the responses of a wide variety of clinically occurring arrhythmias to intravenous or oral calcium channel blockers the role of these compounds appears to be somewhat limited. The reason for the discrepancy between the experimental and clinical results remains uncertain. This is especially so in the case of ventricular tachyarrhythmias.

Considerable doubt also exists regarding the precise relationship between the abnormality of calcium channel activity and the genesis of various types of supraventricular tachyarrhythmias. The subject has been discussed at length by Zipes et al. (1980). As alluded to above, calcium channel-dependent activity has been found to occur in tissues removed from diseased human atria at surgery (Hordof et al., 1976) but whether such regenerative pacemaker potentials mediate ectopic atrial tachycardias remains conjectural. The role of these observed slow response activities in diseased human atria in the genesis of atrial flutter and fibrillation is also unclear. However, the fact that calcium channel blockers rarely produce conversion of such arrhythmias to sinus rhythm (see below) suggests that abnormalities of the calcium channel are unlikely to be a mechanism underlying the development of atrial flutter and fibrillation in man. The most compelling evidence for the role of the calcium channel in the genesis of clinical arrhythmias appears to be in the case of reentrant paroxysmal supraventricular tachycardia, especially that involving the AV and SA nodes.

A. Electrophysiologic Effects of Calcium Channel Blockers

In interpreting the overall actions of a calcium channel blocker in arrhythmias in man, it is worth emphasizing that the normal AV node is calcium channel dependent, and that it may become the site of deranged impulse formation and impulse conduction, the two most significant mechanisms underlying the genesis of cardiac arrhythmias. In the experimental setting, calcium channel blockers block the calcium channel in a concentration or dose-dependent fashion. Thus, in fibers with a fast sodium-dependent depolarization they have no significant effect on atrial, ventricular, or His-Purkinje refractory periods or on conduction velocity. In isolated preparations, they slow phase 4 depolarization in the SA and AV nodes with an associated depression of conduction. Because of their effects on the calcium channel, calcium channel blockers also accelerate repolarization at the plateau phase of the action potential. However, the magnitude of this effect is such that there is little or no change in the voltage-dependent refractoriness. The major effect on the refractory period is in the AV node where both the effective refractory period (ERP-longest coupling interval between basic drive and premature impulse that fails to propagate) and the functional refractory period (FRP-minimal interval between two consecutive conducted impulses) are significantly lengthened in the antegrade as well as in the retrograde directions. Some of the newer agents exert additional effects over and above their propensity to block the myocardial calcium channel. For example, bepridil (Singh et al., 1982) blocks the fast sodium channel and it prolongs

TABLE 1
Clinical Electrophysiologic Effects of Calcium Channel Blockers

Effects	Verapamil[b]	Nifedipine	Diltiazem	Bepridil
Heart rate	↓ ↑	↑ +	↓ +	↓ +
QRS	0	0	0	↑ +
QTc	0	0	0	↑ + +
PR	↑ + +	0	↑ + +	↑ + +
A-H	↑ + + +	0	↑ + + +	↑ + +
H-V	0	0	0	↑ +
Atrial ERP	±	0	±	↑ +
AV node ERP	↑ + + + +	±	↑ + + + +	↑ + + +
AV node FRP	↑ + + + +	±	↑ + + + +	↑ + + +
Ventricular ERP	0	0	0	↑ + +
His-Purkinje ERP	0	0	0	↑ + +
Bypass tract ERP	±	0	±	↑ + +
Sinus node recovery time	0[a]	0	0[a]	↑ +
Ventricular automaticity	0	0	0	↓ +

Note: ↓ = decrease; ↑ = increase; ± = variable effect; + → + + + + = range of change from minimal to large; ERP = effective refractory period; FRP = functional refractory period.

[a] Prolonged in sick sinus syndrome.
[b] Effects of gallopamil and tiapamil are similar to those of verapamil.

the action potential duration. Thus, this compound, unlike the conventional agents, has a wider spectrum of electrophysiologic and antiarrhythmic actions.

The *in vivo* and clinical effects of calcium channel blockers represent a balance of their direct actions and those that result from reflex activation of the sympathetic nervous system engendered by their often potent vasodilator actions. In the case of the dihydropyridines the reflex actions either nullify or reverse the intrinsic actions on the sinus node. Thus, these agents do not exert measurable effects on the electrophysiologic parameters in man and appear to be devoid of antiarrhythmic actions. In general, the clinical electrophysiologic effects of various calcium channel blockers are in accord with their *in vitro* effects. In the case of diltiazem, verapamil and bepridil, the net effects are also modified by the noncompetitive sympatholytic effects of these compounds.

The major clinical electrophysiologic actions of various calcium channel blockers are summarized in Table 1. In general, their antiarrhythmic actions can be explained in terms of their overall electrophysiological effects. Undoubtedly the most significant action whereby calcium channel blockers exert their salutary effects in arrhythmias is by modulating the electrophysiologic function of the AV node. This is an example of their "direct" antiarrhythmic actions. Calcium channel blockers prolong the intranodal conduction time (the AH interval) and lengthen the antegrade and retrograde effective and functional refractory periods. The effect on conduction is utilized in the termination of acute episodes and for the prevention of recurrences of paroxysmal supraventricular tachycardia (PSVT) and for slowing the ventricular response in atrial flutter and fibrillation (see below). As indicated in Table 1, these agents have no measurable effects on intraatrial, intraventricular or His-Purkinje conduction or refractoriness (Singh et al., 1982). It should also be noted that the effects on the AV node and the SA node *in vivo* among different calcium channel blockers differ in terms of the reflex responses engendered by their peripheral vasodilator actions relative to the absence or presence of associated noncompetitive anti-sympathetic effects. Most calcium channel blockers have little or no electrophysiologic effects on ventricular muscle, the exception being bepridil (see Table 1). Thus, they are unlikely to be potent antiarrhythmic

TABLE 2
Mechanisms of Paroxysmal Supraventricular Tachycardias

Enhanced automaticity
  (a) Paroxysmal and acute
  (b) Chronic
Reentry without bypass tracts
  (a) AV nodal reentry: slow-fast
                       fast-slow
  (b) Sinoatrial nodal reentry
  (c) Intra-atrial reentry
Reentry in association with bypass tracts
  (a) Reentry with antegrade AV conduction (orthodromic)
     (i) with evidence of pre-excitation on 12-lead ECG
     (ii) concealed WPW (bypass tract conducting only retrogradely)
  (b) Reentry with antegrade conduction over bypass tract (antidromic) during tachycardia

---

agents in most types of ventricular tachyarrhythmias although recent experimental and clinical evidence has suggested a role for these drugs in the control of arrhythmias due to triggered automaticity (Brugada and Wellens, 1984). There is also evidence that calcium channel blockers may prevent the development of such arrhythmias by primarily influencing the course of myocardial ischemia either by altering demand or supply or both in patients with coronary artery disease, especially those prone to coronary vasospasm. The beneficial actions of these drugs in this context may therefore be considered an "indirect" antiarrhythmic effect.

IV. CALCIUM CHANNEL BLOCKERS IN SUPRAVENTRICULAR ARRHYTHMIAS

A. Electrophysiologic Mechanisms

The experimental and clinical observations that most arrhythmias arise on the basis of automaticity or reentry apply reasonably well in the case of PSVT (Josephson and Kastor, 1976, 1977; Josephson, 1978; Anderson et al., 1982). A simplified classification of the electrophysiologic mechanisms for the clinically occurring PSVTs is presented in Table 2.

The least common variety is the automatic form (not shown in Figure 3); many types of atrial fibers are capable of spontaneous diastolic depolarization, and when enhanced, such spontaneous diastolic depolarizations may form the basis for PSVT in some patients. As indicated above, the exact ionic basis for the automaticity in this setting is not known, although it is tempting to suggest that calcium channel-dependent activity, as has been demonstrated in certain supraventricular tissues, may constitute the basis for the arrhythmia. However, it must be emphasized that the so called automatic PSVT is incompletely characterized and may not be easily distinguished from sinus nodal or intraatrial reentry.

The idea that PSVT could occur as a result of reentrant mechanism is not new (Iliescu and Sebastiani, 1923) but it is only in the last decade or two that electrophysiologic evidence in experimental animals and man has supported the idea that impulse conduction through the AV node can undergo functional longitudinal dissociation and may provide the basis for reentrant PSVT (Mendez and Moe, 1966; Denes et al., 1973). Intranodal reentry is perhaps the most common form of the clinically encountered PSVT (see Table 2); however, the exact pathways used are still not completely defined. The presence of more than one AV nodal pathway in many patients with recurrent AV nodal reentrant PSVT is suggested (albeit without conclusive proof) by the demonstration of a "discontinuous" AV nodal curve (Denes et al., 1973); here, a marked increase in the interval between successive His bundle responses occurring with little or no shortening of the premature atrial intervals, suggests conduction over two pathways, one having a shorter refractory period and a longer conduction time than the other. Thus, when there is antegrade block in the fast pathway with ventricular activation occurring through the slow pathway, the fast pathway is then invaded retrogradely to produce atrial echoes or sustained PSVT ("slow-fast" form). In the much rarer form of AV nodal reentrant PSVT ("fast-slow"), the antegrade conduction occurs over the fast pathway with retrograde conduction (to the atria) over the slow.

The electrophysiologic evidence for reentry as the basis for PSVT is perhaps most compelling for this arrhythmia when it occurs in association with pre-excitation (Heng et al., 1975). Most frequently, ventricular depolarization occurs via an antegrade conduction (orthodromic) through the normal AV node—His Purkinje system with the retrograde conduction over the accessory pathway. Thus, the ventricular activation has a normal sequence so that narrow QRS complexes are generated during the tachycardia. In many patients, the accessory pathway capable of conducting only in the retrograde direction (concealed WPW syndrome) constitutes the retrograde limb of the reentrant loop. In the uncommon form of the PSVT occurring in relation to the accessory pathway, ventricular excitation during the tachycardia occurs in an antegrade direction via the accessory pathway (antidromic tachycardia), the AV node forming the retrograde limb of the circus movement. Examples of the different forms of reentrant PSVT discussed above are shown in Figure 4. PSVT due to sinus nodal or intraatrial reentry are uncommon and are often extremely difficult to distinguish from other types of reentrant PSVT. Calcium channel blockers have a role in the acute termination of PSVT when given intravenously as well as in preventing recurrences when they are administered orally.

B. Acute Termination of PSVT

Most experience in the use of calcium channel blockers in termination of acute paroxysms of SVT has been with intravenous verapamil but preliminary data are available with Type 1 calcium channel blockers such as tiapamil, gallopamil and diltiazem. However, their precise potencies relative to that of verapamil remain to be established.

In the largest numbers of patients with PSVT, the antegrade limb of the tachycardia circuit is the AV node, the "weakest link" of the reentrant loop; it is most susceptible to the blocking action of intravenous verapamil. This may account for the prompt and predictable reversion of 80 to 100% (Krikler, 1974; Heng et al., 1975; Singh et al., 1978; Kuhn, 1981; Anderson and Reiser, 1981; Dargie et al., 1981; Krikler, 1980; Singh et al., 1983) of cases of PSVT given 3 to 5 mg (in children) to 10 to 15 mg (in adults) of verapamil. The earlier studies focused essentially on the issue of efficacy of verapamil in PSVT (Brichard and Zimmerman, 1970; Gotsman et al., 1972; Schamroth et al., 1972; Heng et al., 1975; Hartel and Hartikainen, 1976; Hagemeijer, 1978); the more recent ones have dealt with potential mechanisms whereby the drug exerts its salutary actions in the termination of PSVT (Sung, Elser and McAllister, 1980; Sung, Waxman et al., 1980; Rinkenberger et al., 1980). There has been an increasing concensus that for the acute conversion of PSVT, verapamil is the drug of choice for terminating PSVT with narrow QRS complexes due to reentry involving antegrade conduction over the AV node. As might be expected, Type II calcium channel blockers (i.e., dihydropyridines) are ineffective in this regard. Recently, it has been shown that diltiazem given intravenously or orally may also terminate narrow QRS complex PSVT (Rozansky et al., 1982; Betriu et al., 1983; Rowland et al., 1983; Hung et al., 1984; Yeh et al., 1985), with a potency probably comparable to that of verapamil in studies with a smaller number of patients than in those in which verpamil has been evaluated. As in the case of verapamil, the termination of AV nodal reentrant tachycardia with diltiazem may

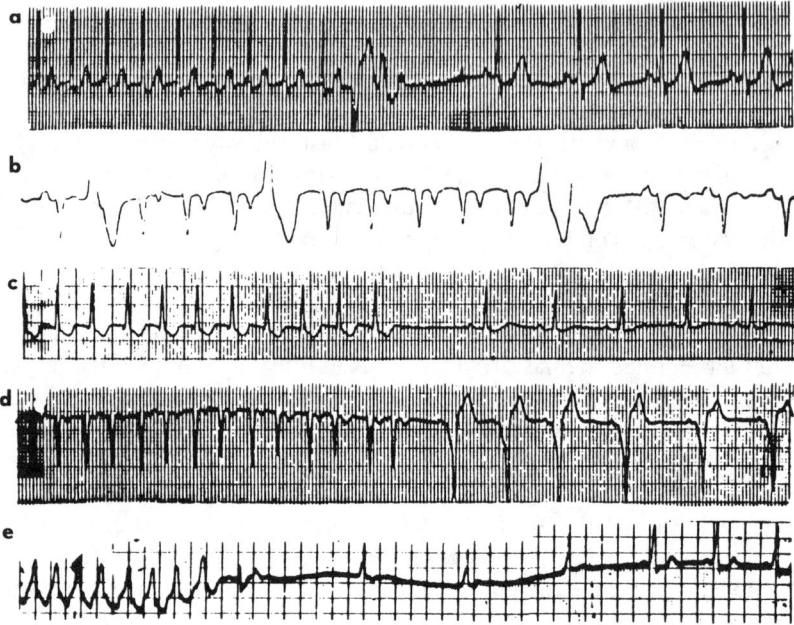

Figure 4. Different forms of paroxysmal supraventricular tachycardia (PSVT). (a) "Slow-fast" intranodal PSVT. Atria and ventricles are activated almost simultaneously. No P-wave activity is identifiable in the surface record during the tachycardia. (b) "Fast-slow" intranodal PSVT. The tachycardia is initiated by one premature ventricular contraction (PVC) and is terminated not by a single PVC but by PVCs occurring in a pair. Note the delayed retrograde activation of the atria (over the slow pathway). (c) Reentrant PSVT with retrograde atrial activation via concealed bypass tract. The antegrade limit of the tachycardia is constituted by the AV node. Note the retrograde P waves immediately following the QRS complexes during the tachycardia. The termination of the tachycardia reveals no evidence of pre-excitation, since conduction over the bypass tract occurs solely in the retrograde direction. (d) Circus movement tachycardia with the WPW Syndrome. During the tachycardia, the ventricular activation occurs via the AV node with the retrograde invasion of the atria over the Kent bundle; there is a retrograde P wave succeeding each QRS complex. The termination of the tachycardia is preceded by cycle length alternation. (e) Antidromic tachycardia. The reentrant loop is formed by antegrade conduction over the bypass tract and retrograde conduction through the A-V node. (From Singh et al., 1983, with the permission of the authors and of the editor of *Drugs*.)

be due to antegrade block at the AV node. Hung et al. (1984) found in 6 patients with inducible AV nodal reentrant PSVT and in 15 of 24 patients with orthodromic PSVT (with the WPW syndrome) that the tachycardia terminated within 60 s of the I.V. injection of 0.25 mg/kg of diltiazem. In three of the 24 patients, termination of the arrhythmia occurred after 1 min, and the cycle length of the remaining 6 was lengthened by slowing of the AH interval. Thus, it is likely that the overall effect of I.V. diltiazem is comparable to that of I.V. verapamil in the acute termination of narrow QRS complex PSVT. The experience with other Type 1 calcium channel blockers (e.g., tiapamil) is limited. However, the preliminary experience and the knowledge of their electrophysiologic effects (Seipel et al., 1980; Russell, 1981; Henderson and Lewis, 1981; Seipel and Breithardt, 1982) suggest that they are likely to be less effective than verapamil in converting acute paroxysms of SVT. Table 3 lists the compounds that are effective in terminating PSVT when administered intravenously. It should be emphasized that beta blockers and calcium channel blockers in general terminate AV nodal reentrant PSVT by blocking the antegrade tachycardia limb of the circuit and Class I agents are also effective in blocking the retrograde limb. This is illustrated for verapamil,

TABLE 3
Comparative Efficacy of Verapamil and Other Antiarrhythmic Agents in the Acute Conversion of Reentrant Supraventricular Tachycardia

Agent (Ref.)	Percentage conversion	Mode of termination	Ref.
Intravenous propranolol	41—44	Antegrade AV block	Bauernfeind et al., 1980; Wu et al., 1974
Intravenous ouabain	44—54	Antegrade AV block	Bauernfeind et al., 1980; Wu et al., 1974
Intravenous ouabain plus intravenous propranolol	58	Antegrade AV block	Ross et al., 1982
Intravenous procainamide	65	Retrograde AV block	Wu et al., 1978
Oral quinidine	78	Retrograde AV block	Wu et al., 1981
Oral disopyramide	67	Retrograde AV block	Swiryn et al., 1981
Intravenous disopyramide	60	Retrograde AV block	Sethi et al., 1983
Intravenous verapamil (see text)	80—100	Antegrade AV block (see text)	
Intravenous ajmaline	80—100	Retrograde AV block	
Intravenous bepridil[a]	80—100	Antegrade AV block	Sethi et al., 1981
Intravenous diltiazem (62)[a]	over 80	Antegrade AV block	
Intravenous tiapamil[a]	over 50	Antegrade AV block	
Intravenous gallopamil[a]	over 80	Antegrade AV block	
Intravenous ATP (89—91)	80—100	Antegrade AV block	
Intravenous adenosine (68)	90—100	Antegrade AV block	
Intravenous amiodarone[a]	? over 50	? Antegrade AV block	

[a] Based essentially on unpublished data. It is also emphasized that the figures cited for individual agents are based on nonstandardized and uncontrolled protocols but they provide an indication of the relative potencies of various agents used for the termination of PSVT. Because of the enormous variability of technique, the overall effects of carotid sinus massage have not been estimated. Similarly, the data on the effects of edrophonium (Tensilon) are essentially anecdotal and variable.

propranolol, and procainamide in Figure 5. Although not extensively studied, the effect of intravenous amiodarone is similar to the action of a beta blocker and that of verapamil or diltiazem with the block occurring in the antegrade limb as it does in the case of vagal maneuvers, adenosine (DiMarco et al., 1985) and ATP (Favale et al., 1985). In the case of amiodarone, the major effect most likely stems from the drug's calcium channel blocking effect manifested acutely (Singh, 1983).

When verapamil is given intravenously during an episode of PSVT, the onset of its action is rapid. Sinus rhythm is restored in most cases in 2 to 3 min, and in some cases within 10 min of drug administration. The success rate of conversion of PSVT to sinus rhythm by verapamil may be improved to nearly 100% by carotid sinus massage or the addition of 5 to 10 mg of edrophonium (Tensilon) given in rapid succession after verapamil. Although less experience is available in children, the antiarrhythmic spectrum of verapamil appears similar in children and adults (Solar-Soler et al., 1979; Porter et al., 1981; Casta, 1981; Sapine et al., 1981; Shakibi, 1981). The data in the elderly is also comparable (Midtbo, 1981a; Midtbo, 1981b). The spectrum of clinical antiarrhythmic effects of other calcium channel blockers relative to age is at present uncertain.

C. Modes of Conversion of PSVT with Calcium Channel Blockers

The continuous monitoring of the ECG with or without intracavitary recording of the AV activation sequence during the tachycardia has shown that the mode of conversion of PSVT to sinus rhythm is variable. The most common is an abrupt termination of the arrhythmia as might occur with carotid sinus massage. In many cases, the cycle length of

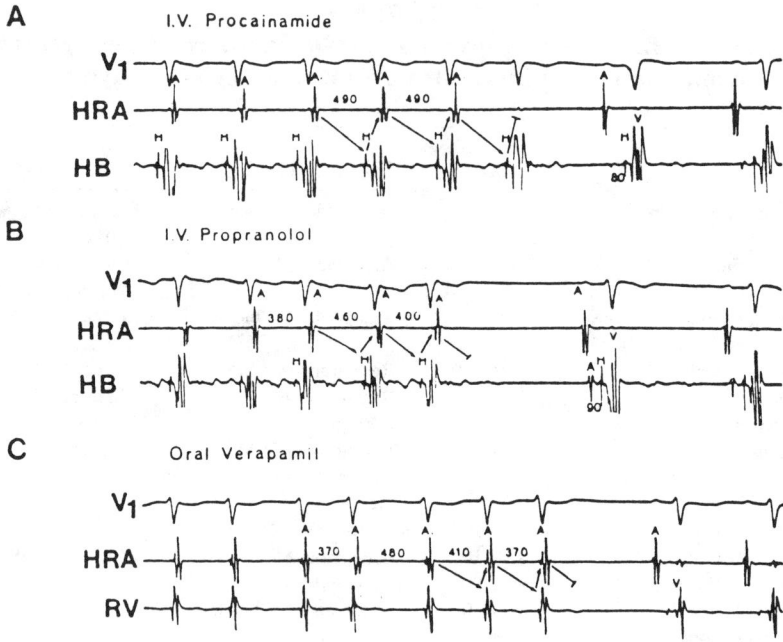

Figure 5. Recordings of site of action in the common type of atrioventricular (AV) nodal reentry, that is, slow antegrade conduction (A-H) and fast retrograde (H-A) conduction terminations are shown after antiarrhythmic agents. In panel A, retrograde block is seen after intravenous procainamide, whereas antegrade block occurs with propranolol (panel B) that is, atrial response (Ac) is not followed by H deflection or QRS complex. Similar termination is seen in panel C with verapamil. V_1, surface ECG lead; HR + 7 high right atrial electrogram; HB, His Bundle electrogram. Same abbreviations are used in subsequent traces. (Reproduced with permission from Akhtar M, in Josephson ME, Wellens HJJ (eds): *Tachycardias: Mechanisms, Diagnosis and Treatment.* Philadelphia, Lea & Febiger, 1984; 0137.)

the tachycardia may lengthen somewhat due to the delayed AV conduction (as evidenced by lengthening of the A-H interval) followed by a short pause before reversion to sinus rhythm. There is some disagreement as to whether verapamil and other calcium channel blockers terminate the intranodal reentrant PSVT by blocking antegrade conduction or retrograde conduction or both (Sung, Elser and McAllister, 1980; Sung, Waxman et al., 1980). The overall experience suggests that the antegrade block is the more common but is not the sole focus of action of these compounds in every case of AV nodal reentrant tachycardia. It is generally agreed, however, that these agents are equally effective in terminating fast-slow as well as slow-fast forms of PSVT.

In some cases of PSVT, given intravenous verapamil (or other Type 1 calcium channel blockers), there is frank AV dissociation followed by a junctional escape rhythm before sinus rhythm is restored. In a few patients, reversion to sinus rhythm is preceded by the occurrence of transient atrial fibrillation, the mechanism of which is unclear since calcium channel blockers (with the possible exception of bepridil) do not reduce atrial flutter rate (Heng et al., 1975). Two other types of response have also been reported. The first is characterized by the occurrence of premature ventricular contractions, but this is also seen during PSVT conversion by the use of beta-adrenoceptor blocking drugs or carotid sinus massage. Neither the mechanism nor the precise clinical significance of such premature ventricular contractions is known. Short runs of ventricular tachycardia have also been reported in this context. Again, the mechanism for their occurrence during the conversion of PSVT to sinus rhythm is obscure. The final type of response during PSVT termination is the development of alternating cycle length before conversion to sinus rhythm (Singh et al., 1978). Intracavitary recordings (Sung, Elser and McAllister, 1980; Rinkenberger et al., 1980) have shown that the alternation in cycle length is due to alternations in the A-H or H-A intervals. The precise mechanism for this phenomenon is not well understood but it is not a unique feature of the action of calcium channel blockers. It has been found to occur with nearly all interventions including ajmaline, which has recently been shown to be extremely effective in terminating PSVT (Sethi et al., 1984). The precise mechanism underlying the occurrence of cycle length alternation in this setting is uncertain. However, a number of possibilities have been considered. For example, it has been suggested to represent the unmasking of a third intranodal pathway conducting the longer cycles (Vohra et al., 1974) or the occurrence of a form of resonance related to the effects of alterations in the cycle length on the AV nodal conduction time (Sung, Waxman et al., 1980). However, recent analysis by the use of a computer model (Ross et al., 1982) has demonstrated that cycle length alternation during a circus-movement tachycardia can be explained by the characteristics of a single curve of AV nodal function without the need to postulate the presence of an additional antegrade accessory tract in the heart.

D. PSVT Conversion by Calcium Channel Blockers Relative to Plasma Drug Levels

At present the data refer essentially to the effects of verapamil. In the acute reversion of PSVT by verapamil, the early prompt response appears to be related to high initial plasma concentration of the drug which aborts the tachycardia. Sung, Elser and McAllister (1980) gave I.V. verapamil to 17 patients with PSVT refractory to propranolol or digoxin. A dose of 0.075 mg/kg converted all patients to sinus rhythm (n = 9) who had plasma levels exceeding 72 ng/ml; the nonresponders had levels of 0 to 62 ng/ml. Following a second bolus injection (0.15 mg/kg), 6 others converted (plasma levels 98 to 1320 ng/ml), only 2 patients with plasma levels above 100 ng/ml having no response. Thus, the data indicate the dependence of drug response on plasma levels which may vary widely in individual patients given an identical dose of verapamil. However, it must be emphasized that the measurement of plasma drug levels for the acute intravenous administration of verapamil is neither practical nor necessary. Rather, the data emphasize the need for titrating the dose of the drug to the desired response on an individual basis rather than adhering to a rigidly predetermined dosage schedule. At present, concentration-response relationships for other calcium channel blockers are not known. It is of interest that little data is available on the issue of whether orally administered calcium channel blockers may lead to the conversion of PSVT. Yeh et al. (1985) recently reported that single doses of 120 mg diltiazem and 160 mg propranolol given orally in combination led to conversion to sinus rhythm in 14 of 15 patients with induced PSVT in 39 ± 49 min; the placebo conversion rate in 164 ± 89 min was noted in 4 of 15 patients. The conversion was related to the attainment of certain serum levels of diltiazem and propranolol. Experience with single oral doses of calcium channel blockers is not available.

E. Acute Conversion of PSVT by Calcium Channel Blockers vs. Other Antiarrhythmic Agents

No direct blind studies have been reported to determine the precise comparative potency of calcium channel blockers and other antiarrhythmic compounds in the termination of PSVT. Approximate figures compiled from the experience with different agents reported in the literature are shown in Table 3. It appears that intravenous beta-blocking drugs produce a conversion rate of 40 to 60% of all cases of reentrant PSVT with narrow QRS complexes (Singh et al., 1978). For example, in one study, with a cross-over design (Hartel and

Hartikainen, 1976), 5 mg of practolol (a cardioselective beta blocker) converted 8 out of 20 cases, whereas verapamil (5 mg) produced sinus rhythm in 19 out of 20 cases; subsequently, verapamil produced conversions in 9 out of 12 cases that had failed on the beta blocker.

Surprisingly, little systematic data have been published on the overall efficacy of vagal maneuvers, intravenous edrophonium, alpha agonists (phenylephrine, metaraminol), or digoxin given either alone or in combination, in acutely terminating PSVT. Thus, a direct comparison of these "standard" measures with I.V. verapamil is not possible. However, at least in comparison to beta blockers (Wu et al., 1974; Bauernfeind et al., 1980) and cardiac glycosides (Wellens et al., 1975; Bauernfeind et al., 1980), verapamil and other Type 1 calcium channel blockers offer clear advantages. For example, verapamil is preferable where urgent termination of PSVT is desirable since the effects of digoxin are often not apparent for 15 to 60 min or longer and the dosage regimen is less clearly defined. Furthermore, in the rare event of failure of reversion to sinus rhythm with verapamil, direct current (DC) conversion can be attempted reasonably promptly because of the short elimination of the I.V. verapamil half-life (30 min). Apart from the issue of the higher conversion rate, verapamil also offers significant advantages over beta blockers in patients whose tachyarrhythmias are associated with chronic obstructive airway disease, diabetes mellitus or peripheral vascular disease.

At present it is difficult to evaluate the significance of the findings from uncontrolled studies that have indicated that Class I agents such as quinidine (Bauernfeind et al., 1980; Wu, et al., 1978), procainamide (Bauernfeind et al., 1980; Wu et al., 1981), and disopyramide (Swiryn et al., 1981; Sethi et al., 1983) may also convert PSVT acutely when given either orally or intravenously (see Table 3). It is, however, of electrophysiologic interest that all of these three compounds, which also retard conduction and prolong refractoriness over the bypass tracts in the WPW syndrome (Sethi et al., 1983), prevent AV nodal reentry by the selective inhibition of the retrograde fast pathway (Wu et al., 1978; Sethi et al., 1983). This has also been reported for intravenous ajmaline, which was found to be effective in terminating inducible PSVT in all 10 patients during electrophysiologic studies (Sethi et al., 1984). Although data are not available regarding the efficacy of the drug in spontaneously occurring PSVT, it is clear that its mode of action is similar to that of other Class I antiarrhythmics and in sharp contrast to that of calcium channel blockers.

F. Calcium Channel Blockers vs. Adenosine and ATP in the Acute Conversion of PSVT

Although it is now generally accepted that verapamil and possibly some other Type 1 calcium channel blockers are the agents of choice in the acute pharmacologic conversion of reentrant PSVT, recent reports (Greco et al. 1982; DiMarco et al., 1984; Belhassen and Pelleg, 1984) have suggested that intravenously administered adenosine or ATP may not only be as effective but may also offer other important advantages (Belhassen and Pelleg, 1984), at least in certain subsets of patients.

A clear electrophysiologic rationale exists for the use of adenosine and ATP in this setting. As with the calcium channel blockers, both adenosine and ATP have a potent depressant effect on the SA and AV nodes (Belhassen and Pelleg, 1984). In experimental preparations, adenosine depresses automaticity of the SA node and Purkinje fibers, shortens the action potential plateau of atrial myocytes and depresses AV nodal conduction; it also hyperpolarizes the membrane (Belhassen and Pelleg, 1984). This aggregate of effects is consistent with slow-channel blockade and a vagomimetic action: overall effects which are expected to be conducive to a prompt termination of AV nodal reentrant tachycardia. Numerous studies with intravenous ATP (3 to 15 mg in pediatric patients and 10 to 20 mg in adults) and adenosine (mean dose of 180 µg/kg) given as bolus injections have indicated that between 90 to 95% of PSVTs are terminated within 40 s. The conversion may be attended by transient (lasting a few seconds) sinus arrest, AV block, bradycardia, facial flushing, bronchospasm, and even convulsions in the case of ATP. The side effects with adenosine are considered to be short-lived (Belhassen and Pelleg, 1984) although further studies are needed.

The short half-life of ATP and particularly of adenosine may offer significant advantages over verapamil and other calcium channel blockers especially in patients with hypotension, heart failure, of pre-treatment with beta blockers. The available data suggest that if the efficacy of adenosine is confirmed to be comparable to that of verapamil, it would be an effective alternative to calcium channel blockers as a drug of first choice for the acute management of PSVT (Belhassen and Pelleg, 1984) in certain clinical situations. There is also increasing belief among cardiologists that adenosine might be preferable to calcium channel blockers in the acute termination of narrow QRS reentrant PSVTs.

G. Response in Other Forms of Reentrant or Ectopic Supraventricular Tachycardia

Although experience is limited, it is possible that verapamil and other Type 1 calcium channel blockers may be of value in other forms of reentrant and automatic supraventricular tachycardia (Sung et al., 1980a; Sung et al., 1980b; Rinkenberger et al., 1980). For example, in two patients with sinus nodal tachycardia (due to reentry within the SA node or its adjacent tissue) given I.V. verapamil, the tachycardia terminated promptly (Sung et al., 1980b). In another study (Rinkenberger et al., 1980), there was no effect of I.V. verapamil in PSVT due to intraatrial reentry. Clearly, further work is needed to define the role of I.V. verapamil in these settings. Similarly, there are increasing data to suggest that I.V. verapamil is less effective in converting ectopic atrial tachycardia (Rinkenberger et al., 1980) although paroxysmal atrial tachycardia with block due to digoxin toxicity as well as to other causes may convert to sinus rhythm following the oral doses of verapamil. The exact mechanism mediating such a beneficial response is not known; it is conceivable that it may be due to the drug's effect on digitalis-induced afterdepolarization but experience with calcium channel blockers in arrhythmias due to digitalis intoxication is limited. It must be emphasized however, that these agents should be used in this setting with the knowledge that the effects of toxic levels of digoxin and verapamil may summate and AV block supervene in some cases. An endocardial pacing system should be available in cases in which verapamil is contemplated in the setting of variable AV block.

H. Calcium Channel Blockers and Multifocal Atrial Tachycardia (MAT)

The role of calcium channel blockers in MAT when used for acute treatment or prophylactically, is uncertain. However, in one study (Rabkin et al., 1980) beneficial effects in supraventricular tachyarrhythmias associated with chronic pulmonary disease were reported. Less encouraging results have been reported by Aronow et al. (1980). For example, they compared responses of verapamil in reentrant PSVT to those in multifocal atrial tachycardia; of the five patients with reentrant PSVT, all converted to sinus rhythm after I.V. verapamil, whereas only one of five with MAT responded. In contrast, much more favorable results have recently been reported by Levin et al. (1985). For example, all 6 patients with MAT given I.V. verapamil responded; the drug decreased the mean atrial rate from 138 to 120 beats per minute ($p < 0.01$) and the ventricular rate from 130 to 109 ($p < 0.001$). The drug also decreased the number of different P-wave forms from 11 to 7 ($p < 0.01$). In three patients verapamil converted MAT to sinus rhythm. In one patient in whom the arrhythmia recurred chronically, complete control was attained by oral therapy with 80 mg qid of verapamil.

The overall response indicates that verapamil may be effective in controlling MAT and that this disorder of rhythm may be a triggered arrhythmia. However, further studies are needed to enlarge the experience with verapamil in MAT and to define the role of other Type 1 calcium channel blockers in this setting.

I. Chronic Prophylaxis of PSVT with Calcium Channel Blockers

Again, the bulk of experience has been with verapamil but other calcium channel blockers (e.g., diltiazem, gallopamil and tiapamil), which also prolong AV nodal conduction and refractoriness, are likely to be variably effective in this context.

Even in the case of verapamil, available data (Rinkenberger et al., 1980; Mauritson et al., 1982; Sakurai et al., 1983; Pritchett et al., 1983) suggest that, in contrast to the extreme potency of the I.V. drug, the oral formulation (80 to 120 mg tid or qid) may not have comparable efficacy in preventing recurrences of PSVT during chronic administration. However, Mauritson et al. (1982) found that during long-term therapy of PSVT, oral verapamil was superior to placebo. Rinkenberger et al. (1980) found that when the drug was given orally (180 to 480 mg/day) to patients who had responded to I.V. verapamil during acute episodes of PSVT, 10 of 19 patients discontinued therapy within a month, either because of side effects or lack of response. While no correlation was sought with plasma level of the drug the development of side effects in many patients suggested that failure of response to the drug could not be attributed to low plasma drug levels in every instance.

It must be emphasized, however, that because of the extremely variable nature of the pattern of recurrences of PSVT in different patients and in the same patient at different times, the precise efficacy of a therapeutic agent in the chronic prophylaxis of PSVT is not easily established. Nevertheless, in the case of verapamil, there are reports that indicate that this compound may be of prophylactic value in PSVT (Winniford et al., 1984; Gonzalez and Scheinman, 1981). For example, Gonzalez and Scheinman (1981) found that patients who had responded to the intravenously administered drug also obtained significant benefit during continuous oral therapy. Of particular interest are the data from Mauritson et al. (1982) who, in a double-blind randomized comparison of digoxin, propranolol and verapamil, found that the three agents were of comparable efficacy in preventing recurrences of PSVT.

An approach that appears to be effective in predicting the efficacy of oral calcium channel blockers in the prophylaxis of PSVT is the use of electrophysiologic studies by programmed electrical stimulation (Tonkin et al., 1980; Klein et al., 1982). Tonkin et al. (1980) made observations before and after I.V. verapamil, 0.15 mg/kg, in 13 patients, of whom 12 were previously resistant to other antiarrhythmic drugs. The drug increased the AV nodal conduction time as well as the functional and the effective refractory periods. After the drug, reentrant PSVT could not be initiated in 5 of the 13 patients and was nonsustained in 3 other patients. The tachycardia rate was decreased in those in whom PSVT was re-inducible after I.V. verapamil. Over a mean follow-up period of 16 months on oral verapamil, 11 of the 13 patients had definite symptomatic improvement, with a decrease in the frequency, duration and/or associated symptoms of their tachyarrhythmia. In only one patient were side effects severe enough to necessitate withdrawal of the drug. Very similar results have recently been reported in 8 patients by Klein et al. (1982); they found that it was not necessary to attain the high serum drug levels observed during acute I.V. infusions to obtain comparable clinical results with oral verapamil. Their data suggested that the response to I.V. verapamil could be extrapolated to predict the electrophysiologic results and the outcome of long-term therapy with the oral drug. A similar predictive outcome between the acute intravenous effect and the chronic clinical response in 18 patients given diltiazem has been reported by Hung et al. (1984). However, these are limited data in a small series of patients and more extensive controlled experiences will be needed to clearly define the role of oral verapamil and diltiazem in the prophylaxis of PSVT. The role of the other calcium channel blockers in the prophylaxis of PSVT remains to be defined.

There is a rational basis for modulating the electrophysiological properties of the AV node by combination therapy with beta-adrenoceptor blocking drugs or digoxin; such an approach may enhance the clinical utility of verapamil in the oral prophylaxis of PSVT. In our own unpublished experience, the most potent agent for the chronic prophylaxis of PSVT is amiodarone and therefore the evaluation of the comparative efficacy and clinical utility of the two agents in resistant PSVT will be of practical importance.

A report by Storstein and Landmark (1975) about the use of oral verapamil in paroxysmal atrial tachycardia with AV block deserves mention. Paroxysmal atrial tachycardia with block is a relatively uncommon arrhythmia, believed, in most cases to be due to digitalis intoxication. However, in the experience of Storstein and Rasmussen (1976) and Storstein and Landmark (1975), digitalis could not be incriminated as a cause in many of their cases. Of particular interest was the finding that verapamil, given orally in a dose of 40 to 80 mg q3h to 14 patients with paroxysmal atrial tachycardia with block, led to reversion to sinus rhythm in 10 initially, with subsequent relapse occurring in 4. The best effect was found in patients with a short duration of the arrhythmia. The mechanism of action of verapamil in this setting is uncertain but the data suggest that the drug has a depressant effect on the ectopic pacemaker that may have been mediated through the slow response.

J. Calcium Channel Blockers in Pre-Excitation Syndromes

The role of these drugs in pre-excitation syndromes is relatively easily defined on the basis of the knowledge of their electrophysiologic effects and of the mechanism of the arrhythmias that complicate pre-excitation. The major action of these compounds is on the AV node. There is a minimal effect on the bypass tracts (presumably "fast response" fibers). For example, in the study of Krikler and Spurrell (1974), in 8 patients the antegrade refractory period of the bypass tract was slightly shortened, was unchanged in one, and was lengthened minimally by verapamil in 4. The shortening was minimal, much less than that effected by digitalis (Wellens et al., 1975). Similar data have been obtained with other calcium channel blockers. For example, Gmeiner et al. (1982) found that 2 mg/kg I.V. tiapamil had no effect on the antegrade or retrograde refractory period of the accessory pathway (nor of the atria or ventricles) but, as might be expected, prolonged AV nodal refractoriness. Although the data are minimal concerning the effects of the drug on the electrophysiologic properties of the heart (Singh et al., 1982; Rozansky, et al., 1982) one may infer that the actions of diltiazem on the anomalous tracts in the WPW syndrome are likely to be qualitatively similar to those of verapamil and its congeners. Thus, all these compounds (verapamil, tiapamil, gallopamil, and diltiazem) are likely to be variably effective in terminating the acute paroxysms of the narrow QRS (orthodromic) SVT using the AV node for antegrade conduction in the setting of the bypass tract syndrome. An example of such a conversion is shown in Figure 5. As in the case of AV nodal reentrant PSVT, the role of calcium channel blockers in the oral prophylaxis of orthodromic PSVT remains to be defined but since they prolong AV nodal refractoriness, they are likely to be of value at least in some patients. In contrast, since they have no significant effect on the refractoriness of the bypass tract, they are unlikely to terminate antidromic PSVT; nevertheless, the theoretical possibility exists that they may, in large doses, terminate the arrhythmia by inhibiting retrograde AV nodal conduction (Singh et al., 1982).

It must be emphasized that in patients with atrial flutter and fibrillation complicating pre-excitation, agents that shorten the effective refractory period of the bypass tracts (e.g.,

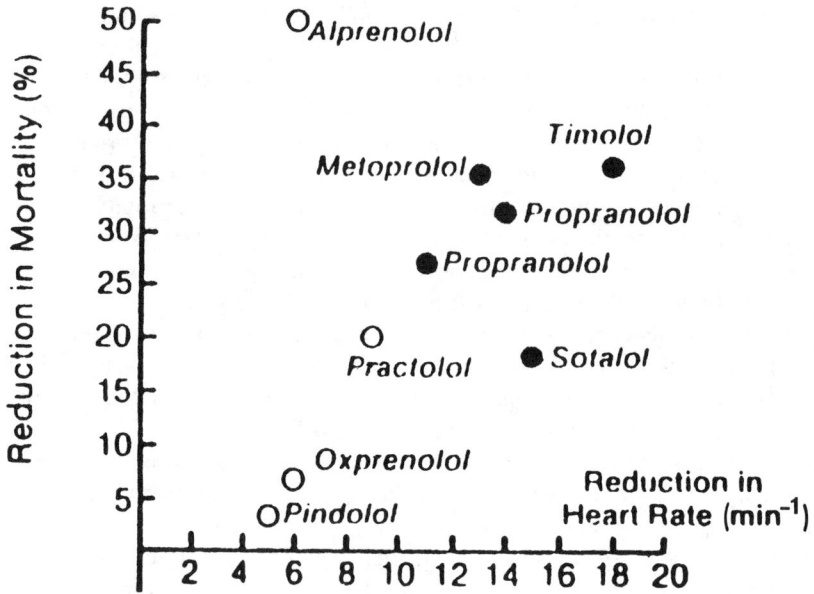

Figure 6. Relationship between heart rate reduction and reduction in mortality induced by beta blockers in survivors of acute infarction. Note that beta blockers that have a significant intrinsic sympathomimetic activity with little or no bradycardia effects exert little or no effect on mortality. (From Kjekshus (1976) with the permission of the author and of the editor of the *American Journal of Cardiology*.)

digitalis) and those that lengthen it over the AV node (digitalis and beta-adrenoceptor blocking drugs) will augment the ventricular response and may possibly precipitate ventricular fibrillation. Calcium antagonists which increase AV nodal refractoriness (e.g., verapamil, diltiazem and other Type 1 agents) fall into the latter category. Published data however, pertain only to verapamil. One such case was reported by Rinkenberger et al. (1980). Significant acceleration of the ventricular response in atrial fibrillation in WPW syndrome was also documented in 8 patients by Gulamhusein et al. (1981). An example from our own experience is shown in Figure 6. Clearly, verapamil and similar calcium channel blockers should not be given to patients in whom atrial fibrillation or flutter complicate the WPW syndrome with anterograde conduction over the bypass tracts. These conditions are best treated acutely by cardioversion or Class I antiarrhythmic drugs (procainamide, disopyramide) and prophylactically by amiodarone or quinidine or possibly surgery. It should be emphasized that when doubt exists regarding the nature of wide QRS tachycardia (irregular or regular), it is prudent not to use calcium channel blockers for acute conversion. As in the case of beta blockers, calcium channel blockers are ineffective in patients with antidromic tachycardia; here, Class I agents are generally preferred.

K. Treatment of Other Atrial Tachyarrhythmias by Calcium Channel Blockers

The fact that certain calcium antagonists impede conduction over the AV node has also been used to control the ventricular response in atrial flutter and fibrillation; however, reliable data exist only for verapamil, the effects of other compounds in this setting being essentially inferential based on the knowledge of the electrophysiologic properties of these drugs. There have been no convincing reports on the potential efficacy of this class of agents in chronic ectopic atrial tachycardia; in such a case, transient AV block converts the ectopic atrial tachycardia to one with variable block without effect on the frequency of the ectopic pacemaker (Rinkenberger et al., 1980). Barring PSVT, most experience has been with the I.V. and oral verapamil in atrial fibrillation and flutter. It should be emphasized that the effects of other Type 1 calcium channel blockers (gallopamil, tiapamil and diltiazem) are likely to be qualitatively similar to those of verapamil and bepridil.

1. Atrial Fibrillation

In the case of atrial fibrillation, three types of response to I.V. verapamil have been reported. The first, perhaps the most common, is the inhibition of AV conduction with subsequent slowing of the ventricular response (Aronow et al., 1979; Talano and Feerst, 1980; Aronow and Ferlinz, 1980; Bender, 1980; Plumb et al., 1980). The effect is of brief duration so that after 30 min the ventricular response gradually accelerates, unless an infusion is commenced after the completion of the injection. The second type of response with verapamil in atrial fibrillation is what has been designated "regularization" (Schamroth, 1971) of the ventricular response. This phenomenon was noted in 25% of our own cases (Heng et al., 1975) but in a higher proportion of those reported by Schamroth (1971). The precise mechanism of this phenomenon is not understood but the regularization of the ventricular response without conversion to sinus rhythm appears to be of little hemodynamic benefit (Singh et al., 1978). The final type of response of atrial fibrillation to I.V. verapamil is reversion to sinus rhythm. This is, however, rare in the experience of most investigators (Sung et al., 1980b; Rinkenberger et al., 1980) and our limited unpublished data suggest that the patients who do convert, are those in whom the arrhythmia is of recent onset associated with a normal or only mild to moderately enlarged left atrium. This observation is in line with the known findings that verapamil and similar drugs have no significant effect on atrial refractoriness (Singh, 1982; Singh et al., 1982) nor do they exert a potent effect on conduction as do Class Ic agents that are known to convert a significant number of cases of atrial flutter to sinus rhythm following intravenous administration.

The clinical value of orally administered verapamil in controlling the ventricular response in atrial fibrillation has recently been investigated by Klein et al. (1979) and by Khalsa and Olsson (1979). In the first study, the effects of 160 to 240 mg daily of verapamil on resting heart rate and that during mild exercise were studied in 23 digitalized patients with chronic atrial fibrillation of various causes. Verapamil significantly reduced the heart rate response during exercise. In a smaller number of patients, the study by Khalsa and Olsson (1979) also confirmed the efficacy of verapamil in attenuating the ventricular response in atrial fibrillation. However, in this study the higher doses were associated with significant side effects. Very similar results have more recently been reported by Morganroth et al. (1982). It should be noted that verapamil tends to reduce both the resting as well as the exercise-induced increases in heart rate in the setting of atrial fibrillation whereas the major effect of digoxin is on the resting rate. Perhaps it should also be emphasized that when digoxin and verapamil are used together, their depressant effect on the AV node may summate. Since a significant interaction occurs between the two compounds, the net effect may be unexpectedly exaggerated, necessitating careful monitoring of serum drug concentration vs. response. On the other hand, recent data have suggested that in many patients verapamil may replace digitalis for the control of the ventricular response in atrial flutter and fibrillation (Lang et al., 1983a; Lang et al., 1983b; Klein and Kaplinsky, 1986). For example, it was found that digoxin 0.25 mg and 0.5 mg reduced the increases in heart rate induced by exercise by 3 and 8%; by 29% with the combination of digoxin and verapamil (240 mg) and 23% with verapamil alone. Such a beneficial effect is undoubtedly due to the direct action of the calcium channel blocker on the AV node in addition to the sympatholytic action of the compound. Clearly, the use of verapamil in this context will be imprudent if there is a significant impairment of ventricular function or a history of clinical heart failure.

At present, little data are available for the other calcium channel blockers in the control of the ventricular response in atrial fibrillation and flutter. However, a recent study (Roth et al., 1986) has clearly established the role of diltiazem (240 to 360 mg/day) when combined with digoxin in effectively controlling the ventricular response in atrial flutter and fibrillation. A significant reduction in resting, as well as exercise heart rate, was attained by both doses of the drug but there was a high incidence of side effects at the higher dose. The data indicated that the medium dose of diltiazem, when combined with digoxin was an effective and safe regimen in the control of the ventricular response in atrial flutter and fibrillation. At present, there is little data on the effects of diltiazem used alone in this setting.

The particular value of agents such as verapamil, diltiazem, tiapamil, gallopamil and bepridil in the chronic control of atrial flutter and fibrillation needs emphasis, although data for most of these compounds need to be obtained. It is nevertheless clear that since verapamil and similar compounds do not exacerbate bronchospastic or vasospastic phenomena, if effective in atrial tachyarrhythmias in patients with obstructive airways or peripheral vascular disease, these drugs will have distinct advantages over beta antagonists for the purposes of chronic prophylaxis in the above subsets of patients. They have advantages over cardiac glycosides (with which they can be used in combination for a greater effect) since calcium channel blockers reduce the exercise-induced increases in heart rate more effectively (Klein and Kaplinsky, 1986).

a. Maintenance of Sinus Rhythm After Conversion

The question has been raised whether calcium antagonists either alone or in combination with other compounds such as quinidine might be of value in preventing relapses in the case of atrial fibrillation treated by electroversion. Controlled data in this context are scant. It is of interest, however, that Basu et al. (1977) reported that patients with atrial fibrillation treated with verapamil before and after DC cardioversion, had a low rate of conversion (one out of 24) *before* electroconversion was attempted. In contrast, on verapamil low energy levels were needed for conversion to sinus rhythm in the remainder, 40% of whom remained in sinus rhythm on the drug at the end of one year. The reported success rate with quinidine is 18 to 29%. In contrast, a recent controlled clinical trial (Rasmussen et al., 1981) comparing verapamil (80 mg tid) with 0.4 mg bid of quinidine demonstrated that quinidine was significantly more effective in producing conversion to sinus rhythm, and in maintaining sinus rhythm for 3 months post-conversion. These data nevertheless need to be confirmed by further direct comparisons between quinidine and verapamil relative to chronicity of atrial fibrillation and left atrial size using a stringently controlled protocol with plasma drug level monitoring. It is known that verapamil does not lengthen the atrial effective refractory period and may induce atrial fibrillation in cases of PSVT or flutter indicating a potential fibrillatory propensity for the compound (see below). It is possible that the drug might act by depressing the abnormal slow response in atrial muscle (El-Sherif and Lazzara, 1979). Its observed salutary effect might be mediated by its stabilizing action on AV conduction with a secondary effect on the fibrillating atria. Whether other calcium channel blockers might have qualitatively or quantitatively different effects in this setting as compared to verapamil at present remains uncertain. Electrophysiologic considerations suggest that lidoflazine and especially bepridil might be of value in the prevention of relapses of atrial fibrillation to sinus rhythm after cardioversion. Both of these agents have a complex electrophysiologic profile which includes lengthening of repolarization and a propensity to induce torsades de pointes under certain circumstances.

2. Atrial Flutter

Again, the bulk of the clinical experience has been with verapamil. The immediate effect of I.V. verapamil in atrial flutter is reasonably consistent; the increased AV block slows the ventricular response, with an occasional patient converting to sinus rhythm. In some, atrial flutter alters to atrial fibrillation before sinus rhythm is restored. Again, as in the case of atrial fibrillation, the rate of conversion to sinus rhythm is low, which is consistent with the known electrophysiologic effects of the drug.

A single intravenous dose of verapamil has been found to be of diagnostic value in differentiating atrial flutter with 2-1 AV conduction from paroxysmal supraventricular tachycardia when these two arrhythmias are not readily distinguished electrocardiographically. If the rhythm is atrial flutter, the AV block increases immediately after I.V. verapamil, thus revealing the true nature of the arrhythmias (Singh et al., 1978; Heng et al., 1975). Other calcium channel blockers that inhibit AV conduction may also be used in this way.

A significant number of patients with atrial flutter develop atrial fibrillation when given verapamil (Heng et al., 1975; Singh et al., 1978). Atrial fibrillation also develops occasionally with supraventricular tachycardia after verapamil administration (Singh et al., 1978). The mechanism of this action is obscure. The drug may enhance conduction velocity in isolated rabbit atria (Singh et al., 1972) but this effect was not observed with clinically relevant drug concentrations. Moreover, in the studies by Heng et al. (1975), no increase in the flutter rate was found after verapamil, including those in whom conversion from atrial flutter to atrial fibrillation subsequently occurred.

There are limited data indicating that orally administered verapamil or other calcium antagonists in combination with other antiarrhythmic compounds may be of value in controlling the ventricular response in atrial flutter (Singh et al., 1978).

V. VENTRICULAR ARRHYTHMIAS

Unlike the situation with supraventricular tachyarrhythmias as outlined above, the role of calcium channel blockers in ventricular arrhythmias is poorly defined (Krikler and Goodwin, 1976; Belhassen and Horowitz, 1984; Stewart et al., 1986; Akhtar et al., 1989). In general, these agents are of limited value in the largest number of patients with life-threatening ventricular arrhythmias. The subject has recently been placed in perspective by Belhassen and Horowitz (1984). Future work is likely to delineate more clearly which subsets of patients with ventricular tachyarrhythmias might respond to calcium channel blockers.

A. Premature Ventricular Contractions

There are no significant systematic data to support the belief that calcium antagonists are potent agents for the suppression of ventricular premature contractions in the setting of chronic ischemic heart disease. Thus, if they were to prevent the incidence of sudden death in such patients, the mechanism of the salutary effect is unlikely to be mediated through the suppression of premature ventricular contractions. Similarly, while reasonably sound theoretical bases have been suggested for the role of the calcium channel in the genesis of ventricular arrhythmias in obstructive cardiomyopathies (Krikler and Goodwin, 1976) and for the mitral valve syndrome (Singh et al., 1978), clinical experience with calcium channel inhibitors in these conditions has not confirmed their efficacy (Singh et al., 1978; McKenna et al., 1980). Moreover, there are few systematic data to indicate that they may be effective in reducing the frequency of PVCs in patients without structural heart disease.

B. Chronic Recurrent Ventricular Tachycardia

There have been few systematic studies with calcium channel blockers in ventricular tachycardias complicating organic heart disease with reduced ventricular ejection fraction. The limited experience available suggests that verapamil and other calcium channel blockers

TABLE 4
Antiarrhythmic Agents and their Effects on Inducibility of Ventricular Tachycardia and Fibrillation and on PVCs[a]

Test Drug by Class		Prevention of inducibility of VT/VF by PES (% of patients)	Suppression of PVCs by >75% (% of patients)	Suppression of VT beats by >90% (% of patients)
Class IA	Quinidine	15—20	60	? 70
	Procainamide	15—25	60	? 70
	Disopyramide	10—20	60	? 70
Class IB	Mexiletine	15	60	>70
	Tocainide	10—15	60	? >70
	Ethmozine	15—20	60	? >70
Class IC	Encainide	10—15	80	90—100
	Flecainide	10—15	80	90—100
	Propafenone	10—20	70	90—100
	Indecainide	10—15	70	90—100
Class II	Beta-blockers (without ISA)	? 1—2	50	? 70—80
Class III	Amiodarone	8—40	80	90—100
	Sotalol	40—45	50—60	80
Class IV	Verapamil	1—2	≤10	≤10
	Diltiazem	?	≤10	≤10

[a] Data included herein are crude estimates from controlled and uncontrolled studies reported in the literature. However, they provide a reasonable approximation of the overall effects of various antiarrhythmic agents in affecting two major end points used in determining the efficacy of pharmacologic therapy of (life-threatening) ventricular arrhythmias. Note the lack of correlation between PVC suppression and the suppression of inducible VT/VF. ISA = intrinsic sympathomimetic action; PES = programmed electrical stimulation.

have little or no effect on chronically occurring ventricular tachyarrhythmias (Wellens et al., 1977; Coumel et al., 1979; Wellens et al., 1980). Table 4 summarizes the overall controlled and uncontrolled data on the effects of various classes of antiarrhythmic agents on the suppression of inducible ventricular tachycardia/fibrillation and premature ventricular contractions. Calcium channel blockers have no significant effect on either parameter. The data are consistent with the knowledge that neither conduction nor refractoriness of the ventricle or His-Purkinje system is lengthened by calcium channel blockade (Wellens et al., 1980). Wellens et al. (1980) studied the effects of I.V. verapamil on ventricular tachycardia (VT) in patients in whom VT could be induced and terminated by programmed electrical stimulation of the heart. Intravenous (10 mg) verapamil given during the tachycardia had no effect in 7 of 8 patients, and the arrhythmia could be readily induced before as well as after verapamil administration. In only one patient could the induced VT be terminated by verapamil administration. The experience of Sung et al. (1983) appears to be similar. They administered intravenous verapamil to 10 patients with inducible ventricular tachycardia (8 with programmed electrical stimulation and 2 with isoproterenol). The drug had no effect on VT induction in any of the 10 patients. Cardioversion was required in two patients because of verapamil-induced hypotension. Mason et al. (1983) found verapamil to be effective in preventing VT induction in only 3 of 16 patients with chronic recurrent VT. A similar lack of effect of calcium channel blockers in patients with nonsustained ventricular tachycardia has recently been reported. For instance, Buxton et al. (1984) found that 10 mg of intravenous verapamil had no effect on nonsustained VT induced by programmed electrical stimulation of the heart in 13 patients. The drug did not prevent or alter the mode of VT induction, nor did it slow the rate of the inducible tachycardia. On the other hand, in 3 of the patients, the induced VT accelerated, became more prolonged and deteriorated into ventricular fibrillation in 2 patients. The authors concluded that the mechanism of the tachyarrhythmias was unlikely to be accountable in terms of triggered activity due to afterdepolarizations.

TABLE 5
An Overview of Mortality in Trials Evaluating Calcium Channel Blockers Following Myocardial Infarction

Agent (No. studies)		Deaths/Patients		Observed expected	Variance
		Calcium blockers	Controls		
Acute, short-term studies					
Verapamil	(1)	0/8	2/9	−0.9	0.5
Nifedipine	(4)	171/2509	157/2521	+7.5	76.7
Diltiazem	(1)	11/287	9/289	+1.0	4.8
Subtotal	(6)	182/2802	168/2819	+7.6	82.0
Long-term studies					
Nifedipine	(1)	66/1140	64/1139	+1.0	30.7
Lidoflazine	(1)	177/896	168/896	+4.5	69.7
Subtotal	(2)	243/2036	232/2035	+5.5	100.4
Acute and long-term studies					
Verapamil	(1)	149/1729	145/1718	+1.5	67.2
Total	(9)	574/6567	545/6572	+14.6	249.6

Typical odds ratio of 1.06; 95% confidence interval (CI) of 0.94 to 1.2, N.S.

Reproduced from Yusuf and Furberg. See text for further details.

Thus, it is unlikely that further intensive and systematic study will reveal a higher rate of success with verapamil (or other calcium channel blockers with a similar electrophysiologic profile) in most patients with life-threatening ventricular tachyarrhythmias but it might identify the subset(s) of patients who may respond more predictably to the drug. The experiences summarized herein suggest that the type of the ventricular tachycardia occurring in patients with organic heart disease is probably not related to a calcium channel abnormality. In contrast, a number of recent reports (Belhassen and Horowitz, 1984) have suggested that there are, indeed, subsets of patients whose ventricular tachyarrhythmias might respond favorably to calcium channel blockers. The experience is limited almost completely to verapamil. For the usual forms of ventricular tachycardias seen in clinical practice, calcium channel blockers exert little or no suppressant effect on premature ventricular contractions, nor do they affect the inducibility of ventricular tachycardia using programmed electrical stimulation of the heart. The overall effects of these drugs, with respect to these parameters, are compared to those of various electrophysiologic classes of antiarrhythmic agents in Table 5.

C. Ischemic Ventricular Arrhythmias

The experimental basis for the possibility that certain ventricular arrhythmias complicating myocardial ischemia or acute infarction (Kaumann and Aramendia, 1968; Fondacaro et al., 1978) might be sensitive to calcium channel blockers has been alluded to above. The clinical data, essentially of an anecdotal type, is conflicting. Perhaps the earliest report was that of Heng et al. (1975) who gave 10 mg I.V. verapamil to patients with sustained ventricular tachycardia in the context of a recent myocardial infarction. Conversion to sinus rhythm occurred in only one but it was uncertain whether this was related to a drug effect or whether the conversion was merely spontaneous and fortuitous. It is of interest, however, that Filias (1974) reported a reduction in the number of premature ventricular contractions (PVCs) complicating acute infarction in patients given an intermittent or continuous intravenous infusion of verapamil. Similar results have been reported by Fazzini et al. (1978) who noted that 0.10 mg/kg of intravenous verapamil given over 2 min resulted in complete abolition of PVCs in 7 of 8 patients who exhibited persistent frequent PVCs during the first

48 h of acute myocardial infarction. Favorable responses have also been reported in sustained arrhythmias in this setting. For example, Sclarovsky et al. (1983) administered 3 to 5 mg of intravenous verapamil to 8 patients with multiform accelerated idioventricular rhythm occurring during the first 12 h of acute myocardial infarction. The arrhythmia was abolished in six, slowed in one and was without effect in the remaining patient. Successful control of polymorphous ventricular tachycardia by verapamil treatment was also recently reported by Grenadier et al. (1984) in 3 of 4 patients during acute myocardial infarction. These 3 patients had failed on lidocaine, Class I agents, overdrive pacing and repeated cardioversion. However, these are uncontrolled studies that raise the possibility but do not provide conclusive proof that calcium channel blockade by verapamil might be effective in controlling ventricular tachyarrhythmias. Consideration should also be given to the possibility that the apparent salutary effect might have resulted from reduced ischemia and improved myocardial conduction (Elharrar et al., 1977) rather than as a consequence of a direct antiarrhythmic effect of the drug in ventricular muscle or the Purkinje fibers. Amelioration of ischemia in this situation might be effective in reversing reentry, and triggered as well as enhanced automaticity. Moreover, it must be emphasized that in the context of myocardial ischemia, verapamil and other calcium channel blockers may exert a potent effect in reversing coronary artery spasm or vasoconstriction and thus relieve ischemia and ischemia-induced ventricular tachyarrhythmias. Such an effect must be regarded as a "secondary" antiarrhythmic effect that is likely to be most readily apparent in the setting of Prinzmetal variant angina (Kimura et al., 1977).

D. Exercise-Triggered Ventricular Tachycardia

There have been several recent reports that have suggested that exercise-triggered ventricular tachycardia, with the morphologic pattern of left bundle branch block and right axis deviation, might respond predictably and promptly to intravenous verapamil. Such an arrhythmia may occur in patients without identifiable cardiac disease (Palileo et al., 1982). Wu et al. (1981) described three such patients; in none was the tachycardia inducible by programmed electrical stimulation, but it could be provoked by isoproterenol infusion. Intravenous verapamil terminated the tachycardia in all three and prevented its occurrence during exercise in two. It is of interest that propranolol and lidocaine also prevented the occurrence of the arrhythmia. Three similar cases have been reported by Sung et al. (1983) although in this series intravenous verapamil was not effective in preventing the induction of ventricular tachycardia by isoproterenol. Unfortunately the QRS axis in these patients during the tachycardia was not stated. Thus, it is unclear whether the nature of the arrhythmia was identical to that in those reported by Wu et al. (1981). At present, it is not clear whether the ventricular tachycardia in question is due to an autonomous focus or to triggered automaticity (Wu, 1984); nor is it known whether the observed response to intravenous verapamil in a small series of patients is a property unique to this drug or a common property of all Type 1 calcium channel blockers. In this context, it is also of interest that Coumel and Attuel (1983) recently reported a group of patients without demonstrable heart disease who experienced syncopal spells due to torsades de pointes in association with a normal QT and an exceptionally short coupling interval of the beat initiating the tachycardia. In the author's experience, the arrhythmia in these patients selectively responded to intravenous and oral verapamil, being resistant to other antiarrhythmic agents including amiodarone and beta blocking drugs.

E. Idiopathic Ventricular Tachycardia with Right Bundle Branch Block and Left Axis Deviation Morphology

It is now increasingly appreciated that, in young patients without demonstrable cardiac disease, the sustained ventricular tachycardia that occurs in a few has the morphologic pattern of right bundle branch block with left axis deviation (Belhassen et al., 1981; Lin et al., 1983; German et al., 1983). This arrhythmia appears to constitute a distinct clinical entity (Belhassen et al., 1984); it is usually possible to initiate the tachycardia by both atrial as well as ventricular stimulation.

It is of interest that in most cases the ventricular tachycardia in these patients responds to verapamil given intravenously (Hordoff et al., 1976; German et al., 1983; Belhassen et al., 1984) whereas beta blockers or Class I antiarrhythmic agents either fail or exert only a partial response (Belhassen et al., 1981; German et al., 1983; Belhassen et al., 1984). The precise mechanism for the arrhythmia is not known; electrophysiologic studies have suggested a reentrant mechanism in the hands of some investigators (Lin et al., 1983; German et al., 1983; Belhassen et al., 1984; Zipes et al., 1979) and triggered automaticity in the hands of others (Wu et al., 1981) but the difficulties of such an approach in the clinical laboratory are well known (Brugada and Wellens, 1984). Further work is needed to define the nature of the arrhythmia and to determine the basis for the apparent unique selectivity of its response to calcium channel blockers.

F. Calcium Channel Blockers and Torsades De Pointes

As indicated above, although the precise ionic basis for every case of torsades de pointes that occurs in the setting of prolonged myocardial repolarization is not known, there is little doubt that there are situations in which the slow current (I_{Ca}) does provide the basis for the genesis of this type of polymorphic ventricular tachycardia (Brachman et al., 1983; Takanaka and Singh, 1990). This is supported by the clinical observation that in hypothyroidism, chronic amiodarone administration and severe hypocalcemia—conditions in which there is considerable QT prolongation—torsades de pointes is rare unless there are associated electrolyte perturbations of a marked degree. In all three, calcium channel activity is markedly attenuated. Furthermore, Hiramasu et al. (1988) have shown that early afterdepolarizations induced by marked lengthening of the action potential duration in canine Purkinje fibers is abolished by verapamil. Similarly, Takanaka and Singh (1990) showed that triggered activity of a similar kind induced by Ba^{2+} was eliminated by verapamil and amiodarone (a potent calcium channel blocker). Finally, Aliot et al. (1985), in a series of preliminary experiments, found that it was possible to prevent the development of torsades de pointes by calcium channel blockade in an *in vivo* animal model of torsades. These experimental observations provide a clear rationale for the critical evaluation of calcium channel blockade in the prevention and control of torsades de pointes in man.

VI. CALCIUM CHANNEL BLOCKERS AND SUDDEN DEATH

Perhaps the most intriguing observation that has emerged with calcium channel blocking drugs is the one that relates to their potential to prevent sudden cardiac death in patients with coronary artery disease especially those surviving acute myocardial infarction. In the case of beta-adrenoceptor blocking drugs, there is increasingly compelling evidence that sudden cardiac death can be reduced significantly by the chronic prophylactic treatment in the survivors of acute myocardial infarction (Norwegian Multicenter Study Group, 1981; Beta Blocker Heart Attack Trial, 1982; Yusuf et al., 1985; ISIS-1 Collaborative Group, 1986). There is little to suggest that such a beneficial effect is mediated through a direct influence on arrhythmias; the presumption is that the primary action is on ischemia with a secondary effect on electrical instability. Calcium channel blockers are potent antianginal drugs with the additional propensity to reverse coronary vasospasm. These drugs also have the property of minimizing myocardial ischemic damage following experimental coronary occlusion (Smith et al., 1975; Henry et al., 1977; Reimer et al., 1977; Henry et al., 1978; Nayler, 1980; Nayler et al., 1980; Bourdillon and Poole-Wilson, 1982; Thandroyen, 1982; Nayler, 1982; Hamm and Opie, 1983; Marban et al., 1989; Singh and Nayler, 1989). There is, therefore, a compelling reason to expect calcium channel blockers to exert a favorable effect on sudden death by ameliorating ischemia in the survivors of acute myocardial infarction in the event the data obtained to date have been either negative, inconclusive or have demonstrated a possibly deleterious effect (Yusuf and Furberg, 1987). For example, in the case of verapamil, the first calcium channel blocker that was investigated with the aim of reducing mortality and reinfarction rate in infarct survivors, the results were found to be inconclusive (Hansen et al., 1980; Danish Multicenter Study Group, 1982, 1984). In this 1436 patient study, 717 were treated with verapamil and 719 with placebo. At the end of the six months, the mortality in the treated group was 12.8 and 13.9% in the placebo series. At 12 months the corresponding figures were 15.2 vs. 16.4%. Although a trend in mortality reduction was evident in this study, it is clear that the drug, given early, did not significantly influence early mortality resulting from acute myocardial infarction.

Yusuf and Furberg (1987) have recently reviewed most of the randomized trials of calcium channel blockers following acute myocardial infarction. They combined all the mortality data by the Mantel-Haenszel method (Yusuf and Furberg, 1987) for each of the calcium channel blockers for which data were available from short-term and long-term trials and for all the trials in total. They calculated a pooled odds ratio and its 98% confidence limits in an effort to provide a typical estimate of the effects of the individual agents on mortality.

The data summarized in Table 5 indicated that 13,139 patients had been studied in 9 randomized controlled trials (Yusuf and Furberg, 1987). It was noteworthy that 8 of the 9 trials showed *a small excess of mortality* in the treated group although the difference from the effects of placebo did not reach a statistical significance. There were 574 deaths among the 6567 patients randomized to the active treatment group (8.7%); in the control group there were 545 deaths in the control population of 6572 (8.3%). This gave an odds ratio of 1.06 with a 95% confidence interval of 0.94 to 1.20. The results were similar when the data were split according to whether the trials evaluated early treatment or when treatment was delayed 1 to 2 weeks following myocardial infarction. It was of interest that a nonsignificant increase in mortality was consistently observed with all of the above agents (Crea et al. 1975; de Geest et al., 1979; Norwegian Nifedipine Group, 1983; Gibson et al., 1986; The Israeli Sprint Study Group, 1988). However, while the data on verapamil (one trial with 3464 patients), lidoflazine (one trial with 1792 patients), and nifedipine (5 trials with 7309 patients) were derived from reasonably large population samples, the data on diltiazem are limited (only 576 in one trial). Thus, the potential beneficial effect of diltiazem on mortality in patients with myocardial infarction could not be excluded. This possibility is of particular interest as diltiazem is the calcium channel blocker that has perhaps the highest potency for reducing heart rate. The issue of the heart rate reduction is of particular importance in this context especially in the case of diltiazem in light of more recent trials involving the compound (see later). It so happens that reduction in heart rate appears to be the common denominator in the case of beta blockers having a salutary effect on sudden death and reinfarction rate in patients surviving infarction (Kjekshus, 1986). Kjekshus (1986) carried out an analysis of the most significant beta-blocker trials in the survivors of acute infarction; he found that a linear correlation existed between the degree of heart rate reduction effected by a beta blocker and the percentage reduction in sudden death that resulted during chronic prophylaxis with the compound (Figure 6). It was of interest that only those beta blockers that reduced the heart rate significantly were effective. In contrast, those (e.g., oxprenolol or pindolol) which had a marked intrinsic sympathomimetic action and little or no bradycardic response at rest either had no effect on sudden death or tended to *increase* its incidence.

In a subsequent meta-analysis to assess the effects of calcium channel blockers on the development of infarcts, reinfarction and mortality, Held, Yusuf and Furberg (1989) performed a systematic overview of all randomized trials of the drugs not only in myocardial infarction but also unstable angina. The pooled data included those from 19,000 patients and 28 trials. In the trials of myocardial infarction, 873 deaths occurred among 8870 patients randomized to active therapy compared with 825 deaths among 8889 control patients. This gave an odds ratio of 1.06, 95% confidence interval of 0.96 to 1.18. There was no evidence of a beneficial effect on the development and the size of infarcts or rate of reinfarction. These results were similar for unstable angina trials—110 out of 561 treated with calcium channel blockers and 104 out of 548 controls developed infarction; 14 out of 591 treated compared with 9 out of 578 controls died. The authors concluded that calcium channel blockers do not reduce the risk of initial or recurrent infarction or death when these agents are given routinely to patients with acute myocardial infarction or unstable angina. The issue might be linked intimately with the question of an inconsistent effect on heart rate produced by calcium channel blockers when compared to beta blockers without intrinsic sympathomimetic actions.

Calcium channel blockers exert variable effects on the 24-h profile of heart rate. The dihydropyridines either have no effect or tend to increase heart rate over 24-h while preserving its circadian periodicity (Figure 7). In contrast, both verapamil and diltiazem tend to produce a modest decrease but their effects do not approach those of beta blockers without intrinsic sympathomimetic actions (Figure 8). A rate-lowering beta blocker consistently decreases heart rate markedly and attenuates its circadian periodicity (Figure 8). Thus, as stressed by Yusuf and Furberg (1987), the variable effect of calcium channel blockers on heart rate in man may account for the unexpectedly disappointing results on mortality. Mortality rates following administration of calcium channel blockers are consistent with the failure of this class of drugs to exhibit unequivocally beneficial effect on infarct size, development of myocardial infarction or reinfarction in trials of myocardial infarction or unstable angina (Holland Interuniversity Nifedipine/Metoprolol Trial, Research Group, 1986; Held et al., 1989). However, the positive data from the recent trial of diltiazem in patients with non-Q wave infarction merits consideration (Gibson et al., 1986). This multi-center randomized double-blind trial (Gibson et al., 1986) involved 576 patients recovering from non-Q wave infarction. Diltiazem (360 mg/day) was given to 278 patients, placebo to 289 for 14 days after acute myocardial infarction. The two groups were identical. The 14-day mortality was similar in both groups but reinfarction occurred in 26 patients on placebo and 11 in the diltiazem group. The difference of 42% was significant (p <0.04). It is striking that there was a very low incidence of mortality (3%) in both the treated as well as the placebo groups but one important feature of the patients entering the study was that 61% of them were treated concurrently with beta blockers. It may well be that the demonstrated salutary effect of diltiazem on reinfarction may be greatest in the presence of beta blockade, a finding also noted in trials involving patients with unstable angina (Holland Interuniversity Nifedipine/Metoprolol Trial, Research Group, 1986) in which a calcium channel blocker was combined with a beta blocker. A major reduction in heart rate is likely during a combination regimen of diltiazem and a beta blocker. An analysis of changes in heart rate in patients taking both beta blockers and diltiazem vs. diltiazem alone relative to the beneficial effect might have been of much interest.

An intriguing observation was made in another larger trial (Multicenter Diltiazem Post-Infarction Trial Research Group, 1988) in which 2466 patients surviving Q and non-Q wave acute infarction and being treated with or without beta blockers were randomized in a double-blind fashion into a group that was given diltiazem and the other, placebo. The data showed that the cumulative mortality and reinfarction rates were similar in diltiazem-treated and placebo-treated patients who were followed for a mean period of 25 ± 8 months. Figure 4

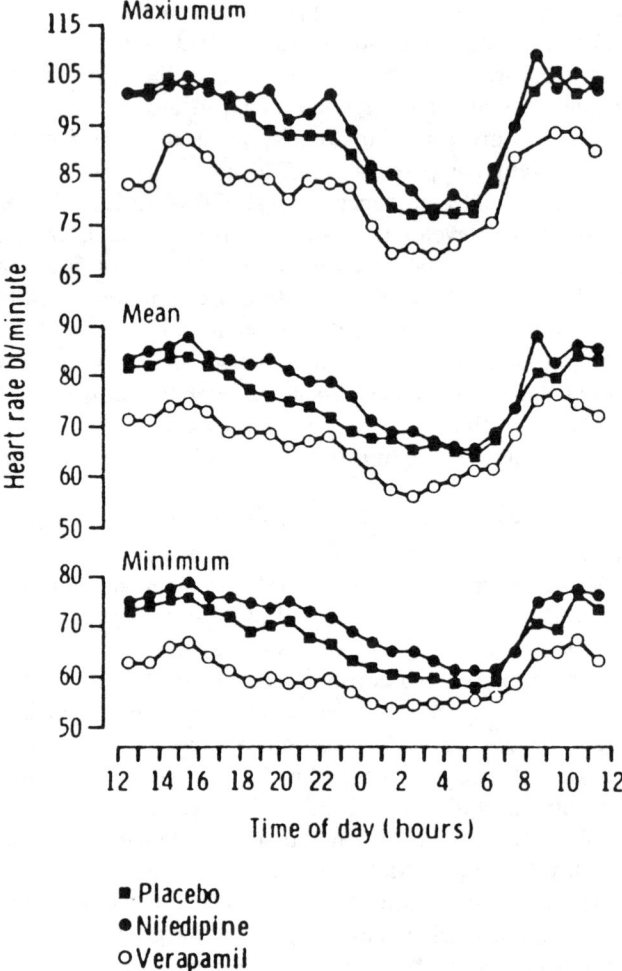

Figure 7. The hourly maximum, mean and minimum heart rate during double-blind trial comparing verapamil 120 mg three times a day with nifedipine, 20 mg three times a day, and placebo in 28 patients. Note that the circadian rhythmicity in heart rate is maintained; nifedipine tends to *increase* the heart rate and verapamil tends to slow it but the mean and maximum heart rates are still high and do not reach the levels induced by beta-blockade (see Figure 8). (From Balasubramanian with the permission of the author and of the editor of the *American Journal of Cardiology*.)

shows the overall data. There was no significant effect of diltiazem on mortality for the entire group. However, when a subset analysis was performed, it was clear that in patients who had a history of pulmonary congestion reflecting a lower ejection fraction, mortality over the follow-up period was *increased*; in the group without pulmonary congestion, there was a significant reduction in mortality (Figure 9). The reason for the difference is not clear. However, the data are in marked contrast to those with beta blockers in which benefit was evident at all levels of function and appeared to be greater in those with cardiomegaly and pulmonary congestion (Chadda et al., 1986).

The fact that calcium channel blockers do not exert a significant effect in reducing sudden death in the survivors of acute infarction does not negate the possibility that the efficacy of beta blockers in this regard is mediated at least in part through an action on myocardial ischemia. There are experimental and theoretical reasons for thinking that slowing

19. Control of Cardiac Arrhythmias by Modulation

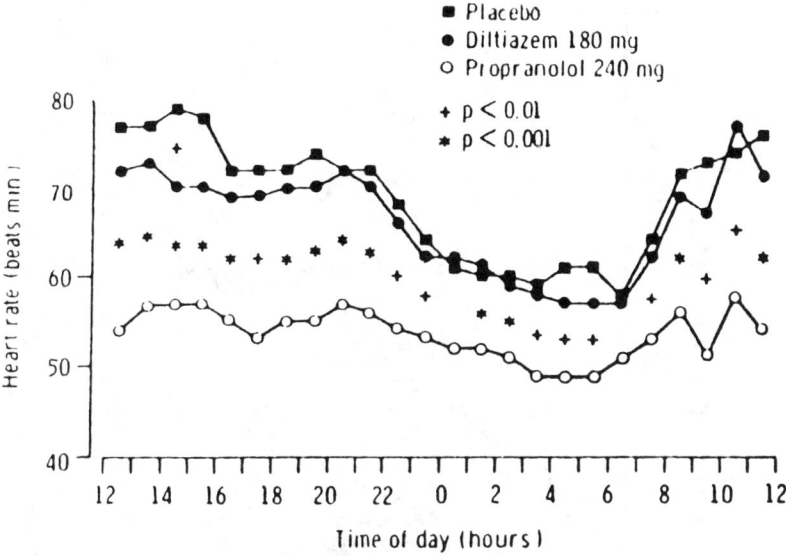

Figure 8. The 24-h ambulatory minimum heart rate on placebo, diltiazem, 180 mg/day, in 14 patients with stable angina. Note that diltiazem hardly affects the minimum heart rate during the 24-h monitoring while propranolol (240 mg/day) consistently and significantly reduces it. (From Balasubramanian (1986) with the permission of the author and of the editor of *American Journal of Cardiology*.)

of heart rate by whatever means might be beneficial in ischemia. As emphasized by Clusin (1987), perhaps the most potent means of modulating the severity of ischemia is through the regulation of heart rate (Blake et al., 1986). The major effect of heart rate on ischemia is generally attributed to alterations in myocardial oxygen demand. However, recent experimental data indicate that changes in oxygen consumption may be less critical (Clusin et al., 1984) than heretofore believed. The major effect of heart rate may stem from the modulation of calcium influx and retention by the ischemic myocardium (Clusin et al., 1984). It is known that myocardial calcium channels activate and inactivate during each action potential (see Chapters 3 and 4, Section I). Therefore, the number of action potentials per unit time will exert a marked influence on calcium influx. Since Ca^{2+}-Na^+ exchange over the plateau range of potentials is voltage dependent, an increase in time at the plateau voltages will have a significant effect on overall cell Ca^{2+} load (Clusin, 1987). An increase in subendocardial blood flow per minute and per beat in the ischemic zone may also relate, in part, to a reduced number of systolic contractions with a longer period for diastolic perfusion (Bache and Cobb, 1977). This may be a mechanism to account for increased subendocardial flow per beat after beta blockade; the longer diastolic period may also allow for recovery of metabolism per beat and over successive beats may contribute to the increased regional contraction in the ischemic zone (Guth et al., 1987). These considerations raise the issue whether future development of calcium channel blockers might focus on compounds that reduce heart rate.

A. Choice Among Calcium Channel Blockers and Combination with Other Antiarrhythmic Agents

The appropriate selection of a calcium channel blocker for a particular antiarrhythmic indication for monotherapy or for combination therapy is based on the knowledge of its electropharmacologic effects and pharmacokinetic properties relative to those of the com-

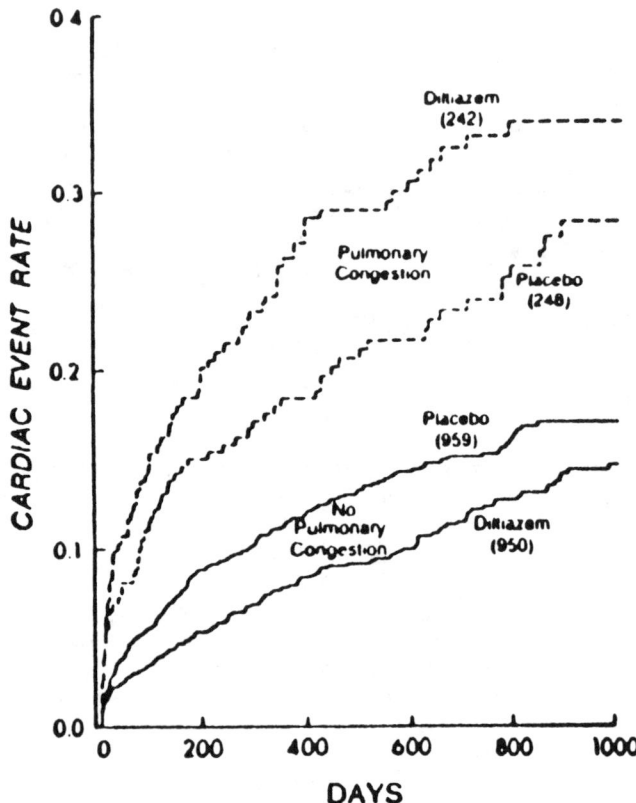

Figure 9. Effects of diltiazem on cumulative coronary incidence according to treatment in patients with and without pulmonary congestion in survivors of acute myocardial infarction. (From the Multicenter Diltiazem Post-Infarction Trial Research Group and the editor of *New England Journal of Medicine* (1988), with permission.)

pound to be used in combination. The dosage regimens of calcium channel blockers commonly employed in the control of cardiac arrhythmias are shown in Table 6.

Although a number of newer agents have recently been added to the growing list of calcium channel blockers, only the effects of verapamil and diltiazem on the AV node remain consistent with a predictable effect on the rate of conversion of paroxysmal supraventricular tachycardia and for the slowing of ventricular response in atrial flutter and fibrillation. Thus, for all practical purposes the choice of a calcium channel blocker for antiarrhythmic indications is between verapamil and diltiazem, both orally and intravenously. However, for the acute PSVT conversion, it is likely that I.V. adenosine might supplant I.V. verapamil or diltiazem because of the ultrashort action of the drug, its safety in hypotension and in patients with congestive heart failure, coupled with a high degree of success in terminating PSVT. The role of flunanzine in ventricular arrhythmias of triggered origin is under investigation and appears promising (Vos et al., 1990).

In recent years, there has been an increasing trend to use combination therapy in the treatment of arrhythmias in general. In the case of calcium channel blockers, combination therapy is particularly applicable for oral prophylactic therapy of recurrent supraventricular tachyarrhythmias and for the control of ventricular response in atrial flutter and fibrillation (Long et al., 1983; Roth et al., 1986). In either case, cardiac glycosides and beta blockers can be combined with diltiazem or verapamil in appropriate doses. The combination regimen has a summated effect on AV nodal conduction and refractoriness. Such a combination of

TABLE 6
Dosage Regimens of Calcium Channel Blocking Agents Employed in the Control of Cardiac Arrhythmias[a]

Agent	I.V. dose	Oral dose	Comments
Verapamil	0.145 mg/kg (bolus) 0.005 mg/kg/min (infusion)	240—480 mg/day	Oral regimen given in 3 or 4 divided doses
Diltiazem	0.15 to 0.25 mg/kg (bolus)	90—360 mg/day	Oral regimen administered in 3 or 4 divided doses
Tiapamil	1 to 2 mg/kg (bolus)	Not established	—
Gallopamil (D600)	0.03 to 0.05 mg/kg (bolus)	Not established	—
Bepridil	2 to 4 mg/kg	300—400 mg/daily	Oral regimen administered in one single daily dose

[a] The dosages cited are well established for verapamil (the "reference" calcium channel blocker). For the other compounds they are tentative and given to serve as an approximate indication.

electrophysiologic effects is likely to prevent the establishment of reentrant circuits involving the AV node and to slow the ventricular response in atrial flutter and fibrillation. The addition of a beta blocker is of particular value in situations in which heart rate control following exercise is suboptimal with a calcium channel blocker alone. However, in certain subsets of patients this may induce excessive effects in terms of AV nodal conduction and/or bradycardia. This is especially a concern when digoxin or beta blockers are combined with verapamil or diltiazem in the elderly and in patients with pre-existing sick sinus syndrome and AV blocks. Recent data have suggested that calcium channel blockers may need to be combined with other antiarrhythmic agents in situations when the drug effects are reversed with catecholamines (as, for example, with exercise). This has been found in patients with supraventricular ventricular tachycardias treated with Class I agents (Dongas et al., 1985; Akhtar et al., 1988; Morady et al., 1988; Vos et al., 1990). Both beta blockers and calcium channel blockers with anti-sympathetic actions may be of value.

VII. SIDE EFFECTS, CONTRAINDICATIONS AND PRECAUTIONS IN THE USE OF CALCIUM ANTAGONISTS AS ANTIARRHYTHMIC DRUGS

The most common side effect during I.V. drug administration is the transient fall in blood pressure (Schamroth, 1971; Schamroth et al., 1972; Heng et al., 1975). More serious side effects have however, been reported: persistent hypotension, bradycardia and rarely ventricular asystole (Benaim, 1972; Boothby et al., 1972; Sacks and Kennelly, 1972; Donnelly and Scaps, 1973). In most such cases, the patients had been on chronic beta-blocking therapy before verapamil was given. Since the negative inotropic actions and depressant effects on impulse generation of beta antagonists and of calcium channel blockers (especially verapamil and diltiazem) are additive, hypotension, bradycardia and asystole are predictable side effects when these drugs are administered in combination. However, in general, side effects that occur in this setting may be reversed by intravenous atropine (only partially effective), isoprenaline and particularly intravenous calcium (10 to 20 ml of 10% solution); temporary ventricular pacemaker therapy may also be used in recalcitrant cases. The adverse side effect profile during oral therapy is perhaps most benign in the case of orally administered diltiazem although experience in arrhythmias is limited and the knowledge of the side effects of the drugs is derived from studies in angina. Dizziness, headache, fatigue, blurred vision, flushing and minor degrees of AV block may occur when the diltiazem is given in a daily dosage between 240 and 360 mg and verapamil 320 to 480 mg. In the case of the latter, constipation is the dominant side effect.

Prolongation of first degree AV block occurs in a proportion of patients given chronic oral therapy with verapamil, its congeners or diltiazem. Advanced grades of heart block are unusual unless antecedent conduction system disease is present. The development of clinically evident cardiac failure in patients with normal ventricular function is very uncommon; however, if the left ventricular ejection fraction is severely depressed or when the drug is given concomitantly with beta-blocking drugs in patients with more than mild impairment of myocardial performance, manifest cardiac failure may ensue.

In general, the main contraindications to the use of calcium channel blockers as antiarrhythmic drugs are the presence of advanced heart failure, unstable AV block, disease of the conduction system including the sick sinus syndrome and low blood pressure states such as cardiogenic shock. However, in situations in which heart failure is related to a persistent rapid atrial tachyarrhythmia, a prompt reversion to sinus rhythm by calcium channel blockers may lead to an improvement in the degree of cardiac decompensation. This is particularly likely to occur in patients with supraventricular tachycardia. It is emphasized that in patients with wide QRS tachycardias (supraventricular vs. ventricular in origin) it is, in general, imprudent to use I.V. verapamil or diltiazem in attempts to revert the arrhythmia. In practice, wide QRS tachycardias are only uncommonly of supraventricular origin and those that are of ventricular origin rarely convert to sinus rhythm in a predictable manner (Heng et al., 1975). If conversion does not occur, an alarming fall in blood pressure may follow. Similarly, in cases of atrial flutter or fibrillation complicating the WPW syndrome, I.V. verapamil and diltiazem are contraindicated. In this setting, AV block that results from the use of these compounds may increase the ventricular response over the bypass tract (Gulamhusein et al., 1981; Rozansky et al., 1982) to dangerous rates with the likelihood of the development of ventricular fibrillation.

Of particular importance is the necessity to avoid the combined use of verapamil (its congeners and diltiazem) and beta-adrenoceptor blocking drugs in patients with overt or marginal hemodynamic dysfunction. For example, I.V. verapamil (0.1 mg/kg) and practolol (0.1 mg/kg) produced minor hemodynamic changes when given individually, but when the combination regimen of the two drugs was employed and heart rate fixed by atrial pacing, a pronounced reduction in left ventricular contractility occurred independently of changes in preload and afterload (Palileo et al., 1982). Thus, precautionary action must be recommended when the combined use of calcium channel blockers and beta blockers in patients with impaired myocardial function is contemplated. This consideration also holds for their use individually or in combination in patients with the sick sinus syndrome and impaired AV conduction. In these cases, beta blockers and calcium channel blockers may, however, be used for the prophylactic treatment of tachyarrhythmias if a demand ventricular pacemaker is first inserted. The same precaution clearly also holds for the combination therapy with digoxin. However, unless there is evidence of impaired AV conduction, prior digitalization is not a contraindication to the use of I.V. verapamil (or diltiazem), since 61% of patients with supraventricular tachyarrhythmias in one series (Heng et al., 1975) and 79% in another (Schamroth et al., 1972) were on digitalis at the time verapamil was given. In neither series was there a significant untoward reaction attributable to the combination of I.V. verapamil and a steady state oral digitalis administration. Nevertheless, caution and careful surveillance are advised during the concomitant oral therapy with the two drugs in view of the significant interaction between certain calcium channel blockers (Klein et al., 1982) and digoxin.

VIII. CONCLUSIONS

Over the last 20 years it has become clear that certain cardiac arrhythmias may arise on the basis of slowed conduction or of abnormalities of impulse initiation resulting from deranged slow calcium channel function in the heart. Such arrhythmias respond reasonably predictably to those calcium channel blockers which exert distinctive electrophysiologic properties especially with respect to nodal tissues. The major clinical utility of these agents is in the elective and prophylactic control of reentrant paroxysmal supraventricular tachycardia, atrial flutter and fibrillation, and possibly in multifocal atrial tachycardia. The role of calcium channel blockers in ventricular arrhythmias is limited to certain triggered tachyarrhythmias (e.g., exercise-induced ventricular arrhythmias) and to tachyarrhythmias resulting from ischemia due especially to coronary artery spasm. However, the role of calcium channel blockers in the reduction of sudden death in the survivors of acute myocardial infarction and in unstable angina remains to be defined.

ACKNOWLEDGMENTS

I am much indebted to Francesca Frederick and Lawrence Kimble in the preparation of this manuscript.

Chapter
20

The Effects of Calcium Channel Blockers on Platelets and their Application in the Management of Vascular Diseases

Wenche Jy and Yeon S. Ahn

TABLE OF CONTENTS

I.	Calcium Channel Blockers and Platelets	364
	A. Platelet Activation	364
	B. Calcium as a Second Messenger	365
	C. Calcium Homeostasis in Platelets	366
	D. Effects of Organic Ca^{2+} Channel Blockers on Platelet Functions	367
	E. Mechanism of Action of Calcium Channel Blockers	368
II.	Calcium Channel Blockers and Blood Vessels	370
	A. Vasodilation	370
	B. Calcium Channel Blockers and Atherosclerosis	371
III.	Use of Calcium Channel Blockers in Vascular Diseases	372
	A. Intermittent Claudication	373
	B. Raynaud's Phenomenon	373
	C. Platelet Related Disorders	374
IV.	Conclusions	375

I. CALCIUM CHANNEL BLOCKERS AND PLATELETS

Calcium channel blockers have multiple actions on platelets and blood vessels that affect blood flow, modify clinical manifestations and effect pathologic processes associated with vascular disorders. This review will focus mainly on their actions on platelets and blood vessels, and on their clinical application in various vascular and related disorders.

A. Platelet Activation

Platelets circulate freely in the blood stream as smooth surfaced discoid particles. Upon activation by agonists such as collagen, adenosine diphosphate (ADP), epinephrine, thrombin, or other external stimuli, platelets change their shape from discoid to spiny spheres, secrete their constituents, adhere to blood vessel walls, and aggregate (Mustard and Packham, 1970).

The participation of platelets in physiologic and pathologic processes requires platelet activation, the complex process which accompanies morphologic, biochemical and functional alterations. Morphologically, the platelet is surrounded by the exterior coat glycocalyx and the plasma membrane. The glycocalyx is a dense layer 15 to 20 nm in thickness covering the external surface of the platelet. The glycocalyx of platelets contains many glycoproteins, some of which are essential for platelet adhesion and aggregation. For instance, an epitope in glycoprotein IIb-IIIa has been identified as the fibrinogen binding site responsible for the formation of the primary platelet aggregation (Plow and Ginsberg, 1989). Under resting conditions, this epitope is not available to fibrinogen. However, upon platelet activation, the conformational change of glycoprotein IIb-IIIa causes the expression of fibrinogen binding site leading to platelet aggregation. The significance of glycoprotein IIb-IIIa is best demonstrated by the effect of its deficiency in platelets from patients with thrombasthenia (Nurden and Caen, 1974).

The plasma membrane maintains the internal cellular millieu by actively or passively transporting various molecules through the membrane. The binding of agonists to the cell surface triggers transduction into biochemical signals across the membrane. During platelet activation, membrane phospholipids are utilized as precursors of active messengers such as thromboxane A_2 (Samuelsson et al., 1978), inositol trisphosphate (IP_3), and diacylglycerol (DAG) (Berridge, 1984). The calcium permeability of the plasma membrane is greatly increased, promoting a rapid influx of Ca^{2+} into the cytoplasm (Rink et al., 1982; Jy and Haynes, 1987).

The cytoplasm of the platelet contains many organelles such as alpha-granules, dense granules, lysosomes, mitochondria, the dense tubular system and microfilaments, microtubule and glycogens. The microfilaments and microtubules play a major role in maintaining the discoid shape in the resting state and in providing the contractile strength for shape change during platelet activation.

Alpha-granules contain many proteins (Holmsen and Weiss 1979), some of which are specific to the platelet. These include platelet factor 4, beta-thromboglobulin, platelet-derived growth factor, thrombospondin, etc. A number of other proteins that are not specific for the platelets are also found in alpha-granules, including fibrinogen, Factor V, Factor VIII, fibronectin, albumin, etc. Among these proteins, several are of special importance for hemostasis and atherosclerosis. Platelet-derived growth factor (PDGF) is a potent mitogenic factor for the proliferation of vascular smooth muscle, one of the major features of atherosclerosis (Ross and Glomset, 1976). The secreted thrombospondin can bind to its receptor on the surface of platelets to provide strong cross-linking between platelets, eventually leading to irreversible aggregation.

The dense granules contain serotonin, calcium, ATP, ADP and pyrophosphate (Holmsen and Weiss, 1979). Some of these substances such as serotonin and ADP are platelet stimulating agents by themselves. The secretion of dense granules is believed to induce a positive feedback on platelet activation.

The dense tubular system, similar to the sarcoplasmic reticulum in muscle cells, is the major intracellular Ca^{2+} uptake and storage site. Upon platelet activation, the sequestered Ca^{2+} inside the lumen of the dense tubular system can be triggered to release by IP_3, an enzymatic product of phosphatidyl inositol diphosphate (Berridge, 1984).

Platelets are activated by various stimuli. Physiological stimuli include collagen, thrombin, ADP, serotonin and epinephrine. Other stimuli in pathological conditions are antigen-antibody complexes, aggregated gamma globulins (Nachman and Weksler, 1980), and platelet-activating factor (PAF) which is an O-ether-phospholipid produced by activated leukocyte and stimulates platelets at nM concentrations (Marcus et al., 1981). Foreign platelet activating agents include calcium ionophore (Massini and Luscher, 1974) and phorbol ester, an activator of protein kinase C (Chiang et al., 1981).

It is remarkable that the platelet responses to this wide variety of stimuli lead into the same final pathway of functional and morphologic alterations: platelets change their shape, then become "sticky" and begin to aggregate, and at the same time the contents of granules are secreted. The latter process induces further aggregation leading to larger and irreversible aggregates.

The shape change is initiated within a few seconds after the addition of the stimulating agents. This change involves first the formation of very fine (0.1 μm diameter) pseudopodia from the rim of the disc followed by a general "rounding up" of the platelet so that it becomes a spiny sphere. With lower concentrations of a weak agonist such as ADP, this initial shape change is followed by a small and reversible aggregation ("primary aggregation") without secretion. Higher concentrations of ADP can induce further secretion accompanied by more extensive and irreversible aggregation ("secondary aggregation"). With strong agonists such as thrombin or collagen, the two phases of aggregation are not usually observed; instead, an irreversible aggregation with secretion are observed (Charo et al., 1977). Recent studies showed that the major factor determining the shape changes, primary and secondary aggregation is the level of cytoplasmic free Ca^{2+} (Rink et al., 1982). The Ca^{2+} threshold required for shape change and primary aggregation is lower than that for secretion and secondary aggregation (Rink et al., 1982). Once the Ca^{2+} level exceeds the threshold for secretion, thrombospondin as well as many other granular substances are released to the external medium. Thrombospondin can further stabilize the platelet aggregate by forming the cross-linkage between platelets (Asch and Nachman, 1989).

B. Calcium as a Second Messenger

Most physiologic agonists as described, do not penetrate the plasma membrane. They bind to specific receptors which initiate intracellular transduction through specific messengers, leading into shape change, secretion and aggregation (Mustard and Packham, 1971).

Calcium is a major second messenger that mediates platelet activations when induced by various stimuli (Feinstein, 1978, 1982; Rink, 1988; Salzman and Ware, 1989). In the resting state, platelets maintain their cytoplasmic calcium at 0.1 μM against 2 mM in the extracellular fluid. An increase of cytoplasmic calcium to about 1 μM is sufficient to induce both shape changes and the release reaction (Rink et al., 1982; Rink and Hallam, 1984). Aggregation requires the presence of both external and intracellular calcium (Feinstein, 1982).

The contraction of microfilaments, the process involved in platelet shape changes (White, 1974) is associated with the phosphorylation of platelet myosin light chain. This reaction is mediated by the calcium/calmodulin dependent enzyme and myosin light chain kinase (Hathaway and Adelstein, 1979; Daniel et al., 1984).

An increase in cytoplasmic free calcium $[Ca^{2+}]_i$ also triggers Ca^{2+} secretion from platelet internal storage granules. This promotes contractile reactions, during which the secretory granules move into the center of the cell and fuse with the membranes of open canalicular systems to release their content to the outside of the cells (White, 1974; Stenberg et al., 1984). The process of this fusion also appears to depend on cytoplasmic calcium (Douglas, 1968; Pollard et al., 1979).

In the process of aggregation, platelets change their shapes, become sticky, and adhere to each other to form aggregates. This requires fibrinogen that bridges platelets by binding to the fibrinogen receptors on platelet surfaces. External calcium is needed to stabilize the bridging of fibrinogen to the receptor (Marguerie et al., 1980; Plow and Ginsberg, 1989).

It has been shown that platelet stickiness results from either the conformational expression of fibrinogen receptors (Marguerie et al., 1980; Plow and Ginsberg, 1989) or release of a lectin-like protein, thrombospondin following the fusion of secretory granules with plasma membranes (Jaffe et al., 1982; Gartner et al., 1984). Both of these processes have also been demonstrated to be calcium dependent (Asch et al., 1987; Shattil and Brass, 1987).

Further evidences of calcium as a secondary messenger in platelet activation have been provided by the following observations. A calcium ionophore, which increases cytoplasmic Ca^{2+} by increasing Ca^{2+} influx from external media, can induce shape changes, secretion, and aggregation of platelets, that are essentially similar to those induced by thrombin (Massini and Luscher, 1974). The important role of calcium in platelet activation can be demonstrated with permeabilized platelets prepared by the application of high voltage electric discharge (Knight and Scrutton, 1980). When the permeabilized platelets are suspended in Ca^{2+}-EGTA buffer at defined concentrations of free Ca^{2+}, the extent of secretion is dependent on the increase of free Ca^{2+} concentration (Knight and Scrutton, 1980).

With techniques employing fluorescent Ca^{2+} probes such as chlortetracycline (CTC) (LeBreton et al., 1976; Jy and Haynes, 1984), Quin 2 (Tsien et al., 1982), and Fura 2 (Grynkiewicz et al., 1985), it is now possible to monitor the release of Ca^{2+} from dense tubules and the elevation of cytoplasmic free Ca^{2+} during platelet shape change, secretion, and aggregation. It has been shown that internal Ca^{2+} release precedes secretion and does so without external Ca^{2+} (Feinstein, 1980). The apparent threshold for $[Ca^{2+}]_i$ for shape change was found to be lower than that for secretion and aggregation (Rink et al., 1982). This is consistent with the observation that shape change is the most difficult of the platelet functions to be inhibited (Kinbough-Rathbone et al., 1970).

C. Calcium Homeostasis in Platelets

Calcium homeostasis in the resting platelets is maintained by two forces: the influx and extrusion of calcium through the plasma membrane, and the sequestration and release of calcium to and from internal organelles (Brass, 1984; Jy and Haynes, 1984, 1988; Johanson and Haynes, 1988). Upon stimulation with agonists, the $[Ca^{2+}]_i$ rises rapidly from 0.1 μM up to about 1 μM and declines slowly to the initial resting level (Rink et al., 1982; Sage et al., 1986). The initial rapid rise of cytoplasmic calcium results from the summation of the agonist induced Ca^{2+} influx and the release of internal Ca^{2+} from the dense tubular system (Jy and Haynes, 1987). Ca^{2+} influx precedes internal Ca^{2+} release by at least 40 to 50 ms (Sage and Rink, 1986). The agonist-induced Ca^{2+} influx is believed to be mediated through receptor-operated Ca^{2+} channels rather than voltage-dependent Ca^{2+} channels (Doyle and Ruegg, 1985; Zschauer et al., 1988) and can be inhibited by a cAMP-dependent protein kinase system (MacIntyre et al., 1985; Sage and Rink, 1985). On the other hand, release of internal Ca^{2+} from the storage organelles is induced by IP_3 (O'Rourke et al., 1985), produced from phosphatidyl inositol diphosphate by phospholipase C (Berridge, 1984). The slow decline of cytoplasmic Ca^{2+} following its initial rise to a peak level is mediated by the inactivation of both Ca^{2+} influx and release from storage organelles which results in a cessation of any further increase in $[Ca^{2+}]_i$ and by the enhancement of both Ca^{2+} extrusion and sequestration of Ca^{2+} to internal storage organelles (Jy and Haynes, 1987, 1988; Johanson and Haynes, 1988). Two components of Ca^{2+} extrusion have been identified: a Ca^{2+} pump and a Na^+-Ca^{2+} exchange system at the plasma membrane (Johanson and Haynes, 1988; Schaeffer and Blaustein, 1989).

The fluorescent probes, CTC and Quin 2, have been utilized to measure abnormal calcium homeostasis of platelets in patients with hypertension, thrombotic disorders, diabetes, and other disorders in which platelet hyperactivation plays a contributing role in vascular complications. In these disorders, the cytoplasmic free calcium and the Ca^{2+} sequestered in dense tubular systems are significantly elevated compared to those of normal controls (Ahn et al., 1987; Jy et al., 1987). This may be the result of high Ca^{2+} influx at the resting state in these disorders.

Further studies revealed a close correlation between calcium homeostasis and platelet aggregability: high Ca^{2+} levels in platelets were associated with hyperaggregability of platelets (Jy et al., 1987; Shanbaky et al., 1987).

Attempts to identify receptor-operated calcium channels in platelets have not met with success. Radiolabeled ligands successfully employed in the study of muscle cells (Glossman et al., 1982), failed to detect the high affinity binding site on platelets. The study of low affinity binding sites is hampered by the high background of nonspecific binding (Motulsky et al., 1983; Erne et al., 1984). However, some circumstantial data suggest that it may be linked to glycoprotein IIb-IIIa complex (Brass, 1985; Dowling and Hardisty, 1985; Rybak et al., 1988).

D. Effects of Organic Ca^{2+} Channel Blockers on Platelet Functions

Extensive data from *in vitro* and *in vivo* studies have demonstrated that organic Ca^{2+} channel blockers (see Chapter 13, Section III) such as verapamil, diltiazem and nifedipine inhibit platelet functions (Han et al., 1980; Ikeda et al., 1981; Ono et al., 1981; Dale et al., 1983; Takahara et al., 1985; Addonizio et al., 1986; Walley et al., 1989). In general, all prototypes of Ca^{2+} channel blockers can reduce platelet aggregability and secretion *in vitro*. The potency (IC_{50}: concentration of drug which induces 50% inhibition on special platelet functions) of verapamil, diltiazem and nifedipine on platelet functions, however, varies with the experimental conditions. Early studies showed that the IC_{50} of verapamil, diltiazem and nifedipine on ADP-induced platelet aggregation was 500, 500 and 100 μM, respectively (Ono et al., 1981). One may argue that these concentrations are much higher than those attained *in vivo*. However, by using a whole blood aggregation technique, it has been shown that the IC_{50} of verapamil or diltiazem on ADP-induced platelets is only 15 and 8 μM, respectively (Pannocchia et al., 1987). The effect of Ca^{2+} channel blockers on platelet functions also varies depending on the agonists used. For instance, epinephrine-induced platelet aggregation is inhibited by relatively low concentrations of verapamil ($IC_{50} = 0.3$ $\mu M - 3$ μM) (Addonizio et al., 1986; Pannocchia et al., 1987) while higher concentrations of nifedipine or other calcium channel blockers are required to inhibit ADP or collagen-induced aggregation. Nifedipine is a very potent inhibitor of U46619; a stable TXA_2 agonist-induced platelet aggregation ($IC_{50} < 3$ μM; Johnson et al., 1988). It would appear, therefore, that in addition to their effect on Ca^{2+} homeostasis, verapamil and nifedipine may also act on the α-adrenergic receptor and the TXA_2 receptor, respectively.

The modification of calcium channel blockers on platelet functions has been investigated in normal donors. Two hours after oral administration of 80 mg of verapamil, platelet aggregation by ADP, epinephrine and collagen was reduced 60 to 80% (Ikeda et al., 1981). When a single dose (20 mg) of nifedipine was administered to patients with cardiovascular disorders, it induced a significant (20 to 26%) inhibition of platelet aggregation induced by ADP and collagen and caused a 12% prolongation of bleeding time (Dale et al., 1983). Nifedipine, at a dose of 10 mg, inhibited enhanced platelet aggregability during exercise *in vivo* (Takahara et al., 1985). Animal experiments demonstrated that nifedipine and verapamil are highly protective against thrombosis experimentally induced by I.V. injection of a mixture of collagen plus epinephrine (Ortega et al., 1987) or hardened red cells (Molinari et al., 1987). These protections were thought to be mediated not only by the inhibition of platelet aggregability but also by the vasodilatation and the prostacycline release from vessel walls produced by calcium channel blockers (Kai et al., 1982; Metha, 1985). When the effect of verapamil and nifedipine on thrombus formation in the Grove-Tex grafted artery in the dog was examined, it was found that the administration of these drugs reduced the platelet deposition on the graft by 40 to 60% (Pumphry et al., 1983). All of the *in vivo* experiments described assessed the short term effect of drugs in a matter of hours or days. The long term effects of calcium channel blockers on platelet function and prevention of thrombosis remain to be further defined. We have reported longitudinal studies on the use of nifedipine in patients with thrombotic and related disorders. Our results indicated that long term treatment with nifedipine not only improves the clinical syndrome in many patients but also corrects the abnormal Ca^{2+} handling defect in platelets (Ahn et al., 1987; Ahn and Harrington, 1987).

E. Mechanism of Action of Calcium Channel Blockers

A review of the literature suggests that calcium channel blockers in addition to (1) inhibiting Ca^{2+} movement, modify platelet function and thrombogenesis through other mechanisms. The possible mechanisms cited include, (2) blockade of agonist-receptor interactions, (3) interference with the arachidonate pathway and (4) inhibition of phosphodiesterase. These actions are summarized in Figure 1.

Inhibition of Ca^{2+} movement by calcium channel blockers in platelets has been studied with the techniques employing Ca^{2+} probes such as Quin 2 (Jy and Haynes, 1987) aequorin (Ware et al., 1986) and $^{45}Ca^{2+}$ influx (Blache et al., 1987; Kribben et al., 1987). The effect of verapamil, diltiazem and nifedipine on the level of $[Ca^{2+}]_i$ was studied in platelets that had been stimulated by various agonists such as epinephrine, arachidonate, thrombin, Ca^{2+} ionophore, A23187, 1-oleoyl-2-acetyl glycerol or ADP (Ware et al., 1986). In calcium-containing media verapamil, diltiazem, and nifedipine inhibit the rise of $[Ca^{2+}]_i$ in a dose dependent manner. In general, a strong inhibition of the aequorin Ca^{2+} signal by verapamil, diltiazem and nifedipine is correlated with inhibition of platelet aggregation although a high concentration of inhibitor is required. A similar correlation between inhibition of Ca^{2+} influx and platelet aggregation by verapamil was also reported by using the Quin 2 technique (Jy and Haynes, 1987). Studies with $^{45}Ca^{2+}$ revealed that thrombin-induced $^{45}Ca^{2+}$ uptake was inhibited by 18 μM nitrendipine. At the same concentration of nifedipine, the thrombin-induced serotonin release can be completely inhibited (Blache et al., 1987). Kribben et al. (1987) studied the effects of nitrendipine and tiapamil on $^{45}Ca^{2+}$ influx in platelets *in vitro*. They found that thrombin stimulated $^{45}Ca^{2+}$ influx could be inhibited by 10 μM nitrendipine or by 100 μM tiapamil. The effect of calcium channel blockers on the release of internal Ca^{2+} has not been clearly elucidated. Various calcium channel blockers have been shown to inhibit IP_3-induced Ca^{2+} release from platelet membrane vesicles (Seiler et al., 1987). Cinnarizine and flunarizine are potent inhibitors of IP_3-induced release of intracellular calcium with IC_{50} below 10^{-6} M. The Ca^{2+} channel blocker bepridil can inhibit both thrombin-induced Ca^{2+} influx and internal Ca^{2+} release but with different IC_{50}: 2 μM for thrombin-induced Ca^{2+} influx; 20 μM for thrombin-induced internal Ca^{2+} release (Jy and Haynes, 1987). It is not known which Ca^{2+} channel blockers are more potent inhibitors of platelet

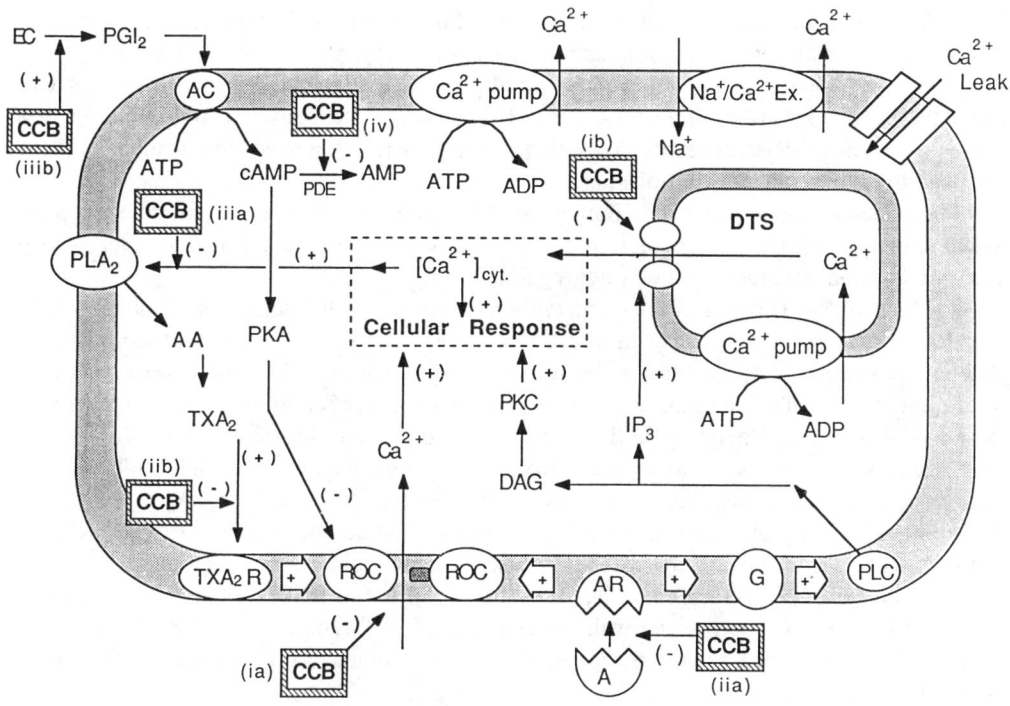

Figure 1. The schematic model illustrates the sites of action for calcium channel blockers on platelets. The (+) symbols indicate stimulation and (−) symbols, inhibition. AA, arachidonic acid; AC, adenylate cyclase; AR, agonist receptor; [Ca^{2+}] cyt, cytoplasmic free calcium; CCB, calcium channel blocker; DAG, diacylglycerol; DTS, dense tubular system; EC, endothelial cell; G, GTP-dependent protein; IP$_3$ inositol trisphosphate; PDE, phosphodiesterase; PGI$_2$, prostacyclin; PKA, protein kinase A; PKC, protein kinase C; PLA$_2$, phospholipase A$_2$; PLC, phospholipase C; ROC, receptor-operated calcium channel; TXA$_2$, thromboxane A$_2$, and TXA$_2$R, thromboxane A$_2$ receptor.

aggregation: those acting on both Ca^{2+} influx and release or those acting on Ca^{2+} influx only.

Ca^{2+} channel blockers interact with three types of platelet-receptors: (1) α-adrenergic receptors, (2) thromboxane A$_2$ (TXA$_2$) receptor and (3) platelet activating factor (PAF) receptor.

Verapamil has been shown to preferentially inhibit epinephrine-induced platelet aggregation through the adrenergic alpha receptor (Barnathan et al., 1983; Motulsky et al., 1983). Ligand binding studies show that verapamil inhibits the binding of α$_2$-adrenergic ligand [^{3}H]-Yohimbine to α$_2$-adrenergic receptor by decreasing the affinity of the α$_2$-ligand to its binding site without altering the total number of binding sites (Johnson et al., 1986). This verapamil-induced inhibition of α$_2$-receptor binding is not shared by other calcium channel blockers.

Bay K8644 is a dihydropyridine derivative that exhibits a partial agonist-antagonist effect on muscle calcium channels (Preuss, 1985). It inhibits, in a concentration-dependent manner, platelet aggregation and serotonin release (Johnson et al., 1988). At the concentration of 1 to 10 μM it inhibits the second phase of platelet aggregation and secretion induced by ADP or epinephrine and prevents aggregation and secretion induced by U46619-TXA$_2$ stable analog. It also blocks all U46619-induced biochemical consequences including the phosphorylation of 40K protein, the formation of TXA$_2$, and the rise in [Ca^{2+}]$_i$. The binding of [^{3}H]-U46619 to intact platelet is also competitively inhibited by Bay K8644 and nifedipine. The Ki for Bay K8644 is estimated to be 1.47 μM. Nifedipine is about ten times less potent than Bay K8644 (Johnson et al., 1988).

Calcium channel blockers interfere with the binding of PAF to its receptor (Valone, 1987). The order of potency among calcium channel blockers is (+)-cis-diltiazem, (+/−) verapamil, (−)-cis-diltiazem followed by nifedipine. The degree of inhibition by four calcium channel blockers of the interaction of [^{3}H]-PAF-acether with high affinity binding sites correlates closely with their respective antiaggregatory activity against PAF-acether-induced response in human platelet rich plasma.

The selective interaction of calcium channel blockers with different membrane receptors would explain in part why one calcium channel blocker is more potent than another against a particular agonist-induced platelet aggregation.

TXA$_2$ and Prostacyclin (PGI$_2$) are two major biologically active metabolites of the arachidonic acid pathway (Moncada and Vane, 1978). The former is a potent stimulator of platelet aggregation, responsible for the secondary aggregation, while the latter is the most potent inhibitor of platelet aggregation. Calcium channel blockers modify the production of these two metabolites, favoring production of PGI$_2$ to prevent thrombosis and inhibition of TXA$_2$ generation to reduce platelet activation. Calcium channel blockers inhibit TXA$_2$ formation induced by various agonists (Metha et al., 1983). This effect may be related to the inhibition of Ca^{2+} mobilization essential for activating membrane bound enzyme, phospholipase A$_2$ (PLA$_2$), which in turn will release arachidonic acid from phospholipids (Rittenhouse-Simmons and Deykin, 1978). On the other hand calcium channel blockers can increase release of PGI$_2$ from blood vessel walls (Kai et al., 1982; Metha, 1985, 1986; Boeynaems et al., 1987). These two known mechanisms may contribute to the antithrombotic effects of calcium channel blockers.

The elevation of intracellular cAMP level in platelets leads to the inhibition of platelet aggregation, partly through its inhibition of Ca^{2+} influx through the receptor-operated Ca^{2+} channel (MacIntyre et al., 1985; Sage and Rink, 1985). In rabbit platelets, cyclic AMP phosphodiesterase is inhibited by the Ca^{2+} channel blockers, verapamil, diltiazem and nifedipine with IC$_{50}$ of 420, 100 and 100 μM, respectively (Moore et al., 1985). Inhibition of phosphodiesterase will lead to an elevation of cAMP in platelets (Zavoico and Feinstein, 1984), which in turn inhibits Ca^{2+} influx through the plasma membrane (see Section I.C).

II. CALCIUM CHANNEL BLOCKERS AND BLOOD VESSELS

A. Vasodilation

One of the major actions of calcium channel blockers on blood vessels is vasodilation. In the vascular smooth muscle cells, intracellular free Ca^{2+} regulates contraction and dilation of blood vessels (see Chapter 6, Section II) (Cauvin et al., 1983; Godfraind et al., 1986). An increase in cytoplasmic calcium triggers vasoconstriction while vasodilation is induced by a decrease in calcium concentration. The opening of calcium channels is the most important step leading to an increase of cytoplasmic Ca^{2+}. Calcium channel blockers inhibit Ca^{2+} influx across the plasma membrane of smooth muscles, decrease cytoplasmic Ca^{2+} and thereby induce vasodilation.

The degree of response of vascular smooth muscles to vasoactive drugs, however, varies considerably between different species and anatomical sites in the same species. A wide range of different sensitivities and differences in potency has been observed among calcium channel blockers (Sjoberg et al., 1987; Van Nueten et al., 1987).

For example, flunarizine and nimodipine are particularly potent in blocking calcium-induced vasoconstriction in cerebral blood vessels (Van Nueten and Vanhoutte, 1981; Towart, 1981; Kazda and Towart, 1982). This may be related to the different configuration of the calcium channel or a different degree of access to the calcium channel at various sites in blood vessels.

Furthermore, the effect of calcium channel blockers also depends on the nature of vasoconstrictor response by different agonists. Calcium agonists are more selective, in general, for inhibiting calcium influx through voltage rather than receptor operated channels. Vasoconstriction mediated by the process of membrane depolarization is particularly sensitive to calcium channel blockers (Cauvin et al., 1983; Janis and Triggle, 1983; Godfraind et al., 1986).

α-Adrenoreceptor induced vasoconstriction gives rise to an increased cytoplasmic Ca^{2+} concentration through enhanced Ca^{2+} influx from the extracellular space and/or by releasing Ca^{2+} from intracellular storage organelles (Van Breemen et al., 1986). In the arteries this is mediated mainly through α_1-adrenoreceptor whereas in veins it is predominantly mediated by α_2-adrenoreceptors (Sjoberg et al., 1987). It has been shown that all calcium channel blockers except nifedipine at high concentrations inhibit noradrenaline induced contraction more effectively in the arteries than in the veins.

The type and size of the artery is another factor in determining the response to calcium channel blockers. The distribution of α_1- and α_2-adrenoreceptors shows a wide variation in blood vessels (Stevens and Moulds, 1981; Steen et al., 1984). In addition, the magnitude of intracellular Ca^{2+} stores varies greatly with the size and type of smooth muscles. Since calcium channel blockers inhibit predominantly Ca^{2+} influx through the plasma membrane, the ratio of Ca^{2+} influx to the calcium release from the storage pool can be an important factor in determining the efficacy of the calcium channel blockers. This ratio is higher in small arteries than in larger ones because of the smaller size of the calcium storage pool in smaller vessels (Van Breemen et al., 1982). Therefore, calcium channel blockers are expected to be more effective in smaller arteries.

Marked differences have been observed in the calcium channel blockers regarding the time course of their actions. For example, the actions of diltiazem, nifedipine and verapamil are rapid and of short duration while cinnarizine, lidoflazine, nicardipine and flunarizine have progressive and sustained actions (Van Nueten et al., 1987).

B. Calcium Channel Blockers and Atherosclerosis

The antiatherogenic properties of calcium channel blockers have been demonstrated (Weinstein and Heider, 1987). The initial event of atherosclerosis appears to be endothelial cell injury, which triggers platelet adhesion and aggregation with subsequent release of platelet constituents. Platelet-derived growth factor stimulates smooth muscle cells to migrate into the intima and to proliferate. Monocytes and macrophages also migrate into the site of injury, releasing factors that stimulate replication not only of smooth muscle cells, but also of endothelial cells. This "repair" of endothelium is followed by development of intimal plaques, which increase in size and protrude into the lumen, restricting blood flow (Ross and Glomset, 1976). The plaques consist primarily of accumulation of smooth muscle cells loaded with cholesterol and cholesterol esters surrounded by lipid, collagen, elastin, and glycosaminoglycans, secretory products of the smooth muscle cells with a high-affinity binding to low-density lipoproteins (Rose, 1985).

Ca^{2+} has been implicated in a number of steps of atherogenesis. When cell injury impairs membrane integrity, the cell is no longer able to maintain a low cytoplasmic calcium concentration against the 20,000-fold higher extracellular concentration. Increased influx of Ca^{2+} across membranes activates various enzymes that may be deleterious to cellular function. Ca^{2+} and cholesterol accumulation are the characteristic features of atherosclerotic plaques. Extensive calcification is common in atheromatous vessel walls. Several studies in animals have demonstrated that an increased Ca^{2+} content of arterial walls enhances atherosclerosis (Morrison et al., 1972), while a decreased content can ameliorate or delay the development of atherosclerosis (Kramsch et al., 1980).

A recent study has demonstrated the inhibitory effects of calcium channel blockers on the myeloproliferative response to balloon catheter injury of the aorta in rabbits without alteration of lipidemia. This suggests a direct protective action of calcium channel blockers on the arterial wall (Jackson et al., 1988). Calcium channel blockers are found to inhibit smooth muscle cell proliferation in vessel walls and to decrease the size of the neo-intima lesion arising from catheter injury *in vivo* (Jackson et al., 1988) and *in vitro* (Nilsson et al., 1985).

The antiatherogenic property of calcium channel blockers has been demonstrated in several animal models (Kramsch et al., 1980, 1981; Henry and Bently, 1981; Ginsburg et al., 1983; Willis et al., 1985; Parmley et al., 1985). When rabbits were fed a high lipid diet, atheromatous plaques formed in the aorta. Animals treated with nifedipine had fewer plaques and fewer cholesterol deposits (Henry and Bently, 1981). However, one study did not demonstrate a protective effect with nicardipine and diltiazem (Naito et al., 1984). Corroborative observations of protective effect were made with diltiazem, verapamil and nicardipine (Willis et al., 1985; Parmley et al., 1985).

Cholesterol accumulation was studied in cultured arterial smooth muscle cells from rabbits fed an egg-supplemented, high cholesterol diet. Addition of nifedipine in the culture did not change the total cholesterol content of smooth muscle cells. However, cells from animals fed high cholesterol diets demonstrated a significant loss in cholesterol and cholesterol ester after treatment with nifedipine. An increase in cholesterol-ester hydrolytic activities was observed, suggesting that nifedipine decreases cholesterol and cholesterol-ester accumulation in arterial smooth muscle cells by increasing arterial cholesterol-ester hydrolysis (Etingin and Hajjar, 1985).

Stein et al. observed that verapamil increases LDL binding and uptake in aortic cells in culture and suggested that the stimulation of LDL uptake might promote the transfer of LDL cholesterol ester from the arterial interstitium to the intracellular catabolism compartment (Stein et al., 1985).

Calcium channel blockers might exert their anti-atherosclerotic activity by counteracting the inhibition of receptor-mediated lipid metabolism altered by calcium deposition in the cellular components of the arterial wall (Paoletti et al., 1987).

The clinical relevance of these findings is uncertain because there is no information on the prevention or treatment of atherosclerosis in man by calcium channel blockers.

III. USE OF CALCIUM CHANNEL BLOCKERS IN VASCULAR DISEASES

The calcium channel blockers, nifedipine, verapamil, diltiazem, and flunarazine have been used in the treatment of occlusive peripheral vascular disease, especially in the management of intermittent claudication of lower extremities and Raynaud's phenomenon of the upper extremities. However, it has been argued that calcium channel blockers are unable to dilate blood vessels in the chronic ischemic state from atherosclerosis.

Calcium channel blockers inhibit the platelet activation in acute ischemia that results from either arterial spasms or thrombosis and in chronic ischemia due to atherosclerosis, and normalize or reduce intracellular platelet Ca^{2+} (Ahn et al., 1987). During acute ischemic episodes, blood vessels can sustain permanent damage. Administration of calcium channel blockers in experimental animals during acute ischemic episodes protected blood vessels by inhibiting calcium influx in the ischemic state, and reduced desquamation of endothelial cells and subsequent platelet activation and aggregation in the vessel wall (Balsano, 1985; Vanhoute, 1987).

A. Intermittent Claudication

Intermittent claudication is the most common manifestation of atherosclerosis of blood vessels in the lower extremities. The symptoms of intermittent claudication are manifested after decades of progressive arterial disease. Over 51% of men over the age of 50 develop some symptoms of leg ischemia. The real incidence of atherosclerosis of blood vessels in the legs is even higher since most are asymptomatic.

Calcium channel blockers were studied in those who manifest clinical symptoms and signs as a result of long-standing progressive pathologic processes. Assessing drug therapy in this disorder, however, has been problematic. There is no practical reliable objective criterion by which to evaluate drug therapy. Most studies measured the pain-free walking distance in assessing the efficacy of calcium channel blockers.

The reports on clinical trials have recently been summarized (Dormandy, 1987). This summary includes nine parallel studies (Schuermans et al., 1971; Staessen, 1977; Schetz et al., 1978; Rudofsky et al., 1979; Barber et al., 1980; Loots et al., 1980; Di Ferri et al., 1984; Balsano, 1985; Catalano, 1985) and five cross-over studies (Adriansen, 1969; Verhaegen et al., 1974; Donald, 1975; Perhoniemi et al., 1984; Lapantalo et al., 1985), ten trials studied flunarizine-type calcium channel blockers in doses of 10 to 25 mg/day. Cinarazin was employed in six studies in doses of 75 to 225 mg/day.

Statistically significant improvement as compared to the placebo groups was observed in all studies except three (Rudofsky, 1979; Catalano et al., 1985; Lepantalo et al., 1985). Pain-free walking distance was usually increased at least two-fold compared to the placebo group. In one study, flunarazine was more effective than pentoxifylline (Perhoniemi et al., 1984).

The duration of these studies was limited to 2 to 4 months except for one study in which 12 months of follow-up was provided (Schetz et al., 1978). Therefore, the long-term effect of calcium channel blockers on this disorder has not been adequately addressed. Further study is needed to evaluate whether long term treatment with calcium channel blockers can prevent the progression of disease and avoid amputation or reduce the incidence of arterial diseases in other vessels such as myocardial infarction or strokes.

B. Raynaud's Phenomenon

Raynaud's phenomenon is caused by vasospasm of arteries and arterioles. No underlying cause is identified in primary forms although many diseases such as scleroderma, systemic lupus erythromatosis, rheumatoid diseases, trauma, drugs, etc. are responsible in secondary forms of Raynaud's phenomenon.

The frequency of attacks is extremely variable and provoked by a variety of stimuli. There is no reliable objective method to quantitate the severity of the disease.

To evaluate drug therapy in Raynaud's disorder, use of a diary by patients to record the frequency and severity of their attacks has been commonly used. Eighteen clinical trials of calcium channel blockers (Vayssairat et al., 1981; Kahan et al., 1982, 1983, 1985; Kinney et al., 1982; Smith et al., 1982; Rodeheffey et al., 1983; Winston et al., 1983; Ettinger et al., 1984; Sauza et al., 1984; Finch et al., 1985; Malamet et al., 1985; Aldoori et al., 1986; Gjorup et al., 1986a; Sarkozi et al., 1986; Waller et al., 1986; White et al., 1986; Dorbin et al., 1988) for Raynaud's phenomenon published since 1981 were recently summarized (Dormandy, 1987). All studies except three employed nifedipine: two studies used diltiazem (Vayssairat et al., 1981; Kahn et al., 1983a) and one, verapamil (Kinney et al., 1982). The dose of nifedipine was 30 to 80 mg/day (usually 60 mg/day). All except one of the 16 placebo-controlled trials have shown that nifedipine was always significantly better than placebo in subjective improvement. Additional studies (Porter et al., 1981b; Gjorup et al., 1986b) reported similar results.

Overall, nifedipine decreased the frequency and severity of attacks in 60 to 90% of patients. The benefit may be less pronounced in patients with severe Raynaud's phenomenon (Ettinger et al., 1984). Among nine trials in which the objective criteria were assessed only two found objective improvement while the rest failed to observe it in spite of subjective improvement. Nifedipine has been shown not to change either the mean cold-induced drop in digital arterial systolic blood pressure or the pulse amplitude of digital pressure as measured by photoplethysmography (Rodeheffer et al., 1983). Diltiazem in a dosage of 30 to 40 mg three times a day also has been shown to relieve symptoms of digital vasospasm in two thirds of patients (Vayssairat et al., 1981; Kahn et al., 1983). Nisoldipine was also found to be effective and may have less side effects (Porter et al., 1981). Verapamil 40 to 80 mg four times a day was not found to be beneficial subjectively or objectively when digital arterial systolic pressure, finger blood flow and skin temperature were monitored (Kinney et al., 1982).

None of the studies addressed the question of the long-term benefits of calcium channel blockers. The duration of the studies was limited to two to four months except for one study that was extended to 12 months. There are no data indicating whether or not treatment with calcium channel blockers will prevent or delay amputation of affected extremities or reduce the incidence of additional arterial occlusion such as cerebral ischemia or myocardial infarction in patients with Raynaud's disease.

C. Platelet Related Disorders

Platelets interact with the vessel wall to participate in various physiologic and pathologic processes. Calcium channel blockers have been applied to inhibit platelet activation and prevent deleterious effects arising from platelet hyperactivation in vascular disorders.

Arterial and venous thrombosis—The benefit of calcium channel blockers in arterial thrombosis has been reported especially in the management of transient ischemic attacks (Ahn and Harrington, 1987), acute stroke (Gelmers et al., 1988) and other arterial thrombosis and related disorders (Ahn et al., 1987). We have observed abnormally high Ca^{2+} levels in platelets from patients with thrombotic disorders and that administration of nifedipine normalized abnormal Ca^{2+} homeostasis (Ahn et al., 1987; Ahn and Harrington, 1987).

High levels of platelet Ca^{2+} were also observed in patients with deep vein thrombus and pulmonary emboli, lupus anticoagulation associated with collagen vascular diseases and Trousseau syndrome (recurrent venous thrombosis associated with underlying neoplasm). Administration of calcium channel blockers corrected high Ca^{2+} levels of platelets to or near normal levels, and improved the clinical courses and prevented recurrences of thrombosis (Ahn and Harrington, 1987). This suggests an important role of platelet activation in the pathogenesis of these related disorders (Ahn et al., 1987). Clinical improvements were associated with the correction of abnormal calcium homeostasis (Ahn et al., 1987).

Vasculitis—Since calcium channel blockers are expected to exert their maximum vasodilation in smaller blood vessels, in which the ratio of Ca^{2+} influx to Ca^{2+} release is higher, their application in the treatment of vasculitis is of particular interest. We have studied patients with vasculitis associated with collagen vascular diseases. The study included patients with collagen vascular disorders manifesting skin vasculitis characterized by erythematous papular eruptions with or without thrombocytopenia (Ahn et al., 1987).

Subjective and objective improvements were followed after nifedipine therapy and this was associated with a rise in platelet counts among thrombocytopenic patients and a fall in platelet CTC ratio (a measure of intracytoplasmic Ca^{2+} level) to or near normal range. The value of calcium channel blockers in small vessel disorders deserves further investigation (Ahn et al., 1987).

Platelet related disorders—In thrombotic thrombocytopenic purpura (TTP), platelets aggregate in the microvasculature and disturb blood flow in microcirculation. Elevated platelet Ca^{2+} levels were observed in patients with TTP and some patients with ITP (idiopathic thrombocytopenic purpura) in which platelets coated with autoantibody are destroyed by macrophages of mononuclear phagocyte system. Platelet calcium homeostasis may be altered in these settings due to platelet aggregating factors in TTP or autoantibodies in ITP, which alter platelet membrane and thus lead to abnormal calcium homeostasis. Therapy with calcium channel blockers prevented recurrence of TTP and improved thrombocytopenia in some patients with immune thrombocytopenia in the stable phase of their disease and tended to correct the abnormality of platelet calcium homeostasis, even though their correction was not as complete (Ahn et al., 1987).

Thrombocythemia—Thromboembolic complications are common in patients with thrombocythemias, especially those associated with myeloproliferative disorders. Various intrinsic platelet abnormalities have been described. We observed extreme variation in CTC ratio, some with high and others with subnormal values indicating aberrant calcium homeostasis of platelets. The symptoms and signs of thrombocythemia such as numbness, tingling, pain and discoloration of digits often improved with administration of nifedipine and other calcium channel blockers. CTC ratios returned toward the normal range with administration of calcium channel blockers (Ahn et al., 1987; Ahn and Harrington, 1987).

IV. CONCLUSION

Calcium channel blockers have multiple actions on platelets, blood vessels and blood flow. Promising short-term clinical benefits of calcium channel blockers have been described in the treatment of vascular disease, especially intermittent claudication, Raynaud's phenomenon and vasculitis. However, long-term benefit in these disorders has not been studied systematically. The important areas of future research of calcium channel blockers in vascular disorders are to further explore their antiatherogenic property in human diseases and antiplatelet actions in vascular diseases and platelet related disorders.

Chapter 21

Calcium Channel Blockers in Neurological Disorders

Gary A. Rosenberg

TABLE OF CONTENTS

I. Introduction 378

II. Normal Calcium Transport and Metabolism 378

III. Pathological Calcium Accumulation 381

IV. Calcium Channel Entry Blockers 383

V. Conclusions 384

I. INTRODUCTION

Calcium is crucial in many aspects of neuronal and glial cell function (Siesjo and Bengtsson, 1989). It is maintained at low concentrations within the cell, where it links membrane activity to cellular metabolism (Berridge, 1988). Four major factors control its movement between the extracellular and intracellular compartments: the membrane potential; the extracellular concentration of excitatory amino acid transmitters; specific membrane receptors linked to second messengers; and the Na^+ concentration. In addition, there are cellular mechanisms that lead to its release from storage sites within the endoplasmic reticulum and the mitochondria. Calcium signaling is a normal aspect of cell function. However, when increases in Ca^{2+} concentration are excessive within the cell, pathological changes are brought about that can lead to the cell's destruction (Farber et al., 1981a; Cheung et al., 1986; Deshpande et al., 1987).

Calcium channels are generally thought of as voltage-regulated, highly selective channels (see Section I). However, in the nervous system another important class of channels capable of passing Ca^{2+} ions are ligand-gated channels sensitive to excitatory amino acids. These channels are important in regulating intracellular Ca^{2+} concentration in injury. Receptor-operated mechanisms linked to phosphoinositol metabolism also are involved in control of intracellular Ca^{2+} concentrations. Since both excitatory amino acid-gated channels and second-messenger systems have a potential role in cell injury, they will be discussed here.

Calcium channels are an important route through which Ca^{2+} can move into cells, and through which Ca^{2+} is shifted into the cytoplasm from sequestered sites in the endoplasmic reticulum and mitochondria. Calcium channels are regulated either by membrane potential or by ligands bound to receptors. In the voltage-operated channel gates are opened by changes in membrane potential and thus permit the movement of Ca^{2+} across the membrane. In the receptor-operated channel gates are responsive to extracellular levels of the excitatory amino acids, glutamate, and aspartate. In addition to voltage and receptor-regulated channels, there are mechanisms to increase intracellular Ca^{2+} that are linked to second messengers (see Chapter 18, Section III). Substance P, histamine, and vasopressin are examples of substrates that probably act by activation of a second messenger-operated mechanism (Berridge, 1988).

Under normal circumstances, intracellular Ca^{2+} levels in CNS cells stay within a narrow range. Pathological changes occur when intracellular levels become elevated (Figure 1). Injury to brain cells leads to a loss of potential across the membrane with a resultant influx of Ca^{2+} via the voltage-operated channels. Injury also leads to increases of excitatory amino acids in the extracellular space with a consequent opening of receptor-operated channels. As a result of activation of second-messenger systems linked to phosphoinositol metabolism further increases occur in the level of free Ca^{2+} in the cell. Accumulation of excessive amounts of Ca^{2+} within cells leads to their death (Farber, 1981b). Calcium influx into cells appears to be a common feature in a number of pathological processes, including stroke, trauma, hypoglycemia, and epilepsy (Siesjo and Bengtsson, 1989). Because of the central role of Ca^{2+}, there is a great deal of interest in the normal function of Ca^{2+} in the brain, in the calcium-related events associated with neurological diseases, and in potential therapeutic agents that can act to block increases of free cytoplasmic Ca^{2+} in the neuron.

The following discussion will describe the normal role of Ca^{2+} ions in cellular metabolism, the changes that occur in cell injury, and the pharmacologic attempts to block Ca^{2+} concentration changes within the cell.

II. NORMAL CALCIUM TRANSPORT AND METABOLISM

The ionic balance in the brain is maintained by a series of interfaces between the blood and the brain collectively referred to as the blood-brain barrier (Bradbury, 1979). The primary

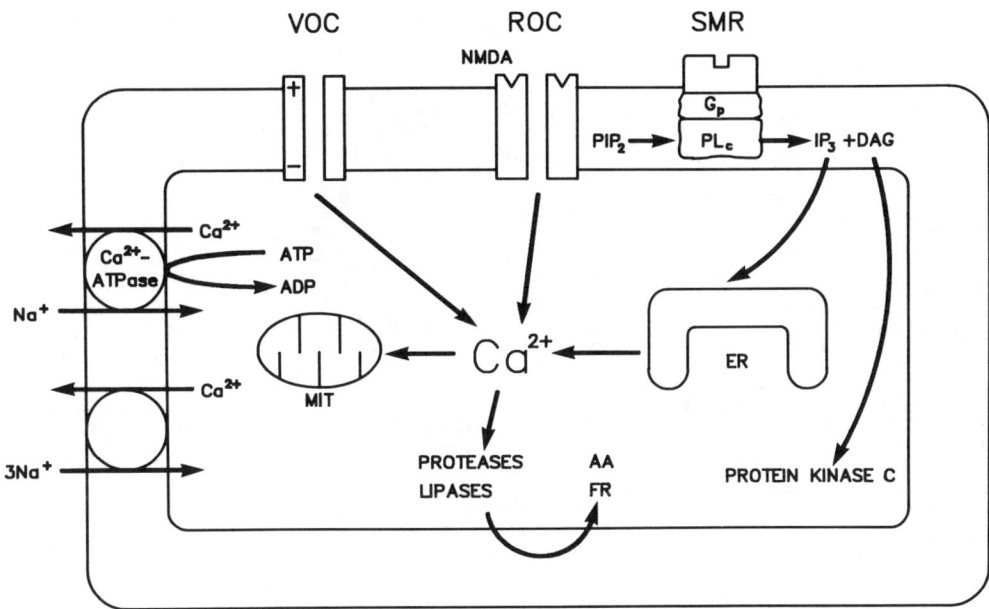

Figure 1. Diagram of calcium-regulating mechanisms in neurons. Voltage-operated channels (VOC) allow Ca^{2+} to enter when membranes are depolarized. One type of receptor-operated channel (ROC) is responsive to glutamate (NMDA). Second messenger-receptor operated channels (SMR) are linked to phosphoinositol metabolism by phospholipase C (PLc); metabolism of phosphatidylinositol 4,5-bisphosphate (PIP_2) indirectly increases intracellular calcium by generation of inositol 1,4,5-triphosphate (IP_3) and diacylglycerol (DAG). IP_3 releases Ca^{2+} from the endoplasmic reticulum (ER) and DAG activates protein kinase C. Ca^{2+} is removed by Ca^{2+}-ATPase pump and Na^+/Ca^{2+} exchange. Excess Ca^{2+} can activate proteases and lipases that release arachidonic acid (AA) from membrane fatty acids and form free radicals (FR). Ca^{2+} is sequestered in the mitochondria (MIT).

function of these interfaces is to prevent ions, proteins, and other lipid insoluble substances from entering the brain (Rosenberg, 1990). Tight junctions between cells are found at each of the interfaces, providing the anatomical basis for the blood-brain barrier (Reese and Karnovsky, 1967). The first interface is between the blood and the cerebrospinal fluid (CSF) at the choroid plexus (Figure 2). Epithelial cells of the choroid plexus secrete the major portion of the CSF. These cells are joined together by tight junctions. The tight junctions at the interface between the CSF and the venous blood are formed by arachnoid cells, which line the surface of the brain and protrude into the venous sinuses as the arachnoid granulations. Tight junctions are located between the endothelial cells of cerebral capillaries (Goldstein and Betz, 1983). Because of their large number and surface area, they form the main site for the blood-brain barrier. In addition to the barrier formed by tight junctions, a complex series of transport functions and enzymes act to provide selective transport between the blood and the CSF. The capillary performs these complex transport functions at the expense of energy. Cerebral capillaries, as opposed to those found systemically, have a large number of mitochondria, few pinocytotic vesicles, and a series of transport molecules that bring glucose and amino acids into the brain (Oldendorf et al., 1977).

Ca^{2+}, Mg^{2+}, and K^+ are exchanged freely between the CSF and brain interstitial fluid, and are normally kept at a level in the brain interstitial fluid lower than that found in the blood (Woodbury et al., 1968). Ca^{2+} levels in the CSF normally range between 2 and 3 mEq/l compared to plasma levels of 4 to 5.5 mEq/l (Katzman and Pappius, 1973). The levels in the CSF are thought to represent those found in the interstitial fluid of brain, since the two fluids are in continuity. Ca^{2+} is actively secreted into the CSF by the choroid plexus and the rate of Ca^{2+} entry from blood to CSF is relatively independent of the serum Ca^{2+}. The ratio of CSF to serum Ca^{2+} in man is around 0.50 (Woodbury et al., 1968). The low

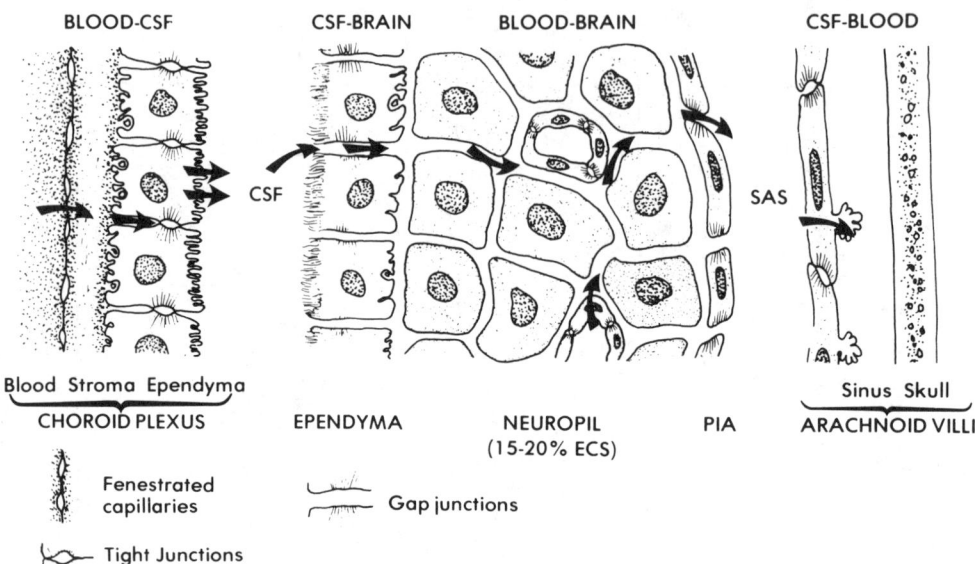

Figure 2. Various interfaces between blood, brain, and CSF. Tight and gap junctions are shown schematically. Arrows indicate formation of CSF at the choroid plexus and interstitial fluid (ISF) at the brain capillary and transport of CSF-ISF through the extracellular space for eventual absorption back into blood at the arachnoid villi. ECS is extracellular space (Rosenberg et al., 1983, with permission).

levels of CSF Ca^{2+} are maintained by the inability of the charged ion to cross the blood-brain barrier, and by transport mechanisms that move Ca^{2+} from the CSF to the blood.

Ca^{2+} concentrations in the brain are insensitive to acute or chronic changes in plasma levels. Fluctuations of plasma Ca^{2+} from 1 to 7 mM/l in dogs changed CSF calcium only from 1 to 2 mM/l (Katzman and Pappius, 1973); similarly, the concentration of the ion in brain tissue remains constant during acute changes in the blood (Bradbury and Wong, 1972). Diet has little effect on brain levels of the ion; young rats fed diets low or high in Ca^{2+} showed a 40% fall or a 30% rise, respectively, in total plasma Ca^{2+}, but brain levels remained within 10% of those in controls (Murphy et al., 1986). Ca^{2+} transport can occur across either the choroid plexuses or the cerebral capillaries. Recent studies suggest that transport across the choroid plexus is the dominant route for Ca^{2+} transport from blood to brain (Tai et al., 1986). Regulation of Ca^{2+} levels in the brain is essential for normal brain function. Elevated levels of the ion in the brain interstitial fluid produce confusion and coma, while low levels produce seizures (Katzman and Pappius, 1973).

Extracellular levels of free Ca^{2+} are higher than those found in the cell. Ca^{2+} within the cell is sequestered in mitochondria and smooth endoplasmic reticulum and is buffered by intracellular proteins. Ca^{2+} enters cells through either voltage- or receptor-operated channels. Three types of channels sensitive to voltage changes have been identified in brain; they are termed, L or long-lasting, T or transient, and N for neither L or T (see Chapter 3, Section I). The L-type channel is blocked by the group of agents termed calcium-channel entry blockers (Siesjo and Bengtsson, 1989). Under normal conditions, levels of free Ca^{2+} in the cytosol are in the nanomolar range. When a membrane channel is opened by a voltage change, the ion enters the cell from the extracellular compartment, down its concentration gradient.

The second mechanism by which calcium ions are moved across the membrane involves the receptor-operated channel that is sensitive to the excitatory amino acids (Choi, 1985). Several types of post-synaptic Ca^{2+}-passing channels that are sensitive to glutamate have been identified. Although they are all activated by glutamate, they have been identified by

TABLE 1
Neurotransmitter Receptors Linked to Phosphoinositide Metabolism in Neuronal Tissue

Neurotransmitter	Receptor
Noradrenaline	α_1-Adrenoceptor
Acetylcholine	Muscarinic
5-Hydroxytryptamine	5-HT
Histamine	H_1
Glutamate	Quisqualate-like
Bradykinin	B_2
Neurotensin	—
Vasopressin	V_1
Substance P and other tachykinins	NK_1, NK_2, NK_2

Modified from Nahorski, 1988.

the substance to which they are most sensitive. Three subtypes have been found, namely, N-methyl-D-aspartate (NMDA), kainate, and quisqualate (Nowak et al., 1984). At present the NMDA receptor has attracted the most attention, because it is the most prevalent type. It is blocked by Mg^{2+} in a voltage-dependent manner, and antagonists with therapeutic potential have been developed. Glutamate and aspartate are the normal excitatory amino acid neurotransmitters that are found in high concentrations in brain tissue. Glutamate is the most abundant amino acid in brain, but only a small proportion is involved in neurotransmission. Cell depolarization results in a rapid increase in the amount of glutamate in the extracellular space (Benveniste et al., 1984).

The third mechanism for elevating free cytoplasmic Ca^{2+} does not involve entry of the divalent ion from the extracellular space. Rather a second messenger is involved in the release of Ca^{2+} from intracellular organelles. Several different receptors are linked to second messengers through the phosphoinositol pathways (Berridge, 1988). Phosphoinositol forms only a small percentage of the membrane phospholipid. It is attached to the membrane as a fatty acid in the triglyceride molecule. When an agonist attaches to the membrane receptor linked to phosphoinositol, a G-protein is activated and triggers phospholipase C in the membrane to release $Ins(1,4,5)P_3$, (D-myoinositol(1,4,5)-trisphosphate), and DAG, (sn(1,2)-diacylglycerol). $Ins(1,4,5)P_3$ releases Ca^{2+} from endoplasmic reticulum, while DAG activates protein kinase C. Other neurotransmitters may act through a similar mechanism to increase intracellular Ca^{2+} (Table 1).

III. PATHOLOGICAL CALCIUM ACCUMULATION

In a number of tissues, the toxic role of Ca^{2+} in cell death is well established (Farber, 1981b; Cheung et al., 1986). The main factor is an excessive increase in the level of the free divalent ion in the cytoplasm. Increased Ca^{2+} ion levels cause a series of damaging intracellular events. Therefore, treatment has been aimed at reducing the influx of the ion into the cell. In brain tissue, Ca^{2+} has been associated with damage in ischemia, hypoglycemia and epileptic seizures (Simon et al., 1984a; Simon et al., 1984b; Deshpande et al., 1987; Benveniste et al., 1988).

Intracerebral microdialysis revealed an increase in the levels of glutamate in the extracellular space during ischemia (Benveniste et al., 1984). Anoxia produced a rise in extracellular K^+ causing a depolarization of the cell and a marked fall in extracellular Ca^{2+} concentration (Nicholson et al., 1977). Removal of excess intracellular Ca^{2+} from the cell

TABLE 2
Cellular Events Triggered by Excess Intracellular Calcium

Activation of proteases and lipases
Arachidonic acid release from membrane
Free radical formation
Energy loss from ATPase use

From Cheung et al., 1986.

is accomplished by uptake into the mitochondria and endoplasmic reticulum, Ca^{2+}-Na^+ exchange, and Ca^{2+}-ATPase-mediated transport. When Ca^{2+} accumulates in cells, a dangerous situation is created with activation of lipases that are active in cellular degradative processes (Table 2).

While the main group of calcium channel blockers studied acts on the voltage-sensitive channels, another group of important agents has recently been studied extensively. These agents block the calcium channel associated with the excitatory amino acids. The Ca^{2+}-passing channel associated with the NMDA receptor is blocked by Mg^{2+}, ketamine, phencyclidine, and drugs, such as, dizocilipine maleate (MK801; Merck, Sharp and Dohme Laboratories). The NMDA channel is strongly antagonized by Mg^{2+} which may be important in its therapeutic action in eclampsia, alcohol withdrawal, and spinal cord injury (Mayer et al., 1984; Nowak et al., 1984). Although Ca^{2+} entry is a normal consequence of cell excitation by glutamate, excessive glutamate stimulation, as is thought to occur in pathological situations, allows more Ca^{2+} into the cell than can normally be handled by the uptake and extrusion processes. There is a resultant activation of cellular processes that are detrimental to the cell with subsequent breakdown of cellular membranes and the formation of products of inflammation (Rothman and Olney, 1986).

Cell damage from excess Ca^{2+} is thought to occur in several stages. Initially the entry of Ca^{2+} causes the cell to swell, probably from the movement of Na^+ into the cell in exchange for Ca^{2+} (Barcenas Ruiz et al., 1987). Thus, by exchanging with Na^+, Ca^{2+} is removed to some extent from the cytoplasm of the cell. The influx of Ca^{2+}, Na^+ and Cl^- results in the osmotic movement of water. The initial phase of injury, therefore, appears to be related to osmotic changes in the cell. More than likely this is a reversible phase.

Later stages in the cell injury process involve the accumulation of Ca^{2+} within the cell and the inability of cellular mechanisms to remove the excess ion. In a recent study of cerebral ischemia in a rat model, Ca^{2+} accumulation was associated with neuronal necrosis (Deshpande et al., 1987); after 10 min of forebrain ischemia in rat, Ca^{2+} concentrations in CA1 pyramidal cells of the hippocampus were significantly increased at 48 h, while cell necrosis was most prominent at 72 h. Therefore, the delayed cellular death followed the Ca^{2+} accumulation. In another study using the same model of 10 min forebrain ischemia, the accumulation of Ca^{2+} in CA1 neurons was blocked by sectioning the glutaminergic input to CA1 from CA3 (Benveniste et al., 1988). The deafferentation of CA3 neurons completely blocked the ischemic nerve damage, supporting the concept that Ca^{2+} influx via glutamate-operated channels was important in cell death.

The mechanism of calcium-mediated injury to cells is only partially understood. Calcium ion mobilizes lipases and proteases that initiate the attack on cell membranes, leading to irreversible damage (Cheung et al., 1986). One important consequence of Ca^{2+} accumulation in the cell is that phospholipase A_2 frees fatty acids from membrane lipids. The free fatty acids, primarily arachidonic acid, injure cells by interfering with the action of Na^+, K^+-ATPase and by forming free radicals (Siesjo, 1984).

TABLE 3
Major Classes of Calcium Entry Blockers

Class	Drug
Dihydropyridines	Nifedipine, nimodipine
Phenylalkylamines	Verapamil, (S)-emopamil
Benzothiazepines	Diltiazem
Diphenylalkylamines	Flunarizine

IV. CALCIUM CHANNEL ENTRY BLOCKERS

The major types of calcium channel blockers that work at the voltage-operated channels have been discussed in Chapter 13, Section III and are shown in Table 3. The use of calcium channel blockers has shown promising results in the treatment of ischemia and hypoxia. Many of the calcium channel blockers currently used for treatment of heart disease are ineffective in the brain, because they cross the blood-brain barrier too slowly. There are, however, several agents that enter brain tissue when given systemically. Two such agents are the experimental agent, PN 200-110 (Sandoz Corporation), and an agent already released for clinical use, nimodipine. PN 200-110 was shown to reduce the lesion evolution as seen on magnetic resonance imaging in a model of stroke in which the middle cerebral artery was occluded in spontaneously hypertensive rats that are prone to develop strokes (Sauter and Rudin, 1987). Nicardipine had a similar effect (Kucharczky et al., 1989). Treatment either before or after cerebral artery occlusion, with both antagonists significantly decreased the infarct size and improved the behavioral performance. PN 200-110 slowed the loss of high energy phosphates as monitored by ^{31}P-NMR in a rat model of global ischemia (Sauter and Rudin, 1987). Although the effect reached statistical significance, it was small and the clinical relevance remains to be determined. The potential for these agents to produce hypotension and the need to give them prior to, or shortly after, the onset of the infarct may limit their clinical efficacy.

Another calcium channel blocker, flunarizine, reduced infarct size in rats given the drug orally 3 h prior to the insult (Van Reempts and Borgers, 1985). The mechanism of flunarizine action may be to delay the Ca^{2+} influx in neuronal and astrocytic cells. Flunarizine readily penetrates the blood brain barrier, where it would have access to brain cells. The drug has multiple other actions on endothelial cells, platelets, and red blood cells; which action is responsible for the protective effect remains to be determined.

Clinical trials of nimodipine in acute ischemic stroke showed a beneficial effect of early treatment in humans (Gelmers et al., 1988). A prospective, placebo-controlled trial of nimodipine (30 mg every 6 h) was done in 186 patients. Treatment was begun within 24 h of the acute ischemic stroke. Fewer deaths occurred in the patients given nimodipine than in those given placebo. The studies are currently being repeated in England and the United States.

In subarachnoid hemorrhage, nimodipine was shown to reduce the incidence of vasospasm and to improve clinical outcome (Petruk et al., 1988). Although the drug is thought to prevent vasospasm, angiographic evidence for this is lacking, and its mechanism of action is unclear. However, the number of strokes in patients receiving the drug was less than in a control group. The drug has been approved by the FDA for use in patients with subarachnoid hemorrhage.

The agents in the phenylalkylamine class have a protective effect in experimentally induced ischemia in rats. A new agent, (S)-emopamil, is highly lipid soluble and readily crosses the blood-brain barrier. In a controlled trial, it was shown to reduce infarct size in rats with middle cerebral artery occlusion; the agent was effective when given prior to treatment and within 1 h after infarction (Nakayama et al., 1988). (S)-emopamil blocks serotonin S_2 receptors. Serotonin is a vasoconstrictor which appears to constrict vessels in subarachnoid hemorrhage. The combination of a calcium channel blocking action with high lipid solubility and a serotonin S_2 receptor antagonist action makes this drug an attractive agent for clinical trials.

Blockade of the influx of Ca^{2+} with the use of NMDA channel blockers has been well studied. The experimental agent, MK801, blocks the NMDA channel, and is thought to prevent the inflow of Ca^{2+}. In gerbils, with bilateral carotid occlusion for 5 min, the injection of MK801 1 h prior to occlusion protected from CA1 damage (Gill et al., 1987). In an experimental study of ischemia in the rat the direct injection of this NMDA channel blocker into the hippocampus was protective against ischemic injury (Simon et al., 1984b). Although not all investigators have found the same protective effects in different stroke models, there is a growing consensus that blockade of the NMDA channel may be protective against stroke and trauma.

Traumatic brain injury has recently been shown to increase extracellular concentrations of glutamate and aspartate in the hippocampus (Faden et al., 1989). When the NMDA receptor antagonists, dextrorphan or 3-(2-carboxypiperazin-4-yl)propyl-1-phosphoric acid, were given shortly after the injury, there was improvement in bioenergic state and behavior. Free Mg^{2+} content was increased. NMDA receptor blockers may be useful in treatment of traumatic injury to brain.

V. CONCLUSIONS

Ca^{2+} appears to play an important role in cell injury. Normally maintained at very low intracellular levels, it enters cells in small amounts during normal stimulation, and in massive amounts in various types of injury. Several Ca^{2+}-passing channels are active in the brain cell membranes. Voltage-operated channels can be blocked with drugs of the dihydropyridine group. Receptor-operated channels sensitive to glutamate and aspartate are blocked with drugs, such as, MK801 and dextrorphan. Calcium channel blocking agents have been shown to be useful in the treatment of ischemia, hypoxia, trauma and subarachnoid hemorrhage. Further study of this diverse and interesting group of agents is underway in a large number of laboratories, and promising agents for use in treatment of patients should continue to be produced.

Chapter 22

Use of Calcium Channel Blockers in Psychiatric Disorders

M. Dose and H. M. Emrich

TABLE OF CONTENTS

I.	Introduction	386
	A. Association of Psychiatric Disorders with Altered Calcium Function	386
II.	Psychiatric Patients	387
	A. Serum and Cerebrospinal Fluid (CSF) Ca^{2+} Levels Related to Behavioral and Mood Changes	387
	B. Calcium-Transport Enzyme Activities in Erythrocytes and Platelets	388
III.	Clinical Effects of Calcium Channel Blockers in Psychiatric Patients	390
IV.	Possible Mechanism of Action	391
	A. Lithium-Calcium Interactions	391
	B. Calcium Channel Blockers and Anticonvulsants	392
	C. Calcium Channel Blockers and Neuroleptics	393
V.	Summary and Conclusions	393

I. INTRODUCTION

A. Association of Psychiatric Disorders with Altered Calcium Function

The occurrence of psychopathological alterations in the course of abnormal Ca^{2+} regulation was first thoroughly investigated by Frankl-Hochwart (Austria) in 1907. His observations, along with those of Barrett were reviewed in 1920 (Barrett). Since then several case reports of psychiatric disturbances during abnormal Ca^{2+} regulation (Popea and Demetresco, 1930a, 1930b; Barr et al., 1938; Knospe, 1938) have been published without any information on the frequency of the psychiatric symptoms. In 1941 Greene and Swanson reported the occurrence of psychiatric symptoms in 5 of 18 patients with hypoparathyroidism during a 9 year period of observation. The psychopathological features in these patients were described as anxiety, depression, a sense of impending disaster, delusions and hallucinations. It was demonstrated that the appearance of psychopathologic symptoms often preceded other clinical signs of parathyroid malfunction and that their disappearance took place one to four weeks after normalization of serum Ca^{2+} levels was achieved by substitution therapy. One patient, after undergoing a subtotal thyroidectomy, developed symptoms of delusional depression, which subsided after normalization of serum Ca^{2+}, during the course of substitution therapy with Ca^{2+} and dihydrotachysterol. However, the symptoms recurred when hypercalcemia developed during continued treatment. Similar psychopathological symptoms associated with altered serum Ca^{2+} levels have also been reported by other authors (see below), and may express the lack of specificity of the effects of hypo- and hypercalcemia.

In one case, 12 years after a total thyroidectomy, the patient began exhibiting intermittent paranoid projections and delusions accompanied by depressive or excited mood shifts, in addition to tetanic and epileptiform seizures (Scarlett and Houghtling, 1944). These symptoms subsided within one week after parathyroid extract was administered. In a review by Bartter (1953), hypersensitivity, apprehensiveness, fatigability, fear, eroticism, extreme irritability, apathy, sluggish speech, exhilaration and talkativeness are summed up as psychiatric features of hypoparathyroidism. The improvement of these symptoms after normalization of serum Ca^{2+} is considered to be indicative of a close relationship between hypocalcemia and psychopathology (Bartter, 1953).

The same review points out that hyperparathyroidism is far less frequently accompanied by psychotic features and that, if they do occur, they can hardly be distinguished from the symptoms of uremia following renal damage that may occur in hyperparathyroidism. However, cases have been described in which psychotic symptoms (e.g., schizophreniform) accompanied hyperparathyroidism without any renal damage, and underwent complete remission after removal of a parathyroid adenoma (Fitz and Hallman, 1952).

An extensive review of psychopathology in hyperparathyroidism and disturbances of Ca^{2+} metabolism, including his own observations, has been published by Petersen from Bleuler's clinic in Switzerland, the "Burghoelzli" (Petersen, 1967). In 1417 cases of hyperparathyroidism that were reviewed he found psychiatric disturbances (loss of drive, fatigue, affective and psychotic symptoms) in 2 to 51% of the patients. His own investigations in 416 patients with hyperparathyroidism revealed abnormalities including acute psychoses in 47% of the patients.

Other studies (Kind, 1959; Karpati and Frame, 1964) have shown that 42 to 67% of patients with hyperparathyroidism exhibit psychiatric symptoms. Again, the point was made that the uniformity of endocrine disorders stands in sharp contrast to the variety of psychopathological changes these disorders can produce (Petersen, 1967). Since in Petersen's sample the psychiatric symptoms of 6 patients with nonendocrine hypercalcemia closely paralleled those of hypercalcemic patients with hyperparathyroidism, he concluded that a disturbance of calcium metabolism, rather than the endocrine dysfunction was the common denominator in these disorders. These findings have recently been confirmed in patients with primary hyperparathyroidism who underwent psychiatric examination prior to and one year after surgery. Compared to healthy volunteers, patients with hyperparathyroidism scored statistically significantly higher in the CPRS (Comprehensive-Psychopathological-Rating-Scale; Äsberg et al., 1978), a scale covering psychopathological items with operationally defined steps between 0 (absence of the particular symptom) and 3 (extreme degree of the symptom). Pre-operatively 18 of 32 patients spontaneously reported psychiatric problems, 8 received psychotropic medication (5 antidepressant, 2 antipsychotic, 1 nootropic, medication that increases intellectual capability). Among the psychic symptoms, anxiety, mild depression, fatigue and psychosomatic complaints were most prominent. At follow-up (one year after operation) there was a statistically significant improvement of psychopathology in 27 of 32 patients with hyperparathyroidism (Jaborn et al., 1988).

II. PSYCHIATRIC PATIENTS

A. Serum and Cerebrospinal Fluid (CSF) Ca^{2+} Levels Related to Behavioral and Mood Changes

Although the evaluation of psychopathology in patients with abnormal Ca^{2+} regulation led to consistent conclusions that there *is* a link between hypo- and (to a minor extent) hypercalcemic states and a number of nonspecific psychiatric symptoms, the reverse approach, in which Ca^{2+} metabolism was investigated in psychiatric patients, has revealed only contradictory and conflicting results. This is not too surprising if one bears in mind the complex mechanisms involved in maintaining constant extra- and intracellular calcium concentrations. In order to link psychiatric symptoms to calcium metabolism, one should be able to determine free (not protein bound) intracellular Ca^{2+} concentration in nerve cells. The free intracellular Ca^{2+} serves as a "second messenger" that contributes to neuronal modulation. Many factors, such as the large difference between extra- and intracellular calcium concentration (up to 10^4-fold), the control of transmembranal Ca^{2+} fluxes by voltage- or transmitter-regulated calcium channels, and active transport and sequestration of intracellular Ca^{2+} serve to diminish the likelihood that correlations will be found between changes of calcium in the serum or CSF and psychiatric disorders. In addition, serum Ca^{2+} (especially the ionized "free" fraction that is physiologically active) is influenced by the concentration of plasma proteins, ingestion, resorption and excretion of Ca^{2+}, and endocrine factors (Crammer, 1977).

In 1922 Weston and Howard analyzed serum- and CSF-Ca^{2+} levels in 17 manic and 10 depressed patients (Weston and Howard, 1922). Although several authors have since related their work to the "positive" findings of these authors (Carman and Wyatt, 1979; Jimerson et al., 1979; Dubovsky and Frank, 1983), Weston and Howard reported that there were no differences in the serum- and CSF-Ca^{2+} ion concentrations that could be related to psychiatric symptomatology. Wagner-Jauregg who had received the Nobel prize in 1927 for his work on "malaria-therapy" for the treatment of progressive paralysis, initiated an investigation of serum Ca^{2+} in psychiatric patients from which lowered calcium levels in depressed and elevated Ca^{2+} levels in manic patients were reported (Klemperer, 1927). While increased serum calcium in depression has been found in subsequent studies (Petersen, 1968; Carman and Wyatt, 1979; Jimerson et al., 1979) it could not be confirmed by others (Bjørum, 1972).

Hypocalcemia was observed in some manic patients (Aronoff et al., 1971) during cyclic mood shifts and hyperactivity (Fourman et al., 1967). Small reductions in serum Ca^{2+} accompanying recovery from depression have been reported following treatment with tricyclic antidepressants, lithium and electroconvulsive therapy (Carman et al., 1977; Carman and Wyatt, 1979; Jimerson et al., 1979).

Some investigators have found CSF-Ca^{2+} to be elevated in depressed patients (Harris and Beauchemin, 1956; Hakim et al., 1975) although other investigators (Bjørum, 1972; Bech et al., 1978) have found it to be within normal range. In one study (Jimerson et al., 1979) the authors found a positive correlation between CSF-Ca^{2+} concentration and symptom severity in hospitalized depressed patients and a decrease of CSF-Ca^{2+} with improvement of symptoms, although mean CSF-Ca^{2+} for the depressed patients as a group did not differ significantly from neurological controls or other psychiatric patients. In the same study, CSF-Ca^{2+} was significantly increased during remission of schizophrenic patients from acute schizophrenic psychoses. The finding that a decrease in CSF-Ca^{2+} accompanied the successful treatment of depressed patients with tricyclic antidepressants or electroconvulsive therapy (Carman et al., 1977; Jimerson et al., 1979) could not be replicated in a study carried out by Mellerup et al., (1979) using a similar design. The immediate decrease in serum- and CSF-Ca^{2+} after electroconvulsive treatment of depressed patients (and in patients suffering from epileptic seizures) was attributed to two factors by these authors.

1. Within 5 min after seizures of any kind there is an increase of plasma proteins followed by a slow decrease to normal values. Thus the serum Ca^{2+} concentration needs to be corrected for the differing concentrations of buffering proteins. If this were done, only a decrease of serum Ca^{2+} at least 5 min after electroconvulsive treatment remained significant.
2. The differences in CSF-Ca^{2+} before and after electroconvulsive treatment may be due in large measure to seizure-induced increase in vascular space.

Beyond this one must, with regard to the contradictory results of CSF studies, consider the high stability of CSF-Ca^{2+} under pathological conditions. Although accompanied by an increase of serum Ca^{2+}, no changes of CSF-Ca^{2+} concentrations could be found in patients with parathyroidal disorders (Balabanov et al., 1984) as well as in patients with acute or chronic hypercalcemia of different etiology (Goldstein et al., 1980; Balabanov et al., 1984).

B. Calcium-Transport Enzyme Activities in Erythrocytes and Platelets

In order to overcome the methodological problems of relating psychiatric disorders to calcium serum- or CSF-Ca^{2+} levels, attempts have been made to determine intracellular Ca^{2+} concentration by measuring the activity of calcium transport enzymes, especially the adenosinetriphosphatases (ATPases). This is important because of the role of Ca^{2+} as an intracellular messenger critical to the release of neurotransmitters (see Section II, Chapter 7). These ATPases are the biochemical correlates of an active transmembrane transport of ions against their electrochemical gradient by "ion pumps". Regulated in their activity by the concentration of Na^+, K^+, and Mg^{2+} in the case of sodium-potassium transport (Na,K-ATPase) or Ca^{2+} and Mg^{2+} in the case of Ca,Mg-ATPase, the ATPases promote the hydrolysis of ATP (adenosinetriphosphate) in order to supply energy for active ion transport (for more detailed information see Section IV, Chapter 21). Most of the studies referred to in this chapter were done on red blood cells, in which a calcium-magnesium dependent ATPase actively removes Ca^{2+} from the cytosol. Following a suggestion by Naylor (Naylor et al., 1973) that depression involves a general reduction in Na,K-ATPase activity, Hesketh and co-workers (Hesketh et al., 1977) compared activities of several ATPases in erythrocyte membranes of patients suffering from depressive illness and those of controls. They defined ATPase activity in the presence of magnesium (6 mM) as the only cation as "Mg-ATPase" activity, "Na,K-ATPase" activity as the difference between activity in the presence of Mg^{2+} alone and the activity in the presence of Mg^{2+}, Na^+ (100 mM), and K^+ (5 mM), and "Ca,Mg-ATPase" activity as the activity in the presence of Mg^{2+} and Ca^{2+} (0.15 mM). Although they could confirm a significantly lower level of Na,K-ATPase-specific activity in the group of patients when compared to controls, no difference in magnesium- or calcium-magnesium ATPase-specific activity between depressed patients and controls could be found. At the same time Choi et al., 1977, reported that the mean erythrocyte Ca^{2+}-ATPase level of depressed patients was 72% of that of a control group prior to electroconvulsive therapy and rose to the levels of the control group after treatment. These findings, however, have been criticized because only single blood samples were drawn by the investigators and the values for different ATPases were 10 times higher than those usually reported in the literature (Linnoila et al., 1983). In their own longitudinal study of red blood cell Na,K-ATPase and calcium-ATPase activities in 8 patients with affective disorders and 12 healthy volunteers, Linnoila and his co-workers found higher mean Ca-ATPase activity in patients with affective disorders but no consistent correlation between mood and ATPase activities. Within the group of patients, Ca-ATPase was higher during mania and euthymia (state without affective disturbance) than during depression (Linnoila et al., 1983). This observation possibly parallels the findings that an increase in Ca-ATPase activity occurs after electroconvulsive treatment (Choi et al., 1977) or after treatment with the tricyclic antidepressant imipramine or nomifensine (Rybakowski et al., 1981) in depressed patients. Differences between the activity of calmodulin-activated Ca-ATPase of erythrocyte membranes in solutions containing LiCl (50 mM) in manic-depressive patients (independent of whether or not treated with lithium) and controls have been interpreted as indicating an abnormality of calmodulin-activated Ca-ATPase in these patients (Meltzer and Kassir, 1983). However, the Li^+ concentrations of 50 mM that were used to activate Ca-ATPase in this study exceed therapeutic levels in patients by approximately 100-fold and only single blood samples were drawn from an outpatient population. In studies using single blood probes, the determination of Ca-ATPase activity in red blood cells of large populations uncovered a bimodal distribution of enzyme activities that may express phenotype-dependent differences between enzymes and their activities as well as diurnal fluctuations (Reinila et al., 1982). Additionally, red blood cells cannot be compared to cells of the CNS because they lack a Ca^{2+}-dependent neurotransmitter release mechanism and may have different ATPases than do neurons (e.g., a direct calcium pump). Therefore, attempts have recently been made (Bowden et al., 1988) to use platelets as a tool to investigate Ca^{2+} metabolism, because of their greater similarity to neurons. According to some investigators, platelets possess metabolic functions virtually identical to those in nerve cells and these functions account for the uptake and storage of the neurotransmitter serotonin. Platelets and nerve cells also have identical intracellular Ca^{2+} concentrations, which increase to 1 to 10 mM when the cells are activated. The elevation in the intracellular Ca^{2+} ion concentration serves to activate many cellular processes including neurotransmitter release (Bowden et al., 1988). The sequestration of free cytosolic Ca^{2+} within the smooth endoplasmic reticulum of nerve endings by a Ca-ATPase also takes place in similar organelles (the dense tubular system) in platelets. Based on these similarities between nerve cells and platelets Bowden and co-workers initiated an investigation in which data obtained from 48 affectively disordered patients were compared to those obtained from 10 age- and sex-matched controls. They found lower plasma Ca^{2+} concentration in the unipolar and manic patients than in controls, lower Ca^{2+} stimulated ATPase in the red blood cells of the unipolar patients and the controls than in the bipolar and manic patients but no differences in values for CSF-Ca^{2+}, platelet Ca-ATPase, and platelet Ca^{2+} uptake. Of statistical significance (p <0.05) was a lower platelet Ca^{2+} concentration in unipolar compared to bipolar depressed patients and in manic compared to depressed patients. Although it is claimed that these results provide further evidence that calcium metabolism is altered in affective disorders, they, as the authors point out, should be viewed with caution. Not all systems that regulate Ca^{2+} metabolism have been included in this study. Since platelets lack voltage-dependent calcium channels and ion exchange mechanisms similar to the ion pumps of nerve cells that maintain a low intracellular Ca^{2+} concentration (Doyle and Ruegg,

1985), measurements of platelet calcium metabolism are probably not the key that will lead to a better understanding of alterations of calcium metabolism in psychiatric disorders. To solve this problem, future studies of Ca^{2+} regulation in nerve cells and its modulatory effects upon neurotransmission will be necessary (also see Section II, Chapter 7). These studies must include the dynamics of modifications in the free intracellular Ca^{2+} ion concentration and a better understanding of the complex role of Ca^{2+} in neurotransmission.

III. CLINICAL EFFECTS OF CALCIUM CHANNEL BLOCKERS IN PSYCHIATRIC PATIENTS

The first case report of the therapeutic effects of the organic calcium channel blocking substance verapamil in a manic patient was presented by Dubovsky and co-workers in 1982. Based on speculations of a link between a disturbance in calcium metabolism and affective disorders and on the similarity between some pharmacological properties of lithium salts and those of calcium channel blockers, they administered 160 mg/day verapamil to a 53-year-old female patient who had suffered from manic episodes for 23 years and had developed tremor during treatment with lithium. Five days after initiation of the verapamil treatment the manic syndrome improved and it deteriorated again under placebo after 3 weeks. At the same time, a French group (Caillard and Massé, 1982) reported on 5 manic patients who had been successfully treated with another calcium channel blocking compound (diltiazem 180 to 240 mg/day). Based on a psychopathometric scale, these patients demonstrated 75% improvement within 2 weeks after admission. In this study which was also published in English (Caillard, 1985), only patients with a manic syndrome in the course of an endogenous affective psychosis benefited from diltiazem while manic patients with organic psychoses did not. Following these preliminary open trials, the first controlled studies (Giannini et al., 1984; Dose et al., 1985; Dubovsky et al., 1986) on antimanic effects of 320 to 480 mg/day verapamil revealed an improvement of manic symptomatology after 5 to 7 days of treatment without signs of sedation or other side effects. In addition a recurrence of manic symptoms occurred under placebo. Similar results were reported in an open study in six manic patients in Czechoslovakia (Höschl et al., 1986). This study also included 23 depressed patients of which 9 showed an improvement with 240 to 400 mg/day verapamil. The results in depressed patients, however, could not be replicated under controlled conditions (Höschl and Kožený, 1988).

Prophylactic effects of verapamil were first reported in a manic-depressive patient who had relapsed frequently under lithium and was stable for one year under constant verapamil medication (160 to 240 mg/day) (Gitlin and Weiss, 1984). Four cases of depression have been reported following the use of the calcium channel blocker nifedipine in cardiac patients who had no previous history of psychiatric disorders. In each case the depression resolved after discontinuation of the nifedipine (Hullett et al., 1988).

After the calcium channel blocking effects of diphenylbutylpiperidin neuroleptics had been described (Gould et al., 1983), investigations of the psychotropic effects of calcium blockers in schizophrenic patients were undertaken. However, no therapeutic effects of verapamil on the symptoms of schizophrenic psychoses could be demonstrated in any of these studies (Grebb et al., 1986; Schepelern and Køster, 1987; Pickar et al., 1987; Uhr et al., 1988) although several different study protocols were used: in one study (Pickar et al., 1987), treatment with neuroleptics in severely disordered chronic schizophrenic patients was stopped and verapamil was given as the only medication. In the other study, chronic schizophrenic patients who had not benefited from high dose neuroleptics even when combined with lithium, received verapamil in addition to the aforementioned medication. Thus, 4 studies in 37 severely ill, chronic patients, who had not responded to neuroleptics, did not demonstrate therapeutic effects of calcium blockers in schizophrenic psychoses. In affectively disordered patients, antimanic effects of verapamil have been demonstrated in 9 (4 controlled) studies in which a total of 60 manic patients were treated.

The question as to whether verapamil can cross the blood-brain barrier (see also Section IV, Chapter 22) has been addressed in two studies (Doran et al., 1985; Dose and Emrich, 1986) in which concentrations of verapamil in the CSF were compared to those in the serum. In patients who received both verapamil and neuroleptics, verapamil concentrations in the serum were found to be of 143.3 ± 65.2 ng/ml and verapamil concentrations in the CSF were found to be 9.2 ± 3.3 ng/ml (6.4%) following daily doses of 480 mg verapamil. In our own study a patient received 480 mg/day verapamil for one week leading to a serum concentration of 587.3 ng/ml and a CSF concentration of 8.8 ng/ml (1.5%) paralleling the findings made on another class of calcium channel blockers, the dihydropyridines, that are supposed to cross the human blood-brain barrier (Rämsch et al., 1984).

IV. POSSIBLE MECHANISM OF ACTION

A. Lithium-Calcium Interactions

In all studies on antimanic effects of calcium channel blockers, especially verapamil, a delay of onset of therapeutic effects (3 to 7 days) and the absence of sedation or other unwanted side effects such as extrapyramidal symptoms have been reported. These observations parallel quite closely those seen during the treatment of acute manic syndromes with lithium salts, that were first described by Cade in Australia (Cade, 1949). Lithium, on the other hand, interacts with such peripheral calcium-dependent processes in humans as glucose metabolism, renal tubular ion transport, cardiac repolarization, and the release of thyroid hormones (Reisberg and Gershon, 1979). With regard to the CNS, lithium has been shown to cause deterioration of dark adaptation in humans, a calcium-dependent function of the retina (Ullrich et al., 1985; Kaschka et al., 1988). Lithium crosses cell membranes via sodium channels (Richelson, 1977) and shares common physicochemical properties with Ca^{2+} (Eisenmann, 1976). Lithium has been shown to slow the intracellular reequilibration of Ca^{2+} during depolarization-induced Ca^{2+} transients in snail neurons at therapeutically relevant concentrations of 1 mM (Aldenhoff and Lux, 1985) leading to an increase of intracellular Ca^{2+} concentration. Since intracellular Ca^{2+} contributes to calcium channel inactivation (see Section I, Chapter 4) and can thereby reduce inward calcium currents (Hagiwara and Byerly, 1981) lithium induced alterations of Ca^{2+} reequilibration could result in a reduction of inward calcium currents similar to Ca^{2+} entry blockade by calcium channel blockers. Biochemically 1 mM Li$^+$ leads to an acute intracellular accumulation of inositol triphosphate, which promotes the release of Ca^{2+} from intracellular stores (van Calkar and Greil, 1988). The consequent changes of intracellular homeostasis could be a factor in the accumulation of intracellular Ca^{2+} observed in snail neurons with lithium (Aldenhoff and Lux, 1985). Since inositol is synthesized intracellularly, the lithium-induced accumulation of its triphosphate inhibits its *de novo* synthesis and thereby slows various calcium-dependent processes in response to external stimuli (Shears, 1988). Similar effects on cellular responses to external stimulation are to be expected from an inhibition of transmembrane inward calcium and/or sodium currents in neurons with calcium channels that respond to channel blockers such as verapamil. One type of calcium channel, found in sensory neurons, discussed in Section I of this book is activated at higher membrane potentials, has a small conductance, and exhibits an incomplete time-dependent inactivation that is controlled by voltage and intracellular Ca^{2+} (Boll and Lux, 1986). Another type of calcium channel in the same preparation has been shown to be activated at low membrane potentials, to be inactivated very rapidly and completely, and to be fully blocked by antagonistic cations but not calcium channel blockers (Carbone and Lux, 1984). Contradictory effects of the calcium channel blocking action of verapamil in neurons possibly relate to these two types of calcium channels: low concentrations of calcium channel blockers do not inhibit noradrenalin-release from adrenergic neurons (Haeusler, 1972), whereas higher concentrations do (Starke and Schümann, 1973; Göthert et al., 1979). In vertebrate slice preparations and synaptosomes high concentrations of verapamil have been shown to inhibit depolarization-induced release of GABA (Vargas et al., 1977; Haycock et al., 1978) and serotonin (Murakami et al., 1978), whereas low concentrations do not affect dopamine- and acetylcholine-release in slices of rabbit nucleus caudatus (Starke et al., 1984). Binding sites for dihydropyridines (Snyder, 1984) and for verapamil (Reynolds et al., 1983) that have been localized in brain tissue give further evidence for a role of Ca^{2+} and calcium channel blockers in neuronal modulation.

Pharmacological studies have revealed that ligand binding to dihydropyridine binding sites is allosterically inhibited by verapamil (Yamamura et al., 1982) but augmented by diltiazem, another calcium channel blocker (Murphy et al., 1983). These effects are probably achieved by accelerating or slowing ligand dissociation from its binding site.

B. Calcium Channel Blockers and Anticonvulsants

An involvement of Ca^{2+} in epileptogenesis is inferred from data showing that extracellular Ca^{2+} concentrations decrease during epileptiform activity in mammals (Heinemann et al., 1977) and that rhythmic discharges, which resemble epileptic activity are induced in hippocampal neurons at low extracellular Ca^{2+} concentrations (Yaari et al., 1983). The appearance of epileptiform activity at low extracellular Ca^{2+} concentrations is thought to be due to changes of surface charges of cell membranes which lower the threshold of Na^+ channels and thereby increase membrane excitability. The prevention of this epileptiform activity at low extracellular Ca^{2+} concentrations by anticonvulsants like carbamazepine and valproic acid as well as by inorganic and organic Ca^{2+} channel blockers suggests a common, possibly membrane stabilizing mechanism of action. Carbamazepine and valproic acid in addition to their anticonvulsive effects, exert acute antimanic as well as prophylactic effects in affective psychoses (Emrich, 1984). Verapamil has been shown to prevent pentylenetetrazol- (Walden et al., 1986) and penicillin-induced (Speckmann et al., 1986) epileptic activity in animals, while, in a pilot study, nifedipine has been shown to be anticonvulsive in humans (Larkin et al., 1988). In cultures of mouse spinal and cortical neurons an inhibition of repetitive firing by carbamazepine in micromolar concentrations was attributed to a blockade of sodium channels (McLean and MacDonald, 1986), although additional effects on calcium channels cannot be ruled out. Such effects were suggested from the inhibition of epileptiform activity in rat hippocampal slices at micromolar carbamazepine concentrations (Heinemann et al., 1985). Indirect effects on Ca^{2+} that are not mediated by direct action upon calcium channels but interactions with adenosine (Skeritt et al., 1982), GABA-B and certain benzodiazepine receptors (Terrence et al., 1983; Mestre et al., 1985) which decrease inward calcium currents into neurons have been suggested for carbamazepine. Direct effects of carbamazepine on calcium inward currents have been described in Helix neurons (Winkel and Lux, 1985) and rat hippocampal slice (Gasser et al., 1988).

Valproic acid is generally thought to exert its anticonvulsive activity by augmenting GABAergic transmission (Löscher, 1985) although it is not yet clear whether its anticonvulsant effects are due to direct pre- or postsynaptic modulation of the GABA system. Recent data suggest that therapeutic concentrations of valproic acid augment GABAergic transmission without increasing presynaptic GABA release (Wolf et al., 1988). Although no direct effects of valproic acid upon calcium and chloride channels have been demonstrated (Nosek, 1981; Ferrendelli and Daniels-McQueen, 1982) the inhibition of repetitive firing without any effect on spontaneous activity in mouse cortical and spinal neurons by 10 mM valproic acid (McLean and MacDonald, 1986) has been interpreted as indicative of an ion channel blockade similar to that induced by carbamazepine.

C. Calcium Channel Blockers and Neuroleptics

Interactions of neuroleptics, such as the phenothiazine derivative trifluoperazine, with the Ca^{2+}-calmodulin system were described shortly after the discovery of this calcium binding protein (Levin and Weiss, 1977). These interactions may be due either to the binding of certain neuroleptics to calmodulin with the subsequent inhibition of calmodulin-dependent enzymes (Levin and Weiss, 1977) or to the blockade of voltage-dependent calcium channels (Sand et al., 1983). According to the calmodulin hypothesis of neurotransmission (De-Lorenzo, 1981) such a channel blockade would hinder the initiation of cellular processes such as the release of neurotransmitters that requires the intracellular entrance of Ca^{2+} ions during depolarization and the formation of a Ca^{2+}-calmodulin complex in order to be activated. Binding studies with drugs have revealed that neuroleptics of the diphenylbutylpiperidine type have, by producing allosteric effects at verapamil binding sites, a strong influence on ligand binding at dihydropyridine binding sites (Gould et al., 1983). These dihydropyridine binding sites are concentrated in brain tissue with their highest density at synaptic zones of the limbic system, especially at the synaptic zones of the hippocampus (Murphy et al., 1982). Their distribution pattern parallels that of other neurotransmitter receptors to which calcium blocker binding sites are probably coupled and exert neuromodulatory and in consequence psychotropic effects. Thus, besides their well-known dopamine-antagonistic effects, some neuroleptics may act on calcium channels and by the same token some calcium channel blocking agents may produce effects similar to those of some neuroleptics.

V. SUMMARY AND CONCLUSIONS

Although there are numerous reports that deal with psychiatric symptoms in patients with hypo- and hypercalcemia of varying etiology, conflicting observations have been made regarding the occurrence of disturbances of calcium metabolism in psychiatric patients and the various concepts that have been advanced are still speculative. Preliminary studies on possible psychotropic effects of calcium channel blocking agents, however, have led to the suggestion that, at least the antimanic effects of the organic calcium channel blocker verapamil are similar to those of lithium salts and some anticonvulsants. There is supporting evidence from clinical observations for a possible common mechanism of action of some anticonvulsants, lithium, calcium channel blockers, and, to some extent, of some neuroleptics as previously mentioned from experimental work. The combination of lithium with verapamil (Price and Giannini, 1986; Dubovsky et al., 1987) and diltiazem (Valdiserri, 1985) may increase synergistically the well-known neurotoxicity of lithium. Neurotoxic side effects of carbamazepine are potentiated by combining it with verapamil (MacPhee et al., 1986; Price and DiMarzio, 1988) or diltiazem (Brodie and MacPhee, 1986).

Similar than commonly seen with neuroleptics, a female patient developed hyperprolactinemia and galactorrhea during verapamil monotherapy for atrial tachycardia (Gluskin et al., 1981). Moreover, the development of "neuroleptic-specific" side effects such as Parkinsonism, akathisia, and tardive dyskinesia have been reported in patients who were treated with calcium channel blockers without any concomitant neuroleptic or psychotropic medication (Chouza et al., 1986; López Alemany, 1986).

A possible common mechanism of action of Ca^{2+} channel blockers, anticonvulsants, and some neuroleptics is an inhibition of transmembranal Ca^{2+}- and/or Na^+-fluxes. Such inhibition of Ca^{2+}- and/or Na^+-fluxes during depolarization of cell membranes slows down many cellular responses to external stimuli. Since an elevated intracellular Ca^{2+} inhibits Ca^{2+} influx during membrane depolarization, the same effect will be brought about by agents that increase intracellular Ca^{2+} as it has been demonstrated for Li^+ (Aldenhoff and Lux,

1985). If one assumes, that psychiatric disorders result from a disturbed homeostasis of central nervous inhibition and excitation by overshoot stimulation or undershoot inhibition, a generalized "slowing" of neuronal responses could reequilibrate this homeostasis and thus modulate affective states. In contrast to neuroleptics and antidepressants that exert their therapeutic as well as their side effects by acting on dopamine receptors (neuroleptics) or the re-uptake of catecholamines (antidepressants), drugs that interact with intracellular Ca^{2+} or calcium channels modulate the activity of neurons or neuronal networks by inhibiting membrane excitability. Possibly, such modulation takes more time compared to the rapid sedative action of neuroleptics and comparable drugs, before behavioral consequences can be observed, but it may also produce fewer unwanted side effects. In addition to continuing investigations of classical neurotransmitter-models of psychotropic drug action, future research strategies should also be directed towards the effects of psychotropic substances on neuronal calcium-dependent processes and calcium channel blocking agents in psychiatric syndromes.

References

REFERENCES

Aaronson, P.I., T.B. Bolton, R.J. Lang and I. MacKenzie: Calcium Currents in Single Isolated Smooth Muscle Cells from the Rabbit Ear Artery in Normal Calcium and High Barium Solutions. *J.Physiol.* **405**:57-(2/10) (1988).

Abrahamsson, H., P.O. Berggren and P. Rorsman: Direct Measurements of Increased Free Cytoplasmic Ca^{2+} in Mouse Pancreatic Beta-Cells Following Stimulation by Hypoglycemic Sulfonylureas. *FEBS Lett.* **190**:21-(2/7) (1985).

Adams, D.J. and P.W. Gage: Calcium Channel in *Aplysia* Nerve Cell Membrane: Ionic and Gating Current Kinetics. *Proc.Austr.Physiol.Pharmacol.Soc.* **8**:115P-(1/3,1/4) (1977).

Adams, D.J. and P.W. Gage: Characteristics of Sodium and Calcium Conductance Changes Produced by Membrane Depolarization in an *Aplysia* Neurone. *J.Physiol.* **289**:143-(2/9) (1979).

Adams, W.B. and I.B. Levitan: Voltage and Ion Dependences of the Slow Currents which Mediate Bursting in *Aplysia* Neuron Rl5. *J.Physiol.* **360**:69-(2/9) (1985).

Addonizio, P.V., C.A. Fisher, J.F. Strauss, Y.T. Wachtfogel, R.W. Lolman and M.E. Josephson: Effects of Verapamil and Diltiazem on Human Platelet Function. *Am.J.Physiol.* **250**:H366-(4/20) (1986).

Adler, E.M., G.J. Augustine, S.N. Duffy and M.P. Charlton: Alien Intracellular Calcium Chelators Attenuate Neurotransmitter Release at the Squid Giant Synapse. *J. Neurosci.* (in press) (2/7) (1990).

Adriaensen, H.: The Treatment of Intermittent Claudication. *Clin.Trials J.* **191**-(4/20) (1969).

Affolter H. R. and R. Coronado: Sideness of Reconstituted Calcium Channels from Muscle Transverse Tubules as Determined by D600 and D890 Blockade. *Biophys.J.* **49**:767-(3/13) (1986).

Ahn, Y.S. and W.J. Harrington: Calcium Channel Blockers in Thrombotic Diseases. *Adv.Int.Med.* **32**:137-(4/15,4/20) (1987).

Ahn, Y.S., W. Jy, W.J. Harrington, N. Shanbaky, L.F. Fernandez and D.H. Haynes: Increased Platelet Calcium in Thrombosis and Related Disorders and its Correction by Nifedipine. *Thrombos.Res.* **45**:135-(4/20) (1987).

Akabas, M.H., J. Dodd and Q. Al-Awqati: A Bitter Substance Induces a Raise in Intracellular Calcium in a Subpopulation of Rat Taste Cells. *Science* **242**:1047-(2/8) (1988).

Akaike, N., H. Kanaide, T. Kuga, M. Nakamura, J.I. Sadoshima and H. Tomoike: Low-Voltage-Activated Calcium Current in Rat Aorta Smooth Muscle Cells in Primary Culture. *J. Physiol.* **416**:141-(1/4) (1989a).

Akaike, N., P.G. Kostyuk and I.V. Osipchuk: Dihydopyridine Sensitive Low Threshold Calcium Channels in Isolated Rat Hypothalamic Neurones. *J.Physiol.* **412**:181-(3/13,3/14) (1989b).

Akaike, N., K.S. Lee and A.M. Brown: The Calcium Currents of *Helix* Neuron. *J.Gen.Physiol.* **71**:509-(1/2,1/3,3/12) (1978).

Akaike, N., K. Nishi and Y. Oyama: Characteristics of Manganese Current and its Comparison with Currents Carried by other Divalent Cations in Snail Soma Membranes. *J.Membr.Biol.* **76**:289-(1/2) (1983).

Akaike, N., Y. Tsuda and Y. Oyama: Separation of Current- and Voltage-Dependent Inactivation of Calcium Current in Frog Sensory Neuron. *Neurosci.Lett.* **84**:46-(1/4) (1988).

Akhtar, M., I. Niazi, G. Naccarelli, P. Tchon, R. Rinkenburger, A. Dougherty and M. Jazayeri: Role of Adrenergic Stimulation by Isoproterenol in Reversal of Drug Effects of Encainide in Supraventricular Tachycardia. *Am.J.Cardiol.* **62**:45L-(4/19) (1988).

Akhtar, M., P. Tochou and M. Jazayeri: Use of Calcium Channel Entry Blockers in the Treatment of Cardiac Arrhythmias. *Circulation* **80**:31-(3/13,4/19) (1989).

Akiyama, S., M.M. Cornwell, M. Kuwano, I. Pastan and M. Gottesman: Most Drugs that Reverse Multidrug Resistance also Inhibit Photoaffinity Labeling of P-glycoprotein by a Vinblastine Analog. *Mol.Pharmacol.* **33**:144-(3/13) (1988).

Al-Dahan, M.I. and R.H. Thalmann: Effects of Dihydropyridine Calcium Channel Ligands on Rat Brain γ-Amino Butyric Acid B Receptors. *J.Neurochem.* **53**:982-(3/13) (1989).

Aldenhoff, J.B. and H.D. Lux: Lithium Slows Neuronal Calcium Regulation in the Snail *Helix pomatia*. *Neurosci.Lett.* **54**:103-(4/22) (1985).

Alderson, B.H. and J.J. Feher: The Interaction of Calcium and Ryanodine with Cardiac Sarcoplasmic Reticulum. *Biochim.Biophys.Acta* **900**:221-(2/10) (1987).

Aldoori, M.N., W.B. Campbell and P.A. Dieppe: Nifedipine in the Treatment of Raynaud's Syndrome. *Cardiovasc.Res.* **20**:466-(4/20) (1986).

Aldrich, R.W., D.P. Corey and C.F. Stevens: A Reinterpretation of Mammalian Sodium Channel Gating Based on Single Channel Recording. *Nature* **306**:436-(1/4) (1983).

Aliot, E., B. Szabo, R. Sweidan and R. Lazarra: Prevention of Torsades de Pointes with Calcium-Channel Blockade in an Animal Model. *J.Am.Coll.Cardiol.* **5**:492-(4/19) (1985).

Almers, W., R. Fink and P.T. Palade: Calcium Depletion in Frog Muscle Tubules: The Decline of Calcium Current Under Maintained Depolarization. *J.Physiol.* **312**:177-(2/10) (1981).

Almers, W. and E.W. McCleskey: Nonselective Conductance in Calcium Channels of Frog Muscle: Calcium Selectivity in a Single File Pore. *J.Physiol.* **353**:585-(1/2,3/12,3/13) (1984).

Almers, W., E.W. McCleskey and P.T. Palade: A Nonselective Cation Conductance in Frog Muscle Membrane Blocked by Micro-Molar External Calcium Ions. *J.Physiol.* **353**:565-(1/12) (1984).

Almers, W. and P.T. Palade: Slow Calcium and Potassium Currents Across Frog Muscle Membrane: Measurements with a Vaseline-Gap Technique. *J.Physiol.* **312**:159-(2/10) (1981).

Alnaes, E. and R. Rahamimoff: On the Role of Mitochondria in Transmitter Release from Motor Nerve Terminals. *J.Physiol.* **248**:285-(2/7) (1975).

Alps, B.J., C. Calder, W.K. Hass and A.D. Wilson: Comparative Protective Effects of Nicardipine, Flunarizine, Lidoflazine and Nimodipine against Ischaemic Injury in the Hippocampus of the Mongolian Gerbil. *Br.J.Pharmacol.* **93**:877-(3/13) (1988).

Altman, R., A. Scazziota and C. Dujovne: Diltiazem Potentiates the Inhibitory Effect of Aspirin on Platelet Aggregation. *Clin.Pharmacol.Ther.* **44**:320-(3/13) (1988).

Amery, W.K.: Prophylactic and Curative Treatment of Migraine with Calcium Antagonists. *Drug Design Delivery* **4**:197-(3/13) (1989).

Ammar, E.M., D. Hutson and T. Scratcherd: Absence of a Relationship between Arterial pH and Pancreatic Bicarbonate Secretion in the Isolated Perfused Cat Pancreas. *J.Physiol.* **388**:495-(2/7) (1987).

Andersen, C.S. and K. Dunlap: Single L-Type Ca Channels in Dorsal Root Ganglion Neurons can be Modulated by Norepinephrine. *Soc.Neurosci.Abs.* **14**:644-(3/14) (1988).

Anderson, C.J. and J. Reiser: Calcium Antagonists and Cardiac Arrhythmias. *Am.Fam.Physician* **23**:214-(4/19) (1981).

Anderson, J.L., J.M. Wagner, F.L. Datz, P.E. Christian, B.E. Bray and A.T. Taylor: Comparative Effects of Diltiazem, Propranolol, and Placebo on Exercise Performance using Radionuclide Ventriculography in Patients with Symptomatic Coronary Artery Disease: Results of a Double-Blind, Randomized, Crossover Study. *Am.Heart J.* **107**:698-(4/16) (1984).

Anderson, K.P., E.B. Stenson and J.W. Mason: Surgical Exclusion of Focal Paroxysmal Atrial Tachycardia. *Am.J.Cardiol.* **49**:869-(4/19) (1982).

Anderson, M.: Manganese Passes through Calcium Channels in Myoepithelial Cells. *J.Exp.Biol.* **82**:227-(3/12) (1979).

Anderson, P.A.V.: Properties and Pharmacology of a TTX-Insensitive Na$^+$ Current in Neurones of the Jellyfish *Cyanea capillata*. *J.Exp.Biol.* **133**:231-(4/17) (1987).

Andersson, K.E.: Calcium Antagonists and Dysmenorrhea. *Ann.NY Acad. Sci.* **522**:747-(4/15) (1988).

Andre-Fouet, X., J.P. Usdin, C. Gayet, C. Wilner, J.F. Thizy, M. Viallet, E. Apoil, P. Vernant and M. Pont: Comparison of Short-term Efficacy of Diltiazem and Propranolol in Unstable Angina at Rest - A Randomized Trial in 70 Patients. *Eur.Heart J.* **4**:691-(4/16) (1983).

Andreassen, F., E. Boye and E. Christoffersen: Assessment of Verapamil in the Treatment of Angina Pectoris. *Eur.J.Cardiol.* **24**:443-(4/16) (1975).

Angelides, K.J., L.W. Elmer, D. Loftus and E. Elson: Distribution and Lateral Mobility of Voltage Dependent Sodium Channels in Neurons. *J.Cell Biol.* **106**:1911-(2/9) (1988).

Anglister, L., I.C. Farber, A. Shahar and A. Grinvald: Localization of Voltage-Sensitive Calcium Channels along Developing Neurites: Their Possible Role in Regulating Neurite Elongation. *Dev.Biol.* **94**:351-(2/10) (1982).

Antman, E., J. Muller, S. Goldberg, R. MacAlpin, M. Rubenfire, B. Tabatznik, C.S. Liang, F. Heupler, S. Achuff, N. Rerchek, E. Geltman, N.Z. Kerin, R.K. Neff and E. Braunwald: Nifedipine Therapy for Coronary Artery Spasm. Experience in 127 Patients. *N.Engl.J.Med.* **19**:119-(4/16) (1980).

Aosaki, T. and H. Kasai: Characterization of Two Kinds of High-Voltage-Activated Ca-Channel Currents in Chick Sensory Neurons: Differential Sensitivity to Dihydropyridines and ω-conotoxin GVIA. *Pflügers Arch.* **414**:150-(1/4,3/13,3/14) (1989).

Apstein, C.S. and W. Grossman: Opposite Initial Effects of Supply and Demand Ischemia on Left Ventricular Diastolic Compliance: The Ischemia-Diastolic Paradox. *J.Mol.Cell.Cardiol.* **19**:119-(4/16) (1987).

Ardissino, D., S. de Servi, J.A. Salerno, G. Specchia, M. Previtali, A. Mussini and P. Bobba: Efficacy, Duration and Mechanism of Action of Nifedipine in Stable Exercise-Induced Angina Pectoris. *Eur.Heart J.* **4**:873-(4/16) (1983).

Arellano, R.O., P. Ramon, A. Rivera and G.A. Zampighi: Calmodulin Acts as an Intermediary for the Effects of Calcium on Gap Junctions from Crayfish Lateral Axons. *J.Membr.Biol.* **101**:119-(2/10) (1988).

Argibay, J.A., R. Fischmeister and H.C. Hartzell: Inactivation, Reactivation and Pacing Dependence of Calcium Current in Frog Cardiocytes: Correlation with Current Density. *J.Physiol.* **401**:201-(1/4) (1988).

Armstrong, C.M.: Evidence for Ionic Pores in Excitable Membranes. *Biophys.J.* **15**:932-(3/12) (1975a).

Armstrong, C.M.: Potassium Pores of Nerve and Muscle Membranes. **In** *Membranes* G. Eisenman, ed. Marcel Dekker, New York **3**:325-(3/12) (1975b).

Armstrong, C.M. and S.R. Taylor: Interaction of Barium Ions with Potassium Channels in Squid Giant Axons. *Biophys.J.* **30**:473-(1/3) (1980).

Armstrong, C.M. and F. Bezanilla: Currents Related to Movements of the Gating Particles of the Sodium Channel. *Nature* **242**:459-(1/4) (1973).

Armstrong, C.M. and F. Bezanilla: Inactivation of the Sodium Channel. II. Gating Current Experiments. *J.Gen.Physiol.* **70**:567-(1/4) (1977).

Armstrong, C.M., F.M. Bezanilla and P. Horowicz: Twitches in the Presence of Ethylene Glycol bis (β-aminoethyl ether)-N,N'-Tetraacetic Acid. *Biochim Biophys.Acta* **267**:605-(2/10) (1972).

Armstrong, C.M. and D.R. Matteson: The Role of Calcium Ions in the Closing of K Channels. *J.Gen.Physiol.* **87**:817-(1/3) (1986).

Armstrong, C.M., R.P. Swenson and S.R. Taylor: Block of Squid Axon K Channels by Internally and Externally Applied Barium Ions. *J.Gen.Physiol.* **80**:663-(1/3) (1982).

Armstrong, D. and R. Eckert: Voltage Activated Calcium Channels must be Phosphorylated to Respond to Membrane Depolarization. *U.S.A.Sci.Natl.Acad.Sci.U.S.A.* **84**:2518-(1/4,3/14) (1987).

Armstrong, D. and D. Kalman: The Role of Protein Phosphorylation in the Response of Dihydropyridine-Sensitive Calcium Channels to Membrane Depolarization in Mammalian Pituitary Tumor Cells. In *Calcium and Ion Channel Modulation* A. Grinnell, C. Armstrong, M. Jackson, eds. Plenum Press, New York. 215-(1/4) (1988).

Armstrong, D. and D. Kalman: Dihydropyridines Modulate Ca^{2+} Channels by Altering their Availability to Protein Phosphorylation and its Removal. *Biophys.J.* **57**:516a-(3/13) (1990).

Arnman, K. and L. Ryden: Comparison of Metoprolol and Verapamil in the Treatment of Angina Pectoris. *Am.J.Cardiol.* **49**:821-(4/16) (1982).

Aronoff, M.S., R.G. Evans and J. Durell: Effect of Lithium Salts on Electrolyte Metabolism. *J.Psychiatr.Res.* **8**:139-(4/22) (1971).

Aronow, W.S. and J. Ferlinz: Verapamil Versus Placebo in Atrial Fibrillation and Flutter. *Clin.Invest.Med.* **3**:35-(4/19) (1980).

Aronow, W.S., D. Landa and G. Plasencia: Verapamil in Atrial Fibrillation and Atrial Flutter. *Clin.Pharm.Ther.* **26**:578-(4/19) (1979).

Aronow, W.S., G. Plasencia and R. Wong: Effect of Verapamil Versus Placebo in PAT and MAT (Paroxysmal Atrial Tachycardia and Multifocal Atrial Tachycardia). *Current Ther.Res.* 25:823-(4/19) (1980).

Arreola, J., J. Calvo, M.C. Garcia and J.A. Sanchez: Modulation of Calcium Channels of Twitch Muscle Fibers of the Frog by Adrenaline and Cyclic Adenosine Monophosphate. *J.Physiol.* **393**:307-(3/14) (1987).

Art, J.J. and R. Fettiplace: Variation of Membrane Properties in Hair Cells Isolated from the Turtle Cochlea. *J.Physiol.* **385**:207-(2/8) (1987).

Artalejo, C.R., A.G. Garcia and D. Aunis: Chromaffin Cell Calcium Channel Kinetics Measured Isotopically through Fast Calcium, Strontium, and Barium Fluxes. *J.Biol.Chem.* **262**:915-(2/7) (1987).

Asberg, M., S.A. Montgomery, C. Perris, D. Schalling and G. Sedvall: A Comprehensive Psychopathological Rating Scale. *Acta Psychiatr.Scand.Suppl.* **271**:5-(4/22) (1978).

Asch, A.S., J. Barnwell, R.L. Silverstein and R.L. Nachman: Isolation of the Thrombospondin Membrane Receptor. *J.Clin.Invest.* **79**:1054-(4/20) (1987).

Asch, A.S., R.L. Nachman: Thrombospondin: Phenomenology to Function. *Prog. Thrombo.* **9**:157-(4/20) (1989).

Ascher, P. and L. Nowak: Calcium Permeability of the Channels Activated by N-methyl-D-Aspartate (NMDA) in Mouse Central Neurons. *J.Physiol.* **377**:35P-(2/9) (1986).

Ashcroft, F.M. and P.R. Stanfield: Calcium and Potassium Currents in Muscle Fibres from Insect: *Carausius morosus*. *J.Physiol.* **323**:93-(3/12) (1982a).

Ashcroft, F.M. and P.R. Stanfield: Calcium Inactivation in Skeletal Muscle Fibres of the Stick Insect, *Carausius morosus*. *J.Physiol.* **330**:349-(1/4) (1982b).

Ashmore, J.F.: Frequency Tuning in a Frog Vestibular Organ. *Nature* **304**:536-(2/8) (1983).

Ashmore, J.F. and D. Attwell: Models for Electrical Tuning in Hair Cells. *Proc.R.Soc.London B* **226**:325-(2/8) (1985).

Atchison, W.D., L. Adgate and C.M. Beaman: Effects of Antibiotics on Uptake of Calcium into Isolated Nerve Terminals. *J.Pharmacol.Exp.Ther.* **245**:394-(3/13) (1988).

Atchison, W.D.: Dihydropyridine-Sensitive and -Insensitive Components of Acetylcholine Release from Rat Motor Nerve Terminals. *J.Pharmacol.Exp.Ther.* **251**:672-(3/13) (1989).

Atterhog, J.H., L.G. Ekelund and A.L. Melin: Effect of Nifedipine on Exercise Tolerance in Patients with Angina Pectoris. *Eur.J.Clin.Pharmacol.* **8**:125-(4/16) (1975).

Attwell, D., S. Werblin and M. Wilson: The Properties of Single Cones Isolated from the Tiger Salamander Retina. *J.Physiol.* **328**:259-(2/8) (1982).

Atwater, I. and P.M. Beigelman: Dynamic Characteristics of Electrical Activity in Pancreatic Beta-Cells 1. Effects of Calcium and Magnesium Removal. *J.Physiol.(Paris)* **72**:769-(2/7) (1976).

Augustine, G.J., J. Buchanan, M.P. Charlton, L.R. Osses and S.J. Smith: Fingering the Trigger for Neurotransmitter Secretion: Studies on the Calcium Channels of Squid Giant Presynaptic Terminals. In *Secretion and its Control*. Rockefeller University Press, New York. 203-(2/7) (1989).

Augustine, G.J., M.P. Charlton and S.J. Smith: Calcium Entry and Transmitter Release at Voltage-Clamped Nerve Terminals of Squid. *J.Physiol.* **367**:163-(2/7) (1985b).

Augustine, G.J., M.P. Charlton and S.J. Smith: Calcium Entry into Voltage-Clamped Presynaptic Terminals of Squid. *J.Physiol.* **367**:143-(2/7) (1985a).

Augustine, G.J., M.P. Charlton and S.J. Smith: Calcium Action in Synaptic Transmitter Release. *Ann.Rev.Neurosci.* **10**:633-(2/7,3/14) (1987).

Auld, V.J., A.L. Goldin, D.S. Krafte, J. Marshall, J.M. Dunn, W.A. Catterall, H.A. Lester, N. Davidson and R.J. Dunn: A Rat Brain Na^+ Channel Alpha-Subunit with Novel Gating Properties. *Neuron* **1**:449-(3/14) (1988).

Autrum, H., R. Jung, W.R. Lowenstein, D.M. MacKay and H.L. Teubersin: In *Handbook of Sensory Physiology*. Springer-Verlag, Berlin. (2/8) (1971).

Avila-Sakar, A.J., G. Cota, J. Gamboa-Aldeco, J. Garcia, M. Huerta, J. Muniz and E. Stefani: Skeletal Muscle Ca^{2+} Channels. *J.Mus.Res. and Cell Motil.* **7**:291-(2/10) (1986).

Ayoub, G. and D.M.K. Lam: The Release of Gamma-Aminobutyric Acid from Horizontal Cells of the Goldfish (*Carassius auratus*). *J.Physiol.* **355**:191-(2/8) (1984).

Azmitia, E.C.: Nimodipine Attenuates NMDA- and MDMA-Induced Toxicity of Cultured Fetal Serotonergic Neurons: Evidence for a Generic Model of Calcium Toxicity. In *Nimodipine and Central Nervous System Function*. J. Traber and W.H. Gispen, eds. Schattauer, Stuttgart and New York, 257-(3/13) (1989).

Baccaglini, P.I. and N.C. Spitzer: Developmental Changes in the Inward Current of the Action Potential of Rohon-Beard Neurones. *J.Physiol.* **271**:93-(2/10) (1977).

Bache, R.J. and F.R. Cobb: Effect of Maximal Coronary Vasodilation on Transmural Perfusion during Tachycardia in the Awake Drug. *Circ.Res.* **41**:648-(4/19) (1977).

Bacon, K.B., J. Westwick and R.D. Camp: Potent and Specific Inhibition of IL-8-, IL-1 a- and IL-1 b-Induced *In Vitro* Human Lymphocyte Migration by Calcium Channel Antagonists. *Biochem.Biophys.Res.Commun.* **165**:349-(3/13) (1989).

Bader, C.R. and D. Bertrand: Effects of Changes in Intra- and Extracellular Sodium on the Inward (Anomalous) Rectification in Salamander Photoreceptors. *J.Physiol.* **331**:253-(2/8) (1982).

Bader, C.R., P.R. MacLeish and E.A. Schwartz: A Voltage-Clamp Study of the Light Response in Solitary Rods of the Tiger Salamander. *J.Physiol.* **296**:1-(2/8) (1979).

Baeyens, J.M.: Interactions between Calcium Channel Blockers and Non-Cardiovascular Drugs: Interactions with Anticancer Drugs. *Pharmacol.Toxicol.* **63**:1-(3/13) (1988).

Baker, P.F.: Transport and Metabolism of Calcium Ions in Nerve. *Prog.Biophys.Mol.Biol.* **24**:177-(2/7) (1972).

Baker, P.F. and H.G. Glitsch: Voltage Dependent Changes in the Permeability of Nerve Membranes to Calcium and other Divalent Cations. *Phil.Trans.R.Soc.* B**270**:389-(3/12) (1975).

Baker, P.F., A.L. Hodgkin and T.I. Shaw: Replacement of the Protoplasm of a Giant Nerve Fibre with Artificial Solutions. *Nature* **190**:885-(3/12) (1961).

Baker, P.F. and D.E. Knight: Calcium Control of Exocytosis in Bovine Adrenal Medullary Cells. *TINS* **7**:120-(2/7) (1984).

Baker, P.F. and D.E. Knight: Theme and Variation in the Control of Exocytosis. In *Cell Calcium and the Control of Membrane Transport*. Rockefeller University Press, New York. 1-(2/7) (1987).

Balabanov, S., U. Tollner, H.P. Richter, F. Pohlandt, G. Gaedicke and W.M. Teller: Immunoreactive Parathyroid Hormone, Calcium, and Magnesium in Human Cerebrospinal Fluid. *Acta Endocrinology* **106**:227-(4/22) (1984).

Balasubramanian, B.V.: Clinical and Research Applications of Ambulatory Holter ST-segment and Heart Rate Monitoring. *Am.J.Cardiol.* **58**:11-(4/19) (1986).

Balazs, R., N. Hack, O.S. Jorgensen and C.W. Cotman: N-Methyl-D-Aspartate Promotes the Survival of Cerebellar Granule Cells: Pharmacological Characterization. *Neurosci.Lett.* **101**:241-(2/10) (1989).

Balazs, R., O.S. Jorgensen and N. Hack: N-Methyl-D-Aspartate Promotes the Survival of Cerebellar Granule Cells in Culture. *Neuroscience* **27**:437-(2/10) (1988).

Baldwin, J.J., D.A. Claremon, P.K. Lumma, D.E. McClure, S.A. Rosenthal, R.J. Winquist, E.P. Faison, G.J. Kaczorowski, M.J. Trumble and G.M. Smith: Diethyl-3,6-Dihydro-2,4-Dimethyl-2,6-Methano-1,3-Benzothiazocine-5,11-Dicarboxylates as Calcium Entry Blockers: New Conformationally Restricted Analogs of Hantzsch 1,4-Dihydropyridines Related to Nitrendipine as Probes of Receptor Conformation. *J.Med.Chem.* **30**:690-(3/13) (1987).

Baldwin, T.J., C.M. Yoshihara, K. Blackmer, C.R. Kintner and S.J. Burden: Regulation of Acetylcholine Receptor Transcript Expression During Development in *Xenopus laevis*. *J.Cell Biol.* **106**:469-(2/10) (1988).

Ballesta, J.J., M. Palermo, M.J. Hidalgo, L.M. Gutierrez and J.A. Reig: Separate Binding and Functional Sites for ω-conotoxin and Nitrendipine Suggest 2 Types of Calcium Channels in Bovine Chromaffin Cells. *J.Neurochem.* **53**:1050-(3/13) (1989).

Balsano, F.: Chronic Treatment of "Claudication". In *Calcium Entry Blockers Tissue Protection*. T. Godfraind, P. Vanhoutte, S. Giovani, and R. Paoletti, eds. Raven Press, New York. 215-(4/20) (1985).

Balwierczak, J.L., C.L. Johnson and A. Schwartz: The Relationship between the Binding Site of [^{3}H]-d-cis-Diltiazem and that of other Non-Dihydropyridine Calcium Entry Blockers in Cardiac Sarcolemma. *Mol.Pharmacol.* **31**:175-(3/13) (1987).

Barber, J.H., C.A. Reuter, A.H.M. Jageneau and W. Loots: Intermittent Claudication: A Controlled Study in Parallel Time of the Short-Term and Long-Term Effects of Cinnarizine. *Pharmatherapeutica* **2**:401-(4/20) (1980).

Barcenas, R.L., D.J. Beuckelmann and W.G. Wier: Sodium-Calcium Exchange in Heart: Membrane Currents and Changes in Ca^{2+}_i. *Science* **238**:1720-(4/21) (1987).

Barhanin, J., A. Schmid and M. Lazdunski: Properties of Structure and Interaction of the Receptor for ω-Conotoxin, a Polypeptide Active on Ca^{2+} Channels. *Biochem.Biophys.Res.Commun.* **150**:1051-(3/13) (1988).

Barish, M.E.: Differentiation of Voltage-Gated Potassium Current and Modulation of Excitability in Cultured Amphibian Spinal Neurones. *J.Physiol.* **375**:229-(2/10) (1986).

Barish, M.E.: Multiple Calcium Currents in Cultured Embryonic Amphibian Spinal Neurons. *Soc.Neurosci.Abs.* **14**:900-(2/10) (1988).

Barnathan, E.S., V.P. Addonizio and S.J. Shattil: Interaction of Verapamil with Human Platelet-Adrenergic Receptors. *Am.J.Physiol.* **242**:Hl9-(4/20) (1983).

Barnes, S. and J.A. Davies: The Effects of Calcium Channel Agonists and Antagonists on the Release of Endogenous Glutamate from Cerebellar Slices. *Neurosci.Lett.* **92**:58-(3/13) (1988).

Barr, D.P., C.M. McBryde and T.E. Sanders: Tetany with Increased Intracranial Pressure and Papilledema Results from Treatment with Dihydrotachysterol. *Trans.Assoc.Am.Physician* **53**:227-(4/22) (1938).

Barres, B.M., L.L.Y. Chan and D.P. Corey: Ion Channel Expression by White Matter Glia I. Type 2 Astrocytes and Oligodendrocytes. *Glia* 1:10-(3/14) (1988).

Barrett, A.M.: Psychosis Associated with Tetany. *Am.J.Insanity* **76**:373-(4/22) (1919).

Barrett, E.F., J.N. Barrett and W.E. Crill: Voltage-Sensitive Outward Currents in Cat Motoneurones. *J.Physiol.* **304**:251-(2/9) (1980).

Barrett, J.N., K.L. Magleby and B.S. Pallotta: Properties of Single Calcium-Activated Potassium Channels in Cultured Rat Muscle. *J.Physiol.* **331**:211-(2/9) (1982).

Barrett, P.Q., C. Isales, W. Bollag, R.T. McCarthy and H. Rasmussen: Role of Calcium in Angiotensin II Mediated Aldosterone Secretion. *Endocrinol.Rev.* **4**:496-(3/13) (1989).

Bartter, F.C.: The Parathyroid Gland and its Relationship to Diseases of the Nervous System. In *Metabolic and Toxic Diseases of the Nervous System.* H. Merritt and C. Hare, eds. Williams & Williams, Baltimore. (4/22) (1953).

Bastian, B.L. and G.L. Fain: Light Adaptation in Toad Rods: Requirements for an Internal Messenger which is not Calcium. *J.Physiol.* **297**:493-(2/8) (1979).

Basu, B., G. Cherian and K.A. Abraham: Oral Verapamil as Maintenance Therapy after Cardioversion for Atrial Fibrillation. *Indian J.Chest Dis.Allied Sci.* **19**:170-(4/19) (1977).

Bauer, J.H., S. Sobha and G. Reams: Effects of Calcium Entry Blockers on Renin-Angiotensin-Aldosterone System, Renal Function and Hemodynamics, Salt and Water Excretion and Body Fluid Composition. *Am.J.Cardiol.* **56**:62H-(4/18) (1985).

Bauernfeind, R.A., C.R.C. Wyndham, R. Dhingra, S.P. Swiryn, E. Palateo, B. Strasberg and K.M. Rosewn: Serial Electrophysiologic Testing of Multiple Drugs in Patients with Atrioventricular Nodal Re-Entrant Paroxysmal Tachycardia. *Circulation* **62**:1341-(4/19) (1980).

Baughman, K.L.: Calcium Channel Blocking Agents in Congestive Heart Failure. *Am.J.Med.* **80**:46-(4/17) (1986).

Baumgold, J., J.B. Parent and I. Spector: Development of Sodium Channels During Differentiation of Chick Skeletal Muscle in Culture. I. Binding Studies. *J. Neurosci.* **3**:995-(2/10) (1983a).

Baumgold, J., J.B. Parent and I. Spector: Development of Sodium Channels During Differentiation of Chick Skeletal Muscle in Culture II. ^{22}Na$^+$ Uptake and Electrophysiological Studies. *J.Neurosci.* **3**:1004-(2/10) (1983b).

Bayer, R. and R. Mannhold: Fendiline: A Review of its Basic Pharmacological and Clinical Properties. *Pharmatherapeutica* **5**:103-(3/13). (1987).

Baylor, D.A.: Photoreceptor Signals and Vision. *Invest.Ophthalmol.Vis.Sci.* **28**:34-(2/8) (1987).

Baylor, D.A. and M.G.F. Fuortes: Electrical Responses of Single Cones in the Retina of the Turtle. *J.Physiol.* **207**:77-(2/8) (1970).

Baylor, D.A., M.G.F. Fuortes and P.M. O'Bryan: Receptive Fields of Single Cones in the Turtle of the Retina. *J.Physiol.* **214**:265-(2/8) (1973).

Baylor, D.A. and A.L. Hodgkin: Detection and Resolution of Visual Stimuli by Turtle Photoreceptors. *J.Physiol.* **234**:165-(2/8) (1973).

Baylor, D.A., A.L. Hodgkin and T.D. Lamb: The Electrical Response of Turtle Cones to Flashes and Steps of Light. *J.Physiol.* **242**:685-(2/8) (1974).

Baylor, D.A. and T.D. Lamb: Local Effects of Bleaching in Retinal Rods of the Toad. *J.Physiol.* **328**:49-(2/8) (1982).

Baylor, D.A., T.D. Lamb and K.W. Yau: Response of Retinal Rods to Single Photons. *J.Physiol.* **288**:613-(2/8) (1979b).

Baylor, D.A., T.D. Lamb and K.W. Yau: The Membrane Current of Single Rod Outer Segments. *J.Physiol.* **288**:589-(2/8) (1979a).

Baylor, D.A., G. Matthews and K.W. Yau: Two Components of Electrical Dark Noise in Toad Retinal Rod Outer Segments. *J.Physiol.* **309**:591-(2/8) (1980).

Baylor, D.A. and B.J. Nunn: Electrical Properties of the Light-sensitive Conductance of Rods of the Salamander *Ambystoma tigrinum*. *J.Physiol.* **371**:115-(2/8) (1986).

Beam, K.G., J.H. Caldwell and D.T. Campbell: Na Channels in Skeletal Muscle Concentrated Near the Neuromuscular Junction. *Nature* **313**:588-(2/10) (1985).

Beam, K.G. and C.M. Knudson: Calcium Currents in Embryonic and Neonatal Mammalian Skeletal Muscle. *J.Gen.Physiol.* **91**:781-(2/10) (1988a).

Beam, K.G. and C.M. Knudson: Effect of Postnatal Development on Calcium Currents and Slow Charge Movement in Mammalian Skeletal Muscle. *J.Gen.Physiol.* **91**:799-(2/10) (1988b).

Beam, K.G., C.M. Knudson and J.A. Powell: A Lethal Mutation in Mice Eliminates the Slow Calcium Current in Skeletal Muscle Cells. *Nature* **320**:168-(2/10) (1986).

Beam, K.G., J.A. Powell amd S. Numa: Restoration of Excitation Contraction Coupling and Slow Calcium Current in Dysgenic Mice by Dihydropyridine Receptor Complementary DNA. *Nature* **336**:134-(3/13) (1988).

Bean, B.P.: Nitrendipine Block of Cardiac Calcium Channels: High Affinity Binding to the Inactivated State. *Ann.N.Y.Acad.Sci.* **81**:6388-(3/13) (1984).

Bean, B.P.: Two Kinds of Calcium Channels in Canine Atrial Cells. Differences in Kinetics, Selectivity, and Pharmacology. *J.Gen.Physiol.* **86**:1-(1/2,1/4,2/10,2/9) (1985).

Bean, B.P.: Classes of Calcium Channels in Vertebrate Cells. *Ann.Rev.Physiol.* **51**:367-(1/3,1/4,2/7,3/13,3/14) (1989a).

Bean, B.P.: Neurotransmitter Inhibition of Neuronal Calcium Currents by Changes in Channel Voltage Dependence. *Nature* **340**:153-(1/3,3/14) (1989b).

Bean, B.P.: Multiple Types of Calcium Channels in Heart Muscle and Neurons: Modulation by Drugs and Neurotransmitters *Ann. N.Y.Acad.Sci.* **560**:334-(1/2) (1989c).

Bean, B.P., M.C. Nowycky and R.W. Tsien: Beta-Adrenergic Modulation of Calcium Channels in Frog Ventricular Heart Cells. *Nature* **307**:371-(1/4) (1984).

Bean, B.P. and E. Rios: Asymmetric Charge Movement in Mammalian Cardiac Muscle Cells: Na and Ca Channel Components. *Biophys.J.*(abstr.) **53**:158a-(2/10) (1988).

Bean, B.P. and E. Rios: Nonlinear Charge Movement in Mammalian Cardiac Ventricular Cells. *J.Gen.Physiol.* **94**:65-(2/10) (1989).

Bean, B.P., M. Sturek, A. Puga and K. Hermsmeyer: Calcium Channels in Muscle Cells Isolated from Rat Mesenteric Arteries: Modulation by Dihydropyridine Drugs. *Cir.Res.* **59**:229-(2/10,3/13) (1986).

Beaty, G.N. and E. Stefani: Calcium Dependent Electrical Activity in Twitch Muscle Fibres of the Frog. *Proc.R.Soc.London B* **194**:141-(2/10) (1976).

Bech, P., C. Kirkegaard, E. Bock, M. Johannesen and O.J. Rafaelsen: Hormones, Electrolytes, and Cerebrospinal Fluid Proteins in Manic-Melancholic Patients. *Neuropsychobiology* **4**:99-(4/22) (1978).

Bechem, M., R. Gross, S. Hebisch and M. Schramm: Ca-agonists: A New Class of Inotropic Drugs. *Basic Res.Cardiol.* **84**:105-(3/13) (1989).

Bechem, M., S. Hebisch and M. Schramm: Ca^{2+} Agonists: New, Sensitive Probes for Ca^{2+} Channels. *TIPS* **99**:257-(3/13) (1988).

Bechem, M. and L. Pott: Removal of Ca Current Inactivation in Dialysed Guinea Pig Atrial Cardioballs by Ca Chelators. *Pflügers Arch.* **404**:10-(1/4) (1985).

Bechem, M. and M. Schramm: Electrophysiology of Dihydropyridine Calcium Agonists. In *The Calcium Channel: Structure, Function and Implications*. M. Morad, S. Nayler, S. Kazda and M. Schramm, eds. Springer-Verlag, Berlin and Heidelberg. 63-(3/13) (1988).

Beech, D.J., I. MacKenzie, T.B. Bolton and M.O. Chisten: Effects of Pinaverium on Voltage-Activated Calcium Channel Currents of Single Smooth Muscle Cells Isolated from the Longitudinal Muscle of the Rabbit Jejunum. *Br.J.Pharmacol.* **999**:374-(3/13) (1990).

Beirao, P.S. and N. Lakshminarayanaiah: Calcium Carrying System in the Giant Muscle Fibre of the Barnacle Species, *Balanus nubilus. J.Physiol.* **293**:319-(3/12) (1979).

Belhassen, B. and L.N. Horowitz: Use of Intravenous Verapamil for Ventricular Tachycardia. *Am.J.Cardiol.* **54**:1141-(4/19) (1984).

Belhassen, B. and M. Pelleg: Acute Management of Paroxysmal Supraventricular Tachycardia: Verapamil, Adenosine Triphosphate or Adenosine. *Am.J.Cardiol.* **54**:225-(4/19) (1984).

Belhassen, B., H.H. Rotmensch and S. Laniado: Response of Recurrent Sustained Ventricular Tachycardia to Verapamil. *Br.Heart J.* **46**:679-(4/19) (1981).

Belhassen, B., I. Shapira, A. Pelleg, I. Cooperman, N. Kauli and S. Laniado: Idiopathic Recurrent Sustained Ventricular Tachycardia Responsive to Verapamil: An ECG-Electrophysiologic Entity. *Am.Heart J.* **108**:1034-(4/19) (1984).

Bellemann, P., A. Schade and R. Towart: Dihydropyridine Receptor in Rat Brain Labeled with [^{3}H]Nimodipine. *Proc.Natl.Acad.Sci.U.S.A.* **80**:2356-(3/13) (1983).

Bem, J.L., A.M. Breckenridge, R.D. Mann and M.D. Rawlins: Review of Yellow Cards (1986): Report to the Committee on the Safety of Medicines. *Br.J.Clin.Pharmacol.* **26**:679-(3/13) (1988).

Benaim, M.E.: Asystole after Verapamil. *Br.Med.J.* **2**:169-(4/19) (1972).

Bender, F.: Clinical Uses of Calcium Ion Antagonists in Arrhythmias. *Clin.Invest.Med.* **3**:21-(4/19) (1980).

Benham, C.D.: ATP-Activated Channels Gate Calcium Entry in Single Smooth Muscle Cells Dissociated from Rabbit Ear Artery. *J.Physiol.* **419**:689-(3/13) (1989).

Benham, C.D. and T.B. Bolton: Spontaneous Transient Outward Currents in Single Visceral and Vascular Smooth Muscle Cells of the Rabbit. *J.Physiol.* **381**:385-(2/10) (1986).

Benham, C.D., P. Hess and R.W. Tsien: Two Types of Calcium Channels in Single Smooth Muscle Cells from Rabbit Ear Artery Studied with Whole-Cell and Single-Channel Recordings. *Circ.Res.Suppl.I.* **61**:1-(2/10) (1987).

Benham, C.D. and R.W. Tsien: Calcium-Permeable Channels in Vascular Smooth Muscle: Voltage-Activated, Receptor-Operated, and Leak Channels, In *Cell Calcium and Membrane Transport*. L.J. Mandel and D.C. Eaton, eds. Rockefeller Unversity Press, New York. 45-(2/10) (1986).

Benham, C.D. and R.W. Tsien: Noradrenaline Increases L-Type Calcium Current in Smooth Muscle Cells of Rabbit Ear Artery Independent of Beta-Adrenoceptors. *J.Physiol.* **390**:85-(3/14) (1987a).

Benham, C.D. and R.W. Tsien: A Novel Receptor-operated Ca^{2+}-permeable Channel Activated by ATP in Smooth Muscle. *Nature* **328**:275-(2/10,2/9,3/13) (1987b).

Benham, C.D. and R.W. Tsien: Noradrenaline Modulation of Calcium Channels in Single Smooth Muscle Cells from Rabbit Ear Artery. *J.Physiol.* **404**:767-(2/10) (1988).

Bennett, J.P., S. Cockcroft and B.D. Gomperts: Rat Mast Cells Permeabilized with ATP Secrete Histamine in Response to Calcium Ions Buffered in the Micromolar Range. *J.Physiol.* **317**:335-(2/7) (1981).

Benotti, J.R., W. Grossman and P.F. Cohn: Clinical Profile of Restrictive Cardiomyopathy. *Circulation* **61**:1206-(4/17) (1980).

Benveniste, H., J. Drejer, A. Schousboe and N.H. Diemer: Elevation of the Extracellular Concentrations of Glutamate and Aspartate in Rat Hippocampus during Transient Cerebral Ischemia Monitored by Intracerebral Microdialysis. *J.Neurochem.* **43**:1369-(4/21) (1984).

Benveniste, H., M.B. Jorgensen, N.H. Diemer and A.J. Hansen: Calcium Accumulation by Glutamate Receptor Activation is Involved in Hippocampal Cell Damage after Ischemia. *Acta Neurol.Scand.* **78**:529-(4/21) (1988).

Bergener, M. and B. Reisberg: *Diagnosis and Treatment of Senile Dementia.* Springer-Verlag, Berlin and Heidelberg. (3/13) (1989).

Bernal, M.J., J.R. Fernandez and A.J. Alvarez-Leefmans: Trifluoperazine Blocks Ca^{2+} Inward Currents in *Helix* Neurons in a Reversible and Dose Dependent Manner. *Biophys.J.* **51**:33a-(3/13) (1987).

Bernard, C.: Establishment of Ionic Permeabilities of the Myocardial Membrane during Embryonic Development of the Rat. **In** *Developmental and Physiological Correlates of Cardiac Muscle.* M. Lieberman and T. Sano, eds. Raven Press, New York. **1**:169-(4/17) (1976).

Bernini, F., A.L. Catapano, A. Corsini, R. Fumagalli and R. Paoletti: Effects of Calcium Antagonists on Lipids and Atherosclerosis. *Am.J.Cardiol.* 64:1291-(3/13) (1989)

Bernstein, J.: Untersuchungen zur Thermodynamik der Biolelktrischen Strome Erster Theil. *Pflügers Arch.* **82**:521-(1/1) (1902).

Berridge, M.J.: Inositol Trisphosphate and Diacylglycerol as Second Messengers. *Biochem.J.* **220**:345-(4/17,4/20) (1984).

Berridge, M.J.: Inositol Lipids and Calcium Signalling. *Proc.R.Soc.London B* **234**:359-(4/21) (1988).

Berridge, M.J. and R.F. Irvine: Inositol Trisphosphate, a Novel Second Messenger in Cellular Signal Transduction. *Nature* **312**:315-(2/7) (1984).

Berridge, M.J. and R.F. Irvine: Inositol Phosphates and Cell Signalling. *Nature* **341**:197-(2/7) (1989).

Berridge, M.J. and A. Galione: Cytosolic Calcium Oscillators. *FASEB J.* **2**:3076-(3/14) (1988).

Bers, D.M.: Ca Influx and Sarcoplasmic Reticulum Ca Release in Cardiac Muslce Activation During Postrest Recovery. *Am.J.Physiol.* **248**:H366-(2/10) (1985).

Berwe, D., G. Gottschalk and C. H. Luttgau: Effects of the Calcium Antagonist Gallopamil (D600) upon Excitation-Contraction Coupling in Toe Muscle Fibres of the Frog. *J.Physiol.* **385**:693-(2/10,3/13) (1987).

Beta-Blocker Heart Attack Trial Research Group: A Randomized Trial of Propranolol in Patients with Acute Myocardial Infarction. I. Mortality Results. *JAMA* **247**:1707-(4/19) (1982).

Betriu, A., B.R. Chaitman, M.G. Bourassa, G. Breuess, J.M. Scholl, P. Bruneau, P. Gague and M. Chabot: Beneficial Effect of Intravenous Diltiazem in the Acute Management of Supraventricular Tachyarrhythmias. *Circulation* **67**:88-(4/19) (1983).

Betz, H.: Regulation of a-Bungarotoxin Receptor Accumulation in Chick Retina Cultures: Effects of Membrane Depolarization, Cyclic Nucleotide Derivatives, and Ca^{2+}. *J.Neurosci.* **3**:1333-(2/10) (1983).

Betz, H. and J.P. Changeaux: Regulation of Muscle Acetylcholine Receptor Synthesis *In Vitro* by Cyclic Nucleotide Derivatives. *Nature* **278**:749-(2/10) (1979).

Beuckelmann, D.J. and W.G. Wier: Mechanism of Release of Calcium from Sarcoplasmic Reticulum of Guinea Pig Cardiac Cells. *J.Physiol.* **405**:233-(2/10) (1988).

Bezanilla, F. and C.M. Armstrong: Inactivation of the Sodium Channel. I. Sodium Current Experiments. *J.Gen.Physiol.* **70**:549-(1/4) (1977).

Bianchi, C.P. and A.M. Shanes: Calcium Influx in Skeletal Muscle at Rest, during Activity, and during Potassium Contracture. *J.Gen.Physiol.* **42**:803-(2/10) (1959).

Bicknell, R.J.: Optimizing Release from Peptide Hormone Secretory Nerve Terminals. *J.Exp.Biol.* **139**:51-(2/7) (1988).

Bielefeld, D.R., R.W. Hadley, P.M. Vassilev and A.R. Hume: Membrane Electrical Properties of Vesicular Na-Ca Exchange Inhibitors in Single Atrial Myocytes. *Circ.Res.* **59**:381-(3/13) (1986).

Bilinski, M., H. Plattner and H. Matt: Secretory Protein Decondensation as a Distinct, Ca^{2+} Mediated Event during the Final Steps of Exocytosis in Paramecium Cells. *J.Cell Biol.* **88**:179-(2/7) (1981).

Bindokas, V.P. and M.E. Adams: ω-Aga-I: A Presynaptic Calcium Channel Antagonist from Venom of the Funnel Web Spider, *Agelenopsis aperta. J.Neurobiol.* **20**:171-(3/13) (1989).

Bingmann, D. and E.J. Speckmann: Specific Suppression of Pentylenetetrazol-Induced Epileptiform Discharges in CA3 Neurons (Hippocampal Slice, Guinea Pig) by the Organic Calcium Antagonists Flunarizine and Verapamil. *Exp.Brain Res.* **239**:(1989).

Binnie, C.D.: Potential Antiepileptic Drugs Flunarizine and other Calcium Entry Blockers. In *Antiepileptic Drugs*. R. Levy, R. Mattson, B. Meldrum, J.K. Penry and F.D. Dreifuss, eds. Raven Press, New York. 971-(3/13) (1989).

Birge, R.: Photophysics of Light Transduction in Rhodopsin and Bacteriorhodopsin. *Ann.Rev.Biophys.Bioeng.* **10**:315-(2/8) (1981).

Birnbaum, M., M.A. Reis and A. Shainberg: Role of Calcium in the Regulation of ACh Receptor Synthesis in Cultured Muscle Cells. *Pflügers Arch.* **385**:37-(2/10) (1980).

Bixby, J.L. and N.C. Spitzer: The Appearance and Development of Chemosensitivity in Rohon-Beard Neurones of the *Xenopus* Spinal Cord. *J.Physiol.* **330**:513-(2/10) (1982).

Bixby, J.L. and N.C. Spitzer: The Appearance and Development of Neurotransmitter Sensitivity in *Xenopus* Embryonic Spinal Neurones In Vitro. *J.Physiol.* **353**:143-(2/10) (1984a).

Bixby, J.L. and N.C. Spitzer: Early Differentiation of Vertebrate Spinal Neurons in the Absence of Voltage-Dependent Ca^{2+} and Na^+ Influx. *Dev.Biol.* **106**:89-(2/10) (1984b).

Bjorum, N.: Electrolytes in Blood in Endogenous Depression. *Acta Psychiatr. Scand.* **48**:59-(4/22) (1972).

Bkaily, G., D. Jacques, A. Sculptoreanu, T. Yamamoto, D. Carrier, D. Vigneault and N. Sperelakis: Apamin Highly Potent Blocker of the TTX- and Mn^{2+}Insensitive Fast Transient Na^+ Current in Young Embryonic Heart. *J. Mol. Cell. Cardiol.* (in press) (4/17) (1991).

Bkaily, G., D. Jacques, T. Yamamoto, A. Sculptoreanu, M.D. Payet and N. Sperelakis: Three Types of Slow Inward Currents as Distinguished by Melitine in 3-Day-old Embryonic Heart. *Can.J.Physiol.Pharmacol.* **66**:1017-(4/17) (1988a).

Bkaily, G., G. Jasmin, C. Tautu, L. Proschek, T. Yamamoto, A. Sculptoreanu, M. Peyrow and D. Jacques: A Tetrodoxin- and Mn^{2+}-Insensitive Current in Duchenne Muscle Dystrophy. *Muscle Nerve* **13**:936-(4/17) (1990).

Bkaily, G., M. Peyrow, T. Yamamoto, A. Sculptoreanu, D. Jacques and N. Sperelakis: Macroscopic Ca^{2+} - Na^+ and K^+ Currents in Single Heart and Aortic Cells. *Mol.Cell. Bioch.* **80**:59-(4/17) (1988).

Blache, D., M. Ciavatti and C. Ojeda: The Effect of Calcium Channel Blockers on Blood Platelet Function, Especially Calcium Uptake. *Biochim.Biophys.Acta* **923**:401-(3/13,4/20) (1987).

Black, I.B., J.E. Adler, C.F. Dreyfus, W.F. Friedman, E.F. LaGamma and A.H. Roach: Biochemistry of Information Storage in the Nervous System. *Science* **236**:1263-(2/10) (1987).

Black, I.B., D.M. Chikaraishi and E.J. Lewis: Trans-Synaptic Increase in RNA Coding for Tyrosine Hydroxylase in a Rat Sympathetic Ganglion. *Brain Res.* **339**:151-(2/10) (1985).

Blair, L.A.C.: The Timing of Protein Synthesis Required for the Development of the Sodium Action Potential in Embryonic Spinal Neurons. *J.Neurosci.* **3**:1430-(2/10) (1983).

Blake, K., N.A. Smith and W.T. Clusin: Rate-Dependence of Ischemic Myocardial Depolarization-Evidence for a Novel Membrane Current. *Cardiovasc.Res.* **20**:557-(4/19) (1986).

Blazynski, C. and A.I. Cohen: Rapid Declines in Cyclic GMP of Rod Outer Segments of Intact Frog Photoreceptors after Illumination. *J.Biol.Chem.* **261**:12142-(2/8) (1986).

Bloch, R.J.: Dispersal and Reformation of Acetylcholine Receptor Clusters of Cultured Rat Myotubes Treated with Inhibitors of Energy Metabolism. *J.Cell Biol.* **82**:626-(2/10) (1979).

Bloch, R.J.: Acetylcholine Receptor Clustering in Rat Myotubes: Requirement for Ca^{2+} and Effects of Drugs which Depolymerize Microtubules. *J.Neurosci.* **3**:2670-(2/10) (1983).

Bloch, R.J. and J.H. Steinbach: Reversible Loss of Acetylcholine Receptor Clusters at the Developing Rat Neuromuscular Junction. *Dev.Biol.* **81**:386-(2/10) (1981).

Block, B.A., T. Imagawa, K.P. Campbell and C. Franzini-Armstrong: Structural Evidence for Direct Interaction Between the Molecular Components of the Transverse Tubule/Sarcoplasmic Reticulum Junction in Skeletal Muscle. *J.Cell Biol.* **107**:2587-(2/10) (1988).

Blumlein, S.L., R. Sievers, P. Kidd and W.W. Parmley: Mechanism of Protection from Atherosclerosis by Verapamil in the Cholesterol-Fed Rabbit. *Am.J.Cardiol.* **54**:884-(4/18) (1984).

Boddeke, H.W., B. Wilffert, J.B. Heynis and P.A. Zwieten: Investigation of the Mechanism of Negative Inotropic Activity of some Calcium Antagonists. *J.Cardiovasc.Pharmacol.* **11**:321-(3/13) (1988).

Bodoia, R.D. and P.B. Detwiler: Patch-clamp Recordings of the Light-Sensitive Dark Noise in Retinal Rods from the Lizard and Frog. *J.Physiol.* 367:183-(2/8) (1985).

Boeynaems, J.M., A. Van-Coevorden and D. DeMolle: Stimulation of Prostacyclin Production in Blood Vessels by the Antithrombotic Drug Suloctidil. *Biochem.Pharmacol.* **36**:1629-(4/20) (1987).

Boland, L.M. and R. Dingledine: Multiple Components of both Transient and Sustained Barium Currents in a Rat Dorsal Root Ganglion Cell Line. *J.Physiol.* **420**:223-(3/13) (1990).

Bolger, G.T., P.J. Gengo, E.M. Luchowski, H. Siegel, R.A. Janis, A.M. Triggle and D.J. Triggle: Characterization of the Binding of the Ca^{2+} Channel Antagonist, [^{3}H]Nitrendipine to Guinea Pig Ileal Smooth Muscle. *J.Pharmacol.Exp.Ther.* **225**:291-(3/13) (1983).

Bolger, G.T., B.A. Weisman and P. Skolnick: The Behavioral Effects of the Calcium Agonist BAY K 8644 in the Mouse: Antagonism by the Calcium Antagonist Nifedipine. *Naunyn Schmied. Arch.Pharmacol.* **328**:373-(3/13) (1985).

Boll, W. and H.D. Lux: Action of Organic Antagonists on Neuronal Calcium Currents. *Neurosci.Lett.* **56**:335-(1/4,3/13) (1985).

Boll, W. and H.D. Lux: Blockade of Neuronal Calcium Conductances by Organic Antagonists. In *Calcium Electrogenesis and Neuronal Functioning*. U. Heinmann, M. Klee, E. Neher and W. Singer, eds. Springer, Berlin. 104-(4/22) (1986).
Bolton, T.B.: Mechanisms of Action of Transmitters and Other Substances on Smooth Muscle. *Physiol.Rev.* **59**:606-(2/10) (1979).
Bolton, T.B. and W.A. Large: Are Junction Potentials Essential? Dual Mechanism of Smooth Muscle Cell Activation by Transmitter Released from Autonomic Nerves. *Q. J.Exp.Physiol.* **71**:1-(2/10) (1986).
Bond, M., A.R. Jaraki, C.H. Disch and B.P. Healy: Subcellular Content in Cardiomyopathic Hamster Hearts *In Vivo*: An Electron Probe Study. *Circ.Res.* **64**:1001-(3/13) (1989).
Bond, M., T. Kitazawa, A.P. Somlyo and A.V. Somlyo: Release and Recycling of Calcium by the Sarcoplasmic Reticulum in Guinea Pig Portal Vein Smooth Muscle. *J.Physiol.* **355**:677-(2/10) (1984).
Bonow, R.O., V. Dilsizian, D.R. Rosing, B.J. Maron, S.L. Bacharach and M.V. Green: Verapamil-Induced Improvement in Left Ventricular Diastolic Filling and Increased Exercise Tolerance in Patients with Hypertrophic Cardiomyopathy Short and Long-Term Effects. *Circulation* **72**:853-(4/17) (1985).
Bonow, R.O., M.B. Leon, D.R. Rosing, K.M. Kent, L.C. Lipson, S.L. Bacharach, M.V. Green and S.E. Epstein: Effects of Verapamil and Propranolol on Left Ventricular Systolic Function and Diastolic filling in Patients with Coronary Artery Disease: Radionuclide Angiographic Studies at Rest and during Exercise. *Circulation* **65**:1337-(4/16) (1981).
Bonow, R.O., H.G. Ostrow, D.R. Rosing, R.O. Cannon, L.C. Lipson, B.J. Maron, K.M. Kent, S.L. Bacharach and M.V. Green: Effects of Verapamil on Left Ventricular Systolic and Diastolic Function in Patients with Hypertrophic Cardiomyopathy: Pressure-Volume Analysis with a Nonimaging Scintillation Probe. *Circulation* **68**:1062-(4/17) (1983).
Bonow, R.O., D.F. Vitale, B.J. Maron, S.L. Bacharach, T.M. Frederick and M.V. Green: Regional Left Ventricular Asynchrony and Impaired Global Left Ventricular Filling in Hypertrophic Cardiomyopathy: Effect of Verapamil. *J.Am.Coll.Cardiol.* **9**:1108-(4/17) (1987).
Bonvallet, R.: A Low Threshold Calcium Current Recorded at Physiological Ca Concentrations in Single Frog Atrial Cells. *Pflügers Arch.* **408**:540-(2/10) (1987).
Bookman, R.J. and F. Schweizer: Fast and Slow Phases in Chromaffin Cell Exocytosis Depend on Ca Entry and Ca Buffering. *J.Gen.Physiol.* **92**:4a-(2/7) (1988).
Boothby, C.B., C.S. Garrard and D. Pickering: Intravenous Verapamil in Cardiac Arrhythmias. *Br.Med.J.* **2**:349-(4/19) (1972).
Borsotto, M., J. Barhanin, M. Fosset and M. Lazdunski: The 1,4-Dihydropyridine Receptor Associated with the Skeletal Muscle Voltage-Dependent Ca^{2+} Channel: Purification and Subunit Composition. *J.Biol.Chem.* **260**:14255-(2/10) (1985).
Bosma, M. and N. Sidell: Retinoic Acid Inhibits Ca^{2+} Currents and Cell Proliferation in a B-Lymphocyte Cell Line. *J.Cell Physiol.* **135**:317-(2/7) (1988).
Bossert, F. and W. Vater: 1,4-Dihydropyridines - A Basis for Developing New Drugs *Medicinal Res.Rev.* **9**:291-(3/13) (1989).
Bossu, J. and A. Fletz: Patch-clamp Study of the Tetrodotoxin-Resistant Sodium Current in Group C Sensory Neurones. *Neurosci.Lett.* **51**:241-(4/17) (1984).
Bossu, J.L., J.L. Dupont and A. Feltz: Calcium Currents in Rat Cerebellar Purkinje Cells Maintained in Culture. *Neuroscience* **30**:605-(2/9) (1989).
Bossu, J.L. and A. Feltz: Inactivation of the Low-threshold Transient Calcium Current in Rat Sensory Neurones: Evidence for a Dual Process. *J.Physiol.* **376**:341-(1/4) (1986).
Bossu, J.L., A. Feltz and J.M. Thomann: Depolarization Elicits Two Distinct Calcium Currents in Vertebrate Sensory Neurons. *Pflügers Arch.* **403**:360-(1/4,2/9) (1985).
Bouchard, R.A., L.V. Hryshko, J.K. Saha and D. Bose: Effects of Caffeine and Ryanodine on Depression of Post-Rest Tension Development Produced by BAY K 8644 in Canine Ventricular Muscle. *Br.J.Pharmacol.* **97**:1279-(3/13) (1989).
Bourdillon, P.D. and P.A. Poole-Wilson: The Effects of Verapamil, Quiescence and Cardioplegia on Calcium Exchange and Mechanical Function in Ischemic Rabbit Myocardium. *Am.J.Cardiol.* **46**:242-(4/19) (1980).
Bournaud, R. and A. Mallart: An Electrophysiological Study of Skeletal Muscle Fibres in the "Muscular Dysgenesis" Mutation of the Mouse. *Pflügers Arch.* **409**:468-(2/10) (1987).
Bourque, C.W. and L.P. Renaud: Calcium Dependent Action Potential in Rat Supraoptic Neurosecretory Neurones recorded *In Vitro*. *J.Physiol.* **363**:419-(3/12) (1985).
Bourson, A., P.C. Moser, A.J. Gower and A.K. Mir: Central and Peripheral Effects of the Dihydropyridine Calcium Channel Activator BAY K 8644 in the Rat. *Eur.J.Pharmacol.* **160**:339-(3/13) (1989).
Bowden, C.L., L.G. Huang, M.A. Javors, J.M. Johnson, E. Seleshi, K. McIntyre, S. Contreras and J.W. Maas: Calcium Function in Affective Disorders and Healthy Controls. *Biol.Psychiatr.* **23**:367-(4/22) (1988).
Boyd, R.A., J.C. Giacomini and K.M. Giacomini: Species Differences in the Negative Inotropic Response of 1,4-Dihydropyridine Calcium Channel Blockers in Myocardium. *J.Cardiovasc.Pharmacol.* **912**:650-(3/13) (1988).
Boyett, M.R., M.S. Kirby and C.H. Orchard: Rapid Regulation of the "Second Inward Current" by Intracellular Calcium in Isolated Rat and Ferret Ventricular Myocytes. *J.Physiol.* **407**:77-(2/10) (1988).

Brachmann, J., B.J. Scherlag, V. Rosenstraukh and R. Lazzara: Bradycardia-Dependent Triggered Activity: Relevance to Drug-Induced Multiform Ventricular Tachycardia. *Circulation* **68**:846-(4/19) (1983).

Bradbury, M.W. and R.P. Wong: Entry of ^{45}Calcium from Blood into Brain and Cerebrospinal Fluid of Normal and Adrenalectomized Rats. *J.Physiol.* **225**:65P-(4/21) (1972).

Bradbury, M.W.B.: *The Concept of a Blood-Brain Barrier.* John Wiley and Sons, Chichester. (4/21) (1979).

Braha, O., M. Klein and E.R. Kandel: Phorbol Ester Increases Ca^{2+} Current in *Aplysia* Sensory Neurons. *Soc.Neurosci.Abs.* **14**:644-(3/14) (1988).

Brass, L.F.: Ca^{2+} Homeostasis in Unstimulated Platelets. *J.Biol.Chem.* **259**:12563-(4/20) (1984).

Brass, L.F.: Ca^{2+} Transport Across the Platelet Plasma Membrane: A Role for Glycoproteins IIb and IIIa. *J.Biol.Chem.* **260**:2231-(4/20) (1985).

Braunwald, E.: Mechanism of Action of Calcium Channel Blocking Agents. *N.Engl.J.Med.* **307**:1618-(4/17) (1982).

Braunwald, E.: A Symposium. Calcium Antagonists - Emerging Clinical Opportunities. *Am.J.Cardiol.* **59**:1B-(4/17) (1987).

Braunwald, E., C.T. Lambrew, S.D. Rockoff, J. Ross and A.G. Morrow: Idiopathic Hypertrophic Subaortic Stenosis. I. A Description of the Disease Based Upon an Analysis of 64 Patients. *Circulation* **30**:3-(4/17) (1964).

Brehm, P. and R. Eckert: Calcium Entry Leads to Inactivation of Calcium Channels in Paramecium. *Science* **202**:1203-(1/4) (1978).

Brehm, P. and L.P. Henderson: Regulation of Acetylcholine Receptor Channel Function During Development of Skeletal Muscle. *Dev.Biol.* **129**:1-(2/10) (1988).

Breitweiser, G.E. and G. Szabo: Uncoupling of Cardiac Muscarinic and Beta-Adrenergic Receptors from Ion Channels by a Guanine Nucleotide Analogue. *Nature* **317**:538-(3/14) (1985).

Bression, D., P.H. Chaumet-Riffaud, A.M. Brandi, A. Comte, F. Peillon and J.R. Kiechel: Binding of (+)-PN200-110 to Rat Pituitaries and to Normal and Adnomatous Human Pituitaries. *Mol.Cell Endocrinol.* **50**:255-(3/13) (1987).

Brethes, D., G. Dayanithi, L. Letellier and J.J. Nordmann: Depolarization-Induced Ca^{2+} Increase in Isolated Neurosecretory Nerve Terminals Measured with Fura-2. *Proc.Natl.Acad.Sci.U.S.A.* **84**:1439-(2/7) (1987).

Brice, A., S. Bernard, B. Raynaud, S. Ansieau, T. Coppola, M.J. Weber and J. Mallet: Complete Sequence of a cDNA Encoding an Active Rat Choline Acetyltransferase: A Tool to Investigate the Plasticity of Cholinergic Phenotype Expression. *J.Neurosci.Res.* **23**:266-(2/10) (1989).

Brichard, G. and P.G. Zimmerman: Verapamil in Cardiac Dysrhythmias During Anesthesia. *Br.J.Anesth.* **42**:1005-(4/19) (1970).

Brigant, J.L. and A. Mallart: Presynaptic Currents in Mouse Motor Endings. *J.Physiol.* **333**:619-(2/7) (1982).

Brinley, F.J.: Calcium Buffering in Squid Axons. *Ann.Rev.Biophys.Bioeng.* **7**:363-(2/7) (1978).

Brodie, M.J. and G.J.A. MacPhee: Carbamazepine Neurotoxicity Precipitated by Diltiazem. *Br.Med.J.* **3**:1170-(4/22) (1986).

Brodin, L. and S. Grillner: Effects of Magnesium on Fictive Locomotion Induced by Activation of N-Methyl-D-Aspartate (NMDA) Receptors in the Lamprey Spinal Cord *In Vitro*. *Brain Res.* **380**:244-(2/9) (1986).

Brown, A.M. and L. Birnbaumer: Direct G-Protein Gating of Ion Channels. *Am.J.Physiol.* **254**:H401-(3/13) (1988).

Brown, A.M., H. Camerer, D.L. Kunze and H.D. Lux: Similarity of Unitary Ca^{2+} Currents in Three Different Species. *Nature* **299**:156-(1/4) (1982).

Brown, A.M., D.L. Kunze and A. Yatani: The Agonistic Effect of Dihydropyridines on Ca Channels. *Nature* **311**:570-(1/4) (1984b).

Brown, A.M., D.L. Kunze and A. Yatani: Dual Effect of Dihydropyridines on Whole Cell and Unitary Calcium Currents in Single Ventricular Cells of Guinea Pig. *J.Physiol.* **379**:495-(3/13) (1989).

Brown, A.M., H.D. Lux and D.L. Wilson: Activation and Inactivation of Single Calcium Channels in Snail Neurons. *J.Gen.Physiol.* **83**:751-(1/3) (1984c).

Brown, A.M., K. Morimoto, Y. Tsuda and D.L. Wilson: Calcium Current-Dependent and Voltage-Dependent Inactivation of Calcium Channels in *Helix aspersa*. *J.Physiol.* **320**:193-(1/4,3/14) (1981).

Brown, A.M., Y. Tsuda and D.L. Wilson: A Description of Activation and Conduction in Calcium Channels Based on Tail and Turn-on Current Measurements in the Snail. *J.Physiol.* **344**:549-(1/3) (1983).

Brown, A.M., D.L. Wilson and H.D. Lux: Activation of Calcium Channels. *Biophys.J.* **45**:125-(1/3) (1984).

Brown, D.A. and W.H. Griffith: Persistent Slow Inward Calcium Current in Voltage Clamped Hippocampal Neurones of the Guinea Pig. *J.Physiol.* **337**:303-(2/9,3/12) (1983).

Brown, E.M., D.G. Gardner, R.A. Windeck and G.D. Aurbach: Relationship of Intracellular 3′,5′-adenosine Monophosphate Accumulation to Parathyroid Hormone Release from Dispersed Bovine Parathyroid Cells. *Endocrinology* **103**:2323-(2/7) (1978).

Brown, J.E. and L.H. Pinto: Ionic Mechanism for the Photoreceptor Potential of the Retina of *Bufo marinus*. *J.Physiol.* **236**:575-(2/8) (1974).

Brugada, P. and H.J.J. Wellens: The Role of Triggered Activity in Clinical Ventricular Arrhythmias. *Pace* **7**:260-(4/19) (1984).

Brum, G., R. Fitts, G. Pizzaro and E. Rios: Voltage Sensors of the Frog Skeletal Muscle Membrane Require Calcium to Function in Excitation-Contraction Coupling. *J.Physiol.* **398**:475-(2/10,3/13) (1988b).

Brum, G., V. Flockerzi, F. Hoffmann, W. Osterrieder and W. Trautwein: Injection of Catalytic Subunit of cAMP-Dependent Protein Kinase into Isolated Cardiac Myocytes. *Pflügers Arch.* **398**:147-(2/10) (1983).
Brum, G., W. Osterrieder and W. Trautwein: Beta-Adrenergic Increase in the Calcium Conductance of Cardiac Myocytes Studied with the Patch Clamp. *Pflügers Arch.* **401**:111-(1/4) (1984).
Brum, G., E. Rios and E. Stefani: Effects of Extracellular Calcium on Calcium Movements of Excitation-Contraction Coupling in Frog Skeletal Muscle Fibres. *J.Physiol.* **398**:441-(2/10) (1988a).
Brush, K.L., M. Perez, M.J. Hawkes, D.R. Pratt and S.L. Hamilton: Low Affinity Binding Sites for 1,4-Dihydropyridines in Mitochondria and in Guinea Pig Ventricular Membranes. *Biochem.Pharmacol.* **36**:4153-(3/13) (1987).
Buhler, F.R., U.L. Hulthen, W. Kowski and P. Bolli: Greater Antihypertensive Efficacy of the Calcium Channel Inhibitor Verapamil in Older and Low Renin Patients. *Clin.Sci.* **63**:439-(4/18) (1982).
Bulbring, E. and T. Tomita: Catecholamine Action on Smooth Muscle. *Pharmacological.Rev.* **39**:49-(3/13) (1987).
Burges, R.A. and M.G. Dodd: Amlodipine. *Cardiovasc.Drug Rev.* **8**:25-(3/13) (1990).
Burgess, T.L. and R.B. Kelly: Constitutive and Regulated Secretion of Proteins. *Ann.Rev.Cell Biol.* **3**:243-(2/7) (1987).
Burgoyne, R.D.: Mechanisms of Secretion from Adrenal Chromaffin Cells. *Biochim.Biophys.Acta* **779**:201-(2/7) (1984).
Burkart, F.E.: Proceedings of a Symposium-Circadian Variation in Myocardial Ischemia: The 24-hour antianginal Effect of Amlodipine. *Am.Heart J..* **118**:1083-(3/13) (1989).
Burris, J.F., M.R. Weir, S. Oparil, M.W. Weber, W.J. Cady and W.H. Stewart: Application of Factorial Trial Design to a Multicenter Clinical Trial of Combination Therapy: An Assessment of Diltiazem and Hydrocholorothiazide in Hypertension. *JAMA* (in press) (4/18) (1990).
Bursztajn, S., J.L. McManaman and S.H. Appel: Organization of Acetylcholine Receptor Clusters in Cultured Rat Myotubes is Calcium Dependent. *J.Cell Biol.* **98**:507-(2/10) (1984).
Bussmann, W.D., W. Scher and M. Grungras: Reduktion der CK und CKMB-Infarktgrosse durch Verapamil. *Deutsche Med.Wochschr.* **108**:1047-(4/16) (1983).
Buxton, A.E., H.L. Waxman, F.E. Marchlinski and M.E. Josephson: Electropharmacology of Non-Sustained Ventricular Tachycardia: Effects of Class I Antiarrhythmic Agents, Verapamil and Propranolol. *Am.J.Cardiol.* **53**:738-(4/19) (1984).
Byerly, L., P.B. Chase and J.R. Stimers: Calcium Current Activation Kinetics in Neurones of the Snail *Lymnaea stagnalis*. *J.Physiol.* **348**:187-(1/3) (1984).
Byerly, L., P.B. Chase and J.R. Stimers: Permeation and Interaction of Divalent Cations in Calcium Channels of Snail Neurons. *J.Gen.Physiol.* **85**:491-(1/2,3/12) (1985).
Byerly, L. and S. Hagiwara: Calcium Currents in Internally Perfused Nerve Cell Bodies of *Lymnaea stagnalis*. *J.Physiol.* **322**:503-(1/3,1/4,3/12) (1982).
Byerly, L. and W.J. Moody: Membrane Currents of Internally Perfused Neurones of the Snail, *Lymnaea stagnalis* at low Intracellular pH. *J.Physiol.* **376**:477-(3/12) (1986).
Cachelin, A.B., J.E. DePeyer, S. Kokubun and H. Reuter: Calcium Channel Modulation by 8-Bromo-Cyclic AMP in Cultured Heart Cells. *Nature* **304**:462-(1/4) (1983).
Cade, J.F.J.: Lithium Salts in the Treatment of Psychotic Excitement. *Med.J.Aust.* **2**:349-(4/22) (1949).
Caffrey, J.M. and M.C. Farach-Carson: Vitamin D3 Metabolites Modulate Dihydropyridine-sensitive Calcium Currents in Clonal Rat Osteosarcoma Cells. *J.Biol.Chem.* **264**:20265-(3/13) (1989).
Caillard, V.: Treatment of Mania Using a Calcium Antagonist - Preliminary Trial. *Neuropsychobiology* **14**:23-(4/22) (1985).
Caillard, V. and G. Masse: Traitement de la Manie par un Inhibiteur Calcique. Etude Preliminaire. *L'Encephale* **8**:587-(4/22) (1982).
Calker van, D. and W. Greil: Effects of Lithium Ions on the Accumulation of Inositolphosphates in PC-12 Cells and Human Granulocytes. In *Inorganic Pharmacology and Psychiatric Use.* N.J. Birch, ed. IRL Press, Oxford. 209-(4/22) (1988).
Callewaert, G., L. Cleemann and M. Morad: Epinephrine Enhances Ca^{2+} Current-Regulated Ca^{2+} Release and Ca^{2+} Reuptake in Rat Ventricular Myocytes. *Proc.Natl.Acad.Sci.U.S.A.* **85**:2009-(2/10) (1988).
Callewaert, G., I. Hanbauer and M. Morad: Modulation of Calcium Channels in Cardiac and Neuronal Cells by an Endogenous Peptide. *Science* **243**:663-(3/13,3/14) (1989).
Cambier, J.C. and J.T. Ransom: Molecular Mechanisms of Transmembrane Signalling in B Lymphocytes. *Ann.Rev.Immunol.* **5**:175-(2/7) (1987).
Campbell, D.L., W.R. Giles, J.R. Hume and E.F. Shibata: Inactivation of Calcium Current in Bullfrog Atrial Myocytes. *J.Physiol.* **403**:287-(1/4) (1988a).
Campbell, K.P., C.M. Knudson, T. Imagawa, A.T. Leung, J.L. Sutko, S.D. Kahl, C.R. Raab and L. Madson: Identification and Characterization of the High Affinity [^{3}H]Ryanodine Receptor of the Junctional Sarcoplasmic Reticulum Ca^{2+} Release Channel. *J.Biol.Chem.* **262**:6460-(2/10) (1987).
Campbell, K.P., A.T. Leung and A.H. Sharp: The Biochemistry and Molecular Biology of the Dihydropyridine-Sensitive Calcium Channel. *TINS* **11**:425-(1/3,3/13) (1988b).

Campbell, L.W., S.Y. Hao and P.W. Landfield: Aging-Related Increases in L-Like Calcium Currents in Rat Hippocampal Slices. *Soc.Neurosci.Abs.* **15**:260-(3/13) (1989).

Campbell, N.C., C.F. Ekerot and G. Hesslow: Interactions Between Responses in Purkinje Cells Evoked by Climbing Fibre Impulses and Parallel Fibre Volleys in the Cat. *J.Physiol.* **340**:225-(2/9) (1983b).

Campbell, N.C., C.F. Ekerot, G. Hesslow and O. Oscarsson: Dendritic Plateau Potentials Evoked in Purkinje Cells by Parallel Fibre Volleys in the Cat. *J.Physiol.* **340**:209-(2/9) (1983a).

Canfield, D.R. and K. Dunlap: Pharmacological Characterization of Amine Receptors on Embryonic Chick Sensory Neurones. *Br.J.Pharmacol.* **82**:557-(3/14) (1984).

Cannell, M.B., J.R. Berlin and W.J. Lederer: Effect of Membrane Potential Changes on the Calcium Transient in Single Rat Cardiac Muscle Cells. *Science* **238**:1419-(2/10) (1987).

Cannon, R.O., D.R. Rosing, B.J. Maron, M.B. Leon, R.O. Bonow, R. Watson and S.E. Epstein: Myocardial Ischemia in Patients with Hypertrophic Cardiomyopathy: Contribution of Inadequate Vasodilatator Reserve and Elevated Left Ventricular Filling Pressures. *Circulation* **71**:231-(4/17) (1985).

Capella, D., J.R. Laporte, J.M. Castel, C. Tristan, A. Cos and F.J. Morales-Olivas: Parkinsonism, Tremor, and Depression Induced by Cinnarizine and Flunarizine. *Br.Med.J.* **297**:722-(3/13) (1988).

Capovilla, M.L., L. Cervetto, E. Pasino and V. Torre: The Sodium Current Underlying the Response of Toad Rods to Light. *J.Physiol.* **317**:223-(2/8) (1981).

Capucci, A., L. Bassein, D. Bracchetti, G. Carini, A. Maresta and B. Magnani: Propranolol v. Verapamil in the Treatment of Unstable Angina. A Double-Blind Cross-Over Study. *Eur.Heart J.* **4**:148-(4/16) (1983).

Caputo, C. and P. Bolanos: Contractile Inactivation in Frog Skeletal Muscle Fibers. The Effects of Low Calcium, Tetracaine, Dantrolene, D-600, and Nifedipine. *J.Gen.Physiol.* **89**:421-(2/10) (1987).

Carafoli, E.: Intracellular Calcium Homeostasis. *Annu.Rev.Biochem.* **56**:395-(3/14) (1987).

Carafoli, E. and T.J. Penniston: The Calcium Signal. *Sci.Am.* **253**:70-(3/14) (1985).

Carbone, E., A. Formenti and A. Pollo: Multiple Actions of Bay K 8644 on High-Threshold Ca Channels in Adult Rat Sensory Neurons. *Neurosci.Lett.*(in press) (1/4) (1990a).

Carbone, E. and H.D. Lux: A Low Voltage-Activated, Fully Inactivating Ca Channel in Vertebrate Sensory Neurones. *Nature* **310**:501-(1/2,1/4,2/9,4/22) (1984b).

Carbone, E. and H.D. Lux: A Low Voltage-Activated Calcium Conductance in Embryonic Chick Sensory Neurons. *Biophys.J.* **46**:413-(1/2,1/4) (1984a).

Carbone, E. and H.D. Lux: Low- and High-Voltage Activated Ca Channels in Vertebrate Cultured Neurons: Properties and Functions. *Exp.Brain Res.Suppl.* **14**:1-(1/4) (1986b).

Carbone, E. and H.D. Lux: Sodium Channels in Cultured Chick Dorsal Root Ganglion Neurons. *Eur.Biophys.J.* **13**:259-(1/4) (1986a).

Carbone, E. and H.D. Lux: External Ca^{2+} Ions Block Unitary Na^+ Currents Through Ca^{2+} Channels of Cultured Chick Sensory Neurones by Favouring Prolonged Closures. *J.Physiol.* **382**:124P-(1/2) (1987c).

Carbone, E. and H.D. Lux: Kinetics and Selectivity of a Low-Voltage Activated Calcium Current in Chick and Rat Sensory Neurones. *J.Physiol.* **386**:547-(1/2,1/4,2/9,3/12) (1987a).

Carbone, E. and H.D. Lux: Single Low Voltage Activated Calcium Channels in Chick and Rat Sensory Neurones. *J.Physiol.* **386**:571-(1/2,1/3,1/4,2/9,3/12) (1987b).

Carbone, E. and H.D. Lux: ω-Conotoxin Blockade Distinguishes Ca from Na Permeable States in Neuronal Calcium Channels. *Pflügers Arch.* **413**:14-(1/4,3/13) (1988a).

Carbone, E. and H.D. Lux: Sodium Currents Through Neuronal Calcium Channels: Kinetics and Sensitivity to Calcium Antagonists. In *The Calcium Channel: Structure, Function and Implications.* M. Morad, W. Nayler, S. Kazda, and M. Schramm, eds. Springer-Verlag, Berlin. 115-(1/4) (1988b).

Carbone, E., M. Morad and H.D. Lux: External Ni^{2+} Selectively Blocks the Low-Threshold Ca^{2+} Current of Chick Sensory Neurons. *Pflügers Arch.* **408**:R60-(1/4) (1987).

Carbone, E., E. Sher and F. Clementi: Ca Currents in Human Neuroblastoma IMR32 Cells: Kinetics, Permeability and Pharmacology. *Pflügers Arch.*(in press) (1/4) (1990b).

Carbone, E. and D. Swandulla: Neuronal Calcium Channels: Kinetics, Blockade and Modulation. *Prog. Biophys. Mol. Biol.* **54**:31-(1/3) (1989).

Carboni, E., W.J. Wojcik and E. Costa: Dihydropyridines Change the Uptake of Calcium Induced by Depolarization into Primary Cultures of Cerebellar Granule Cells. *Neuropharmacology* **24**:1123-(3/13) (1985).

Care, A.D., L.M. Sherwood, J.T. Potts, and G.D. Aurbach: Perfusion of the Isolated Parathyroid Gland of the Goat and Sheep. *Nature* **209**:55-(2/7) (1966).

Carman, J.S., R.M. Post, F.K. Goodwin and W.E. Bunney: Calcium and Electroconvulsive Therapy of Severe Depressive Illness. *Biol.Psychiatr.* **12**:5-(4/22) (1977).

Carman, J.S. and R.J. Wyatt: Calcium: Bivalent Cation in the Bivalent Psychoses *Biol. Psychiatr.* 14:295-(4/22) (1979).

Caro, G., M. Barrios and J.M. Baeywna: Dose-Dependent and Stereoselective Antagonism by Diltiazem of Naloxone-Precipitated Morphine Abstinence after Acute Morphine- Dependence *In Vivo* and *In Vitro. Life Sci.* **43**:1523-(3/13) (1988).

Carvalho, C.M., C.R. Oliveira, M.P. Lima, J.E. Leysen and A.P. Carvalho: Partition of Ca^{2+} Antagonists in Brain Plasma Membranes. *Biochem.Pharmacol.* **38**:2121-(3/13) (1989).

Casta, A.: Verapamil in Patients with WPW. *J.Pediatr.* **99**:502-(4/19) (1981).
Castillo, C.J., R.I. Fonterez, M.G. Lopez, K. Rosenheck and A.J. Garcia: (+)-PN200-110 and Ouabain Binding Sites in Purified Bovine Adrenomedullary Plasma Membranes and Chromaffin Cells. *J.Neurochem.* **53**:1442-(3/13) (1989).
Castle, N.A., D.G. Haylett and D.H. Jenkinson: Toxins in the Characterization of Potassium Channels. *TINS* **12**:59-(2/9) (1989).
Catalano, M., A. Aronica, S. Belletti, E. Sacchi, M. Tommasini and A. Libretti: Flunarizine in the Treatment of Peripheral Vascular Disease. *Intern.Angiol.* **4**:73-(4/20) (1985).
Catterall, W.A.: Structure and Function of Voltage Sensitive Ion Channels. *Science* **242**:50-(1/3,3/14) (1988).
Caubet, J.F.: C-fos Protooncogene Expression in the Nervous System During Mouse Development. *Mol.Cell Biol.* **9**:2269-(2/10) (1989).
Cauvin, C., R. Loutzehiser and C. Van Breemen: Mechanisms of Calcium Antagonist Induced Vasodilation. *Ann.Rev.Pharmacol.Toxicol.* **23**:373-(2/10,4/20) (1983).
Cavalie, A., R. Ochi, D. Pelzer and W. Trautwein: Elementary Currents through Calcium Channels in Guinea Pig Myocytes. *Pflügers Arch.* **398**:284-(1/4,3/12) (1983).
Cavero, I., N. Shepperson, F. Lefevre-Borg and S.Z. Langer: Differential Inhibition of Vascular Smooth Muscle Responses to Alpha-1 Alpha-2 Adrenoreceptor Agonists by Diltiazem and Verapamil. *Circ.Res.* **52**:69-(4/17) (1983).
Cazalis, M., G. Dayanithi and J. Nordmann: Hormone Release from Isolated Nerve Endings of the Rat Neurohypophysis. *J.Physiol.* **390**:55-(2/7,3/13) (1987a).
Cazalis, M., G. Dayanithi and J.J. Nordmann: The Role of Patterned Burst and Interburst Interval on the Excitation-Coupling Mechanism in the Isolated Rat Neural Lobe. *J.Physiol.* **369**:45-(2/7) (1985).
Cazalis, M., G. Dayanithi and J.J. Nordmann: Requirements for Hormone Release from Permeabilized Nerve Endings Isolated from the Rat Neurohypophysis. *J.Physiol.* **390**:71-(2/7) (1987b).
Celentano, A., M. Galderisi, M. Garofalo, G.F. Mureddu, P. Tammaro, M. Petitto, S. Di Somma and O. de Divitiis: Blood Pressure and Cardiac Morphology in Young Children of Hypertensive Subjects. *J.Hypertens.* **6**:l07-(4/18) (1988).
Cervetto, L., L. Lagnado, R.J. Perry, D.W. Robinson and P.A. McNaughton: Extrusion of Calcium from Rod Outer Segment is Driven by Both Sodium and Potassium Gradients. *Nature* **337**:740-(2/8) (1989).
Cervetto, L., A. Menini, G. Rispoli and V. Torre: The Modulation of the Ionic Selectivity of the Light-Sensitive Current in Isolated Rods of the Tiger Salamander. *J.Physiol.* **406**:181-(2/8) (1988).
Chad, J.: Inactivation of Calcium Channels. *Comp.Biochem.Physiol.* **93**:95-(1/4) (1989).
Chad, J.E. and R. Eckert: Calcium Domains Associated with Individual Channels can Account for Anomalous Voltage Relations of Ca-Dependent Responses. *Biophys.J.* **45**:993-(1/4) (1984).
Chad, J.E. and R. Eckert: An Enzymatic Mechanism for Calcium Current Inactivation in Dialysed *Helix* Neurones. *J.Physiol.* **378**:31-(1/4,2/9,3/14) (1986).
Chadda, K., S. Goldstein, R. Cyington and J.D. Curb: Effect of Propranolol after Acute Myocardial Infarction in Patients with Heart Failure. *Circulation* **73**:503-(4/19) (1986).
Chaffman, M. and R.N. Brogden: Diltiazem: A Review of its Pharmacological Properties and Therapeutic Efficacy. *Drugs* **29**:387-(3/13) (1985).
Chaitman, B.R., P. Wagniart, A. Pasternac, G. Brevers, J. School, J. Lam, M. Methe, R.J. Furguson and M.G. Bourossa: Improved Exercise Tolerance after Propranolol, Diltiazem or Nifedipine in Angina Pectoris: Comparison at l, 3 and 8 Hours and Correlation with Plasma Drug Concentration. *Am.J.Cardiol.* **53**:1-(4/16) (1984).
Chalazonitis, A. and G.D. Fischbach: Elevated Potassium Induced Morphological Differentiation of Dorsal Root Ganglionic Neurons in Dissociated Cell Culture. *Dev.Biol.* **78**:173-(2/10) (1980).
Chambers, E.L., B.C. Pressman and B. Rose: The Activation of Sea Urchin Eggs by the Divalent Ionophores A23l87 and X-537A. *Biochem.Biophys.Res.Commun.* **60**:126-(2/7) (1974).
Chandler, W.K. and H. Meves: Slow Changes in Membrane Permeability and Long-lasting Action Potentials in Axons Perfused with Fluoride Solutions. *J.Physiol.* **211**:707-(1/4) (1970).
Chandler, W.K., R.F. Rakowski and M.F. Schneider: A Non-Linear Voltage Dependent Charge Movement in Frog Skeletal Muscle. *J.Physiol.* **254**:245-(2/10) (1976a).
Chandler, W.K., R.F. Rakowski and M.F. Schneider: Effects of Glycerol Treatment and Maintained Depolarization on Charge Movement in Skeletal Muscle. *J.Physiol.* **254**:285-(2/10) (1976b).
Chang, F.C. and M.M. Hosey: Dihydropyridine and Phenylalkyamine Receptors Associated with Cardiac and Skeletal Muscle Calcium Channels are Structurally Different. *J. Biol. Chem.* **263**:18929-(3/13) (1988).
Changelian, P.S., P. Feng, T.C. King and J. Milbrandt: Structure of the NGFI-A Gene and Detection of Upstream Sequences Responsible for its Transcriptional Induction by Nerve Growth Factor. *Proc.Natl.Acad.Sci.U.S.A.* **86**:377-(2/10) (1989).
Charlton, M.P., S.J. Smith and R.S. Zucker: Role of Presynaptic Calcium Ions and Channels in Synaptic Facilitation and Depression at the Squid Giant Synapse. *J.Physiol.* **323**:173-(2/7) (1982).
Charo, I.F., R.D. Feinman and T.C. Detrviler: Interrelationships of Platelet Aggregation and Secretion. *J.Clin.Invest.* **60**:866-(4/20) (1977).
Chatelain, P., D. Demol and J. Roba: Comparison of [^{3}H]Nitrendipine Binding to Heart Membranes of Normotensive and Spontaneously Hypertensive Rats. *J. Cardiovasc. Pharmacol.* **6**:222-(3/13) (1984).

Chatterjee, K.: Calcium Antagonist Agents in Hypertrophic Cardiomyopathy. *Am.J.Cardiol.* **59**:146B-(4/17) (1987).

Chaudhari, N. and K.G. Beam: The Muscular Dysgenesis Mutation in Mice Leads to the Arrest of the Genetic Program for Muscle Differentiation. *Dev.Biol.* **133**:456-(2/10) (1989).

Chen, C., M.J. Corbley, T.M. Roberts and P. Hess: Voltage-sensitive Calcium Channels in Normal and Transformed 3T3 Fibroblasts. *Science* **239**:1024-(3/13,3/14) (1988).

Cherksey, B., J.W. Lin, M. Sugimori and R. Llinas: Synthetic FTX-Like Blocks P-Type Calcium Currents. *Biophys.J.* **57**:305a-(3/13) (1990).

Chesnoy-Marchai, D. and J. Fritsch: Voltage Gated Sodium and Calcium Currents in Rat Osteoblasts. *J.Physiol.* **398**:291-(3/14) (1988).

Chester, D.W., L.G. Herbette, R.P. Mason, A.F. Joslyn, D.J. Triggle and D.E. Koppel: Diffusion of Dihydropyridine Calcium Channel Antagonists in Cardiac Sarcolemmal Lipid Multilayers *Biophys. J.* **52**:1021-(3/13) (1987).

Cheung, D.G., J.L. Gasster, J.M. Neutel and M.A. Weber: Acute Pharmacokinetic and Hemodynamic Effects of Intravenous Bolus Dosing of Nicardipine. *Am.Heart J.* **119**:438-(4/18) (1990).

Cheung, J.Y., J.V. Bonventre, C.D. Malis and A. Leaf: Calcium and Ischemic Injury. *N.Engl.J.Med.* **314**:1670-(4/21) (1986).

Chew, C.Y.C., B.G. Brown, B.N. Singh, M.M. Wong, C. Pierce and R. Petersen: Effects of Verapamil on Coronary Hemodynamic Function and Vasomobility Relative to its Mechanism of Antianginal Action. *Am.J.Cardiol.* **51**:699-(4/16) (1983).

Chiang, T.M., L.M. Cagen and A.H. Kang: Effects of 1-0-Tetradecanoyl Phorbol 13-Acetate on Platelet Aggregation. *Thromb.Res.* **21**:611-(4/20) (1981).

Chiarandini, D.J. and E. Stefani: Calcium Action Potentials in Rat Fast-Twitch and Slow-Twitch Muscle Fibres. *J.Physiol.* **335**:29-(3/12) (1983).

Chiu, S.Y., H.E. Mrose and J.M. Ritchie: Anomalous Temperature Dependence of the Sodium Conductance in Rabbit Nerve Compared with Frog Nerve. *Nature* **279**:327-(1/4) (1979).

Cho, K.O., W.C. Skarnes, B. Minsk, S. Palmieri, L. Jackson-Grusby and J.A. Wagner: Nerve Growth Factor Regulates Gene Expression by Several Distinct Mechanisms. *Mol.Cell Biol.* **9**:135-(2/10) (1989).

Choi, D.W.: Glutamate Neurotoxicity in Cortical Cell Culture is Calcium Dependent. *Neurosci.Lett.* **58**:293-(4/21) (1985).

Choi, O.H., W.L. Padgett, Y. Nishizawa, F. Gusovsky, T. Yasumoto and J.W. Daly: Effects on Calcium Channels, Phosphoinositide Breakdown, and Arachidonate Release in Pheochromocytoma PC12 Cells. *Mol. Pharmacol.* **37**:222-(3/13) (1990).

Choi, S.M., M.A. Taylor and R. Abrams: Depression, ECT, and Erythrocyte Adenosinetriphosphatase Activity. *Biol. Psychiatr.* **12**:75-(4/22) (1977).

Chouza, C., J.L. Caamano, R. Aljanati, A. Scaramelli, O. DeMedina and S. Romero: Parkinsonsim, Tardive Dyskinesia, Akathisia, and Depression Induced by Flunarizine. *Lancet* **8493**:1303-(4/22) (1986).

Chow, I. and M.W. Cohen: Developmental Changes in the Distribution of Acetylcholine Receptors in the Myotomes of *Xenopus laevis*. *J.Physiol.* **339**:553-(2/10) (1983).

Cicotera, P., D.J. McConkey, D.P. Jones and S. Orrenius: ATP Stimulates Ca^{2+} Uptake and Increases the Free Ca^{2+} Concentration in Isolated Rat Liver Nuclei. *Proc.Natl.Acad.Sci.U.S.A.* **86**:453-(2/10) (1989).

Clapham, D.E. and E. Neher: Trifluoperazine Reduces Inward Ionic Currents and Secretion by Separate Mechanisms in Bovine Chromaffin Cells. *J.Physiol.* **353**:541-(2/7) (1984).

Clarke, B., D. Grant, L. Patmore and R.L. Whiting: Comparative Calcium Entry Blocking Properties of Nicardipine, Nifedipine, and PY-l08068 on Cardiac and Vascular Smooth Muscle. *Br.J.Pharmacol.* **79**:333P-(4/18) (1983).

Clasbrummel, B., H. Osswald and P. Illes: Inhibition of Noradrenaline Release by ω-Conotoxin GVIA in the Rat Tail artery. *Br.J.Pharmacol.* **96**:101-(3/13) (1989).

Clusin, W.T.: What is the Solution to Sudden Cardiac Death: Calcium Modulation or Arrhythmia Clinics? *Cardiovasc.Drugs Ther.* **1**:335-(4/19) (1987).

Clusin, W.T., M.R. Bristow, D.S. Baim, J.S. Schroeder, P. Jaillon, P. Brett and D.C. Harrison: The Effects of Diltiazem and Reduced Serum Ionized Calcium on Ischemic Ventricular Fibrillation in the Dog. *Circ.Res.* **50**:518-(4/16) (1982).

Clusin, W.T., L.H. Opie and F. Thandroyen: Drastic Reduction of Cardiac Work Fails to Mimic the Antifibrallatory Effect of Low Calcium. *Circulation* **70**:II-(4/19) (1984).

Cobbs, W.H. and E.N. Pugh: Cyclic GMP can Increase Rod Outer Segment Light-sensitive Current l0 Fold without Delay of Excitation. *Nature* **313**:585-(2/8) (1985).

Cobbs, W.H. and E.N. Pugh: Kinetics and Components of the Flash Photocurrent of Isolated Retinal Rods of the Larval Salamander, *Ambystoma tigrinum*. *J.Physiol.* **394**:529-(2/8) (1987).

Coburn, R.A., M. Wierzba, M.J. Suto, A.J. Solo, A.M. Triggle and D.J. Triggle: 1,4-Dihydropyridine Antagonist Activities at the Calcium Channel: A Quantitative Structure-Activity Relationship Approach. *J.Med.Chem.* **31**:2103-(3/13) (1988).

Cocco, G., C. Strozzi, D. Chu, R. Amrein and E. Castagnoli: Therapeutic Effects of Pindolol and Nifedipine in Patients with Stable Angina Pectoris and Asymptomatic Resting Ischemia. *Eur.J.Cardiol.* **10**:59-(4/16) (1979).

Cochrane, D.E. and W.W. Douglas: Calcium-Induced Extrusion of Secretory Granules (Exocytosis) in Mast Cells Exposed to 48/80 or the Ionophores A-23l87 and X-537A. *Proc.Natl.Acad.Sci.U.S.A.* **71**:408-(2/7) (1974).

Coetzee, W.A., S.C. Dennis, L.H. Opie and C.A. Muller: Calcium Channel Blockers and Early Ischemic Ventricular Arrhythmias: Electrophysiological versus Anti-Ischemic Effects. *J.Mol.Cell Cardiol.* **19**:77-(4/16) (1987).

Coetzee, W.A. and L.H. Opie: Effects of Components of Ischemia and Metabolic Inhibition on Delayed After Depolarizations in Guinea Pig Papillary Muscle. *Circ.Res.* **61**:157-(4/16) (1987).

Cognard, C.A., G. Romey, J.P. Galizzi, M. Fosset and M. Lazdunski: Dihydropyridine-sensitive Ca^{2+} Channels in Mammalian Skeletal Muscle Cell in Culture: Electrophysiological Properties and Interaction with Ca^{2+} Channel Activator (BAY K 8644) and Inhibitor (PN 200-110). *Proc.Natl.Acad.Sci.U.S.A.* **83**:1518-(3/13) (1986).

Cohan, C.S., J.A. Connor and S.B. Kater: Electrically and Chemically Mediated Increases in Intracellular Calcium in Neuronal Growth Cones. *J.Neurosci.* **7**:3588-(2/10) (1987).

Cohen, A.I.: Rods and Cones. In *Handbook of Sensory Physiology*. Springer-Verlag, New York. **7**-(2/8) (1972).

Cohen, A.I. and C. Blazynski: Light-Induced Losses and Dark Recovering Rates of Guanosine 3′, 5′-Cyclic Monophosphate in Rod Outer Segments of Intact Amphibian Photorecptors. *J.Gen.Physiol.* **92**:731-(2/8) (1989).

Cohen, A.I., I.A. Hall and J.A. Ferendelli: Calcium and Cyclic Nucleotide Regulation in Incubated Mouse Retinas. *J.Gen.Physiol.* **71**:595-(2/8) (1978).

Cohen, C.J., T. Bale and M.D. Leibowitz: Calcium Agonist and Antagonist Effects of Fatty Acids in Guinea Pig Atrial Myocytes *Biophys. J.* **57**:307a-(3/13) (1990).

Cohen, C.J. and R.T. McCarthy: Nimodipine Block of Calcium Channels in Rat Anterior Pituitary Cells. *J.Physiol.* **387**:195-(3/13,3/14) (1987).

Cohen, C.J., R.T. McCarthy, P.Q. Barrett and H. Rasmussen: Ca Channels in Adrenal Glomerulosa Cells: K^+ and Angiotensin II Increase T-Type Ca Channel Current. *Proc.Natl.Acad.Sci.U.S.A.* **85**:2412-(3/13,3/14) (1988).

Cohen, C.J., S. Spires and D. Van Skiver: Modulation of L- and T-Type Calcium Channels in Guinea Pig Atrial Cells by 1,4-Dihydropyridines. In *Molecular and Cellular Mechanisms of Antiarrhythmic Agents*. L. Hondeghem and B.G. Katzung, eds. Futura Press. Ch. 9 (3/13) (1990).

Cohen, M.W., M. Greschner and M. Tucci: *In Vivo* Development of Cholinesterase at a Neuromuscular Junction in the Absence of Motor Activity in *Xenopus laevis*. *J.Physiol.* **348**:57-(2/10) (1984).

Colatsky, T.J.: Voltage Clamp Measurements of Sodium Channel Properties in Rabbit Cardiac Purkinje Fibres. *J.Physiol.* **305**:215-(1/4) (1980).

Cole, K.S. and H. Curtis: Electrical Impedance of *Nitella* during Activity. *J.Gen.Physiol.* **22**:37-(1/1) (1938).

Collingridge, G.L., C.E. Herron and R.A.J. Lester: Frequency-Dependent N-Methyl-D-Aspartate Receptor-Mediated Synaptic Transmission in Rat Hippocampus. *J.Physiol.* **399**:301-(2/9) (1988).

Collins, F. and J.D. Lile: The Role of Dihydropyridine Sensitive Voltage-Gated Channels in Potassium-Mediated Neuronal Survival. *Brain Res.* **502**:99-(2/10) (1989).

Colucci, W.S.: Usefulness of Calcium Antagonists for Congestive Heart Failure. *Am.J.Cardiol.* **59**:52B-(4/17) (1987).

Colvin, R.A., R.A. Allen, R.G. Williams, D.T. Eagle, J.A. Oibo and G.D. Miner: [125I] ω-Conotoxin Binding to Human Frontal Cortex from Normal, Alzheimer's and Non-Alzheimer's Dementia Patients. *Neurobiol.Aging* **11**:1-(3/13) (1990).

Colvin, R.A., T.F. Ashavaid and L.G. Herbette: Structure-Function Studies of Canine Cardiac Sarcolemmal Membranes. I. Estimation of Receptor Site Densities. *Biochim. Biophys. Acta* **812**:601-(3/13) (1985).

Colvin, R.A., R.G. Williams, D.T. Eagle, R.A. Allen, J.A. Oibo, I. Ibok and G.D. Miner: Evidence for Altered Neuronal Binding of [^{3}H]Nitrendipine in Dementia. In *Familial Alzheimer's Disease, Molecular Genetics and Clinical Perspectives*. G.D. Miner, R.W. Richter, J.P. Blass, J.L. Valentine and L.A. Winters-Miner, eds. Mercel Dekker, New York and Basel, 325-(3/13) (1990).

Connor, J.A.: Calcium Localization Associated with c-Fos Induction in Acutely Isolated CA1 Neurons. *Soc.Neurosci.Abst.* **14**:82-(2/10) (1988).

Connor, J.A., M.C. Cornwall and G.H. Williams: Spatially Resolved Cytosolic Calcium Reponse to Angiotensin II and Potassium in Rat Glomerulosa Cells Measured by Digital Imaging Techniques. *J.Biol.Chem.* **262**:2919-(2/7) (1987).

Connor, J.A., W.J. Wadman, P.E. Hockberger and R.K.S. Wong: Sustained Dendritic Gradients of Ca^{2+} Induced by Excitatory Amino Acids in CA1 Hippocampal Neurons. *Science* **240**:649-(2/9) (1988).

Cook, D.I., G.B. Gard, M. Champion and J.A. Young: Patch-clamp Studies of the Electrolyte Secretory Mechanism of Rat Mandibular Gland Cells Stimulated with Acetylcholine or Isoproterenol. In *Molecular Mechanisms in Secretion*. 25th Alfred Benzon Symposium, Munksgaard, Copenhagen. 133-(2/7) (1988).

Cook, N.J., W. Hanke and U.B. Kaupp: Identification, Purification, and Functional Reconstitution of the Cyclic GMP-Dependent Channel from Rod Photoreceptors. *Proc.Natl.Acad.Sci.U.S.A.* **84**:585-(2/8) (1987).

Cook, N.J., L.L. Molday, D. Reid, U.B. Kaupp and R.S. Molday: The cGMP-gated Channel of Bovine Rod Photoreceptors is Localized Exclusively in the Plasma Membrane. *J.Biol.Chem.* **264**:6996-(2/8) (1989).

Cook, N.J., C. Zeilinger, K.W. Koch and U.B. Kaupp: Solubilization and Functional Reconstitution of the cGMP-Dependent Cation Channel from Bovine Rod Outer Segment. *J.Biol.Chem.* **261**:17033-(2/8) (1986).

Cooke, J.P., T.J. Rimele, N.A. Flavahan and P.M. Vanhoutte: Nimodipine and Inhibition of Alpha Adrenergic Activation of the Isolated Canine Saphenous Vein. *J.Pharmacol.Exp.Ther.* **234**:598-(4/18) (1985).

Cooperman, S.S., R.L. Barchi, R.H. Goodman and G. Mandel: Modulation of Sodium-Channel mRNA Levels in Rat Skeletal Muscle. *Proc.Natl.Acad.Sci.U.S.A.* **84**:8721-(2/10) (1987).

Corey, D.P., J.M. Dubinsky and E.A. Schwartz: The Ca Current in Inner Segments of Rods from Salamander (*Ambystoma tigrinum*) Retina. *J.Physiol.* **254**:557-(2/8) (1984).

Corey, D.P. and A.J. Hudspeth: Ionic Basis of the Receptor Potential in a Vertebrate Hair Cell. *Nature* **281**:675-(2/8) (1979).

Cota, G. and E. Stefani: Saturation of Calcium Channels and Surface Charge Effects in Skeletal Muscle Fibres of the Frog. *J.Physiol.* **351**:135-(3/12) (1984).

Cota, G. and E. Stefani: Effects of External Calcium Reduction on the Kinetics of Potassium Contractures in Frog Twitch Muscle Fibers *J. Physiol.* **317**:303-(2/10) (1981)

Cota, G. and E. Stefani: A Fast-Activated Inward Calcium Current in Twitch Muscle Fibres of the Frog (*Rana montezume*). *J.Physiol.* **370**:151-(2/10) (1986).

Cote, R.H., M.S. Biernbaum, G.D. Nicol and M.D. Bownds: Light-Induced Decreases in cGMP Concentration Precede Changes in Membrane Permeability in Frog Rod Photoreceptors. *J.Biol.Chem.* **259**:9635-(2/8) (1984).

Cote, R.H., G.D. Nicol, S.A. Burke and M.D. Bownds: Changes in cGMP Concentration Correlate with Some, but not All, Aspects of the Light-Regulated Conductance of Frog Rod Photoreceptors. *J.Biol.Chem.* **261**:12965-(2/8) (1986).

Coulter, D.A., J.R. Huguenard and D.A. Prince: Specific Petit Mal Anticonvulsants Reduce Calcium Currents in Thalamic Neurons. *Neurosci.Lett.* **98**:74-(3/13) (1989b).

Coulter, D.A., J.R. Huguenard and D.A. Prince: Anticonvulsants Depress Calcium Spikes and Calcium Currents of Mammalian Thalamic Neurons *In Vitro*. *Neurobiology* **39**:412-(3/13) (1989a).

Coulter, D.A., J.R. Huguenard and D.A. Prince: Calcium Currents in Rat Thalamocortical Relay Neurones: Kinetic Properties of the Transient, Low-Threshold Current. *J.Physiol.* **414**:587-(1/4) (1989c).

Coumel, P. and P. Attuel: Which Arrhythmias are Specifically Susceptible to Calcium Antagonists? In *Frontiers of Cardiac Electrophysiology*. M.B. Rosenbaum, M.V Elizari, eds. Martinus Nijhoff, Boston. 341-(4/19) (1983).

Coumel, P., J.F. Leclercy and P. Attuel: Drug-resistant Paroxysmal Ventricular Tachycardia: Approach to Drug Management by Intracardiac Electrocardiography and Ambulatory ECG Monitoring. In *Management of Ventricular Tachycardia: Role of Mexiletine*. E. Sandoe, D.J. Julian and J. W. Bell, eds. Excerpta Medica, Holland. 433-(4/19) (1979).

Crammer, J.L.: Calcium Metabolism and Mental Disorder. *Psycholog. Med.* **7**:557-(4/22) (1977).

Cranefield, P.F.: *The Conduction of the Cardiac Impulse*. Futura, Mount Kisco, New York. 1-(3/12) (1975).

Cranefield, P.F.: Action Potentials, Afterpotentials and Cardiac Arrhythmias. *Cir.Res.* **41**:415-(4/19) (1977).

Cranefield, P.F., R.L.S. Aronson and A.L. Wit: Effects of Verapamil on the Normal Action Potential and on the Calcium-Dependent Slow Response of Canine Purkinje Fibers. *Circ.Res.* **34**:204-(4/19) (1974).

Crawford, A.C. and R. Fettiplace: An Electrical Tuning Mechanism in Turtle Cochlear Hair Cells. *J.Physiol.* **312**:377-(2/8) (1981).

Crea, F., J. Deanfield, P. Crean, M. Sharom, G. Davies and A. Maseri: Effects of Verapamil in Preventing Early Postinfarction Angina and Reinfarction. *Am.J.Cardiol.* **55**:900-(4/16,4/19) (1985).

Creager, M.A., K.M. Pariser, E.M. Winston, H.M. Rasmussen, K.B. Miller and J.D. Coffman: Nifedipine-Induced Fingertip Vasodilation in Patients with Raynaud's Phenomenon. *Am.Heart J.* **108**:370-(4/20) (1984).

Crepel, F., S.S. Dhanjal and J. Garthwaite: Morphological and Electrophysiological Characteristics of Rat Cerebellar Slices Maintained *In Vitro*. *J.Physiol.* **316**:127-(2/9) (1981).

Crunelli, V., S. Lightowler and C.E. Pollard: A T-Type Ca^{2+} Current Underlies Low-Threshold Ca^{2+} Potentials in Cells of the Cat and Rat Lateral Geniculate Nucleus. *J.Physiol.* **413**:543-(1/4) (1989).

Curran, T. and B.R. Franza Jr.: Fos and Jun: The AP-1 Connection. *Cell* **55**:395-(2/10) (1988).

Curran, T. and J.I. Morgan: Barium Modulates c-Fos Expression and Post-Translational Modification. *Proc.Natl.Acad.Sci.U.S.A.* **83**:8521-(2/10) (1986).

Curry, D.L., L.L. Bennett and G.M. Grodsky: Requirement for Calcium Ion in Insulin Secretion by the Perfused Rat Pancreas. *Am.J.Physiol.* **214**:174-(2/7) (1968).

Curtis, B.A.: Ca Fluxes in Single Twitch Muscle Fibers. *J.Gen.Physiol.* **50**:255-(2/10) (1966).

Curtis, B.M. and W.A. Catterall: Purification of the Calcium Antagonist Receptor of the Voltage-Sensitive Calcium Channel from Skeletal Muscle Transverse Tubules. *Biochem.* **23**:2113-(2/10) (1984).

Curtis, B.M. and W.A. Catterall: Phosphorylation of the Calcium Antagonist Receptor of the Voltage Sensitive Calcium Channel by cAMP Dependent Protein Kinase. *Proc.Natl.Acad.Sci.U.S.A.* **82**:2528-(3/14) (1985).

D'Brot, J., W.M. Abraham and T. Ahmed: Effect of Calcium Antagonist Gallopamil on Antigen-Induced Early and Late Bronchoconstrictor Responses in Allergic Sheep. *Am.Rev.Respir.Dis.* **139**:915-(3/13) (1989).

D'Orleans-Juste, P., G. DeNucci and J.R. Vane: Endothelin-1 Contracts Isolated Vessels Independently of Dihydropyridine-Sensitive Ca^{2+} Channel Activation. *Eur.J.Pharmacol.* **165**:289-(3/13) (1989).

Dacquet, C., P. Pacaud, G. Loirand, C. Mironneau and J. Mironneau: Comparison of Binding Affinities and Calcium Current Inhibitory Effects of a 1,4-Dihydropyridine Derivative (PN 200-110) in Vascular Smooth Muscle. *Biochem.Biophys.Res.Commun.* **152**:1165-(3/13) (1988).

Daemen, F.: Vertebrate Rod Outer Segment Membranes. *Biochim.Biophys.Acta* **300**:255-(2/8) (1973).

Dale, J., K.H. Landmark and E. Myhre: The Effect of Nifedipine, a Calcium Antagonist on Platelet Function. *Am.Heart J.* **105**:103-(4/20) (1983).

Dale, N. and A. Roberts: Dual-Component Amino-Acid-Mediated Synaptic Potentials: Excitatory Drive for Swimming in *Xenopus* Embryos. *J.Physiol.* **363**:35-(2/9) (1985).

Dalton, W.G., T.M. Grogan, P.S. Meltzer, R.J. Scheper, B.G.M. Durie, C.W. Taylor, T.P. Miller and S.E. Salmon: Drug-resistance in Multiple Myeloma and Non-Hodgkin's Lymphoma: Detection of P-Glycoprotein and Potential Circumvention by Addition of Verapamil to Chemotherapy. *J.Clin.Oncol.* **7**:415-(3/13) (1989).

Damiano, B.V. and M.R. Rosen: Effects of Pacing on Triggered Activity Induced by Early Afterdepolarizations. *Circulation* **69**:977-(4/19) (1984).

Daniel, J.L., I.R. Molish, C.C. Rigmaiden and G. Stewart: Evidence for a Role of Myosin Phosphorylation in the Initiation of the Platelet Shape Change Responses. *J. Biol.Chem.* **259**:9826-(4/20) (1984).

Danish Multicenter Study Group: Verapamil in Myocardial Infarction. *Clin.Pharmacol.Exp.Physiol.* **6**:89-(4/19) (1982).

Danish Multicenter Study Group: Verapamil in Myocardial Infarction. *Am.J.Cardiol.* **54**:24E-(4/19) (1984).

Danish Multicenter Study Group: Verapamil in Myocardial Infarction. *Eur.Heart J.* **4**:516-(4/19) (1984).

Danish Study Group on Verapamil in Myocardial Infarction: The Danish Studies on Verapamil in Acute Myocardial Infarction. *Br.J.Clin.Pharmacol.* **21**:197S-(4/16) (1986).

Danish Study Group on Verapamil in Myocardial Infarction. Effects of Verapamil on Mortality and Major Events After Acute Myocardial Infarction (The Danish Verapamil Infarction Trial II—DAVIT II). *Am. J. Cardiol.* **66**:779-(4/16) (1990).

Dargie, H., E. Rowland and D. Kirkler: Role of Calcium Antagonists in Cardiovascular Disease. *Br.Heart J.* **46**:8-(4/19) (1981).

Davey, D.F. and M.W. Cohen: Localization of Acetycholine Receptors and Cholinesterase on Nerve-Contacted and Noncontacted Muscle Cells Grown in the Presence of Agents that Block Action Potentials. *J.Neurosci.* **6**:673-(2/10) (1986).

David, J.D., W.M. See and C.A. Higginbotham: Fusion of Chick Embryo Skeletal Myoblasts: Role of Calcium Influx Preceding Membrane Union. *Dev.Biol.* **82**:308-(2/10) (1981).

Dawson, J.R., N.H.G. Whitaker and G.C. Sutton: Calcium Antagonists in Chronic Stable Angina. Comparison of Verapamil and Nifedipine. *Br.Heart J.* **46**:508-(4/16) (1981).

Dayanithi, G., N. Martin-Moutot, S. Barlier, D.A. Colin, M. Kretz-Zaepfel, F. Couraud and J.J. Nordmann: The Calcium Channel Antagonist ω-Conotoxin Inhibits Secretion from Peptidergic Nerve Terminals. *Biochem.Biophys.Res.Commun.* **156**:255-(3/13) (1988).

de Boland, A.R., I. Nemere and A.W. Norman: Ca^{2+}-channel Agonist BAY K 8644 Mimics 1,25(OH)2-Vitamin D3 Rapid Enhancement of Ca^{2+} Transport in Chick Perfused Duodenum. *Biochem.Biophys.Res.Commun.* **166**:217-(3/13) (1990).

De Jong, J.W., E. Harmsen, P.P. De Tombe and E. Keijzer: Nifedipine Reduces Adenine Nucleotide Breakdown in Ischemic Rat Heart. *Eur.J.Pharmacol.* **81**:89-(4/16) (1982).

De Leiris, J., V. Richard and S. Pestre: Calcium Antagonists and Experimental Myocardial Ischemia and Infarction. In *Calcium Antagonists and Cardiovascular Disease.* L.H. Opie, ed. Raven Press, New York. 105-(4/16) (1984).

De Lorme, E.M., C.S. Rabe and R. McGee: Regulation of the Number of Functional Voltage-Sensitive Ca^{2+} Channels in PC12 Cells by Chronic Changes in Membrane Potential. *J.Pharmacol.Exp.Ther.* **244**:838-(3/13) (1988).

de Geest, R., H. Kesteloot and J. Piessens: Secondary Prevention Ischemic Heart Disease: A Long-Term Controlled Lidoflazine Study. *Acta Cardiol.* **24**:7-(4/19) (1979).

Deanfield, J., C. Wright and K. Fox: Treatment of Angina Pectoris with Nifedipine: Importance of Dose Titration. *Br.Med.J.* **286**:1467-(4/16) (1983).

Deanfield, J., C. Wright, S. Krikler, P. Rebeiro and K. Fox: Cigarette Smoking and the Treatment of Angina with Propranolol, Atenolol and Nifedipine. *N.Engl.J.Med.* **310**:951-(4/16) (1984).

DeAzeredo, F.A.M., W.D. Lust and J.V. Passoneau: Guanine Nucleotide Concentrations *In Vivo* in Outer Segments of Dark and Light Adapted Frog Retina. *Biochem.Biophys.Res.Commun.* **85**:293-(2/8) (1978).

DeCamilli, P. and F. Navone: Regulated Secretory Pathways of Neurons and their Relation to the Regulated Pathway of Endocrine Cells. *Ann.N.Y Acad.Sci.* **493**:461-(2/7) (1987).

DeCino, P. and Y. Kidokoro: Development and Subsequent Neural Tube Effects on the Excitability of Cultured *Xenopus* Myocytes. *J.Neurosci.* **5**:1471-(2/10) (1985).

Deisz, R. and H.D. Lux: γ-Aminobutyric Acid-Induced Depression of Calcium Currents of Chick Sensory Neurons. *Neurosci.Lett.* **56**:205-(1/3) (1985).

Deitmer, J.: Evidence for Two Voltage Dependent Calcium Currents in the Membrane of Ciliate *Stylonchia*. *J.Physiol.* **355**:137-(3/12) (1984).

del Castillo, J. and B. Katz: Quantal Components of the End-plate Potential. *J.Physiol.* **124**:560-(2/7) (1954).

DeLorenzo, R.J.: The Calmodulin Hypothesis of Neurotransmission. *Cell Calcium* **2**:365-(4/22) (1981).

Deneris, E.S., J. Connolly, J. Boulter, E. Wada, K. Wada, L.W. Swanson, J. Patrick and S. Heinemann: Primary Structure and Expression of Beta-2: A Novel Subunit of Neuronal Nicotinic Acetylcholine Receptors. *Neuron* **1**:45-(3/14) (1988).

Denes, P., D. Wu, R. Dhingra, R. Chuquimia and K.M. Rosen: Demonstration of Dual AV Nodal Pathways in Patients with Paroxysmal Supraventricular Tachycardia. *Circulation* **48**:549-(4/19) (1973).

Dennis, M.J.: Development of the Neuromuscular Junction: Inductive Interactions Between Cells. *Ann.Rev.Neuroscience* **4**:43-(2/10) (1981).

DeRiemer, S.A. and B. Sakmann: Two Calcium Currents in Normal Rat Anterior Pituitary Cells Identified by a Plaque Assay. In *Experimental Brain Research*, Springer-Verlag Berlin, Heidelberg. **14**:139-(2/7) (1986).

DeRiemer, S.A., J.A. Strong, K.A. Albert, P. Greengard and L.K. Kaczmarek: Enhancement of Calcium Current in *Aplysia* Neurones by Phorbol Ester and Protein Kinase C. *Nature* **313**:313-(3/14) (1985).

Deshpande, J.K., B.K. Siesjo and T. Wieloch: Calcium Accumulation and Neuronal Damage in the Rat Hippocampus Following Cerebral Ischemia. *J.Cereb.Blood Flow Metab.* **7**:89-(4/21) (1987).

Detwiler, P.B., J.D. Conner and R.D. Bodoia: Gigaseal Patch Clamp from Outer Segments of Intact Retinal Rods. *Nature* **300**:59-(2/8) (1982).

Deyo, R.A., K.T. Straube and J.F. Disterhoft: Nimodipine Facilitates Associative Learning in Aging Rabbits. *Science* **243**:809-(3/13) (1989).

Di Perri, T., F. Laghi Pasini and M. Guerrini: Calcium Entry Blockers in the Treatment of Peripheral Obliterative Arterial Disease. In *Calcium Entry Blockers Cardiovascular and Cerebral Dysfunction*. T. Godfraind, A. Herman, D. Wellen, eds. Martinus Nijhoff. The Hague, Holland. 103-(4/20) (1984).

Diebler, H., G. Manfred, G. Ilgenfritz, G. Maab and R. Winkler: Kinetics and Mechanism of Reactions of Main Group Metal Ions with Biological Carriers. *Pure Appl.Chem.* **20**:93-(1/2) (1969).

Dillon, J.S., X.H. Gu and W.G. Nayler: Effect of Age and of Hypertrophy on Cardiac Ca^{2+} Antagonist Binding Sites. *J.Cardiovasc.Pharmacol.* **14**:233-(3/13) (1989).

DiMarco, J.P., T. Duncan Sellers, R.M. Beone, G.A. West and L. Belardinelli: Adenosine: Electrophysiologic Effects and Therapeutic Use for Terminating Paroxysmal Supraventricular Tachycardia. *Circulation* **68**:1254-(4/19) (1984).

DiMarco, J.P., J.D. Sellers, B.B. Lerman, M.L. Greenberg, R.M. Berne and L. Belardinelli: Diagnostic and Therapeutic Use of Adenosine in Patients with Supraventricular Tachyarrhythmias. *J.Am.Coll.Cardiol.* **6**:630-(4/19) (1985).

Docherty, R.J.: Gadolinium Selectively Blocks a Component of Calcium Current in Rodent Neuroblastoma x Glioma Hybrid (NG108-15) Cells. *J.Physiol.* **398**:33-(3/14) (1988).

Docherty, R.J. and I. McFadzean: Noradrenaline-Induced Inhibition of Voltage-Sensitive Calcium Currents in NG108-15 Cells. *Eur.J.Neurosci.* **1**:132-(1/3) (1989).

Dodge, F.A. Jr. and R. Rahamimoff: Cooperative Action of Calcium Ions in Transmitter Release at the Neuromuscular Junction. *J.Physiol.* **193**:419-(2/7) (1967).

Doerner, D. and B.E. Alger: Cyclic GMP Depresses Hippocampal Ca^{2+} Current through a Mechanism Independent of cGMP Dependent Protein Kinase. *Neuron* **1**:693-(3/14) (1988).

Doerner, D., T.A. Pittler and B.E. Alger: Protein Kinase C Activators Block Specific Calcium and Potassium Current Components in Isolated Hippocampal Neurons. *J.Neurosci.* **8**:4069-(2/9,3/14) (1988).

Dolin, S.J., M.J. Halsey and H. Little: Effects of the Calcium Channel Activator BAY K 8644 on General Anesthetic Potency in Mice. *Br.J.Pharmacol.* **94**:413-(3/13) (1988b).

Dolin, S.J., A.B. Hunter, M.H. Halsey and H.J. Little: Anticonvulsant Profile of the Dihydropyridine Calcium Channel Antagonists, Nitrendipine and Nimodipine. *Eur.J.Pharmacol.* **152**:19-(3/13) (1988a).

Dolin, S.J. and H.V. Little: Are Changes in Neuronal Calcium Channels Involved in Ethanol Tolerance. *J. Pharmacol. Exp. Ther.* 250:985-(3/13) (1989).

Dolin, S.J., H. Little, M. Hudspith, C. Pagonis, and J. Littleton: Increased Dihydropyridine-Sensitive Calcium Channels in Rat Brain may Underlie Ethanol Physical Dependence. *Neuropharmacol.* **26**:275-(3/13) (1987).

Dolphin, A.C., S.R. Forda and R.H. Scott: Calcium-Dependent Currents in Cultured Rat Dorsal Root Ganglion Neurones are Inhibited by an Adenosine Analogue. *J.Physiol.* **373**:47-(1/3,) (1986).

Dolphin, A.C. and R.H. Scott: A Guanine Nucleotide Analogue Blocks a Calcium Current in Rat Sensory Neurones and Enhances Inhibition by Baclofen. *J.Physiol.* **372**:81-(3/14) (1986).

Dolphin, A.C. and R.H. Scott: Calcium Channel Currents and Their Inhibition by (-)Baclofen in Rat Sensory Neurones: Modulation by Guanine Nucleotides. *J.Physiol.* **386**:1-(1/3,3/13,3/14) (1987).

Dolphin, A.C. and R.H. Scott: Interaction Between Calcium Channel Ligands and Guanine Nucleotides in Cultured Rat Sensory and Sympathetic Neurones. *J.Physiol.* **413**:271-(1/3,1/4,3/13) (1989).

Dolphin, A.C., J.F. Wootton, R.H. Scott and D.R. Trentham: Photoactivation of Intracellular Guanosine Triphosphate Analogues Reduces the Amplitude and Slows the Kinetics of Voltage Activated Calcium Channel Currents in Sensory Neurons. *Pflügers Arch.* **411**:428-(3/14) (1988).

Donald, J.F.: A Multicentre General Practice Study of Cinnarizine in the Treatment of Peripheral Vascular Disease. *J.Int.Med.Res.* **7**:502-(4/20) (1979).

Donaldson, P.L. and K.G. Beam: Calcium Currents in a Fast-Twitch Skeletal Muscle of the Rat. *J.Gen.Physiol.* **82**:449-(2/10) (1983).

Donaldson, S.K., R. Dunn and D. Huetteman: Peeled Mammalian Skeletal Muscle Fibers: Reversible Block of Cl⁻ Induced Tension Transients by D600 and D890. *Biophys.J.*(abstr.) **45**:46A-(2/10) (1984).

Dongas, J., P. Tchou, R. Mahmud, M.H. Lehman, S. Denker and M. Akhtar: Catecholamine-Mediated Reversal of Procainamide-Induced Retrograde Block in Paroxysmal Supraventricular Tachycardias: Possible Cause of Treatment Failure. *Circulation Abstr.* **72**:126-(4/19) (1985).

Donnelly, G.L. and P. Scaps: Intravenous Verapamil in Cardiac Arrhythmias. *Aust.N.Z.J.Med.Abstr.* **3**:63-(4/19) (1973).

Doran, A.R., P.K. Narang, C.Y. Meigs, W.M. Wolkowitz, A. Roy, A. Breier and D. Pickar: Verapamil Concentrations in Cerebrospinal Fluid after Oral Administration. *N. Engl.J.Med.* **312**:1261-(4/22) (1985).

Dorbin, D.O.C., D.A. Wood, C.C.A. Macintyre and E. Housley: A Randomized Double Blind Cross-over Trial of Nifedipine in the Treatment of Primary Raynaud's Phenomenon. *Eur.Heart J..***7**:165-(4/20) (1987).

Dormandy, J.: Clinical Use of Calcium Antagonists in Peripheral Circulatory Disease. *Ann.N.Y.Acad.Sci.* **522**:611-(4/20) (1987).

Doroshenko, P.A., P.G. Kostyuk and E.A. Luk'Yanetz: Modulation of Calcium Current by Calmodulin Antagonists. *Neuroscience* **27**:1073-(3/14) (1988).

Doroshenko, P.A., P.G. Kostyuk and A.E. Martynyuk: Intracellular Metabolism of Adenosine 3′,5′-Cyclic Monophosphate and Calcium Inward Current in Perfused Neurons of *Helix pomatia. Neuroscience* **7**:2125-(1/4) (1982).

Doroshenko, P.A., P.G. Kostyuk, A.I. Martynyuk, M.D. Kursky and Z.D. Vorobetz: Intracellular Protein Kinase and Calcium Inward Current in Perfused Neurones of the Snail *Helix pomatia. Neuroscience* **11**:263-(3/14) (1986).

Dose, M. and H.M. Emrich: Calcium Antagonists in Mania: Clinical Experiences and Theoretical Considerations. *Clin. Neuropharmacol.* **4**:556-(4/22) (1986).

Dose, M., H.M. Emrich, C. Cording-Tommel and D. von Zerssen: Calcium Antagonists in Mania: A Preliminary Clinical Report. **In** *Psychiatry: The State of the Art*. P. Pichot, P. Berner, and R. Wolf, eds. Plenum Press, New York. **3**:501-(4/22) (1985).

Douglas, W.W.: Stimulus-Secretion Coupling: The Concept and Clues from Chromaffin and other Cells. *Br.J.Pharmacol.* **34**:451-(2/7,4/20) (1968).

Douglas, W.W. and R.P. Rubin: The Role of Calcium in the Secretory Response of the Adrenal Medulla to Acetycholine. *J.Physiol.* **159**:40-(2/7) (1961).

Dowling, M.J. and R.M. Hardisty: Glycoprotein IIb-IIIa Complex and Ca^{2+} Influx into Stimulated Platelets. *Blood* **66**:731-(4/20) (1985).

Doyle, V.M. and U.T. Ruegg: Lack of Evidence for Voltage Dependent Calcium Channels on Platelets. *Biochem.Biophys.Res.Commun.* **127**:161-(4/20,4/22) (1985).

Drayer, J.I.M., W.D. Hall, V.E. Smith, M.A. Weber, G.L. Wollam and W. White: Effects of the Calcium Channel Blocker, Nitrendipine, on LV Mass in Hypertensive Patients. *Clin.Pharmacol.Ther.* **40**:679-(4/18) (1986).

Droogmans, G., I. Declerck and R. Casteels: Effect of Adrenergic Agonists on Ca^{2+} Channel Currents in Single Vascular Smooth Muscle Cells. *Pflügers Arch.* **409**:7-(2/10) (1987).

Droogmans, G. and B. Nilius: Kinetic Properties of the Cardiac T-Type Calcium Channel in the Guinea Pig. *J.Physiol.* **419**:627-(1/4) (1989).

Droogmans, G., L. Raeymaekers and R. Casteels: Electro- and Pharmacomechanical Coupling in the Smooth Muscle Cells of the Rabbit Ear Artery. *J.Gen.Physiol.* **70**:129-(2/10) (1977).

Dubinsky, J.M. and G.S. Oxford: Ionic Currents in Two Strains of Rat Pituitary Tumor Cells. *J. Gen. Physiol.* **83**:309-(2/7) (1984).

Dubovsky, S.L. and R.D. Franks: Intracellular Calcium Ions in Affective Disorders: A Review and an Hypothesis. *Biol.Psychiatry.* **18**:781-(4/22) (1983).

Dubovsky, S.l., R.D. Franks, S. Allen and J. Murphy: Calcium Antagonists in Mania: A Double-Blind Study of Verapamil. *Psychiatr.Res.* **18**:309-(4/22) (1986).

Dubovsky, S.L., R.D. Franks, M. Lifschitz and P. Coen: Effectiveness of Verapamil in the Treatment of a Manic Patient. *Am.J.Psychiatry.* **139**:502-(4/22) (1982).

Dulhunty, A.F. and P.W. Gage: Effects of Extracellular Calcium Concentration and Dihydropyridines on Contraction in Mammalian Skeletal Muscle. *J.Physiol.* **399**:63-(2/10) (1988).

Dunlap, E.D., J.S. Plowden, D.A. Lathrop and R.W. Millard: Hemodynamic and Electrophysiologic Effects of Amlodipine, a New Calcium Channel Blocker. *Am.J.Cardiol.* **64**:711-(3/13) (1989).

Dunlap, K.: Localization of Calcium Channels in *Paramecium caudatum. J.Physiol.* **271**:119-(2/7) (1977).

Dunlap, K. and G.D. Fischbach: Neurotransmitters Decrease the Calcium Component of Sensory Neuron Action Potentials. *Nature* **276**:837-(3/14) (1978).

Dunlap, K. and G.D. Fischbach: Neurotransmitters Decrease the Calcium Conductance Activated by Depolarization of Embryonic Chick Sensory Neurones. *J.Physiol.* **317**:519-(2/9,3/14) (1981).

Dunlap, K., G.G. Holz and S.G. Rane: G Proteins as Regulators of Ion Channel Function. *TINS* **10**:241-(1/4) (1987).

Dunn, L.A. and R.W. Holz: Catecholamine Secretion from Digitonin-Treated Adrenal Medullary Chromaffin Cells. *J.Biol.Chem.* **258**:4989-(2/7) (1983).

Durroux, T., N. Gallo-Peyet and M.D. Payet: Three Components of the Calcium Current in Cultured Glomerulosa Cells from Rat Adrenal Gland. *J.Physiol.* **404**:713-(2/7,3/13,3/14) (1988).

Ebashi, S., M. Endo and I. Ohtsuki: Control of Muscle Contraction. *Q.Rev.Biophys.* **2**:351-(2/5) (1969).

Ebersole, B.J., Z.L. Cajary and P.B. Molinoff: Endogenous Modulators of Binding of [^{3}H]Nitrendipine in Extracts of Rat Brain. *J.Pharmacol.Exp.Ther.* **244**:971-(3/13) (1988).

Eccles, J.C., R. Llinas and K. Sasaki: Parallel Fibre Stimulation and the Responses Induced thereby in the Purkinje Cells of the Cerebellum. *Exp.Brain Res.* **1**:17-(2/9) (1966a).

Eccles, J.C., R. Llinas and K. Sasaki: The Excitatory Synaptic Action of Climbing Fibers of the Purkinje Cells of the Cerebellum. *J.Physiol.* **182**:268-(2/9) (1966b).

Eckert, R. and P. Brehm: Ionic Mechanisms of Excitation in Paramecium. *Ann.Rev.Biophys.Bioeng.* **8**:358-(2/5,2/7) (1979).

Eckert, R. and J.E. Chad: Inactivation of Ca Channels. *Prog.Biophys.Mol.Biol.* **44**:215-(1/4,3/14) (1984).

Eckert, R. and D. Ewald: Calcium Tail Currents in Voltage-Clamped Intact Nerve Cell Bodies of *Aplysia californica*. *J.Physiol.* **345**:533-(1/3) (1983).

Eckert, R. and H.D. Lux: A Voltage Dependent Persistent Calcium Current in Neuronal Somata of *Helix*. *J.Physiol.* **254**:129-(2/9) (1976).

Eckert, R. and D. Tillotson: Calcium Mediated Inactivation of the Calcium Conductance in Caesium Loaded Giant Neurones of *Aplysia californica*. *J.Physiol.* **254**:265-(1/4,2/9) (1981).

Ehrlich, B.E., A. Finkelstein, M. Forte and C. Kung: Voltage-Dependent Calcium Channels from Paramecium Cilia Incorporated into Planar Lipid Bilayers. *Science* **225**:427-(2/7) (1984).

Ehrlich, B.E. and J. Watras: Inositol 1,4,5-Trisphosphate Activates a Channel from Smooth Muscle Sarcoplasmic Reticulum. *Nature* **336**:583-(2/7) (1988).

Eisen, A., D.P. Kiehart, S.J. Wieland and G.T. Reynolds: Temporal Sequence and Spatial Distribution of Early Events of Fertilization in Single Sea Urchin Eggs. *J.Cell Biol.* **99**:1647-(2/7) (1984).

Eisenberg, R.S., R.T. McCarthy and R.L. Milton: Paralysis of Frog Skeletal Muscle Fibres by the Calcium Antagonist D-600. *J.Physiol.* **341**:495-(2/10) (1983).

Eisenmann, G.: The Molecular Basis for Ion Selectivity and its Possible Bearing on the Neurobiology of Lithium. In *The Neurobiology of Lithium*. W.E. Bunney, D.L. Murphy, eds. Boston. 154-(4/22) (1976).

El Sherif, N., R. Zeiter, W. Craelius and R. Henkin: QTu Prolongation and Polymorphic Ventricular Tachyarrhythmias due to Bradycardia-Dependent Early Afterdepolarizations: Afterdepolarizations and Ventricular Arrhythmias. *Circ.Res.* **63**:286-(4/19) (1988).

El-Sherif, N. and R. Lazzara: Re-Entrant Ventricular Arrhythmias in the Late Myocardial Infarction Period. Effect of Verapamil and D600 and the Role of the Slow Conduction. *Circulation* **60**:605-(4/19) (1979).

Elharrar, V.J., W.E. Gaum and D.P. Zipes: Effect of Drugs on Conduction Delay and Incidence of Ventricular Arrhythmias Induced by Acute Coronary Occlusion in Dogs. *Am.J.Cardiol.* **39**:544-(4/19) (1977).

Ellis, S.B., M.E. Williams, N.R. Ways, R. Brenner, A.H. Sharp, A.T. Leung, K.P. Campbell, E. McKenna, W.J. Koch, A. Hui, A. Schwartz and M.M. Harpold: Sequence and Expression of mRNAs Encoding the $\alpha-1$ and $\alpha-2$ Subunits of a DHP-Sensitive Calcium Channel. *Science* **241**:1661-(3/13) (1988).

Ellrodt, G., C.Y.C. Chew and B.N. Singh: Therapeutic Implications of Slow Channel Blockade in Cardiocirculatory Disorders. *Circulation* **62**:669-(4/19) (1980).

Emrich, H.M., T. Okuma and A.A. Muller: *Anticonvulsants in Affective Disorders*. Excerpta Medica, Amsterdam. (4/22) (1984).

Endicott, J.A. and V. Ling: The Biochemistry of P-Glycoprotein-Mediated Multidrug Resistance. *Annu.Rev.Biochem.* **58**:137-(3/13) (1989).

Endo, M., M. Tanaka and Y. Ogawa: Calcium Induced Release of Calcium from the Sarcoplasmic Reticulum of Skinned Skeletal Muscle Fibers. *Nature* **228**:34-(2/10) (1970).

Endo, M.: Calcium Release from Sarcoplasmic Reticulum. In *Current Topics in Membranes and Transport*. Academic Press, New York. **25**:181-(2/10) (1985).

Endo, T. and B. Nadal-Ginard: Three Types of Muscle-Specific Gene Expression in Fusion-Blocked Rat Skeletal Muscle Cells: Translational Control in EGTA-Treated Cells. *Cell* **49**:515-(2/10) (1987).

Entwistle, A., R.J. Zalin, S. Bevan and A.E. Warner: The Control of Chick Myoblast Fusion by Ion Channels Operated by Prostaglandins and Acetylcholine. *J.Cell Biol.* **106**:1693-(2/10) (1988).

Entwistle, A., R.J. Zalin, A.E. Warner and S. Bevan: A Role for Acetylcholine Receptors in the Fusion of Chick Myoblasts. *J.Cell Biol.* **106**:1703-(2/10) (1988).

Enyeart, J.J., S.S. Sheu and P.M. Hinkle: Dihydropyridine Modulators of Voltage-Sensitive Ca^{2+} Channels Specifically Regulate Prolactin Production by GH4C1 Pituitary Tumor Cells. *J.Biol.Chem.* **262**:3154-(3/13) (1987).

Epstein, M. and R.D. Loutzenhiser: *Calcium Antagonists and the Kidney*. Hanley & Belfus, Inc., Philadelphia. (3/13) (1990).

Epstein, S.E. and D.R. Rosing: Verapamil: Its Potential for Causing Serious Complication in Patients with Hypertrophic Cardiomyopathy. *Circulation* **64**:437-(4/17) (1990).

Erdmann, R. and H.C. Littgau: The Effect of the Phenylalkylamine D888 (Devapamil) on Force and Ca^{2+} Current in Isolated Frog Skeletal Muscle Fibres. *J.Physiol.* **413**:521-(3/13) (1989).

Erne, P., E. Burgisser, F.R. Buhler, B. Dubach, H. Kuhnis, M. Meier and H. Rogg: Enhancement of Calcium Influx in Human Platelets by CGP 28392, a Novel Dihydropyridine. *Biochem.Biophys.Res.Commun.* **118**:842-(4/20) (1984).

Ervasti, J.M., M.T. Claessens, J.R. Mickelson and C.F. Lois: Altered Transverse Tubule Dihydropyridine Receptor Binding in Malignant Hyperthermia. *J.Biol.Chem.* **264**:2711-(3/13) (1989).

Eskinder, H., F.D. Supan, N.J. Rusch, J.P. Kampine and Z.J. Bosnjak: Effects of Halothane (HAL) on Ca^{2+} Channel Currents in Canine Cardiac Purkinje Cells. *Biophys.J.* **57**:518a-(3/13) (1990).

Espluges, J.V., R. Brage, J. Cortijo, M. Marti-Cabera, G. Hernandez, B. Sarria and J. Espluges: Differential Effects of Verapamil on Various Gastric Lesions in Rats. *Pharmacology* **36**:69-(3/13) (1988).

Etingin, O.R. and D.P. Hajjar: Nifedipine Increases Cholesteryl Ester Hydrolytic Activity in Lipid Laden Rabbit Arterial Smooth Muscle Cells. *J.Clin.Invest.* **75**:1554-(4/20) (1985).

Etingin, O.R. and D.P. Hajjar: Calcium Channel Blockers Enhance Cholesteryl Ester Hydrolysis and Decrease Total Cholesterol Accumulation in Human Aortic Tissue. *Circ.Res.* **66**:185-(3/13) (1990).

Ettinger, W.H., R.A. Wise, D. Schaffhauser and F.M. Wigley: Controlled Double-Blinded Trial of Dazoxiben and Nifedipine in the Treatment of Raynaud's Phenomenon. *Am.J.Med.* **77**:451-(4/20) (1984).

Evans, S., D. Goldman, S. Heinemann and J. Patrick: Muscle Acetylcholine Receptor Biosynthesis - Regulation by Transcript Availability. *J.Biol.Chem.* **262**:4911-(2/10) (1987).

Ewald, D.A., H.J.G. Matthies, T.M. Perney, M.W. Walker and R.J. Miller: The Effect of Protein Kinase C Down Regulation of the Inhibitory Modulation of Dorsal Root Ganglion Neuron Ca^{2+} Currents by Neuropeptide Y. *J.Neurosci.* **8**:2447-(3/14) (1988b).

Ewald, D.A., I.H. Pang, P.C. Sternweis and R.J. Miller: Differential G Protein-Mediated Coupling of Neurotransmitter Receptors to Ca^{2+} Channels in Rat Dorsal Root Ganglion Neurons In Vitro. *Neuron* **2**:1185-(2/9,3/13,3/14) (1989).

Ewald, D.A., P.C. Sternweis and R.J. Miller: G_o Induced Coupling of NPY Receptors to Calcium Channels in Sensory Neurons. *Proc.Natl.Acad.U.S.A.* **85**:3633-(3/14) (1988a).

Fabiato, A.: Calcium-Induced Release of Calcium from the Cardiac Sarcoplasmic Reticulum. *Am.J.Physiol.* **245**:Cl-(2/10) (1983).

Fabiato, A.: Simulated Calcium Current Can Both Cause Calcium Loading in and Trigger Calcium Release from the Sarcoplasmic Reticulum of a Skinned Canine Cardiac Purkinje Cell. *J.Gen.Physiol.* **85**:291-(2/10) (1985b).

Fabiato, A.: Time and Calcium Dependence of Activation and Inactivation of Calcium-Induced Release of Calcium from the Sarcoplasmic Reticulum of a Skinned Canine Cardiac Purkinje Cell. *J.Gen.Physiol.* **85**:247-(2/10,2/7) (1985a).

Faden, A.I., P. Demediuk, S.S. Panter and R. Vink: The Role of Excitatory Amino Acids and NMDA Receptors in Traumatic Brain Injury. *Science* **244**:798-(4/21) (1989).

Fain, G.L.: Sensitivity of Toad Rods: Dependence on Wavelength and Background Illumination. *J.Physiol.* **261**:71-(2/8) (1976).

Fain, G.L. and W.H. Schroder: Calcium Content and Calcium Exchange in Dark-Adapted Toad Rods. *J.Physiol.* **368**:641-(2/8) (1985).

Fain, G.L. and W.H. Schroder: Calcium in Dark-Adapted Toad Rods: Evidence for Pooling and Cyclic Guanosine-3'-5' Monophosphate-Dependent Release. *J.Physiol.* **389**:361-(2/8) (1987).

Farber, J.L.: The Role of Calcium in Cell Death. *Life Sci.* **29**:1289-(4/21) (1981b).

Farber, J.L., K.R. Chien and S. Mittnacht: The Pathogenesis of Irreversible Cell Injury in Ischemia. *Am.J.Pathol.* **102**:271-(4/21) (1981a).

Fatt, P. and B.L. Ginsborg: Ionic Requirements for the Production of Action Potentials in Crustacean Muscle Fibres. *J.Physiol.* **142**:516-(3/12) (1958).

Fatt, P. and B. Katz: The Electrical Properties of Crustacean Muscle Fibers. *J.Physiol.* **120**:171-(2/9,3/12) (1953).

Favale, S., M. DiBiase, U. Rizzo, L. Bellardinelli and P. Rizzou: Effect of Adenosine and Adenosine-5'-triphosphate on Atrioventricular Conduction in Patients. *JACC* **5**:230-(4/19) (1985).

Fazzini, P.F., F. Marchi and P. Pucci: Effects of Verapamil on Ventricular Premature Beats of Acute Myocardial Infarction. *Acta Cardiol.* **33**:25-(4/19) (1978).

Fedulova, S.A., P.G. Kostyuk and N.S. Veselovsky: Two Types of Calcium Channels in the Somatic Membrane of New-Born Rat Dorsal Root Ganglion Neurons. *J.Physiol.* **359**:431-(1/2,1/4,3/12) (1985).

Fedulova, S.A., P.G. Kostyuk and N.S. Veselovsky: Changes in Ionic Mechanisms of Electrical Excitability of the Somatic Membrane of Rat Dorsal Root Ganglion Neurones during Ontogenesis: Correlation Between Inward Current Densities. *Neurophysiol.Kiev.* **18**:820-(2/9) (1986).

Feigenbaum, P., M.L. Garcia and G.J. Kaczorowski: Evidence for Distinct Sites Coupled to High Affinity ω-Conotoxin Receptors in Rat Brain Synaptic Plasma Membrane Vesicles. *Biochem.Biophys.Res.Commun.* **154**:298-(3/13) (1988).

Feinstein, M.B.: Role of Ca^{2+}-Ions in the Regulation of Platelet Function. *Rec.Prog.Cell Biol.* **45**:1-(4/20) (1978).

Feinstein, M.B.: Release of Intracellular Membrane-Bound Ca^{2+} Precedes the Onset of Stimulus-Induced Exocytosis in Platelets. *Biochem.Biophys.Res.Commun.* **93**:593-(4/20) (1980).

Feinstein, M.B.: The Role of Calmodulin in Hemostasis. *Prog.Hemostas.Thrombos.* **6**:25-(4/20) (1982).

Fenwick, E.M., A. Marty and E. Neher: Sodium and Calcium Channels in Bovine Chromaffin Cells. *J.Physiol.* **331**:519-(1/2,1/3,1/4,2/7,3/12,3/14) (1982).

Ferguson, D.G., H.W. Schwartz and C. Franzini-Armstrong: Subunit Structure of Junctional Feet in Traids of Skeletal Muscle: A Freeze-Drying, Rotary-Shadowing Study. *J.Cell Biol.* **99**:1735-(2/10) (1984).

Fernandez, J.M., A.P. Fox and S. Krasne: Membrane Patches and Whole-Cell Membranes: A Comparison of Electrical Properties in Rat Clonal Pituitary (GH3) Cells. *J.Physiol.* **356**:565-(1/4) (1984).

Fernandez-Alfonso, M., J. Alonso, I. Rico, M. Salaices, C.F. Sanchez-Ferrer and J. Maron: Effects of the Ca^{2+} Agonists, BAY K 8644 and CGP 28392, on Isolated Cat Cerebral and Peripheral Arteries. *Brain Res.* **474**:147-(3/13) (1988).

Ferrante, J., E. Luchowski, A. Rutledge and D.J. Triggle: Binding of the 1,4-Dihydropyridine Calcium Channel Activator, (-)S-BAY K 8644, to Cardiac Preparations. *Biochem.Biophys.Res.Commun.* **158**:149-(3/13) (1989).

Ferrante, J. and D.J. Triggle: Drug- and Disease-Induced Regulation of Voltage-Dependent Calcium Channels. *Rev.Pharmacol.* **42**:29-(3/13) (1990).

Ferrendelli, J.A. and S. Daniels-McQueen: Comparative Actions of Phenytoin and Other Anticonvulsant Drugs on Potassium- and Veratridine-Stimulated Calcium Uptake in Synaptosomes. *J.Pharmacol.Exp.Ther.* **220**:29-(4/22) (1982).

Ferrier, G.R.: Digitalis Arrhythmias: Role of Oscillatory After Potentials. *Prog.Cardiovasc.Dis.* **19**:459-(4/19) (1977).

Fesenko, E.E., S.S. Kolesnikov and A.L. Lyubarsky: Induction by Cyclic GMP of Cationic Conductance in Plasma Membrane of Retinal Rod Outer Segment. *Nature* **313**:310-(2/8) (1985).

Fesenko, E.E., S.S. Kolesnikov and A.L. Lyubarsky: Direct Action of cGMP on the Conductance of the Retinal Rod Plasma Membrane. *Biochim.Biophys.Acta* **856**:661-(2/8) (1986).

Fetter, R.D. and J.M. Corless: Morphological Components Associated wtih Frog Cone Outer Segment Disc Martins. *Invest.Ophthalmol.Vis.Sci.* **28**:646-(2/8) (1987).

Field, A.C., C. Hill and G.D. Lamb: Asymmetric Charge Movement and Calcium Currents in Ventricular Myocytes of Neonatal Rat. *J.Physiol.* **406**:277-(2/10) (1988).

Figulla, H.R., J.V. Rechenberg, V. Wiegand, R. Sobolla and H. Kreuzer: Beneficial Effects of Long-Term Diltiazem Treatment in Dilated Cardiomyopathy. *J.Am.Coll.Cardiol.* **13**:653-(4/17) (1989).

Filias, N.: Verapamil-Behandlung bei Herzerhythmusstorungen. *Scheveiz.Rundschan.Med.* **3**:66-(4/19) (1974).

Fill, M.D. and P.M. Best: Block of Contracture in Skinned Frog Skeletal Muscle Fibers by Calcium Antagonists. *J.Gen.Physiol.* **93**:429-(2/10) (1989).

Finch, M.B., G.D. Johnston and J. Dawson: The Peripheral Vascular Effects of Nifedipine in Raynaud's Disease. *Br.J.Clin.Pharmacol.* **21**:100-(4/20) (1985).

Findlay, T. and M.J. Dunne: Voltage-Activated Ca^{2+} Currents in Insulin-Secreting Cells. *FEBS Lett.,* **189**:281-(2/7) (1985).

Findlay, I.N., K. MacLeod, G. Gillen, A.T. Elliot, T. Aitchison and H.J. Dargie: A Double-Blind Placebo-Controlled Comparison of Verapamil, Attenolol, and their Combination in Patients with Chronic Stable Angina Pectoris. *Br.Heart J.* **57**:336-(4/16) (1987).

Fine, D.G., I.P. Clements and M.J. Callahan: Myocardial Stunning in Hypertrophic Cardiomyopathy: Recovery Predicted by Single Photon Emission Computed Tomographic Thallium-20l Scintigraphy. *J.Am.Coll.Cardiol.* **13**:1415-(4/17) (1989).

Finger, S. and S.B. Dunnett: Nimodipine Enhances Growth and Vascularization of Neural Grafts. *Exp.Neurol.* **104**:1-(3/13) (1989).

Finkel, M.S., R.E. Patterson, W.C. Roberts, T.D. Smith and H.R. Keiser: Calcium Channel Binding Characteristics in the Human Heart. *Am.J.Cardiol.* **62**:1281-(3/13) (1988).

Fisch, T.M., R. Prywes and R.G. Roeder: An AP1-Binding Site in the c-Fos Gene can Mediate Induction by Epidermal Growth Factor and l2-O-Tetradecanoyl Phorbol-l3-acetate. *Mol.Cell Biol.* **9**:1327-(2/10) (1989).

Fischbach, G.D. and S. Schuetze: A Post-Natal Decrease in Acetylcholine Channel Open Time at Rat End-Plates. *J.Physiol.* **303**:125-(2/10) (1980).

Fischl, S.J., M.V. Herman and R. Gorlin: The Intermediate Coronary Syndrome. Clinical, Angiographic and Therapeutic Aspects. *N.Engl.J.Med.* **288**:1193-(4/16) (1973).

Fischmeister, R., D. Mentrand and G. Vassort: Slow Inward Current Inactivation in Frog Heart Atrium. *J.Physiol.* **320**:27-(1/4) (1981).

Fish, R.D., G. Sperti, W.S. Colucci and D.E. Clapham: Phorbol Ester Increases the Dihydropyridine Sensitive Calcium Conductance in a Vascular Smooth Muscle Cell Line. *Circ.Res.* **62**:1049-(3/13,3/14) (1988).

Fitz, T.E. and B.L. Hallman: Mental Changes Associated with Hyperparathyroidism; Report of 2 Cases. *Arch.Int.Med.* **89**:547-(4/22) (1952).

Fitzpatrick, L.A. and H. Chin: Inhibition of Parathyroid Hormone Secretion by Calcium: The Role of Calcium Channels. In *The Calcium Channel: Structure, Function and Implications.* M. Morad, W. Nayler, S. Kazda, M. Schramm, eds. Springer-Verlag, Berlin, Heidelberg. 418-(3/13) (1988).

Flaim, S.F. and D.M. Cummings: Bepridil Hydrochloride: A Review of its Pharmacological Properties. *Curr.Ther.Res.* **39**:568-(3/13) (1986).

Flatman, J.A., P.C. Schwindt and W.E. Crill: The Induction and Modification of Voltage-Sensitive Responses in Cat Neocortical Neurons by N-Methyl-D-Aspartate. *Brain Res.* **363**:62-(2/9) (1986).

Fleckenstein, A.: Specific Pharmacology of Calcium in Myocardium. Cardiac Pacemakers and Vascular Smooth Muscle. *Ann.Rev.Pharmacol.Toxicol.* **17**:149-(4/17) (1977).

Fleckenstein, A.: History of Calcium Antagonists. *Circ.Res.* **52**:1-(4/17) (1983).

Fleckenstein, A.: *Calcium Antagonism In Heart and Smooth Muscle. Experimental Facts and Therapeutic Prospects.* John Wiley & Sons, New York. (3/13) (1983b).

Fleckenstein, A.: Calcium Antagonism in Heart and Vascular Smooth Muscle. *Medicinal Res. Rev.* **5**:395-(4/15) (1985).
Fleckenstein, A., G. Fleckenstein-Grun, M. Frey and J. Zorn: The Fundamental Role of Arterial Calcium Overload in Atherogenesis. *Abstracts of the 4th International Symposium on Calcium Antagonists: Pharmacology and Clinical Research, Florence* 89-(3/13) (1989).
Fleckenstein, A., M. Frey, J. Zorn and G. Fleckenstein-Grun: The Role of Calcium in the Pathogenesis of Experimental Arteriosclerosis. *TIPS* **8**:496-(3/13) (1987).
Fleckenstein, A., M. Frey, J. Zorn and G. Fleckenstein-Grun: Calcium, a Neglected Key Factor in Hypertension and Arteriosclerosis. In *Hypertension: Physiology, Diagnosis and Management* . J.H. Laragh, B.M. Brenner, eds. Raven Press, New York. 471-(3/13) (1990).
Fleischer, J.E., W.L. Lanier, J.H. Milde and J.D. Michenfelder: Lidoflazine does not Improve Neurologic Outcome when Administered after Complete Cerebral Ischemia in Primates. *J.Cereb.Blood Flow Metab.* **7**:366-(3/13) (1987).
Fleischer, S., E.M. Ogunbunmi, M.C. Dixon and E.A.M. Fleer: Localization of Ca^{2+} Release Channels with Ryanodine in Junctional Terminal Cisternae of Sarcoplasmic Reticulum of Fast Skeletal Muscle. *Proc.Natl.Acad.Sci.U.S.A.* **82**:7256-(2/10) (1985).
Flockerzi, V., H.J. Oeken, F. Hofmann, D. Pelzer, A. Cavalie and W. Trautwein: Purified Dihydropyridine-Binding Site from Skeletal Muscle t-Tubules is a Functional Calcium Channel. *Nature* **323**:66-(2/10,3/13). (1986).
Fondacaro, J.D., J.H. Han and M.S. Joon: Effect of Verapamil on Ventricular Rhythm During Acute Coronary Occlusion. *Am.Heart J.* **96**:81-(4/19) (1978).
Fontaine, B., A. Klarsfeld and J.P. Changeux: Calcitonin Gene-Related Peptide and Muscle Activity Regulate Acetylcholine Receptor Alpha-Subunit mRNA Levels by Distinct Intracellular Pathways. *J.Cell Biol.* **105**:1337-(2/10) (1987).
Ford, J.M., W.C. Prozialeck and W.N. Hait: Structural Features Determining Activity of Phenothiazines and Related Drugs for Inhibition of Cell Growth and Reversal of Multidrug Resistance. *Mol.Pharmacol.* **35**:105-(3/13) (1989).
Ford, L.E. and R.J. Podolsky: Regenerative Calcium Release Within Muscle Cells. *Science* **167**:58-(2/10) (1970).
Forder, J., A. Scriabine and H. Rasmussen: Plasma Membrane Ca^{2+} Flux, Protein Kinase C Activation and Smooth Muscle Concentration. *J.Pharmacol.Exp.Ther.* **234**:267-(3/14) (1985).
Foreman, J.C. and J.L. Mongar: The Role of Alkaline Earth Ions in Anaphylactic Histamine Secretion. *J.Physiol.* **224**:753-(2/7) (1972).
Foreman, J.C., J.L. Mongar and B.D. Gomperts: Calcium Ionophores and Movement of Calcium Ions Following the Physiological Stimulus to a Secretory Process. *Nature* **245**:249-(2/7) (1973).
Forrester, J.S., F. Litvack, W. Grunfest and A. Hickey: A Perspective of Coronary Disease seen Through the Arteries of Living Man. *Circulation* **75**:505-(4/16) (1987).
Forscher, P., G. Oxford and D. Schultz: Noradrenaline Modulates Calcium Channels through Tight Receptor Channel Coupling. *J.Physiol.* **379**:131-(3/14) (1986).
Forscher, P. and G.S. Oxford: Modulation of Calcium Channels by Norepinephrine in Internally Dialyzed Avian Sensory Neurons. *J.Gen.Physiol.* **85**:743-(1/3,3/14) (1985).
Fourman, P., K. Rawnsley, R.H. Davis, K.H. Jones and D.B. Morgan: Effect of Calcium on Mental Symptoms in Partial Parathyroid Insufficiency. *Lancet* **7522**:914-(4/22) (1967).
Fox, A.P. and S. Krasne: Two Calcium Currents in Egg Cells. *Biophys.J.* **33**:145a-(3/14) (1981).
Fox, A.P. and S. Krasne: Two Calcium Currents in *Neanthes arenaceodentata* Egg Cell Membranes. *J.Physiol.* **356**:491-(3/14) (1984).
Fox, A.P., M. Nowycky and R.W. Tsien: Single Channel Recordings of Three Types of Calcium Channels in Chick Sensory Neurones. *J.Physiol.* **394**:173-(1/2,1/4,2/9,3,12,3/14) (1987b).
Fox, A.P., M.C. Nowycky and R.W. Tsien: Kinetic and Pharmacological Properties Distinguishing Three Types of Calcium Currents in Chick Sensory Neurones. *J.Physiol.* **394**:149-(1/4,2/9,3/12,3/13,3/14) (1987a).
Frank, G.B.: Blockade of Ca^{2+} Channels Inhibit K^+ Contractures but not Twitches in Skeletal Muscle. *Can.J.Physiol.Pharmacol.* **62**:374-(3/13) (1984).
Franzini-Armstrong, C.: Studies of Triad: I. Structure of the Junction in Frog Twitch Fibers. *J.Cell Biol.* **47**:488-(2/10) (1970).
Franzini-Armstrong, C. and G. Nunzi: Junctional Feet and Particles in the Triads of a Fast-Twitch Muscle Fiber. *J.Mus.Res. Cell Motil.* **4**:233-(2/10) (1983).
Fraser, S.E., C.R. Green, H.R. Bode and N.B. Gilula: Selective Disruption of Gap Junctional Communication Interferes with a Patterning Process in Hydra. *Science* **237**:49-(2/10) (1987).
Friedel, H.A. and E.M. Sorkin: Nisoldipine: A Preliminary Review of its Pharmacodynamic and Pharmacokinetic Properties, and Therapeutic Efficacy in the Treatment of Angina Pectoris, Hypertension and Related Cardiovascular Disorders. *Drugs* **36**:682-(3/13) (1988).
Frishman, W.H., N.A. Klein, J.A. Strom, H. Willens, T.H. LeJemtel, J. Jentzer, L. Siegel, P. Klein, N. Kirschen, R. Silverman, S. Pollack, R. Doyle, E. Kirsten and E.H. Sonnenblick: Superiority of Verapamil to Propranolol in Stable Angina Pectoris: A Double-Blind Randomized Crossover Trial. *Circulation* **65**:51-(4/16) (1982).
Frishman, W.H., A.E. Skolnick and J.A. Strom: Effects of Calcium Entry Blockade on Hypertension-Induced Left Ventricular Hypertrophy. *Circulation* **80**:151-(3/13) (1989).

Fritschka, E., A. Kribben, A. Distler and T. Philipp: Inhibition of Aggregation and Calcium Influx of Human Platelets by Nitrendipine. *J.Cardiovasc.Pharmacol.* **9**:S85-(3/13) (1987).

Fry, H.K. and R.T. McCarthy: Nimodipine Block of L-Type Calcium Channels in Freshly Dispersed Rabbit DRG Neurons. *Soc.NeuroscienceAbs.* **14**:134-(3/13) (1988).

Fujino, M., T. Yamaguchi and K. Suzuki: "Glycerol Effect" and the Mechanism Linking Excitation of the Plasma Membrane with Contraction. *Nature* **192**:1159-(2/10) (1961).

Fukuda, J. and K. Kawa: Permeation of Manganese, Cadmium, Zinc and Beryllium through Calcium Channels of an Insect Muscle Membrane. *Science* **196**:309-(3/12) (1977).

Fukushima, Y. and S. Hagiwara: Currents Carried by Monovalent Cations Through Calcium Channels in Mouse Neoplastic B Lymphocytes. *J.Physiol.* **358**:255-(1/2,1/4,2/7,3/12) (1985).

Fukushima, Y. and S. Hagiwara: Voltage Gated Ca^{2+} Channel in Mouse Myeloma Cells. *Proc.Natl.Acad.Sci.U.S.A.* **80**:2240-(3/12,3/14) (1983).

Fukushima, Y.S., S. Hagiwara and R.E. Saxton: Variation of Calcium Current during the Cell Growth Cycle in Mouse Hybridoma Lines Secreting Immunoglobulins. *J.Physiol.* **355**:313-(2/7) (1984).

Fuortes, M.G., E.A. Schwartz and E.J. Simon: Colour Dependence of Cone Responses in the Turtle Retina. *J.Physiol.* **234**:199-(2/8) (1973).

Furchgott, R.F.: Role of Endothelium in Responses of Vascular Smooth Muscle. *Circ.Res.* **53**:557-(4/16) (1983).

Fuster, V. and J.H. Chesebro: Current Concepts of Thrombogenesis: Role of Platelets. *Mayo Clin.Proc.* **56**:172-(4/20) (1981).

Fuster, V. and J.H. Chesebro: Pathogenesis of Atherosclerosis: The Role of Platelets and Thrombosis. In*Thrombosis*. H.C. Kwaan and H.C. Bowie, eds. Saunders, Philadelphia. 57-(4/20) (1982).

Gainer, H., S.A. Wolfe, A.L. Obaid and B.M. Salzberg: Action Potentials and Frequency-Dependent Secretion in the Mouse Neurohypophysis. *Neuroendocrinology* **43**:557-(2/7) (1986).

Galizzi, J.P., M. Borsotto, J. Barhanin, M. Fosset and M. Lazdunski: Characterization and Photoaffinity Labeling of Receptor Sites for the Ca^{2+} Channel Inhibitors d-cis-Diltiazem, ($\pm$)-Bepridil, Desmethoxyverapamil, and (+)-PN 200-110 in Skeletal Muscle Tranverse Tubule Membranes. *J.Biol.Chem.* **261**:1393-(3/13) (1986).

Galizzi, J.P., M. Fosset and M. Lazdunski: Characerization of the Ca^{2+} Coordination Site Regulating Binding of Ca^{2+} Channel Inhibitors d-dis-Diltiazem, (+/-)bepridil and (-) desmethoxyverapamil to their Receptor Site in Skeletal Muscle Transverse Tubule Membranes. *Biochem.Biophys.Res.Commun.* **132**:49-(3/13) (1985).

Galizzi, J.P., M. Fosset, G. Romey, P. Laduron and M. Lazdunski: Neuroleptics of the Diphenylbutylpiperidine Series are Potent Calcium Channel Inhibitors. *Proc.Natl.Acad.Sci.U.S.A.* **83**:7513-(3/13) (1986b).

Galizzi, J.P., J. Qar, M. Fosset, C. van Reterghen and M. Lazdunski: Regulation of Calcium Channels in Aortic Muscle Cells by Protein Kinase C Activators (Diacylglycerol and Phorbol Esters) and by Peptides (Vasopressin and Bombesin) that Stimulate Phosphoinositide Breakdown. *J.Biol.Chem.* **262**:6974-(3/14) (1987).

Gallo, V., A. Kingsbury, R. Balazs and O.S. Jorgensen: The Role of Depolarization in the Survival and Differentiation of Cerebellar Granule Cells in Culture. *J.Neurosci.* **7**:2203-(2/10) (1987).

Gandia, L., M.G. Lopez, R.I. Fonteriz, C.R. Artalejo and A.G. Garcia: Relative Sensitivities of Chromaffin Cell Calcium Channels to Organic and Inorganic Calcium Antagonists. *Neurosci.Lett.* **77**:333-(2/7) (1987).

Ganitkevich, V.Y., M.F. Shuba and S.V. Smirnov: Saturation of Calcium Channels in Single isolated Smooth Muscle Cells of Guinea Pig, *Taenia caeci. J.Physiol.* **399**:419-(3/12) (1988).

Garcia, A.G., F. Sala, J.A. Reig, S. Viniegra, J. Frias, R.I. Fonteriz and L. Gandia: Dihydropyridine BAY-K-8644 Activates Chromaffin Cell Calcium Channels. *Nature* **309**:69-(2/7) (1984).

Garcia, M.L., V.F. King, J.L. Shevell, J.S. Slaughter, G. Suarez-Kurtz, R.J. Winquist and G.J. Kaczorowski: Amiloride Analogs Inhibit L-Type Calcium Channels and Display Calcium Entry Blocker Activity. *J.Biol.Chem.* **265**:3763-(3/13) (1990).

Garcia, M.L., V.F. King, P.K. Siegl, J.P. Reuben and G.J. Kaczorowski: Binding of Ca^{2+} Entry Blockers to Cardiac Sarcolemmal Membrane Vesicles. *J.Biol.Chem.* **261**:8146-(3/13) (1986).

Garcia, M.L., M.J. Trumble, J.P. Reuben and J. Kaczorowski: Characteristics of Verapamil Binding Sites in Cardiac Membrane Vesicles. *J.Biol.Chem.* **259**:15013-(3/13) (1984).

Garfield, R.E. and A.P. Somlyo: Structure of Smooth Muscle. In *Calcium and Contractility-Smooth Muscle*. A.K. Grover and E.E. Daniels, eds. Humana Press, Clifton, New Jersey. 1-(2/10) (1985).

Garofalo, R.S. and B.H. Satir: Paramecium Secretory Granule Contant: Quantitative Studies on *In Vitro* Expansion and its Regulation by Calcium and pH. *J.Cell Biol.* **99**:2193-(2/7) (1984).

Garrett, J.S., J. Wikman-Coffelt, R. Sievers, W.E. Finkbeiner and W.W. Parmley: Verapamil Prevents the Development of Alcoholic Dysfunction in Hamster Myocardium. *J.Am.Coll.Cardiol.* **9**:1326-(3/13) (1987).

Garthoff, B. and P. Bellemann: Effects of Salt Loading and Nitrendipine on Dihydropyridine Receptors in Hypertensive Rats. *J.Cardiovasc.Pharmacol.* **10**:S36-(3/13) (1987).

Garthoff, B., C. Hirth, A. Federmann, S. Kazda and J.P. Stasch: Renal Effects of 1,4-Dihydropyridines in Animal Models of Hypertension and Renal Failure. *J.Cardiovasc.Pharmacol.* **9**:S8-(3/13) (1987).

Garthoff, B., S. Kazda, A. Knorr and G. Thomas: Factors Involved in the Antihypertensive Action of Calcium Antagonists. *Hypertension* **4**:34-(3/13) (1983).

Gartner, T.K., M.J. Doyle and D.F. Mosher: Effect of Anti-Thrombospondin Antibodies on the Hemagglutination Activities of the Endogenous Platelet Lectin and Thrombospodin. *Thromb.Haemost.* **52**:354-(4/20) (1984).

Gasser, T., M. Reddington and P. Schubert: Effects of Carbamasepine on Stimulus-Evoked Ca^{2+} Fluxes in Rat Hippocampal Slices and its Interaction with A_1-Adenosine Receptors. *Neurosci.Lett.* **91**:189-(4/22) (1988).

Gazes, P.C., E.M. Mobley, H.M. Faris, R.C. Duncan and G.B. Humphries: Preinfarctional (Unstable) Angina - A Prospective Study - Ten Year Follow-Up. Prognostic Significance of Electrocardiographic Changes. *Circulation* **48**:331-(4/16) (1973).

Gelmers, H.J., K. Gorter, C.J. de Weerdt and H.J. Wiezer: A Controlled Trial of Nimodipine in Acute Ischemic Stroke. *N.Engl.J.Med.* **318**:203-(4/21) (1988).

Gengo, P., A. Skattebol, J.F. Moran, S. Gallant, M. Hawthorn and D.J. Triggle: Regulation by Chronic Drug Administration of Neuronal and Cardiac Calcium Channel, Beta-Adrenoceptor and Muscarinic Receptor Levels. *Biochem.Pharmacol.* **37**:627-(3/13) (1988).

German, L.D., D.L. Packer, G.H. Brady and J.J. Gallagher: Ventricular Tachycardia Induced by Atrial Stimulation in Patients without Symptomatic Cardiac Disease. *Am.J.Cardiol.* **52**:1202-(4/19) (1983).

Germano, I.M., H.M. Bartkowski, M.E. Cassel and L.H. Pitts: The Therapeutic Value of Nimodipine in Experimental Focal Cerebral Ischemia. *J.Neurosurg.* **67**:81-(3/13) (1987).

Gerschenfeld, H.M. and M. Piccolino: Sustained Feedback Effect of L-Horizontal Cells on Turtle Cones. *Proc.Natl.Acad.Sci.U.S.A.* **83**:1150-(2/8) (1986).

Gerschenfeld, H.M., M. Piccolin and J. Neyton: Feedback Modulation of Cone Synapses by L-Horizontal Cells of Turtle Retinas. *J.Exp.Biol.* **89**:177-(2/8) (1980).

Gershengorn, M.C.: Role of Inositol Lipid Second Messengers in Regulation of Secretion: Studies of Thyrotropin-Releasing Hormone Action in Pituitary Cells. *Secretion and its Control.* G.S. Oxford and C.M. Armstrong, eds. Rockefeller University Press. 1-(2/7) (1989).

Ghanayem, B.I., H.B. Matthews and R.R. Maronpot: Calcium Channel Blockers Protect Against Ethanol- and Indomethacin-Induced Gastric Lesions in Rats. *Gastroenterology* **92**:106-(3/13) (1987).

Giannini, A.J., W.L. Houser, R.H. Loiselle, M.C. Giannini and W.A. Price: Antimanic Effects of Verapamil. *Am.J.Psychiatry* **141**:1602-(4/22) (1984).

Gibson, R.S., W.E. Boden, P. Theroux, H.D. Strauss, C.M. Pratt, M. Gheorghiade, R.J. Capone, M.H. Crawford, R.C. Schlant, R.E. Kleiger, P.M. Young, K. Schectman, M.B. Perryman and R. Roberts: Diltiazem and Reinfarction in Patients with Non-Q Wave Myocardial Infarction. *N.Engl.J.Med.* **315**:423-(4/16,4/19) (1986).

Giebisch, G., A. Scriabine and B. Garthoff: Symposium on Renal Effects of Dihydropyridine-Type Calcium Antagonists. *J.Cardiovasc.Pharmacol.* **9**:1-(3/13) (1987).

Gill, R., A.C. Foster and G.N. Woodruff: Systemic Administration of MK-801 Protects Against Ischemia-Induced Hippocampal Neurodegeneration in the Gerbil. *J.Neurosci.* **7**:3343-(4/21) (1987).

Gilligan, D.M. and B.H. Satir: Stimulation and Inhibition of Secretion in Paramecium: Role of Divalent Cations. *J.Cell Biol.* **97**:224-(2/7) (1983).

Ginsberg, R., I.H. Lamb, J.S. Schroeder, M. Hu and D.C. Harrison: Randomized Double-Blind Comparison of Nifedipine and Isosorbide Dinitrate Therapy in Variant Angina Pectoris due to Coronary Artery Spasm. *Am.Heart J.* **103**:44-(4/16) (1982).

Ginsburg, R., K. David, M.R. Bristow, K. McKennett, S.R. Kodsi, M.E. Billingham and J.S. Schroeder: Calcium Antagonists Suppress Atherogenesis in Aorta but not in the Intramural Coronary Arteries of Cholesterol Fed Rabbits. *Lab.Invest.* **49**:154-(4/20) (1983).

Girsch, S.J. and C. Peracchia: Lens Cell-to-Cell Channel Proteins: I. Self-Assembly into Liposomes and Permeability Regulation by Calmodulin. *J.Membr.Biol.* **83**:217-(2/10) (1985).

Gispen, W.H., T. Schuurman and J. Traber: Nimodipine and Neural Plasticity in the Peripheral Nervous System of Adult and Aged Rats. In *The Calcium Channel: Structure, Function and Implications.* M. Morad, W. Nayler, S. Kazda, M. Schramm, eds. Springer-Verlag, Berlin, Heidelberg. 491-(3/13) (1988).

Gitlin, M.J. and J. Weiss: Verapamil as Maintenance Treatment in Bipolar Illness: A Case Report. *J.Clin.Psychopharmacol.* **4**:341-(4/22) (1984).

Gjorup, T., O.J. Hartling, H. Kelbaek and S.L. Nielson: Controlled Double Blind Trial of Nisolipine in the Treatment of Idiopathic Raynaud's Phenomenon. *Eur.J.Clin.Pharmacol.* **31**:387-(4/20) (1986b).

Gjorup, T., H. Kelbaek, O.J. Hartling and S.L. Nielson: Controlled Double Blind Trail of the Clinical Effect of Nifedipine in the Treatment of Idiopathic Raynaud's Phenomenon. *Am.Heart J..* **111**:742-(4/20) (1986a).

Glagoleva, I.M., Y.A. Liberman and Z.KM. Khashayev: Effect of Uncoupling Agents of Oxidative Phosphorylation on the Release of Acetycholine from Nerve Endings. *Biophys.J.* **15**:74-(2/7) (1970).

Glasstone, S.: *Text-Book of Physical Chemistry.* Macmillan & Co., London. 872-(3/12) (1940).

Glavin, G.B.: Calcium Channel Modulators: Effects on Gastric Function. *Eur.J.Pharmacol.* **160**:323-(3/13) (1989).

Glazier, J.J., S. Chierchia, M.J. Brown and A. Maseri: Importance of Generalized Defective Perception of Painful Stimuli as a Cause of Silent Myocardial Ischemia in Chronic Stable Angina Pectoris. *Am.J.Cardiol.* **58**:667-(4/16) (1986).

Glossman, H., D.R. Ferry, F. Lubbecke, R. Mewes and F. Hoffman: Calcium Channels: Direct Identification with Radioligand Binding Studies. *TIPS* **3**:431-(4/20) (1982).

Glossmann, H. and J. Striessnig: Calcium Channels. *Vitamins and Hormones* **49**:155-(3/13) (1988).

Glossmann, H., G. Zernig, I. Graziadei and T. Moshammer: Non L-Type Ca^{2+} Channel Linked Receptors for 1,4-Dihydropyridines and Phenylalkylamines. In *Nimodipine and Central Nervous System Functions: New Vistas*. J. Traber and W.H. Gispen, eds. Schattauer, Stuttgart. 51-(3/13) (1989).

Gluskin, L.E., B. Strasberg and J.H. Shah: Verapamil-Induced Hyperprolactinemia and Galactorrhea. *Ann.Intern.Med.* **95**:66-(4/22) (1981).

Gmeiner, R., C.K. Ng and M. Hochleitner: Effect of Tiapamil in the Pre-Excitation Syndrome. *Cardiology* **69**:149-(4/19) (1982).

Go, M. and M. Hollenberg: Improved Efficacy of High-Dose versus Medium- and Low Dose Diltiazem Therapy for Chronic Stable Angina Pectoris. *Am.J.Cardiol.* **53**:669-(4/16) (1984).

Goa, K.L. and E.M. Sorkin: Nitrendipine: A Review of its Pharmacodynamic and Pharmacokinetic Properties and Therapeutic Efficacy in the Treatment of Hypertension. *Drugs* **33**:123-(3/13) (1987).

Godfraind, T., R. Miller and M. Wibo: Calcium Antagonism and Calcium Entry Blockade. *Pharmacol.Rev.* **38**:321-(3/13,4/15,4/20) (1986).

Godfraind, T., D. Mennig, G. Bravo, C. Chalant, and P. Javmin: Inhibition by Amlodipine of Activity Evoked in Isolated Human Coronary Arteries by Endothelin, Prostaglandin F2 and Depolarization. *Am.J.Cardiol.* **64**:58I-(4/18) (1989).

Gold, G.H.: Plasma Membrane Calcium Fluxes in Intact Rods are Inconsistent with the Calcium-Hypothesis. *Proc.Natl.Acad.Sci.U.S.A.* **83**:1150-(2/8) (1986).

Gold, G.H. and J.I. Korenbrot: The Regulation of Ca in Intact Retinal Rods: A Study of Light-Induced Calcium Release by the Outer Segment. In *Current Topics in Membranes and Transport*. **15**:307-(2/8) (1981).

Goldberg, N.D., A. Ames, J.E. Gander and T.F. Walseth: Magnitude of Increase in Retinal cGMP Metabolic Flux Determined by 0(l8) Incorporation into Nucleotide Alpha-Phosphoryls Corresponds with Intensity of Photic Stimulation. *J.Biol.Chem.* **258**:9213-(2/8) (1983).

Goldman, D., J. Boulter, S. Heinemann and J. Patrick: Muscle Denervation Increases the Levels of Two mRNAs Coding for the Acetylcholine Receptor Alpha-Subunit. *J.Neurosci.* **5**:2553-(2/10) (1985).

Goldman, D.E.: Potential, Impedance, and Recitification in Membranes. *J.Gen.Physiol.* **27**:37-(1/3) (1943).

Goldstein, D.A. and S.G. Massry: Divalent Ions in Blood and Cerebrospinal Fluid: Effect of Hypercalcemia, Hyperphosphatemia, Renal Failure and Parathyroid Hormone. *Adv.Exp.Med.Biol.* **128**:289-(4/22) (1980).

Goldstein, G.W. and A.L. Betz: Recent Advances in Understanding Brain Capillary Function. *Ann.Neurol.* **14**:389-(4/21) (1983).

Goll, A., H. Glossmann and R. Mannhold: Correlation Between the Negative Inotropic Potency and Binding Parameters of 1,4-Dihydropyridine and Phenylalkylamine Calcium Channel Blockers in Cat Heart. *Naunyn Schmied Arch.Pharmacol.* **334**:303-(3/13) (1986).

Gomperts, B.D.: Calcium Shares the Limelight in Stimulus-Secretion Coupling. *Trends Biochem.Sci.* **11**:290-(2/7) (1986).

Gonoi, T. and S. Hasegawa: Post-Natal Disappearance of Transient Calcium Channels in Mouse Skeletal Muscle: Effects of Denervation and Culture. *J.Physiol.* **401**:617-(2/10) (1988).

Gonzalez, R. and M.M. Scheinman: Treatment of Supraventricular Arrhythmias with Intravenous and Oral Verapamil. *Chest* **80**:465-(4/19) (1981).

Gonzalez-Rudo, R., J.B. Patlak and W.R. Gibbons: A Single Calcium Current Type in Rabbit Ventricular Myoctyes. *Biophys.J.* **55**:306a-(2/10) (1989).

Gonzalez-Serratos, H., R. Valle-Aguilera, D.A. Lathrop and M. del Carmen Garcia: Slow Inward Calcium Curents have no Obvious Role in Muscle Excitation-Contraction Coupling. *Nature* **298**:292-(2/10) (1982).

Gordon, G.D., T.A. Mabin, S. Issacs, E.A. Lloyd, H.G. Eichler and L.H. Opie: Hemodynamic Effects of Sublingual Nifedipine in Acute Myocardial Infarction. *Am.J.Cardiol.* **53**:1228-(4/16) (1984).

Gordon, G.D., T.A. Mabin and E.A. Lloyd: Nifedipine in Acute Myocardial Infarction. In *6th International Adalat Symposium.* P.R. Lichtlen, ed. Excerpta Medica, Amsterdam. 271-(4/16) (1986).

Gorin, M.B., S.B. Yancey, J. Cline, J.P. Revel and J. Horwitz: The Major Intrinsic Protein (MIP) of the Bovine Lens Fiber Membrane: Characterization and Structure Based Upon cDNA Cloning. *Cell* **39**:49-(2/10) (1984).

Gorman, A.L.F., A. Hermann and M.V. Thomas: Ionic Requirements for Membrane Oscillations and Their Dependence on the Calcium Concentration in a Molluscan Pacemaker Neurone. *J.Physiol.* **327**:185-(2/9,3/12) (1982).

Gorman, A.L.F. and M.V. Thomas: Changes in the Intracellular Concentration of Free Calcium Ions in a Pacemaker Neuron, Measured with the Metallochromic Indicator Dye Arsenazo III. *J.Physiol.* **275**:357-(1/4) (1978).

Gothert, M., P. Nawroth and H. Neumeyer: Inhibitory Effects of Verapamil, Prenylamine and D 600 on Ca^{2+}-Dependent Noradrenaline Release from the Sympathetic Nerves of Isolated Rabbit Hearts. *Naunyn Schmied Arch.Pharmacol.* **310**:11-(4/22) (1979).

Gotsman, M.S., A. Bakst and A.S. Mitha: Verapamil in Life-Threatening Tachyarrhythmias. *S.Afr.Med.J.* **46**:2017-(4/19) (1972).

Gottlieb, S.O., P. Ouyang, S.C. Achuff, K.I. Baughman, T.A. Traill, E.D. Mellitis, M.R. Weisfeldt and G. Gerstenblith: Acute Nifedipine Withdrawal: Consequences of Preoperative and Late Cessation of Therapy in Patients with Prior Unstable Angina. *J.Am.Coll.Cardiol.* **4**:382-(4/16) (1984).

Gottlieb, S.O., M.L. Weisfeldt, P. Ouyang, E.D. Mellitis and G. Gerstenblith: Silent Ischemia as a Marker for Early Unfavorable Outcomes in Patients with Unstable Angina. *N.Engl.J.Med.* **314**:1214-(4/16) (1986).

Gottmann, K., I.D. Dietzel, H.D. Lux, S. Huck and H. Rohrer: Development of Inward Currents in Chick Sensory and Autonomic Neuronal Precursor Cells in Culture. *J.Neurosci.* **8**:3722-(2/10) (1988).

Gould, R.J., K.M.M. Murphy, I.G. Reynolds and S.H. Snyder: Antischizophrenic Drugs of the Diphenylbutylpiperidine Type Act as Calcium Channel Antagonists. *Proc.Natl.Acad.Sci.U.S.A.* **80**:5122-(4/22) (1983).

Govardovskii, V.I. and A.L. Berman: Light-Induced Changes of Cyclic GMP Content in Frog Retinal Rod Outer Segments Measured with Rapid Freezing and Microdissection. *Biophys.Struct.Mech.* **7**:125-(2/8) (1981).

Govoni, S., F. Battaini, S.M. Magnoni, L. Lucchi, R.A. Rius and M. Trabucchi: Plasticity of Neuronal L-Type Calcium Channels. *Ann.N.Y.Acad.Sci.* **522**:187-(3/13) (1988).

Graettinger, W.F., M.A. Weber, J.M. Gardin and M.L. Knoll: Diastolic Blood Pressure as a Determinant of Doppler Left Ventricular Filling Indices in Normotensive Adolescents. *JACC* **10**:1280-(4/18) (1987).

Grassi, F. and H.D. Lux: Voltage-Dependent GABA-Induced Modulation of Calcium Currents in Chick Sensory Neurons. *Neurosci.Lett.* **105**:113-(1/3) (1989).

Gray, P. and D. Attwell: Kinetics of Light-Sensitive Changes in Vertebrate Photoreceptors. *Proc.R.Soc.Lond.* 379-(2/8) (1985).

Gray, R. and D. Johnston: Noradrenaline and Beta-Adrenoreceptor Agonists Increase Activity of Voltage Dependent Calcium Channels in Hippocampal Neurons. *Nature* **327**:620-(2/9,3/14) (1987).

Gray, W.R., B.M. Olivera and L.J. Cruz: Peptide Toxins from Venomous Conus Snails. *Ann.Rev.Biochem.* **57**:665-(3/13) (1988).

Grebb, J.A., R.C. Shelton, E.H. Taylor and L.B. Bigelow: A Negative Double-Bind, Placebo-Controlled, Clinical Trial of Verapamil in Chronic Schizophrenia. *Biol.Psychiatr.* **21**:691-(4/22) (1986).

Greco, R., B. Musto, V. Arieuzo, A. Alborino, S. Garofalo and F. Marsico: Treatment of Paroxysmal Supraventricular Tachycardia in Infancy with Digitalis, Adenosine 5-l-Triphosphate and Verapamil: A Comparative Study. *Circulation* **66**:504-(4/19) (1982).

Green, K.A. and G.A. Cottrell: Actions of Baclofen on Components of the Ca^{2+} Currents in Rat and Mouse Dorsal Root Ganglion Neurons in Culture. *Br.J.Pharmacol.* **94**:235-(3/14) (1988).

Greenberg, D.A., R.O. Messing, S.S. Marks, C.L. Carpenter and D.L. Wilson: Alcohol, Neurodegenerative Disorders, and Calcium Channel Antagonist Receptors. In *The Calcium Channel: Structure, Function and Implications*. M. Morad, W. Nayler, S. Kazda and M. Schramm, eds. Springer-Verlag, Berlin, Heidelberg. 541-(3/13) (1988).

Greenberg, M.E., E.B. Ziff and L.A. Greene: Stimulation of Neuronal Acetylcholine Receptors Induces Rapid Gene Transcription. *Science* **234**:80-(2/10) (1986).

Greene, L.A. and A.S. Tischler: Establishment of a Noradrenergic Clonal Line of Rat Adrenal Pheochromocytoma Cells which Respond to Nerve Growth Factor. *Proc.Natl.Acad.Sci.U.S.A.* **73**:2424-(2/10) (1976).

Greene, J.A. and L.W. Swanson: Psychosis in Hypoparathyroidism; with a Report of Five Cases. *Ann.Int.Med.* **14**:1233-(4/22) (1941).

Grenadier, E., G. Alpan, N. Maor, S. Keidar, C. Binenboim, T. Marguiles and A. Palant: Polymorphous Ventricular Tachycardia in Acute Myocardial Infarction. *Am.J.Cardiol.* **53**:1280-(4/19) (1984).

Griffith, T.M., M.J. Lewis, A.C. Newby and A.H. Henderson: Endothelium-Derived Relaxing Factor. *J.Am.Coll.Cardiol.* **12**:797-(4/16) (1988).

Grinstein, S., S. MacDougall, R.K. Cheung and E.W. Gelfand: Role and Properties of Ligand-Induced Calcium Fluxes in Lymphocytes. In *Cell Calcium Metabolism*. G. Fiskum, ed. Academic Press, New York. 283-(2/7) (1990).

Gross, R.A., K.M. Kelly and R.L. MacDonald: Ethosuximide and Dimethadione Selectively Reduce Calcium Currents in Cultured Neurons by Different Mechanisms. *Neurology* **39**:412-(3/13) (1989).

Gross, R.A. and R.L. MacDonald: Reduction of the Same Calcium Current Component by A and C Kinases: Differential Pertussis Toxin Sensitivity. *Neurosci.Lett.* **88**:50-(3/14) (1988).

Gross, R.A. and R.L. MacDonald: Dynorphin A Selectively Reduces a Large Transient (N-Type) Calcium Current of Mouse Dorsal Root Ganglion Neurons in Cell Culture. *Proc.Natl.Acad.Sci.U.S.A.* **84**:5469-(3/13) (1987).

Gross, R.A. and R.L. MacDonald: Activators of Protein Kinase C Selectivity Enhance Inactivation of a Calcium Current Component of Cultured Sensory Neurons in a Pertussis Toxin-Sensitive Manner. *J.Neurophysiol.* **6**:1259-(3/13) (1989).

Grygorczyk, C., R. Grygorczyk and J. Ferrier: Osteoblastic Cells have L-Type Calcium Channels. *Bone Min.* **7**:137-(3/13) (1989).

Grynkiewicz, G., M. Poenie and R.Y. Tsien: A New Generation of Ca^{2+} Indicators with Greatly Improved Fluorescence Properties. *J.Biol.Chem.* **260**:3440-(2/9,4/20) (1985).

Guggino, S.E., D. Lajeunesse, J.A. Wagner and S.H. Snyder: Bone Remodeling Signaled by a Dihydropyridine- and Phenylalkylamine-Sensitive Calcium Channel. *Proc.Natl.Acad.Sci.U.S.A.* **86**:2957-(3/13) (1989).

Guggino, S.E., J.A. Wagner, A.M. Snowman, L.D. Hester and B. Sacktor: Phenylalkylamine-Sensitive Calcium Channels in Osteoblast-Like Osteosarcoma Cells. Characterization by Ligand Binding and Single Channel Recordings. *J.Biol.Chem.* **263**:10155-(3/13) (1988).

Gulamhussein, S., P. Ko and S. Carruthers: Acceleration of the Ventricular Response during Atrial Fibrillation in the Wolff-Parkinson-White Syndrome after Verapamil. *Circulation* **65**:348-(4/19) (1982).

Gundersen, C.B., B. Katz and R. Miledi: The Antagonism Between Botulinum Toxin and Calcium in Motor Nerve Terminals. *Proc.Roy.Soc.London B* **216**:369-(2/7) (1982).

Gusovsky, F., T. Yasamuto and J.W. Daly: Maitotoxin, a Potent, General Activator of Phosphoinositide Breakdown. *FEBS Lett.*, **243**:307-(3/13) (1989).

Guth, B.D., G. Heusch, R. Seiteberger and J. Ross: Mechanisms of Benefical Effect of Beta-Adrenergic Blockade on Exercise-Induced Ischemia in Conscious Dogs. *Circ.Res.* **80**:738-(4/19) (1987).

Gutnick, M.J., H.D. Lux, D. Swandulla and H. Zucker: Voltage-Dependent and Calcium Dependent Inactivation of Calcium Channel Current in Identified Snail Neurones. *J.Physiol.* **412**:197-(1/4) (1989).

Haag-Weber, M., P. Schollmeyer and W.H. Hörl: Granulocyte Activation During Haemodialysis in the Absence of Complement Activation: Inhibition by Calcium Channel Blockers. *Eur.J.Clin.Invest.* **18**:380-(3/13) (1988).

Hadley, R.W. and J.R. Hume: An Intrinsic Potential-Dependent Inactivation Mechanism Associated Calcium Channels in Guinea Pig Myocytes. *J.Physiol.* **389**:205-(1/4,2/10) (1987).

Hadley, H.N., J.M. Zabranski, R.F. Spetzler, D. Rigamonti, M.S. Fifield and P.C. Johnson: The Efficacy of Intravenous Nimodipine in the Treatment of Focal Cerebral Ischemia in a Primate Model. *Neurosurgery* **25**:63-(3/13) (1989).

Hadley, R.W. and W.J. Lederer: Description of Charge Movement in Isolated Guinea Pig and Rat Ventricular Myocytes. *Biophys.J.* **55**:46a-(2/10) (1989).

Haeusler, G.: Differential Effect of Verapamil on Excitation-Contraction Coupling in Smooth Muscle and on Excitation-Secretion Coupling in Adrenergic Nerve Terminals. *J.Pharmacol.Exp.Ther.* **180**:672-(4/22) (1972).

Hagemeijer, F.: Verapamil in the Management of Supraventricular Tachyarrhythmias Occurring After a Recent Myocardial Infarction. *Circulation* **57**:751-(4/19) (1978).

Hagins, W.A.: The Visual Process: Excitatory Mechanisms in the Primary Receptor Cells. *Ann.Rev.Biophys.Bioeng.* **1**:131-(2/8) (1972).

Hagins, W.A., R.D. Penn and S. Yoshikami: Dark Current and Photocurrent in Retinal Rods. *Biophys.J.* **10**:380-(2/8) (1970).

Hagins, W.A. and S. Yoshikami: Ionic Mechanisms in the Excitation of Photoreceptors. *Neuroscience* **264**:314-(2/8) (1975).

Hagiwara, S., H. Irisawa and M. Kameyama: Contribution of Two Types of Calcium Currents to the Pacemaker Potentials of Rabbit Sino-atrial Node Cells. *J.Physiol.* **395**:233-(1/2,2/10,3/12,3/13,4/19) (1988).

Hagiwara, S.: Ca Spike. In *Advances in Biophysics*. M. Kotani, ed. University of Tokyo Press. **4**:71-(3/12) (1973).

Hagiwara, S.: Calcium Dependent Action Potential. In *Membranes*. G. Eisenman, ed. Marcel Dekker, New York. **3**:359-(3/12) (1975).

Hagiwara, S.: *Membrane Potential Dependent Ion Channels in Cell Membranes*. New York, Raven Press. (1983).

Hagiwara, S. and L. Byerly: Calcium Channel. *Ann.Rev.Neuroscience* **4**:69-(2/7,2/9,3/12,3/14,4/22) (1981).

Hagiwara, S. and L. Byerly: The Calcium Channel. *TINS* **6**:189-(3/12) (1983).

Hagiwara, S., J. Fukuda and D.C. Eaton: Membrane Currents Carried by Calcium, Strontium and Barium in Barnacle Muscle Fiber during Voltage Clamp. *J.Gen.Physiol.* **63**:564-(3/12) (1974).

Hagiwara, S. and L.A. Jaffe: Electrical Properties of Egg Cell Membranes. *Ann.Rev.Biophys.Bioeng.* **8**:358-(2/5) (1979).

Hagiwara, S., H. Irisawa and M. Kameyama: Contribution of Two Types of Calcium Channels to the Pacemaker Potentials of Rabbit Sino Atrial Node Cells. *J. Physiol.* **395**:233-(3/12) (1988).

Hagiwara, S. and K. Naka: The Initiation of Spike Potential In Barnacle Muscle Fibers under Low Intracellular Calcium. *J.Gen.Physiol.* **48**:141-(3/12) (1964).

Hagiwara, S. and H. Ohmori: Studies of Calcium Channels in Rat Clonal Pituitary Cells with Patch Electrode Voltage-clamp. *J.Physiol.* **331**:231-(1/3,2/7,3/12) (1982).

Hagiwara, S. and H. Ohmori: Studies of Single Calcium Channel Currents in Rat Clonal Pituitary Cells. *J.Physiol.* **336**:649-(1/2,1/4,3/12) (1983).

Hagiwara, S., S. Ozawa and O. Sand: Voltage Clamp Analysis of Two Inward Current Mechanisms in the Egg Cell Membrane of a Starfish. *J.Gen.Physiol.* **65**:617-(2/9,3/12) (1975).

Hagiwara, S. and K. Takahashi: Surface Density of Calcium Ions and Calcium Spikes in the Barnacle Muscle Fiber Membrane. *J.Gen.Physiol.* **50**:583-(1/2,3/12) (1967).

Hagiwara, S. and I. Tasaki: A Study on the Mechanism of Impulse Transmission Across the Giant Synapse of the Squid. *J.Physiol.* **143**:114-(2/7) (1958).

Hakim, A.H., B.S. Bomb, S.K. Pandey and S.V. Singh: A Study of Cerebrospinal Fluid Calcium and Magnesium in Depression. *J.Assoc.Physicians* **23**:311-(4/22) (1975).

Hakim, A.M., M. Hogan, L. Berger and A. Gjedde: Nimodipine Binding in Cerebral Ischemia. *Abstracts of the 4th International Symposium on Calcium Antagonists: Pharmacology and Clinical Research* 101-(3/13) (1989).

Halvorsen, S.W. and K.K. Berg: Affinity Labeling of Neuronal Acetylcholine Receptor Subunits with an α-Neurotoxin that Blocks Receptor Function. *J.Neurosci.* **7**:2547-(2/10) (1987).

Hamaguchi, Y. and Y. Hiramoto: Activation of Sea Urchin Eggs by Microinjection of Calcium Buffers. *Exp.Cell Res.* **134**:171-(2/7) (1981).

Hamill, O.P., A. Marty, E. Neher, B. Sakmann and F.J. Sigworth: Improved Patch-clamp Techniques for High-Resolution Current Recording from Cells and Cell-Free Membrane Patches. *Pflügers Arch.* **391**:85-(1/1,1/2,1/4,4/17) (1981).

Hamilton, S.L., A. Yatani, K. Brush, A. Schwartz and A.M. Brown: A Comparison Between the Binding and Electrophysiological Effects of Dihydropyridines on Cardiac Membranes. *Mol.Pharmacol.* **31**:221-(3/13) (1987).

Hamlyn, J.M., D.W. Harris and J.H. Ludens: Digitalis-Like Activity in Human Plasma, Purification, Affinity and Mechanism. *J.Biol.Chem.* **264**:7395-(3/13) (1989).

Hamm, C.W. and L.H. Opie: Protection of Infarcting Myocardium by Slow Channel Inhibitors. Comparative Effects of Verapamil, Nifedipine, and Diltiazem in the Coronary-Ligated, Isolated Working Rat Heart. *Circ.Res.* **52**:129-(4/16,4/19) (1983).

Han, P., G. Boatwright and N.G. Ardlie: Effect of the Calcium-Entry-Blocking Agent Nifedipine on Activation of Human Platelets and Comparison with Verapamil. *Thromb.Haemast.* **50**:513-(4/20) (1980).

Handa, J., H. Matsuda, Y. Nakasu, A. Shiino, T. Kanazawa and K. Harada: KB-2796, a New Ca^{2+} Antagonist. *Cardiovasc.Drug Rev.*(in press) (1990).

Hanke, W., N.J. Cook and U.B. Kaupp: cGMP-Dependent Channel Protein from Photoreceptor Membranes: Single-Channel Activity of the Purified and Reconstituted Protein. *Proc.Natl.Acad.Sci.U.S.A.* **85**:94-(2/8) (1988).

Hanrath, P., P. Kremer and W. Bleifeld: Influence of Nifedipine on Left Ventricular Dysfunction at Rest and During Exercise. *Eur.Heart J.* **3**:325-(4/17) (1982).

Hanrath, P., D.G. Mathey, P. Kremer, F. Sonntag and W. Bleifeld: Effect of Verapamil on Left Ventricular Isovolumic Relaxation Time and Regional Left Ventricular Filling in Hypertrophic Cardiomyopathy. *Am.J.Cardiol.* **45**:1258-(4/17) (1980).

Hansen, J.F., B. Sigurd, K. Mellemgaard and I. Lynby: Verapamil in Acute Myocardial Infarction. *Dan.Med.Bull.* **27**:105-(4/19) (1980).

Hansson, L.: What are we Really Achieving with Long-Term Drug Therapy. *Am.J.Hypertens.* **1**:414-(4/18) (1988).

Harder, D.R.: Membrane Electrical Activation of Arterial Smooth Muscle. In *Vascular Smooth Muscle: Metabolic, Ionic and Contractile Mechanisms.* M.F. Crass and C.D. Barnes, eds. Academic Press, New York. 71-(2/10) (1982).

Harris, G.L., L.P. Henderson and N.C. Spitzer: Changes in Densities and Kinetics of Delayed Rectifier Potassium Channels During Neuronal Differentiation. *Neuron* **1**:739-(2/10) (1988).

Harris, G.L. and N.C. Spitzer: Spatial Distribution of GABA Receptors at the Time of Their Appearance on Spinal Neurons Differentiating in Culture. *Soc.NeuroscienceAbst.* **13**:151-(2/10) (1987).

Harris, K.M., S. Kongsamut and R.J. Miller: Protein Kinase C Mediated Regulation of Calcium Channels in PC-12 Pheochromocytoma Cells. *Biochem.Biophys.Res.Commun.* **134**:1298-(3/14) (1986).

Harris, W.H. and J.A. Beauchemin: Cerebrospinal Fluid Calcium, Magnesium, and their Ratio in Psychoses of Organic and Functional Origin. *Yale J.Biol.Med.* **29**:117-(4/22) (1956).

Harris-Warwick, R.M., C. Hammond, D. Paupardin-Tritsch, V. Homburger, B. Rouot, J. Bockaert and H.M. Gerschenfeld: An α-40 Subunit of a GTP Binding Protein Immunologically Related to G_o Mediates a Dopamine Induced Decrease of Ca^{2+} Current in Snail Neurons. *Neuron* **1**:27-(3/14) (1988).

Hartel, G. and M. Hartikainen: Comparison of Verapamil and Practolol in Paroxysmal Supraventricular Tachycardia. *Eur.J.Cardiol.* **4**:87-(4/19) (1976).

Hartzell, H.C.: Regulation of Cardiac Ion Channels by Catecholamines, Acetylcholine and Second Messenger Systems. *Proc.Biophys.Molec.Biol.* **52**:165-(1/4) (1988).

Hartzell, H.C. and R.E. White: Effects of Magnesium on Inactivation of the Voltage-Gated Calcium Current in Cardiac Myocytes. *J.Gen.Physiol.* **94**:745-(1/4) (1989).

Hathaway, B.R. and R.S. Adelstein: Human Platelet Myosin Light Chain Kinase Requires the Calcium-Binding Protein Calmodulin for Reactivity. *Proc.Natl.Acad.Sci.U.S.A.* **76**:1653-(4/20) (1979).

Hatta, K., T.S. Okada and M. Takeichi: A Monoclonal Antibody Disrupting Calcium-Dependent Cell-Cell Adhesion of Brain Tissues. Possible Role of its Target Antigen in Animal Pattern Formation. *Proc.Natl.Acad.Sci.U.S.A.* **82**:2789-(2/10) (1985).

Hauswirth, O. and B.N. Singh: Ionic Mechanisms in Heart Muscle in Relation to the Genesis and Pharmacologic Control of Cardiac Arrhythmias. *Pharmacol.Rev.* **30**:5-(4/19) (1979).

Hawthorn, M., P. Gengo, X.Y. Wei, A. Rutledge, J.F. Moran, S. Gallant and D.J. Triggle: Effect of Thyroid Status on Beta-Adrenoceptors and Calcium Channels in Rat Cardiac and Vascular Tissues. *Naunyn Schmied Arch.Pharmacol.* **337**:539-(3/13) (1988).

Haycock, J.W., W.F. White and C.W. Cotman: Differences in Alkaline Earth Stimulation of Neurotransmitter Release from Isolated Brain Synaptosomes. *Naunyn Schmied Arch.Pharmacol.* **301**:175-(4/22) (1978).

Haydon, P.G. and H. Man-Son-Hing: Low and High Voltage Activated Calcium Currents: Their Relationship to the Site of Neurotransmitter Release in an Identified Neuron of Helisoma. *Neuron* **1**:919-(2/9,3/14) (1988).

Haynes, L.W., A.R. Kay and K.W. Yau: Single cGMP-Activated Channel Activity in Excised Patches of Rod Outer Segment Membranes. *Nature* **321**:66-(2/8) (1986).

Haynes, L.W. and K.W. Yau: Cyclic GMP-Sensitive Conductance in Outer Segment Membrane of Catfish Cones. *Nature* **317**:61-(2/8) (1985).

Hearse, D.J.: Release of Enzymes from Ischemic Myocardium. In *Degradation Processes in Heart and Skeletal Muscle* K. Wildenthal, ed. Research Monographs in Cell and Tissue Physiology. Elsevier, North Holland. **3**:419-(4/17) (1980).

Hecht, H.S., C.Y.C. Chew, M.H. Burnam, J. Hopkins, S. Schnugg and B.N. Singh: Verapamil in Chronic Stable Angina: Amelioration of Pacing-Induced Abnormalities of Left Ventricular Ejection Fraction, Regional Wall Motion, Lactate Metabolism and Hemodynamics. *Am.J.Cardiol.* **48**:536-(4/16) (1981).

Hedberg, A., F. Kempf, M.E. Josephson and P. Molinoff: Coexistence of Beta-1 and Beta-2 Adrenergic Receptors in the Human Heart: Effects of Treatment with Receptor Antagonists and Calcium Entry Blockers. *J.Pharmacol.Exp.Ther.* **234**:561-(3/13) (1985).

Hefti, F., H. Gnahn, M.E. Schwab and H. Thoenen: Induction of Tyrosine Hydroxylase by Nerve Growth Factor and by Elevated K^+ Concentrations in Cultures of Dissociated Sympathetic Neurons. *J.Neurosci.* **2**:1554-(2/10) (1982).

Heider, J.G., D.B. Weinstein, C.E. Pickens, S. Lan and C.M. Su: Antiatherogenic Activity of the Calcium Channel Blocker Isradipine (PN 200-110): A Novel Effect on Matrix Synthesis Independent of Calcium Channel Blockade. *Transplant Proc.* **29**:96-(3/13) (1987).

Heidland, A., K. Klutsch and A. Oebeck: Myogenbedingte Vasodilation bei Nierenischaemie. *Muench.Med.Wschr.* **104**:1636-(4/18) (1962).

Heinemann, U., S. Franceschetti, B. Hamon, A. Konnerth and Y. Yaari: Effects of Anticonvulsants on Spontaneous Epileptiform Activity which Develops in the Absence of Chemical Synaptic Transmission in Hippocampal Slices. *Brain Res.* **325**:349-(4/22) (1985).

Heinemann, U., H.D. Lux and M.J. Gutnick: Extracellular Free Calcium and Potassium During Paroxysmal Activity in Cerebral Cortex of the Cat. *Exp.Brain Res.* **27**:237-(4/22) (1977).

Held, P.H., S. Yusuf and C.D. Furberg: Calcium-Channel Blockers in Acute Myocardial Infarction and Unstable Angina: An Overview. *Br.Med.J.* **299**:1187-(4/19) (1989).

Helfferich, F.: *Ion Exchange*. McGraw Hill Book Co., New York. 304-(3/12) (1962).

Hellman, B. and E. Gylfe: Calcium and the Control of Secretion. In *Calcium and Cell Function*. W.Y. Cheung, ed. Academic Press, New York. **6**:253-(2/7) (1986).

Henderson, A.H. and M.J. Lewis: Basic Mechanisms of Calcium Antagonists: (Slow-channel Blockers). *Scot.Med.J.* **26**:156-(4/19) (1981).

Henderson, L.P., M.A. Smith and N.C. Spitzer: The Absence of Calcium Blocks Impulse-evoked Release of Acetylcholine but not *de novo* Formation of Functional Neuromuscular Synaptic Contacts in Culture. *J.Neurosci.* **4**:3140-(2/10) (1984).

Henderson, L.P. and N.C. Spitzer: Autonomous Early Differentiation of Neurons and Muscle Cells in Single Cell Cultures. *Dev.Biol.* **113**:381-(2/10) (1986).

Hendry, S.H.C. and E.G. Jones: Activity-Dependent Regulation of GABA Expression in the Visual Cortex of Adult Monkeys. *Neuron* **1**:701-(2/10) (1988).

Heng, M.K., B.N. Singh, A.H.G. Roche, R.M. Norris and C.J. Mercer: Effects of Intravenous Verapamil on Cardiac Arrhythmias and on the Electrocardiogram. *Am.Heart J.* **90**:487-(4/19) (1975).

Henkart, P., M. Henkart, R. Hodes and M. Taplits: Secretory Processes in Lymphocyte Function. *Biosci.Rep.* **7**:345-(2/7) (1987).

Henry, P.D.: Calcium Antagonists as Antiatherogenic Agents. *Ann.N.Y.Acad.Sci.* **522**:411-(3/13) (1988).

Henry, P.D. and K.I. Bentley: Suppression of Atherogenesis in Cholesterol-fed Rabbit Treated with Nifedipine. *J.Clin.Invest.* **68**:1366-(4/18,4/20) (1981).

Henry, P.D., R. Shuchleib, L.J. Broda, R. Roberts, J.R. Williams and B.E. Sobel: Effects of Nifedipine on Myocardial Perfusion and Ischemic Injury in Dogs. *Circ.Res.* **43**:372-(4/19) (1978).

Henry, P.D., R. Shuchleib, J. Davis, E.S. Weiss and B.E. Sobel: Myocardial Contracture and Accumulation of Mitochondrial Calcium in Ischemic rabbit heart. *Am.J.Physiol.* **233**:H677-(4/19) (1977).

Herbette, L.G., Y.M.H. Vant Erve and D.G. Rhodes: Interaction of 1,4-Dihydropyridine Calcium Channel Antagonists with Biological Membranes: Lipid Bilayer Partitioning could Occur before Drug Binding to Receptors. *J.Mol.Cell Cardiol.* **21**:187-(3/13) (1989).

Herdon, H. and S.R. Nahorski: Investigations of the Roles of Dihydropyridine and ω-Conotoxin-Sensitive Calcium Channels in Mediating Depolarization-Evoked Endogenous Dopamine Release from Striatal Slices. *Naunyn Schmied Arch.Pharmacol.* **340**:36-(3/13) (1989).

Hering, S., D.J. Beech, T.B. Bolton and S.P. Lim: Action of Nifedipine or BAY K 8644 is Dependent on Calcium Channel State in Single Smooth Muscle Cells from Rabbit Ear Artery. *Pflügers Arch.* **411**:590-(3/13) (1988).

Hering, S., T.B. Bolton, D.J. Beech and S.P. Lim: Mechanism of Calcium Channel Block by D600 in Single Smooth Muscle Cells from Rabbit Ear Artery. *Circ.Res.* **64**:928-(3/13) (1989).

Herling, A.W. and M. Ljungstrom: Effects of Verapamil on Gastric Acid *In Vitro* and *In Vivo*. *Eur.J.Pharmacol.* **156**:341-(3/13) (1988).

Hermsmeyer, K., M. Sturek, W. Marvin, R. Mason and A. Puga: Vascular Muscle Calcium Channel Modulation in Hypertension. *J.Cardiovasc.Pharmacol.* **14**:S45-(3/13) (1989).

Hertz, F. and A. Cloarec: Comparative Antiulcer and Antisecretory Effects of Various Calcium Antagonists. *Gen.Pharmacol.* **20**:635-(3/13) (1989).

Hertz, L., A.S. Bender, D.M. Woodbury and H.S. White: Potassium-Stimulated Calcium Uptake in Astrocytes and its Potent Inhibition by Nimodipine. *J.Neurosci.Res.* **22**:209-(3/13) (1989).

Hertzberg, E.L. and N.B. Gilula: Liver Gap Junctions and Lens Fiber Junctions: Comparative Analysis and Calmodulin Interaction. *Cold Spring Harbor Symp.Quant.Biol.* **46**:639-(2/10) (1981).

Hertzberg, E.L. and R.G. Johnson: In *Gap Junctions in Modern Cell Biology*. Alan R..Liss, Inc., New York. **7**:(2/10) (1988).

Hescheler, J., M. Kameyama and W. Trautwein: On the Mechanism of Muscarinic Inhibition of the Cardiac Ca^{2+} Current. *Pflügers Arch.* **402**:182-(3/14) (1986).

Hescheler, J., M. Kameyama, W. Trautwein, G. Miesks and H.D. Soling: Regulation of the Cardiac Calcium Channel by Protein Phosphatases. *Eur.J.Biochem.* **165**:261-(3/13,3/14) (1987a).

Hescheler, J., H. Nawrath, M. Tang and W. Trautwein: Adrenoceptor-Mediated Changes of Excitation and Contraction in Ventricular Heart Muscle from Guinea Pigs and Rabbits. *J.Physiol.* **397**:657-(2/10) (1988a).

Hescheler, J., W. Rosenthal, K.D. Hinsch, M. Wulfern, W. Trautwein and G. Schultz: Angiotensin II Induced Stimulation of Voltage Dependent Ca^{2+} Currents in Adrenal Cortical Cell Line. *EMBO J.* **7**:619-(3/13,3/14) (1988b).

Hescheler, J., W. Rosenthal, W. Trautwein and G. Schultz: The GTP Binding Protein G_o Regulates Neuronal Calcium Channels. *Nature* **325**:445-(3/14) (1987b).

Hescheler, J. and W. Trautwein: Modification of L-Type Calcium Current by Intracellular Applied Trypsin in Guinea Pig Ventricular Myocytes. *J.Physiol.* **404**:259-(1/4) (1988).

Hesketh, J.E., A.I.M. Glen and H.W. Reading: Membrane ATPase Activities in Depressive Illness. *J.Neurochem.* **28**:1401-(4/22) (1977).

Hess, O.M., J. Grimm and H.P. Krayenbuehl: Diastolic Function in Hypertrophic Cardiomyopathy: Effects of Propranolol and Verapamil on Diastolic Stiffness. *Eur.Heart J.* **4**:47-(4/17) (1983).

Hess, P.: Elementary Properties of Cardiac Calcium Channels: A Brief Review. *Can.J.Physiol.Pharmacol.* **66**:1218-(2/10,4/17) (1988).

Hess, P.: Calcium Channels in Vertebrate Cells. *Ann.Rev.Neurosci.* **13**:337-(3/13) (1990).

Hess, P., C. Chen, M. Plummer and D. Logothetis: Elementary Properties of Calcium Channels. *4th Intl. Symp. on Calcium Antagonists: Pharmacology and Clinical Research* 1-(3/13) (1990).

Hess, P., J.B. Lansman and R.W. Tsien: Different Modes of Ca Channel Gating Behaviours Favoured by Dihydropyridine Ca Agonists and Antagonists. *Nature* **311**:538-(1/2,1/4,3/13,3/14) (1984).

Hess, P., J.B. Lansman and R.W. Tsien: Calcium Channel Selectivity for Divalent and Monovalent Cations. Voltage and Concentration Dependence of Single Channel Current in Ventricular Heart Cells. *J.Gen.Physiol.* **88**:293-(1/2,3/12) (1986).

Hess, P. and R.W. Tsien: Mechanism of Ion Permeation through Calcium Channels. *Nature* **309**:453-(1/2,3/12) (1984).

Hestrin, S.: The Properties and Function of Inward Rectification in Rod Photoreceptors of the Tiger Salamander. *J.Physiol.* **390**:319-(2/8) (1987).

Hestrin, S. and J.I. Korenbrot: Effects of Cyclic GMP on the Kinetics of the Photocurrent in Rods and in Detached Rod Outer Segments. *J.Gen.Physiol.* **90**:527-(2/8) (1987).

Hestrin, S. and J.I. Korenbrot: Kinetics of Activation of Phototransduction in Retinal Rods and Cones. *J.Neurosci.*(submitted) (2/8) (1989).

Heuser, J.E., T.S. Reese, M.J. Dennis, Y. Jan, L. Jan and L. Evans: Synaptic Vesicle Exocytosis Captured by Quick Freezing and Correlated with Quantal Transmitter Release. *J.Cell Biol.* **81**:275-(2/7) (1979).

Higo, K., H. Saito and N. Matsuki: Characteristics of [^{3}H]Nimodipine Binding to Sarcolemmal Membranes from Rat Vas Deferens and its Regulation by Guanine Nucleotide. *Jpn.J.Pharmacol.* **48**:213-(3/13,3/14) (1988).

Hill, J.A., R.L. Feldman, C.J. Pepine and C.R. Conti: Randomized Double-blind Comparison of Nifedipine and Isosorbide Dinitrate in Patients with Coronary Arterial Spasm. *Am.J.Cardiol.* **49**:431-(4/16) (1982).

Hille, B.: In *Ionic Channels of Excitable Membranes*. Sinauer Assoc. Sunderland, MA. (1/1,1/2, 3/14) (1984).

Hille, B.: Ionic Selectivity of Na and K Channels of Nerve Membranes. In *Membranes*. G. Eisenman, ed. Marcel Dekker, New York. (1/1,3/12) (1975).

Hille, B. Ionic Channels in Excitable Membranes: Current Problems and Biophysical Approaches. *Biophys. J.* **22**:283-(3/12) (1978).

Hille, B., A.M. Woodhull and B.I. Shapiro: Negative Surface Charge near Sodium Channels of Nerve. Divalent Ions, Monovalent Ions and pH. *Phil.Trans.R.Soc.* **270**:301-(3/12) (1975).

HINT Research, Group (Holland Interuniversity Nifedipine/Metoprolol Trial) Early Treatment of Unstable Angina in the Coronary Care: A Randomized Double-Blind Placebo-Controlled Comparison of Recurrent Ischaemia in Patients Treated with Nifedipine or Metoprolol or Both. *Br.Heart J*. **56**:400-(4/16,4/19) (1986).

Hiramasu, S., H. Coto, Z.Y. Li, C. Maldonado and J. Kupersmith: Dextro Rotary Isomer of Sotalol Electrophysiologic Effects and Interaction with Verapamil. *Am.Heart J.* **116**:1552-(4/19) (1988).

Hirano, T. and S. Hagiwara: Kinetics and Distribution of Voltage Gated Ca, Na and K Channels on the Somata of Rat Cerebellar Cells. *Pflügers Arch.* **413**:463-(2/9) (1989).

Hirano, Y., H.A. Fozzard and C.T. January: Inactivation Properties of T-Type Calcium Current in Canine Cardiac Purkinje Cells. *Biophys.J.* **56**:1007-(1/4) (1989).

Hirning, L.D., A.P. Fox, E.W. McCleskey, B.M. Olivera, S.A. Thayer, R.J. Miller and R.W. Tsien: Dominant Role of N-Type Ca^{2+} Channels in Evoked Release of Norepinephrine from Sympathetic Neurons. *Science* **239**:57-(1/4,3/13,3/14) (1988).

Hirst, G.D.S., S.M. Johnson and D.F. van Helden: The Calcium Current in a Myenteric Neurone of the Guinea Pig Ileum. *J.Physiol.* **361**:297-(3/12) (1985).

Hirst, G.D.S. and E.M. McLachlan: Development of Dendritic Calcium Currents in Ganglion Cells of the Rat Lower Lumbar Sympathetic Chain. *J.Physiol.* **377**:349-(2/9) (1986).

Hirst, G.D.S. and T.O. Neild: Evidence for Two Populations of Excitatory Receptors for Noradrenaline on Anteriolar Smooth Muscle. *Nature* **283**:767-(2/10) (1980).

Hirst, G.D.S. and T.O. Neild: Localization of Specialized Noradrenaline Receptors at Neuromuscular Junctions on Arterioles of the Guinea Pig. *J.Physiol.* **313**:343-(2/10) (1981).

Hirst, G.D.S., G.D. Silverberg and D.F. van Helden: The Action Potential and Underlying Ionic Currents in Proximal Rat Middle Cerebral Arterioles. *J.Physiol.* **371**:298-(2/10) (1986).

Hisayama, T. and I. Takayanagi: Ryanodine: Its Possible Mechanism of Action in the Caffeine-Sensitive Calcium Store of Smooth Muscle. *Pflügers Arch.* **412**:376-(2/10) (1988).

Hockberger, P., M. Toselli, D. Swandulla and H.D. Lux: A Diacylglycerol Analogue Reduces Neuronal Calcium Currents Independently of Protein Kinase C Activation. *Nature* **338**:340-(3/13,3/14) (1989).

Hockberger, P.E., H.Y. Tseng and J.A. Connor: Fura-2 Measurements of Cultured Rat Purkinje Neurons Show Dendritic Localization of Ca^{2+} Influx. *J.Neurosci.* **9**:2272-(2/9) (1989).

Hodgkin, A.L. and P. Horowicz: Potassium Contractures in Single Muscle Fibers. *J.Physiol.* **153**:386-(2/10) (1960).

Hodgkin, A.L. and A.F. Huxley: A Quantitative Description of Membrane Current and its Application to Conduction and Excitation in Nerve. *J.Physiol.* **117**:500-(1/1,1/3,1/4) (1952b).

Hodgkin, A.L. and A.F. Huxley: Current Carried by Sodium and Potassium Ions through the Membranes of the Giant Squid *Loligo*. *J.Physiol.* **116**:449-(3/12) (1952a).

Hodgkin, A.L. and B. Katz: The Effect of Sodium Ions on the Electric Activity of the Giant Axon of the Squid. *J.Physiol.* **108**:37-(1/3) (1949).

Hodgkin, A.L., P.A. McNaughton and B.J. Nunn: The Ionic Selectivity and Calcium Dependence of the Light-Sensitive Pathway in Toad Rods. *J.Physiol.* **358**:447-(2/8) (1985).

Hodgkin, A.L., P.A. McNaugton, B.J. Nunn and K.W. Yau: Effects of Ions on Retinal Rods from *Bufo marinus*. *J.Physiol.* **350**:649-(2/8) (1984).

Hodgkin, A.L. and B.J. Nunn: Control of Light-Sensitive Current in Salamander Rods. *J.Physiol.* **403**:439-(2/8) (1988).

Hof, R.P. and A. Hof: Vasoconstrictor and Vasodilator Effects in Normal and Atherosclerotic Conscious Rabbits. *Br.J.Pharmacol.* **95**:1075-(3/13) (1988).

Hoffman, B.F. and K.H. Dangman: Mechanisms for Cardiac Arrhythmias. *Experientia* **43**:1049-(4/19) (1987).

Hoffman, B.F. and M.R. Rosen: Cellular Mechanism for Cardiac Arrhythmias. *Circ.Res.* **49**:1-(4/19) (1981).

Holck, M. and W. Osterrieder: The Peripheral High Affinity Benzodiazepine Binding Site is not Coupled to the Cardiac Ca^{2+} Channel. *Eur.J.Pharmacol.* **118**:293-(3/13) (1985).

Holliday, J. and N.C. Spitzer: Role of Calcium Dependent Influx in Neuronal Differentiation: A Period of Spontaneously Elevated Internal Calcium is Correlated with a Period of Sensitivity to External Calcium. *Soc.Neurosci.Abst.* **14**:1040-(2/10) (1988).

Holliday, J. and N.C. Spitzer: Spontaneous Calcium Influx and its Roles in Differentiation of Spinal Neurons in Culture. *Dev.Biol.* **141**:13-(2/10) (1990).

Hollifield, J.W., K. Sherman and P. Slaton: Age, Race and Sex as a Determinant of Successful Antihypertensive Therapy. *Prev.Med.* **7**:88-(4/18) (1978).

Holmsen, H. and H.J. Weiss: Secretable Storage Pools in Platelets. *Ann.Rev.Med.* **30**:119-(4/20) (1979).

Holz, G.G., K. Dunlap and R.M. Kream: Characterization of the Electrically Evoked Release of Substance P from Dorsal Root Ganglion Neurons: Methods and Dihydropyridine Sensitivity. *J.Neurosci.* **8**:463-(3/13) (1988).

Holz, G.G., S.G. Rane and K. Dunlap: GTP-Binding Proteins Mediate Transmitter Inhibition of Voltage-Dependent Calcium Channels. *Nature* **319**:670-(3/13) (1986).

Hondeghem, L.M. and B.G. Katzung: Antiarrhythmic Agents: the Modulated Receptor Mechanism of Sodium and Calcium Channel Blocking Drugs. *Ann.Rev.Pharmacol.Toxicol.* **24**:387-(3/13) (1984).

Hopf, R. and M. Kaltenbach: Effects of Nifedipine and Propranolol Combined Therapy in Patients with Hypertrophic Cardiomyopathy. *Z.Kardiol.* **76**:105-(4/17) (1987).

Hordof, A.J., R. Edie, J.R. Malm, B.F. Hoffman and M.R. Rosen: Electrophysiologic Properties and Response to Pharmacologic Agents of Fibers in Diseased Atria. *Circulation* **54**:776-(4/19) (1976).

Horne, W.A., M. Abdel-Ghany, R. Racker, G.A. Weiland, R.E. Oswald and R.A. Cerione: Functional Reconstitution of Skeletal Muscle Ca Channel: Separation of Regulatory and Channel Components. *Proc.Natl.Acad.Sci.U.S.A.* **85**:3718-(3/13) (1988).

Horwitz, B.A. and G.E. Hanes: Isoproterenol Induced Calorigenesis of Dystrophic and Normal Hamsters. *Proc.Soc.Exp.Biol.Med.* **147**:392-(4/17) (1974).

Höschl, C., J. Blahos and J. Kabes: The Use of Calcium Channel Blockers in Psychiatry. In *Biological Psychiatry*. C. Shagass, R.C. Josiassen, W.H. Bridger, K.J. Weiss, D. Stoff, G.M. Simpson, eds. Elsevier, New York. (4/22) (1986).

Höschl, C. and J. Kozeny: Verapamil in Depression: A Controlled, Double-Blind Study. *Psychopharmacol.Suppl.* **96**:212-(4/22) (1988).

Hosey, M.M., J. Barhanin, A. Schmid, S. Vandaele, J. Ptasienski, C. O'Callahan, C. Cooper and M. Lazdunski: Photoaffinity Labelling and Phosphorylation of a l65 Kilodalton Peptide Associated with Dihydropyridine and Phenylalkylamine Sensitive Calcium Channels. *Biochem.Biophys.Res.Commun.* **147**:1137-(3/14) (1987).

Hosey, M.M., M. Borsotto and M. Lazdunski: Phosphorylation and Dephosphorylation of Dihydropyridine Sensitive Voltage Dependent Ca^{2+} Channel in Skeletal Muscle Membranes by cAMP and Ca^{2+} Dependent Processes. *Proc.Natl.Acad.Sci.U.S.A.* **83**:3733-(3/14) (1986).

Hosey, M.M., F.C. Chang, C.M. O'Callahan and J. Ptasienski: L-Type Calcium Channels in Cardiac and Skeletal Muscle. *Ann.N.Y.Acad.Sci.* **560**:27-(1989) (1989).

Hosey, M.M. and M. Lazdunski: Calcium Channels: Molecular Pharmacology, Structure and Regulation. *J.Membr.Biol.* **104**:81-(3/13,3/14) (1988).

Hoshi, T. and S.J. Smith: Large Depolarization Induces Long Openings of Voltage-Dependent Calcium Channels in Adrenal Chromaffin Cells. *J.Neurosci.* **7**:571-(2/7) (1987).

Hossack, K.F., P.E. Pool, P. Steele, M.H. Crawford, A.N. DeMaria, L.S. Cohen and T.A. Ports: Efficacy of Diltiazem in Angina of Effort: A Multicenter Trial. *Am.J.Cardiol.* **49**:567-(4/16) (1982).

Hossmann, K.A.: Calcium Antagonists for the Treatment of Brain Ischemia: A Critical Appraisal. In *Pharmacology of Cerebral Ischemia.* K. Krieglstein, ed. CRC Press, Boca Raton, FL **53** -(3/13) (1989).

Hounsgaard, J. and J. Midtgaard: Intrinsic Determinants of Firing Pattern in Purkinje Cells of the Turtle Cerebellum *In Vitro. J.Physiol.* **402**:731-(2/9) (1988).

Hounsgaard, J. and J. Midtgaard: Synaptic Control of Excitability in Turtle Cerebellar Purkinje Cells. *J.Physiol.* **409**:157-(2/9) (1989).

Howard, J. and A.J. Hudspeth: Compliance of the Hair Bundle Associated with Gating of Mechanoelectrical Transduction Channels in the Bullfrog's Saccular Hair Cell. *Neuron* **1**:189-(2/8) (1988).

Howard, J., W.M. Roberts and A.J. Hudspeth: Mechanoelectrical Transduction by Hair Cells. *Ann.Rev.Biophys.Chem.* **17**:99-(2/8) (1988).

Howard, P.: Vasodilator Drugs in Chronic Obstructive Airways Disease. *Eur.Respir.* **2**:6785-(3/13) (1989).

Howlett, S.E., V.F. Rafuse and T. Gordon: [^{3}H]Nitrendipine Binding Sites in Normal and Cardiomyopathic Hamsters: Absence of a Selective Increase in Putative Calcium Channels in Cardiomyopathic Hearts. *Cardiovasc.Res.* **22**:840-(3/13) (1988).

Hoyle, G., P.A. McNeill and A.I. Selverston: Ultrastructure of Barnacle Muscle Fibers. *J.Cell Biol.* **56**:74-(3/12) (1973).

Hryshko, L.V., R. Bouchard, T. Chau and D. Bose: Inhibition of Rest Potentiation in Canine Ventricular Muscle by BAY K 8644: Comparison with Caffeine. *Am.J.Physiol.* **257**:H399-(3/13) (1989).

Hu, P.S., E. Lindgren, K.A. Jacobson and B.B. Fredholm: Interaction of Dihydropyridine Calcium Channel Agonists and Antagonists with Adenosine Receptors. *Pharmacol.Toxicol.* **61**:121-(3/13) (1987).

Huang, L.Y.M.: Calcium Channels in Isolated Rat Dorsal Horn Neurons, Including Labeled Spinothalamic and Trigemninothalamic Cells. *J.Physiol.* **411**:161-(2/9) (1989).

Hudspeth, A.J. and R.S. Lewis: A Model for Electrical Resonance and Frequency Tuning in Saccular Hair Cells of the Bullfrog, *Rana catesbiana. J.Physiol.* **400**:275-(2/8) (1988b).

Hudspeth, A.J. and R.S. Lewis: Kinetic Analysis of Voltage- and Ion-Dependent Conductances in Saccular Hair Cells of the Bullfrog, *Rana catesbiana. J.Physiol.* **400**:237-(2/8) (1988a).

Hugenholtz, P.G., P.W. Serruys, A. Fleckenstein and W. Nayler: Why Ca^{2+} Antagonists will be Most Useful Before or During Early Myocardial Ischaemia and not after Infarction has been Established. *Eur.Heart J..***7**:270-(4/16) (1986).

Hugtenburg, J.G.: *The Effects of Calcium Antagonists on the Normoxic and the Ischaemic Myocardium.* Uitgeverij University of Amsterdam Press. (3/13) (1989).

Huguenard, J.R., O.P. Hammill and D.A. Prince: Sodium Channels in Dendrites of Rat Cortical Neurons. *Proc.Natl.Acad.Sci.U.S.A.* **86**:2473-(2/9) (1989).

Hui, C.S. and R.L. Milton: Suppression of Charge Movement in Frog Skeletal Muscle by D600. *J.Mus.Res. Cell Motil.* **8**:195-(2/10) (1987).

Hui, C.S., R.L. Milton and R.S. Eisenberg: Charge Movement in Skeletal Muscle Fibers Paralyzed by the Calcium-Entry Blocker D600. *Proc.Natl.Acad.Sci.U.S.A.* **81**:2582-(2/10) (1984).

Hullet, F.J., S.G. Potkin, A.B. Levy and R. Ciasca: Depression Associated with Nifedipine-Induced Calcium Channel Blockade. *Am.J.Psychiatr.* **145**:10-(4/22) (1988).

Hulthen, U., P. Bolli, F.W. Amann, W. Kiowski and F.R. Buhler: Enhanced Vasodilatation in Essential Hypertension by Calcium Channel Blockade with Verapamil. *Hypertension* **4**:26-(4/18) (1982).

Hung, J., I.H. Lamb, S.J. Connolly, K.R. Jutzy, M.I. Goris and J.S. Schroeder: The Effect of Diltiazem and Propranolol, Alone and in Combination, on Exercise Performance and Left Ventricular Function in Patients with Stable Effort Angina: A Double-Blind, Randomized, and Placebo Controlled Study. *Circulation* **68**:560-(4/16) (1983).

Hung, J.S., S.J. Yeh, F.C. Liu, M. Fu, Y.S. Lee and D. Wu: Usefulness of Intravenous Diltiazem in Predicting Subsequent Electrophysiologic and Clinical Responses to Oral Diltiazem. *Am.J.Cardiol.* **54**:1259-(4/19) (1984).

Hunt, B.A., T.H. Self, R.L. Lalonde and M.B. Bottorff: Calcium Channel Blockers as Inhibitors of Drug Metabolism. *Chest* **96**:393-(3/13) (1989).

Hurley, J.B.: Molecular Properties of the cGMP Cascade of Vertebrate Photoreceptors. *Ann.Rev.Physiol.* **49**:793-(2/8) (1987).

Hwang, K.S. and C. Van Breemen: Ryanodine Modulation of ^{45}Ca Efflux and Tension in Rabbit Aortic Smooth Muscle. *Pflügers Arch.* **408**:343-(2/10) (1987).

Hymel, L., M. Inui, S. Fleischer and H. Schindler: Purified Ryanodine Receptor of Skeletal Muscle Sarcoplasmic Reticulum Forms Ca^{2+}-Activated Oligomeric Ca^{2+} Channels in Planar Bilayers. *Proc.Natl.Acad.Sci.U.S.A.* **85**:441-(2/10) (1988a).

Hymel, L., J. Striessnig, H. Glossmann and H. Schindler: Purified Skeletal Muscle 1,4-Dihydropyridine Receptor Forms Phosphorylation-Dependent Oligomeric Calcium Channels in Planar Bilayers. *Proc.Natl.Acad.Sci.U.S.A.* **85**:4290-(3/13) (1988b).

Iijima, T., S. Ciani and S. Hagiwara: Effects of External pH on Ca Channels: Experimental Studies and Theoretical Considerations using a Two Site, Two-Ion Model. *Proc.Natl.Acad.Sci.U.S.A.* **83**:654-(3/12) (1986).

Ikeda, S.R. and G.G. Schofield: Somatostatin Blocks a Calcium Current in Rat Sympathetic Neurones. *J.Physiol.* **409**:221-(1/3) (1989).

Ikeda, Y., M. Kikuchi, K. Toyama, K. Watanabe and Y. Ando: Inhibition of Human Platelet Functions by Verapamil. *Thromb.Haemostas.* **45**:158-(4/20) (1981).

Iliescu, C.C. and A. Sebastiani: Notes on the Effects of Quinidine Upon Paroxysms of Tachycardia. *Heart* **10**:223-(4/19) (1923).

Imagawa, T., J.S. Smith, R. Coronado and K.P. Campbell: Purified Ryanodine Receptor from Skeletal Muscle Sarcoplasmic Reticulum is the Ca^{2+}-Permeable Pore of the Calcium Release Channel. *J.Biol.Chem.* **262**:16636-(2/10) (1987).

Inui, M., A. Saito and S. Fleischer: Purification of the Ryanodine Receptor and Identity with Feet Structures of Junctional Terminal Cisternae of Sarcoplasmic Reticulum from Fast Skeletal Muscle. *J.Biol.Chem.* **262**:1740-(2/10) (1987a).

Inui, M., A. Saito and S. Fleischer: Isolation of the Ryanodine Receptor from Cardiac Sarcoplasmic Reticulum and Identity with the Feet Structures. *J.Biol.Chem.* **262**:15637-(2/10) (1987b).

Isawe, M., I. Sotobata, S. Takagi, S. Miyaguchi, H.X. Jing and M. Yokota: Effects of Diltiazem on Left Ventricular Diastolic Behavior in Patients with Hypertrophic Cardiomyopathy: Evaluation with Exercise Pulsed Doppler Echocardiography. *J.Am.Coll.Cardiol.* **9**:1099-(4/17) (1987).

Isenberg, G. and U. Klöckner: Calcium Currents of Isolated Bovine Ventricular Myocytes are Fast and of Large Amplitude. *Pflügers Arch.* **395**:30 (1/2) (1982).

Ishida, I. and T. Deguchi: Effect of Depolarizing Agents on Choline Acetyltransferase and Acetylcholinesterase Activities in Primary Cell Cultures of Spinal Cord. *J.Neurosci.* **3**:1818-(2/10) (1983).

Ishii, K., T. Kano, Y. Kurobe and J. Ando: Binding of [^{3}H]Nitrendipine to Heart and Brain Membranes from Normotensive and Spontaneously Hypertensive Rats. *Eur.J.Pharmacol.* **88**:277-(3/13) (1983).

ISIS-1 Collaborative Group.: Randomized Trial of Intravenous Atenolol Among 16,027 Cases of Suspected Acute Myocardial Infarction. *Lancet* **2**:57-(4/19) (1986).

Ito, M., N. Sakurai and P. Tongroach: Climbing Fibre Induced Depression of Both Mossy Fibre Responsiveness and Glutamate Sensitivity of Cerebellar Purkinje Fibres. *J.Physiol.* **324**:113-(2/9) (1982).

Itoh, T., H. Kuriyama and H. Suzuki: Differences and Similarities in the Noradrenaline- and Caffeine-Induced Mechanical Responses in the Rabbit Mesenteric Artery. *J.Physiol.* **337**:609-(2/10) (1983).

Ivens, I.: Different Properties of Two Voltage Dependent Inward Currents of the Ciliate *Stylonychia mytilus*. *J.Physiol.* **381**:1-(3/12) (1986).

Iwatsuki, N., Y. Maruyama, O. Matsumoto and A. Nishiyama: Activation of Ca^{2+}-Dependent Cl$^-$ and K$^+$ Conductances in Rat and Mouse Parotid Acinar Cells. *Jpn.J.Physiol.* **35**:933-(2/7) (1985).

Jackson, C.L., R.C. Bush and D.E. Bowyer: Inhibitory Effect of Calcium Antagonists on Balloon Catheter-Induced Arterial Smooth Muscle Cell Proliferation and Lesion Size. *Atherosclerosis* **69**:115-(4/20) (1988).

Jackson, H. and T.N. Parks: Spider Toxins: Recent Applications in Neurobiology. *Ann.Rev.Neuroscience* **12**:405-(3/13) (1989).

Jaffe, E.A., L.L. Leung, R.L. Nachman, R.I. Levine and D.F. Mosher: Thrombospondin is the Endogenous Lectin of Human Platelets. *Nature* **295**:246-(4/20) (1982).

Jaffe, L.A.: Regulation of Cortical Vesicle Exocytosis in Sea Urchin Eggs. In *Calcium and Ion Channel Modulation*. A.D. Grinnell, D. Armstrong and M.B. Jackson, eds. Plenum Press, New York. 305-(2/7) (1988).

Jahnsen, H. and R. Llinas: Ionic Basis for the Electroresponsiveness and Oscillatory Properties of Guinea Pig Thalamic Neurones In Vitro. *J.Physiol.* **349**:227-(2/9) (1984).

Jahr, C.E. and C.F. Stevens: Glutamate Activates Multiple Single Channel Conductances in Hippocampal Neurons. *Nature* **325**:522-(2/9) (1987).

Janis, R.A., D.E. Johnson, A.V. Shrikhande, R.T. McCarthy, A.D. Howard, R. Greguski and A. Scriabine: Endogenous 1,4-Dihydropyridine-Displacing Substances Acting on L-Type Ca^{2+} Channels: Isolation and Characterization of Fractions from Brain and Stomach. In *The Calcium Channel: Structure, Function and Implications*. M. Morad, W. Nayler, S. Kazda, M. Schramm, eds. Springer-Verlag, Berlin, Heidelberg. 564-(3/13) (1988b).

Janis, R.A., S.C. Maurer, J.G. Sarmiento, G.T. Bolger and D.J. Triggle: Binding of [^{3}H]Nimodipine to Cardiac and Smooth Muscle Membranes. *Eur.J.Pharmacol.* **82**:191-(3/13) (1982).

Janis, R.A., D. Rampe, J.G. Sarmiento and D.J. Triggle: Specific Binding of a Calcium Channel Activator, [^{3}H]BAY K 8644, to Membranes from Cardiac Muscle and Brain. *Biochem.Biophys.Res.Commun.* **121**:317,-(3/13) (1984b).

Janis, R.A., J.G. Sarmiento, S.C. Maurer, G.T. Bolger and D.J. Triggle: Characteristics of the Binding of [^{3}H]Nitrendipine to Rabbit Ventricular Membranes: Modification by other Calcium Channel Antagonists and by the Calcium Channel Agonist BAY K 8644. *J.Pharmacol.Exp.Ther.* **231**:8-(3/13) (1984a).

Janis, R.A. and A. Scriabine: Sites of Action of Ca^{2+} Channel Inhibitors. *Biochem.Pharmacol.* **32**:3499-(4/17) (1983).

Janis, R.A., A.V. Shrikhande, D.E. Johnson, R.T. McCarthy, A.D. Howard, R. Greguski and A. Scriabine: Isolation and Characterization of a Fraction from Brain that Inhibits 1,4-[^{3}H]Dihydropyridine Binding and L-Type Calcium Channel Current. *FEBS Lett.*, **239**:233-(3/13) (1988a).

Janis, R.A., P. Silver and D.J. Triggle: Drug Action and Cellular Calcium Regulation. *Adv.Drug Res.* **16**:309-(3/13) (1987).

Janis, R.A. and D.J. Triggle: New Developments in Ca^{2+} Channel Antagonists. *J.Med.Chem.* **26**:775-(4/20) (1983).

Janssens, W. and R. Verhaege: Sources of Calcium Used during Alpha-l and Alpha-l Adrenergic Contractions in Canine Saphenous Veins. *J.Physiol.* **347**:525-(4/18) (1984).

January, C.T. and H.A. Fozzard: Delayed Afterdepolarizations in Heart Muscle: Mechanisms and Relevance. *Pharmacol.Rev.* **40**:219-(4/19) (1988).

January, C.T. and J.M. Riddle: Early Afterdepolarizations: Mechanism of Induction and Block A Role for L-Type Ca^{2+}-Current. *Circ.Res.* **64**:977-(4/19) (1989).

January, C.T., J.M. Riddle and J.J. Salata: Model for Early Afterdepolarizations: Induction with the Ca^{2+} Channel Agonist Ba K8644. *Circ.Res.* **28**:61-(4/19) (1988).

Jaramillo, R., S. Vicini, S.M. Schuetze, M.F. Johnston and F. Ramone: Embryonic Acetylcholine Receptors Guarantee Spontaneous Contractions in Rat Developing Muscle. *Nature* **335**:66-(2/10) (1988).

Jariwalla, A.G. and E.G. Anderson: Side Effects of Drugs. Production of Ischaemic Cardiac Pain by Nifedipine. *Br.Med.J.* **1**:1181-(4/16) (1978).

Jasmin, G. and E. Bajusz: Polymyopathie et Cardiomyopathie Hereditaire Chez le Hamster de Syrie Inhibtion Selective des Lesions du Myocarde. *Ann.Anat.Pathol.* **18**:49-(4/17) (1973).

Jasmin, G. and L. Proschek: The Paradoxical Effect of Isoproterenol on Hamster Hereditary Polymopathy. *Muscle Nerve* **6**:408-(4/17) (1983).

Jasmin, G. and L. Proschek: Calcium and Myocardial Cell Injury. An Appraisal in the Cardiomyopathic Hamster. *Can.J.Physiol.Pharmacol.* **62**:891-(4/17) (1984).

Jasmin, G. and L. Proschek: Pathogenesis of the Hamster Hereditary Cardiomyopathy. A Pharmacologic Appraisal. In *Cardiomyopathy.* C. Kawai, and W.H. Abelman, eds. University of Tokyo Press. 79-(4/17) (1987b).

Jasmin, G., L. Proschek, G. Brisson and N.S. Dhalla: The Hypothyroid State in Cardiomyopathic Hamsters. In *Pathophysiology of Heart Disease*. N.S. Dhalla, P.K. Singal, R.E. Beamish, eds. Martinus Nijhoff Publishing, Boston **311**-4/17) (1987a).

Jasmin, G. and B. Solymoss: Prevention of Hereditary Cardiomyopathy in the Hamster by Verapamil and other Agents. *Proc.Soc.Exp.Biol.Med.* **149**:193-(4/17) (1975).

Jimerson, D.C., R.M. Post, J.S. Carman, D.P. van Kammen, J.H. Wood, F.K. Goodwin and W.E. Bunney Jr: CSF Calcium: Clinical Correlates in Affective Illness and Schizophrenia. *Biol.Psychiatr.* **14**:37-(4/22) (1979).

Jmari, K., C. Mironneau and J. Mironneau: Selectivity of Calcium Channels in Rat Uterine Smooth Muscle: Interaction between Sodium, Calcium and Barium Ions. *J.Physiol.* **384**:247-(3/12) (1987).

Joborn, C., J. Hetta, J. Rastad, H. Agren, G. Akerstrom and S. Ljunghall: Psychiatric Symptoms and Cerebrospinal Fluid Monoamine Metabolites in Primary Hyperparathyroidism. *Biol.Psychiatr.* **23**:149-(4/22) (1988).

Johanson, J.S. and D.H. Haynes: Deliberate Quin 2 Overload as a Method for in situ Characterization of Active Calcium Extrusion System and Cytoplasmic Calcium Binding: Application to the Human Platelet. *J.Membr.Biol.* **104**:147-(4/20) (1988).

Johansson, B. and A.P. Somlyo: Electrophysiology and Excitation-Contraction Coupling. In *Handbook of Physiology—The Cardiovascular System.* D.F. Bohr, P. Somlyo and H.V. Sparks, eds. Am.Physiol.Soc.Washington, DC. **7**:301-(2/10) (1980).

Johns, E.J. and J. Manitius: The Renal Actions of Nitrendipine and its Influence on the Neural Regulation of Calcium and Sodium Reabsorption in the Rat. *J.Cardiovasc.Pharmacol.* **9**:S49-(3/13) (1987).

Johnson, G.J., P.C. Dunlap and L.A. Leis: From AHL: Dihydropyridine Agonist Bay K 8644 Inhibits Platelet Activation by Competitive Antagonism of Thromboxane A$_2$-Prostaglandin H$_2$ Receptor. *Circ.Res.* **62**:494-(3/13,4/20) (1988).

Johnson, G.J., L.A. Leis and G.S. Francis: Disparate Effects of the Calcium Channel Blockers Nifedipine Verapamil on Adrenergic Receptors and Thromboxane A$_2$-Induced Aggregation of Human Platelets. *Circulation* **73**:847-(4/20) (1986).

Johnston, D.: Voltage Clamp Reveals Basis for Calcium Regulation of Bursting Pacemaker Potentials in *Aplysia* Neurons. *Brain Res.* **107**:418-(2/9) (1976).

Johnston, D., J.J. Hablitz and W.A. Wilson: Voltage Clamp Discloses Slow Inward Current in Hippocampal Burst Firing Neurons. *Nature* **286**:391-(2/9) (1980).

Johnston, D.L., R. Lesoway, D.P. Humen and W.J. Kostuk: Clinical and Hemodynamic Evaluation of Propranolol in Combination with Verapamil, Nifedipine and Diltiazem in Exertional Angina Pectoris: A Placebo-Controlled, Double-Blind, Randomized, Crossover Study. *Am.J.Cardiol.* **55**:680-(4/16) (1985).

Johnston, M.F.and F. Ramon: Electronic Coupling in Internally Perfused Crayfish Segmented Axons. *J.Physiol.* **317**:509-(2/10) (1981).

Jolly, S.R. and G.J. Gross: Improvement in Ischemic Myocardial Blood Flow Following a New Calcium Antagonist. *Am.J.Physiol.* **239**:Hl63-(4/17) (1980).

Jones, J.G., S. Cho and S.M. Factor: The Anatomy and Pathophysiology of the Microvasculature in Different Organs: Relationship to Vasculogenic Necrosis and Tissue Damage. In *Meth.Achiev.Exp.Pathol.* G. Jasmin, ed. Karger, Basel. **13**:114-(4/17) (1988).

Jones, K.A. and R.W. Baughman: NMDA- and non-NMDA-Receptor Components of Excitatory Synaptic Potentials Recorded from Cells in Layer V of Rat Visual Cortex. *J.Neurosci.* **8**:3522-(2/9) (1988).

Jones, O.T., D.L. Kinze and K.J. Angelides: Localization and Cellular Mobility of ω-Conotoxin-sensitive Ca^{2+} Channels in Hippocampal CA1 Neurons. *Science* **244**:1189-(2/9,3/13) (1989).

Jones, P.M., J. Stutchfield and S.L. Howell: Effects of Ca^{2+} and a Phorbol Ester on Insulin Secretion from Islets of Langerhans Permeabilized by High-voltage Discharge. *FEBS Lett.,* **191**:102-(2/7) (1985).

Jones, S.W. and T.N. Marks: Calcium Currents in Bullfrog Sympathetic Neurons. I. Activation Kinetics and Pharmacology. *J.Gen.Physiol.* **94**:151-(1/3,1/4) (1989).

Josephson, M.E.: Paroxysmal Supraventricular Tachycardia: An Electrophysiologic Approach. *Am.J.Cardiol.* **41**:1123-(4/19) (1978).

Josephson, M.E. and J.A. Kastor: Paroxysmal Supraventricular Tachycardia. Is the Atrium a Necessary Link. *Circulation* **54**:430-(4/19) (1976).

Joyal, M., K. Cremer, J. Pieper, R.L. Feldman and C.J. Pepine: Effects of Diltiazem During Tachycardia-Induced Angina Pectoris. *Am.J.Cardiol.* **57**:10-(4/16) (1986).

Juematsu, E., M. Hirata, T. Hashimoto and H. Kuriyama: Inositol 1,4,5-Triphosphate Releases Ca^{2+} from Intracellular Store Sites in Skinned Single Cells of Porcine Coronary Artery. *Biochem.Biophys.Res.Commun.* **120**:481-(2/10) (1984).

Jy, W., Y.S. Ahn, N. Shanbaky, L. Fernandez, W.J. Harrington and D.H. Haynes: Abnormal Calcium Handling by Platelets in Thrombotic Disorders. *Circ.Res.* **60**:346-(4/20) (1987).

Jy, W. and D.H. Haynes: Intracellular Calcium Storage and Release in the Human Platlet: Chloro-tetracycline as a Continuous Monitor. *Circ.Res.* **55**:594-(4/20) (1984).

Jy, W. and D.H. Haynes: Thrombin-Induced Calcium Movements in Platelet Activation. *Biochem.Biophys.Acta* **929**:88-(4/20) (1987).

Jy, W. and D.H. Haynes: Calcium Uptake and Release. Characteristics of the Dense Tubules of Digtonin-Permeabilized Human Platelets. *Biochim.Biophys.Acta* **944**:374-(4/20) (1988).

Kahan, A., J.M. Foult, S. Webber, B. Amor, C.J. Menkes and M. Degeorges: Nifedipine and Alpha1-Adrenergic Blockade in Raynaud's Phenomenon. *Eur.Heart J.* **6**:702-(4/20) (1985).

Kahan, A., S. Webber, B. Amor, L. Saporta, M. Hodara and M. Degeorges: Etude Controlee de la Nifedipine dans le Traitment du Phenomene de Raynaud. *Rev.Rhum.Mal Osteoarticulares* **49**:337-(4/20) (1982).

Kahan, A., S. Webber, C. Amor, C.J. Menkes, L. Saporta, M. Hordara, F. Guerin and M. Degeorges: Calcium Entry Blocking Agents in Digital Vasospasm (Raynaud's Phenomenon). *Eur.Heart J.* **4**:123-(4/20) (1983).

Kai, I., K. Ogawa and T. Ita: Effects of Peripheral Vasolidation Caused by Verapamil, Nifedipine, and Nitroglycerin on Plasma Prostaglandins and Thromboxane Concentration. *Jpn.Heart J.* **23**:941-(4/20) (1982).

Kaibara, M. and M. Kameyama: Inhibition of the Calcium Channel by Intracellular Protons in Single Ventricular Myocytes of the Guinea Pig. *J.Physiol.* **403**:621-(3/12) (1988).

Kalman, D. and D.L. Armstrong: Protein Phosphorylation and the Inactivation of Dihydropyridine Sensitive Calcium Channels in Mammalian Pituitary Tumor Cells. In *The Calcium Channel: Structure, Function, and Implications.* M. Morad, W. Nayler, S. Kazda, M. Schramm, eds. Springer-Verlag, Berlin, Hiedelberg. **103**-(3/14) (1988).

Kaltenbach, M., R. Hopf, G. Kober, W.D. Bussmann, M. Keller and Y. Ptersen: Treatment of Hypertrophic Obstructive Cardiomyopathy with Verapamil. *Br.Heart J.* **42**:35-(4/17) (1979).

Kameyama, M., J. Hescheler, F. Hofmann and W. Trautwein: Modulation of Ca Current During the Phosphorylation Cycle in the Guinea Pig Heart. *Pflügers Arch.* **407**:123-(1/4,3/14) (1986).

Kameyama, M., J. Hescheler, F. Hofmann and W. Trautwein: Protein Phosphorylation Regulate the Ca^{2+} Channel in Guinea Pig Heart. *Pflügers Arch.* **407**:461-(3/14) (1988).

Kameyama, M., F. Hofmann and W. Trautwein: On the Mechanism of Beta-adrenergic Regulation of the Ca^{2+} Channel in Guinea Pig Heart. *Pflügers Arch.* **405**:285-(1/4,2/10,3/14) (1985).

Kamp, T.J. and R.J. Miller: Voltage-Dependent Nitrendipine Binding to Cardiac Sarcolemmal Vesicles. *Mol.Pharmacol.* **32**:278-(3/13) (1987).

Kamp, T.J., M.C. Sanguinetti and R.J. Miller: Voltage- and Use-Dependent Modulation of Cardiac Calcium Channels by the Dihydropyridine (+)-202-791. *Circ.Res.* **64**:338-(3/13) (1989).

Kanaide, H., T. Matsumoto and M. Nakamura: Inhibition of Calcium Transients in Cultured Vascular Smooth Muscle Cells by Pertussis Toxin. *Biochem.Biophys.Res.Commun.* **140**:195-(2/10) (1986).

Kaneko, A. and M. Tachibana: Effects of Gamma-Aminobutyric Acid on Isolated Cone Photoreceptors of Turtle Retina. *J.Physiol.* **373**:443-(2/8) (1986).

Kano, M.: Development of Excitability in Embryonic Chick Skeletal Muscle Cells. *J.Cell Physiol.* **86**:503-(2/10) (1975).

Kano, M. and M. Kato: Quisqualate Receptors are Specifically Involved in Cerebellar Synaptic Plasticity. *Nature* **325**:276-(2/9) (1987).

Kano, M. and Y. Shimada: Tetrodotoxin-Resistant Electric Activity in Chick Skeletal Muscle Cells Differentiated *In Vitro*. *J.Cell Physiol.* **81**:85-(2/10) (1973).

Kano, M., Y. Shimada and K. Ishikawa: Electrogenesis of Embryonic Chick Skeletal Muscle Cells Differentiated *In Vitro*. *J.Cell Physiol.* **79**:363-(2/10) (1972).

Kano, M. and M. Yamamoto: Development of Spike Potentials in Skeletal Muscle Cells Differentiated *In Vitro* from Chick Embryo. *J.Cell Physiol.* **90**:431-(2/10) (1977).

Kaplan, N.M.: Calcium Entry Blockers in the Treatment of Hypertension. *JAMA* **262**:817-(3/13) (1989).

Kappagoda, C.T. and A.B.R. Thomson: Nisoldipine in Experimental Atherosclerosis in Rabbits. *4th Intl. Symp. on Calcium Antagonists: Pharmacology and Clinical Research* 121-(3/13) (1989).

Karasch, E.D., A.M. Mellow and E.M. Silinsky: Intracellular Magnesium does not Antagonize Calcium-Dependent Acetylcholine Secretion. *J.Physiol.* **314**:255-(2/7) (1981).

Karliner, J.S., H. Alabster, P. Stephens, P. Barnes and C. Dollery: Enhaced Noradrenaline Response in Cardiomyopathic Hamsters: Possible Relation to Changes in Adrenoceptors Studied by Radioligand Binding. *Cardiovasc.Res.* **15**:296-(4/17) (1981).

Karpati, G. and B. Frame: Neuropsychiatric Disorders in Primary Hyperparathyroidism. *Arch.Neurol.* **10**:387-(4/22) (1964).

Karpen, J.W., A.L. Zimmerman, L. Stryer and D.A. Baylor: Gating Kinetics of the Cyclic GMP-activated Channel of Retinal Rods: Flash Photolysis and Voltage-Jump Studies. *Proc.Natl.Acad.Sci.U.S.A.* **85**:1287-(2/8) (1988).

Kasai, H., T. Aosaki and J. Fukuda: Presynaptic Calcium Antagonist ω-Conotoxin Irreversibly Blocks N-Type Calcium Channels in Chick Sensory Neurons. *NeuroscienceRes.* **4**:228-(1/4,3/13,3/14) (1987).

Kasai, H. and T. Aosaki: Divalent Cation Dependent Inactivation of the High-Voltage-Activated Ca Channel Current in Chick Sensory Neurons. *Pflügers Arch.* **411**:695-(1/4) (1988).

Kasai, H. and T. Aosaki: Modulation of Ca-Channel Current by an Adenosine Analog Mediated by a GTP-Binding Protein in Chick Sensory Neurons. *Pflügers Arch.* **414**:145-(1/3) (1989).

Kaschka, W.P., N. Thurauf, T.H. Mokrusch and M. Korth: Electrooculography and Dark Adaptation in Patients with Affective and Schizoaffective Psychoses: Early Physiological Effects of Carbamazepine and Lithium. *Pharmacopsychiatry* **21**:404-(4/22) (1988).

Kass, R.S.: Voltage-Dependent Modulation of Cardiac Calcium Channel Current by Optical Isomers of Bay K 8644:Implications for Channel Gating. *Circ.Res.* **61**:1-(3/13) (1987).

Kass, R.S. and J.P. Arena: Influence of pH_o on Calcium Channel Block by Amlodipine, a Charged Dihydropyridine Receptor. *J.Gen.Physiol.*(in press) (3/13) (1989).

Kass, R.S., J.P. Arena and D. DiManno: Block of Heart Calcium Channels by Amlodipine: Influence of Drug Charge on Blocking Activity. *J.Cardiovasc.Pharmacol.* **12**:S45-(3/13) (1988).

Kass, R.S. and M.C. Sanguinetti: Inactivation of Calcium Channel Current in the Calf Cardiac Purkinje Fiber: Evidence for Voltage- and Calcium-Mediated Mechanisms. *J.Gen.Physiol.* **84**:705-(1/4) (1984).

Kass, R.S., R.W. Tsien and R. Weingart: Ionic Basis of Transient Inward Current Induced by Strophanthidin in Cardiac Purkinje Fibers. *J.Physiol.* **281**:209-(4/19) (1978).

Kasuya, Y., T. Ishikawa, M. Yanigasawa, S. Kimura, K. Goto and T. Masaki: Mechanism of Contraction to Endothelin in Isolated Porcine Coronary Artery. *Am.J.Physiol.* **257**:H1828-(3/13) (1989).

Katz, B. and R. Miledi: The Effect of Calcium on Acetylcholine Release from Motor Nerve Terminals. *Proc.R.Soc.London B* **161**:496-(2/9) (1965).

Katz, B. and R. Miledi: A Study of Synaptic Transmission in the Absence of Nerve Impulses. *J.Physiol.* **192**:407-(2/7) (1967).

Katz, B. and R. Miledi: Tetrodotoxin-Resistant Electrical Activity in Presynaptic Terminals. *J.Physiol.* **203**:459-(2/7) (1969).

Katz, G., L. Roy-Contancin, T. Bale and J.P. Reuben: Arachidonic, Linoleic and Other Unsaturated Fatty Acids Enhance K^+ and Depress Na^+ and Ca^{2+} Channel Activity. *Biophys.J.* **57**:506a-(3/13) (1990).

Katzka, D.A. and M. Morad: Properties of Calcium Channels in Guinea Pig Gastric Myocytes. *J.Physiol.* **413**:175-(3/13) (1989).

Katzman, R. and H.M. Pappius: In *Brain Electrolytes and Fluid Metabolism*. Williams & Wilkins Co., Baltimore. (4/21) (1973).

Kauker, M.L., D.W. Zeigler and E.T. Zawada: Renal Tubular Effect of Nisoldipine, a Calcium Channel Blocker, in Rats. *J.Cardiovasc.Pharmacol.* **9**:S32-(3/13) (1987).

Kaumann, A.J. and P. Aramendia: Prevention of Ventricular Fibrillation Induced by Coronary Ligation. *J.Pharmacol.Exp.Ther.* **166**:326-(4/19) (1968).

Kawa, K.: Zinc Dependent Action Potentials in Giant Neurons of Snail *Euhadra quaestia*. *J.Membr.Biol.* **49**:325-(3/12) (1979).

Kawa, K.: Voltage-Gated Sodium and Potassium Currents and Their Variation in Calcitonin-Secreting Cells of the Chick. *J.Physiol.* **399**:93-(2/7) (1988).

Kawamura, S. and M. Murakami: In situ cGMP Phosphodiesterase and Photoreceptor Potential in Gecko Retina. *J.Gen.Physiol.* **87**:737-(2/8) (1986).

Kay, A.R. and R.K.S. Wong: Calcium Current Activation Kinetics in Isolated Pyramidal Neurones of the CA1 Region of the Mature Guinea Pig Hippocampus. *J.Physiol.* **392**:603-(1/3,2/9) (1987).

Kazda, S.: The Calcium Channel and Vascular Injury. In *The Calcium Channel: Structure, Function and Implications*. M. Morad, W. Nayler, S. Kazda, M. Schramm, eds. Springer-Verlag, Berlin, Heidelberg. **326**-(3/13) (1988).

Kazda, S., C. Hirth and J.P. Stasch: Diuretic Effect of Nitrendipine Contributes to its Antihypertensive Efficacy: A Review. *J.Cardiovasc.Pharmacol.* **12**:1-(3/13) (1988).

Kazda, S., J.P. Stasch and C. Hirth: Nitrendipine in Experimental Hypertension: Effects on Cardiac Hypertrophy, Heart Failure and Atrial Natriuretic Peptides. *J.Cardiovasc.Pharmacol.* **9**:S90-(3/13) (1987).

Kazda, S. and R. Towart: Nimodipine: A New Calcium Antagonistic Drug with a Preferential Cerebrovascular Action. *Acta Neurochir.* **63**:259-(4/20) (1982).

Kelly, R.A. and T.W. Smith: The Search for the Endogenous Digitalis: an Alternative Hypothesis. *Am.J.Physiol.* **256**:C937-(3/13) (1989).

Kelly, R.B.: Pathways of Protein Secretion in Eukaryotes. *Science* **230**:25-(2/7) (1985).

Kelly, R.B.: The Cell Biology of the Nerve Terminal. *Neuron* **1**:431-(2/7) (1988).

Keynes, R.D., E. Rojas, R.E. Taylor and J. Vergara: Calcium and Potassium Systems of a Giant Barnacle Muscle Fibre under Membrane Potential Control. *J.Physiol.* **229**:409-(3/12) (1973).

Khalsa, A. and S.B. Olssen: Verapamil-Induced Ventricular Regularity in Atrial Fibrillation. *Acta Med.Scand.* **205**:509-(4/19) (1979).

Khurmi, N.S., M.J. Bowles, M.J. O'Hara, V.B. Subramanian and E.B. Raftery: Long-Term Efficacy of Diltiazem Assessed with Multistage Graded Exercise Tests in Patients with Chronic Stable Angina Pectoris. *Am.J.Cardiol.* **54**:738-(4/16) (1984).

Khurmi, N.S. and E.B. Raftery: Comparative Effects of Prolonged Therapy with Four Calcium Ion Antagonists (Diltiazem, Nicardipine, Tiapamil and Verapamil) in Patients with Chronic Stable Angina Pectoris. *Cardiovasc.Drugs Ther.* **1**:81-(4/16) (1987).

Khurmi, N.S. and E.B. Raftery: A Comparison of Nine Calcium Ion Antagonists and Propranolol: Exercise Tolerance, Heart Rate and ST-Segment Changes in Patients with Chronic Stable Angina Pectoris. *Eur.J.Clin.Pharmacol.* **32**:539-(4/16) (1987).

Kidokoro, Y.: Spontaneous Calcium Action Potentials in a Clonal Pituitary Cell Line and Their Relationship to Prolactin Secretion. *Nature* **258**:741-(2/7) (1976).

Kidokoro, Y. and R. Gruener: Distribution and Density of Alpha-Bungarotoxin Binding Sites on Innervated and Noninnervated *Xenopus* Muscle Cells in Culture. *Dev.Biol.* **91**:78-(2/10) (1982).

Kidokoro, Y. and A.K. Ritchie: Chromaffin Cell Action Potentials and Their Possible Role in Adrenaline Secretion from Rat Adrenal Medulla. *J.Physiol.* **307**:199-(2/7) (1980).

Kidokoro, Y. and M. Saito: Early Cross-Striation Formation in Twitching *Xenopus* Myocytes in Culture. *Proc.Natl.Acad.Sci.U.S.A.* **85**:1978-(2/10) (1988).

Killbride, P.: Ca Effects on Frog Retinal cGMP Levels and Their Light-Initiated Rate of Decay. *J.Gen.Physiol.* **75**:457-(2/8) (1980).

Kim, D.H., S.T. Ohnishi and N. Ikemoto: Kinetic Studies of Calcium Release from Sarcoplasmic Reticulum *In Vitro*. *J.Biol.Chem.* **258**:9662-(2/10) (1983).

Kimura, E. and H. Kishida: Treatment of Variant Angina with Drugs: A Survey of 11 Cardiology Institutes in Japan. *Circulation* **63**:844-(4/16) (1981).

Kimura, E., K. Tanaka, K. Mizuno, Y. Handa and H. Hashimoto: Suppression of Repeatedly Occurring Ventricular Fibrillation with Nifedipine in Variant Form of Angina Pectoris. *Jpn.Heart J.* **18**:736-(4/19) (1977).

Kimura, J.E. and H. Meves: The Effect of Temperature on the Asymmetrical Charge Movement in Squid Giant Axons. *J.Physiol.* **289**:479-(1/4) (1979).

Kinbough-Rathbone, R.L., M.A. Packham and J.F. Mustard: The Effect of Prostaglandin E1 on Platelet Function *In Vitro* and *In Vivo*. *Br.J.Haematol.* **19**:570-(4/20) (1970).

Kind, H.: Psychische Störungen bei Hyperparathyreoidismus. Archiv für Psychiatrie und Zeitschrift fur die ges. *Neurologie*. **200**:1-(4/22) (1959).

King, V.F., M.L. Garcia, D. Himmel, J.P. Reuben, Y.K.T. Lam, J.X. Pan, G.Q. Han and G.J. Kaczorowski: Interaction of Tetrandine with Slowly Inactivating Calcium Channels. *J.Biol.Chem.* **263**:2238-(3/13) (1988).

King, V.F., M.L. Garcia, J.L. Shevell, R.S. Slaughter and G.J. Kaczorowski: Substituted Diphenylbutylpiperidines Bind to a Unique High Affinity Site on the L-Type Calcium Channel. *J.Biol.Chem.* **264**:5633-(3/13) (1989).

Kingma, J.G. and D.M. Yellon: Limitation of Myocardial Necrosis with Verapamil During Sustained Coronary Occlusion in the Closed-Chest Dog. *Cardiovasc.Drugs Ther.* **2**:313-(4/16) (1988).

Kinney, E.L., G. Nicholas, J. Gallo, C. Pontoriero and R. Zelia: The Treatment of Severe Raynaud's Phenomenon with Verapamil. *J.Clin.Pharmacol.* **22**:74-(4/20) (1982).

Kiowski, W., F.R. Buhler, M.O. Fadayomi, P. Erne, F.G. Muller, U.L. Hulthen and P. Bolli: Age, Race, Blood Pressure and Renin: Predictors for Antihypertensive Treatment with Calcium Antagonists. *Am.J.Cardiol.* **56**:81-(4/18) (1985).

Kiowski, W., P. Erne, O. Bertel, P. Bolli and F. Buhler: Acute and Chronic Sympathetic Reflex Activation and Antihypertensive Response to Nifedipine. *J.Am.Coll.Cardiol.* **7**:344-(4/16) (1986).

Kirch, W., C.H. Kleinbloesem and G.G. Belz: Drug Interactions with Calcium Antagonists. *Pharmacol.Ther.* **45**:109-(3/13) (1990).

Kita, H. and W. Van der Kloot: Effects of the Ionophore X-537A on Acetylcholine Release at the Frog Neuromuscular Junction. *J.Physiol.* **259**:177-(2/7) (1976).

Kitamura, Y., X.H. Zhao, T. Ohnuki and Y. Nomura: Ligand-Binding Characteristics of [^{3}H]QNB, [^{3}H]Prazosin, [^{3}H]Rauwolscine, [^{3}H]TCP and [^{3}H]Nitrendipine to Cerebral Cortical and Hippocampal Membranes of Senescence Accelerated Mouse. *Neurosci.Lett.* **106**:334-(3/13) (1989).

Kitazawa, T.: Effect of Exracellular Calcium on Contractile Activation in Guinea Pig Ventricular Muscle. *J.Physiol.* **355**:635-(2/10) (1984).

Kjekshus, J.K.: Importance of Heart Rate in Determining Beta-Blocking Efficacy in Acute and Long Term Myocardial Infarction Trials. *Am.J.Cardiol.* **57**:43F-(4/19) (1986).

Kjeldsen, K. and S. Stender: Calcium Antagonists and Experimental Atherosclerosis. *Proc.Soc.Exp.Biol.Med.* **190**:219-(3/13) (1989).

Klarsfeld, A. and J.P. Changeux: Activity Regulates the Levels of Acetylcholine Receptor Alpha-Subunit mRNA in Cultured Chicken Myotybes. *Proc.Natl.Acad.Sci.U.S.A.* **82**:4558-(2/10) (1985).

Klarsfeld, A., R. Laufer, B. Fontaine, A. Devillers-Thiery, C. Dubreuil and J.P. Changeux: Regulation of Muscle AChR Alpha-Subunit Gene Expression by Electrical Activity: Involvement of Protein Kinase C and Ca^{2+}. *Neuron* **2**:1229-(2/10) (1989).

Klein, G.J., S. Gulamhusein, S.G. Carruthers, A.P. Donner and P.T. Ko: Comparison of the Electrophysiologic Effects of Intravenous and Oral Verapamil in Patients with Paroxysmal Supraventricular Tachycardia. *Am.J.Cardiol.* **49**:117-(4/19) (1982).

Klein, H.O. and E. Kaplinsky: Digitalis and Verapamil in Atrial Fibrillation and Flutter. Is Verapamil Now the Preferred Agent. *Drugs* **31**:185-(4/19) (1986).

Klein, H.O., R. Long, E. Weiss, C. Desegni, J.C. Libhaber, J. Guerrero and E. Kaplinsky: The Influence of Verapamil on Serum Digoxin. *Circulation* **65**:998-(4/19) (1982).

Klein, H.O., H. Pauzner, E. DiSegni, D. David and E. Kaplinsky: The Beneficial Effects of Verapamil in Chronic Atrial Fibrillation. *Ann.Int.Med.* **139**:747-(4/19) (1979).

Klein, W., D. Brandt, K. Vrecko and M. Harringer: Role of Calcium Antagonists in the Treatment of Essential Hypertension. *Circ.Res.* **52**:74-(4/18) (1983).

Klemperer, E.: Untersuchungen über den Stoffwechsel bei manischen und depressiven Zustandsbildern. Veranderungen des Kalzium- und Kaliumspiegels des Gesamtblutes. In *Jahrbücher für Psychiatrie und Neurologie*. F. Mautman, C. Mayer, O. Pötzl and J. Wagner-Jauregg, eds. Verlag Julius Springer, Wien. 62-(4/22) (1927).

Kloner, R.A. and E. Braunwald: Effects of Calcium Antagonists on Infarcting Myocardium. *Am.J.Cardiol.* **59**:84B-(4/16) (1987).

Klugmann, S., A. Salvi and F. Camerini: Hemodynamic Effects of Nifedipine in Heart Failure. *Br.Heart J.* **43**:440-(4/17) (1980).

Knauss, H.G., J. Striessnig, A. Koza and H. Glossmann: Neurotoxic Aminoglycoside Antibiotics are Potent Inhibitors of [125I]ω-Conotoxin GVIA Binding to Guinea Pig Cerebral Cortex Membranes. *Naunyn Schmied Arch.Pharmacol.* **336**:583-(3/13) (1987).

Knight, D.E.: Intracellular Factors Controlling Exocytosis. *Molecular Mechanisms in Secretion 25th Alfred Benzan Symposium*, Munksgaard, Copenhagen. 192-(2/7) (1988).

Knight, D.E. and P.F. Baker: Calcium-Dependence of Catecholamine Release from Bovine Adrenal Medullary Cells after Exposure to Intense Electric Fields. *J.Membr.Biol.* **68**:107-(2/7) (1982).

Knight, D.E. and N.T. Kesteven: Evoked Transient Intracellular Free Ca^{2+} Changes and Secretion in Isolated Bovine Adrenal Medullary Cells. *Proc.R.Soc.London B* **218**:177-(2/7) (1983).

Knight, D.E. and M.C. Scrutton: Direct Evidence for a Role for Ca^{2+} in Amin Storage Granule Secretion by Human Platelets. *Thrombos. Res.* **20**:437-(4/20) (1980).

Knight, D.E., H. von Grafenstein and C.M. Athyade: Calcium-Dependent 2nd Calcium-Independent Exocytosis. *TINS* **12**:451-(3/13) (1989).

Knoch, G., M. Schlepper and E. Witzleb: Isoptin - A Clinical Study Using Normal Subjects and Patients with Coronary Disease. *Med.Klin.* **58**:1485-(4/16) (1963).

Knorr, A.: The Pharmacology of Nisoldipine. *Cardiovasc.Drugs Ther.* **1**:393-(3/13) (1987).

Knospe, H.: Tetany Psychoses. *Monatschr. Psychiatr.Neurol.* **99**:503-(4/22) (1938).

Knowles, A. and H.J.A. Dartnall: *The Photobiology of Vision in "The Eye"*. Academic Press, New York. H. Davson, ed. **2**-(2/8) (1977).

Knudson, C.M., N. Chaudhauri, A.H. Sharp, J.A. Powell, K.G. Beam and K.P. Campbell: Specific Absence of the Alpha 1 Subunit of the Dihydropyridine Receptor in Mice with Muscular Dysgenesis. *J.Biol.Chem.* **264**:1345-(2/10,3/13) (1989).

Kobayashi, S., H. Kanaide and M. Nakamura: Complete Overlap of Caffeine- and K^+ Depolarization-Sensitive Intracellular Calcium Storage Site in Cultured Rat Arterial Smooth Muscle Cells. *J.Biol.Chem.* **261**:15709-(2/10) (1986).

Kobayashi, S., A.P. Somlyo and A.V. Somlyo: Guanine Nucleotide- and Inositol 1,4,5-Trisphosphate-Induced Calcium Release in Rabbit Main Pulmonary Artery. *J.Physiol.* **403**:601-(2/10) (1988).

Kober, G., R. Hopf, G. Biamino, P. Bubenheimer, K. Forster, K.H. Kuck, P. Hanrath, K.E. Olshausen, M. Schlepper and M. Kaltenbach: Long-Term Treatment of Hypertrophic Cardiomyopathy with Verapamil or Propranolol in Matched Pairs of Patients: Results of Multicenter Study. *Kardiology* **76**:113-(4/17) (1987).

Koch, K.W. and L. Stryer: Highly Cooperative Feed-back Control of Retinal Rod Guanylate Cyclase by Calcium Ion. *Nature* **334**:64-(2/8) (1988).

Koch, W.J., A. Hui, G.E. Shull, P. Ellinor and A. Schwartz: Characterization of cDNA Clones Encoding Two Putative Isoforms of the Alpha-1 Subunit of the Dihydropyridine-Sensitive Voltage-Dependent Calcium Channel Isolated from Rat Brain and Rat Aorta. *FEBS Lett.* **250**:386-(3/13) (1989).

Kohlhardt, M., H. Fichtner and J.W. Herzig: The Response of Single Cardiac Sodium Channels in Neonatal Rats to the Dihydropyridines CGP 28392 and (-)BAY K 8644. *Naunyn Schmied Arch.Pharmacol.* **340**:210-(3/13) (1989).

Kohlhardt, M. and K. Haap: The Blockade of V_{max} of the Atrioventricular Action Potential Produced by the Slow Channel Inhibitors Verapamil and Nifedipine. *Naunyn Schmied Arch.Pharmacol.* **316**:178-(4/17) (1981).

Koike, T., D.P. Martin and E.M. Johnson: Role of Ca^{2+} Channels in the Ability of Membrane Depolarization to Prevent Neuronal Death Induced by Trophic-Factor Deprivation: Evidence that Levels of Internal Ca^{2+} Determine Nerve Growth Factor Dependence of Sympathetic Ganglion Cells. *Proc.Natl.Acad.Sci.U.S.A.* **86**:6421-(2/10) (1989).

Kojima, M. and N. Sperelakis: Calcium Antagonistic Drugs Differ in Ability to Block the Slow Na^+ Channels of Young Embryonic Chick Hearts. *Eur.J.Pharmacol.* **94**:9-(4/17) (1983).

Kokubun, S., B. Prod'hom, C. Becker, H. Porzig and H. Reuter: Studies on Ca Channels in Intact Cardiac Cells: Voltage-Dependent Effects and Cooperative Interactions of Dihydropyridine Enantiomers. *Mol.Pharmacol.* **30**:571-(3/13) (1986).

Kokubun, S. and H. Reuter: Dihydropyridine Derivatives Prolong the Open State of Ca Channels in Cultured Cardiac Cells. *Proc.Natl.Acad.Sci.U.S.A.* **81**:4824-(3/14) (1984).

Kongsamut, S., D. Lipscombe and R.W. Tsien: The N-Type Ca Channel in Frog Sympathetic Neurons and its Role in α-adrenergic Modulation of Transmitter Release. *Ann.N.Y.Acad.Sci.* **560**:312-(1/4) (1989).

Konnerth, A., H.D. Lux and M. Morad: Proton Induced Transformation of Calcium Channel in Chick Dorsal Root Ganglion Cells. *J.Physiol.* **386**:603-(3/12) (1987).

Korenbrot, J.I. and R.A. Cone: Dark Ionic Flux and the Effects of Light in Isolated Rod Outer Segments. *J.Gen.Physiol.* **60**:20-(2/8) (1972).

Korenbrot, J.I. and D.L. Miller: Calcium Ions Act as Modulators of Intracellular Information Flow in Retinal Rod Phototransduction. *Neurosci.Res.* **4**:Sll-(2/8) (1986).

Korenbrot, J.I. and D.L. Miller: Cytoplasmic Free Calcium Concentration in Dark-Adapted Retinal Rod Outer Segments. *Vision Res.*(in press) (2/8) (1989).

Korenbrot, J.I., D.L. Ochs, J. Williams, D.L. Miller and J.E. Brown: The Use of Tetracarboxylate Indicators in the Measure and Control of Cytoplasmic Calcium. In *Optical Methods in Cell Physiology*. P. De Weer and B.M. Salzberg, eds. John Wiley & Sons, New York. 347-(2/8) (1986).

Kostyuk, P., N. Akaike, Y. Osipchuk, A. Savchenko and Y. Shuba: Gating and Permeation of Different Types of Ca Channels. *Ann.N.Y.Acad.Sci.* **560**:63-(3/13) (1989).

Kostyuk, P.G.: Calcium Ionic Channels in Electrically Excitable Membrane *Neuroscience* **5**:945-(3/12) (1980)

Kostyuk, P.G.: Calcium Channels in Neuronal Membrane. *Biochim.Biophys.Acta* **650**:128-(3/12) (1981).

Kostyuk, P.G.: Diversity of Calcium Ion Channels in Cellular Membranes. *Neuroscience* **28**:253-(3/13) (1989).

Kostyuk, P.G. and O.A. Krishtal: Effects of Calcium and Calcium-Chelating Agents on the Inward and Outward Current in the Membrane of Mollusc Neurones. *J.Physiol.* **270**:569-(1/2) (1977b).

Kostyuk, P.G., O.A. Krishtal and V.I. Pidoplichko: Asymmetrical Displacement Currents in Nerve Cell Membrane and Effect of Internal Fluoride. *Nature* **267**:70-(1/4,3/12) (1977).

Kostyuk, P.G., O.A. Krishtal and V.I. Pidoplichko: Effect of Internal Fluoride and Phosphate on Membrane Currents during Intracellular Dialysis of Nerve Cell. *Nature* **257**:691-(3/12) (1975).

Kostyuk, P.G., O.A. Krishtal and Y.A. Shakhovalov: Separation of Sodium and Calcium Currents in the Somatic Membrane of Mollusc Neurones. *J.Physiol.* **270**:545-(1/3,2/9,3/12) (1977a).

Kostyuk, P.G., S.L. Mironov and M. Shuba Ya: Two Ion-Selective Filters in the Calcium Channel of the Somatic Membrane of Mollusc Neurons. *J.Membr.Biol.* **76**:83-(1/2,3/12) (1983).

Kostyuk, P.G., N.S. Veselovsky and S.A. Fedulova: Ionic Currents in the Somatic Membrane of Rat Dorsal Root Ganglion Neurons II. Calcium Current. *Neuroscience* **6**:2431-(1/3, 1/4) (1981).

Kowarski, D., H. Shuman, A.P. Somlyo and A.V. Somlyo: Calcium Release by Noradrenaline from Central Sarcoplasmic Reticulum in Rabbit Main Pulmonary Artery Smooth Muscle. *J.Physiol.* **366**:153-(2/10) (1985).

Krafte, D.S. and R.S. Kass: Hydrogen Ion Modulation of Ca Channel Current in Cardiac Ventricular Cell. *J.Gen.Physiol.* **91**:641-(3/12) (1988).

Kramer, R.H., E.S. Levitan, G.M. Carrow and I.B. Levitan: Modulation of Subthreshold Calcium Current by the Neuropeptide FMR-amide in *Aplysia* Neuron R15. *J.Neurophysiol.* **60**:1728-(3/14) (1988).

Kramer, R.H. and R.S. Zucker: Calcium Induced Inactivation of Calcium Current Causes the Inter Burst Hyperpolarization of *Aplysia* Bursting Pacemaker Neurones. *J.Physiol.* **362**:131-(2/9) (1985b).

Kramer, R.H. and R.S. Zucker: Calcium Dependent Inward Current in *Aplysia* Bursting Pacemaker Neurones. *J.Physiol.* **362**:107-(2/9) (1985a).

Kramsch, D.M., A.J. Aspen and C.S. Apstein: Suppression of Experimental Atherosclerosis by the Ca^{2+}-Antagonist Lanthanum. *J.Clin.Invest.* **65**:967-(4/20) (1987).

Kramsch, D.M., A.J. Aspen and L.J. Rozler: Atherosclerosis: Prevention by Agents not Affecting Abnormal Levels of Blood Lipids. *Science* **213**:1511-(4/20) (1981).

Kribben, A., E. Fritschka, D. Senger, M. Sibold, A. Distler and T. Philipp: Effects of Nitrendipine and Tiapamil on $^{45}Ca^{2+}$ Influx and on Platelet Aggregation. *J.Cardiovasc.Pharmacol.* **10**:S68-(3/13,4/20) (1987).

Krieglstein, J.: *Pharmacology of Cerebral Ischemia.* CRC Press, Boca Raton, Florida 1-(3/13) (1988).

Krikler, D.M.: Verapamil in Cardiology. *Eur.J.Cardiol.* **2**:3-(4/19) (1974).

Krikler, D.M., and J.F. Goodwin: Arrhythmia as a Cause of Sudden Death in Hypertrophic Cardiomyopathy. *Lancet* **2**:937-(4/19) (1976).

Krikler, D.M.: Calcium Ion Antagonists: Mechanisms of Action in Arrhythmias. *Clin.Invest.Med.* **3**:29-(4/19) (1980).

Krikler, D.M. and R.A.J. Spurrell: Verapamil in the Treatment of Paroxysmal Supraventricular Tachycardia. *Postgrad.Med.J.* **50**:447-(4/19) (1974).

Kril, J.J., A.L. Gundlach, P.R. Dodd, G.A.R. Johnston and C.G. Harper: Cortical Dihydropyridine Binding Sites are Unaltered in Human Alcoholic Brain. *Ann.Neurol.* **26**:395-(3/13) (1989).

Krishtal, O.A., A.V. Petrov, S.V. Smirnov and M.C. Nowycky: Hippocampal Synaptic Plasticity Induced by Excitatory Amino Acids Includes Changes in Sensitivity to the Calcium Channel Blocker, ω-Conotoxin. *Neurosci.Lett.* **102**:197-(3/13) (1989).

Kruijer, W., D. Schubert and I.M. Verma: Induction of the Proto-Oncogene fos by Nerve Growth Factor. *Proc.Natl.Acad.Sci.U.S.A.* **82**:7330-(2/10) (1985).

Krukoff, T.L. and T.M. Scott: the Development of Two Subnuclei of the Nucleus Tractus Solitarius in Spontaneously Hypertensive Rats. *Brain Res.* **314**:39-(3/13) (1984).

Kucharczky, J., W. Chew, N. Derugin, M. Moseley, C. Rollin, I. Berry and D. Norman: Nicardipine Reduces Ischemic Brain Injury: Magnetic Resonance Imaging/Spectroscopy Study in Cats. *Stroke* **20**:268-(4/21) (1989).

Kuhn, M.: Verapamil in the Treatment of PSVT. *Ann.Emerg.Med.* **10**:538-(4/19) (1981).

Kumar, N. and N.B. Gilula: Cloning and Characterization of Human and Rat Liver cDNA's Coding for a Gap Junction Protein. *J.Cell Biol.* **103**:767-(2/10) (1986).

Kuno, M. and P. Gardner: Ion Channels Activated by Inositol 1,4,5-Trisphosphate in Plasma Membrane of Human T-Lymphocytes. *Nature* **326**:301-(2/7) (1987).

Kunze, D.L., S.L. Hamilton, M.J. Hawkes and A.M. Brown: Dihydropyridine Binding and Calcium Channel Function in Clonal Rat Adrenal Medullary Tumor Cells. *Mol.Pharmacol.* **31**:401-(3/13) (1987).

Kunze, D.L., A.E. Lacerda, D.L. Wilson and A.M. Brown: Cardiac Na^+ Currents and the Inactivating, Reopening, and Waiting Properties of Single Cardiac Na^+ Channels. *J.Gen.Physiol.* **86**:691-(1/4) (1985).

Kusano, K., D.R. Livengood and R. Werman: Correlation of Transmitter Release with Membrane Properties of the Presynaptic Fiber of the Squid Giant Synapse. *J.Gen.Physiol.* **50**:2579-(2/7) (1967).

Kwon, Y.W., G. Franckowiak, D.A. Langs, M. Hawthorn, A. Joslyn and D.J. Triggle: Pharmacologic and Radioligand Binding Analysis of the Actions of 1,4-Dihydropyridine Activators Related to BAY K 8644 in Smooth Muscle, Cardiac Muscle and Neuronal Preparations. *Naunyn Schmied Arch.Pharmacol.* **339**:19-(3/13) (1989).

Kwon, Y.W., Q. Zhong, X.Y. Wei, W. Zheng and D.J. Triggle: The Interactions of 1,4-Dihydropyridines Bearing a 2-(2- aminoethylthiomethyl) Substituent at Voltage-Dependent Ca^{2+} Channels of Smooth Muscle, Cardiac Muscle and Neuronal Tissue. *Naunyn Schmied Arch.Pharmacol.* **341**:128-(3/13) (1990).

Lacerda, A.E. and A.M. Brown: Nonmodal Gating of Cardiac Calcium Channels as Revealed by Dihydropyridines. *J.Gen.Physiol.* **93**:1243-(1/4,3/13) (1989).

Lacerda, A.E., D. Rampe and A.M. Brown: Effects of Protein Kinase C Activation on Cardiac Calcium Channels. *Nature* **335**:249-(3/14) (1988).

Laher, I. and J.A. Bevan: Stretch of Vascular Smooth Muscle Activates Tone and $^{45}Ca^{2+}$ Influx. *J.Hypertens.* **7**:S17-(3/13) (1989).

Lahiri, A., P. Dasgupta and E.A. Rodrigues: Acute Drug Withdrawal in Patients with Stable Angina on Long-Term Treatment with Verapamil. In *Verapamil.* E.B. Raftery, ed. ADIS Press International Inc. 18-(4/16) (1987).

Lai, F.A., H.P. Erickson, E. Rousseau, Q.Y. Liu and G. Meissner: Purification and Reconstitution of the Calcium Release Channel from Skeletal Muscle. *Nature* **331**:315-(2/10) (1988).

Laidler, K.J. and P.S. Bunting: *The Chemical Kinetics of Enzyme Action.* Clarendon Press, Oxford. **2**:89-(3/12) (1973).

Lakshminarayanaiah, N.: *Transport Phenomena in Membranes.* Academic Press, New York. 274-(3/12) (1969).

Lakshminarayanaiah, N.: Transport Processes in Membranes: A Consideration of Membrane Potential Across Thick and Thin Membranes. In *Subcellular Biochemistry.* D.B. Roodyn, ed. Plenum Press, New York. 449-(3/12) (1979).
Lakshminarayanaiah, N.: Calcium Channels in Barnacle Muscle Membrane. In *New Perspectives on Calcium Antagonists.* G.B. Weiss, ed. American Physiological Society, Bethesda, Maryland. 19-(3/12) (1981).
Lakshminarayanaiah, N.: *Equations of Membrane Biophysics.* Academic Press, New York. 87-(3/12) (1984).
Lakshminarayanaiah, N. and P.S. Beirao: Calcium System of a Giant Barnacle Muscle Fiber. *Proc.W.Pharmacol.Soc.* **22**:301-(3/12) (1979).
Lakshminarayanaiah, N. and E. Rojas: Effects of Anions and Cations on the Resting Membrane Potential of Internally Perfused Barnacle Muscle Fibres. *J.Physiol.* **233**:613-(3/12) (1973).
Lamb, G.D.: Components of Charge Movement in Rabbit Skeletal Muscle: The Effect of Tetracaine and Nifedipine. *J.Physiol.* **376**:85-(2/10) (1986).
Lamb, G.D. and T. Walsh: Calcium Currents, Charge Movement and Dihydropyridine Binding in Fast- and Slow-Twitch Muscles of Rat and Rabbit. *J.Physiol.* **393**:595-(3/13) (1987).
Lamb, T.D. and H.R. Matthews: External and Internal Actions on the Response of Salamander Retinal Rods to Altered External Calcium Concentration. *J.Physiol.* **403**:473-(2/8) (1988).
Lamb, T.D., H.R. Matthews and V. Torre: Incorporation of Calcium Buffers into Salamander Retinal Rods: A Rejection of the Calcium Hypothesis of Phototransduction. *J.Physiol.* **372**:315-(2/8) (1986).
Lamb, T.D., P.A. McNaughton and K.W. Yau: Spatial Spread of Activation and Background Desensitization in Toad Rod Outer Segments. *J.Physiol.* **319**:463-(2/8) (1981).
Lamborghini, J.E.: Rohon-Beard Cells and Other Large Neurons in *Xenopus* Embryos Originate During Gastrulation. *J.Comp.Neurol.* **189**:323-(2/10) (1980).
Lamborghini, J.E. and A. Iles: Development of a High-Affinity GABA Uptake System in Embryonic Amphibian Spinal Neurons. *Dev.Biol.* **112**:167-(2/10) (1985).
Lancet, D.: Vertebrate Olfactory Reception. *Ann.Rev.Neurosci.* **9**:329-(2/8) (1986).
Landfield, P.: Calcium Homeostasis in Brain Aging and Alzheimer's Disease. In *Diagnosis and Treatment of Senile Dementia* M. Bergener and B. Reisberg, eds. Springer-Verlag, Berlin and Heidelberg 276-(3/13) (1989).
Landfield, P.W., D.G. Fleenor, J.C. Eldridge and B.S. McEwen: Effects of Aging on [^{3}H]Nimodipine Binding in Rat Brain. *Soc.Neurosci.Abs.* (1989).
Landmark, K.: Antihypertensive and Metabolic Effects of Long-term Therapy with Nifedipine Slow-Release Tablets. *J.Cardiovasc.Pharmacol.* **7**:12-(4/16) (1985).
Landreth, G., P. Cohen and E.M. Shooter: Ca^{2+} Transmembrane Fluxes and Nerve Growth Factor Action on a Clonal Cell Line of Rat Pheochromocytoma. *Nature* **283**:202-(2/10) (1980).
Lang, R., H.O. Klein, E. DiSegni, J. Gefen and P. Sareli: Verapamil Improves Exercise Capacity in Chronic Atrial Fibrillation: Double-Blind Cross-Over Study. *Am.Heart J..* **105**:820-(4/19) (1983a).
Lang, R., H.O. Klein, E. Weiss, D. David and P. Sareli: Superiority of Verapamil Therapy to Digoxin in the Treatment of Chronic Atrial Fibrillation. *Chest* **83**:491-(4/19) (1983b).
Langley, M.S. and E.M. Sorkin: Nimodipine: A Review of its Pharmacodynamic and Pharmacokinetic Properties, and Therapeutic Potential in Cerebrovascular Disease. *Drugs* **37**:669-(3/13) (1989).
Lankford, K.L. and P.C. Letourneau: Evidence that Calcium May Control Neurite Outgrowth by Regulating Stability of Actin Filaments. *J.Cell Biol.* **109**:1229-(2/10) (1989).
Lansman, J.B., T.J. Hallam and T.J. Rink: Single Stretch-Activated Ion Channels in Vascular Endothelial Cells as Mechanotransducers. *Nature* **325**:811-(1/2) (1987).
Lansman, J.B., P. Hess and R.W. Tsien: Blockage of Current through Single Calcium Channels by Cadmium, Magnesium and Calcium: Voltage and Concentration Dependence of Calcium Entry into the Pore. *J.Gen.Physiol.* **88**:321-(1/2,3/12) (1986).
Lappin, R.I. and L.L. Rubin: Molecular Forms of Acetylcholinesterase in *Xenopus* Muscle. *Dev.Biol.* **110**:269-(2/10) (1985).
Laragh, J.H.: Vasoconstriction-Volume Analysis for Understanding and Treating Hypertension: The Use of Renin and Aldosterone Profiles. *Am.J.Med.* **55**:261-(4/18) (1973).
Lasansky, A.: Synaptic Action Mediating Cone Responses to Annular Stimulation in the Retina of the Larval Tiger Salamander. *J.Physiol.* **310**:205-(2/8) (1981).
Lasher, R.S. and I.S. Zagon: The Effect of Potassium on Neuronal Differentiation in Cultures of Dissociated Newborn Rat Cerebellum. *Brain Res.* **41**:482-(2/10) (1972).
Läuger, P.: Conformational Transitions of Ionic Channels. In *Single Channel Recording.* B. Sakmann and E. Neher, eds. Plenum Press, New York. 177-(1/2) (1983).
Lazdunski, M., C. Frelin and P. Vigne: The Sodium/Hydrogen Exchange System in Cardiac Cells: Its Biochemical and Pharmacological Properties and its Role in Regulating Internal Concentrations of Sodium and Internal pH. *J.Mol.Cell Cardiol.* **17**:1029-(4/17) (1985).
Lazzara, R., N. El-Sherif, R.R. Hope and B.J. Scherlag: Ventricular Arrhythmias and Electrophysiologic Consequences of Myocardial Ischemia and Infarction. *Circ.Res.* **42**:740-(4/19) (1978).
Lazzara, R. and B. Scherlag: The Treatment of Arrhythmias by Blocking Slow Current. *Ann.Intern.Med.* **93**:919-(4/19) (1980).

Leblanc, N. and J.R. Hume: D600 Block of L-Type Ca^{2+} Channel in Vascular Smooth Muscle Cells: Comparison with Permanently Charged Derivative, D 890. *Am.J.Physiol.* **257**:C689-(3/13) (1989).

LeBreton, G.C., R.J. Dimerstein, L.J. Roth and H. Feinberg: Direct Evidence for Intracellular Divalent Cation Redistribution Associated with Platelet Shape Change. *Biochem.Biophys.Res.Commun.* **71**:362-(4/20) (1976).

Lederballe Pederson, O.: Calcium Blockade as a Therapeutic Principle in Arterial Hypertension. *Acta. Pharmacol. Toxicol. (Suppl. II)* **49**:5-(4/15) (1981).

Lee, H.C., N. Smith, R. Mohabir and W.T. Clusin: Cytosolic Calcium Transients from the Beating Mammalian Heart. *Proc.Natl.Acad.Sci.U.S.A.* **84**:7793-(4/16) (1987).

Lee, H.R., M. Watson, H.I. Yamamura and W.R. Roeske: Decreased [^{3}H]Nitrendipine Binding in the Brain Stem of Deoxycorticosterone-NaCl Hypertensive Rats. *Life Sci.* **37**:971-(3/13) (1985).

Lee, K.S., N. Akaike and A.M. Brown: Properties of Internally Perfused, Voltage Clamped, Isolated Nerve Cell Bodies. *J.Gen.Physiol.* **71**:489-(3/12) (1978).

Lee, K.S., E. Marban and R.W. Tsien: Inactivation of Calcium Channels in Mammalian Heart Cells: Joint Dependence on Membrane Potential and Intracellular Calcium. *J.Physiol.* **364**:395-(1/4,2/10,3/14,4/17) (1985).

Lee, K.S. and R.W. Tsien: Reversal of Current Through Calcium Channels in Dialysed Single Heart Cells. *Nature* **297**:498-(1/2,3/12) (1982).

Lee, K.S. and R.W. Tsien: High Selectivity of Calcium Channels in Single Dialysed Heart Cells of the Guinea Pig. *J.Physiol.* **354**:253-(3/12) (1984).

Lee, W., P. Mitchell and R. Tjian: Purified Transcription Factor AP-1 Interacts with TPA-Inducible Enhancer Elements. *Cell* **49**:741-(2/10) (1987).

Leibowitz, M.D., T. Bale and C.G. Cohen: Tissue Specific Regulation of T-Type Ca Channels by Fatty Acids. *Biophys.J.* **57**:307a-(3/13) (1990).

Lemos, J.R., J.J. Nordmann, I.M. Cooke and E.L. Stuenkel: Single Channels and Ionic Currents in Peptidergic Nerve Terminals. *Nature* **319**:410-(2/7) (1986).

Lemos, J.R. and M.C. Nowycky: Two Types of Calcium Channels Coexist in Peptide-Releasing Vertebrate Nerve Terminals. *Neuron* **2**:1419-(2/7) (1989).

Leon, M.B., D.R. Rosing, R.O. Bonow, L.C. Lipson and S.E. Epstein: Clinical Efficacy of Verapamil Alone and Combined with Propranolol in Treating Patients with Chronic Stable Angina Pectoris. *Am.J.Cardiol.* **48**:131-(4/16) (1981).

Leonard, D.G.B., E.B. Ziff and L.A. Green: Identification and Characterization of mRNAs Regulated by Nerve Growth Factor in PC12 Cells. *Mol.Cell Biol.* **7**:3156-(2/10) (1987).

Leonard, J.P., J. Nargeot, T.P. Snutch, N. Davidson and H.A. Lester: Ca^{2+} Channels Induced in *Xenopus* Oocytes by Rat Brain mRNA. *J.Neurosci.* **7**:875-(3/14) (1987).

Leong, D.A.: A Complex Mechanism of Facilitation in Pituitary ACTH Cells: Recent Single-Cell Studies. *J.Exp.Biol.* **139**:151-(2/7) (1988).

Lepantalo, M., S. Sundberg and A. Gordin: Lack of Effect of Flunarizine in Intermittent Claudication. *Intern.Angiol.* **4**:75-(4/20) (1985).

Lester, H.A., T.P. Snutch, J.P. Leonard, J. Nargeot, B.M. Curtis and N. Davidson: Expression of mRNA Encoding Voltage-Dependent Ca Channels in *Xenopus* Oocytes. Review and Progress Report. *Ann.N.Y.Acad.Sci.* **560**:174-(3/13,3/14) (1989).

Letourneau, P.C.: Growth of Neurites without Filopodial or Lamellipodial Activity in the Presence of Cytochalasin B. *J.Cell Biol.* **99**:2041-(2/10) (1984).

Lette, J., R.M. Gagnon, J.G. Lemire and M. Morissette: Rebound of Vasospastic Angina after Cessation of Long-Term Treatment with Nifedipine. *Can.Med.Assoc.J.* **130**:1169-(4/16) (1984).

Leung, A.T., T. Imagawa and K.P. Campbell: Structural Characterization of the 1,4-Dihydropyridine Receptor of the Voltage-Dependent Ca^{2+} Channel from Rabbit Skeletal Muscle: Evidence for Two Distinct High Molecular Weight Subunits. *J.Biol.Chem.* **262**:7943-(2/10) (1987).

Levey, G.S., C.L. Skelton and S.E. Epstein: Decreased Myocardial Adenyl Cyclase. *J.Clin.Invest.* **48**:2244-(4/17) (1969).

Levin, R.M. and B. Weiss: Binding of Trifluoperazine to the Calcium-Dependent Activator of Cyclic Nucleotide Phosphodiesterase. *Mol.Pharmacol.* **13**:690-(4/22) (1977).

Levine, J.H., J.R. Michael and T. Guarnieri: Treatment of Multifocal Atrial Tachycardia with Verapamil. *N.Engl.J.Med.* **312**:21-(4/19) (1985b).

Levine, J.H., J.F. Spear, T. Guarnieri and M. Weisfeldt: Cesium Chloride-Induced Long OT Syndrome: Demonstration of Afterdepolarizations and Triggered Activity In Vivo. *Circulation* **72**:1092-(4/19) (1985a).

Levitan, E.S., R.H. Kramer and I.B. Levitan: Augmentation of Bursting Pacemaker Activity in *Aplysia* Neuron R15 by Egg Laying Hormone is Mediated by a Cyclic AMP Dependent Increase in Ca^{2+} and K$^+$ Currents. *Proc.Natl.Acad.Sci.U.S.A.* **84**:6307-(3/14) (1987).

Lewis, D., F.F. Weight and A. Luini: A Guanine Nucleotide-Binding Protein Mediates the Inhibition of Voltage-dependent Calcium Current by Somatostatin in a Pituitary Cell Line. *Proc.Natl.Acad.Sci.U.S.A.* **83**:9035-(1/3,3/13) (1986).

Lewis, D.V.: Spike Aftercurrents in R15 of *Aplysia*: Their Relationship to Slow Inward Current and Calcium Influx. *J.Neurophysiol.* **51**:387-(2/9) (1984).

Lewis, R.S. and M.D. Cahalan: Mitogen-Induced Oscillations of Cytosolic Ca^{2+} and Transmembrane Ca^{2+} Current in Human Leukemic T Cells. *Cell Regulation* **1**:99-(2/7) (1989).

Lewis, R.S. and M.D. Cahalan: Ion Channels and Signal Transduction in Lymphocytes. *Ann. Rev. Physiol.* **52**:415-(2/7) (1990).

Lewis, R.S. and A.J. Hudspeth: Voltage-and Ion-Dependent Conductances in Solitary Vertebrate Hair Cells. *Nature* **304**:538-(2/8) (1983).

Lichtlen, P.R.: New Therapy of Ischemic Heart Disease and Hypertension. *6th.Intl.Adalat.[R.] Symp.*, Excerpta Medica, Amsterdam. (3/13) (1986).

Lichtlen, P.R. and P.G. Hugenholtz: *Recent Aspects in Calcium Antagonism: Nisoldipine*. Schattauer, Stuttgart, New York. (3/13) (1988).

Liebman, P.A., K.R. Parker and E.A. Dratz: The Molecular Mechanism of Visual Excitation and its Relation to the Structure and Composition of the Rod Outer Segment. *Ann.Rev.Physiol.* **49**:765-(2/8) (1987).

Liebman, P.A. and E.N. Pugh: The Control of Phosphodiesterase in Rod Disk Membranes: Kinetics, Possible Mechanisms and Significance for Vision. *Vision Res.* **19**:375-(2/8) (1979).

Liebman, P.A. and E.N. Pugh: Gain, Speed and Sensitivity of GTP Binding vs PDE Activation in Visual Excitation. *Vision Res.* **22**:1475-(2/8) (1982).

Lin, A.Y., S.C. Chang and A.S. Lee: A Calcium Ionophore-Inducible Cellular Promoter is Highly Active and has Enhancer Like Properties. *Mol.Cell Biol.* **6**:1235-(2/10) (1986).

Lin, F.C., D. Finley, S.H. Rahimtoola and D. Wu: Idiopathic Paroxysmal Ventricular Tachycardia with a QRS Pattern of Right Bundle Branch Block and Left Axis Deviation: A Unique Clinical Entity with Specific Properties. *Am.J.Cardiol.* **52**:96-(4/19) (1983).

Lindau, M., J.M. Fernandez: IgE-Mediated Degranulation of Mast Cells does not Require Opening of Ion Channels. *Nature* **319**:150-(2/7) (1986).

Linden, D.J. and A. Routtenberg: Cis-Fatty Acids, Which Activate Protein Kinase C, Attenuate Na^+ and Ca^{2+} Currents in Mouse Neuroblastoma Cells. *J.Physiol.* **419**:95-(3/13) (1989).

Lindenberg, B.S., D.A. Weiner, C.H. McCabe, S.S. Cutler, T.J. Ryan and M.D. Klein: Efficacy and Safety of Incremental Doses of Diltiazem for the Treatment of Stable Angina Pectoris. *J.Am.Coll.Cardiol.* **2**:1129-(4/16) (1983).

Lindgren, C.A. and J.W. Moore: Identification of Ionic Currents at Presynaptic Nerve Endings of the Lizard. *J.Physiol.* **414**:201-(2/7) (1989).

Lineweaver, H. and D. Burk: The Determination of Enzyme Dissociation Constants. *J.Am.Chem.Soc.* **56**:658-(3/12) (1934).

Linnoila, M., E. MacDonald, M. Reinila, A. Leroy, D.R. Rubinow and F.K. Goodwin: RBC Membrane Adenosine Triphosphatase Activities in Patients with Major Affective Disorders. *Arch.Gen.Psychiatr.* **40**:1021-(4/22) (1983).

Lipkin, D.P. and P.A. Poole-Wilson: Treatment of Chronic Heart Failure: A Review of Recent Drug Trials. *Br.Med.J.* **291**:993-(4/17) (1985).

Lipp, P., S. Mechmann and L. Pott: Effects of Calcium Release from Sarcoplasmic Reticulum on Membrane Currents in Guinea Pig Atrial Cardioballs. *Pflügers Arch.* **410**:121-(2/10) (1987).

Lipscombe, D., K. Bley and R.W. Tsien: Modulation of Neuronal Ca^{2+} Channels by cAMP and Phorbol Esters. *Soc.Neurosci.Abs.* **14**:153-(3/14) (1988a).

Lipscombe, D., S. Kongsamut and R.W. Tsien: α-Adrenergic Inhibition of Sympathetic Neurotransmitter Release Mediated by Modulation of N-Type Calcium-Channel Gating. *Nature* **340**:639-(1/3,3/13,3/14) (1989).

Lipscombe, D., D.V. Madison, M. Poenie, H. Reuter, R.Y. Tsien and R.W. Tsien: Imaging of Cytosolic Ca^{2+} Transients Arising from Ca^{2+} Stores and Ca^{2+} Channels in Sympathetic Neurons. *Neuron* **1**:355-(3/14) (1988c).

Lipscombe, D., D.V. Madison, M. Poenie, H. Reuter, R.Y. Tsien and R.W. Tsien: Spatial Distribution of Calcium Channels and Cytosolic Calcium Transients in Growth Cones and Cell Bodies of Sympathetic Neurons. *Proc.Natl.Acad.Sci.U.S.A.* **85**:2398-(2/9,3/14) (1988b).

Lipton, S.A., M.P. Frosch, M.D. Phillips, D.L. Tauck and E. Aizenman: Nicotinic Antagonists Enhance Process Outgrowth by Rat Retinal Ganglion Cells in Culture. *Science* **239**:1293-(2/10) (1988).

Lisman, J.: A Mechanism for the Hebb and the Anti-Hebb Processes Underlying Learning and Memory. *Proc.Natl.Acad.Sci.U.S.A.* **86**:9574-(3/13) (1989).

Livesley, B., P.F. Catley and R.C. Campbell: Double-Blind Evaluation of Verapamil, Propranolol and Isosorbide Dinitrate Against a Placebo in the Treatment of Angina Pectoris. *Br.Med.J.* **1**:375-(4/16) (1973).

Llinas, R.: The Intrinsic Electrophysiological Properties of Mammalian Neurons: Insights into Central Nervous System Function. *Science* **242**:1654-(3/13,3/14) (1988).

Llinas, R., J.R. Blinks and C. Nicholson: Calcium Transient in Presynaptic Terminal of Squid Giant Synapse: Detection with Aequorin. *Science* **176**:1127-(2/7) (1972).

Llinas, R. and C. Nicholson: Electrophysiological Properties of Dendrites and Somata in Alligator Purkinje Cells. *J.Neurophysiol.* **34**:534-(2/9) (1971).

Llinas, R. and C. Nicholson: Calcium Role in Depolarization-Secretion Coupling: An Aequorin Study in Squid Giant Synapse. *Proc.Natl.Acad.Sci.U.S.A.* **72**:187-(2/7) (1975).

Llinas, R., I.Z. Steinberg and K. Walton: Presynaptic Calcium Currents and Their Relation to Synaptic Transmission: Voltage Clamp Study in Squid Giant Synapse, and Theoretical Model for the Calcium Gate. *Proc.Natl.Acad.Sci.U.S.A.* **73**:2918-(2/7) (1976).

Llinas, R. and M. Sugimori: Electrophysiological Properties of *In Vitro* Purkinje Cell Somata in Mammalian Cerebellar Slices. *J.Physiol.* **305**:171-(2/9,3/14) (1980a).

Llinas, R. and M. Sugimori: Electrophysiological Properties of *In Vitro* Purkinje Cell Dendrites in Mammalian Cerebellar Slices. *J.Physiol.* **305**:197-(2/9) (1980b).

Llinas, R., M. Sugimori and B. Cherksey: Voltage-Dependent Calcium Conductances in Mammalian Neurons. The P Channel. *Ann.N.Y.Acad.Sci.* **560**:103-(3/13,3/14) (1989b).

Llinas, R., M. Sugimori, J.W. Lin and B. Cherksey: Blocking and Isolation of a Calcium Channel from Neurons in Mammals and Cephalopods Utilizing a Toxin Fraction (FTX) from Funnel-web Spider Poison. *Proc.Natl.Acad.Sci.U.S.A.* **86**:1689-(2/9,3/13) (1989a).

Llinas, R. and Y. Yarom: Properties and Distribution of Ionic Conductances Generating Electroresponsiveness of Mammalian Inferior Olivary Neurones *In Vitro*. *J.Physiol.* **315**:569-(1/4,2/9) (1981b).

Llinas, R. and Y. Yarom: Electrophysiology of Mammalian Inferior Olivary Neurons *In Vitro*: Different Types of Voltage-Dependent Ionic Conductances. *J.Physiol.* **315**:549-(2/9,3/14) (1981a).

Llinas, R. and Y. Yarom: Oscillatory Properties of Guinea Pig Inferior Olivary Neurones and Their Pharmacological Modulation: An *In Vitro* Study. *J.Physiol.* **376**:163-(2/9) (1986b).

Llinas, R. and Y. Yarom: Specific Blockage of the Low Threshold Calcium Channel by High Molecular Weight Alcohols. *Soc.Neurosci.Abs.* **12**:49-(3/13,3/14) (1986a).

Llinas, R.R.: Calcium in Synaptic Transmission. *Sci.Am.* **247(4)**:56-(2/7) (1982).

Llinas, R.R.: The Squid Giant Synapse. *Curr.Top.Membr.Transp.* **22**:519-(2/7) (1984).

Llinas, R.R., I.Z. Steinberg and K. Walton: Presynaptic Calcium Currents in Squid Giant Synapse. *Biophys.J.* **33**:289-(1/3,2/7) (1981a).

Llinas, R.R., I.Z. Steinberg and K. Walton: Relationship Between Presynaptic Calcium Current and Postsynaptic Potential in Squid Giant Synapse. *Biophys.J.* **33**:323-(2/7) (1981b).

Loirand, G., C. Mironneau, J. Mironneau and P. Pacaud: Two Types of Calcium Currents in Single Smooth Muscle Cells from Rat Portal Vein. *J.Physiol.* **412**:333-(2/10,3/13) (1989).

Loirand, G., P. Pacaud, C. Mironneau and J. Mironneau: Evidence for Two Distinct Calcium Channels in Rat Vascular Smooth Muscle Cells in Short-Term Primary Culture. *Pflügers Arch.* **407**:566-(2/10) (1986).

Lolley, R.N. and E. Racz: Calcium Modulation of Cyclic GMP Synthesis in Rat Visual Cells. *Vision Res.* **22**:1481-(2/8) (1982).

Lomo, T. and J. Rosenthal: Control of ACh Sensitivity by Muscle Activity in the Rat. *J.Physiol.* **221**:493-(2/10) (1972).

London, B. and J.W. Krueger: Contraction in Voltage-Clamped, Internally Perfused Single Heart Cells. *J.Gen.Physiol.* **88**:475-(2/10) (1986).

Loots, W., J. Dom and C. Horig: Intermittent Claudication Attempts to Make this Disease Objective on Account of the Results of Treatment Obtainable by Means of Sibelium. *Med.Welt.* **31**:189-(4/20) (1980).

Lopez Alemany, M.: Parkinsonism as Late Toxic Effect of Dihydropyridines. *Lancet* (March 29) 737-(4/22) (1986).

Lopez-Barneo, J. and C.M. Armstrong: Depolarizing Response of Rat Parathyroid Cells to Divalent Cations. *J.Gen.Physiol.* **82**:269-(2/7) (1983).

Lorell, B.H., W.J. Paulus, W. Grossman, J. Wynne and P.F. Cohn: Modification of Abnormal Left Ventricular Diastolic Properties by Nifedipine in Patients with Hypertrophic Cardiomyopathy. *Circulation* **65**:499-(4/17) (1982).

Loscher, W.: Valproic Acid. In *Handbook of Experimental Pharmacology, Antiepileptic Drugs*. H.H. Frey and D. Janz, eds. Springer, Berlin. **74**:508-(4/22) (1985).

Loutzenhiser, R., M. Epstein, C. Horton and P. Sonke: Reversal by the Calcium Antagonist Nisoldipine of Norepinephrine-Induced Reduction of GFR: Evidence for Preferential Antagonism of Preglomerular Vasoconstriction. *J.Pharmacol.Exp.Ther.* **232**:382-(4/18) (1984).

Loutzenhiser, R.D., M. Epstein, and K. Hayashi: Renal Hemodynamic Effects of Calcium Antagonists. *Am.J.Cardiol.* **64**:41F-(3/13) (1989b).

Loutzenhizer, R., P. Leyton, K. Saida and C. Van Breemen: Calcium Compartments and Mobilization during Contraction of Smooth Muscle, In *Calcium and Contractility Smooth Muscle*. A.K. Grover and E.E. Daniel, eds. Humana Press, Clifton, New Jersey. 61-(2/10) (1985).

Loutzenhizer, R.D., M. Epstein, F. Fischetti and C. Horton: Effects of Amlodipine on Renal Hemodynamics. *Am.J.Cardiol.* **64**:1221-(3/13) (1989a).

Lowry, O.H., N.J. Rosebrough, A.L. Farr and R.J. Randall: Protein Measurement with the Folin Phenol Reagent. *J.Biol.Chem.* **193**:265-(4/17) (1951).

Lubbe, W.F., T. Podzuweit, P.S. Daries and L.H. Opie: The Role of Cyclic Adenosine Monophosphate in Adrenergic Effects on Vulnerability to Fibrillation in the Isolated Perfused Rat Heart. *J.Clin.Invest.* **61**:1260-(4/16) (1978).

Luchowski, E.M., F. Yousif, D.J. Triggle, S.C. Maurer, J.G. Sarmiento, and R.A. Janis: The Effects of Metal Cations and Calmodulin Antagonists on [^{3}H]Nitrendipine Binding in Smooth and Cardiac Muscle. *J.Pharmacol.Exp.Ther.* **230**:607-(3/13) (1984).

Luft, F.C., G. Aronoff, R. Sloan, N. Fineberg and M. Weinberger: Calcium Channel Blockade with Nitrendipine. *Hypertension* **7**:438-(4/18) (1985).

Lugaresi, A., P. Montagna, R. Gallassi and E. Lagaresi: Extrapyramidal Syndrome and Depression Induced by Flunarizine. *Eur.Neurol.* **28**:208-(3/13) (1988).

Luini, A., D. Lewis, S. Guild, G. Schofield and F. Weight: Somatostatin, an Inhibitor of ACTH Secretion, Decreases Cytosolic Free Calcium and Voltage-Dependent Calcium Current in a Pituitary Cell Line. *J.Neurosci.* **6**:3128(1/3,3/13) (1986).

Luttgau, H., C.H. and W. Spiecker: The Effects of Calcium Deprivation Upon Mechanical and Electrophysiological Parameters in Skeletal Muscle Fibres of the Frog. *J.Physiol.* **296**:411-(2/10) (1979).

Lux, H.D., E. Carbone and H. Zucker: Na^+ Currents through Low-Voltage-Activated Ca^{2+} Channels of Chick Sensory Neurones: Block by External Ca^{2+} and Mg^{2+}. *J.Physiol.* **430**:159-(1/2) (1990).

Lux, H.D.: Studies of the Development of Voltage-Activated Calcium Channels in Vertebrate Neurons. *Puerto Rican Health Sci.J.* **7**:116-(2/10) (1988).

Lux, H.D. and A.M. Brown: Single Channel Studies on Inactivation of Calcium Currents. *Science* **225**:432-(1/4) (1984a).

Lux, H.D. and A.M. Brown: Patch and Whole Cell Calcium Currents Recorded Simultaneously in Snail Neurons. *J.Gen.Physiol.* **83**:727-(1/3,1/4) (1984b).

Lux, H.D. and A.M. Brown: Activation of Single Neuronal Calcium Channels. *Proceedings of the 16th FEBS Congress* 407-(1/3) (1985).

Lux, H.D., E. Carbone and H. Zucker: Block of Na^+ Ion Permeation and Selectivity of Ca Channels. *Ann.N.Y.Acad.Sci.* **560**:94-(1/2,1/4) (1989).

Lux, H.D. and K. Nagy: Single Channel Ca^{2+} Currents in *Helix pomatia* Neurons. *Pflügers Arch.* **391**:252-(1/2,1/3,1/4) (1981).

Lynch, P., H. Dargie, S. Krikler and D. Krikler: Objective Assessment of Antianginal Treatment: A Double-Blind Comparison of Propranolol, Nifedipine and their Combination. *Br.Med.J.* **281**:184-(4/16) (1980).

MacDermott, A.B., M.L. Mayer, G.L. Westbrook, S.J. Smith and J.L. Barker: NMDA-Receptor Activation Increases Cytoplasmic Calcium Concentration in Cultured Spinal Cord Neurones. *Nature* **321**:519-(2/9) (1986).

MacDonald, J.F., A.V. Porietis and J.M. Wojtowicz: L-Aspartic Acid Induces a Region of Negative Slope Conductance in the Current-Voltage Relationship of Cultured Spinal Cord Neurons. *Brain Res.* **237**:248-(2/9) (1982).

MacDonald, R.L. and B.S. Meldrum: Principles of Antiepileptic Drug Action. In *Antiepileptic Drugs.* R. Levy, R. Mattson, B. Meldrum, J.K. Henry and F.E. Dryfuss, eds. Raven Press, New York. 59-(3/13) (1989).

MacDonald, T., D. Pelzer and W. Trautwein: Dual Action (Stimulation, Inhibition) of D600 on Contractility and Calcium Channels in Guinea Pig and Cat Heart Cells. *J.Physiol.* **414**:569-(3/13) (1989).

MacDougall, S.L., S. Grinstein and E.W. Gelfand: Detection of Ligand-Activated Conductive Ca^{2+} Channels in Human B Lymphotytes. *Cell* **54**:229-(2/7) (1988).

MacIntyre, D.E., M. Buchfield and A.M. Shaw: Regulation of Platelet Cytosolic Free Calcium by Cyclic Nucleotides and Protein Kinase C. *FEBS Lett.* **188**:383-(4/20) (1985).

MacLeish, P.R., E.A. Schwartz and M. Tachibana: Control of the Generator Current in Solitary Rods of the *Ambystoma tigrinum* Retina. *J.Physiol.* **348**:645-(2/8) (1984).

Madison, D.V.: Phorbol Esters Increase Unitary Calcium Channel Activity in Cultured Hippocampal Neurons. *Soc.Neurosci.Abs.* **15**:16-(3/14) (1989).

Madison, D.V. and R.A. Nicoll: Control of the Repetitive Discharge of Rat CA1 Pyramidal Neurones *In Vitro*. *J.Physiol.* **354**:319-(2/9) (1984).

Maeda, N., K. Wada, M. Yuzaki and K. Mikoshiba: Autoradiographic Visualization of a Calcium Channel Antagonist, [125I]- ω-Conotoxin GVIA, Binding Site in the Brains of Normal and Cerebellar Mutant Mice (pcd and Weaver). *Brain Res.* **489**:21-(3/13) (1989).

Magnoni, M.S., S. Govoni, F. Battaini and M. Trabucchi: L-Type Calcium Channels are Modified in Rat Hippocampus by Short-Term Experimental Ischemia. *J.Cereb.Blood Flow Metab.* **8**:96-(3/13) (1988).

Makino, N., G. Jasmin, R.E. Beamish and N.S. Dhalla: Sarcolemmal Na^+ - Ca^{2+} Exchange during the Development of Genetically Determined Cardiomyopathy. *Biochem.Biophys.Res.Commun.* **133**:491-(4/17) (1985).

Malaisse-Lagae, F., P.C.F. Mathias and W.J. Malaisse: Gating and Blocking of Calcium Channels by Dihydropyridines in the Pancreatic Beta-cell. *Biochem.Biophys.Res.Commun.* **123**:1062-(2/7) (1984).

Malaret, R., R.A. Wise, W.H. Egginger and F.M. Wigley: Nifedipine in the Treatment of Raynaud's Phenomenon. *Am.J.Med.* **78**:602-(4/20) (1985).

Malécot, C.O., P. Feindt and W. Trautwein: Intracellular N-Methyl-D-Glucamine Modifies the Kinetics and Voltage-Dependence of the Calcium Current in Guinea Pig Ventricular Heart Cells. *Pflügers Arch.* **411**:235-(1/4) (1988).

Malhotra, R.K., S.V. Bhave, T.D. Wakade and A.R. Wakade: Protein Kinase C of Sympathetic Neuronal Membrane is Activated by Phorbol Esters - Correlation Between Transmitter Release, $^{45}Ca^{2+}$ Uptake and the Enzyme Activity. *J.Neurosci.* **51**:967-(3/14) (1988).

Marban, E. and R.W. Tsien: Effects of Nystatin-Mediated Intracellular Ion Substitution on Membrane Currents in Calf Purkinje Fibres. *J.Physiol.* **329**:569-(1/4) (1982).

Marbau, E., Y. Koretsune, M. Corretti, V.P. Chacko and H. Kusuoka: Calcium and its Role in Myocardial Cell Injury during Ischemia and Reperfusion. *Circulation* **80**:17-(4/19) (1989).

Marchetti, C. and A.M. Brown: The Protein Kinase C Activator l-oleyl-2-acetyl-sn-glycerol inhibits two Types of Calcium Currents in GH_3 Cells. *Am.J.Physiol.* **254**:C207-(3/14) (1988).

Marchetti, C., E. Carbone and H.D. Lux: Effects of Dopamine and Noradrenaline on Ca Channels of Cultured Sensory and Sympathetic Neurons of Chick. *Pflügers Arch.* **406**:104-(1/3,3/14) (1986).

Marchetti, C. and M. Robello: Guanosine-5′-O-(30 Thiotriphosphate) Modifies Kinetics of Voltage-Dependent Calcium Current in Chick Sensory Neurons. *Biophys.J.* **56**:1267-(1/3) (1989).

Marcus, A.J., L.B. Safier and H.L. Ulllman: Effects of Acetylglyceryl Ether Phosphonylcholine on Human Platelet Function *In Vitro*. *Blood* **58**:1027-(4/20) (1981).

Margiotta, J.F., D.K. Berg and V.E. Dionne: Cyclic AMP Regulates the Proportion of Functional Acetylcholine Receptors on Chicken Ciliary Ganglion Neurons. *Proc.Natl.Acad.Sci.U.S.A.* **84**:8155-(2/10) (1987a).

Margiotta, J.F., D.K. Berg and V.E. Dionne: The Properties and Regulation of Functional Acetylcholine Receptors on Chick Ciliary Ganglion Neurons. *J.Neurosci.* **7**:3612-(2/10) (1987b).

Marguerie, G.A., T.S. Edgington and E.F. Plow: Interaction of Fibrinogen with its Platelet Receptor as Part of a Multi-step Reaction in ADP-Induced Platelet Aggregation. *J.Biol.Chem.* **255**:154-(4/20) (1980).

Maricq, A.V. and J.I. Korenbrot: Calcium and Calcium-Dependent Chloride Currents Generate Action Potentials in Solitary Cone Photoreceptors. *Neuron* **1**:503-(2/8) (1988).

Marion Laboratories - Data on File, M. Moser, J. Lunn and B.J. Materson: Comparative Effects of Diltiazem and Hydrocholorothiazide in Blacks with Systemic Hypertension. *Am.J.Cardiol.* **56**:101-(4/18) (1985).

Marks, S.S., D.L. Watson, C.L. Carpenter, R.O. Messing and D.A. Greenberg: Comparative Effects of Chronic Exposure to Ethanol and Calcium Channel Antagonists on Calcium Channel Antagonist Receptors in Cultured Neural (PC-12) Cells. *J.Neurochem.* **53**:168-(3/13) (1989).

Markwardt, F., R. Albitz, T. Franke and B. Nilius: Thrombin Stimulates Ca^{2+} Channel Currents in Isolated Frog Ventricular Cells. *Pflügers Arch.* **412**:668-(3/14) (1988).

Markwartd, F. and B. Nilius: Modulation of Calcium Channel Currents in Guinea Pig Single Ventricular Heart Cells by the Dihydropyridine Bay K 8644. *J.Physiol.* **399**:559-(1/4) (1988).

Maron, B.J., S.E. Epstein and W.C. Roberts: Hypertrophic Cardiomyopathy and Transmural Myocardial Infarction without Significant Atherosclerosis of the Extramural Coronary Arteries. *Am.J.Cardiol.* **43**:1086-(4/17) (1979).

Marsh, J.D.: Coregulation of Calcium Channels and β-Adrenergic Receptors in Cultured Chick Embryo Ventricular Cells. *J.Clin.Invest.* **84**:817-(3/13) (1989).

Martin, A.R.: **In** Junctional Transmission. II. Presynaptic Mechanisms. 1: *Handbook of Physiology Section 1 The Nervous* System. J.M. Brookhart, V.B. Mountcastle and E.R. Kandel, eds. The American Physiological Society, Bethesda, MD. 329-(2/7) (1977).

Martin, S.K., A.M. Oduola and W.K. Milhous: Reversal of Chloroquine Resistance in Plasmodium Faciparum by Verapamil. *Science* **235**:899-(3/13) (1987).

Marty, A., Y.P. Tan and A. Trautmann: Three Types of Calcium-Dependent Channels in Rat Lacrimal Glands. *J.Physiol.* **357**:293-(2/7) (1984).

Maseri, A. and S. Chierchia: Coronary Artery Spasm: Demonstration, Definition, Diagnosis and Consequences. *Prog. Cardiovasc. Dis.* **25**:169-(4/15) (1982).

Maseri, A., S. Chierchia and J.C. Kaski: Mixed Angina Pectoris. *Am.J.Cardiol.* **56**:30E-(4/16) (1985).

Maseri, A., S. Severi, M. DeNes, A. L'Abbate, S. Chierchia, M. Marzelli, A.M. Ballestra, O. Parodi, A. Biagini and A. Distante: "Variant" Angina: One Aspect of a Continuous Spectrum of Vasospastic Myocardial Ischemia. Pathogenetic Mechanisms, Estimated Incidence and Clinical and Coronary Arteriographic Findings in 138 Patients. *Am.J.Cardiol.* **42**:1019-(4/16) (1978).

Masiakowski, P. and E.M. Shooter: Nerve Growth Factor Induces the Genes for Two Proteins Related to a Family of Calcium-Binding Proteins in PCl2 Cells. *Proc.Natl.Acad.Sci.U.S.A.* **85**:1277-(2/10) (1988).

Mason, J.W., C.D. Swerdlow and L.B. Mitchell: Efficacy of Verapamil in Chronic, Recurrent Ventricular Tachycardia. *Am.J.Cardiol.* **51**:1614-(4/19) (1983).

Mason, R.P., S.F. Campbell, S.D. Wang and L.G. Herbette: Comparison of Location and Binding for the Positively Charged 1,4-Dihydropyridine Calcium Channel Antagonist Amlodipine with Uncharged Drugs of this Class in Cardiac Membranes. *Mol.Pharmacol.* **36**:634-(3/13) (1989b).

Mason, R.P., G.E. Gonye, D.W. Chester and L.G. Herbette: Partitioning and Location of BAY K 8644, 1,4-Dihydropyridine Calcium Channel Agonist, in Model and Biological Membranes. *Biophys.J.* **55**:769-(3/13) (1989a).

Mason, R.P., J. Moring and L.G. Herbette: Cholesterol/Drug Molecular Interactions with Model and Native Membranes. *Biophys.J.* **57**:523a-(3/13) (1990b).

Mason, R.P., D.G. Rhodes and L.G. Herbette: Reevaluation of Dissociation Constants for 1,4-Dihydropyridines (DHP) Based on Membrane Drug Concentration. *Biophys.J.* **57**:302a-(3/13) (1990a).

Mason, W.T., S.R. Rawlings, P. Cobbett, S.K. Sikdar, R. Zorec, S.N. Akerman, C.D. Benham, M.J. Berridge, T. Cheek and R.B. Moreton: Control of Secretion in Anterior Pituitary Cells - Linking Ion Channels, Messengers and Exocytosis. *J.Exp.Biol.* **139**:287-(2/7) (1988).

Massey, S.C. and D.A. Redburn: Transmitter Circuits in the Vertebrate Retina. *Prog.Neurobiol.* **28**:55-(2/8) (1987).

Massieu, L. and R. Tapia: Relationship of Dihydropyridine Binding Sites with Calcium-Dependent Neurotransmitter Release in Synaptosomes. *J.Neurochem.* **51**:1184-(3/13) (1988).

Massini, P. and E.P. Luscher: Some Effects of Ionophores for Divalent Cations on Blood Platelets - Comparison with the Effects of Thrombin. *Biochim.Biophys.Acta* **372**:109-(4/20) (1974).

Massumi, R.A., J.C. Rios, A.S. Gooch, D. Nutter, V.T. de Vita and D.W. Datlow: Primary Myocardial Disease. Report of Fifty Cases and Review of the Subject. *Circulation* **31**:19-(4/17) (1965).

Matlib, M.A.: Relaxation of Potassium Chloride-Induced Contractions by Amlodipine and its Interaction with the 1,4-Dihydropyridine-Binding Site in Pig Coronary Artery. *Am.J.Cardiol.* **64**:51-(3/13) (1989).

Matlib, M.A., J.F. French, I.L. Grupp, P.L. Vaghy, G. Grupp and A. Schwartz: Pharmacological Effects of Amlodipine, a New Slow-Acting 1,4-Dihydropyridine. *J.Cardiovasc.Pharmacol.* **12**:S14-(3/13) (1988).

Matsuda, H.: Sodium Conductance in Calcium Channels of Guinea Pig Ventricular Cells Induced by Removal of External Calcium Ions. *Pflügers Arch.* **407**:465-(1/2) (1986).

Matsuda, H. and A. Noma: Isolation of Calcium Current and its Sensitivity to Monovalent Cations in Dialysed Ventricular Cells of Guinea Pig. *J.Physiol.* **357**:553-(1/2) (1984).

Matsudaira, P. and P. Janmey: Pieces in the Actin-Severing Protein Puzzle. *Cell* **54**:139-(2/10) (1988).

Matteson, D.R. and C.M. Armstrong: Properties of Two Types of Calcium Channels in Clonal Pituitary Cells. *J.Gen.Physiol.* **87**:161-(2/7) (1986).

Matthews, G.: Comparison of the Light-Sensitive and Cyclic GMP-Sensitive Conductances of the Rod Photoreceptor: Noise Characteristics. *J.Neurosci.* **6**:2521-(2/8) (1986).

Matthews, G.: Single-Channel Recordings Demonstrate that cGMP Opens the Light-Sensitive Ion Channel of the Rod Photoreceptor. *Proc.Natl.Acad.Sci.U.S.A.* **84**:299-(2/8) (1987).

Matthews, G. and S.I. Watanabe: Activation of Single Ion Channels from Toad Retinal Rod Inner Segments by Cyclic GMP: Concentration Dependence. *J.Physiol.* **403**:389-(2/8) (1988).

Matthews, H.R., R.L.W. Murphy, G.L. Fain and T.D. Lamb: Photoreceptor Light Adaptation is Mediated by Cytoplasmic Ca^{2+} Concentration. *Nature* **334**:67-(2/8) (1988).

Mattson, M.P., P.B. Guthrie and S.B. Kater: Intracellular Messengers in the Generation and Degeneration of Hippocampal Neuroarchitecture. *J.Neurosci.Res.* **21**:447-(2/10) (1988b).

Mattson, M.P. and S.B. Kater: Calcium Regulation of Neurite Elongation and Growth Cone Motility. *J.Neurosci.* **7**:4034-(2/10) (1987).

Mattson, M.P., A. Taylor-Hunter and S.B. Kater: Neurite Outgrowth in Individual Neurons of a Neuronal Population is Differentially Regulated by Calcium and Cyclic AMP. *J.Neurosci.* **8**:1704-(2/10) (1988a).

Mauritson, D.R., M.D. Winniford, W.S. Walker, R.E. Rude, J.R. Cary and L.D. Hillis: Oral Verapamil for Paroxysmal Supraventricular Tachycardia. A Long-term Double-blind Randomized Trial. *Ann.Int.Med.* **96**:409-(4/19) (1982).

Mayer, M.L., A.B. MacDermott, G.L. Westbrook, S.J. Smith and J.L. Barker: Agonist- and Voltage-Gated Calcium Entry in Mouse Spinal Cord Neurons Under Voltage Clamp Measured Using Arsenazo III. *J.Neurosci.* **7**:3230-(2/9) (1987a).

Mayer, M.L. and G.L. Westbrook: Permeation and Block of n-Methyl-D-Aspartic Acid Receptor Channels by Divalent Cations in Mouse Cultured Central Neurones. *J.Physiol.* **394**:501-(2/9) (1987b).

Mayer, M.L. and G.L. Westbrook: Cellular Mechanisms Underlying Excitotoxicity. *TINS* **10**:59-(2/10) (1987c).

Mayer, M.L., G.L. Westbrook and P.B. Guthrie: Voltage-Dependent Block by Mg^{2+} of NMDA Responses in Spinal Cord Neurones. *Nature* **309**:261-(2/9,4/21) (1984).

McBurney, R.N. and S.J. Kehl: Electrophysiology of Neurosecretory Cells from the Pituitary Intermediate Lobe. *J.Exp.Biol.* **139**:317-(2/7) (1988).

McCaig, C.D.: Myoblasts and Myoblast-Conditioned Medium Attract the Earliest Spinal Neurites from Frog Embryos. *J.Physiol.* **375**:39-(2/10) (1986).

McCarthy, R.T.: Nimodipine Block of L-Type Calcium Channels in Dorsal Root Ganglion Cells. In *Nimodipine and Central Nervous System Function: New Vistas*. J. Traber and W. Gispen, eds. Schattauer, Stuttgart and New York. 35-(3/13) (1989).

McCarthy, R.T. and C.J. Cohen: The Enantiomers of BAY K 8644 have Different Effects on Ca Channel Gating in Rat Anterior Pituitary Cells. *Biophys.J.* **49**:432a-(3/13) (1986).

McCarthy, R.T. and C.J. Cohen: Nimodipine Block of Calcium Channels in Rat Vascular Smooth Muscle Cell Lines. *J.Gen.Physiol.* **94**:669-(3/13) (1989).

McCarthy, R.T. and H.K. Fry: Nitrendipine Block of Calcium Channel Currents in Vascular Smooth Muscle and Adrenal Glomerulosa Cells. *J.Cardiovasc.Pharmacol.* **12**:S98-(3/13) (1988).

McCleskey, E.W.: Calcium Channels and Intracellular Calcium Release are Pharmacologically Different in Frog Skeletal Muscle. *J.Physiol.* **361**:231-(2/10) (1985).

McCleskey, E.W., A.P. Fox, D. Feldman and R.W. Tsien: Different Types of Calcium Channels. *J.Exp.Biol.* **124**:177-(3/13) (1987a).

McCleskey, E.W., A.P. Fox, D.H. Feldman, L.J. Cruz, B.M. Olivera, R.W. Tsien and D. Yoshikama: ω-Conotoxin: Direct and Persistent Blockade of Specific Types of Calcium Channels in Neurons but not in Muscle. *Proc.Natl.Acad.Sci.U.S.A.* **84**:4327-(1/4,3/13,3/14) (1987b).

McClure, W.O., B.C. Abbott, D.E. Baxter, T.H. Hsiao, L.S. Satin, A. Siger and J.E. Yoshino: Leptinotarsin: A Presynaptic Neurotoxin that Stimulates Release of Acetylcholine. *Proc.Natl.Acad.Sci.U.S.A.* **77**:1219-(2/7) (1980).

McCobb, D.P., P.M. Best and K.G. Beam: Development of the Expression of Calcium Currents in Chick Limb Motoneurons. *Neuron* **2**:1633-(2/10,3/13) (1989).

McCobb, D.P. and S.B. Kater: Membrane Voltage and Neurotransmitter Regulation of Neuronal Growth Cone Motility. *Dev.Biol.* **130**:599-(2/10) (1988).

McCormick, D.A. and D.A. Prince: Acetylcholine Induces Burst Firing in Thalamic Reticular Neurons by Activing a Potassium Conductance. *Nature* **319**:402-(2/9) (1986).

McDonagh, P.F. and D.J. Roberts: Prevention of Transcoronary Macromolecular Leakage After Ischemia-Reperfusion by the Calcium Entry Blocker Nisoldipine. Direct Observations in Isolated Rat Hearts. *Circ.Res.* **58**:127-(3/13) (1986).

McKenna, W.J., L. Harris and G. Perez: Hypertrophic Cardiomyopathy: Comparison of Verapamil and Amiodarone in the Treatment of Arrhythmia. *Br.Heart J.* **45**:354-(4/19) (1980).

McLean, M.J. and R.L. MacDonald: Carbamazepine and 10,11-Epoxycarbamazepine Produce Use- and Voltage-Dependent Limitation of Rapidly Firing Action Potential of Mouse Central Neurons in Cell Culture. *J.Pharmacol.Exp.Ther.* **238**:727-(4/22) (1986).

McManaman, J.L., J.C. Blosser and S.H. Appel: The Effect of Calcium on Acetylcholine Receptor Synthesis. *J.Neurosci.* **1**:771-(2/10) (1981).

McNaughton, P.A., L. Cervetto and B.J. Nunn: Measurement of the Intracellular Free Calcium Concentration in Salamander Rods. *Nature* **322**:261-(2/8) (1986).

McTavish, D. and E.M. Sorkin: Verapamil - An Updated Review of its Pharmacodynamic and Pharmacokinetic Properties, and Therapeutic use in Hypertension. *Drugs* **38**:19-(3/13) (1989).

Means, G.E. and R.E. Feeney: *Chemical Modification of Proteins*. Holden-Day Inc. San Francisco, California. (3/12) (1971).

Meech, R.W.: The Sensitivity of *Helix aspersa* Neurones to Injected Calcium Ions. *J.Physiol.* **237**:259-(2/5) (1974).

Meech, R.W. and N.B. Standen: Potassium Activation in *Helix aspersa* Under Voltage Clamp: A Component Mediated by Calcium Influx. *J.Physiol.* **249**:211-(2/9) (1975).

Mehta, J.: Influence of Calcium-Channel Blockers on Platelet Function and Arachidonic Acid Metabolism. *Am..J.Cardiol.* **55**:158B-(4/20) (1985).

Mehta, J., P. Mehta and N. Ostrowski: Calcium Blocker Diltiazem Inhibits Platelet Activation and Stimulates Vascular Prostacyclin Synthesis. *Am.J.Med.Sci.* **29**:20-(3/13,4/20) (1986).

Mehta, J., P. Mehta, N. Ostrowski, and F. Crews: Effects of Verapamil on Platelet Aggregation, ATP Release and Thromboxane Formation. *Thromb. Res.* **30**:469-(4/20) (1983).

Meisheri, K.D., O. Hwang and C. Van Breemen: Evidence for Two Separate Ca^{2+} Pathways in Smooth Muscle Plasmalemma. *J.Membr.Biol.* **59**:19-(2/10) (1981).

Meissner, G., E. Darling and J. Eveleth: Kinetics of Rapid Ca^{2+} Release by Sarcoplasmic Reticulum. Effects of Ca^{2+}, Mg^{2+}, and Adenine Nucleotides. *Biochem.* **25**:236-(2/10) (1986).

Meldolesi, J., P. Volpe and T. Pozzan: The Intracellular Distribution of Calcium. *TINS* **11**:449-(2/7) (1988).

Mellerup, E.T., P. Bech, T. Sorensen, A. Fuglsang-Frederiksen and O.J. Rafaelsen: Calcium and Electroconvulsive Therapy of Depressed Patients. *Biol.Psychiatr.* **14**:711-(4/22) (1979).

Meltzer, H.L. and S. Kassir: Abnormal Calmodulin-Activated Ca ATPase in Manic-Depressive Subjects. *J.Psychiatr.Res.* **17**:29-(4/22) (1983).

Mendez, G.C. and G.K. Moe: Demonstration of Dual AV Conduction System in the Isolated Heart. *Circulation* **19**:378-(4/19) (1966).

Menini, A., G. Rispoli and V. Torre: The Ionic Selectivity of the Light-Sensitive Current in Isolated Rods of the Tiger Salamander. *J.Physiol.* **402**:279-(2/8) (1988).

Merlie, J.P., K.E. Esenberg, S.D. Russel and J.R. Sanes: Denervation Supersensitivity in Skeletal Muscle: Analysis with a Cloned cDNA Probe. *J.Cell Biol.* **99**:332-(2/10) (1984).

Messerli, F.H., U.R. Kaesser and C.J. Losem: Effects of Antihypertensive Therapy on Hypertensive Heart Disease. *Circulation* **80**:145-(3/13) (1989).

Mestre, M., T. Carriot, C. Belin, A. Uzan, C. Renault, M.C. Dubroeucq, C. Gueremy, A. Doble and G. Le Fur: Electrophysiological and Pharmacological Evidence that Peripheral Type Benzodiazepine Receptors are Coupled to Calcium Channels in the Heart. *Life Sci.* **36**:391-(4/22) (1985).

Meyer, F.B.: Calcium, Neuronal Hyperexcitability and Ischemic Injury. *Brain Res.Rev.* **14**:227-(3/13) (1989).

Meyer, F.B., R.E. Anderson and T.M. Sundt: Anticonvulsant Properties of Dihydropyridine Calcium Antagonists. In *The Calcium Channel: Structure, Function, and Implications*. M. Morad, W. Nayler, S. Kazda, and M. Schramm, eds. Springer-Verlag, Berlin and Heidelberg. 503-(3/13) (1988).

Meyer, F.B., R.E. Anderson, T.L. Yaksh and T.M. Sundt: Effect of Nimodipine on Intracellular Brain pH, Cortical Blood Flow and EEG in Experimental Focal Cerebral Ischemia. *J.Neurosurg.* **64**:617-(3/13) (1986).

Meyer, H., E. Wehinger, F. Bossert, H. Boshagen, G. Franckowiak, S. Goldman, W. Seidel and J. Stoltefuss: Chemistry of Dihydropyridines [1-4]. In *Cardiovascular Effects of Dihydropyridine-Type Calcium Antagonists and Agonists* (Bayer Symposium IX. A. Fleckenstein, C. Van Breemen, R. Gross, and F. Hoffmeister, eds. Springer-Verlag, Berlin and Heidelberg. 90-(3/13) (1985).

Meyers, D.E.R. and J.L. Barker: Whole-Cell Patch-Clamp Analysis of Voltage-Dependent Calcium Conductances in Cultured Embryonic Rat Hippocampal Neurons. *J.Neurophysiol.* **61**:467-(2/9) (1989).

Midtbo, K.: Verapamil in the Prophylactic Treatment of Paroxysmal Supraventricular Tachycardia. *Curr. Ther.Res.* **30**:372-(4/19) (1981b).

Midtbo, K.: Intravenous Verapamil in the Treatment of Supraventricular Tachyarrhythmias in the Elderly. *Curr.Ther.Res.* **30**:378-(4/19) (1981a).

Mikami, A., K. Imoto, T. Tanabe, T. Niidome, Y. Mori, H. Takeshima, S. Narumiya and S. Numa: Primary Structure and Functional Expression of the Cardiac Dihydropyridine-sensitive Calcium Channel. *Nature* **340**:230-(1/3,3/13,3/14) (1989).

Milbrandt, J.: Nerve Growth Factor Rapidly Induces c-Fos mRNA in PC12 Rat Pheochromocytoma Cells. *Proc.Natl.Acad.Sci.U.S.A.* **83**:4789-(2/10) (1986).

Milbrandt, J.: A Nerve Growth Factor-Induced Gene Encodes a Possible Transcriptional Regulatory Factor. *Science* **238**:797-(2/10) (1987).

Milbrandt, J.: Nerve Growth Factor Induces a Gene Homologous to the Glucocorticoid Receptor Gene. *Neuron* **1**:183-(2/10) (1988).

Miledi, R.: Transmitter Release Induced by Injection of Calcium Ions into Nerve Terminals. *Proc.R.Soc.London B* **183**:421-(2/7) (1973).

Miledi, R. and I. Parker: Calcium Transients Recorded with Arsenazo III in the Presynaptic Terminal of the Squid Giant Synapse. *Proc.R.Soc.London B* **212**:197-(2/7) (1981).

Miledi, R., I. Parker and G. Shalow: Transmitter Induced Calcium Entry Across the Post-Synaptic Membrane at Frog End-Plates Measured Using Arsenazo III. *J.Physiol.* **300**:197-(2/9) (1980).

Miledi, R., I. Parker and P.H. Zhu: Calcium Transients Studied under Voltage-Clamp Control in Frog Twitch Muscle Fibres. *J.Physiol.* **340**:649-(2/10) (1983).

Miller, D.L. and J.I. Korenbrot: Kinetics of Light-Dependent Ca Fluxes Across the Plasma Membrane of Rod Outer Segments. A Dynamic Model of the Regulation of Cytoplasmic Ca Concentration. *J.Gen.Physiol.* **90**:397-(2/8) (1987).

Miller, J.L. and B.J. Litman: The Amplification of Light Initiated Cyclic GMP Hydrolysis in Preparations Derived from Bovine Rod Outer Segments is Modulated by Calcium and Magnesium Ions. *J.Biol.Chem.*(submitted) (1989).

Miller, R.J.: Multiple Calcium Channels and Neuronal Function. *Science* **235**:46-(1/4,2/7,2/9,3/13,3/14) (1987).

Miller, R.J.: Calcium Channels in Neurones. In *Structure and Physiology of the Slow Inward Calcium Channel*. D.J. Triggle and J.C. Venter, eds. Alan R. Liss, New York. 161-(3/13) (1986).

Miller, R.J. and A.P. Fox: Voltage-Sensitive Calcium Channels. In *Intracellular Calcium Regulation*. F. Bronner, ed. (in press) (3/14) (1989).

Mines, G.R.: On Functional Analysis by the Action of Electrolytes. *J.Physiol.* **46**:188-(2/10) (1913).

Mishina, M., T. Takai, D. Imoto, M. Noda, T. Takahashi, S. Numa, C. Methfessel and B. Sakmann: Molecular Distinction Between Fetal and Adult Forms of Muscle Acetylcholine Receptor. *Nature* **321**:406-(2/10) (1986).

Mitchell, M.R., T. Powell, D.A. Terrar and V.W. Twist: Influence of a Change in Stimulation Rate on Action Potentials, Currents, and Contractions in Rat Ventricular Cells. *J.Physiol.* **364**:113-(2/10) (1985).

Mitra, R. and M. Morad: Two Types of Calcium Channels in Guinea Pig Ventricular Myocytes. *Proc.Natl.Acad.Sci.U.S.A.* **83**:5340-(2/10) (1986).

Mitsuhashi, T., R.C. Morris and H.E. Ives: Endothelin-Induced Increases in Vascular Smooth Muscle Ca^{2+} do not Depend on Dihydropyridine-Sensitive Ca^{2+} Channels. *J.Clin.Invest.* **84**:635-(3/13) (1989).

Moczydlowski, E., B.M. Olivera, W.R. Gray and G.R. Strichartz: Discrimination of Muscle and Neuronal Na^+-Channel Subtypes by Binding Competition between ^{3}H-saxotoxin and ω-Conotoxins. *Proc.Natl.Acad.Sci.U.S.A.* **83**:5321-(3/14) (1986).

Mogul, D.J. and A.P. Fox: Nimodipine Exhibits Voltage-Dependent Agonist and Antagonist Effects on Ca Current in Hippocampal CA3 Neurons. *Biophys.J.* **57**:304a-(3/13) (1990).

Molinari, A., L. Guarneri, E. Pacei, F. De Marchi, G. Cerletti and G. De Gaetano: Mouse Antithrombotic Assay: The Effects of Ca^{2+} Channel Blockers are Platelet-Independent. *J.Pharmacol.Exp.Ther.* **240**:623-(4/20) (1987).

Moll, M.G., J.M. Dominguez and D. Obrador: Nifedipine (N) vs. Propranolol (P) in Unstable Angina (UA): A Prospective Randomized Study. *Eur.Heart J.* **5**:238-(4/16) (1984).

Moncada, S. and J.R. Vane: Unstable Metabolites of Arachidonic Acid and Their Role in Haemostasis and Thrombosis. *Br.Med.Bullet.* **34**:129-(4/20) (1978).

Montminy, M., K. Sevarino, J. Wagner, G. Mandel and R. Goodman: Identification of a Cyclic AMP-Responsive Element Within the Rat Somatostatin Gene. *Proc.Natl.Acad.Sci.U.S.A.* **83**:6682-(2/10) (1986).

Moody-Corbett, F. and M.W. Cohen: Localization of Cholinesterase at Sites of High Acetylcholine Receptor Density on Embryonic Amphibian Muscle Cells Cultured without Nerve. *J.Neurosci.* **1**:596-(2/10) (1981).

Moody-Corbett, F. and M.W. Cohen: Influence of Nerve on the Formation and Survival of Acetylcholine Receptor and Cholinesterase Patches on Embryonic *Xenopus* Muscle Cells in Culture. *J. Neurosci.* **2**:633-(2/10) (1982).

Moore, J.B., B.L. Fuller, R. Falotico and E.L. Tolman: Inhibition of Rabbit Platelet Phosphodiesterase Activity and Aggregation by Calcium Channel Blockers. *Thromb.Res.* **40**:401-(4/20) (1985).

Morad, M., N.W. Davies, J.H. Kaplan and H.D. Lux: Inactivation and Block of Calcium Channels by Photo-released Ca^{2+} in Dorsal Root Ganglion Neurons. *Sci. U.S.A.* **241**:842-(1/4) (1988).

Morad, M., Y.E. Goldman and D.R. Trentham: Rapid Photochemical Inactivation of Ca^{2+}-Antagonists Shows that Ca^{2+} Entry Directly Activates Contraction in Frog Heart. *Nature* **304**:635-(2/10) (1983).

Morad, M., W. Nayler, S. Kazda and M. Schramm: *The Calcium Channel: Structure, Function and Implications.* Springer-Verlag, Berlin and Heidelberg. (1/4,3/13) (1988a).

Morady, F., W.H. Kou, A.H. Kadish, S.D. Nelson, L.K. Toivonen, J.A. Kishner, S. Schalmtz and M. DeBuitleir: Antagonism of Quinidine's Electrophysiologic Effect by Epinephrine in Patients with Ventricular Tachycardia. *J.Am.Coll.Cardiol.* **12**:388-(4/19) (1988).

Moran, J. and A.J. Patel: Stimulation of the N-Methyl-D-Aspartate Receptor Promotes the Biochemical Differentiation of Cerebellar Granule Neurons and Not Astrocytes. *Brain Res.* **486**:15-(2/10) (1989a).

Moran, J. and A.J. Patel: Effect of Potassium Depolarization on Phosphate-activated Glutaminase Activity in Primary Cultures of Cerebellar Granule Neurons and Astroglial Cells During Development. *Dev.Brain Res.* **46**:97-(2/10) (1989b).

Morgan, J.I. and T. Curran: Role of Ion Flux in the Control of c-fos Expression. *Nature* **322**:552-(2/10) (1986).

Morgan, J.I. and T. Curran: Calcium as a Modulator of the Immediate-Early Gene Cascade in Neurons. *Cell Calcium* **9**:303-(2/10) (1988).

Morgan, J.P. and J.R. Blinks: Intracellular Ca^{2+} Transients in the Cat Papillary Muscle. *Can.J.Physiol.Pharmacol.* **60**:524-(2/10) (1982).

Morgan, J.P., W.G. Wier, P. Hess and J.R. Blinks: Influence of Ca^{2+}-Channel Blocking Agents on Calcium Transients and Tension Development in Isolated Mammalian Heart Muscle. *Circ.Res.* **52**:47-(2/10) (1983).

Morganroth, J., C.C. Chen, S. Sturn and L.S. Dreifus: Oral Verapamil in the Treatment of Atrial Filbrillation/Flutter. *Am.J.Cardiol.* **49**:981-(4/19) (1982).

Moriyama, T., S. Takamura, S. Narita, K. Tanaka, T. Matsuura and M. Kito: Dihydropyridine-Sensitive and Insensitive Ca^{2+} Channels in Human Platelets. *J.Biochem.* **104**:875-(3/13) (1988).

Morrison, I.M., G.S. Bajwa and R.B. Alfin-Slater: Prevention of Vascular Lesions in the Coronary Artery and Aorta of Rat Induced by Hypervitaminosis D and Cholesterol-Containing Diet. *Atherosclerosis* **16**:105-(4/20) (1987).

Moron, M.A., C.W. Stevens and T.L. Yaksh: Diltiazem Enhances and Flunarizine Inhibits Nimodipine's Antiseizure Effects. *Eur.J.Pharmacol.* **163**:299-(3/13) (1989).

Motulsky, H.J., M.D. Snavely, R.J. Hughes and P.A. Insel: Interaction of Verapamil and Other Calcium Channel Blockers with 1- and 2-Adrenergic Receptors. *Circ.Res.* **52**:226-(4/20) (1983).

Moy, G.W., G.S. Kopf, C. Gache and V.D. Vacquier: Calcium-Mediated Release of Glucanase Activity from Cortical Granules of Sea Urchin Eggs. *Dev.Biol.* **100**:267-(2/7) (1983).

Mueller, H.S. and R.A. Chahine: Interim Report of Multicenter Double-Blind, Placebo-Controlled Studies of Nifedipine in Chronic Stable Angina. *Am.J.Med.* **7**:645-(4/16) (1981).

Muiesan, G., E. Agabiti-Rosei, G. Romanelli, M.L. Muiesan, M. Castellano and M. Beschi: Adrenergic Activity and Left Ventricular Function During Treatment of Essential Hypertension with Calcium Antagonists. *Am.J.Cardiol.* **57**:44D-(4/16) (1986).

Muller, J.E., J. Morrison, P.H. Stone, R.E. Rude, B. Rosner, R. Roberts, D.L. Pearle, Z.G. Turi, J.F. Schneider, D.H. Serfas, C. Tate, E. Scheiner, B.E. Sobel, C.H. Hennekens and E. Braunwald: Nifedipine Therapy for Patients with Threatened and Acute Myocardial Infarction: A Randomized Double-Blind Placebo-Controlled Comparison. *Circulation* **69**:740-(4/16) (1984).

Multicenter Diltiazem Post-Infarction Trial Research Group: The Effect of Diltiazem on Mortality and Reinfarction after Myocardial Infarction. *N.Engl.J.Med.* **319**:385-(4/16) (1988).

Murakami, H., E. Kaji and T. Segawa: Influence of Verapamil, X-537A, A-23187 and Adenosine 3', 5'-Cyclic Monophosphate on Release of 5-Hydroxytryptamine from Rat Brain Slices. *Jpn.J.Pharmacol.* **28**:589-(4/22) (1978).

Murakami, M., Y. Shimoda, K. Nakatani, E. Miyachi and S. Watanabe: GABA-Mediated Negative Feedback from Horizontal Cells to Cones in Carp Retina. *Jpn.J.Physiol.* **32**:911-(2/8) (1982).

Murakami, T., O.M. Hess and H.P. Krayenbuehl: Left Ventricular Function Before and After Diltiazem in Patients with Coronary Artery Disease. *J.Am.Coll.Cardiol.* **5**:723-(4/16) (1985).

Murase, K. and M. Randic: Electrophysiological Properties of Rat Spinal Dorsal Horn Neurones *In Vitro*: Calcium Dependent Action Potentials. *J.Physiol.* **334**:141-(2/9) (1983).

Murayama, K. and N. Lakshminarayanaiah: Some Electrical Properties of the Membrane of the Barnacle Muscle Fibers under Internal Perfusion. *J.Membr.Biol.* **35**:257-(3/12) (1977).

Murphy, B.J., C.A. Rogers, R.K. Sunahara, S. Lemaire and B.S. Tuana: Identification, Characterization, and Photoaffinity Labeling of the Dihydropyridine Receptor Associated with the L-Type Calcium Channel from Bovine Adrenal Medulla. *Mol.Pharmacol.* **37**:173-(3/13) (1990).

Murphy, K.M.M., R.J. Gould and S.H. Snyder: Autoradiographic Visualization of [^{3}H]Nitrendipine Binding Sites in Rat Brain: Localization to Synaptic Zones. *Eur.J.Pharmacol.* **81**:517-(4/22) (1982).

Murphy, V.A., Q.R. Smith and S.I. Rapoport: Homeostasis of Brain and Cerebrospinal Fluid Calcium Concentrations During Chronic Hypo- and Hypercalcemia. *J.Neurochem.* **47**:1735-(4/21) (1986).

Mustard, J.F. and M.A. Packham: Factors Influencing Platelet Function: Adhesion, Release and Aggregation. *Pharmacol.Rev.* **22**:97-(4/20) (1970).

Mutoh, S. and I. Yamaguchi: Nilvadipine, a Potent Calcium Antagonist, Exerts Antiatherosclerotic Activity Through Inhibition of Vascular Smooth Muscle Cell Migration. *J.Pharmacodyn.* **12**:5-(3/13) (1989).

Nabauer, M., G. Callewaert, L. Cleemann and M. Morad: Depolarization-Induced Ca^{2+} Release in Single Mammalian Cardiac Myocytes Requires Ca^{2+} Influx Through the Ca^{2+} Channel. *Biophys.J.* **55**:488a-(2/10) (1989).

Nachman, R.L. and B.B. Weksler: The Platelet as an Inflammatory Cell. In *The Cell Biology of Inflammation*. G. Weissman, ed. Elsevier, Amsterdam. 145-(4/20) (1980).

Nachshen, D.A.: Selectivity of the Ca Binding Site in Synaptosome Ca Channels Inhibition of Ca Influx by Multivalent Metal Cations. *J.Gen.Physiol.* **83**:941-(3/12) (1984).

Nachshen, D.A. and M.P. Blaustein: Regulation of Nerve Terminal Calcium Channel Selectivity by a Weak Acid Site. *Biophys.J.* **26**:329-(3/12) (1979).

Nachshen, D.A. and M.P. Blaustein: Some Properties of Potassium Stimulated Calcium Influx in Presynaptic Nerve Endings. *J.Gen.Physiol.* **76**:709-(3/12) (1980).

Nachshen, D.A. and M.P. Blaustein: Influx of Calcium, Strontium and Barium in Presynaptic Nerve Endings. *J.Gen.Physiol.* **79**:1065-(2/7,3/12) (1982).

Nahas, G., R. Trouve, C. Latour, M. Sitbon and J.F. Demus: Nitrendipine: An Antidote to the Lethal Toxicity of Cocaine. *Ann.N.Y.Acad.Sci.* **522**:796-(3/13,4/15) (1988).

Nahorski, S.R.: Inositol Polyphosphates and Neuronal Calcium Homeostasis. *TINS* **11**:444-(4/21) (1988).

Naito, M., F. Kuzuya and K. Asai: Ineffectiveness of Ca^{2+} Antagonist Nicardipine and Diltiazem on Experimental Atherosclerosis in Cholestereol-Fed Rabbits. *Angiology* **35**:622-(4/20) (1984).

Naito, M. and T. Tsuruo: Competitive Inhibition by Verapamil of ATP-Dependent High Affinity Vincristine Binding to the Plasma Membrane of Multidrug-Resistant K562 Cells without Calcium Ion Involvement. *Cancer Res.* **49**:1452-(3/13) (1989).

Naitoh, Y., R. Eckert and K. Friedman: A Regenerative Calcium Response in Paramecium. *J.Exp.Biol.* **56**:667-(2/7) (1972).

Nakamura, T. and G.H. Gold: A Cyclic Nucleotide-Gated Conductance in Olfactory Receptor Cilia. *Nature* **325**:442-(2/8) (1987).

Nakatani, K. and K.W. Yau: Calcium and Light Adaptation in Retinal Rods and Cones. *Nature* **334**:69-(2/8) (1988b).

Nakatani, K. and K.W. Yau: Calcium and Magnesium Fluxes Across the Plasma Membrane of the Toad Rod Outer Segment. *J.Physiol.* **395**:695-(2/8) (1988a).

Nakayama, H., M.D. Ginsberg and W.D. Dietrich: (S)-Emopamil, a Novel Calcium Channel Blocker and Serotonin S2 Antagonist, Markedly Reduces Infarct Size Following Middle Cerebral Artery Occlusion in the Rat. *Neurology* **38**:1667-(4/21) (1988).

Nakayama, N., T. Kirley, P.L. Vaghy, E. McKenna and A. Schwartz: Purification of a Putative Ca^{2+} Channel Protein from Rabbit Skeletal Muscle: Determination of the Amino-Terminal Sequence. *J.Biol.Chem.* **262**:6572-(2/10) (1987).

Narahashi, T., A. Tsunoo and M. Yoshii: Characterization of Two Types of Calcium Channels in Mouse Neuroblastoma Cells. *J.Physiol.* **383**:231-(1/2,1/4) (1987).

Nastainczyk, W., W. Rohrkasten, M. Sieber, C. Rudolph, C. Schachtele, D. Marme and F. Hofmann: Phosphorylation of the Purified Receptor for Calcium Channel Blockers by cAMP Kinase and Protein Kinase C. *Eur.J.Biochem.* **169**:137-(3/14) (1987).

Naukkarinen, V.A., K. Karppinin and S. Sarma: Comparison of Nicardipine and Propanolol in the Treatment of Mild and Moderate Hypertension. *Eur.J.Clin.Pharmacol.* **33**:119-(4/18) (1987).

Navarette, R. and K.D. Walton: Calcium Conductances Trigger Doublet Firing in Neonatal Rat Motoneurones. *J.Physiol.* **415**:73P-(2/9) (1989).

Nayler, W.G.: The Pharmacological Protection of the Ischemic Heart: The Use of Calcium Antagonists and Beta-Adrenocepter Antagonists. *Eur.Heart J.* **1**:5-(4/19) (1980).

Nayler, W.G.: Protection of the Myocardium against Post-Ischemic Reperfusion Damage. *J.Thorac.Cardiovasc.Surg.* **84**:897-(4/16,4/19) (1982).

Nayler, W.G.: *Calcium Antagonists*. Academic Press, New York. (3/13) (1988).

Nayler, W.G., R. Ferrari and A. Williams: Protective Effect of Pretreatment with Verapamil, Nifedipine and Propranolol on Mitochondrial Function in the Ischemic and Reperfused Myocardium. *Am.J.Cardiol.* **46**:242-(4/16,4/19) (1980).

Nayler, W.G., A. Grau and A. Slade: A Protective Effect of Verapamil on Hypoxic Heart Muscle. *Cardiovasc.Res.* **10**:650-(4/16) (1976).

Nayler, W.G., S. Panagiotopoulos, J.S. Elz and W.J. Sturrock: Fundamental Mechanisms of Action of Calcium Antagonists in Myocardial Ischemia. *Am.J.Cardiol.* **59**:75B-(4/16) (1987).

Nayler, W.G: Review: Calcium Antagonists and the Ischemic Myocardium. *Int.J.Cardiol.* **15**:267-(4/16) (1987).

Naylor, G.J., D.A.T. Dick and E.G. Dick: Erythrocyte Membrane Cation Carrier in Depressive Illness. *Psychol.Med.* **3**:502-(4/22) (1973).

Nedergaard, S., J.P. Bolam and S.A. Greenfield: Facilitation of a Dendritic Calcium Conductance by 5-Hydroxytryptamine in the Substantia Nigra. *Nature* **333**:174-(3/14) (1988).

Neher, E.: Concentration Profiles of Intracellular Calcium in the Presence of a Diffusible Chelator. *Exp.Brain Res.* **14**:80-(1/4) (1986).

Neher, E.: The Influence of Intracellular Calcium Concentration on Degranulation of Dialysed Mast Cells from Rat Peritoneum. *J.Physiol.* **395**:193-(2/7) (1988).

Neher, E. and W. Almers: Fast Calcium Transients in Rat Peritoneal Mast Cells are not Sufficient to Trigger Exocytosis. *EMBO J.* **5**:51-(2/7) (1986).

Nelson, M.T. and Worley J.F.: Dihydropyridine Inhibition of Single Calcium Channels and Contraction in Rabbit Mesenteric Artery Depends on Voltage. *J.Physiol.* **412**:65-(3/13) (1989).

Nelson, M.T.: Interactions of Divalent Cations with Single Calcium Channels from Rat Brain Synaptosomes. *J.Gen.Physiol.* **87**:201-(3/12) (1986).

Nelson, M.T., R.J. French and B.K. Krueger: Voltage Dependent Calcium Channels from Brain Incorporated into Planar Lipid Bilayers. *Nature* **308**:77-(3/12) (1984).

Nelson, M.T., N.B. Standen, J.E. Brayden and J.F. Worley: Noradrenaline Contracts Arteries by Activating Voltage Dependent Calcium Channels. *Nature* **336**:382-(2/10,3/13,3/14) (1988).

Nemeth, E.F. and A. Scarpa: Rapid Mobilization of Cellular Ca^{2+} in Bovine Parathyroid Cells Evoked by Extracellular Divalent Cations; Evidence for a Cell Surface Calcium Receptor. *J.Biol.Chem.* **262**:5188-(2/7) (1987).

New, W. and W. Trautwein: The Ionic Nature of Slow Inward Current and its Relation to Contraction. *Pflügers Arch.* **334**:24-(2/10) (1972).

Nicholson, C., G.T. Bruggencate, R. Steinberg and H. Stockle: Calcium Modulation in Brain Extracellular Microenvironment Demonstrated with Ion-Selective Micropipette. *Proc.Natl.Acad.Sci.U.S.A.* **74**:1287-(4/21) (1977).

Nicholson, J.P., L.M. Resnick and J.H. Laragh: The Antihypertensive Effect of Verapamil at Extremes of Sodium Intake. *Ann.Int.Med.* **107**:329-(4/18) (1987).

Nicol, G.D., D. Atwell and F.S. Werblin: Membrane Potential Affects Photocurrent Kinetics in Salamander Rods and Cones. *Brain Res.* **297**:164-(2/8) (1984).

Nicol, G.D., K.B. Kaupp and M.D. Bownds: Transduction Persists in Rod Photoreceptors after Depletion of Intracellular Calcium. *J.Gen.Physiol.* **89**:297-(2/8) (1987).

Nicola Siri, L., J.A. Sanchez and E. Stefani: Effect of Glycerol Treatment on the Calcium Current of Frog Skeletal Muscle. *J.Physiol.* **305**:87-(2/10) (1980).

Nicotera, P., D.J. McConkey, D.P. Jones, and S. Orrenius: ATP Stimulates Ca^{2+} Uptake and Increases the Free Ca^{2+} Concentration in Insulated Rat Liver Nuclei. *Proc. Natl. Acad. Sci. U.S.A.* **86**:453-(2/10) (1989).

Nieoullon, A., A. Cheramy and J. Glowinski: Release of Dopamine *In Vivo* from Cat Substantia Nigra. *Nature* **266**:375-(2/7) (1977).

Nieuwkoop, P.D. and J. Faber: *Normal Table of Xenopus Laevis (daudin): A Systemmatical and Chronological Survey of the Development from the Fertilized Egg Till the End of Metamorphosis.* 2nd edition, North Holland Publ., Amsterdam. (2/10) (1967).

Nilius, B., P. Hess, J.B. Lansman and R.W. Tsien: A Novel Type of Cardiac Calcium Channel in Ventricular Cells. *Nature* **316**:443-(1/2,1/4,2/10,2/9) (1985).

Nilsson, H., T. Jonasson and I. Ringqvist: Treatment of Digital Vasospastic Disease with the Calcium Entry Blocker Nifedipine. *Acta Med.Scand.* **215**:13-(4/20) (1984).

Nilsson, J., M. Solund, L. Palmberg, A.M. Von Euler, B. Jonzon and J. Thyberg: The Calcium Antagonist Nifedipine Inhibits Arterial Smooth Muscle Cell Proliferation. *Atherosclerosis* **58**:109-(4/20) (1985).

Nishi, R. and D.K. Berg: Effects of High K^+ Concentrations on the Growth and Development of Ciliary Ganglion Neurons in Cell Culture. *Dev.Biol.* **87**:301-(2/10) (1981).

Nishikibe, M.: Effect of the Calcium Entry Blocker NB-818 on Cerebral Blood Flow after Unilateral Carotid Occlusion in the Mongolian Gerbil. *J.Pharmacol.Exp.Ther.* **246**:719-(3/13) (1988).

Nishimura, M., C.M. Follmer and D.H. Singer: Amiodarone Blocks Calcium Current in Single Guinea Pig Ventricular Myocytes. *J.Pharmacol.Exp.Ther.* **251**:650-(3/13) (1989).

Nishino, N., S.A. Noguchi-Kuno, T. Sugiyama and C. Tanaka: [^{3}H]Nitrendipine Binding Sites are Decreased in the Substantia Nigra and Striatum of the Brain from Patients with Parkinson's Disease. *Brain Res.* **377**:186-(3/13) (1986).

Nishizuka, Y.: The Role of Protein Kinase C in Cell Surface Signal Transduction and Tumor Promotion. *Nature* **308**:693-(3/14) (1984).

Nishizuka, Y.: Studies and Perspectives of Protein Kinase C. *Science* **233**:305-(3/14) (1986).

Nobile, M., E. Carbone, H.D. Lux and H. Zucker: Temperature Sensitivity of Ca Currents in Chick Sensory Neurons. *Pflügers Arch.* **415**:658-(1/3,1/4) (1990).

Noble, D.: The Surprising Heart: A Review of Recent Progress in Cardiac Electrophysiology. *J.Physiol.* **353**:1-(2/10) (1984).

Noda, M., S. Shimizu, T. Tanabe, T. Takai, T. Kayano, T. Ikeda, H. Takahashi, H. Nakayama, Y. Kanaoka, N. Minamino, K. Kangawa, H. Matsuo, M. Raftery, T. Hirose, S. Inayama, H. Hayashida, T. Miyata and S. Numa: Primary Structure of *Electrophorus Electricus* Sodium Channel Deduced from cDNA Sequence. *Nature* **312**:121-(1/1) (1984).

Nogae, I., K. Kohno, J. Kikuchi, M. Kuwano, S.I. Akiyama, A. Kiue, K.I. Suzuki, Y. Yoshida, M. Cornwell, I. Pastan and M.M. Gottesman: Analysis of Structural Features of Dihydropyridine Analogs Needed to Reverse Multidrug Resistance and to Inhibit Photoaffinity Labeling of P-Glycoprotein. *Biochem.Pharmacol.* **38**:519-(3/13) (1989).

Nokin, P., M. Clinet, P. Polster, P. Beaufort, L. Meysmans, J. Gougat and P. Chatelain: SR33557, a Novel Calcium-Antagonist: Interaction with [^{3}H]-(-)-Desmethoxy-Verapamil Binding Sites in Cerebral Membranes. *Naunyn Schmied.Arch.Pharmacol.* **339**:31-(3/13) (1989).

Noll, G., H. Stieve and J. Winterhager: Interaction of Bovine Rhodopsin with Calcium Ions II: Calcium Release in Bovine Rod Outer Segments Upon Bleaching. *Biophys.Struct.Mech.* **5**:43-(2/8) (1979).

Nomoto, A., S. Mutoh, H. Hagihara and I. Yamaguchi: Smooth Muscle Cell Migration Induced by Inflammatory Cell Products and its Inhibition by a Potent Calcium Antagonist, Nilvadipine. *Atherosclerosis* **72**:213-(3/13) (1988).

Nordmann, J.J.: Stimulus-Secretion Coupling. *Prog.Brain Res.* **60**:281-(2/7) (1983).

Nordmann, J.J., G. Dayanithi and J.R. Lemos: Isolated Neurosecretory Nerve Endings as Tool for Studying the Mechanism of Stimulus-Secretion Coupling. *Biosci.Rep.* **7**:411-(2/7) (1987).

Norman, R.A. and F.S. Werblin: Control of Retinal Sensitivity. Light and Dark Adaptation of Vertebrate Rods and Cones. *J.Gen.Physiol.* **63**:37-(2/8) (1974).

Norwegian Nifedipine Group: Nifedipine in Acute Myocardial Infarction. No Influence on Enzymatic Infarct Size. *Acta Med.Scan.Abstr.* **68**:22-(4/19) (1983).

Nosek, T.M.: How Valproate and Phenytoin Affect the Ionic Conductances and Active Transport Characteristics of the Crayfish Giant Axon. *Epilepsia* **22**:651-(4/22) (1981).

Nowak, L., P. Bregestovski, P. Ascher, A. Herbet and A. Prochiantz: Magnesium Gates Glutamate-Activated Channels in Mouse Central Neurones. *Nature* **307**:462-(2/9,4/21) (1984).

Nowycky, M.C., A.P. Fox and R.W. Tsien: Long-Opening Mode of Gating of Neuronal Calcium Channels and its Promotion by the Dihydropyridine Calcium Agonist Bay K 8644. *Proc.Natl.Acad.Sci.U.S.A.* **82**:2178-(1/4,3/14) (1985b).

Nowycky, M.C., A.P. Fox and R.W. Tsien: Three Types of Neuronal Calcium Channel with Different Calcium Agonist Sensitivity. *Nature* **316**:440-(1/4,2/9,3/14) (1985a).

Nuglisch, J., C. Karkoutly, D. Sauer, H.D. Mennel, C. Robberg and J. Krieglstein: Protective Effect of Nimodipine Against Neuronal Damage Caused by Ischemia. In *Pharmacology of Cerebral Ischemia*. CRC Press Inc., Boca Raton, Florida. 87-(3/13) (1989).

Nunn, B.J. and D.A. Baylor: Visual Transduction in Retinal Rods of the Monkey *Macaca fascicularis*. *Nature* **299**:726-(2/8) (1982).

Nunoki, K., V. Florio and W.A. Catterall: Activation of Purified Calcium Channels by Stoichiometric Protein Phosphorylation. *Proc.Natl.Acad.Sci.U.S.A.* **86**:6816-(3/13) (1989).

Nurden, A.T. and J.P. Caen: An Abnormal Platelet Plasma Membrane Glycoprotein. *J.Biol.Chem.* **252**:2121-(4/20) (1974).

Nyakas, C., E. Markel, R.J.K. Kramers, E. Gaspar, B. Bohus and P.G.M. Luiten: Nimodipine and Central Nervous System Function. In *New Vistas*. J. Traber and W.H. Gespen, eds. Schattauer, Stuttgart and New York. 175-(3/13) (1989).

O'Bryan, P.M.: Properties of the Depolarizing Synaptic Potential Evoked by Peripheral Illumination in Cones of the Turtle Retina. *J.Physiol.* **235**:207-(2/8) (1973).

O'Dowd, D.K.: Development of Voltage-Dependent Conductances in Cultured Amphibian Spinal Neurons. *Soc.Neurosci.Abs.* **9**:506-(2/10) (1983a).

O'Dowd, D.K.: RNA Synthesis Dependence of Action Potential Development in Spinal Cord Neurones. *Nature* **303**:619-(2/10) (1983b).

O'Dowd, D.K., A.B. Ribera and N.C. Spitzer: Development of Voltage-Dependent Calcium, Sodium and Potassium Currents in *Xenopus* Spinal Neurons. *J.Neurosci.* **8**:792-(2/10) (1988).

O'Gara, P.T., R.O. Bonow and B.J. Maron: Myocardial Perfusion Abnormalities in Patients with Hypertrophic Cardiomyopathy: Assessment with Thallium-201 Emission Computed Tomography. *Circulation* **76**:1214-(4/17) (1987).

O'Neill, S.K. and G.T. Bolger: Phencyclidine and MK-801: A Behavioral and Neurochemical Comparison of their Interactions with Dihydropyridine Calcium Antagonists. *Brain Res.Bull.* **22**:611-(3/13) (1989).

O'Rourke, F.A., S.P. Halenda, G.B. Zavoico and M.B. Feinstein: Inositol 1,4,5-Trisphosphate Releases Ca^{2+} from a Ca^{2+}-Transporting Membrane Vesicles Fraction Derived from Human Platelets. *J.Biol.Chem.* **260**:956-(4/20) (1985).

O'Rourke, R.A.: Calcium Entry Blockers: Usefulness in Ischemic and Hypertensive Heart Disease. *Circulation* **80**:1-(3/13) (1989).

O'Sullivan, A.J., T.R. Cheek, R.B. Moreton, M.J. Berridge and R.D. Burgoyne: Localization and Heterogeneity of Agonist-Induced Changes in Cytosolic Calcium Concentration in Single Bovine Adrenal Chromaffin Cells from Video Imaging of Fura-2. *EMBO J.* **8**:401-(2/7) (1989).

Obaid, A.L., R. Flores and B.M. Salzberg: Calcium Channels that are Required for Secretion from Intact Nerve Terminals of Vertebrates are Sensitive to ω–Conotoxin and Relatively Insensitive to Dihydropyridines. *J.Gen.Physiol.* **93**:715-(2/7,3/13) (1989).

Ochi, R.: The Slow Inward Current and the Action of Manganese Ions in Guinea Pig's Myocardium. *Pflügers Arch.* **316**:81-(1/2,3/12) (1970).

Ochi, R.: Manganese Action Potentials in Mammalian Cardiac Muscle. *Experientia* **31**:1048-(3/12) (1975).

Ochi, R.: Manganese Dependent Propagated Action Potentials and Their Depression by Electrical Stimulation in Guinea Pig Myocardium Perfused by Na-free Media. *J.Physiol.* **263**:139-(3/12) (1976).

Oetting, M., M.S. LeBoff, S. Levy, L. Swiston, J. Preston, C. Chen and E.M. Brown: Permeabilization Reveals Classical Stimulus-Secretion Coupling in Bovine Parathyroid Cells. *Endocrinology* **121**:1571-(2/7) (1987).

Ogata, N. and T. Narahashi: Differential Effects of Chlorpromazine on Sodium Channels and Two Types of Calcium Channels. *Biophys.J.* **51**:430a-(3/13) (1987).

Ogata, N., M. Yoshii and T. Narahashi: Blocking Action of Chlorpromazine on Two Types of Calcium Channels in Neuroblastoma Cells. *Soc.Neurosci.Abs.* **12**:1193-(3/13) (1986).

Ogawa, N., and H. Ono: Influence of: BAY K 8644 on Aequorin-Loaded Human Platelets. *Jpn.J.Pharmacol.* **51**:306-(3/13) (1989).

Ogura, A. and K. Takahashi: Artificial Deciliation Causes Loss of Calcium-Dependent Responses in Paramecium. *Nature* **264**:170-(2/7) (1976).

Ohmori, H.: Studies of Ionic Currents in the Isolated Vestibular Hair Cell of the Chick. *J.Physiol.* **350**:561-(2/8) (1984).

Ohmori, H.: Mechano-electrical Transduction Currents in Isolated Vestibular Hair Cells of the Chick. *J.Physiol.* **359**:189-(2/8) (1985).

Ohmori, H.: Mechanical Stimulation and Fura-2 Fluorescence in the Hair Bundle of Dissociated Hair Cells of the Chick. *J.Physiol.* **399**:115-(2/8) (1988).

Ohnishi, T., K. Saito, K. Matsumoto, M. Sakuda and K. Ishii: Changes in [^{3}H]Nitrendipine Binding and Gamma-Aminobutyric Acid Release in Rat Hippocampus Following Repeated Morphine Administration. *J.Neurochem.* **53**:1507-(3/13) (1989).

Ohtsuka, M., M. Yokota, I. Kodama, K. Yamada and S. Shibata: New Generation Dihydropyridine Calcium Entry Blockers: In Search of Greater Selectivity for one Tissue Subtype. *Gen.Pharmacol.* **20**:539-(3/13) (1989).

Oikawa, T., C.S. Spyropoulos, I. Tasaki and T. Teorell: Methods for Perfusing the Giant Axon of *Loligo pealii*. *Acta Physiol.Scand.* **52**:195-(3/12) (1961).

Okamoto, H., K. Takahashi and N. Yamashita: Ionic Currents through the Membrane of the Mammalian Oocyte and their Comparison with those in the Tunicate and Sea Urchin. *J.Physiol.* **267**:465-(3/12) (1977).

Okamoto, H., M. Takahashi and M. Yoshii: Two Components of the Calcium Current in the Egg Cell Membrane of the Tunicate. *J.Physiol.* **255**:527-(3/12) (1976).

Oldendorf, W.H., M.E. Cornford and W.J. Brown: The Large Apparent Work Capability of the Blood-Brain Barrier: A Study of the Mitochondrial Content of Capillary Endothelial Cells in Brain and other Tissues of the Rat. *Ann.Neurol.* **1**:409-(4/21) (1977).

Olivari, M.T., C. Bartorelli, A. Polese, C. Florentini, P. Moruzzi and M.D. Guazzi: Treatment of Hypertension with Nifedipine, a Calcium Antagonist Agent. *Circulation* **59**:1056-(4/18) (1979).

Olive, J.: The Structural Organization of Mammalian Disc Membranes. *Int.Rev.Cytol.* **64**:107-(2/8) (1980).

Olivera, B.M., W.R. Gray, R. Zeikus, J.M. McIntosh, L. Varga, J. Rivier, V. de Santos and L.J. Cruz: Peptide Neurotoxins from Fish Hunting Cone Snails. *Science* **230**:1338-(3/13) (1985).

Ono, H. and M. Kimura: Effect of Ca^{2+} Antagonistic Vasodilators, Dilitiazem, Nifedipine, Perhexcilin and Verapamil on Platelet Aggregation *In Vitro*. *Arzneim-Forsch.* **31**:1131-(4/20) (1981).

Ono, K., M. Delay, T. Nakajima, H. Irisawa and W. Giles: Calcitonin Gene-Related Peptide Regulates Calcium Current in Heart Muscle. *Nature* **340**:721-(3/14) (1989).

Onoda, J.M., K.K. Nelson, J.D. Taylor and K.V. Honn: *In Vivo* Characterization of Combination Chemotherapy with Calcium Channel Blockers and cis-Diaminedichloroplatinum(II). *Can.Res.* **49**:2844-(3/13) (1989).

Onuma, E.K. and S.W. Hui: Electric Field-Directed Cell Shape Changes, Displacement, and Cytoskeletal Reorganization are Calcium Dependent. *J.Cell Biol.* **106**:2067-(2/10) (1988).

Opie, L.H.: Metabolism of Free Fatty Acids, Glucose and Catecholamines in Acute Myocardial Infarction. Relation to Myocardial Ischemia and Infarct Size. *Am.J.Cardiol.* **36**:938-(4/16) (1975).

Opie, L.H.: Calcium Channel Antagonists. I. Fundamental Properties, Mechanisms, Classification, Sites of Action. *Cardiovasc.Drugs Ther.* **1**:411-(3/13) (1988a).

Opie, L.H.: Calcium Channel Antagonists. Part II. Use and Comparative Properties of the Three Prototypical Calcium Antagonists in Ischemic Heart Disease, Including Recommendations Based on an Analysis of 4l Trials. *Cardiovasc.Drugs Ther.* **1**:461-(4/16) (1988b).

Opie, L.H.: Calcium Channel Antagonists. Part V. Second-Generation Agents. *Cardiovasc.Drugs Ther.* **2**:191-(3/13,4/16) (1988c).

Opie, L.H. and W.A. Coetzee: Are Calcium Ions Involved in the Genesis of Early Ischemic Ventricular Arrhythmias? **In** *Life-Threatening Arrhythmias during Ischemia and Infarction*. D. Hearse, A. Manning, and M. Janse, eds. Raven Press, New York. 63-(4/16) (1987).

Opie, L.H., W.A. Coetzee, S.C. Dennis and F.T. Thandroyen: A Potential Role of Calcium Ions in Early Ischemic and Reperfusion Arrhythmias. *Ann.N.Y Acad.Sci.* **522**:464-(4/19) (1988).

Opie, L.H. and A. Maseri: Vasospastic Angina. **In** *Treatment of Cardiovascular Diseases by Adalat (Nifedipine)*. R. Krebs, ed. Schattauer, Stuttgart. 231-(4/16) (1986).

Opie, L.H., C. Muller, D. Nathan, P. Daries and W.F. Lubbe: Evidence for Role of Cyclic AMP as Second Messenger of Arrhythmogenic Effects of Beta-Stimulation. *Adv.Cycl.Nucleotide Res.* **12**:63-(4/16) (1980).

Opie, L.H., D. Nathan and W.F. Lubbe: Biochemical Aspects of Arrhythmogenesis and Ventricular Fibrillation. *Am.J.Cardiol.* **43**:131-(4/16) (1979).

Orchard, C.H. and E.G. Lakatta: Intracellular Calcium Transients and Developed Tension in Rat Heart Muscle. *J.Gen.Physiol.* **86**:637-(2/10) (1985).

Ortega, M.P., C. Sunkel, J.G. Priego and P.R. Statkow: The Antithrombogenic *In Vivo* Effects of Calcium Channel Blockers in Experimental Thrombosis in Mice. *Thromb.Haemost.* **57**:283-(4/20) (1987).

Owen, D.G., M. Segal and J.L. Barker: A Ca-Dependent Cl- Conductance is Present in Cultured Mouse Spinal Neurones. *Nature* **311**:567-(2/9) (1984).

Oyama, Y., Y. Tsuda, S. Sakakibara and N. Akaike: Synthetic ω-Conotoxin: A Potent Calcium Channel Blocking Neurotoxin. *Brain Res.* **242**:58-(3/13) (1987).

Pacaud, P., G. Loirand, L. Mironneau and J. Mironneau: Opposing Effects of Noradrenaline on the Two Classes of Voltage Dependent Calcium Channels of Single Vascular Smooth Muscle Cells in Short Term Primary Culture. *Pflügers Arch.* **410**:557-(3/14) (1987).

Packham, M.A.: Methods for Detection of Hypersensitive Platelets. *Thromb.Haemost.* **40**:175-(4/20) (1978).

Palade, P.T. and W. Almers: Slow Calcium and Potassium Currents in Frog Skeletal Muscle: Their Relationship and Pharmacologic Properties. *Pflügers Arch.* **405**:91-(1/2) (1985).

Palfreyman, M.G., M.W. Dudley, H.C. Cheng, A.K. Mir and S. Yamada: Lactamides: A Novel Chemical Class of Calcium Antagonists with Diltiazem-Like Properties. *Biochem.Pharmacol.* **38**:2459-(3/13) (1989).

Palileo, E.V., W.W. Ashley, S. Swiryn, R.A. Bauernfeind, B. Strasberg, A.T. Petrogoulos and K.M. Rosen: Exercise Provocable Right Ventricular Outflow Tract Tachycardia. *Am.Heart J.* **104**:185-(4/19) (1982).

Pallotta, B.S., K.L. Magleby and J.N. Barrett: Single Channel Recordings of Ca^{2+}-Activated K^+ Currents in Rat Muscle Cell Culture. *Nature* **293**:471-(3/14) (1981).

Palmer, C.G., E.W. Harris, R.K. Ray, M.L. Stagnitto and R.J. Schmiesing: Are There Correlates Between NMDA Antagonism and Prevention of MES and PTZ Seizures in Mice? *Winter Conference on Brain Research, Snowmass* (3/13) (1990).

Panagia, V., J.N. Singh, M.B. Anand-Srivastava, G.N. Pierce, G. Jasmin and N.S. Dhalla: Sarcolemmal Alterations During the Development of Genetically Determined Cardiomyopathy. *Cardiovasc.Res.* **18**:567-(4/17) (1984).

Pancrazio, J.J., M.P. Viglione and Y.I. Kim: Effects of BAY K 8644 on Spontaneous and Evoked Transmitter Release at the Mouse Neuromuscular Junction. *Neuroscience* **30**:215-(3/13) (1989).

Pandiella-Alonso, A., A. Malgaroli, L.M. Vicentini and J. Meldolesi: Early Rise of Cytosolic Ca^{2+} Induced by NGF in PC12 and Chromaffin Cells. *FEBS Lett.* **208**:48-(2/10) (1986).

Pannocchia, A., N. Praloran, C. Arduino, N. Della Dora, M. Bazzan, P. Schinco, M. Buraglio, A. Pileri and G. Tamponi: Absence of (-)[^{3}H]Desmethoxyverapamil Binding Sites on Human Platelets and Lack of Evidence for Voltage-Dependent Calcium Channels. *Eur.J.Pharmacol.* **142**:83-(3/13,4/20) (1987).

Panza, G., J.A. Grebb, E. Sanna, A.G. Wright and I. Hanbauer: Evidence for Down-Regulation of [^{3}H]Nitrendipine Recognition Sites in Mouse Brain after Long-Term Treatment with Nifedipine or Verapamil. *Neuropharmacology* **24**:1113-(3/13) (1985).

Paoletti, R., F. Bernini, R. Fumagalli, M. Allorio and A. Corsini: Calcium Antagonists and Low Density Lipoprotein Receptors. *Ann.N.Y.Acad.Sci.* 522:390-(4/20) (1988).

Papaioannou, S., S. Panzer-Knodle and P.C. Yang: Calcium Uptake Studies of 1,4-Dihydropyridine Agonists into Rabbit Aortic Smooth Muscle Cells in Culture. *Life Sci.* **44**:1751-(3/13) (1989).

Parenti, F.I., A.D. Angel and P.M. Mannucci: Methods for the Detection of Activated Platelets. *Adv.Exp.Med.Biol.* **104**:155-(4/20) (1984).

Parmley, W.W., S. Blumlein and R. Stevens: Modification of Experimental Atherosclerosis by Calcium-Channel Blocker. *Am.J.Cardiol.* **55**:165-(4/20) (1985).

Parodi, O., A. Maseri and I. Simonetti: Management of Unstable Angina at Rest by Verapamil. A Double-Blind Cross-Over Study in Coronary Care Unit. *Br.Heart J.* **41**:167-(4/16) (1979).

Parodi, O., I. Simonetti, A. L'Abbate and A. Maseri: Verapamil Versus Propranolol for Angina at Rest. *Am.J.Cardiol.* **50**:923-(4/16) (1982).

Parodi, O., I. Simonetti, C. Michelassi, C. Carpeggiani, A. Biagini, A. L'Abbate and A. Maseri: Comparison of Verapamil and Propranolol Therapy for Angina Pectoris at Rest: A Randomized, Multiple Cross-Over, Controlled Trial in the Coronary Care Unit. *Am.J.Cardiol.* **57**:899-(4/16) (1986).

Partridge, L.D.: A Mechanism for the Control of Low-Frequency Repetitive Firing. *Cell Mol.Neurobiol.* **2**:33-(2/9) (1982).

Partridge, L.D. and D. Swandulla: Calcium-Activated Non-Specific Cation Channels. *TINS* **11**:69-(2/9) (1988).

Pasternac, A.: Myocardial Stunning in Hypertrophic Cardiomyopathy. *J.Am.Coll.Cardiol.* **13**:1419-(4/17) (1989).

Pasternac, A., J. Noble, Y. Streulens, R. Elie, C. Henschke and M.G. Bourassa: Pathophysiology of Chest Pain in Patients with Cardiomyopathies and Normal Coronary Arteries. *Circulation* **65**:778-(4/17) (1982).

Paul, D.L.: Molecular Cloning of cDNA for Rat Liver Gap Junction Protein. *J.Cell Biol.* **103**:123-(2/10) (1986).

Paupardin-Tritsch, D., C. Hammond, H.N. Gerschenfeld, A.C. Nairn and P. Greengard: cGMP Dependent Protein Kinase Enhances Ca^{2+} Current and Potentiates the Serotonin Induced Ca^{2+} Current Increase in Snail Neurons. *Nature* **323**:812-(3/14) (1986).

Pauwels, P.J., H.P. Van Assouw and J.E. Leysen: Depolarization of Chick Myotubes Triggers the Appearance of (+)-[^{3}H]PN-200-110-Binding Sites. *Mol.Pharmacol.* **32**:785-(3/13) (1987).

Pauwels, P.J., H.P. Van Assouw, J.E. Leysen and P.A.J. Janssen: Ca^{2+}-Mediated Neuronal Death in Rat Brain Neuronal Cultures by Veratridine: Protection by Flunarizine. *Mol.Pharmacol.* **36**:525-(3/13) (1989).

Pellegrini-Giampietro, D.E., L. Bacciottini, V. Carla and F. Moroni: Morphine Withdrawal in Cortical Slices: Suppression by Ca^{2+}-Channel Inhibitors of Abstinence-Induced [^{3}H]-Noradrenaline Release. *Br.J.Pharmacol.* **93**:535-(3/13) (1988).

Pelzer, D., S. Pelzer and T.F. McDonald: Properties and Regulation of Calcium Channels in Muscle Cells. In *Review of Physiology, Biochemistry, Pharmacology*. Springer-Verlag, Berlin and Heidelberg. 114:107-(1/2) (1990).

Pelzer, S., J. Barhanin, D. Pauron, W. Trautwein, M. Lazdunski and D. Pelzer: Diversity and Novel Pharmacological Properties of Ca^{2+} Channels in Drosophila Brain Membranes. *EMBO J.* **8**:2365-(3/13) (1989).

Peng, H.B.: Participation of Calcium and Calmodulin in the Formation of Acetycholine Receptor Clusters. *J.Cell Biol.* **98**:550-(2/10) (1984).

Peng, H.B., P.C. Cheng and P.W. Luther: Formation of AChR Receptor Clusters Induced by Positively Charged Latex Beads. *Nature* **292**:831-(2/10) (1981).

Peng, H.B., K.X. Gao, M.Z. Xie and D.Y. Zhao: Development of Acetylcholinesterase Induced by Basic Polypeptide-Coated Latex Beads in Cultured *Xenopus* Muscle Cells. *Dev.Biol.* **127**:452-(2/10) (1988).

Penington, N.J., A.P. Fox, J.S. Kelly and R.J. Miller: G-Protein Involvement in the Inhibitory Action of 5-HT on Ca^{2+} Currents in Acutely Isolated Adult Rat Dorsal Raphe Neurons. *Biophys.J.*(in press) (3/14) (1990).

Penn, R.D. and W.A. Hagins: Signal Transmission Along Retinal Rods and the Origin of the Electroretinographic A-Wave. *Nature* **223**:201-(2/8) (1969).

Penn, R.D. and W.A. Hagins: Kinetics of the Photocurrent of Retinal Rods. *Biophys.J.* **12**:1073-(2/8) (1972).

Penner, R. and F. Dreyer: Two Different Presynaptic Calcium Currents in Mouse Motor Nerve Terminals. *Pflügers Arch.* **406**:190-(2/7) (1986).

Penner, R., G. Matthews and E. Neher: Regulation of Calcium Influx by Second Messengers in Rat Mast Cells. *Nature* **334**:499-(2/7) (1988).

Penner, R. and E. Neher: The Role of Calcium in Stimulus-Secretion Coupling in Excitable and Non-excitable Cells. *J.Exp.Biol.* **139**:329-(2/7) (1988b).

Penner, R. and E. Neher: Secretory Responses of Rat Peritoneal Mast Cells to High Intracellular Calcium. *FEBS Lett.* **226**:307-(2/7) (1988a).

Pepe, I.M., A. Boero, L. Vergani, I. Panfoli and C. Cugnoli: Effect of Light and Calcium on Cyclic GMP Synthesis in Rod Outer Segments of Toad Retina. *Biochim.Biophys.Acta* **889**:271-(2/8) (1986a).

Pepe, I.M., I. Panfola and C. Cugnoli: Guanylate Cyclase in Rod Outer Segments of the Toad Retina. *Fed. Eur. Biochem. Soc.* **203**:73-(2/8) (1986b).

Pepine, C.J., R.L. Feldman, J.A. Hill, C.R. Conti, J. Mehta, C. Hill and E. Scott: Clinical Outcome after Treatment of Rest Angina with Calcium Blockers: Comparative Experience during the Initial Year of Therapy with Diltiazem, Nifedipine and Verapamil. *Am.Heart J.* **106**:1341-(4/16) (1983).

Peracchia, C.: Communicating Junctions and Calmodulin: Inhibition of Electrical Uncoupling in *Xenopus* Embryo by Calmidazolium. *J.Membr.Biol.* **91**:49-(2/10) (1984).

Peracchia, C.: Calmodulin-Like Proteins and Communicating Junctions-Electrical Uncoupling of Crayfish Septate Axons is Inhibited by the Calmodulin Inhibitor W7 and is not Affected by Cyclic Nucleotides. *Pflügers Arch.* **408**:379-(2/10) (1987).

Peracchia, C., G. Bernardini and L.L. Peracchia: A Calmodulin Inhibitor Prevents Gap Junction Crystallization and Electrical Uncoupling. *J.Cell Biol.* **91**:124a-(2/10) (1981).

Peracchia, C., G. Bernardini and L.L. Peracchia: Is Calmodulin Involved in the Regulation of Gap Junction Permeability? *Pflügers Arch.* **399**:152-(2/10) (1983).

Peres, A., M. Sturani and R. Zippel: Properties of the Voltage Dependent Calcium Channel of Mouse Swiss 3T3 Fibroblasts. *J.Physiol.* **401**:639-(3/12) (1988).

Perez-Reyes, E., S.H. Kim, A.E. Lacerda, W. Horne, X. Wei, D. Rampe, K.P. Campbell, A.M. Brown and L. Birnbaumer: Induction of Calcium Currents by the Expression of the Alpha-1-Subunit of the Dihydropyridine Receptor from Skeletal Muscle. *Nature* **340**:233-(1/3,3/13,3/14) (1989).

Perhoniemi, V., K. Salmenkivi, S. Sundber, R. Johnson and A. Gordin: Effects of Flunarizine and Pentoxifylline on Walking Distance and Blood Rheology in Claudication. *Angiology* **35**:366-(4/20) (1984).

Perney, T.M., L.D. Hirning, S.E. Leeman and R.J. Miller: Multiple Calcium Channels Mediate Neurotransmitter Release from Peripheral Neurons. *Proc.Natl.Acad.Sci.U.S.A.* **83**:6656-(3/13) (1986).

Perney, T.M. and R.J. Miller: Neuropeptide Y and Bradykinin Activate Inositol Triphosphate Synthesis in the Same Sensory Neurons but Utilize Different G-Proteins. *J.Biol.Chem.*(in press) (1989).

Pernu, H.E., K. Oikarinen, J. Hietanen and M. Knuuttila: Verapamil-Induced Gingival Overgrowth: A Clinical, Histologic, and Biochemic Approach. *J.Oral.Pathol.Med.* **18**:422-(3/13) (1989).

Petersen, O.H. and I. Findlay: Electrophysiology of the Pancreas. *Physiol.Rev.* **67**:1054-(2/7) (1987).

Peterson, P.: Die Psychiatrie des Primaren Hyperparathyreodismus. In *Monographien aus dem Gesamtgebiete der Neurologie und Psychiatrie*. M. Muller, H. Spatz, and P. Vogel, eds. Springer, Berlin. 118-(4/22) (1967).

Petruk, K.C., M. West, G. Mohr, B.K.A. Weir, B.G. Benoit, F. Gentili, L.B. Disney, M.I. Khan, M. Grace, R.O. Holness, M.S. Karwon, R.M. Ford, G.S. Cameron, W.S. Tucker, G.B. Purves, J.D.R. Miller, K.M. Hunter, M.T. Richard, F.A. Durity, R. Chan, L.J. Clein, F.B. Maroun and A. Godon: Nimodipine Treatment in Poor-Grade Aneurysm Patients: Results of a Multicenter Double-Blind Placebo-Controlled Trial. *J.Neurosurg.* **68**:505-(4/21) (1988).

Pfaffinger, P.J., J.M. Martin, D.D. Hunter, N.M. Nathanson and B. Hille: GTP-Binding Proteins Couple Cardiac Muscarinic Receptors to a K Channel. *Nature* **317**:536-(3/14) (1985).

Phillipson, O. and M. Sandler: The Influence of NGF, Potassium Depolarization and Dibutyryl (Cyclic) AMP on Explant Cultures of Chick Sympathetic Ganglia. *Brain Res.* **90**:273-(2/10) (1975).

Piccolino, M. and H.M. Gerschenfeld: Activation of a Regenerative Calcium Conductance in Turtle Cones by Peripheral Stimulation. *Proc.R.Soc.London B* **201**:309-(2/8) (1978).

Piccolino, M. and H.M. Gerschenfeld: Characteristics and Ionic Processes Involved in Feedback Spikes of Turtle Cones. *Proc.R.Soc.London* **206**:439-(2/8) (1980).

Pickar, D., O.M. Wolkowitz, A.R. Doran, R. Labarca, A. Roy, A. Breier and P.K. Narang: Clinical and Biochemical Effects of Verapamil Administration to Schizophrenic Patients. *Arch.Gen.Psychiatr.* **44**:113-(4/22) (1987).

Pickard, J.D., G.D. Murray, R. Illingworth, M.DM. Shaw, G.M. Teasdale, P.M. Foy, P.R.D. Humphrey, D.A. Lang, R. Nelson, P. Richards, J. Sinar, S. Bailey and A. Skene: Effect of Oral Nimodipine on Cerebral Infarction and Outcome after Subarachnoid Haemorrhage: British Aneurysm Nimodipine Trial. *Br.Med.J.* **298**:636-(3/13) (1989).

Piette, J., A. Klarsfeld and J.P. Changeux: Interaction of Nuclear Factors with the Upstream Region of the Alpha-Subunit Gene of Chicken Muscle Acetylcholine Receptor: Variations with Muscle Differentiation and Denervation. *EMBO.J.* **8**:687-(2/10) (1989).

Pincon-Raymond, M., F. Rieger, M. Fosset and M. Lazdunski: Abnormal Transverse Tubule System and Abnormal Amount of Receptors for Ca^{2+} Channel Inhibitors of the Dihydropyridine Family in Skeletal Muscle from Mice with Embryonic Muscular Dysgenesis. *Dev.Biol.* **112**:458-(3/13) (1985).

Pinquier, J.L., S. Urien, P. Chaumet-Riffaud, A. Comte and J.P. Tillement: Binding of [^{3}H]Isradipine (PN 200-110) on Smooth Muscle Cell Membranes from Different Bovine Arteries. *J.Cardiovasc.Pharmacol.* **11**:402-(3/13) (1988).

Plant, T.D.: Properties and Calcium-Dependent Inactivation of Calcium Currents in Cultured Mouse Pancreatic Beta-cells. *J.Physiol.* **404**:731-(2/7) (1988).

Plant, T.D. and N.B. Standen: Calcium Current Inactivation in Identified Neurons of *Helix aspersa*. *J.Physiol.* **321**:273-(1/4) (1981).

Plow, E.F. and M.H. Ginsberg: Cellular Adhesion: GpIIb-IIIa as a Prototypic Adhesion Receptor. *Prog.Hemostat.Thrombos.* **9**:117-(4/20) (1989).

Plumb, V.J., R.B. Karp and N.T. Kouchoukos: Verapamil Treatment of Atrial Fibrillation and Atrial Flutter after Open Heart Surgery. *Circulation* (Abstract) **62**:84-(4/19) (1980).

Plummer, M.R., D.E. Logothetis and P. Hess: Elementary Properties and Pharmacological Sensitivities of Calcium Channels in Mammalian Peripheral Neurons. *Neuron* **2**:1453-(1/4,3/13,3/14) (1989).

Polans, A.S., S. Kawamura and M.D. Bownds: Influence of Calcium on Guanosine 3',5'-Cyclic Monophosphate Levels in Frog Rod Outer Segments. *J.Gen.Physiol.* **77**:41-(2/8) (1981).

Pollard, H.B., C.J. Pazoles, C.E. Creutz and O. Zindo: The Chromaffin Granule and Possible Mechanism of Exocytosis. *Int.Rev.Cytol.* **58**:159-(4/20) (1979).

Pool, P.E., S.C. Seagren, A.F. Salel and L.M. Skalland: Effects of Diltiazem on Serum Lipids, Exercise Performance and Blood Pressure: Randomized, Double-Blind, Placebo Controlled Evaluation for Systemic Hypertension. *Am.J.Cardiol.* **56**:86H-(4/18) (1985).

Popea, A. and T.R. Demetresco: Tetany and Psychosis. *Rev.Stiint.Med.* **19**:531-(4/22) (1930b).

Popea, A. and T.R. Demetresco: Tetany and Psychosis in Nursing Mother. *Bruxelles Med.* **10**:708-(4/22) (1930a).

Porter, C.J., P.C. Gillette and A. Farson: Effects of Verapamil on Supraventricular Tachycardia in Children. *Am.J.Cardiol.* **48**:487-(4/19) (1981a).

Porter, J.M., S.P. Rivers, C.J. Anderson and G.M. Baur: Evaluation and Management of Patients with Raynaud's Syndrome. *Am.J.Surg.* **142**:183-(4/20) (1981b).

Potreau, D. and G. Raymond: Calcium-Dependent Electrical Activity and Contraction of Voltage-Clamped Frog Single Muscle Fibres. *J.Physiol.* **307**:9-(2/10) (1980b).

Potreau, D. and G. Raymond: Slow Inward Barium Current and Contraction on Frog Single Muscle Fibres. *J.Physiol.* **303**:91-(2/10) (1980a).

Powell, J.A. and D.M. Fambrough: Electrical Properties of Normal and Dysgenic Mouse Skeletal Muscle in Culture. *J.Cell Physiol.* **82**:21-(2/10) (1973).

Pozzan, T., P. Volpe, F. Zorzato, M. Bravin, D.P. Lew, K.H. Krause, S. Hashimoto, B. Bruno and J. Meldolesi: Immunological Identification of the Microsomal Calcium Store of Nonmuscle Cells. *J.Cardiol.Pharmacol.Suppl.l5.* **12**:580-(3/14) (1988).

Prentki, M. and F.M. Matschinsky: Ca^{2+}, cAMP, and Phospholipid-Derived Messengers in Coupling Mechanisms of Insulin Secretion. *Physiol.Rev.* **67**:1185-(2/7) (1987).

Preuss, K.C., G.J. Gross, H.L. Brooks and D.C. Warltier: Slow Channel Calcium Activator, a New Group of Pharmacological Agents. *Life Sci.* **37**:1271-(4/20) (1985).

Previtali, M., J.A. Salerno, L. Tavazzi, M. Ray, A. Medici, M. Chimienti, G. Specchia and P. Bobba: Treatment of Angina at Rest with Nifedipine: A Short-Term Controlled Study. *Am.J.Cardiol.* **45**:825-(4/16) (1980).

Price, W.A. and L.R. DiMarzio: Verapamil-Carbamazepine Neurotoxicity. *J.Clin.Psychiatry* **49**:80-(4/22) (1988).

Price, W.A. and A.J. Giannini: Neurotoxicity Caused by Lithium-Verapamil Synergism. *J.Clin.Pharmacol.* **26**:717-(4/22) (1986).

Prida, X.E., J.S. Gelman, R.L. Feldman, J.A. Hill, C.J. Pepine and E. Scott: Comparison of Diltiazem and Nifedipine Alone and in Combination in Patients with Coronary Artery Spasm. *J.Am.Coll.Cardiol.* **9**:412-(4/16) (1987).

Prinzmetal, M., R. Kennamer, R. Merliss, T. Wada and N. Bor: Angina Pectoris. I. A Variant Form of Angina Pectoris. *Am.J.Med.* **27**:375-(4/16) (1959).

Pritchett, E.L.C., S.C. Hammill, M.J. Reiter, K.L. Lee, E.A. McCarthy, J.M. Zimmerman and D.G. Shand: Life-Table Method for Evaluating Antiarrhythmic Drug Efficacy in Patients with Paroxysmal Atrial Tachycardia. *Am.J.Cardiol.* **52**:1007-(4/19) (1983).

Prod'hom, B., D. Pietrobon and P. Hess: Direct Measurement of Proton Transfer Rates to a Group Controlling Dihydropyridine Sensitive Ca Channel. *Nature* **329**:243-(3/12) (1987).

Proschek, L. and G. Jasmin: Hereditary Polymyopathy and Cardiomyopathy in the Syrian Hamster II. Development of Heart Necrotic Changes in Relation to Defective Mitochondrial Function. *Muscle Nerve* **5**:26-(4/17) (1982).

Prosser, C.L., D.L. Kreulen, R.J. Weigel and W. Yau: Prolonged Potentials in Gastrointestinal Muscles Induced by Calcium Chelation. *Am.J.Physiol.* **233**:Cl9-(1/2) (1977).

Pugh, E.N.: The Nature and Identity of the Internal Excitational Transmitter of Vertebrate Phototransduction. *Ann.Rev.Physiol.* **49**:715-(2/8) (1987).

Pugh, E.N. and W.H. Cobbs: Visual Transduction in Vertebrate Rods and Cones: A Tale of Two Transmitters, Calcium and Cyclic GMP. *Vision Res.* **26**:1613-(2/8) (1986).

Pumphrey, C.W., V. Fuster, M.K. Dewanjee, J.H. Chesebro, R.E. Vlietstra and M.P. Kays: Comparison of Antithrombotic Action of Calcium Antagonist Drugs with Dipyridamole in Dogs. *Am.J.Cardiol.* **51**:591-(4/20) (1983).

Pumplin, D.W., T.S. Reese and R. Llinas: Are the Pre-Synaptic Membrane Particles the Calcium Channels. *Proc.Natl.Acad.Sci.U.S.A.* **78**:7210-(2/9,3/14) (1981).

Qar, J., J. Barhanin, G. Romey, R. Henning, U. Lerch, R. Oekonomupolos, H. Urbach and M. Lazdunski: A Novel High Affinity Class of Ca^{2+} Channel Blockers. *Mol.Pharmacol.* **33**:363-(3/13) (1988).

Quirion, R. and N.P.V. Nair: Dihydropyridine and Phenylalkylamine Binding Sites in Alzheimer's Disease and other Neurological Disorders. In *Nimodipine and Central Nervous System Function: New Vistas.* J. Traber W.H. Gispen, eds. Schattauer, Stuttgart and New York. 257-(3/13) (1990).

Quyyumi, A.A., C. Wright, L. Mockus and K.M. Fox: Effect of Partial Agonist Activity in β-Blockers in Severe Angina Pectoris: A Double-Blind Comparison of Pindolol and Atenolol. *Br.Med.J.* **289**:951-(4/16) (1984).

Rabkin, S.A., C. Tomlinson, B.N. Corbett and T.E. Cuddy: Verapamil and Supraventricular Tachyarrhythmias: Beneficial Effects in Patients with Chronic Pulmonary Disease. *Can.Med.Assoc.J.* **122**:64-(4/19) (1980).

Raftery, E.B.: Cardiovascular Drug Withdrawal Syndromes. A Potential Problem with Calcium Antagonists. *Drugs* **28**:371-(3/13) (1984).

Rahamimoff, R., S.D. Erulkar, A. Lev-Tov and H. Meiri: Intracellular and Extracellular Calcium Ions in Transmitter Release at the Neuromuscular Synapse. *Ann.N.Y.Acad.Sci.* **307**:583-(2/7) (1978a).

Rahamimoff, R., H. Meiri, S.D. Erulkar and Y. Barenholz: Changes in Transmitter Release Induced by Ion Containing Liposomes. *Proc.Natl.Acad.Sci.U.S.A.* **75**:5214-(2/7) (1978b).

Rahimtoola, S.H., D. Nunley, G. Grunkemeier, J. Tepley, L. Lambert and A. Starr: Ten Year Survival after Coronary Artery Bypass Surgery for Unstable Angina. *N.Engl.J.Med.* **308**:676-(4/16) (1983).

Rahwan, G.: Mechanism of Action of Membrane Calcium Channel Blockers and Intracellular Calcium Antagonists. *Med.Res.Rev.* **3**:21-(4/18) (1983).

Ramkumar, V. and E.E. El-Fakahany: Prolonged Morphine Treatment Increases Brain Dihydropyridine Binding Sites: Possible Involvement in Development of Morphine Dependence. *Eur.J.Pharmacol.* **146**:73-(3/13) (1988).

Ramon, F., R.O. Arellano, A. Rivera and G.A. Zampighi: Effects of Internally Perfused Calmodulin on the Junctional Resistance of Crayfish Lateral Axons. In *Gap Junctions*. Hertzberg and Johnston, eds. Alan R. Liss, New York. 255-(2/10) (1988).

Rampe, D., E. Luchowski, A. Rutledge, R.A. Janis and D.J. Triggle: Comparative Aspects of [^{3}H]1,4-Dihydropyridine Ca^{2+} Channel Antagonist and Activator Binding to Neuronal and Muscle Membranes. *Can.J.Physiol.Pharmacol.* **65**:1452-(3/13) (1987a).

Rampe, D., T. Poder, Z.Y. Zhao and W.P. Schilling: Calcium Channel Agonist and Antagonist Binding in Highly Enriched Sarcolemma Preparation Obtained from Canine Ventricle. *J.Cardiovasc.Pharmacol.* **13**:547-(3/13) (1989b).

Rampe, D., A. Skattebol, D.J. Triggle and A.M. Brown: Effects of McN 6186-11 on Voltage-Dependent Ca^{2+} Channels in Heart and Pituitary Cells. *J.Pharmacol.Exp.Ther.* **248**:164-(3/13) (1989a).

Rampe, D. and D.J. Triggle: New Ligands for L-Type Ca^{2+} Channels. *Trends Pharmacol.Sci.* **11**:112-(3/13) (1990).

Rampe, D., D.J. Triggle and A.M. Brown: Electrophysiological and Biochemical Studies on the Putative Ca^{2+} Channel Blocker MDL 12,330A in an Endocrine Cell. *J.Pharmacol.Exp.Ther.* **243**:402-(3/13) (1987b).

Rane, S.G. and K. Dunlap: Kinase C Activator 1,2-Oleoylacetylglycerol Attenuates Voltage-Dpendent Calcium Current in Sensory Neurones. *Proc.Natl.Acad.Sci.U.S.A.* **83**:184-(3/14) (1986).

Rane, S.G., G.G. Holtz and K. Dunlap: Dihydropyridine Inhibition of Neuronal Calcium Current and Substance P Release. *Pflügers Archiv.* **409**:361-(3/13,3/14) (1987).

Rane, S.G., M.P. Walsh, J.R. McDonald and K. Dunlap: Specific Inhibitors of Protein Kinase C Block Transmitter-Induced Modulation of Sensory Neuron Calcium Current. *Neuron* **3**:239-(3/13) (1989).

Rapaport, R.M., M.B. Draznin and F. Murad: Endothelium-Dependent Relaxation in Rat Aorta may be Mediated Through Cyclic GMP-Dependent Protein Dephosphorylation. *Nature* **306**:174-(4/16) (1983).

Rapaport, R.M. and F. Murad: Endothelium-Dependent and Nitrovasodilator-Induced Relaxation of Vascular Smooth Muscle: Role of Cyclic GMP. *J.Cycl.Nucl.Prot.Phosphoryl.Res.* **9**:281-(4/16) (1983).

Rapuano, M., A.F. Ross and J. Prives: Opposing Effects of Calcium Entry and Phorbol Esters on Fusion of Chick Muscle Cells. *Dev.Biol.* **134**:271-(2/10) (1989).

Rasmussen, H.: The Calcium Messenger System (1). *N.Engl.J.Med.* **314**:1094-(2/10) (1986a).

Rasmussen, H.: The Calcium Messenger System (2). *N.Engl.J.Med.* **314**:1164-(2/10) (1986b).

Rasmussen, H. and P.Q. Barrett: Calcium Messenger System: An Integrated View. *Physiol.Rev.* **64**:938-(2/10) (1984).

Rasmussen, K., H. Wang and D. Faisa: Comparative Efficacy of Quinidine and Verapamil in the Maintenance of Sinus Rhythm after DC Conversion of Atrial Fibrillation. *Acta Med.Scand.* **645**:23-(4/19) (1981).

Ratto, G.M., R. Payne, R.G. Owen and R.Y. Tsien: The Concentration of Cytosolic Free Calcium in Vertebrate Rod Outer Segment Measured with Fura-2. *J.Neurosci.* **8**:3240-(2/8) (1988).

Rauscher, F.J., III, D.R. Cohen, T. Curran, T.J. Bos, P.K. Vogt, D. Bohmann, R. Tjian and B.R. Franza, Jr.: Fos Associated Protein p39 is the Product of the Jun Proto-Oncogene. *Science* **240**:1010-(2/10) (1988).

Raynaud, B., D. Clarous, S. Vidal, C. Ferrand and M.J. Weber: Comparison of the Effects of Elevated K^+ Ions and Muscle-Conditioned Medium on the Neurotransmitter Phenotype of Cultured Sympathetic Neurons. *Dev.Biol.* **121**:548-(2/10) (1987a).

Raynaud, B., N. Faucon Biguet, S. Vidal, J. Mallet and M.J. Weber: The Use of a Tyrosine-Hydroxylase cDNA Probe to Study the Neurotransmitter Plasticity of Rat Sympathetic Neurons in Culture. *Dev.Biol.* **119**:305-(2/10) (1987b).

Reese, T.S. and M.J. Karnovsky: Fine Structural Localization of a Blood-Brain Barrier to Exogenous Peroxidase. *J.Cell Biol.* **34**:207-(4/21) (1967).

Regan, L.J.: Ca^{2+} Channels in Freshly Dissociated Rat Cerebellar Purkinje Cells. *Biophys.J.(Abst.)* **51**:223-(3/14) (1987).

Regehr, W.G. and J.A. Connor, D.W. Tank: Optical Imaging of Calcium Accumulation in Hippocampal Pyramidal Cells During Synaptic Activation. *Nature* **341**:533-(2/9) (1989).

Reichardt, L.F. and R.B. Kelly: A Molecular Description of Nerve Terminal Function. *Ann.Rev.Biochem.* **52**:871-(2/7) (1983).

Reimer, K.A., R.B. Jennings, F.R. Cobb, R.H. Murdock, J.C. Greenfield, L.C. Becker, B.H. Bulkey, G.M. Hutchins, Jr., R.P. Schwartz and K.R. Bailey: Animal Models for Protecting Ischemic Myocardium: Results of the NHLBI Cooperative Study. Comparison of Unconscious and Conscious Dog Models. *Circ.Res.* **56**:651-(4/16) (1985).

Reimer, K.A., J.E. Lowe and R.B. Jennings: Effect of the Calcium Antagonist Verapamil on Necrosis Following Temporary Coronary Artery Occlusion in Dogs. *Circulation* **55**:581-(4/166) (1977).

Reinila, M., E. MacDonald, N. Salem Jr., M. Linnoila and E.G. Trams: A Standardized Method for the Determination of Human Erythrocyte Membrane Adenosine Triphosphatases. *Anal.Biochem.* **124**:19-(4/22) (1982).

Reisberg, B. and S. Gershon: Side Effects Associated with Lithium Therapy. *Arch.Gen.Psychiatry.* **36**:879-(4/22) (1979).

Report of WHO/ISFC Task Force on the Definition and Classification of Cardiomyopathies. *Br. Heart J.* **44**:672-(4/15) (1980).

Resendez, E., J. Ting, K.S. Kim, S.K. Wooden and A.S. Lee: Calcium Ionophore A23187 as a Regulator of Gene Expression in Mammalian Cells. *J.Cell Biol.* **103**:2145-(2/10) (1986).

Reuter, H.: Divalent Cations as Charge Carriers in Excitable Membranes. *Biophys.Mol.Biol.* **26**:1-(3/12) (1973).

Reuter, H.: Calcium Channel Modulation of Neurotransmitters, Enzymes and Drugs. *Nature* **301**:569-(1/4,2/10,3/14,4/17) (1983).

Reuter, H.: A Variety of Calcium Channels. *Nature* **316**:391-(3/14) (1985).

Reuter, H.: Modulation of Ion Channels Phosphorylation and Second Messengers. *NIPS* **2**:168-(4/17) (1987).

Reuter, H. and H. Scholz: A Study of Ionic Selectivity and Kinetic Properties of Calcium Dependent Slow Inward Current in Mammalian Cardiac Muscle. *J.Physiol.* **264**:17-(1/2,3/12) (1977a).

Reuter, H. and H. Scholz: The Regulation of the Calcium Conductance of Cardiac Muscle by Adrenaline. *J.Physiol.* **264**:49-(1/2,2/10) (1977b).

Reuter, H., C.F. Stevens, R.W. Tsien and G. Yellen: Properties of Single Calcium Channels in Cardiac Cell Culture. *Nature* **297**:501-(1/2,1/3,1/4) (1982).

Reuter, H.: Localization of Beta Adrenergic Receptors, and Effects of Noradrenaline and Cyclic Nucleotides on Action Potentials, Ionic Currents and Tension in Mammalian Cardiac Muscle. *J.Physiol.* **242**:429-(1/4) (1974).

Reverdin, E.C., G.A. Cohen, N.M. Birchall and E.L. Boulpaep: Ca^{2+} Current in Human Epidermal Cells. *Biophys.J.* **55**:603a-(3/13) (1989).

Reynolds, I.J., R.J. Gould and S.H. Snyder: ^{3}H Verapamil Binding Sites in Brain and Skeletal Muscle: Regulation by Calcium. *Eur.J.Pharmacol.* **95**:319-(4/22) (1983).

Reynolds, I.J., A.M. Snowman and S.H. Snyder: (-)-[^{3}H]Desmeth-oxyverapamil Labels Multiple Calcium Channel Modulator receptors in Brain and Skeletal Muscle Membranes: Differentiation by Temperature and Dihydropyridines. *J.Pharmacol.Exp.Ther.* **237**:731-(3/13) (1986a).

Reynolds, I.J., J.A. Wagner, S.H. Snyder, S.A. Thayer, B.M. Olivera and R.J. Miller: Brain Voltage-Sensitive Calcium Channel Subtypes Differentiated by ω-Conotoxin Fraction GVIA. *Proc.Natl.Acad.Sci.U.S.A.* **83**:8804-(3/13) (1986b).

Reynolds, J.N. and P.L. Carlen: Diminished Calcium Currents in Aged Hippocampal Dentate Gyrus Granule Neurons. *Brain Res.* **479**:384-(3/13) (1989).

Rhodes, D.G., J.G. Sarmiento and L.G. Herbette: Kinetics of Binding of Membrane-Active Drugs to Receptor Sites: Diffusion-Limited Rates for a Membrane Bilayer Approach of 1,4-Dihydropyridine Calcium Channel Antagonists to their Active Site. *Mol.Pharmacol.* **27**:612-(3/13) (1985).

Ribalet, B. and P.M. Beigelman: Calcium Action Potentials and Potassium Permeability Activation in Pancreatic β-Cells. *Am.J.Physiol.* **239**:C124-(2/7) (1980).

Ribera, A.B. and N.C. Spitzer: A Critical Period of Transcription Required for Differentiation of the Action Potential of Spinal Neurons. *Neuron* **2**:1055-(2/10) (1989).

Rich, T.L., G.A. Langer and M.G. Klassen: Two Components of Coupling Calcium in Single Ventricular Cell of Rabbits and Rats. *Am.J.Physiol.* **254**:H937-(2/10) (1988).

Richard, S., F. Tiaho, P. Charnet, J. Nargeot and J.M. Nerbonne: Two Components of High Threshold Ca^{2+} Current Modulated Differentially by Physiological Stimuli. *Biophys.J.* **55**:38a-(2/10) (1989).

Richelson, E.: Lithium Entry through the Sodium Channel of Cultured Mouse Neuroblastoma Cells: A Biochemical Study. *Science* **196**:1001-(4/22) (1977).

Ringer, S.: A Further Contribution Regarding the Influence of the Different Constituents of the Blood on the Contraction of the Heart. *J.Physiol.* **4**:29-(2/10) (1883).

Rink, T.J.: A Real Receptor-Operated Calcium Channel. *Nature* **334**:649-(2/10) (1988b).

Rink, T.J.: Cytosolic Calcium in Platelet Activation. *Experimentia* **44**:97-(4/20) (1988a).

Rink, T.J. and T.J. Hallam: What Turns Platelets on? *Trends Biochem.* **9**:215-(4/20) (1984).

Rink, T.J., S.W. Smith and R.Y. Tsien: Cytoplasmic free Ca^{2+} in Human Platelets: Ca^{2+} Threshold and Ca^{2+} Independent Activation for Shape Change and Secretion. *FEBS Lett.* **148**:21-(4/20) (1982).

Rinkenberger, R.I., E.N. Prystowsky, J.J. Heger, P.J. Troup, W.M. Jackman and D.P. Zipes: Effects of Intravenous and Chronic Oral Verapamil Administration in Patients with Supraventricular Tachyarrhythmias. *Circulation* **62**:996-(4/19) (1980).

Rios, E. and G. Brum: Involvement of Dihydropyridine Receptors in Excitation-Contraction Coupling in Skeletal Muscle. *Nature* **325**:717-(2/10,3/14) (1987).

Rios, E. and G. Pizarro: Voltage Sensors and Calcium Channels of Excitation-Contraction Coupling. *News Physiol.Sci.* **3**:223-(2/10) (1988).

Rittenhouse-Simmons, S. and D. Deykin: The Activation of Ca^{2+} of Platelet-Phospholipase A2: Effects of Dibutyryl Cycle Adenosine Monophosphate and 8-(N,N-Diethylamino)-octyl-3,4,5-trimethoxybenzoate. *Biochim.Biophys.Acta* **543**:409-(4/20) (1978).

Rius, R.A., L. Lucchi, S. Govoni and M. Trabucchi: *In Vivo* Chronic Lead Exposure Alters [^{3}H]Nitrendipine Binding in Rat Striatum. *Brain Res.* **322**:180-(3/13) (1984).

Rivier, J., R. Galyean, W.R. Gray, A. Azimi-Zonooz and J.M. McIntosh: Neural Calcium Channel Inhibitors. *J.Biol.Chem.* **262**:1194-(3/13) (1987).

Rizzon, P., D. Scrutinio, S.G. Mangini, R. Lagioia and L. de Toma: Randomized Placebo-Controlled Comparative Study of Nifedipine, Verapamil and Isosorbide Dinitrate in the Treatment of Angina at Rest. *Eur.Heart J.* **7**:67-(4/16) (1986).

Roberts, A., D.O.P. Ottersen and J. Storm-Mathisen: The Early Development of Neurons with GABA Immunoreactivity in the CNS of *Xenopus laevis* Embryos. *J.Comp.Neurol.* **261**:435-(2/10) (1987).

Roberts, W.C. and V.J. Ferrans: Pathologic Anatomy of the Cardiomyopathies. Idiopathic Dilated and Hypertrophic Types, Infiltrative and Endomyocardial Disease with and without Eosinophilia. *Hum.Pathol.* **6**:287-(4/17) (1975).

Roberts, W.M., J. Howard and A.J. Hudspeth: Hair Cells: Transduction, Tuning and Transmission in the Inner Ear. *Ann.Rev.Cell Biol.* **4**:63-(2/8) (1988).

Roberts, W.M., L. Robles and A.J. Hudspeth: Correlation Between the Kinetic Properties of Ionic Channels and the Frequency of Membrane-Potential Resonance in Hair Cells of the Bullfrog. In *Auditory Frequency Selectivity*. B.C.J. Moore and R.D. Patterson, eds. Plenum Press, New York. 89-(2/8) (1986).

Robertson, M.J. and P. Lumley: Effects of Hypoxia, Elevated K^+ and Acidosis on the Potency of Verapamil, Diltiazem and Nifedipine in the Guinea Pig Isolated Papillary Muscle. *Br.J.Pharmacol.* **98**:937-(3/13) (1989).

Robinson, P.R., S. Kawamura, B. Abramson and M.D. Bownds: Control of the Cyclic GMP Phospodiesterase of Frog Photoreceptor Membranes. *J.Gen.Physiol.* **76**:631-(2/8) (1980).

Robinson, R.A. and R.H. Stokes: *Electrolyte Solutions*. Academic Press, New York. (3/12) (1959).

Robishaw, J.D. and K.A. Foster: Role of G Proteins in the Regulation of the Cardiovascular System. *Ann.Rev.Physiol.* **51**:229-(3/13) (1989).

Robson, S.J. and R.D. Burgoyne: L-Type Calcium Channels in the Regulation of Neurite Outgrowth from Rat Dorsal Root Ganglion Neurons in Culture. *Neurosci.Lett.* **104**:110-(2/10) (1989).

Rocci, M.L., P.H. Vlasses, M.E. Lener, R.A. Fruncillo and M.A. Sirgio: Comparative Evaluation of the Effects of Labetalol, Verapamil and Diltiazem on Antipyrine and Indocyanine Green Clearances. *J.Clin.Pharmacol.* **29**:891-(3/13) (1989).

Rodeheffer, R.J., J.A. Rommer, F. Wigley and C.R. Smith: Controlled Double-Blind Trial of Nifedipine in the Treatment of Raynaud's Phenomenon. *N.Engl.J.Med.* **308**:880-(4/20) (1983).

Rodenkirchen, R., R. Bayer, R. Steiner, F. Bossert, H. Meyer and E. Moller: Structure-Activity Studies on Nifedipine in Isolated Cardiac Muscle. *Naunyn Schmied Arch.Phamacol.* **310**:69-(3/13) (1978).

Role, L. and J.H. Schwartz: Cross-Talk Between Signal Transduction Pathways. *TINS* **12**:-(2/10) (1989).

Romey, G., L. Garcia, V. Dimitriadou, M. Pincon-Raymond, F. Rieger and M. Lazdunski: Ontogenesis and Localization of Ca^{2+} Channels in Mammalian Skeletal Muscle in Culture and Role in Excitation-Contraction Coupling. *Proc.Natl.Acad.Sci.U.S.A.* **86**:2933-(3/13) (1989).

Ronning, S.A., G.A. Heatley and T.F.J. Martin: Thyrotropin-Releasing Hormone Mobilizes Ca^{2+} from Endoplasmic Reticulum and Mitochondria of GH3 Pituitary Cells: Characterization of Cellular Ca^{2+} Pools by a Method Based on Digitonin Permeabilization. *Proc.Natl.Acad.Sci.U.S.A.* **79**:6294-(2/7) (1982).

Roof, D.J. and J.E. Heuser: Surfaces of Rod Photoreceptor Disk Membranes: Integral Membrane Components. *J.Cell Biol.* **95**:487-(2/8) (1982a).

Roof, D.J., J.I. Korenbrot and J.E. Heuser: Surfaces of Rod Photoreceptor Disk Membranes: Light-Activated Enzymes. *J.Cell Biol.* **95**:501-(2/8) (1982b).

Rorsman, P., F.M. Ashcroft and G. Trube: Single Ca Channel Currents in Mouse Pancreatic β-Cells. *Pflügers Arch.* **412**:597-(2/7) (1988).

Rorsman, P. and G. Trube: Glucose Dependent K^+-Channels in Pancreatic β-Cells are Regulated by Intracellular ATP. *Pflügers Arch.* **405**:305-(2/7) (1985).

Rorsman, P. and G. Trube: Calcium and Delayed Potassium Currents in Mouse Pancreatic β-Cells under Voltage-Clamp Conditions. *J.Physiol.* **374**:531-(2/7) (1986).

Rosario, L.M., I. Atwater and A.M. Scott: Pulsatile Insulin Release and Electrical Activity from Single Ob/ob Mouse Islets of Langerhans. *Adv.Exp.Biol.Med.* **211**:413-(2/7) (1986).

Rosario, L.M., B. Soria, G. Feurstein and H.B. Pollard: Voltage-Sensitive Calcium Flux into Bovine Chromaffin Cells Occurs through Dihydropyridine-Sensitive and Dihydropyridine- and ω-Conotoxin- Sensitive Pathways. *Neuroscience* **29**:735-(3/13) (1989).

Rosen, M.R. and R.F. Reder: Does Triggered Activity have a Role in the Genesis of Cardiac Arrhythmias. *Ann.Intern.Med.* **96**:794-(4/19) (1981).

Rosenberg, G.A.: *Brain Fluids and Metabolism.* Oxford University Press, New York. (4/21) (1990).

Rosenberg, G.A., L.I. Wolfson and R. Katzman: *Disorders of Cerebrospinal Fluid Circulation.* R.N. Rosenberg, ed. Churchill Livingstone, New York. 285-(4/21) (1983).

Rosenberg, R.L., P. Hess and R.W. Tsien: Cardiac Calcium Channels in Planar Lipid Bilayers. *J.Gen.Physiol.* **92**:27-(3/13) (1988).

Rosenthal, W., J. Hescheler, K.D. Hinsch, K. Spicher, W. Trautwein and G. Schultz: Cyclic AMP Independent Dual Regulation of Voltage Dependent Ca^{2+} Currents by LHRH and Somatostatin in a Pituitary Cell Line. *EMBO.J.* **7**:1627-(3/14) (1988).

Rosing, D.R., U. Idanpaan-Heikkila, B.j. Maron, R.O. Bonow and S.E. Epstein: Use of Calcium-Channel Blocking Drugs in Hypertrophic Cardiomyopathy. *Am.J.Cardiol.* **55**:185B-(4/17) (1985).

Rosing, D.R., K.M. Kent, J.S. Borer, S.F. Seides, B.J. Maron and S.E. Epstein: Verapamil Therapy: A New Approach to the Pharmacologic Treatment of Hypertrophic Cardiomyopathy. I. Hemodynamic Effects. *Circulation* **60**:1201-(4/17) (1979).

Ross, R.: Platelets, Platelet-Derived Growth Factor, Growth Control and their Interactions with the Vascular Wall. *J.Cardiovasc.Pharmacol.* **7**:5186-(4/20) (1985).

Ross, R. and J.A. Glomset: The Pathogenesis of Atherosclerosis. *N.Engl.J.Med.* **295**:369-(4/20) (1976).

Ross, R.D., W.R.M. Dassen, E.J. Vanaght, P. Brugada, W.H.M. Barr and H.J.J. Wellens: Cycle Length Alteration in Circus Movement Tachyardia Using an Atrio-Ventricular Accessory Pathway: A Study of the Role of AV Node Using a Computer Model of Tachycardia. *Circulation* **65**:862-(4/19) (1982).

Rossier, J.R., J.A. Cox, E.J. Niesor and C.L. Bentzen: A New Class of Calcium Entry Blockers Defined by 1,3-Diphosphonates Interactions of SR-7037 (Belfosdil) with Receptors for Calcium Channel Ligands. *J.Biol.Chem.* **264**:16598-(3/13) (1989).

Roth, A., E. Harrison, G. Mitani, J. Cohen, S.H. Rahimtoola and U. Elkayam: Efficacy and Safety of Medium- and High-Dose Diltiazem Alone and in Combination with Digoxin for Control of Heart Rate at Rest and During Exercise in Patients with Chronic Atrial Fibrillation. *Circulation* **73**:316-(4/19) (1986).

Rothman, S.M. and J.W. Olney: Glutamate and the Pathophysiology of Hypoxic-Ischemic Brain Damage. *Ann.Neurol.* **19**:105-(4/21) (1986).

Rougier, O., G. Vassort, D. Garnier, Y.M. Gargouil and E. Coraboeuf: Existence and Role of a Slow Inward Current during the Frog Atrial Action Potential. *Pflügers Arch.* **308**:91-(1/2) (1969).

Rousseau, E., J. Ladine, Q.Y. Liu and G. Meissner: Activation of the Ca^{2+} Release Channel of Skeletal Muscle Sarcoplasmic Reticulum by Caffeine and Related Compounds. *Arch.Biochem.Biophys.* **267**:75-(2/10) (1988).

Rousseau, E., J.S. Smith, J.S. Henderson and G. Meissner: Single Channel and $^{45}Ca^{2+}$ Flux Measurements of the Cardiac Sarcoplasmic Reticulum Calcium Channel. *Biophys.J.* **50**:1009-(2/10) (1986).

Rousseau, M.F., C. Hanet, E. Pardonge-Lavenne, G. van den Berghe, F. Van Hoof and H. Pouleur: Changes in Myocardial Metabolism during Therapy in Patients with Chronic Stable Angina: A Comparison of Long-Term Dosing with Propranolol and Nicardipine. *Circulation* **73**:1270-(4/18) (1986).

Rowland, E., W.J. McKenna, H. Gulker and D.M. Krikler: The Comparative Effects of Diltiazem and Verapamil on Atrioventricular Conduction and AV Re-entry Tachycardia. *Circ.Res.* **1**:163-(4/19) (1983).

Rozansky, J.J., L. Zaman and A. Castellanos: Electrophysiologic Effects of Diltiazem Hydrochloride on Supraventricular Tachycardia. *Am.J.Cardiol.* **49**:621-(4/19) (1982).

Rubin, L.: Increases in Muscle Ca^{2+} Mediate Changes in Acetylcholinesterase and Acetylcholine Receptors Caused by Muscle Contraction. *Proc.Natl.Acad.Sci.U.S.A.* **82**:7121-(2/10) (1985).

Rubin, L.L., S.M. Schuetze, C.L. Weill and G.D. Fischbach: Regulation of Acetylcholinesterase Appearance at Neuromuscular Junctions *In Vitro*. *Nature* **283**:264-(2/10) (1980).

Rudolfsky, G., F.E. Brock, M. Ulrich and F. Nobel: Clinical Evaluation of Flunarizine; Walking Distance, Ergometric Performance and Haemodynamic and Biochemical Effects. *Angiology* **30**:470-(4/20) (1979).

Russell, D.C.: Electrophysiology and Antiarrhythmic Effects of Calcium Antagonists. *Scott.Med.J.* **2**:161-(4/19) (1981).

Rybak, M.E., L.A. Renzulli, M.J. Burns and D.P. Cahaly: Platelet Glycoprotein IIb and IIIa as a Calcium Channel in Liposomes. *Blood* **72**:714-(4/20) (1988).

Rybakowski, J., E. Potok and W. Strzyzewski: Erythrocyte Membrane Adenosine Triphosphatase Activities in Patients with Endogenous Depression and Healthy Subjects. *Eur.J.Clin.Invest.* **11**:61-(4/22) (1981).

Ryu, P.D., G. Gerber, K. Murase and M. Randic: Calcitonin Gene Related Peptide Enhances Calcium Current of Rat Dorsal Root Ganglion Neurons and Spinal Excitatory Synaptic Transmitters. *Neurosci.Lett.* **89**:305-(3/14) (1988).

Sacks, H. and B.N. Kennelly: Verapamil in Cardiac Arrhythmias. *Br.Med.J.* **2**:716-(4/19) (1972).

Sage, S.O. and T.J. Rink: Inhibition of Forskolin of Cytosolic Calcium Rise, Shape Change and Aggregation in Quin-2 -Loaded Human Platelets. *FEBS Lett.***188**:135-(4/20) (1985).

Sage, S.O. and T.J. Rink: Kinetic Differences between Thrombin-Induced and ADP-Induced Calcium Influx and Release from Internal Stores in Fura-2-loaded Human Platelets. *Biochem.Biophys.Res.Commun.* **136**:1124-(4/20) (1986).

Sah, D.W.Y., L.J. Regan and B.P. Bean: Calcium Channels in Rat Neurons; High Threshold Channels that are Resistant to Both ω-Conotoxin and Dihydropyridine Blockers. *Soc.Neurosci.Abs.* **15**:823-(1/4,3/13) (1989).

Saha, J.K., L.V. Hryshko, R.A. Bouchard, T. Chau and D. Bose: Analysis of the Effects of (-) and (+) Isomers of the 1,4-Dihydropyridine Calcium Channel Agonist BAY K 8644 on Post-Test Potentiation in the Canine Ventricular Muscle. *Can.J.Physiol.Pharmacol.* **67**:788-(3/13) (1989).

Saida, K. and C. Van Breemen: A Possible Ca^{2+}-Induced Ca^{2+} Release Mechanism Mediated by Norepinephrine in Vascular Smooth Muscle. *Pflügers Arch.* **397**:166-(2/10) (1983).

Saimi, Y. and C.H. Kung: Are Ions Involved in the Gating of Calcium Channels. *Science* **218**:153-(1/3) (1982).

Saito, A., M. Inui, M. Radermacher, J. Frank and S. Fleischer: Ultrastructure of the Calcium Release Channel of Sarcoplasmic Reticulum. *J.Cell Biol.* **107**:211-(2/10) (1988).

Sakmann, B. and H.R. Brenner: Change in Synaptic Channel Gating During Neuromuscular Development. *Nature* **276**:401-(2/10) (1978).

Sakmann, B. and E. Neher: *Single Channel Recording*. Plenum Press, New York. (1/4,3/12) (1984).

Sakurai, M.: Synaptic Modification of Parallel Fibre-Purkinje Cell Transmission in *In Vitro* Guinea Pig Cerebellar Slices. *J.Physiol.* **394**:463-(2/9) (1987).

Sakurai, M., I.T. Yasuda, N. Kato, A. Nomura, M. Fujita, T. Nishino, K. Fujita, Y. Kolke and H. Saito: Acute and Chronic Effects of Verapamil in Patients with Paroxysmal Supraventricular Tachycardia. *Am.Heart J.* **105**:619-(4/19) (1983).

Saleh, M., R.J. Lang and P.F. Bartlett: Thy-1 Mediated Regulation of a Low Threshold Transient Calcium Current in Cultured Sensory Neurons. *Proc.Natl.Acad.Sci.U.S.A.* **85**:4543-(3/14) (1988).

Salkoff, L.B. and M.A. Tanouye: Genetics of Ion Channels. *Physiol.Rev.* **66**:301-(1/1) (1986).

Saltiel, E., A.G. Ellrodt, J.P. Monk and M.S. Langley: Felodipine: A Review of its Pharmacodynamic and Pharmacokinetic Properties and Therapeutic Use in Hypertension. *Drugs* **36**:387-(3/13) (1988).

Salzberg, B.M. and A.L. Obaid: Optical Studies of the Secretory Event of Vertebrate Nerve Terminals. *J.Exp.Biol.* **139**:195-(2/7) (1988).

Salzberg, B.M., A.L. Obaid, K. Staley, J.W. Lin, M. Sugimon, B.D. Cherksey and R. Llinas: FTX, a Low Molecular Weight Fraction of Funnel Web Spider Venom, Blocks Calcium Channels in Nerve Terminals of Vertebrates. *Biophys.J.* **57**:305a-(3/13) (1990).

Salzman, E.W. and J.A. Ware: Ionized Calcium as an Intracellular Messenger in Blood Platelets. *Prog.Hemost.Thrombo.* **9**:177-(4/20) (1989).

Samuelsson, B., M. Goldyne, E. Granstom, M. Hamberg, S. Hammarstorm and C. Malmsten: Prostaglandins and Thromboxanes. *Ann.Rev.Biochem.* **47**:997-(4/20) (1978).

Sanchez, J.A. and E. Stefani: Inward Calcium Current in Twitch Muscle Fibres of the Frog. *J.Physiol.* **283**:197-(2/10) (1978).

Sand, O., K. Sletholt, K.M. Gautvik and E. Haug: Trifluoperazine Blocks Calcium-Dependent Action Potentials and Inhibits Hormone Release from Rat Pituitary Tumor Cells. *Eur.J.Pharmacol.* **86**:177-(4/22) (1983).

Sandler, G., G.A. Clayton and S.G. Thornicroft: Clinical Evaluation of Verapamil in Angina Pectoris. *Br.Med.J.* **3**:224-(4/16) (1968).

Sandow, A.: Excitation-Contraction Coupling in Muscular Response. *Yale J.Biol.Med.* **25**:176-(2/10) (1952).

Sanguinetti, M.C. and R.S. Kass: Regulation of Cardiac Calcium Channel Current and Contractile Activity by the Dihydropyridine Bay K 8644 is Voltage Dependent. *J.Mol.Cell Cardiol.* **16**:667-(2/10) (1984b).

Sanguinetti, M.C. and R.S. Kass: Voltage-Dependent Block of Calcium Channel Current in Calf Cardiac Purkinje Fiber by Dihydropyridine Calcium Channel Antagonists. *Circ.Res.* **55**:36-(3/13) (1984a).

Sanguinetti, M.C., D.S. Krafte and R.S. Kass: Voltage-Dependent Modulation of Ca Channel Current in Heart Cells by Bay K8644. *J.Gen.Physiol.* **88**:369-(1/4,3/13) (1986).

Sapine, D., M. Schdelman and A. O'Riordan: Safety and Efficacy of Short and Long-Term Verapamil Therapy in Children with Tachycardia. *Am.J.Cardiol.* (abstr.) **47**:439-(4/19) (1981).

Sarkozi, J., A.A.M. Bookman, W. Mahon, C. Ramsay, A.S. Detsky and E.C. Keystone: Nifedipine in the Treatment of Idiopathic Raynaud's Syndrome. *J.Rheumatol.* **13**:331-(4/20) (1986).

Sarmiento, J.G., A.V. Shrikhande, R.A. Janis and D.J. Triggle: [^{3}H]BAY K 8644, a 1,4-Dihydropyridine Ca^{2+} Channel Activator: Characteristics of Binding to High and Low Affinity Sites in Cardiac Membranes. *J.Pharmacol.Exp.Ther.* **241**:140-(3/13) (1987).

Sarto, A., M. Inui and S. Fleischer: Ultrastructure of the Ca Release Channel (CRC) from Heart Sarcoplasmic Reticulum. *Biophys.J.* **55**:4-(2/10) (1989).

Sassone-Corsi, P., J.C. Sisson and I.M. Verma: Transcriptional Autoregulation of the Proto-Oncogne Fos. *Nature* **334**:314-(2/10) (1988).

Satin, L.S. and D.L. Cook: Evidence for Two Calcium Currents in Insulin-Secreting Cells. *Pflügers Arch.* **411**:401-(2/7) (1988).

Satin, L.S. and D.L. Cook: Calcium Current Inactivation in Insulin-Secreting Cells is Mediated by Calcium Influx and Membrane Depolarization. *Pflügers Archiv.* **414**:1-(1/4) (1989).

Satir, B.H., G. Busch, A. Vuoso and T.J. Murtaugh: Aspects of Signal Transduction in Stimulus Exocytosis-Coupling in Paramecium. *J.Cell Biochem.* **36**:429-(2/7) (1988).

Satir, B.H. and S.G. Oberg: Paramecium Fusion Rosettes: Possible Function as Ca^{2+} Gates. *Science* **199**:536-(2/7) (1978).

Sauter, A. and M. Rudin: Effects of Calcium Antagonists on High-Energy Phosphates in Ischemic Rat Brain Measured by ^{31}P NMR Spectroscopy. *Magn.Reson.Med.* **4**:1-(4/21) (1987).

Sauter, A., R. Rudin, K.H. Wiederhold and R.P. Hof: Cerebrovascular, Biochemical, and Cytoprotective Effects of Isradipine in Laboratory Animals. *Am.J.Med.* **86**:134-(3/13) (1989).

Sauza, J., A. Kraus, R. Gonzalez-Amaro and D. Alarcon-Segovia: Effect of the Calcium Channel Blocker Nifedipine on Raynaud's Phenomenon A Controlled Double-Blind Trial. *J.Rheumatol.* **11**:362-(4/20) (1984).

Scarlett, E.P. and W.J. Houghtling: Psychosis in Hypoparathyroidism. *Can.M.A.J.* **50**:351-(4/22) (1944).

Schaeffer, J. and M.P. Blaustein: Platelet Free Calcium Concentration Measured with Fura-2 are Influenced by the Transmembrane Sodium Gradient. *Cell Calcium* **10**:101-(4/20) (1989).

Schaeffer, P., C. Lugnier and J.C. Stoclet: Interaction of Calmodulin and Calcium Antagonists with [^{3}H]Nitrendipine Binding Sites. *J.Cardiovasc.Pharmacol.* **12**:S102-(3/13) (1988).

Schamroth, L.: Immediate Effects of Intravenous Verapamil on Atrial Fibrillation. *Cardiovasc.Res.* **5**:419-(4/19) (1971).

Schamroth, L., D.M. Krikler and C. Garrett: Immediate Effects of Intravenous Verapamil in Cardiac Arrhythmias. *Br.Heart J.* **1**:660-(4/19) (1972).

Scheidt, S., W.H. Frishman, M. Packer, J. Mehta, O. Parodi and V.B. Subramanian: Long-Term Effectiveness of Verapamil in Stable and Unstable Angina Pectoris. One-Year Follow-Up of Patients Treated in Placebo-Controlled Double-Blind Randomized Clinical Trials. *Am.J.Cardiol.* **50**:1185-(4/16) (1982).

Schepelern, S. and S. Koster: Verapamil in Treatment of Severe Schizophrenia. *Acta Psychiatr.Scand.* **75**:557-(4/22) (1987).

Schetz, J., H. Bostoen, D. Clement, M. Fornhoff, A. Haerens, P. Roekaerts and A.J. Staessen: Flunerizine in Chronic Obstructive Peripheral Arterial Disease: A Placebo-Controlled, Double Blind, Randomized Multicentre Trial. *Curr.Ther.Res.* **23**:121-(4/20) (1978).

Schilling, W.P. and J.A. Drewe: Voltage-Sensitive Nitrendipine Binding in an Isolated Cardiac Sarcolemma Preparation. *J.Biol.Chem.* **261**:2750-(3/13) (1986).

Schlant, R.C. and S.B. King: Usefulness of Calcium Entry Blockers During and after Percutaneous Transluminal Coronary Artery Angioplasty. *Circulation* **80**:88-(3/13) (1989).

Schlegel, W., B.P. Winiger, P. Mollard, P. Vacher, F. Wuarin, G.R. Zahnd, C.B. Wollheim and B. Dufy: Oscillations of Cytosolic Ca^{2+} in Pituitary Cells Due to Action Potentials. *Nature* **329**:719-(2/7) (1987).

Schmage, N., K. Boehme, J. Dycka and H. Schmitz: Nimodipine for Psychogeriatric Use: Methods, Strategies, and Considerations Based on Experience with Clinical Trials. In *Diagnosis and Treatment of Senile Dementia*. M. Bergener and B. Reisberg, eds. Springer-Verlag, Berlin and Heidelberg. 374-(3/13) (1989).

Schmid, A., T. Kazazoglou, J. Renaud and M. Lazdunski: Comparative Changes of Levels of Nitrendipine Ca^{2+} Channels, of Tetrodotoxin-Sensitive Na^+ Channels and of Ouabain-Sensitive $(Na^+ K^+)$-ATPase Following Denervation of Rat and Chick Skeletal Muscle. *FEBS Lett.* **172**:114-(3/13) (1984).

Schmid, A, G. Romey, J. Barhanin and M. Lazdunski: SR33557, an Indolizinsulfone Blocker of Ca^{2+} Channels: Identification of Receptor Sites and Analysis of its Mode of Action. *Mol.Pharmacol.* **35**:776-(3/13) (1989).

Schneider, M.F. and W.K. Chandler: Voltage Dependent Charge Movement in Skeletal Muscle: A Possible Step in Excitation-Contraction Coupling. *Nature* **242**:244-(2/10) (1973).

Schoemaker, H. and S.Z. Langer: Effects of Ca^{2+} on [^{3}H]Diltiazem Binding and its Allosteric Interaction with Dihydropyridine Calcium Channel Binding Sites in the Rat Cortex. *J.Pharmacol.Exp.Ther.* **248**:710-(3/13) (1989).

Schoen, R.E., W.H. Frishmann and H. Shamoon: Hormonal and Metabolic Effects of Calcium Channel Antagonists in Man. *Am.J.Med.* **84**:492-(3/13) (1988).

Schroeder, J.S., S.D. Walker, M.L. Skalland and J.A. Hemberger: Absence of Rebound from Diltiazem Therapy in Prinzmetal's Variant Angina. *J.Am.Coll.Cardiol.* **6**:174-(4/16) (1985).

Schubert, D., M. LaCorbiere, C. Whitlock and W. Stallcup: Alterations in the Surface Properties of Cells Responsive to Nerve Growth Factor. *Nature* **273**:718-(2/10) (1978).

Schuermans, V., A. Bofrmans, H. Geivers, A. Jageneau and J. Brugmans: Cinnarizine in Peripheral Vascular Insufficiency. *Drug Res.* **21**:154-(4/20) (1971).

Schulz, W., S. Jost, G. Kober and M. Kaltenback: Relation of Antianginal Efficacy of Nifedipine to Degree of Coronary Arterial Narrowing and to Presence of Coronary Collateral Vessels. *Am.J.Cardiol.* **55**:26-(4/16) (1985).

Schurtz, C., J.P. Lesbre, A. Kalisa and J.G. Fardelonne: Interet des Inhibiteurs Calciques Dans L'angor D'effort Stable: Diltiazem Versus Nifedipine. *Ann.Cardiol.Angiol.* **32**:337-(4/16) (1983).

Schuurman, T. and J. Traber: Old Rats as an Animal Model for Senile Dementia: Behavioral Effects of Nimodipine. **In** *Diagnosis and Treatment of Senile Dementia.* M. Bergener and B. Reisberg, eds. Springer-Verlag, Berlin and Heidelberg. 295-(3/13) (1990).

Schwartz, A. and D.J. Triggle: Cellular Action of Calcium Channel Blocking Drugs *Ann.Rev.Med.* **35**:325-(4/17) (1984).

Schwartz, A. and P.A. van Zwieten: A Symposium: Pharmacology and Therapeutic Considerations of Amlodipine, a New 1,4-Dihydropyridine Calcium Antagonist. *Am.J.Cardiol.* **64**:1-(3/13) (1989).

Schwartz, E.A.: Synaptic Transmission in Amphibian Retinae During Conditions Unfavourable for Calcium Entry into Presynaptic Terminals. *J. Physiol.* **376**:411-(2/7) (1986).

Schwartz, J.B. and D.R. Abermethy: Responses to Intravenous and Oral Diltiazem in Elderly and Younger Patients with Systemic Hypertension. *Am.J.Cardiol.* **59**:1111-(4/18) (1987).

Schwartz, L.M., W.E. McCleskey and W. Almers: Dihydropyridine Receptors in Muscle are Voltage-Dependent but most are not Functional Calcium Channels. *Nature* **314**:747-(2/10,3/13) (1985).

Schwarz, T.L., B.L. Tempel, D.M. Papzian, Y.N. Jan and L.Y. Jan: Multiple Potassium Channel Components are Produced by Alternative Splicing at the Shaker Locus in Drosophila. *Nature* **331**:137-(3/14) (1988).

Sclarovsky, S., B. Strasberg, J. Fuchs, R.F. Lewin, E. Arditi, E. Kleinman, O.H. Kracoff and J. Agmon: Multiform Accelerated Idioventricular Rhythm in Acute Myocardial Infarction. *Am.J.Cardiol.* **52**:43-(4/19) (1983).

Scott, B.S.: The Effects of Elevated Potassium on the Time Course of Neuron Survival in Cultures of Dissociated Dorsal Root Ganglia. *J.Cell Physiol.* **91**:305-(2/10) (1977).

Scott, R.H. and A.C. Dolphin: Activation of a G-Protein Promotes Agonist Responses to Calcium Channel Ligands. *Nature* **330**:760-(3/13,3/14) (1987).

Scott, R.H. and A.C. Dolphin: The Agonist Effect of BAY K 8644 on Neuronal Calcium Channel Currents is Promoted by G-protein Activation. *Neurosci.Lett.* **89**:170-(1/4,3/14) (1988).

Scott, R.H. and A.C. Dolphin: G-Protein Regulation of Neuronal Voltage-Activated Calcium Currents. *Gen.Pharmacol.* **20**:715-(3/13) (1989).

Scriabine, A.: Ca^{2+} Channel Ligands: Comparative Pharmacology. **In** *Structure and Physiology of the Slow Inward Calcium Channel.* J.C. Venter and D. Triggle, eds. Alan R. Liss Inc., New York. 51-(3/13) (1987).

Scriabine, A. and S. Kazda: Pharmacological Basis for Use of Calcium Antagonists in Hypertension. *Magnesium* **8**:253-(3/13) (1989a).

Scriabine, A. and M. Pan: Ca^{2+} Antagonists and Inhibitory Effects of Dopamine on Isolated Rabbit Ear Artery. *J.Cardiovasc.Pharmacol.* **12**:S107-(3/13) (1988).

Scriabine, A., T. Schuurman and J. Traber: Pharmacological Basis for the Use of Nimodipine in Central Nervous System Disorders. *FASEB J.* **3**:1799-(3/13) (1989b).

Seabra-Gomes, R., A. Rickard and R. Sutton: Hemodynamic Effects of Verapamil and Practolol in Man. *Eur.J.Cardiol.* **4**:79-(4/19) (1976).

Seiler, S.M., A.J. Arnold and H.C. Stanton: Inhibitors of Inositol Trisphaphate-Induced Ca^{2+} Release from Iolated Platelet Membrane Vesicles. *Biochem.Pharmacol.* **36**:3331-(4/20) (1987).

Seipel, L. and G. Breithardt: Electrophysiological Actions of Calcium Antagonists in the Heart. *Cardiology* **1**:105-(4/19) (1982).

Seipel, L., G. Breithardt, R.R. Abendroth and E. Wiebringhaus: The Electrophysiological Effects of Ca-Antagonists Gallopamil, Dimeditiapramine and Verapamil in Man. *Z.Kardiol.* **68**:551-(4/19) (1980).

Sellers, T.D., R.W.F. Campbell, T.M. Bashore and J.J. Gallagher: Effects of Procainamide and Quinidine Sulphate in the Wolff-Parkinson-White Syndrome. *Circulation* **55**:15-(4/19) (1977).

Sethi, K.K., S. Jaishankar and M.P. Gupta: Salutary Effects of Intravenous Ajmaline in Patients with Paroxysmal Supraventricular Tachycardia Mediated by Dual Atrioventricular Nodal Pathways: Blockade of the Retrograde Fast Pathway. *Circulation* **70**:876-(4/19) (1984).

Sethi, K.K., S. Jaishankar, M. Khallilulah and M.P. Gupta: Selective Blockade of Retrograde Fast Pathway by Intravenous Disopyramide in Paroxysmal Supraventricular Tachycardia Mediated by Dual AV Nodal Pathways. *Br.Heart J.* **49**:532-(4/19) (1983).

Shainberg, A. and M. Burstein: Decrease of Acetylcholine Receptor Synthesis in Muscle Cultures by Electrical Stimulation. *Nature* **264**:368-(2/10) (1976).

Shainberg, A., G. Yagil and D. Yaffe: Control of Myogenesis *In Vitro* by Ca^{2+} Concentration in Nutritional Medium. *Exp.Cell Res.* **58**:163-(2/10) (1969).

Shainberg, A., G. Yagil and D. Yaffe: Alteration of Enzymatic Activities during Muscle Differentiation *In Vitro*. *Dev.Biol.* **25**:1-(2/10) (1971).

Shakibi, J.G.: Arrhythmias in Infants and Children. *Pediatrician* **10**:117-(4/19) (1981).

Shanbaky, N., Y.S. Ahn, W. Jy, W.J. Harrington and D.H. Haynes: Abnormal Aggregation Accompanies Abnormal Platelet Ca^{2+} Handling in Arterial Thrombosis. *Thromb.Haemost.* **57**:1-(4/20) (1987).

Shattil, S.J. and J.S. Bennett: Platelets and their Membranes in Hemostasis: Physiology and Pathophysiology. *Ann.Int.Med.* **94**:108-(4/20) (1981).

Shattil, S.J. and L.F. Brass: Induction of the Fibrinogen Receptor on Human Platelets by Intracellular Mediators. *J.Biol.Chem.* **262**:992-(4/20) (1987).

Shears, S.B.: Lithium and Inositol Lipid Turnover. In *Lithium: Inorganic Pharmacology and Psychiatric Use.* J. Birch, ed. I.R.L. Press, Oxford. 201-(4/22) (1988).

Shen, W.F., G.S. Roubin, C.Y. Choong, B.F. Hutton, P.J. Harris, P.J. Fletcher and D.T. Kelly: Left Ventricular Response to Exercise in Coronary Artery Disease: Relation to Myocardial Ischemia and Effects of Nifedipine. *Eur.Heart J.* **6**:1025-(4/18) (1985).

Sheng, M., S.T. Dougan, G. McFadden and M.E. Greenberg: Calcium and Growth Factor Pathways of c-fos Transcriptional Activation Require Distinct Upstream Regulatory Sequences. *Mol.Cell Biol.* **8**:2787-(2/10) (1988).

Sher, E., N. Canal, G. Piccolo, C. Gotti, C. Scoppetta, A. Evoli and F. Clementi: Specificity of Calcium Channel Autoantibodies in Lambert-Eaton Myasthenic Syndrome. *Lancet* **16**:640-(3/13) (1989).

Sher, E., A. Pandiella and F. Clementi: ω-Conotoxin Binding and Effects on Calcium Channel Function in Human Neuroblastoma and Rat Pheochromocytoma Cell Lines. *FEBS Lett.* **235**:178-(1/4) (1988).

Sherman, L.G. and C.S. Liang: Nifedipine in Chronic Stable Angina: A Double-Blind Placebo-Controlled Crossover Trial. *Am.J.Cardiol.* **51**:706-(4/16) (1983).

Sherman, S.J. and W.A. Catterall: Biphasic Regulation of Development of the High-Affinity Saxitoxin Receptor by Innervation in Rat Skeletal Muscle. *J.Gen.Physiol.* **80**:753-(2/10) (1982).

Sherman, S.J. and W.A. Catterall: Electrical Activity and Cytosolic Calcium Regulate Levels of Tetrodotoxin-Sensitive Sodium Channels in Cultured Rat Muscle Cells. *Proc.Natl.Acad.Sci.U.S.A.* **81**:262-(2/10) (1984).

Sherman, S.J., J. Chrivia and W.A. Catterall: Cyclic Adenosine 3':5'-Monophosphate and Cytosolic Calcium Exert Opposing Effects on Biosynthesis of Tetrodotoxin-Sensitive Sodium Channels in Rat Muscle Cells. *J.Neurosci.* **5**:1570-(2/10) (1985).

Shieh, B.H., M. Ballivet and J. Schmidt: Quantitation of an Alpha Subunit Splicing Intermediate: Evidence for Transcriptional Activation in the Control of Acetylcholine Receptor Expression in Denervated Chick Skeletal Muscle. *J.Cell Biol.* **104**:1337-(2/10) (1987).

Shigenobu, K. and N. Sperelakis. Development of Sensitivity to Tetrodotoxin of Chick Embryonic Hearts with Age. *J.Mol.Cell Cardiol.* **3**:271-(4/17) (1971).

Shoback, D.M., L.A. Membreno and J.G. McGhee: High Calcium and Other Divalent Cations Increase Inositol Triphosphate in Bovine Parathyroid Cells. *Endocrinology* **123**:382-(2/7) (1988).

Shoback, D.M., J. Thatcher, R. Leombruno and E.M. Brown: Relationship Between Parathyroid Hormone Secretion and Cystosolic Calcium Concentration in Dispersed Bovine Parathyroid Cells. *Proc.Natl.Acad.Sci.U.S.A.* **81**:3113-(2/7) (1984).

Shrager, P.: Ionic Conductance Changes in Voltage Clamped Crayfish Axons at Low pH. *J.Gen.Physiol.* **64**:666-(3/12) (1974).

Sia, S.T.B., P.S. MacDonald, B. Triester, L.E. Oliver and J.D. Horowitz: Aggravation of Myocardial Ischemia by Nifedipine. *Med.J.Aust.* **142**:48-(4/16) (1985).

Siegelbaum, S.A., A. Trautmann and J. Konig: Single Acetylcholine-Activated Channel Currents in Developing Muscle Cells. *Dev.Biol.* **104**:366-(2/10) (1980).

Siegl, P.K.S., M.L. Garcia, V.F. King, G. Morgan and G.J. Kaczorowski: Interactions of DPI 201-106, a Novel Cardiotonic Agent, with Cardiac Calcium Channels. *Naunyn Schmied.Arch.Pharmacol.* **338**:684-(3/13) (1988).

Siesjö, B.K.: Cerebral Circulation and Metabolism. *J.Neurosurg.* **60**:883-(4/21) (1984).

Siesjö, B.K. and F. Bengtsson: Calcium Fluxes, Calcium Antagonists, and Calcium-Related Pathology in Brain Ischemia, Hypoglycemia, and Spreading Depression: A Unifying Hypothesis. *J.Cereb.Blood Flow Metab.* **9**:127-(3/13,4/21) (1989).

Sievers, R., J. Wikman-Coffelt, W.W. Parmley and G. Jasmin: Verapamil Preserves Adenine Nucleotide Pool in Cardiomyopathic Syrian Hamsters. *Am.J.Physiol.* 251:H22-(4/17) (1986).

Sievers, R.E., T. Rashid and H. Garrett: Verapamil and Diet Halt Progression of Atherosclerosis in Cholesterol Fed Rabbits. *Cardiovasc.Drugs Ther.* **1**:65-(4/16) (1987).

Sigl, E. and R. Baur: Activation of Protein Kinase C Differentially Modulates Neuronal Na^+, Ca^{2+}, and Gamma-Aminobutyrate Type A Channels. *Proc.Natl.Acad.Sci.U.S.A.* **85**:6192-(3/14) (1988).

Silinsky, E.M.: The Biophysical Pharmacology of Calcium-Dependent Acetylcholine Secretion. *Pharmacol.Rev.* **37**:81-(2/7) (1985).

Silke, B., S.P. Verma, M. Hussain, N.C. Jackson, M. Hafizullah, S.G. Reynolds and S.H. Taylor: Comparative Hemodynamic Effects of Nicardipine and Verapamil in Coronary Artery Disease. *Herz* **10**:112-(4/18) (1985).

Simes, P.A., K. Overskeid, T.R. Pedersen, J. Bathen, A. Drivenes, G.S. Froland, J.K. Kjekshus, K. Landmark, R. Rokseth, K.E. Simes, A. Sundoy, B.R. Torjussen, K.M. Westlund and B.A. Wik: Evolution of Infarct Size During the Early Use of Nifedipine in Patients with Acute Myocardial Infarction: The Norwegian Nifedipine Multicenter Trial. *Circulation* **70**:638-(4/16) (1984).

Simmons, M.A., E.C. Johnson, J.B. Becker, D.G. Todd, V.E. Reichenbecher, W.D. McCumbee and G.L. Wright: An Endogenous "Hypertensive Factor" Enhances the Voltage-Dependent Calcium Current. *FEBS Lett.* **254**:137-(3/13) (1989).

Simon, G. and D.K. Snyder: Altered Pressor Responses in Long-Term Nitrendipine Treatment. *Clin.Pharmacol.Ther.* **36**:315-(4/18) (1984).

Simon, R.P., T. Griffiths, M.C. Evans, J.H. Swan and B.S. Meldrum: Calcium Overload in Selectivity Vulnerable Neurons of the Hippocampus During and After Ischemia: An Electron Microscopy Study in the Rat. *J.Cereb.Blood Flow Metab.* **4**:350-(4/21) (1984a).

Simon, R.P., J.H. Swan, T. Griffiths and B.S. Meldrum: Blockade of N-Methyl-D-Aspartate Receptors may Protect Against Ischemic Damage in the Brain. *Science* **226**:850-(4/21) (1984b).

Simon, S.M. and R.R. Llinas: Compartmentalization of the Submembrane Calcium Activity during Calcium Influx and its Significance in Transmitter Release. *Biophys.J.* **48**:485-(2/7) (1985).

Simoncini, L., M.L. Block and W.J. Moody: Lineage-specific Development of Calcium Currents during Embryogenesis. *Science* **242**:1572-(2/10) (1988).

Singer, S.J. and G.L. Nicolson: The Fluid Mosaic Model of the Structure of Cell Membrane. *Science* **175**:720-(3/12) (1972).

Singh, B.N.: Pharmacological Basis for the Therapeutic Applications of Slow-Channel Blocking Drugs. *Angiology* **33**:492-(4/19) (1982).

Singh, B.N.: Amiodarone: Historical Development and Pharmacologic Profile. *Am.Heart J.* **106**:788-(4/19) (1983).

Singh, B.N., S. Baky and N. Koonlawee: Second-Generation Calcium Antagonists: Search for Greater Selectivity and Versatility. *Am. J. Cardiol.* **55**:214B-(4/15) (1985).

Singh, B.N., G. Ellrodt and C.T. Peter: Verapamil: A Review of its Pharmacological Properties and Therapeutic Use. *Drugs* **15**:169-(4/19) (1978).

Singh, B.N., H.S. Hecht, K. Nademanee and C.Y.C. Chew: Electrophysiological and Hemodynamic Actions of Slow-Channel Blocking Compounds. *Prog.Cardiovasc.Dis.* **25**:103-(4/19) (1982).

Singh, B.N. and M.A. Josephson: Clinical Pharmacology, Pharmacokinetics, and Hemodynamic Effects of Nicardipine. *Am.Heart J.* **119**:427-(3/13) (1990).

Singh, B.N., K. Nademanee and S. Baky: Calcium Antagonists: Clinical Uses in Treating Arrhythmias. *Drugs* **25**:125-(4/19) (1983).

Singh, B.N. and W.G. Nayler: The Role of Calcium Antagonists in Acute Myocardial Infarction. In *Early Interventions in Acute Myocardial Infarction*. E. Rapaport, ed. Kluwer Academic Publications, Boston. 123-(4/19) (1989).

Singh, B.N. and E.M. Vaughan Williams: A Fourth Class of Antiarrhythmic Action? Effects of Verapamil on Ouabain Toxicity on Atrial and Ventricular Intracellular Potentials and Other Features of Cardiac Function. *Cardiovasc.Res.* **6**:109-(4/19) (1972).

Sjoberg, T., K.E. Anderson, L. Norgren and S. Steen: Comparative Effects of Some Calcium-Channel Blockers on Human Peripheral Arteries and Veins. *Acta Physiol.Scand.* **130**:419-(4/20) (1987a).

Sjoberg, T., S. Steen, T. Skarby, L. Norgren and K.E. Anderson: Postjunctional-Adrenoceptors in Human Superficial Epigastric Arteries and Veins. *Pharmacol.Toxicol.* **60**:43-(4/20) (1987b).

Skattebol, A., A.M. Brown and D.J. Triggle: Homologous Regulation of Voltage-Dependent Ca^{2+} Channels by 1,4-Dihydropyridines. *Biochem.Biophys.Res.Commun.* **160**:929-(3/13) (1989).

Skattebol, A., R.E. Hruska, M. Hawthorn and D.J. Triggle: Kainic Acid Lesions Decrease Striatal Dopamine Receptors and 1,4-Dihydropyridine Binding Sites. *Neurosci.Lett.* **89**:85-(3/13) (1988).

Skattebol, A. and R.A. Rabin: Effects of Ethanol on $^{45}Ca^{2+}$ Uptake in Synaptosomes and in PC 12 Cells. *Biochem.Pharmacol.* **36**:2227-(3/13) (1987b).

Skattebol, A. and D.J. Triggle: 6-Hydroxydopamine Treatment Increases Beta-Adrenoceptors and Ca^{2+} Channels in Rat Heart. *Eur.J.Pharmacol.* **127**:287-(3/13) (1986).

Skattebol, A. and D.J. Triggle: Regional Distribution of Calcium Channel Ligand [1,4-Dihydropyridine] Binding Sites and $^{45}Ca^{2+}$-Uptake Processes in Rat Brain. *Biochem.Pharmacol.* **36**:4163-(3/13) (1987).

Skerritt, J.H. and G.A.R. Johnston: Inhibition of Amino Acid Transmitter Release by Diphenylhydantoin and other Anticonvulsants. *Neurosci.Lett.* **8**:79-(4/22) (1982).

Slish, D.F., D.B. Engle, G. Varadi, I. Lotan, D. Singer, N. Dascal and A. Schwartz: Evidence for the Existence of a Cardiac Specific Isoform of the α1 Subunit of the Voltage Dependent Calcium Channel. *FEBS Lett.* **250**:509-(3/13) (1989).

Smith, C.D. and R.J.R. McKendry: Controlled Trial of Nifedipine in the Treatment of Raynaud's Phenomenon. *Lancet* (Dec. 11):1299-(4/20) (1982).

Smith, H.J. and M.G. Briscoe: The Relative Sensitization of Acidosis of Five Calcium Blockers in Cat Papillary Muscles. *J.Mol.Cell Pharmacol.* **17**:709-(3/13) (1985).

Smith, H.J., R.A. Goldstein and J.M. Griffith: Regional Contractility. Selective Depression of Ischemic Myocardium by Verapamil. *Circulation* **70**:638-(4/16) (1984).

Smith, H.J., B.N. Singh, H.D. Nisbet and R.M. Norris: Effects of Verapamil on Infarct Size Following Experimental Coronary Occlusion. *Cardiovasc.Res.* **9**:569-(4/19) (1975).

Smith, J.S., R. Coronado and G. Meissner: Sarcoplasmic Reticulum Contains Adenine Nucleotide-Activated Calcium Channels. *Nature* **316**:446-(2/10) (1985).

Smith, J.S., R. Coronado and G. Meissner: Single Channel Measurements of the Calcium Release Channel from Skeletal Muscle Sarcoplasmic Reticulum. *J.Gen.Physiol.* **88**:573-(2/10) (1986).

Smith, J.S., T. Imagawa, J. Ma, M. Fill, K.P. Campbell and R. Coronado: Purified Ryanodine Receptor from Rabbit Skeletal Muscle is the Calcium Release Channel of Sarcoplasmic Reticulum. *J.Gen.Physiol.* **92**:1-(2/10) (1988).

Smith, J.S., E.J. McKenna, J. Ma, J. Vilven, P.L. Vaghy, A. Schwartz and R. Coronado: Calcium Channel Activity in Purified Dihydropyridine Receptor Preparation of Skeletal Muscle. *Biochemistry* **26**:7182-(3/13) (1987).

Smith, M.A., J.F. Margiotta, A. Franco, Jr., J.M. Lindstrom and D.K. Berg: Cholinergic Modulation of an Acetylcholine Receptor-Like Antigen on the Surface of Chick Ciliary Ganglion Neurons in Cell Culture. *J.Neurosci.* **6**:946-(2/10) (1986).

Smith, P.A., P. Rosman and F.M. Ashcroft: Modulation of Dihydropyridine Sensitive Ca^{2+} Channels by Glucose Metabolism in Mouse Pancreatic Beta-Cells. *Nature* **342**:550-(3/14) (1989).

Smith, S.J.: Neuronal Cytomechanics: The Actin-Based Motility of Growth Cones. *Science* **242**:708-(2/10) (1988).

Smith, S.J. and G.J. Augustine: Calcium Ions, Active Zones, and Synaptic Tansmitter Release. *TINS* **11**:458-(2/5,2/7,2/9,3/14) (1988).

Smith, S.J. and S.H. Thompson: Slow Membrane Currents in Bursting Pacemaker Neurones of Tritonia. *J.Physiol.* **382**:425-(2/9) (1987).

Snyder, S.H.: Drug and Neurotransmitter Receptors in the Brain. *Science* **224**:22-(4/22) (1984).

Sokabe, M., N. Fujitsuka, A.A. Kori and F. Ito: Effects of Cyclic Nucleotides and Calcium on Transduction and Encoding Processes in Frog Muscle Spindle. *Brain Res.* **443**:254-(2/8) (1988).

Solar-Soler, J., J. Sagrista-Asuleda and A. Cabrera: Effect of Verapamil in Infants with Paroxysmal Supraventricular Tachycardia. *Circulation* **59**:876-(4/19) (1979).

Sole, M.J., C.M. Lo, C.W. Laird, E.H. Sonnenblick and R.J. Wurtman: Norepinephrine Turnover in the Heart and Spleen of the Cardiomyopathic Hamster. *Circ.Res.* **37**:855-(4/17) (1975).

Somlyo, A.P.: Excitation-Contraction Coupling and the Ultrastructure of Smooth Muscle. *Circ.Res.* **57**:497-(2/10) (1985a).

Somlyo, A.P. and A.V. Somlyo: Vascular Smooth Muscle. I. Normal Structure, Pathology, Biochemistry, and Biophysics. *Pharmacol.Rev.* **20**:197-(2/10) (1968).

Somlyo, A.P., J.W. Walker, Y.E. Goldman, D.R. Trentham, F.R.S. Kobayashi, T. Kitazawa and A.V. Somlyo: Inositol Trisphosphate, Calcium and Muscle Contraction. *Philos.Trans.R.Soc.Lond.* **320**:399-(2/10) (1988).

Somlyo, A.V., M. Bond, A.P. Somlyo and A. Scarpa: Inositol Trisphosphate-Induced Calcium Release and Contraction in Vascular Smooth Muscle. *Proc.Natl.Acad.Sci.U.S.A.* **82**:5231-(2/10) (1985b).

Somlyo, A.V. and C. Franzini-Armstrong: New Views of Smooth Muscle Structure Using Freezing, Deep-Etching, and Rotary Shadowing. *Experientia* **41**:841-(2/10) (1985c).

Sonnenberg, J.L., F.J. Rausche, J.I. Morgan and T. Curran: Regulation of Proenkephalin by Fos and Jun. *Science* **246**:1622-(2/10) (1989).

Sorkin, E.M. and S.P. Clissold: Nicardipine: A Review of its Pharmacodynamic and Pharmacokinetic Properties, and Therapeutic Efficacy, in the Treatment of Angina Pectoris, Hypertension and Related Cardiovascular Disorders. *Drugs* **33**:296-(3/13) (1987).

Spalding, B.C.: Properties of Toxin-Resistant Sodium Channels Produced by Channel Modification in Frog Skeletal Muscle. *J.Physiol.* **305**:485-(3/12) (1980).

Spear, J.F., L.N. Horowitz, A.B. Hodess, H. MacVaugh and E.N. Moore: Cellular Electrophysiology of Human Myocardial Infarction. Abnormalities of Cellular Activation. *Circulation* **59**:247-(4/19) (1979).

Speckmann, E.J., O.W. Witte and J. Walden: Involvement of Calcium Ions in Focal Epileptic Activity of the Neocortex. In *Calcium Electrogenesis and Neuronal Functioning*. V. Heinemann, M. Klee, E. Neher, W. Singer, eds. Springer, Berlin. (4/22) (1986).
Spedding, M.: Activators and Inactivators of Ca^{2+} Channels: New Perspectives. *J.Pharmacol.* **16**:319-(3/13) (1985).
Spedding, M., A.J. Anderson and L. Patmore: Definition of the Interactions of Acyl Carnitines, BAY K 8644 and of Diphenylalkylamine Calcium-Antagonists. *4th International Symposium on Calcium Antagonists: Pharmacology and Clinical Research* 207-(3/13) (1989).
Spedding, M., M. Gittos and A.K. Mir: Calcium Antagonist Properties of Diclofurime Isomers. I. Functional Aspects. *J.Cardiovasc.Pharmacol.* **9**:461-(3/13) (1987b).
Spedding, M. and A.K. Mir: Direct Activation of Ca^{2+} Channels by Palmitoyl Carnitine, Putative Endogenous Ligand. *Br.J.Pharmacol.* **92**:457-(3/13) (1987a).
Spencer, W.A. and E.R. Kandel: Electrophysiology of Hippocampal Neurons: IV. Fast Prepotentials. *J.Neurophysiol.* **24**:272-(2/9) (1961).
Sperelakis, N.: Regulation of Calcium Slow Channels of Cardiac Muscle by Cyclic Nucleotides and Phosphorylation. *J.Mol.Cell Cardiol.* **20**:75-(4/17) (1988).
Sperelakis, N. and G. Bkaily: Cultured Cell Models for Studying Problems in Cardiac Toxicology. In *In Vitro Methods in Toxicology*. E.K. Atterwill, C.E. Steele, eds. Cambridge University Press, New York. 59-(4/17) (l987).
Sperelakis, N., G. Bkaily, H. Sada and M. Kojima: Development of Conduction and Contraction Systems in the Heart. In *Fetal and Neonatal Development*. C.T. Jones, ed. Perinatology Press, Ithaca, New York. 113-(4/17) (1988).
Sperti, G. and W.S. Colucci: Phorbol Ester Stimulated Bidirectional Transmembrane Calcium Flux in A7r5 Vascular Smooth Muscle Cells. *Mol.Pharmacol.* **32**:37-(3/14) (1987).
Spiegler, K.S. and M.R.J. Wyllie: Electrical Potential Differences. In *Physical Techniques in Biological Research*. G. Oster and A. Pollister, eds. Academic Press, New York. **2**:301-(3/12) (1956).
Spires, S., D.M. Van Skiver, D.J. Plotkin and C.J. Cohen: High Affinity and Tissue Specific Block of Ca Channels by Fluspirilene. *Biophys.J.* **53**:233a-(3/13) (1988).
Spitzer, N.C.: Low pH Selectively Blocks Calcium Action Potentials in Amphibian Neurons Developing in Culture. *Brain Res.* **161**:555-(3/12) (1979).
Spitzer, N.C.: Voltage and Stage-Dependent Uncoupling of Rohon-Beard Neurones during Embryonic Development of *Xenopus* Tadpoles. *J.Physiol.* **330**:145-(2/10) (1982).
Spitzer, N.C.: The Control of Development of Neuronal Excitability. In *Molecular Bases of Neural Development*. G.E. Edelman, W.E. Gall, and W.M. Cowan, eds. Wiley, New York. 67-(2/10) (1985).
Spitzer, N.C.: Reconstruction of Action Potentials of Embryonic Spinal Neurons from Whole Cell Voltage Clamped Currents. *Biophys.J.* **53**:258a-(2/10) (1988).
Spitzer, N.C., R.C. de Baca and J. Holliday: Developmental Acquisition of GABA-Like Immunoreactivity by Amphibian Spinal Neurons Differentiating *In Vitro*. *Soc.Neurosci.Abs.* **14**:163-(2/10) (1988).
Spitzer, N.C. and J.E. Lamborghini: The Development of the Action Potential Mechanism of Amphibian Neurons Isolated in Culture. *Proc.Natl.Acad.Sci.U.S.A.* **73**:1641-(2/10) (1976).
Staessen, A.J.: Treatment of Circulatory Disturbances with Flunarizine and Cinnarizine. *VASA* **6**:59-(4/20) (1977).
Stafstrom, C.E., P.C. Schwindt, M.C. Chubb and E. Crill: Properties of Persistent Sodium Conductance and Calcium Conductance of Layer V Neurons from Cat Sensorimotor Cortex *In Vitro*. *J.Neurophysiol.* **53**:152-(2/9) (1985).
Stanfield, P.R.: A Calcium Dependent Inward Current in Frog Skeletal Muscle Fibres. *Pflügers Arch.* **368**:267-(2/10) (1977).
Starke, K. and H.J. Schümann: Wirkung von Nifedipine auf die Funktion der sympathischen Nerven des Herzens. *Arzneim. Forsch.* **23**:193-(4/22).
Starke, K., L. Spath and T. Wichmann: Effects of Verapamil, Diltiazem and Ryosidine on the Release of Dopamine and Acetylcholine in Rabbit Caudate Nucleus Slices. *Naunyn Schmied.Arch.Pharmacol.* **325**:124-(4/22) (1984).
Steen, S., T. Sjoberg, T.V.C. Skarby, L. Norgren and K.E. Andersson: Postjunctional A1-and A2-Adrenoceptors Mediating Contraction in Isolated Human Groin Arteries and Veins. *Acta Physiol.Scand.* **122**:323-(4/20) (1984).
Steen, S., T.V.C. Skarby, L. Norgren and K.E. Andersson: Pharmacological Characterization of Postjunctional-Adrenoceptors in Isolated Human Omental Arteries and Vein. *Acta Physiol.Scand.* **120**:109-(4/20) (1984).
Stein, O., E. Leitersdorf and Y. Stein: Verapamil Enhances Receptor-Mediated Endocytosis of Low Density Lipoproteins by Aortic Cells in Culture. *Arteriosclerosis* **5**:35-(4/20) (1985).
Steinhardt, R.A. and J.M. Alderton: Calmodulin Confers Calcium Sensitivity on Secretory Exocytosis. *Nature* **295**:154-(2/7) (1982).
Steinhardt, R., R. Zucker and G. Schatten: Intracellular Calcium Release at Fertilization in the Sea Urchin Egg. *Dev.Biol.* **58**:185-(2/7) (1977).
Steinhardt, R.A. and D. Epel: Activation of Sea Urchin Eggs by a Calcium Ionophore. *Proc.Natl.Acad.Sci.U.S.A.* **71**:1915-(2/7) (1974).
Stenberg, P.E., M.A. Shuman, S.P. Levine and D.F. Bainton: Redistribution of Alpha Granules and their Contents in Thrombin Stimulated Platelets. *J.Cell Biol.* **98**:748-(4/20) (1984).
Stern, J.H., U.B. Kaupp and P.R. MacLeish: Control of the Light-Regulated Current in Rod Photoreceptors by Cyclic GMP, Calcium and 1-cis-Diltiazem. *Proc.Natl.Acad.Sci.U.S.A.* **83**:1163-(2/8) (1986).

Stevens, M.J. and R.F.W. Moulds: Heterogeneity of Post-Junctional-Adrenoceptors in Human Vascular Smooth Muscle. *Arch.Int.Pharmacodyn.* **254**:43-(4/20) (1981).

Stewart, R.B., G.H. Bardy and H.L. Greene: Wide Complex Tachycardia: Misdiagnosis and Outcome after Emergency Therapy. *Ann.Int.Med.* **104**:766-(4/19) (1986).

Stirling, C.E. and A. Lee: ^{3}H Ouabain Autoradiography of Frog Retina. *J.Cell Biol.* **85**:313-(2/8) (1980).

Stockbridge, N. and W.N. Ross: Localized Ca^{2+} and Calcium-Activated Potassium Conductances in Terminals of Barnacle Photoreceptor. *Nature* **309**:266-(2/7) (1984).

Stoclet, J.C., C. Boulanger-Saunier, B. Lassegue and C. Lugnier: Cyclic Nucleotides and Calcium Regulation in Heart and Smooth Muscle Cells. *Ann.N.Y.Acad.Sci.* **522**:106-(4/17) (1988).

Stoehr, S.J., J.E. Smolen, R.W. Holz and B.W. Agranoff: Inositol Trisphosphate Mobilizes Intracellular Calcium in Permeabilized Adrenal Chromaffin Cells. *J.Neurochem.* **46**:637-(2/7) (1986).

Stone, P.H.: Calcium Antagonists for Prinzmetal's Variant Angina, Unstable Angina and Silent Myocardial Ischemia. Therapeutic Tool and Probe for Identification of Pathophysiologic Mechanisms. *Am.J.Cardiol.* **59**:101-(4/16) (1987).

Stone, P.H., E.M. Antman, J.E. Muller and E. Braunwald: Calcium-Blocking Agents in the Treatment of Cardiovascular Disorders. Part II. Hemodynamic Effects and Clinical Applications. *Ann.Intern.Med.* **93**:886-(4/19) (1980).

Stone, P.H., J.E. Muller, Z.G. Turi, E. Geltman, A.S. Jaffe and E. Braunwald: Efficacy of Nifedipine Therapy in Patients with Refractory Angina Pectoris: Significance of the Presence of Coronary Vasospasm. *Am.Heart J.* **106**:644-(4/16) (1983).

Storstein, O. and K.H. Landmark: Verapamil in the Treatment of Atrial Tachycardia with Block. *Acta Med.Scand.* **198**:483-(4/19) (1975).

Storstein, O. and K. Rasmussen: Digitalis and Atrial Tachycardia with Block. *Br.Heart J.* . **36**:171-(4/19) (1976).

Strandgaard, S., J. Olesen, E. Skinhoj and N.A. Lassen: Autoregulation of Brain Circulation and Severe Arterial Hypertension. *Br.Med.J.* **1**:507-(4/18) (1973).

Strauss, W.E. and A.F. Parisi: Superiority of Combined Diltiazem and Propranolol Therapy for Angina Pectoris. *Circulation* **71**:951-(4/16) (1985).

Strichartz, G., T. Rando and G.K. Wang: An Integrated View of the Molecular Toxicology of Sodium Channel Gating in Excitable Cells. *Ann.Rev.Neurosci.* **10**:237-(3/13) (1987).

Strickberger, S.A., L.N. Russek and R.D. Phair: Evidence for Increased Aortic Plasma Membrane Calcium Transport Caused by Experimental Atherosclerosis in Rabbits. *Circ.Res.* **62**:75-(3/13) (1988).

Striessnig, J., E. Moosburger, M. Grabner, H.G. Knaus, H. Glossmann, J. Kaiser, B. Scholkens, R. Becker, W. Linz and R. Henning: Evidence for a Distinct Ca^{2+} Antagonist Receptor for the Novel Benzothiazinone Compound HOE 166. *Naunyn Schmied.Arch.Pharmacol.* **337**:331-(3/13) (1988).

Strong, J.A., A.P. Fox, R.W. Tsien and L.K. Kaczmarek: Phorbol Ester Promotes a Large Conductance Ca Channel in *Aplysia* Bag Cell Neurons. *Biophys.J.* **49**:430a-(3/14) (1986).

Stryer, L.: Cyclic GMP Cascade of Vision. *Ann.Rev.Neurosci.* **9**:87-(2/8) (1986).

Stuhmer, W., F. Conti, H. Suzuki and X. Wang: Structural Parts Involved in Activation and Inactivation of the Sodium Channel. *Nature* **339**:597-(1/3) (1989).

Sturek, M. and K. Hermsmeyer: Calcium and Sodium Channels in Spontaneously Contracting Vascular Muscle Cells. *Science* **233**:475-(2/10) (1986).

Suarez-Isla, B.A., V. Irribarra, R. Bull, A. Oberhauser, L. Larralde, E. Jaimovich and C. Hidalgo: Inositol (1,4,5)-Trisphosphate Activates a Calcium Channel in Isolated Sarcoplasmic Reticulum (SR) Membranes. *Biophys.J.*(abstr.) **53**:467a-(2/10) (1988).

Suarez-Isla, B.A., D.J. Pelto, J.M. Thompson and S.I. Rapoport: Blockers of Calcium Permeability Inhibit Neurite Extension and Formation of Neuromuscular Synapses in Cell Culture. *Dev.Brain Res.* **14**:263-(2/10) (1984).

Suarez-Kurtz, G. and G.J. Kaczorowski: Effects of Dichlorobenzamil on Calcium Currents in Clonal GH3 Pituitary Cells. *J.Pharmacol.Exp.Ther.* **247**:248-(3/13) (1988).

Subramaniam, V.B., M.J. Bowles, N.S. Khurmi, A.B. Davies and E.B. Raftery: Rationale for the Choice of Calcium Antagonists in Chronic Stable Angina An Objective Double-Blind Placebo-Controlled Comparison of Nifedipine and Verapamil. *Am.J.Cardiol.* **50**:1173-(4/16) (1982).

Subramanian, V.B.: *Calcium Antagonists in Chronic Stable Angina Pectoris*. Excerpta Medica, Amsterdam 97-(4/16) (1983).

Subramanian, V.B., A. Lahiri, R. Paramasivan and E.B. Raftery: Verapamil in Chronic Stable Angina. *Lancet* **1**:841-(4/16) (1980).

Sugano, M., Y. Nakashima, T. Matsushima, K. Takahara, M. Takasugi, A. Kuroiwa and O. Koide: Suppression of Atherosclerosis in Cholesterol-Fed Rabbits by Diltiazem Injection. *Arteriosclerosis* **6**:237-(4/18) (1986).

Sukhatme, V.P., X. Cao, L.C. Chang, C.-H. Tsai-Morris, D. Stamenkovich, P.C.P. Ferreira, D.R. Cohen, S.A. Edwards, T.B. Shows, T. Curran, M.M. Le Beau and E.D. Adamson: A Zinc Finger-Encoding Gene Coregulated with C-fos during Growth and Differentiation, and after Cellular Depolarization. *Cell* **53**:37-(2/10) (1988).

Sung, R.J., B. Elser and R.G. McAllister: Intravenous Verapamil for Termination of Reentrant Supraventricular Tachycardias. Intracardiac Studies Correlated with Plasma Verapamil Concentrations. *Ann.Int.Med.* **93**:682-(4/19) (1980a).

Sung, R.J., W.A. Shapira, E.W. Shen, F. Morady and J. Davis: Effects of Verapamil on Ventricular Tachycardias Possibly Caused by Re-entry, Automaticity, and Triggered Activity. *J.Clin.Invest.* **72**:350-(4/19) (1983).

Sung, R.J., H.L. Waxman, B. Elser and Z. Juma: Treatment of Paroxysmal Supraventricular Tachycardia and Atrial Flutter/ Fibrillation with Intravenous Verapamil: Efficacy and Mechanism of Action. *Clin.Invest.Med.* **3**:41-(4/19) (1980b).

Sunkel, C.E., M. Fau de Casa-Juana, F.J. Cillero, J.G. Priego and M.P. Ortega: Synthesis, Platelet Aggregation Inhibitory Activity, and *In Vivo* Antithrombotic Activity of New 1,4-Dihydropyridines. *J.Med.Chem.* **31**:1886-(3/13) (1988).

Suszkiw, J.B., M.M. Murawsky and M. Shi: Further Characterization of Phasic Calcium Influx in Rat Cerebrocortical Synaptosomes: Inferences Regarding Calcium Channel Type(s) in Nerve Endings. *J.Neurochem.* **52**:1260-(3/13) (1989).

Sutko, J.L., K. Ito and J.L. Kenyon: Ryanodine: A Modifier of Sarcoplasmic Reticulum Calcium Release in Striated Muscle. *Fed.Proc.* **44**:2984-(2/10) (1985).

Suzuki, N. and T. Yoshioka: Differential Blocking Action of Synthetic ω-Conotoxin on Components of Ca^{2+} Channel Current in Clonal GH3 Cells. *Neurosci.Lett.* **75**:235-(3/13) (1987).

Swandulla, D. and C.M. Armstrong: Fast Deactivating Calcium Channels in Chick Sensory Neurons. *J.Gen.Physiol.* **92**:197-(1/3,1/4) (1988).

Swandulla, D. and C.M. Armstrong: Calcium Channel Block by Cadmium in Chicken Sensory Neurons. *Proc.Natl.Acad.Sci.U.S.A.* **86**:1736-(1/3) (1989).

Swandulla, D., E. Carbone and H.D. Lux: Do Calcium Channel Classifications Account for Neuronal Calcium Channel Diversity? *TINS* **14**:46-(1/3) (1991).

Swandulla, D., E. Carbone, K. Schäfer and H.D. Lux: Effect of Menthol on Two Types of Ca Currents in Cultured Sensory Neurons of Vertebrates. *Pflügers Arch.* **409**:52-(1/4) (1987).

Swandulla, D. and H.D. Lux: Activation of a Nonspecific Cation Conductance by Intracellular Ca Elevation in Bursting Pacemaker Neurons of *Helix pomatia*. *J.Neurophysiol.* **54**:1430-(2/9) (1985).

Swandulla, D., K. Schäfer and H.D. Lux: Ca Channel Current is Selectively Modulated by Menthol. *Neurosci.Lett.* **68**:23-(1/4) (1986).

Swann, K. and M. Whitaker: The Part Played by Inositol Triphosphate and Calcium in the Propagation of the Fertilization Wave in Sea Urchin Eggs. *J.Cell Biol.* **103**:2333-(2/7) (1986).

Swenson, R.P.: Gating Charge Immobilization and Sodium Current Inactivation in Internally Perfused Crayfish Axons. *Nature* **287**:644-(1/4) (1980).

Swiryn, S.O., R.A. Bauernfeind, C.R.C. Wyndham, R. Dhingra, E. Palileo, B. Strasberg and K.M. Rosen: Effects of Oral Disopyramide Phosphate on Induction of Paroxysmal Supraventricular Tachycardia. *Circulation* **64**:169-(4/19) (1981).

Szlacheic, J., A.T. Hirsch, J.F. Tuban, C. Vollmer, S. Henderson and B.M. Massie: Diltiazem versus Propanolol in Essential Hypertension: Responses of Rest and Exercise Blood Pressure and Effects on Exercise Capacity. *Am.J.Cardiol.* **59**:393-(4/18) (1987).

Tai, C.Y., Q.R. Smith and S.I. Rapoport: Calcium Influxes into Brain and Cerebrospinal Fluid are Linearly Related to Plasma Ionized Calcium Concentration. *Brain Res.* **385**:227-(4/21) (1986).

Taira, N.: Differences in Cardiovascular Profile Amount Calcium Antagonists. *Am.J.Cardiol.* **59**:24B-(4/17) (1987).

Takahara, K., A. Kuroiwa, T. Matsushima, Y. Nakashima and M. Takasugi: Effects of Nifedipine of Platelet Function. *Am.Heart J.* **109**:4-(4/20) (1985).

Takahashi, M. and Y. Fujimoto: Identification of a Dihydropyridine-Sensitive Calcium Channel in Chick Brain by a Monoclonal Antibody. *Biochem.Biophys.Res.Commun.* **163**:1182-(3/13) (1989).

Takahashi, M., Y. Ohizumi and T. Yasumoto: Maitotoxin, A Ca^{2+} Channel Activator Candidate. *J.Biol.Chem.* **257**:7287-(2/7) (1982).

Takanaka, C. and B.N. Singh: Barium-Induced Nondriven Action Potential as a Model of Triggered Potentials from Early Afterdepolarizations: Significance of Slow Channel Activity and Differing Effects of Quinidine and Amiodarone. *J.Am.Coll.Cardiol.* **15**:43-(4/19) (1990).

Takayazu, M., J.E. Bassett and R.G. Dacey: Effects of Calcium Antagonists on Intracerebral Penetrating Arterioles in Rats. *J.Neurosurg.* **69**:104-(3/13) (1988).

Takemura, M., J. Kishino, A. Yamatodani and H. Wada: Inhibition of Histamine Release from Rat Hypothalamic Slices by ω-Conotoxin GVIA, but not by Nilvadipine, a Dihydropyridine Derivative. *Brain Res.* **496**:351-(3/13) (1989a).

Takemura, M., H. Kiyama, H. Fukui, M. Tohyma and H. Wada: Distribution of the ω-conotoxin Receptor in Rat Brain. An Autoradiographic Mapping. *Neuroscience* **32**:405-(3/13) (1989b).

Takeuchi, A. and N. Takeuchi: Active Phase of Frog's Endplate Potential. *J.Neurophysiol.* **22**:395-(1/1) (1959).

Talano, J.V. and D. Feerst: Verapamil: A New Class of Antiarrhythmic Agent with a Variety of Beneficial Cardiovascular Effects. *Arch.Int.Med.* **140**:314-(4/19) (1980).

Talvenheimo, J.A., J.F. Worley and M.T. Nelson: Heterogeneity of Calcium Channels from a Purified Dihydropyridine Receptor Preparation. *Biophys.J.* **52**:891-(3/13) (1987).

Tanabe, T., K.G. Beam, J.A. Powell and S. Numa: Restoration of Excitation-Contraction Coupling and Slow Calcium Current in Dysgenic Muscle by Dihydropyridine Receptor Complementary DNA. *Nature* **336**:134-(2/10,2/10,3/13,3/14) (1988).

Tanabe, T., H. Takeshima, A. Mikami, V. Flockerzi, H. Takahashi, K. Kangawa, M. Kojima, H. Matsuo, T. Hirose and S. Numa: Primary Structure of the Receptor for Calcium Channel Blockers from Skeletal Muscle. *Nature* **328**:313-(2/10,3/13,3/14) (1987).

Tanford, C. and J.D. Hauenstein: Hydrogen Ion Equilibria of Ribonuclease. *J.Am.Chem.Soc.* **78**:5287-(3/12) (1956).

Tang, C.M., F. Presser and M. Morad: Amiloride Selectively Blocks the Low Threshold Calcium Channel. *Science* **240**:213-(3/13,3/14) (1988).

Tank, D.W., M. Sugimori, J.A. Connor and R.R. Llinas: Spatially Resolved Calcium Dynamics of Mammalian Purkinje Cells in Cerebellar Slice. *Science* **242**:773-(2/9,3/14) (1988).

Taraskevitch, P.S. and W.W. Douglas: Action Potentials Occur in Cells of the Normal Pituitary Gland and are Stimulated by the Hypophysiotropic Peptide Thyrotropin-Releasing Hormone. *Proc.Natl.Acad.Sci.U.S.A.* **74**:4064-(2/7) (1977).

Tartakoff, A.M. and P. Vassalli: Plasma Cell Immunoglobulin Secretion: Arrest is Accompanied by Alterations of the Golgi Complex. *J.Exp.Med.* **146**:1332-(2/7) (1977).

Taylor, J.S.H. and A. Roberts: The Early Development of the Primary Sensory Neurones in an Amphibian Embryo: A Scanning Electron Microscope Study. *J.Embryol.Exp.Morphol.* **75**:49-(2/10) (1983).

Taylor, W.R.: Two-Suction-Electrode Voltage-Clamp Analysis of the Sustained Calcium Current in Cat Sensory Neurones. *J. Physiol.* **407**:405-(1/3) (1988a).

Taylor, W.R.: Permeation of Barium and Calcium through Slowly Inactivating Calcium Channels in Cat Sensory Neurones. *J.Physiol.* **407**:433-(1/3,3/12) (1988b).

Tegmeier, F., D. Scheller, J. Urenjak, J. Kolb, A. Bock and M. Hiller: Flunarizine Improves Postischemic Energy Metabolism and Flow after Global Ischemia of the Rat Brain. In *Pharmacology of Cerebral Ischemia.* J. Krieglstein, ed. CRC Press, Boca Raton, Florida. 91-(3/13) (1989).

Tencate, F.J., P.W. Serruys, S. Mey and J. Roelandt: Effects of Short-Term Administration of Verapamil of Left Ventricular Relaxation and Filling Dynamics by a Combined Hemodynamic Ultrasonic Technique in Patients with Hypertrophic Cardiomyopathy. *Circulation* **68**:1274-(4/17) (1983).

Terrence, C.F., M. Sax, G.H. Fromm, C.H. Chang and C.S. Yoo: Effect of Baclofen Enantiomorphs on the Spinal Trigeminal Nucleus and Steric Similarities of Carbamazepine. *Pharmacology* **27**:85-(4/22) (1983).

Thandroyen, F.T.: Protective Action of Calcium-Channel Antagonist Agents Against Ventricular Fibrillation in the Isolated Perfused Rat Heart. *J.Mol.Cell Cardiol.* **14**:21-(4/19) (1982).

Thandroyen, F.T., L. Higginson, L.H. Opie and E. Yon: The Influence of Verapamil and its Isomers on Vulnerability to Ventricular Fibrillation during Acute Myocardial Ischemia and Adrenergic Stimulation in Isolated Rat Heart. *J.Mol.Cell Cardiol.* **18**:645-(4/16) (1986).

Thangnipon, W., A. Kingsbury, M. Webb and R. Balazs: Observations on Rat Cerebellar Cells *In Vitro*: Influence of Substratum, Potassium Concentration and Relationship Between Neurones and Astrocytes. *Dev.Brain Res.* **11**:177-(2/10) (1983).

Thate, A. and D.K. Meyer: Effect of ω-conotoxin GVIA on Release of ^{3}H-γ-Aminobutyric Acid from Slices of Rat Neostriatum. *Naunyn Schmied.Arch.Pharmacol.* **339**:359-(3/13) (1989).

Thaulow, E., B.D. Guth and J.R. Ross, Jr.: Role of Calcium Channel Blockers in Experimental Exercise-Induced Ischemia. *Cardiovasc.Drugs Ther.* **1**:503-(4/16) (1988).

Thayer, S.A., L.D. Hirning and R.J. Miller: Distribution of Multiple Types of Ca^{2+} Channels in Rat Sympathetic Neurons *In Vitro*. *Mol.Pharmacol.* **32**:579-(3/13,3/14) (1987).

The Israeli SPRINT Study Group: Secondary Prevention Reinfarction Israeli Nifedipine Trial (SPRINT): A Randomized Intervention Trial of Nifedipine in Patients with Acute Myocardial Infarction. *Eur.Heart J.* **9**:354-(4/19) (1988).

The Multicenter Diltiazem Post-Infarction Trial Research Group: The Effect of Diltiazem on Mortality and Reinfarction after Myocardial Infarction. *N.Engl.J.Med.* **319**:385-(4/19) (1988).

The Norwegian Multicenter Study Group: Timolol-Induced Reduction in Mortality and Reinfarction in Patients Surviving Acute Myocardial Infarction. *N.Engl.J.Med.* **304**:801-(4/19) (1981).

Theroux, P., T. Taeymans, D. Morissette, X. Bosch, G.B. Pelletier and D.D. Waters: A Randomized Study Comparing Propranolol and Diltiazem in the Treatment of Unstable Angina. *J.Am.Coll.Cardiol.* **5**:717-(4/16) (1985).

Thomas, G., R. Gross, G. Pfitzer and J.C. Ruegg: The Positive Inotropic Dihydropyridine Bay K 8644 does not Affect Calcium Sensitivity or Calcium Release of Skinned Cardiac Fibres. *Arch.Pharmacol.* **328**:378-(2/10) (1985).

Thomas, P.G., A.J. Zimmermann and A.J. Verkleij: Prevention of Calcium-Induced Membrane Structural Alterations in Erythrocyte Membranes by Flunarizine. *Biochim.Biophys.Acta* **946**:439-(3/13) (1989).

Thompson, L.T., R.A. Deyo and J.F. Disterhoft: Nimodipine Enhances Spontaneous Activity of Hippocampal Pyramidal Neurons in Aging Rabbits at a Dose that Facilitates Learning. *Brain Res.* (submitted) (3/13) (1990).

Thompson, S. and J. Coombs: Spatial Distribution of Ca^{2+} Currents in Molluscan Neuron Cell Bodies and Regional Differences in the Strength of Inactivation. *J.Neurosci.* **8**:1929-(3/14) (1988).

Thompson, S.H.: Three Pharmacologically Distinct Potassium Channels in Molluscan Neurones. *J.Physiol.* **265**:465-(2/9) (1977).

Thompson, S.H. and S.J. Smith: Depolarizing Afterpotentials and Burst Production in Molluscan Pacemaker Neurons. *J.Neurophysiol.* **39**:153-(2/9) (1976).

Thompson, S.M. and R.K.S. Wong: Low Threshold, Transient Calcium Current is Present in Young, but not Adult, Isolated Rat Hippocampal Pyramidal Cells. *Soc.Neurosci.Abs.* **15**:653-(2/9) (1989).

Thomson, A.M.: A Magnesium Sensitive Postsynaptic Potential in Rat Cerebral Cortex Resembles Neuronal Responses to N-Methyl-Aspartate. *J.Physiol.* **370**:531-(2/9) (1986).

Thuesen, L., J.R. Jorgensen, H.J. Kvistgaard, J.A. Sorensen, M. Vaeth, E.B. Jensen, J.J. Jensen and L. Hagerup: Effect of Verapamil on Enzyme Release after Early Intravenous Administration in Acute Myocardial Infarction: Double-Blind Randomized Trial. *Br.Med.J.* **286**:1107-(4/16) (1983).

Thurman, R.G., E. Apel, M. Badr and J.L. LeMasters: Protection of Liver by Calcium Entry Blockers. *Ann. N.Y.Acad. Sci.* **522**:757-(4/15) (1988.

Tillotson, D.: Inactivation of Ca Conductance Dependent on Entry of Ca Ions in Molluscan Neurons. *Proc.Natl.Acad.Sci.U.S.A.* **77**:1497-(1/4) (1979).

Tillotson, D. and A.L.F. Gorman: Nonuniform Ca^{2+} Buffer Distribution in a Nerve Cell Body. *Nature* **286**:816-(3/14) (1980).

Todd, P.A. and P. Benfield: Flunarizine: Reappraisal of its Pharmacological Properties and Therapeutic use in Neurological Disorders. *Drugs* **38**:481-(3/13) (1989).

Tomaselli, J.J., K.M. Neugebauer, J.L. Bixby, J. Lilien and L.F. Reichardt: N-Cadherin and Integrins: Two Receptor Systems that Mediate Neuronal Process Outgrowth on Astrocyte Surfaces. *Neuron* **1**:33-(2/10) (1988).

Tomita, T.: Electrical Activity of Vertebrate Photoreceptors. *Q.Rev.Biophys.* **3**:179-(2/8) (1970).

Tonkin, A.M., P.E. Aylward and S.E. Joel: Verapamil in Prophylaxis of Paroxysmal Atrioventricular Nodal Re-entrant Tachycardia. *J.Cardiovasc.Pharmacol.* **2**:473-(4/19) (1980).

Tonosaki, K. and M. Funakoshi: Cyclic Nucleotides may Mediate Taste Transduction. *Nature* **331**:354-(2/8) (1988).

Torre, V., H.R. Matthews and T.D. Lamb: The Role of Calcium in Regulating the Cyclic GMP Cascade of Phototransduction in Retinal Rods. *Proc.Natl.Acad.Sci.U.S.A.* **83**:7109-(2/8) (1986).

Torri-Tarelli, F., F. Grohovaz, R. Fesce and B. Ceccarelli: Temporal Coincidence between Synaptic Vesicle Fusion and Quantal Secretion of Acetylcholine. *J.Cell Biol.* **101**:1386-(2/7) (1985).

Towart, R.: The Selective Inhibition of Serotonin-Induced Contractions of Rabbit Cerebral Vascular Smooth Muscle by Calcium-Antagonistic Dihydropyridines An Investigation of the Mechanisms of Action of Nimodipine. *Circ.Res.* **48**:650-(4/20) (1981).

Townley, R.G., J. Cheng, A.K. Bewtra, N. Nair, R. Hopp and D.K. Agrawal: The Role of Calcium Channel Blockers in Reactive Airway Disease. *Ann. N.Y.Acad. Sci.* **522**:732-(4/15) (1988).

Traber, J. and W.H. Gispen: *Nimodipine and Central Nervous System Function. New Vistas.* Schattauer, Stuttgart and New York. (3/13) (1989).

Trautwein, W.: Membrane Currents in Cardiac Muscle Fibers. *Physiol.Rev.* **53**:793-(3/12)(1973).

Trautwein, W. and D. Pelzer: Voltage-Dependent Gating of Single Calcium Channels in Cardiac Cell Membrane and its Modulation by Drugs. In *Calcium and Cell Physiology.* D. Marme, ed. Springer-Verlag, Berlin. (1/4) (1985).

Trautwein, W. and D. Pelzer: Kinetics and Beta-Adrenergic Modulation of Cardiac Ca^{2+} Channels. In *The Calcium Channel: Structure, Function and Implications.* M. Morad, W. Nayler, W. Kazda, S. Schramm, eds. Springer-Verlag, Berlin and New York. 39-(3/14) (1988).

Traynor, A.E.: The Relationship Between Neurite Extension and Phospholipid Metabolism in PC12 Cells. *Dev.Brain Res.* **14**:205-(2/10) (1984).

Treisman, R.: Identification of a Protein-Binding Site that Mediates Transcriptional Response of the c-fos Gene to Serum Factors. *Cell* **46**:567-(2/10) (1986).

Triggle, D.J.: Calcium Antagonists in Atherosclerosis: A Review and Commentary. *Cardiovasc.Drug Rev.* **6**:320-(3/13) (1989).

Triggle, D.J.: Endogenous Ligands for the Calcium Channel: Myths and Realities. In *The Calcium Channel: Structure, Function and Implications.* M. Morad, W. Nayler, S. Kazda, M. Schramm, eds. Springer-Verlag, Berlin and Heidelberg. 549-(3/13) (1988).

Triggle, D.J.: Calcium Antagonists. In *Cardiovascular Pharmacology.* M. Antonaccio, ed. Raven Press, New York. (in press) (3/13) (1990a).

Triggle, D.J., J. Ferrante, Y.-W. Kwon, A. Skattebol, M. Hawthorn, D.A. Langs and R. Bangalore: 1,4-Dihydropyridine-sensitive Neuronal Ca^{2+} Channels: Pharmacology and Regulation. In *Nimodipine and Central Nervous System Function.* J. Traber and W. H. Gispen, eds. Schattauer, Stuttgart. 3-(3/13) (1989).

Triggle, D.J. and R.A. Janis: Calcium Channel Ligands. *Ann.Rev.Pharmacol.Toxicol.* **27**:347-(3/13) (1987).

Triggle, D.J. and R.A. Janis: Calcium Channels and Calcium Channel Ligands. In *Receptor Pharmacology and Function.* M. Williams, R.A. Glennon, P.B.M.W. Timmermans, eds. Marcel Dekker, New York. (3/13) (1989).

Triggle, D.J. and D.A. Langs: Ligand Gated- and Voltage-Gated Ion Channels. *Ann.Rep.Med.Chem.* (1990).

Triggle, D.J., D.A. Langs and R.A. Janis: Ca^{2+} Channel Ligands: Structure-function Relationships of the 1,4-Dihydropyridines. *Med.Res.Revs.* **9**:123-(3/13) (1989a).

Triggle, D.J. and D. Rampe: 1,4-Dihydropyridine Activators and Antagonists: Structural and Functional Distinctions. *TIPS* **10**:507-(3/13) (1989).

Triggle, D.J. and V.C. Swamy: Calcium Antagonists. Some Chemical-Pharmacologic Aspects. *Circ.Res.* **52**:17-(4/17) (1983).

Triggle, D.J., W. Zheng, M. Hawthorn, Y.W. Kwon, X.Y. Wei, A. Joslyn, J. Ferrante and A.M. Triggle: Calcium Channels in Smooth Muscle: Properties and Regulation. *Ann.N.Y.Acad.Sci.* **560**:215-(3/13) (1989b).

Triller, A., F. Cluzeaud, F. Pfeiffer, H. Betz and H. Korn: Distribution of Glycine Receptors at Central Synapses: An Immunoelectron Microscope Study. *J.Cell Biol.* **101**:683-(2/9) (1985).

Trube, G. and P. Rorsman: Calcium and Potassium Currents Recorded from Pancreatic β-cells under Voltage Clamp Control. In *Biophysics of the Pancreatic β-Cell* . Plenum, New York. (2/7) (1990).

Tschirdewahn, B. and H. Klepzig: Clinical Studies on the Effect of Isoptin and Isoptin S in Patients with Coronary Insufficiency (in German). *Deutsche Med.Wochensch.* **88**:1702-(4/16) (1963).

Tseng, G.N. and P.A. Boyden: Multiple Types of Ca^{2+} Currents in Single Canine Purkinje Cells. *Circ.Res.* **65**:1735-(3/13) (1989).

Tsien, R.W.: Adrenaline-Like Effects of Intracellular Iontophoresis of Cyclic AMP in Cardiac Purkinje Fibers. *Nature New Biol.* **245**:120-(3/14) (1973).

Tsien, R.W.: Calcium Channels in Excitable Cell Membranes. *Ann.Rev.Physiol.* **45**:341-(3/12,3/14) (1983).

Tsien, R.W.: Calcium Channels in Heart Cells and Neurons. In *Neuromodulation*. L.K. Kaczmared and I.B. Levitan, eds. Oxford University Press (3/14) (1986).

Tsien, R.W., B.P. Bean, P. Hess, J.B. Lansman, B. Nilius and M.C. Nowycky: Mechanisms of Calcium Channel Modulation by Beta-Adrenergic Agents and Dihydropyridine Calcium Antagonists. *J.Mol.Cell Cardiol.* **18**:691-(2/10,3/14,4/17) (1986).

Tsien, R.W., A.P. Fox, P. Hess, E.W. McCleskey, B. Nilius, M.C. Nowkcky and R.L. Rosenberg: Multiple Types of Calcium Channel in Excitable Cells. In *Protein of Excitable Membranes*. B. Ille and D.M. Fambrough, eds. John Wiley and Sons, Inc. 167-(2/10) (1987).

Tsien, R.W., W. Giles and P. Greengard: Cyclic AMP Mediates the Action of Adrenaline on the Action Potential Plateau of Cardiac Purkinje Fibers. *Nature New Biol.* **240**:181-(3/14) (1972).

Tsien, R.W., P. Hess, E.W. McCleskey and R.L. Rosenberg: Calcium Channels: Mechanisms of Selectivity, Permeation, and Block. *Annu.Rev.Biophys.Chem.* **16**:265-(1/2,3/13) (1987).

Tsien, R.W., D. Lipscombe, D.V. Madison, K.R. Bley and A.P. Fox: Multiple Types of Neuronal Calcium Channels and their Selective Modulation. *TINS* **11**:431-(1/3,1/4,2/7,3/13,3/14) (1988).

Tsien, R.Y., T. Pozzan and J.J. Rink: Calcium Homeostasis in Intact Lymphocytes: Cytoplasmic Free Calcium Monitored with a New Intracellularly Trapped Fluorescent Indicator. *J.Cell Biol.* **94**:325-(4/20) (1982).

Tsoukaris-Kupfer, D., X. Girerd, S. Laurent, M. Legrand, A.M. Huchet-Brisac and H. Schmitt: Opposite Central Cardiovascular Effects of Nifedipine and BAY K 8644 in Anesthetized Rats. *J.Cardiovasc.Pharmacol.* **10**:S40-(3/13) (1987).

Tsuchida, K., M. Muramatsu, K. Kaneko, R. Yamazaki, T. Yamada, N. Ogawa and H. Aihara: CD-349. *Cardiovasc.Drug Rev.* **8**:45-(3/13) (1990).

Tsuji, T. and D.A. Cook: Effect of Nimodipine on Canine Cerebrovascular Responses to 5-Hydroxytryptamine and Potassium Chloride after Exposure to Blood. *Stroke* **20**:105-(3/13) (1989).

Tsunoo, A., Y. Mitsunobu and T. Narahashi: Block of Calcium Channels by Enkephalin and Somatostatin in Neuroblastomaglioma Hybrid NGl08-l5 Cells. *Proc.Natl.Acad.Sci.U.S.A.* **83**:9832-(1/3) (1986).

Tsunoo, A., M. Yoshii and T. Narahashi: Block of Calcium Channels by Enkephalin and Somatostatin in Neuroblastoma Cells. *Biophys.J.* **47**:433a-(3/13) (1985).

Turner, P.R., L.A. Jaffe and A. Fein: Regulation of Cortical Vesicle Exocytosis in Sea Urchin Eggs by Inositol 1,4,5-Triphosphate and GTP-Binding Protein. *J.Cell Biol.* **102**:70-(2/7) (1986).

Twombly, D.A. and T. Narahashi: Phenytoin Block of Low Threshold Calcium Channels is Voltage- and Frequency-Dependent. *Soc.Neurosci.Abs.* **12**:1193-(3/13) (1986).

Twombly, D.A., M. Yoshii and T. Narahashi: Mcchanism of Calcium Channcl Block by Phcnytoin. *J.Pharmacol.Exp.Ther.* **246**:189-(3/13,3/14) (1988).

Tytgat, J., J. Vereecke and E. Carmeliet: Differential Effects of Verapamil and Flunarizine on Cardiac L-Type and T-Type Ca Channels. *Naunyn Schmied.Arch.Pharmacol.* **337**:690-(3/13) (1988).

Uematsu, D., J.W. Greenberg, W.F. Hickey and M. Reivich: Nimodipine Attenuates both Increase in Cytosolic Free Calcium and Histologic Damage Following Focal Cerebral Ischemia and Reperfusion in Cats. *Stroke* **20**:1531-(3/13) (1990).

Uhr, S.B., K. Jackson, P.A. Berger and J.G. Csernansky: Effects of Verapamil Administration on Negative Symptoms of Chronic Schizophrenia. *Psychiatr.Res.* **23**:351-(4/22) (1988).

Ullrich, A., J. Adamczyk, J. Zihl and H.M. Emrich: Lithium Effects on Ophthalmological Electrophysiological Parameters in Young Healthy Volunteers. *Acta Psychiatr.Scand.* **72**:113-(4/22) (1985).

Usowicz, M., H. Becker, H. Porzig and H. Reuter: Regulation of Calcium Channels in PC12 Cells by Nerve Growth Factor: Patch Clamp and Binding Studies. *Eur.J.Neurosci.Suppl.* **2**:276-(1/4) (1989).

Vacquier, V.D.: The Isolation of Intact Cortical Granules from Sea Urchin Eggs: Calcium Ions Trigger Granule Discharge. *Dev.Biol.* **43**:62-(2/7) (1975).

Vaghy, P.L., K. Itagaki, K. Miwa, E. McKenna and A. Schwartz: Mechanism of Action of Calcium Channel Modulator Drugs *Ann.N.Y.Acad.Sci.* **522**:176-(4/17) (1988).

Valdiserri, E.V.: A Possible Interaction between Lithium and Diltiazem: Case Report. *J.Clin.Psychiatry* **46**:540-(4/22) (1985).

Valdivia, H. and R. Coronado: Dihydropyridine Pharmacology of the Reconstituted Calcium Channel of Skeletal Muscle. In *The Calcium Channel: Structure, Function and Implications*. M. Morad, W. Nayler, S. Kazda, M. Schramm, eds. Springer-Verlag, Berlin and Heidelberg. 252-(3/13) (1988a).

Valdivia, H. and R. Coronado: Pharmacological Profile of Skeletal Muscle Calcium Channels in Planar Lipid Bilayers. *Biophys. J.* **53**:555a-(3/13) (1988b).

Valdivia, H.H. and R. Coronado: Internal and External Effects of Dihydropyridines in the Calcium Channel of Skeletal Muscle. *J.Gen.Physiol.* **95**:1-(3/13) (1990).

Valone, F.H.: Inhibition of Platelet-activating Factor Binding to Human Platelets by Calcium Channel Blockers. *Thromb.Res.* **45**:427-(4/20) (1987).

Van Amsterdam, F.T. and J. Zaagsma: Stereoisomers of Calcium Antagonists Discriminate between Coronary Vascular and Myocardial Sites. *Naunyn Schmied.Arch.Pharmacol.* **337**:213-(3/13) (1988).

Van Breemen, C., P.I. Aaronson, C.A. Cauvin, R.D. Loutzenhiser, A.W. Mangel and K. Saida: The Calcium Cycle in Arterial Smooth Muscle. In *Calcium Blockers. Mechanisms of Action and Clinical Applications*. S.F. Flaim and R. Zelis, eds. Urban Schwarzenberg, Baltimore and Munich. 53-(4/20) (1982).

Van Breemen, C., P., Aaronson and R. Loutzenhiser: Sodium-Calcium Interactions in Mammalian Smooth Muscle. *Pharmacol.Rev.* **30**:167-(2/10) (1978).

Van Breemen, C., P. Leijten, H. Yamamoto, P. Aaronson and C. Cauvin: Calcium Activation of Vascular Smooth Muscle. *Hypertension* **8**:1189-(4/20) (1986).

Van Breeman, C. and K. Saida: Cellular Mechanisms Regulating (Ca^{2+}) in Smooth Muscle. *Ann.Rev.Physiol.* **51**:315-(2/10) (1989).

Van den Kerckhoff, W. and L.R. Drewes: Transfer of Nimodipine and Another Calcium Antagonist Across the Blood-Brain Barrier and their Regional Distribution *In Vivo*. In *Diagnosis and Treatment of Senile Dementia*. M. Bergener, B. Reisberg, eds. Springer-Verlag, Berlin and Heidelberg. 308-(3/13) (1989).

van Inwegen, R.G., I. Weinryb, H. Jones and A. Khandwala: Comparative Structure-Activity Relationships for Dihydropyridines as Inhibitors of [^{3}H]-Nitrendipine Binding vs. Cyclic Nucleotide phosphodiesterases. *Res.Commun.Chem.Path.Pharmacol.* **45**:191-(3/13) (1984).

Van Nueten, J.M., W.J. Janssens and Vanhoutte: Calcium Antagonism and Vascular Smooth Muscle. *Ann.N.Y.Acad.Sci.* **522**:234-(4/20) (1987).

Van Nueten, J.M. and P.M. Vanhoutte: Calcium Entry Blockers and Vascular Smooth Muscle Heterogeneity. *Fed.Proc.* **40**:2862-(4/20) (1981).

Van Reempts, J. and M. Borgers: Ischemic Brain Injury and Cell Calcium: Morphologic and Therapeutic Aspects. *Ann.Emerg.Med.* **14**:736-(4/21) (1985).

Van Skiver, D.M., S. Spires and C.J. Cohen: Block of T-Type Ca Channels in Guinea Pig Atrial Cells by Cinnarizine. *Biophys.J.* **53**:233a-(3/13) (1988).

Van Skiver, D.M., S. Spires and C.J. Cohen: High Affinity and Tissue Specific Block of T-Type Ca Channels by Felodipine. *Biophys.J.* **55**:593a-(3/13) (1989).

van Wagoner, D.R.: *Biophysical Chemistry of the Calcium Channel in Barnacle Muscle*. Ph.D. Thesis. Thomas Jefferson University, Philadelphia. (3/12) (1985).

van Zwieten, P.A., J.C.A. van Meek and P.B.W.M. Timmermans: Functional Interaction between Calcium Antagonists and Vasoconstriction Induced by the Stimulation of Postsynaptic A2-receptors. *Cir.Res.* **52**:77-(4/17) (1983).

Vandenberg, C.A. and R. Horn: Inactivation Viewed through Single Sodium Channels. *J.Gen.Physiol.* **84**:535-(1/4) (1984).

Vanhoutte, P.M.: Ca^{2+} Antagonists and Vascular Disease. *Ann.N.Y.Acad.Sci.* **520**:380-(4/20) (1987).

Vanhoutte, P.M.: Vascular Endothelium and Ca^{2+} Antagonists. *J.Cardiovasc.Pharmacol.* **12**:S21-(3/13) (1988).

Vanhoutte, P.M., A. Paoletti and S. Govoni: Calcium Antagonists, Pharmacology and Clinical Research. *Ann.N.Y.Acad.Sci.* **522**:(1989).

Vargas, O., M.C. Doria de Lorenzo, M.C. Saldate and F. Orrego: Potassium-Induced Release of ^{3}H GABA and ^{3}H Noradrenaline from Normal and Reserpinized Rat Brain Cortex Slices. Differences in Calcium-Dependency, and in Sensitivity to Potassium Ions. *J.Neurochem.* **28**:165-(4/22) (1977).

Vassort, G., O. Rougier, D. Garnier, M.P. Sauviat, E. Coraboeuf and Y.M. Gargouil: Effects of Adrenaline on Membrane Inward Currents during the Cardiac Action Potential. *Pflügers Arch.* **390**:70-(3/14) (1969).

Vayssairat, M., L. Capron, J.N. Fiessinger, J.F. Mathieu and E. Housset: Calcium Channel Blockers and Raynaud's Disease. *Ann.Intern.Med.* **95**:243-(4/20) (1981).

Velasco, J.M., J.U.H. Petersen and O.H. Petersen: Single-Channel Ba^{2+} Currents in Insulin-Secreting Cells are Activated by Glyceraldehyde Stimulation. *FEBS Lett.* **231**:366-(2/7) (1988).

Velly, J., M. Grima, N. Decker, E.J. Cragoe, Jr. and J. Schwartz: Effects of Amiloride and its Analogues on [^{3}H]Batrachotoxinin- A 20-Alpha Benzoate Binding, [^{3}H]tetracaine Binding and ^{22}Na Influx. *Eur.J.Pharmacol.* **149**:97-(3/13) (1988).

Vergara, J., R.Y. Tsien and M. Delay: Inositol 1,4,5-Trisphosphate: A Possible Chemical Link in Excitation-Contraction Coupling in Muscle. *Proc.Natl.Acad.Sci.U.S.A.* **82**:6352-(2/10) (1985).

Verhaegen, H., V. Roel, H. Adriaensen, J. Brugmans, W. De Cock, J. Dony, A. Jageneau and V. Schuermans: The Arteriolar Effects of Cinnarizine and Flunarizine. *Angiology* **25**:261-(4/20) (1974).

Verselis, V., A. Campos de Carvalho, R.L. White, M.V.L. Bennett and D.C. Spray: Calmodulin and Gap Junction Regulation. *J. Cell Biol.* **103**:73a-(2/10) (1986).

Vicini, S. and S.M. Schuetze: Gating Properties of Acetylcholine Receptors at Developing Rat Endplates. *J.Neurosci.* **5**:2212-(2/10) (1985).

Vidal, S., B. Raynaud, D. Clarous and M.J. Weber: Neurotransmitter Plasticity of Cultured Sympathetic Neurones. Are the Effects of Muscle-Conditioned Medium Reversible. *Development* **101**:617-(2/10) (1987).

Vidal, S., B. Raynaud and M.J. Weber: The Role of Ca^{2+} Channels of the L-Type in Neurotransmitter Plasticity of Cultured Sympathetic Neurons. *Mol.Brain Res.* **6**:187-(2/10) (1989).

Villereal, M.L. and G.A. Jamieson: Epidermal Growth Factor Stimulates Calcium Influx via Voltage Sensitive Calcium Channels in Cultured Human Fibroblasts. *J.Cell Biochem.* (abstr.) 159-(3/14) (1988).

Vilmart-Seuwen, J., H. Kersken, R. Stüerzl and H. Plattner: ATP Keeps Exocytosis Sites in a Primed State but is not Required for Membrane Fusion: An Analysis with Paramecium Cells *In Vivo* and *In Vitro*. *J.Cell Biol.* **103**:1279-(2/7) (1986).

Vilven, J. and R. Coronado: Opening of Dihydropyridine Calcium Channels in Skeletal Muscle Membranes by Inositol Trisphosphate. *Nature* **356**:587-(3/13) (1988).

Vilven, J., A.T. Leung, T. Imagawa, A.H. Sharp, K.P. Campbell and R. Coronado: Interaction of Calcium Channels of Skeletal Muscle with Monoclonal Antibodies Specific for its Dihydropyridine Receptor. *Biophys.J.* **53**:556a-(3/13) (1988).

Vincent, A., B. Lang and J. Newsom-Davis: Autoimmunity to the Voltage-Gated Calcium Channel Underlies the Lambert-Eaton Myasthenic Syndrome, a Paraneoplastic Disorder. *TINS* **12**:496-(3/13) (1989).

Visvader, J., P.S. Sassone-Corsi and I.M. Verma: Two Adjacent Promoter Elements Mediate Nerve Growth Factor Activation of the c-fos Gene and Bind Distinct Nuclear Complexes. *Proc.Natl.Acad.Sci.U.S.A.* **85**:9574-(2/10) (1988).

Vohra, J., D. Hunt, J. Strickey and G. Sloman: Cycle Length Alterations in Supraventricular Tachycardia after Administration of Verapamil. *Br.Heart J.* **36**:570-(4/19) (1974).

Volpe, P., K.H. Krause, S. Hashimoto, F. Zarzato, T. Pozzan, J. Meldolesi and D.P. Lew: Calciosome, a Cytoplasmic Organelle: The Inositol 1,4,5,- Triphosphate Sensitive Ca^{2+} Store of Non-Muscle Cells. *Proc.Natl.Acad.Sci.U.S.A.* **85**:1091-(3/14) (1988).

Vos, M.A., A.P.M. Gorgels, J.D.M. Lennissen and H.J.J. Wellens: Flunarizine Allows Differentiation Between Mechanisms of Arrhythmias in the Infarct Heart. *Circulation* **81**:343-(4/19) (1990).

Wagner, J.A., I.J. Reynolds, H.F. Weisman, P. Pudeck, M.L. Weisfeldt and S.H. Snyder: Calcium Antagonist Receptors in Cardiomyopathic Hamster: Selective Increases in Heart, Muscle and Brain. *Science* **232**:515-(3/13) (1986).

Wagner, J.A., F.L. Sax, H.F. Weisman, J. Porterfield and C. McIntosh: Calcium-Antagonist Receptors in the Atrial Tissue of Patients with Hypertrophic Cardiomyopathy. *N.Engl.J.Med.* **320**:755-(3/13,4/17) (1989a).

Wagner, J.A., A.M. Snowman, A. Biswass, M.M. Olivera and S.H. Snyder: ω-Conotoxin GVIA Binding to a High-Affinity Receptor in Brain: Characterization, Calcium Sensitivity and Solubilization. *J.Neurosci.* **8**:3354-(3/13) (1988).

Wagner, J.A., A.M. Snowman, B. Olivera and S.H. Snyder: Aminoglycoside Effects on Voltage Sensitive Calcium Channels Predict Neurotoxicity. *N.Engl.J.Med.* **317**:1669-(3/13,3/14) (1987).

Wagner, J.A., H.F. Weisman, A.M. Snowman, I.J. Reynolds, M.L. Weisfeldt and S.H. Snyder: Alterations in Calcium Antagonist Receptors and Sodium-Calcium Exchange in Cardiomyopathic Hamster Tissues. *Circ.Res.* **65**:205-(3/13) (1989b).

Wagniart, P., R.J. Ferguson, B.R. Chaitman, F. Achard, A. Benacerraf, B. Delanguenhagen, B. Morin, A. Pasternac and M.G. Bourassa: Increased Exercise Tolerance and Reduced Electrocardiographic Ischemia with Diltiazem in Patients with Stable Angina Pectoris. *Circulation* **66**:23-(4/16) (1982).

Walden, J., E.J. Speckmann and O.W. Witte: Depression of Focal Interictal Epileptiform Discharges by Intracerebroventricular Perfusion with the Calcium Channel Blocker Verapamil. **In** *Epilepsy and Calcium*. E.J. Speckman, H. Schulze, J. Walden, eds. Urban and Schwarzenberg, Munich. 335-(4/22) (1986).

Walicke, P., R. Campenot and P. Patterson: Determination of Transmitter Function by Neuronal Activity. *Proc.Natl.Acad.Sci.U.S.A.* **74**:5767-(2/10) (1977).

Walicke, P.A. and P.H. Patterson: On the Role of Ca^{2+} in the Transmitter Choice Made by Cultured Sympathetic Neurons. *J.Neurosci.* **1**:343-(2/10) (1981).

Walker, J.W., A.V. Somlyo, Y.E. Goldman, A.P. Somlyo and D.R. Trentham: Kinetics of Smooth and Skeletal Muscle Activation by Laser Pulse Photolysis of Caged Inositol 1,4,5 Trisphosphate. *Nature* **327**:249-(2/10) (1987).

Walker, M.J.A. and S.K.L. Chia: Calcium Channel Blockers as Antiarrhythmics. *Cardiovasc.Drug Rev.* **7**:265-(3/13) (1989).

Walker, M.W., D.A. Ewald, T.M. Perney and R.J. Miller: Neuropeptide Y Modulates Neurotransmitter Release and Ca^{2+} Currents in Rat Neurons. *J.Neurosci.* **8**:2438-(3/14) (1988).

Waller, D.G., V.F. Challenor, D.A. Francis and O.S. Roath: Clinical and Rheological Effects of Nifedipine in Raynaud's Phenomenon. *Br.J.Clin.Pharmacol.* **22**:449-(4/20) (1986).

Walley, T.J., K.L. Woods and D.B. Barnett: The Effects of Intravenous and Oral Nifedipine on *Ex Vivo* Platelet Function. *Eur.J.Pharmacol.* **37**:449-(3/13) (1989b).

Walley, T.J., K.L. Woods and D.B. Barnett: Effects of Calcium Channel Blockers on *In Vitro* Platelet Function in Whole Blood Using Single Platelet Counting. *Thromb.Haemost.* **61**:137-(4/20) (1989a).

Wallin, J.D., E. Fletcher, C.VS. Ram, M.E. Cook, G. Cheung, E.P. MacCarthy, R. Townsend, E. Saunders, W.R. Davis, H.G. Langford, G. DeValut, W. Flamenbaum, G. Ellrodt, B. Hamilton, S. Frank and W. Frishman: Intravenous Nicardipine for the Treatment of Severe Hypertension: A Double-Blind, Placebo Controlled, Multicellular Trial. *Arch.Intern.Med.* **149**:2662-(4/18) (1990).

Walsch, R.W., C.B. Porter, M.R. Starling and R.A. O'Rourke: Beneficial Hemodynamic Effects of Intravenous and Oral Diltiazem in Severe Congestive Heart Failure. *J.Am.Coll.Cardiol.* **3**:1044-(4/17) (1984).

Walsh, K.B., S.H. Bryant and A. Schwartz: Effect of Calcium Antagonist Drugs on Calcium Currents in Mammalian Skeletal Muscle Fibers. *J.Pharmacol.Exp.Ther.* **236**:403-(3/13) (1986).

Wang, R., E. Karpinski and P.KT. Pang: Two Types of Calcium Channels in Isolated Smooth Muscle Cells from Rat Tail Artery. *Am.J.Physiol.* **256**:H136l-(2/10) (1989).

Wanke, E., A. Ferroni, A. Malgaroli, A. Ambrosini, Y. Pozzan and J. Meldolesi: Activation of a Muscarinic Receptor Selectively Inhibits a Rapidly Inactivated Ca^{2+} Current in Rat Sympathetic Neurons. *Proc.Natl.Acad.Sci.U.S.A.* **84**:4313-(1/3,3/13) (1987).

Wanke, E., P.L. Testa, G. Prestipino and E. Carbone: High Intracellular pH Reversibly Prevents Gating-Charge Immobilization in Squid Axons. *Biophys.J.* **44**:281-(1/4) (1983).

Ware, A.J., P.C. Johnson, M. Smith and E.W. Salzman: Inhibition of Human Platelet Aggregation and Cytoplasmic Calcium Response by Calcium Antagonists: Studies with Aequorin and Quin 2. *Circ.Res.* **59**:39-(4/20) (1986).

Warner, A.E., S.C. Guthrie and N.B. Gilula: Antibodies to Gap-Junctional Protein Selectively Disrupt Junctional Communication in the Early Amphibian Embryo. *Nature* **311**:127-(2/10) (1984).

Watabe, S., M. Yoshii, N. Ogata and T. Narahashi: Clonazepam Differs from Diazepam and Nitrazepam in Blocking Two Types of Calcium Channels. *Soc.Neurosci.Abs.* **12**:1193-(3/13) (1986).

Watanabe, A.M. and H.R. Besch, Jr.: Cyclic Adenosine Monophosphate Modulation of Slow Calcium Influx Channels in Guinea Pig Heart. *Circ.Res.* **35**:316-(3/14) (1974).

Watanabe, N., Y. Ishikawa, R. Okamoto, Y. Watanabe and H. Fukuzaki: Nifedipine Suppressed Atherosclerosis in Cholesterol-Fed Rabbits, but not in Wantanabe Heritable Hyperlipidemic Rabbits. *Artery* **14**:283-(4/18) (1987).

Waters, D.D., D.D. Miller and J. Szlachcic: Factors Influencing the Long-Term Prognosis of Treated Patients with Variant Angina. *Circulation* **68**:258-(4/16) (1983).

Waters, D.D., P. Theroux, J. Szlachcic and F. Dauwe: Provocative Testing with Ergonovine to Assess the Efficacy of Treatment with Nifedipine, Diltiazem and Verapamil in Variant Angina. *Am.J.Cardiol.* **48**:123-(4/16) (1981).

Watts, J.A., E.M. Hawes, S.H. Jenkins and T.C. Williams: Effects of Nisoldipine in the No Reflow Phenomenon in Globally Ischemic Rat Hearts. *J. Cardiovasc. Pharmacol.* **16**:487-(3/13) (1990).

Watts, J.A., L.J. Maiorano and P.C. Maiorano: Protection by Verapamil of Globally Ischemic Rat Hearts: Energy Preservation, a Partial Explanation. *J.Mol.Cell Cardiol.* **17**:797-(4/16) (1985).

Watts, J.A., J.P. Whipple and A.A. Hatley: A Low Concentration of Nisoldipine Reduces Ischemic Heart Injury: Enhanced Reflow and Recovery of Contractile Function Without Energy Preservation During Ischemia. *J.Mol.Cell Cardiol.* **19**:809-(3/13) (1987).

Weber, M. and J.I.M. Drayer: The Calcium Channel Blocker, Nitrendipine, in Single and Multiple Agent Antihypertensive Regimens: Preliminary Report of a Multicenter Study. *J.Cardiovasc.Pharmacol.* **6**:S1077-(4/18) (1984).

Weber, M.A., D.G. Cheung, W.F. Graettinger and J.L. Lipson: Characterization of Antihypertensive Therapy by Whole-Day BP Monitoring. *JAMA* **22**:3281-(4/18) (1988).

Weber, M.A., R.E. Purdy, G.L. Stupecky and B.A. Prins: Augmentation of Sympathomimetic Arterial Contraction by Angiotensin II: A Novel Mechanism. *J.Vasc.Med.Biol.* **1**:7-(4/18) (1980).

Weeds, A.: Actin-Binding Proteins; Regulators of Cell Architecture and Motility. *Nature* **296**:811-(2/10) (1982).

Wei, X.Y., E.M. Luchowski, A. Rutledge, C.M. Su and D.J. Triggle: Pharmacologic and Radioligand Binding Analysis of the Actions of 1,4-Dihydropyridine Activator-Antagonist Pairs in Smooth Muscle. *J.Pharmacol.Exp.Ther.* **239**:144-(3/13) (1986).

Wei, X.Y., A. Rutledge and D.J. Triggle: Voltage-Dependent Binding of 1,4-Dihydropyridine Ca^{2+} Channel Antagonists and Activators in Cultured Neonatal Rat Ventricular Myocytes. *Mol.Pharmacol.* **35**:541-(3/13) (1989).

Weinberg, C.B., J.R. Sanes and Z.W. Hall: Formation of Neuromuscular Junctions in Adult Rats: Accumulation of Acetylcholine Receptors, Acetylcholinesterase, and Components of Synaptic Basal Lamina. *Dev.Biol.* **84**:255-(2/10) (1981).

Weiner, D.A., S.S. Cutler and M.D. Klein: Efficacy and Safety of Sustained-Release Diltiazem in Stable Angina Pectoris. *Am.J.Cardiol.* **57**:6-(4/16) (1986).

Weiner, D.A., M.D. Klein, S.S. Cutler: Efficacy of Sustained-Release Verapamil in Chronic Stable Angina Pectoris. *Am. J. Cardiol.* **59**:215-(4/16) (1987).

Weiner, D.A., C.H. McCabe, S.S. Cutler, T.J. Ryan and M.D. Klein: The Efficacy and Safety of High-Dose Verapamil and Diltiazem in the Long-Term Treatment of Stable Exertional Angina. *Clin.Cardiol.* **7**:648-(4/16) (1984).

Weinstein, D.B.: The Antiatherogenic Potential of Calcium Antagonists. *J.Cardiovasc.Pharmacol.* **12**:S29-(3/13) (1988).

Weinstein, D.B. and J.G. Heider: Anti-Atherogenic Properties of Calcium Antagonists. *Am.J.Cardiol.* **59**:163-(4/20) (1987).

Weishaar, R.E. and R.J. Bing: The Beneficial Effect of a Calcium Channel Blocker, Diltiazem, on the Ischemic-Reperfused Heart. *J.Mol.Cell Cardiol.* **12**:993-(4/16) (1980).

Weiss, M.B., K. Ellis, R.R. Sciacca, L.L. Johnson, D.G. Schmidt and P.T. Cannon: Myocardial Blood Flow in Congestive and Hypertrophic Cardiomyopathy: Relationship to Peak Wall Stress and Mean Velocity of Circumferential Fiber Shortening. *Circulation* **54**:484-(4/17) (1976).

Weiss, R.J. and B. Bent: Diltiazem-Induced Left Ventricular Mass Regression in Hypertensive Patients. *J.Clin.Hypertens.* **3**:135-(4/18) (1987).

Wellens, H.J.J., F.W. Bar, K.I. Lie, D.R. Duren and H.J. Dohmen: Effects of Procainamide, Propranolol and Verapamil on Mechanism of Tachycardia in Patients with Chronic Recurrent Ventricular Tachycardia. *Am.J.Cardiol.* **40**:579-(4/19) (1977).

Wellens, H.J.J., D.R. Duren, K.L. Liem and K.L. Lie: Effects of Digitalis in Patients with Paroxysmal Atrioventricular Nodal Tachycardia. *Circulation* **52**:779-(4/19) (1975).

Wellens, H.J.J., J. Farre and W.B. Bar: The Role of the Slow Inward Current in the Genesis of Ventricular Tachyarrhythmias in Man. In *The Slow Inward Current and Cardiac Arrhythmias.* D.P. Zipes, J.C. Bailey, V. Elharrar, eds. Martinus Nijhoff, Boston. 507-(4/19) (1980).

Welsh, M.J., J.C. Aster, M. Ireland, J. Alcala and H. Maisel: Calmodulin Bindings to Chick Lens Gap Junction Protein in a Calcium Independent Manner. *Science* **216**:624-(2/10) (1982).

Werz, M.A. and R.L. MacDonald: Phorbol Esters: Voltage Dependent Effects on Calcium Dependent Action Potentials of Mouse Central and Peripheral Neurons in Cell Culture. *J.Neurosci.* **7**:1639-(3/14) (1987a).

Werz, M.A. and R.L. MacDonald: Dual Action of Protein Kinase C to Decrease Calcium and Potassium Conductances in Mouse Neurons. *Neurosci.Lett.* **78**:101-(3/14) (1987b).

Wesselinovitch, D., J.F. Mullan, R.W. Wissler, H.R. Davis and T. Bridenstine: Carotid Atherogenesis Inhibited by Sympathectomy, Propranolol, and Nifedipine in Rhesus Monkeys. *Arteriosclerosis* (abstr.) **6**:561-(4/18) (1986).

Weston, P.G. and M.D. Howard: The Determination of Sodium Potassium, Calcium and Magnesium in the Blood and Spinal Fluid of Patients Suffering from Manic-Depressive Insanity. *Arch.Neurol.Psychiatr.* **8**:179-(4/22) (1922).

Whitaker, M. and R.F. Irvine: Inositol 1,4,5-Triphosphate Microinjection Activates Sea Urchin Eggs. *Nature* **312**:633-(2/7) (1984).

Whitaker, M.J. and P.F. Baker: Calcium-Dependent Exocytosis in an *In Vitro* Secretory Granule Plasma Membrane Preparation from Sea Urchin Eggs and the Effects of Some Inhibitors of Cytoskeletal Function. *Proc.R.Soc.London B* **218**:397-(2/7) (1983).

White, C.J., W.A. Phillips, L.A. Abrahams, T.D. Watson and P.T. Singlton: Objective Benefit of Nifedipine in the Treatment of Raynaud's Phenomenon. *Am.J.Med.* **80**:623-(4/20) (1986).

White, G., D.M. Lovinger and F.F. Weight: Transient Low-Threshold Ca^{2+} Current Triggers Burst Firing Through an After Depolarizing Potential in an Adult Mammalian Neuron. *Proc.Natl.Acad.Sci.U.S.A.* **86**:6802-(3/13) (1989).

White, J.G.: Electron Microscopic Studies of Platelet Secretion. *Prog.Hemostas.Thrombo.* **2**:49-(4/20) (1974).

White, W.B., J.J. Viadero, T.J. Lane and S. Podesla: Effects of Combination Therapy with Captopril and Nifedipine in Severe or Resistant Hypertension. *Clin.Pharmacol.Ther.* **39**:43-(4/18) (1986).

Whitmer, K.R., J.S. Willams-Lawson, R.F. Highsmith and A. Schwartz: Effect of Calcium Channel Modulators on Isolated Endothelial Cells. *Biochem.Biophys.Res.Commun.* **154**:591-(3/13) (1988).

Wier, W.G. and G. Isenberg: Intracellular (Ca^{2+}) Transients in Voltage Clamped Cardiac Purkinje Fibres. *Pflügers Arch.* **392**:284-(2/10) (1982).

Wier, W.G., D.T. Yue and E. Marban: Effects of Ryanodine on Intracellular Ca^{2+} Transients in Mammalian Cardiac Muscle. *Fed.Proc.* **44**:2989-(2/10) (1985).

Wikman-Coffelt, J., R. Sievers, W.W. Parmley and G. Jasmin: Cardiomyopathic and Healthy Acidotic Hamsters Hearts: Mitochondrial Activity may Regulate Cardiac Performance. *Cardiovasc. Res.* **20**:471-(4/17) (1986a).

Wikman-Coffelt, J., R. Sievers, W.W. Parmley and G. Jasmin: Verapamil Preserves Nucleotide Pool in Cardiomyopathic Syrian Hamsters. *Am. J. Physiol.* **250**:H22-(4/17) (1986).

Wilcox, R.G., J.R. Hampton, D.C. Banks, J.S. Birkhead, I.A.B. Brooksby, C.J. Burns-Cox, M.J. Hayes, M.D. Joy, A.D. Malcolm, H.G. Mather and J.M. Rowley: Trial of Early Nifedipine in Acute Myocardial Infarction: The Trent Study. *Br.Med.J.* **293**:1204-(4/16) (1986).

Williams, D.A., K.E. Fogarty, R.Y. Tsien and F.S. Fay: Calcium Gradients in Single Smooth Muscle Cells Revealed by the Digital Imaging Microscope Using Fura-2. *Nature* **318**:558-(2/10) (1985).

Williamson, J.R. and J.R. Monck: Hormone Effects on Cellular Ca^{2+} Fluxes. *Ann.Rev.Physiol.* **51**:107-(2/7) (1989).

Willis, A.L., B. Nagel and V. Churchioll: Antiatherosclerotic Effects of Nicardipine and Nifedipine in Cholesterol-Fed Rabbits. *Atherosclerosis* **5**:250-(4/20) (1985).

Wilson, D.L., K. Morimoto, Y. Tsuda and A.M. Brown: Interaction between Calcium Ions and Surface Charge as it Relates to Calcium Currents. *J.Membr.Biol.* **72**:117-(1/2) (1983).

Wilson, S.P. and N. Kirshner: Calcium-Evoked Secretion from Digitonin-Permeabilized Adrenal Medullary Chromaffin Cells. *J.Biol.Chem.* **258**:4994-(2/7) (1983).

Wilson, W.A. and H. Wachtel: Negative Resistance Characteristic Essential for the Maintenance of Slow Oscillations in Bursting Neurons. *Science* **186**:932-(2/9) (1974).

Winegar, B.D., E.R. Rosick and R. Schaffer: Calcium and Olfactory transduction. *Comp.Biochem.Physiol.* **91A**:309-(2/8) (1988).

Winkel, R. and H.D. Lux: Carbamazepine Reduces Calcium Currents in *Helix* Neurones. In *Abstracts, 4th World Congress of Biological Psychiatry*. Philadelphia (Abstr). 6125-(4/22) (1985).

Winkler, H., K.D. Apps and R. Fischer-Colbrie: The Molecular Function of Adrenal Chromaffin Granules: Established Facts and Unresolved Topics. *Neuroscience* **18**:261-(2/7) (1986).

Winniford, M.D., K.L. Fulton, J.R. Corbett, C.H. Croft and L.D. Hillis: Propranolol-Verapamil Versus Propranolol-Nifedipine in Severe Angina Pectoris of Effort: A Randomized, Double-Blind Crossover Study. *Am.J.Cardiol.* **55**:281-(4/16) (1985).

Winniford, M.D., M.D. Fulton and D.L. Hillis: Long-Term Therapy of Paroxysmal Supraventricular Tachycardia: A Randomized, Double-Blind Comparison of Digoxin, Propranolol and Verapamil. *Am.J.Cardiol.* **54**:1138-(4/19) (1984b).

Winniford, M.D., G. Gabliani, S.M. Johnson, D.R. Mauritson, K.L. Fulton and L.D. Hillis: Concomitant Calcium Antagonists Plus Isosorbide Dinitrate Therapy for Markedly Active Variant Angina. *Am.Heart J.* **108**:1269-(4/16) (1984a).

Winniford, M.D., D.E. Jansen, G.A. Reynolds, P. Apprill, W.H. Black and L.D. Hillis: Cigarette Smoking-Induced Coronary Vasoconstriction in Atherosclerotic Coronary Artery Disease and Prevention by Calcium Antagonists and Nitroglycerin. *Am.J.Cardiol.* **59**:203-(4/16) (1987).

Winniford, M.D., S.M. Johnson, D.R. Mauritson, J.S. Rellas, G.A. Redish, J.T. Willerson and L.D. Hillis: Verapamil Therapy for Prinzmetal's Variant Angina: Comparison with Placebo and Nifedipine. *Am.J.Cardiol.* **50**:913-(4/16) (1982).

Winslow, E., P. Wright, J.K. Campbell and R.J. Marshall: Comparative Effect of the Isomers of Bepridil on Isolated Coronary and Aortic Arteries. *Eur.J.Pharmacol.* **166**:241-(3/13) (1989).

Winston, E.L., K.M. Pariser, K.B. Miller, D.M. Salem and M.A. Creager: Nifedipine as a Therapeutic Modality for Raynaud's Phenomenon. *J.Arthritis Rheum.* **26**:1177-(4/20) (1983).

Wisden, W., B.J. Morris, M.G. Darlison, S.P. Hunt and E.A. Barnard: Distinct GABA-A Receptor Alpha-Subunit mRNA's Show Differential Patterns of Expresion in Bovine Brain. *Neuron* **1**:937-(3/14) (1988).

Wit, A.L. and P.F. Cranefield: Effects of Verapamil on Sino-Atrial and Atrioventricular Nodes of the Rabbit and the Mechanisms by which it Terminates AV Nodal Re-entrant Tachycardia. *Circ.Res.* **35**:413-(4/19) (1974).

Wohlfart, B. and M.I. Noble: The Cardiac Excitation-Contraction Cycle. *Pharmacol.Ther.* **16**:1-(2/10) (1982).

Wolf, R., F. Habel and E. Witt: Wirkung von Verapamil auf die Hemodynamik und Grosse des Akuten Myokardinfarkts. *Herz* **2**:110-(4/16) (1977).

Wolf, R., U. Tscherne and H.M. Emrich: Suppression of Preoptic GABA Release Caused by Push-Pull-Perfusion with Sodium Valproate. *Naunyn Schmied.Arch.Pharmacol.* **338**:658-(4/22) (1988).

Wolfe, S.E. and M.A. Brostrum: Mechanisms of Action of Inhibitors of Prolactin Secretion in GH3 Pituitary Cells II Blockade of Voltage-Dependent Ca^{2+} Channels. *Mol.Pharmacol.* **29**:420-(3/13) (1986).

Wollheim, C.B. and T.J. Biden: Second Messenger Function of Inositol 1,4,5-Trisphosphate. Early Changes in Inositol Phosphates, Cytosolic Ca^{2+}, and Insulin Release in Carbamylcholine-Stimulated RINm5F Cells. *J.Biol.Chem.* **261**:8314-(2/7) (1986).

Wollheim, C.B. and T. Pozzan: Correlation between Cytosolic Free Ca^{2+} and Insulin Release in an Insulin-Secreting Cell Line. *J.Biol.Chem.* **259**:2262-(2/7) (1984a).

Wollheim, C.B. and G.W.G. Sharp: Regulation of Insulin Release by Calcium. *Physiol.Rev.* **61**:914-(2/7) (1981).

Wollheim, C.B., S. Ullrich, P. Meda and L. Vallar: Regulation of Exocytosis in Electrically Permeabilized Insulin-Secreting Cells. Evidence for Ca^{2+} Dependent and Independent Secretion. *Biosci.Rep.* **7**:443-(2/7) (1987).

Wollheim, C.B., S. Ullrich and T. Pozzan: Glyceraldehyde, but not Cyclic AMP-Stimulated Insulin Release is Preceded by a Rise in Cytosolic Free Ca^{2+}. *FEBS Lett.* **177**:17-(2/7) (1984b).

Wong, R.K.S. and D.A. Prince: Dendritic Mechanisms Underlying Penicillin Induced Epileptiform Activity. *Science* **204**:1228-(2/9) (1979).

Wong, R.K.S., D.A. Prince and A.I. Basbaum: Intradendritic Recordings from Hippocampal Neurons. *Proc.Natl.Acad.Sci.U.S.A.* **76**:986-(2/9) (1979).

Wood, E.H., R.L. Heppner and S. Weidmann: Inotropic Effects of Electric Currents. *Circ.Res.* **24**:409-(2/10) (1969).

Woodbury, J., K. Lyons, R. Carretta, A. Hahn and J.F. Sullivan: Cerebrospinal Fluid and Serum Levels of Magnesium, Zinc, and Calcium in Man. *Neurology* **18**:700-(4/21) (1968).

Wooden, S.K., R.P. Kapur and A.S. Lee: The Organization of the Rat GRP78 Gene and A23l87-Induced Expression of Fusion Gene Products Targeted Intracellularly. *Exp.Cell Res.* **178**:84-(2/10) (1988).

Woodruff, M.L., B.L. Bastian and G.L. Fain: Light-Dependent Ion Influx into Toad Photoreceptors. *J.Gen.Physiol.* **80**:517-(2/8) (1982a).

Woodruff, M.L. and G.L. Fain: Ca Dependent Changes in Cyclic GMP are not Correlated with Opening and Closing of the Light Dependent Permeability of the Toad Photoreceptors. *J.Gen.Physiol.* **80**:537-(2/8) (1982b).

Woodward, J.J., S.M. Rezazadeh and S.W. Leslie: Differential Sensitivity of Synaptosomal Calcium Entry and Endogenous Dopamine Release to ω-Conotoxin. *Brain Res.* **475**:141-(3/13) (1988).

Worley, III, J.F., J.W. Deitmer and M.T. Nelson: Single Nisoldipine-Sensitive Calcium Channels in Smooth Muscle Cells Isolated from Rabbit Mesenteric Artery. *Proc.Natl.Acad.Sci.U.S.A.* **83**:5746-(2/10,3/13) (1986).

Wray, D.W., R.I. Norman and P. Hess: Calcium Channels: Structure and Function. *Ann.N.Y.Acad.Sci.* **560** (1/4,3/13) (1989a).

Wray, W.D., B. Lang, J. Newsom-Davis and C. Peers: Antibodies Against Calcium Channels in the Lambert-Eaton Myasthenic Syndrome. *Ann.N.Y.Acad.Sci.* **560**:269-(3/13) (1989b).

Wu, D.: What is the Value of Calcium Antagonists for Ventricular Tachycardia. *Int.J.Cardiol.* **5**:543-(4/19) (1984b).

Wu, D., P. Denes, R. Bauernfeind, R. Kehoe, Amat-Y-Leo and K.M. Rosen: Effects of Procainamide on Atrioventricular Nodal Re-Entrant Paroxysmal Tachycardia. *Circulation* **57**:1171-(4/19) (1978).

Wu, D., P. Denes, R.C. Dhingra, A. Khan and K.M. Rosen: The Effects of Propranolol on Induction of AV Nodal Re-Entrant Paroxysmal Tachycardia. *Circulation* **50**:665-(4/19) (1974).

Wu, D., J. Hung, C. Kuo, K. Hsu and W. Shieh: Effects of Quinidine on AV Nodal Re-Entrant Paroxysmal Tachycardia. *Circulation* **64**:169-(4/19) (1981a).

Wu, D., H.C. Kou and J.S. Hung: Exercise-Triggered Paroxysmal Ventricular Tachycardia. A Repetitive Rhythmic Activity Possibly Related to Afterdepolarization. *Ann.Int.Med.* **95**:410-(4/19) (1981).

Yaari, Y., B. Hamon and H.D. Lux: Development of Two Types of Calcium Channels in Cultured Mammalian Hippocampal Neurons. *Science* **235**:680-(2/9,2/10,3/13,3/14) (1989).

Yaari, Y., A. Konnerth and U. Heinemann: Spontaneous Epileptiform Activity of CA1 Hippocampal Neurons in Low Extracellular Calcium Solutions. *Exp.Brain Res.* **51**:153-(4/22) (1987).

Yamada, K. and S. Shibata: Recent Advances in Calcium Channels and Calcium Antagonists. *Proceedings of the Japan-U.S.A. Symposium on Cardiovascular Drugs.* Pergamon Press, New York. (3/13) (1990).

Yamaguchi, D.T., J. Green, C.R. Kleeman and S. Maullem: Characterization of Volume-Sensitive, Calcium-Permeating Pathways in the Osteosarcoma Cell Line UMR-106-01. *J.Biol.Chem.* **264**:4383-(3/13) (1989).

Yamamoto, K.D. and H. Washio: Permeation of Sodium Through Calcium Channels of an Insect Membrane. *Can.J.Physiol.Pharmacol.* **57**:220-(1/2) (1979).

Yamamura, H.I., H. Schoemaker, R.G. Boles and W.R. Roeske: Diltiazem Enhancement of ^{3}H Nitrendipine Binding to Calcium Channel Associated Drug Receptor Sites in Rat Brain Synaptosomes. *Biochem.Biophys.Res.Commun.* **108**:640-(4/22) (1982).

Yasue, H., S. Omote, A. Takizawa, M. Nagas, K. Miwa and S. Tanaka: Circadian Variation of Exercise Capacity in Patients with Prinzmetal's Variant Angina: Role of Exercise-Induced Coronary Artery Spasm. *Circulation* **59**:938-(4/16) (1979).

Yasue, H., S. Omote, A. Takizawa, M.N. Nagas, K. Miwa and S. Tanaka: Exertional Angina Pectoris Caused by Coronary Artery Spasm: Effects of Various Drugs. *Am.J.Cardiol.* **43**:647-(4/16) (1979).

Yatani, A. and A.M. Brown: Rapid Beta-Adrenergic Modulation of Cardiac Calcium Channel Currents by Fast G Protein Pathway. *Science* **245**:71-(3/14,4/17) (1989).

Yatani, A., J. Codina, A.M. Brown and L. Birnbaumer: Direct Activation of Mammalian Atrial Muscarinic Potassium Channels by GTP by the Regulatory Protein Gk. *Science* **235**:207-(3/14) (1987a).

Yatani, A., J. Codina, Y. Imoto, J. Reeves, L. Birnbaumer and A.M. Brown: A G Protein Directly Regulates Mammalian Cardiac Calcium Channels. *Science* **238**:1288-(3/14) (1987b).

Yatani, A., Y. Imoto, J. Codina, S.L. Hamilton, A.M. Brown and L. Birnbaumer: The Stimulating G Protein of Adenylyl Cyclase, G_s, also Stimulates Dihydropyridine-Sensitive Ca^{2+} Channels. *J.Biol.Chem.* **263**:9887-(3/13,3/14) (1988).

Yatani, A., D.L. Kunze and A.M. Brown: Effects of Dihydropyridine Calcium Channel Modulators on Cardiac Sodium Channels. *Am.J.Physiol.* **23**:H443-(3/13) (1988b).

Yatani, A., C.L. Seidel, J. Allen and A.M. Brown: Whole Cell and Single-Channel Calcium Currents of Isolated Smooth Muscle Cells from Saphenous Vein. *Circ.Res.* **60**:523-(2/10,3/13) (1987c).

Yatani, A., D.L. Wilson and A.M. Brown: Recovery of Ca Currents from Inactivation: The Roles of Ca Influx, Membrane Potential, and Cellular Metabolism. *Cell.Molec.Neurobiol.* **3**:381-(1/4) (1983).

Yau, K.W. and D.A. Baylor: Cyclic GMP-Activated Conductance of Retinal Photoreceptor Cells. *Ann.Rev.Neurosci.* **12**:289-(2/8) (1989).

Yau, K.W. and K. Nakatani: Cation Selectivity of Light-Sensitive Conductance in Retinal Rods. *Nature* **309**:352-(2/8) (1984).

Yau, K.W. and K. Nakatani: Light-Suppressible, Cyclic GMP-Sensitive Conductance in the Plasma Membrane of a Truncated Rod Outer Segment. *Nature* **317**:252-(2/8) (1985b).

Yau, K.W. and K. Nakatani: Light-Induced Reduction of Cytoplasmic Free Calcium in Retinal Rod Outer Segment. *Nature* **313**:579-(2/8) (1985a).

Ye, Z. and K. Van Dyke: Reversal of Chloroquine Resistance in Falciparum Malaria Independent of Calcium Channels. *Biochem.Biophys.Res.Commun.* **155**:476-(3/13) (1988).

Ye, Z. and K. Van Dyke: Selective Antimalarial Activity of Tetrandine Against Chloroquine Resistant Plasmodium Falciparum. *Biochem.Biophys.Res.Commun.* **159**:242-(3/13) (1989).

Yeh, S.H., F.C. Liu, Y.Y. Chan, J.S. Hung and D. Wu: Termination of Paroxysmal Supraventricular Tachycardia with a Single Oral Dose of Diltiazem and Propranolol. *Circulation* **71**:104-(4/19) (1985).

Yellen, G.: Single Ca^{2+} Activated Nonselective Cation Channels in Neuroblastoma. *Nature* **296**:357-(2/9,3/14) (1982).

Yokoyama, M., T. Koizumi, K. Fujitani, T. Mizutani and H. Fukuzaki: Adverse Response to Nifedipine in Unstable Angina Pectoris. *Chest* **81**:646-(4/16) (1982).

Yoshida, S.: Permeation of Divalent and Monovalent Cations Through the Ovarian Oocyte Membrane of the Mouse. *J.Physiol.* **339**:631-(3/12) (1983).

Yoshii, M., A. Tsunoo and T. Narahashi: Effects of Pyrethroids and Veratridine on Two Types of Ca Channels in Neuroblastoma Cells. *Soc.Neurosci.Abs.* **11**:518(3/13)-518(3/13) (1985).

Yoshikami, D., Bagabaldo and B.M. Olivera: The Inhibitory Effects on ω-Conotoxins on Ca Channels and Synapses. *Ann.N.Y.Acad.Sci.* **560**:230-(3/13) (1989).

Yu, C., M. Jia, M. Litzinger and P.G. Nelson: Calcium Agonist BAY K 8644 Augments Voltage-Sensitive Calcium Currents but not Synaptic Transmission in Cultured Mouse Spinal Cord Neurons. *Exp.Brain Res.* **71**:467-(3/13) (1988).

Yu, Y.M., F. Lermioglu and A. Hassid: Modulation of Ca by Agents Affecting Voltage-Sensitive Ca Channels in Mesangial Cells. *Am.J.Physiol.* **257**:F1094-(3/13) (1989).

Yue, D.T., S. Herzig and E. Marban: β-Adrenergic Stimulation of Calcium Channels Occurs by Potentiation of High-Activity Gating Modes. *Proc.Natl.Acad.Sci.U.S.A.* **87**:753-(3/13) (1990).

Yusuf, S. and C. Furberg: Effect of Acute or Chronic Administration of Calcium Antagonists on Mortality Following Myocardial Infarction. *J.Am.Coll.Cardiol.* (abstr.) **9**:24-(4/16) (1987).

Yusuf, S. and C.D. Furberg: Effects of Calcium-Channel Blockers on Survival after Acute Myocardial Infarction. *Cardiovasc.Drugs Ther.* **1**:343-(4/19) (1987).

Yusuf, S., R. Petro, J. Lewis, R. Collins and P. Sleight: Beta-blockade During and After Myocardial Infarction: An Overview of the Randomized Trials. *Prog.Cardiovasc.Dis.* **27**:235-(4/16,4/19) (1985).

Zamora, J.M., H.L. Pearce and W.T. Beck: Physical-Chemical Properties Shared by Compounds that Modulate Multidrug Resistance in Human Leukemic Cells. *Mol.Pharmacol.* **33**:454-(3/13) (1988).

Zavoico, G.B. and M.B. Feinstein: Cytoplasmic Ca^{2+} in Platelets is Controlled by Cyclic AMP: Antagonism Between Stimulation and Inhibition of Adenylate Cyclase. *Biochem.Biophys.Res.Commun.* **120**:759-(4/20) (1984).

Zernig, G.: Widening Potential for Ca^{2+} Antagonists: Non-L-Type Ca^{2+} Channel Interaction. *Trends Pharmacol.Sci.* **11**:38-(3/13) (1990).

Zernig, G. and H. Glossmann: A Novel 1,4-Dihydropyridine-Binding Site on Mitochondrial Membranes from Guinea Pig Heart, Liver and Kidney. *Biochem.J.* **253**:49-(3/13) (1988).

Zhu, D.L. and H.B. Peng: Increase in Intracellular Calcium Induced by the Polycation-Coated Latex Bead, a Stimulus that Causes Postsynaptic-Type Differentiation in Cultured *Xenopus* Muscle Cells. *Dev.Biol.* **126**:63-(2/10) (1988).

Zimmerberg, J.: Molecular Mechanisms of Membrane Fusion: Steps during Phospholipid and Exocytotic Membrane Fusion. *Biosci.Rep.* **7**:251-(2/7) (1987).

Zimmerman, A.L. and D.A. Baylor: Cyclic GMP-sensitive Conductance of Retinal Rods Consists of Aqueous Pores. *Nature* **321**:70-(2/8) (1986).

Zimmerman, A.L., G. Yamanaka, F. Eckstein, D.A. Baylor and L. Stryer: Interaction of Hydrolysis-Resistant Analogs of Cyclic GMP with the Phosphodiesterase and Light-Sensitive Channel of Retinal Rod Outer Segments. *Proc.Natl.Acad.Sci.U.S.A.* **82**:8813-(2/8) (1985).

Zimmerman, B.G. and P.C. Raich: Renal Hemodynamic Effects of a Selected Calcium Antagonist. *Am.J.Cardiol.* **62**:69G-(3/13) (1988).

Zimmerman, J.J., S.M. Zuk and J.R. Millard: *In Vitro* Modulation of Human Neutrophil Superoxide Anion Generation by Various Calcium Channel Antagonists Used in Ischemia-Reperfusion Resuscitation. *Biochem.Pharmacol.* **36**:3601-(3/13) (1989).

Zipes, D.P., P.R. Foster, P.J. Troup and D.H. Pedersen: Atrial Induction of Ventricular Tachycardia: Re-Entry Versus Triggered Automaticity. *Am.J.Cardiol.* **44**:1-(4/19) (1979).

Zipes, D.P., J.J. Heger and E.N. Prystowsky: Pathophysiology of Arrhythmias: Clinical Electrophysiology. *Am.Heart J.* **106**:812-(4/19) (1983).

Zipes, D.P., R.L. Rinkenberger, J.J. Heger and E.N. Prystowsky: The Role of the Slow Inward Current in the Genesis and Maintenance of Supraventricular Arrhythmias in Man. **In** *The Slow Inward Current in Cardiac Arrhythmias*. D.P. Zipes, C.J. Bailey, and V. Elharrar, eds. Martinus Nijhoff, Boston. 481-(4/19) (1980).

Zschauer, A., C. Van Breemen, F.R. Buhler and M.T. Nelson: Calcium Channel in Thrombin-Activated Human Platelet Membrane. *Nature* **334**:703-(4/20) (1988).

Zucker, R.S. and A.L. Fogelson: Relationship Between Transmitter Release and Presynaptic Calcium Influx when Calcium Enters through Discrete Channels. *Proc.Natl.Acad.Sci.U.S.A.* **83**:3032-(2/7) (1986).

Zucker, R.S. and R.A. Steinhardt: Prevention of the Cortical Reaction in Fertilized Sea Urchin Eggs by Injection of Calcium Chelating Ligands. *Biochim.Biophys.Acta* **541**:459-(2/7) (1978).

Index

INDEX

A

A23187, calcium ionophore, 154
Acetylcholine, and calcium channel activation, 31
Acetylcholinesterase (AChE), and calcium influx, 144
Acetylcholinesterase receptor, 143—145
Action potential, 138, 141, 154
Activation, calcium channel, 7—31, 160
Activation kinetics, 22—26, 30
Activators, calcium channel, 161
Adenosine, 31, 342—343
Adrenal glomerulosa, and calcium-dependent secretion, 99
Adrenergic agents, competitive with gallopamil, 302—303
β-aldrenergic blockade, calcium blockers compared with, 280
β-adrenergic/cAMP mediated stimulation, 258—259
β-adrenoceptor antagonists, 202
Adrenomedullary cells, binding sites in, 218
Affective disorders, 389—390
Age-related behavioral deficits, effect of calcium channel blockers on, 238
Alkaline ion permeation, 26
Alzheimer's patients, 243
Amiloride, in T channels, 204, 209
Amiloride-like drugs, binding site for, 222
Amino acid sequences, 60
Amino acid transmitters, excitatory, 378
Aminoglycoside antibiotics, effect on calcium channels of, 220—221
Amiodarone, 209, 232
Amlodipine
  binding sites for, 221
  in L channels, 210
  partition coefficient of, 206—207
  selectivity of action of, 227—228
  structural formula of, 212
Amplitudes, calcium current, 12
Anesthetics, 202
Angina
  caused by coronary spasm, 282—285
  chronic stable effort, 292
    diltiazem for, 279
    efficacy of calcium blockers in, 280, 282, 292
    nifedipine for, 278—279
    verapamil for, 277—278
    withdrawal of calcium blockers in, 279—280
  effects of calcium channel ligands on, 233
  ergonovine-induced, 284
  "mixed", 286
  pectoris, 268, 275, 296
  at rest, 286—288
  "silent" ischemia in, 285—286
  treatment of, 223
  vasospastic, 285
Angiotensin II, 311
Animal models. See also specific animals
  effects of calcium channel ligands in cardiovascular systems, 232—234
Anomalous mole-fraction effects, 174, 180
Antagonists, calcium channel. See also Blockers, calcium channel
  effects of, 246—248
  types of, 310—313
Antianginal agents, See Blockers, calcium channel
Antiarrhythmia drugs. See also Arrhythmias
  calcium antagonists as, 359—360
  comparison of, 339
Antiatherogenic effects, of calcium channel blockers, 234
Antibodies
  from B lymphocytes, 101—102
  and calcium channel current, 241
Anticompetitive inhibition, 192
Anticonvulsants, calcium channel blockers and, 392
Antidepressants, 202
Antiepileptics, 248
Anti-ischemic effect, of calcium channel blockers, 277
Antimalarial drugs, 247
Aplysia
  burst firing activity in, 133
  bursting pacemaker neuron R-15 of, 171
Arrhythmias. See also Cardiovascular diseases
  calcium channel blockers for, 328
  calcium channel in genesis of, 330—334
  treatment of, 223
Arrhythmias, supraventricular calcium channel blockers in
  acute conversion of PSVT, 341—343
  acute termination of PSVT, 337—339
  chronic prophylaxis of PSVT, 344—345
  in ectopic supraventricular tachycardia, 343
  electrophysiologic mechanisms, 336—337
  modes of conversion of PSVT, 339
  in multifocal atrial tachycardia (MAT), 343—344
  pre-excitation syndromes, 345—346
  PSVT conversion by, 341
  in reentrant supraventricular tachycardia, 343
Arrhythmias, ventricular, calcium channel blockers in, 349—353
Arteriosclerosis, calcium-overload-induced, 248
Artificial lipid bilayers, inactivation curves for ion channels of, 166
Atherosclerosis, calcium channel blockers and, 325, 371—372
ATP, in acute conversion of PSVT, 342—343
Atrial fibrillation, 347
Atrial flutter, 348—349

B

^{33}Ba, K$^+$-stimulated influxes of, 172
Ba^{2+} ions, 24
   compared with Ca^{2+} ions, 17
   effect on calcium channels of, 161
Baclofen, and calcium channel activation, 31
BAPTA, and rod photoresponse, 116
Barbiturates, 204
   calcium blocker properties of, 202
   N channels inhibited by, 198
Barnacle muscle fiber, 182
   determing K$_{Ca}$ and K$_{Mg}$ in, 188—189
   electrical properties of, 175—176
   resting membrane potential of, 165
   value of pK$_a$ for, 194
Barnacle muscle membrane
   ions that competitively inhibit I$_{Ca}$ in, 174
   resting membrane potential across, 193
BAY K 8644, 205, 243—248
   effect on calcium channels of, 161
   and HVA channel inactivation, 58
   inhibited by dihydropyridines, 236—237
   mechanisms of action for, 229—230
   promotion of long channel opening by, 210
   selectivity of action of, 227
Belfosdil, 222
Benextramine, structural formula of, 208
Benzodiazepines, 202, 307. See also Diltiazem
Benzothiazepines, 311, 383
Benzothiazocines, 214, 215
Bepridil. See also Blockers, calcium channel
   binding of, 222
   in cardiomyopathic hamsters, 300—301
   dosage regimen for, 359
   effect on platelets of, 368
   sites of action of, 244
β-blockade, comparison of nifedipines with, 278—279
Beta blockers
   calcium channel blockers combined with, 318, 320
   effects on AV nodes of, 359
   as antiarrhythmic agents, 360
Beta cells, pancreatic, as calcium-dependent process, 96—98
Binding, 221
Binding proteins, Ca^{2+}, 155
Binding sites, calcium channel
   allosterically coupled, 221—223
   chemical nature of, 189—193
   for drugs acting on, 218
   endogenous substances acting on, 239—241
   location of, 223—224
Biophysics, of calcium channels, 3—8
Blockers, calcium channel, 161
   acute administration of, 321—324
   affinity for L channels of, 220
   antianginal effects of, 275—277
   and anticancer drugs, 247
   and anticonvulsants, 392
   antihypertensive actions of, 311
   and atherosclerosis, 371—372
   in atrial tachyarrhythmias, 346—349
   calcium homeostasis, 366—370
   cardiac effects of, 312
   for cardiomyopathies, 296, 301—302
   in combination with other drugs, 318—319, 320, 357—359
   comparison of, 281—282, 292
   contraindications for 359
   in control of arrhythmias, 328, 357—359
   in dilated myopathies, 298—299
   drugs acting as, 208
   effects on AV node of, 358
   effect on blood vessels, 370—372
   effect on drug-resistant cells, 247
   effect on renal function of, 314—315
   efficacy of, 316—318, 325
   in effort angina, 280, 282, 292
   electrophysiological effects of, 334—336
   failure of, 291
   in hypertension, 310
   in hypertrophic cardiomyopathy, 297
   in left ventricular hypertrophy, 313
   mechanisms of, 268, 311—312
   metabolic effects of, 315—316
   and neuroleptics, 393
   and platelets, 366—370
   in post-infarction follow-up, 290
   potential uses for, 202—203
   and prevention of muscle contractures, 72—73
   in psychiatric patients, 390—391
   side effects of, 324
   structural formulae of, 201, 206
   and sudden death, 353—359, 361
   in supraventricular arrhythmias, 314
      acute conversion of PSVT, 341—343
      acute termination of PSVT, 337—339
      chronic prophylaxis of PSVT, 344—345
      in ectopic supraventricular tachycardia, 343
      electrophysiology mechanisms, 336—337
      modes of conversion of PSVT, 339
      in multifocal atrial tachycardia (MAT), 343—344
      vs. other antiarrhythmias agents, 341—343
      pre-excitation syndromes, 345—346
      PSVT conversion by, 341
      in reentrant supraventricular tachycardia, 343
   therapeutic uses of, 202
   for use in brain, 383—384
   in vascular disease, 372—375
   for vasopastic angina, 284—285
   vasoselectivity of, 312—313
   for vasospastic angina, 282—284
   in ventricular arrhythmias, 349—352
   in ventricular tachycardia, 352—353
Blocking ions, 10—12
Blood cells, effect of drugs on calcium channels of, 226
Blood pressures, pre-treatment, 320—321

Blood vessels, 370—372
B lymphocytes, antibodies secreted from, 101—102
Brain. See also CNS disorders
  calcium channel blockers in, 383
  effects of calcium channel ligands on, 234—237
  pathological accumulation of calcium in, 381—382
Bronchial asthma, 270
Buffering mechanisms, 64
Buffering systems, intracellular Ca^{2+}, 256
Burst firing patterns, 134—135

C

^{45}Ca, K^+-stimualted influxes of, 172
Ca^{2+} ions
  binding of, 5
  effect on calcium channels of, 161
  fundamental role of, 252
N-Cadherin, 147
Calcium
  excess intracellular, 382
  interaction with lithium of, 391—392
  metabolism of, 378—381
  pathological accumulation of, 381—382
  as second messenger, 365—366
  as secretory trigger, 103
  sources of trigger, 89—90
  transport of, 378—381
Calcium-antagonists, in hypertrophic cardiopathy, 298. See also Blockers, calcium channel
Calcium channel blockers. See Blockers, calcium channel
Calcium channel. See also specific calcium channels
  cardiac dihydropyridine-sensitive, 31
  chronic regulation of 7, 241—243, 254, 330—334
  ionic selectivity of, 10, 18
  localization of, 129—132
  micro-organization of, 132
  properties of, 164
  selective permeability of, 12
  slow, 306
  types of, 126—129
  voltage-sensitive, 22
Calcium currents, L-type, 71. See also L channels
Calcium trigger, 89—90
Calmodulin, 138, 145, 150, 246, 249
  blockade of, 310
  in gap junction regulation, 151
Cardiac death, prevention of, 353, 361
Cardiac muscle
  effects of drugs on calcium channels in, 224
  excitation-contraction coupling in, 77—81
  L-channel function in, 262
Cardiac necrosis, assessment of, 301
Cardiomyopathies
  abnormalities of myocardial perfusion in, 297
  calcium channel blockers in, 268
    cardiac necrosis, 301
    dilated cardiomyopathies, 298—299
  comparative effects of, 301—302
    hypertrophic cardiomyopathy, 297
    restrictive cardiomyopathy, 299
  hamster hereditary, 299
Cardiovascular diseases. See also Arrhythmias
  channel blockers used in, 235, 268—269
Caroverine, 200
Ca^{2+} spikes, in larval muscle fibers of beetle, 177
Catecholamine, secretion of, 96—97
Cd^{2+}, effect on calcium channels of, 161
Cell growth, 64
Cells, visualization of Ca^{2+} influx in, 130—131
Cellular interactions, 150—152
Cellular processes, regulation of, 64—65
Central nervous system. See also CNS disorders
Cerebrospinal fluid (CSF), 379—380
C-fos, induction of, 153
cGMP
  light-sensitive channels activated by, 119
  photocurrent altered by, 116
Channel gating, voltage-dependence of, 172. See also Gating mechanisms
Chelators, internal Ca^{2+}, 51
Chick ciliary ganglion, choline acetyltransferase activity in, 149
Chick DRG cell, calcium channel currents of, 44
Chick sensory neuron
  single LVA channels in, 44—45
  sodium channel currents in, 40—41
Cholesterol
  in atherosclerosis, 325
  as cardiovascular risk factor, 315
Chromaffin cells, 89
  binding sites in, 218
  catecholamine secretion from, 96—97
Cinnarizine
  binding site for, 222
  effect on platelets of, 368
  sites of action of, 244
  structural formula, 209
  T channels inhibited by, 232
Cisplatin, 247
Claudication, intermittent, 373
CNS disorders, 237—239
Co^{2+}, effect on calcium channels of, 161
Competitive inhibition, 185, 187—188, 194
Conductance, defined, 5
ω-conotoxin GVIA (ω-CgTX), 197
ω-conotoxins, 202, 215, 248
Constipation, 324
Contraindications, for calcium channel blockers, 359
Conus toxins, sequences of, 208
Converting enzyme inhibitors, and calcium blockers, 319
Coronary artery disease, 268. See also Cardiovascular disease
Crustacean muscle fibers, electrical activity generated in, 175
CSF, see Cerebrospinal Fluid

Cyclic 3'-5' guanosine monophosphate (cGMP), 109
Cyproheptadine, structural formula of, 208

D

D600, 72, 73, 80, 200, 296. See also Gallopamil
Dark adaption, 114
Deactivation, of calcium channels, 160
Deactivation kinetics, 25—26, 30—31
Dementia, 243
Dendrites, 129—130. See also Neurons
Depression. See also Psychiatric disorders
   Ca-AtPase in, 389
   increased serum calcium in, 387
Desmenthoxyverapamil, partition coefficient of, 207
Devapamil, 217—218
DHP, see Dihydropyridine
Diazepam, 204, 208
Dihydropyridine binding, 239—240, 242
Dihydropyridine blockers, 213
Dihydropyridine derivatives. See Nifedipine; Nimodipine; Nisoldipine; Nitrodipine
Dihydropyridine (DHP) receptor
   alpha 1 subunit of, 74
   calcium channels, 254
   cAMP-dependent phosphorylation and, 258
   in excitation-contraction coupling, 73
Dihydropyridines (DHP), 72, 80, 154, 244, 307
   antiarrhythmiac action of, 329
   binding sites of, 246
   as calcium entry blocker, 383
   in cat papillary muscle, 216
   channel activation, 212
   charged, 200
   effect on heart rate of, 355
   effect on nucleoside transporter of, 245
   mechanisms for selectivity for, 235—236
   partition coefficients of, 244
   quantitative structure activity studies for, 212—213
   structural requirements in, 211
   voltage-dependent interactions, 213—214
   as vasodilators, 312
Diltiazem, 200, 311. See also Blockers; calcium channel
   for acute myocardial infarction, 290, 292
   antianginal potency of, 275
   in atherosclerosis, 325
   binding of, 222
   as calcium entry blocker, 383
   in cardiomyopathic hamsters, 300—301
   in cardiovascular disease, 293
      antiarrhythmic action of, 329, 339, 360
      in dilated cardiomyopathies, 298
      in chronic effort angina, 281
      effect on AV node of, 358
      for effort angina, 279
      in left ventricular hypertrophy, 313
      in vasospastic angina, 283
   dosage regimens for, 359
   and drug metabolism, 246
   effect on calcium channels of, 161
   effect on heart rate of, 355
   effects on renal function of, 314
   in hypertensive patients, 317, 322—323
   in hypertrophic cardiomyopathy, 297
   in peripheral vascular disease, 372
   in platelet aggregation, 246
   structural formula for, 201
   vasoselectivity of, 312
Dimethadione, in T channels, 209, 248—249
Diphenylbutylpiperidines, 202
Diphosphonates, binding site for, 222
Dipicolinic acid (DPA), HVA channel inactivation affected by, 59
Disease. See also Cardiovascular disease
   calcium channel alternation of, 243—244
   effect of calcium blockers in, 234
Disopyramide, as antiarrhythmic agent, 339
Dissociation constant, 167—169, 188
Distribution of, calcium channels, 255—256
Diuretics, calcium channel blockers and, 318
Diversity, of calcium channels, 253
Dizziness, 324
Dopamine, and calcium channel activation, 31
DPI 201-106, binding site for, 222
Drug binding sites, for organic ligands, 207
Drug-lipid interactions, 208
Drugs. See also specific drugs
   acting on calcium channels, 200
      binding sites for, 223—224
      modes and selectivity of action, 227—232
      sites of action, 217—227
      structure-activity relationships, 206—217
      T channels, 230
Dysmenorrhea, 270

E

Edema, peripheral, 324
EGTA, and rod photoresponse, 116
Egg fertilization, cortical granule discharge during, 100
Electrophysiology, of calcium channel blockers, 334
Electrophysiological studies, in CM hamsters, 303
Embryonic rat DRG, whole-cell clamp LVA calcium currents from, 37
(S)-Emopamil, as calcium entry block, 383
Endocrine cells, effect of drugs on calcium channels, 224
Endothelial cells, effect of drugs on calcium channels in, 226
Endothelin, 239, 311
Epilepsy, age-associated, 223. See also Brain
Epinephrine, calcium channel activity modulated by, 257
Erythrocytes, 245—246, 388—390
Ethosuximide, 248
Excitation-contraction coupling
   in cardiac muscle, 78—81
   mechanisms of, 65, 70

Index

in skeletal muscle, 72—77
in vascular smooth muscle, 82—84
Excitation-secretion coupling, 66, 96
Exercise tolerance, 324
Exocrine cells, effect of drugs on calcium channels in, 226
Exocytosis, 88—89
   Ca^{2+} as trigger for antibody secretion from B lymphocytes, 101—102
   and catecholamine secretion, 96
   cortical granule discharge, 100
   criteria for identifying calcium-dependent secretion, 95
   and discharge of trichocysts from paramecium, 99—100
   histamine secretion, 100—101
   hormone secretion, 98—99
   neuronal secretion, 90—95
   pancreatic beta cells, 96—98
   parathyroid hormone secretion, 102—103

F

Fatty acids, dihydropyridine binding inhibited by, 240
Felodipine, 209, 231
Fendiline, 200
Filtering systems, ion-selecting, 180—183
Flunarizine. See also Blockers, calcium channel
   in brain tissue, 383
   as calcium entry blocker, 383
   effects of, 246
   effect on platelets of, 368
   partition coefficient of, 207
   in peripheral vascular disease, 372
   sites of action of, 244
Fluspirilene, 202, 222
Frog muscle fibers
   anomalous mole-fraction effect in, 170
   fast activating calcium channels in, 72
   membrane permeation in, 176
Functional states, of calcium channels, 159

G

GABA, see Gamma Amino Butyric Acid
Gallopamil (D600), 218, 296. See also Blockers, calcium channel; Phenylalkylamines
   activities of enantiomers of, 217
   in cardiomyopathic hamsters, 300—301
   competition with adrenergic agents of, 302—303
   dosage regimen for, 359
   efficacy as antiarrhythmic agent, 339
   interaction of adrenergic agonists and antagonists with, 301
Gamma amino butyric acid (GABA)
   in amphibian spinal cord neurons, 139
   calcium currents inhibited by, 261—262, 263
   and calcium regulation, 149
   in cone photoreceptors, 121
Gap junctions, and Ca^{2+} influx, 150—152
Gating currents, 38—39
Gating mechanisms, 26—27, 160
Gene regulation, role of calcium in, 152—154
Glomerular filteration rate, 325
Glucose metabolism, effects of calcium channel blockers on, 316
GMP-dependent kinase, 262
Gouy-Chapman theory, 166, 170
G proteins, 249
GTP binding proteins, 222—223, 239
Guinea pig, 171, 173
Guinea pig myocytes
   calcium channels in, 172
   single LVA channels in, 44—45

H

Hair cells, 108, 121—122
Hamsters. See also Syrian hamsters, cardiomyopathic, electrophysiological studies in, 303
Headache, 324
Heart block, 324
Heart cells, 51—52, 58
Heart muscle cells, anomalous mole-fraction effect in, 170
Helix aspera neuron
   anions of, 194
   calcium channels in, 170
Helix pomatia neuron
   calcium channels in, 182
   inactivation calcium channel current in, 48
Hepatic disorders, 270
Heterogeneity, calcium channel, 253
High-threshold calcium currents, characteristics of, 127
High-threshold calcium channels
   Ca^{2+} buffering and inactivation gating in, 53
   model for inactivation in, 50—53
   single, 54—55
   voltage- and Ca^{2+}-dependent inactivation of, 47—50, 53
High voltage-activated (HVA) calcium channels
   activation kinetics of, 22—26
   Ca^{2+}-buffering and inactivation gating in, 53
   and cellular development, 142
   and inactivation kinetics, 55
   and inactivation mechanisms, 42
   models for inactivation in, 50—53
   recovery from inactivation in, 50
   voltage- and Ca^{2+}-dependent inactivation of, 47—50, 53
Hippocampal pyramidal cells, 132—133
Histamine, secretion of, 100—101
Hodgkin and Huxley formalism, 28
HOE, 166, 202
Hormones, effect on calcium channels of, 161
Hormone secretion, calcium-dependent, 98—99
Huntington's disease, 243
HVA, see High Voltage-Activated Calcium Channels
Hybridoma cell line, 174
6-Hydroxydopamine, 243

Hyperparathyroidism, 386
Hypertension
  calcium channel blockers, for, 310
  calcium-overload-induced, 248
  effects of calcium channel ligands in, 232—233
  importance of pre-treatment blood pressure in, 320—321
  and left ventricular hypertrophy, 313
  treatment of, 223
Hypertensive factor, 240
Hypoparathyroidism, 386

I

I_{Ca}, cations as competitive inhibitors of, 174—175
IgG antibodies, 241
Imipramine, structural formula of, 208
Inactivation
  Ca^{2+} dependent, 39
  of calcium channels, 160
  controlled by voltage, 37—38
  through gating current measurements, 38—39
  ion channel, 36—37
  mechanisms of, 41
    calcium channel types, 41—42
    high-threshold channels, 47—55
    low-threshold channel, 42—47
  modulation of, 55—59
  through single channel measurements, 37—41
  time-dependent, 44
Inactivation gates, 5
Information, of sensory receptor cells, 109
Inhibition, types of, 185
Inorganic cations
  and calcium channels, 165—166
    cations carrying current, 166—174
    cations as competitive inhibitors of I_{Ca}, 174—175
    cations exhibiting different actions, 175—178
    effect of pH on, 178—179
  dissociation constants for, 167—169
  effect on calcium channels of, 161
  and structural integrity of calcium channels, 164—165
Insulin secretion, calcium dependence of, 97
Ion channels, 4—8, 36—37
Ion selectivity, of light-sensitive channels in rods, 117—118
Ions, calcium channels blocked by, 92
Ischemic episodes, silent, 285
Ischemic heart disease
  calcium channel blockers in, 268
  and modulation of calcium channels, 274
Ischemic myocardium, effects of calcium channel ligands in, 233
Ischemic stroke, 270
Isoproterenol, in cardiomyopathies, 302
Isradipine
  L channels blocked by, 224—225
  structural formula of, 212
I-V-relationship
  for calcium channels, 25
  for peak calcium currents, 49

K

Kainate, 381
Kainic acid, 242
KB 944, binding site for, 222
KB-2796, T channels inhibited by, 232
Kinetic models, of low-threshold channels, 45—47
Kinetics
  activation, 30
  deactivation, 30—31
  inactivation, 42, 55
  tail current, 27

L

La^{3+}, effect on calcium channels of, 161
Lambert-Eaton myasthenic syndrome, 241
Langmuir's isotherm, 182
L channel antagonists, 212. See also Amlodipine
L channels
  allosterically coupled binding sites associated with 221—223
  allosteric sites on, 201—202
  characteristics of, 54
  drugs active at, 210—215
  and inactivation mechanisms, 42
  low affinity inhibitors of, 230
  in platelets, 245—246
  properties of, 197—198, 248—249
  subclasses of, 216
  vs. T channels, 254
  types of, 199
  voltage-dependent kinetics of, 172
L-channel drugs, therapeutic uses of, 200
Learning deficits, age-associated, 223
Left ventricular hypertrophy, 313
Leucine-encephalin, and calcium channel activation, 31
Lidocaine, 246
Ligand binding data studies, and physiology data, 217—220
Ligand-gated calcium channels, 104
Light, electrical response to, ionic mechanisms of, 111
Lipid bilayers, and divalent cation movement, 175
Lipid concentration, 316
Lithium, interaction with calcium, 391—392
Loperamide, structural formula of, 208
Low-threshold (LVA) channels, 42—47, 105. See also L-type calcium channels
Low-threshold currents, 127, 135
Low-voltage-activated (LVA or T) channels, 10, 14, 18, 22
  Ca^{2+}-block of sodium currents through, 15—17
  cardiac, 46
  and cellular development, 142
  and inactivation mechanisms, 42
  kinetic models of, 45—47

neuronal, 46
permeability of, 10
single, 44—45
voltage-dependent inactivation of, 42
voltage-dependent recovery from ianctivation of, 43
L-type calcium channels
in arrhythmias, 329
blocked by dihydropyridines, 154
characteristics of, 254
epinephrine modulation of, 257, 263
inactivation properties of, 79
in neuronal survival, 147
in patch clamp ventricular myocytes, 179
Low-voltage activated (LVA) sodium currents, 45. See also Sodium channels
LVA, see Low Voltage-activated
Lymnaea neurons, anomalous mole-fraction effects in, 183

M

Maitotoxin, action of, 216
Manic-depressive patients, 389. See also Psychiatric disorders; Psychiatric patients
Mast cells, secretion of histamine from, 100
McN 6186-11
in L-type channels, 202
structural formula for, 206
Membranes, selective permeability of, 3, 8
Menthol, and HVA channel inactivation, 56—58
Messenger systems, intracellular, 138
Mesenteric arteries, binding sites in, 218
MD1 12, 330A, 202
binding site for, 222
structural formula of, 206
Mediaster aequalis, calcium channel currents in, 171
Memory-enhancing drugs, and dihydropyridine binding sites, 243—244
Mg^{2+} ion, 10, 13—15
Michaelis-Menten constant, 184
"Mixed" angina, 286
Mn spikes, 177
Models of calcium channels, 179—188. See also Animal models
Modulation, calcium channel, 252, 256—257
and cardiac responses to protein kinase A, 258, 263
with epinephrine, 257
mechanisms of, 253
phosphorylation in, 257—258
by protein kinase, 259, 263
and response to β-adrenergic stimulation, 258—259
Mollusk neurons
calcium inactivation in, 53
inactivation in, 47
Mortality, cardiac, prevention of, 353—354, 361
Mouse neoplastic lymphocytes, calcium channel models for, 181

Muscular dysgenesis, 74, 243
Muscle. See Cardiac muscle; Frog muscle fiber; Smooth muscle
Muscle cells, denervation of, 146. See also Barnacle muscle membrane; Guinea pig myocytes; Myocytes
Muscle contraction, 65
Muscle spindles, transduction in, 108
Myocardial hypertrophy, effects of calcium channel ligands in, 232—233
Myocardiac infarction. See also Cardiac disease
calcium channel blockers following, 351
diltiazem for, 290
failure of calcium blockers to benefit, 291
follow-up, 290
infarct size, 289
nifedipine for, 289
ventricular fibrillation, 288—289
verapamil for, 289—290
Myocardial ischemia
and modulation of calcium channels, 274
pathogenesis of, 268
Myocytes, 138—143
Xenopus, 140

N

Na^+ ions, effect on calcium channels of, 161. See also Sodium channels
N channels, 254—255
characteristics of, 54
drugs acting on, 202
drugs active at, 215—216
and inactivation mechanism, 42
properties of, 198
regulation of, 204
types of, 199
voltage-dependent kinetics of, 172
Naturally occurring compounds, effect on calcium channels of, 161
Neomycin
N channels inhibited by, 198
potency of, 205
Nernst equation, 164—165
Neurite outgrowth, Ca^{2+} influx effects on, 147—148
Neuroleptics, and calcium channel blockers, 393
Neurologic disorders, calcium channel blockers in, 270
Neuromuscular interactions, and Ca^{2+} influx, 152
Neuronal cells, effect of drug on calcium channels in, 224. See also Neurons
Neuronal survival, Ca^{2+}-influx and, 145—147
Neurons. See also Mollusk neurons; Purkinje cells
amphibian spinal cord, 139—142
calcium conductances in differentiation of, 138—143
calcium currents in burst firing of, 133—134
calcium-regulating mechanisms of, 379

distribution of calcium channels in, 256
dendrite calcium channels in, 131—132
effects of calcium channel ligands on, 234—237
effects of protein kinase C activation on, 260
and ligand-binding data, 220—221
NMDA channels in, 127—128
process outgrowth, 149
rhythmic activity in, 134—135
Neuro peptide Y (NPY), calcium currents inhibited by, 261—263
Neurotransmitter metabolism, Ca^{2+} effect on, 148—150
Neurotransmitter receptors, 244—245
Neurotransmitters
effect on calcium channels of, 161, 261
and GTP-binding proteins, 261—262
phenotype, 148—150
Ca^{2+} regulation of sensitivity, 105
and voltage-sensitive calcium channels, 148
Nerve growth factor (NGF), 152
Nicardipine, 205, 311. See also Dihydropyridines
in brain tissue, 383
metabolism effects of, 315
structural formula for, 201
vasoselectivity of, 312—313
Nickel, antiarrhythmias action of, 329
Nicotinic channel, relative permeability in Ca^{2+} of, 128
Nifedipine, 200, 311. See also Dihydropyridines
for acute myocardial infarction, 289
antianginal potency of, 275
as calcium entry blocker, 383
in cardiomyopathic hamsters, 300—301
in cardiovascular disease, 293
in chronic effort angina, 281
in dilated cardiomyopathies, 298
and drug metabolism, 246
effect on calcium channels of, 161
effect on platelets of, 368
for effort angina, 278—279
in L channels, 210
and left ventricular hypertrophy, 313
in peripheral vascular disease, 372
structural formula for, 201
vasoselectivity of, 312
in vasospastic angina, 283
vs. verapamil, 280, 281
Nimodipine. See also blockers, calcium channel
in acute ischemic stoke, 383
antiseizure activity of, 244
in brain tissue, 383
as calcium entry blocker, 383
direct neuronal effects of, 237—239
effects on neuronal L channels of, 236—237
L channels blocked by, 224—225
for neurological defects, 200
partition coefficient of, 207
selectivity of action of, 227
structural formula for, 201
in subarachnoid hemorrhage, 200, 237—238, 383
Nisoldipine

L channels blocked by, 224—225
partition coefficient of, 207
in rabbit model of atherosclerosis, 234
selectivity of action of, 227
structural formula of, 212
Nitrendipine, 205
effect on platelets of, 368
effects on renal function of, 314—315
in hypertensive patients, 319—320
inhibition of T channels by, 232
L channels blocked by, 224—225
partition coefficient of, 207
potential uses for, 271
structural formula of, 212
N-methyl-D-asparate (NMDA), 381
N-methyl-D-asparate (NMDA) channels, 127—128
N-methyl-D-asparate (NMDA) receptors, 132, 145—146
N-methyl-D-glucamine (NMG), HVA channel activation affected by, 59
Noncompetitive inhibition, 186—187
Noradrenaline, 83
Norepinephrine, calcium currents inhibitied by, 261—263
NPY, see Neuro Peptide Y
Nucleic acid hybridization studies, 199

O

Octanil/buffer distribution, 207
Olfactory cells, transduction in, 108
Open channels, 173
Open state, of Ca^{2+} channels, 40
Open state block, 224
Open time, probability of, 161
Organic ions, and structural integrity of calcium channels, 164—165. See also specific ions
Osteoblast-like cells, 226
Ovarian oocytes membrane, Ca spikes in, 177
Oxiracetam, and dihydropyridine binding site, 243

P

Pancreas, beta cells of, 96—98
Paramecium cells, trichocysts from, 99
Parathyroid hormone, secretion of, 102—103
Parkinson's disease, 243
Partition coefficients, 206
Patch clamp experiments, 4, 171
of macroscopic inward currents, 303—307
technique, 22, 172
P channels, properties of, 198. See also P-type calcium channels
PDE, see Phosphodiesterase
PDGF, see Platelet-derived Growth Factor
Peptides, dihydropyridines binding inhibited by, 239—240
Percutaneous Transluminal Coronary Artery Angioplasty, see PTCA
Peripheral vascular disorders, calcium channel blockers in, 269—270. See also Vascular diseases

Index

Permeability
  and ion block, 10—19
  selective, 8
pH, ion channel conductances as function of, 178
Pharmacomechanical coupling, in vascular smooth muscle, 70
Phenoxybenzamine, structural formula of, 208
Phenylalkylamines, 307. See also Gallopamil; Verapamil
  binding of, 222
  in brain tissue, 383
  as calcium entry blockers, 383
  selectivity of action of, 228
Phenylalkylamine group, 311
Phenytoin, 204, 209
Phosphodiesterase (PDE)
  cGMP, 115
  inhibitors of, 249
Phosphoinositide, metabolism of, 381
Phosphorylation
  of calcium channels, 58
  of calcium channel proteins, 307
  selectivity of action of, 227
Photoreceptors, 108—112, 120—121
Phototransduction, 111—116. See also Transduction
Physiological data. Ligand binding and, 217—220
Physiology, of calcium channels, 63—67, 197
Pimozide, 202
Piracetam, and dihydropyridine binding sites, 243
Pituitary gland, and calcium-dependent hormone secretion, 99
pKa, 190—194
Planar bilayer techniques, 22
Platelet-derived growth factor (PDGF), 364
Platelet related disorders, 374—375
Platelets
  and calcium channel blockers, 366—370
  platelet activation, 364—365
  role as second messenger, 365—366
  calcium-transport enzyme activities in, 388—390
  L channels in, 245—246
  sites of action for calcium channel blockers on, 369
PN 200-110, 383
Point mutations, 32
Potassium channels, 27
Potassium currents, 129
  in amphibian spinal cord neurons, 139
  voltage-dependent, 134
Precautions, for calcium channel blockers, 359
Prenylamine, 200—201
Presynaptic terminals, calcium channels in, 91—93
Prinzmetal's angina, 282
Propranolol, 278. See also Beta blockers
  in cardiomyopathies, 302
  in hypertensive patients, 320
  metabolic effects of, 315
  vasoselectivity of, 313
Protein kinase
  activation of, 138
  phorbol ester stimulation of, 153

Protein kinase A, cardiac response to activation of, 258, 263
Protein Kinase C
  calcium channels modulated by, 259—261, 263
  in receptor regulation, 144
Proteins
  calcium binding, 138
  molecular identity of channels, 120
Protein synthesis, in amphibian spinal cord neurons, 140—141
Psychiatric disorders
  with altered calcium function, 386—387
  calcium channel blockers in, 270
Psychiatric patients, 387—391
Psychosomatic complaints, 387
PTCA, 233, 291
P-type calcium channels, 263
  characteristics of, 255
  vs. T channels, 253
Pumping mechanisms, 64
Purkinje cells, 129—132, 253
Purkinje cell dendrites, 130—132

Q

Quaternary structures, 60
Quinidine
  efficacy as antiarrhythnic agent, 339
  therapeutic index of, 246
Quisqualate, 381

R

Rat uterine smooth muscle cells, anomalous mole-fraction effect in, 170
Raynaud's phenomenon, 373—374
Renal effects, of calcium channel ligands, 232—233
Reperfusion, 291, 233—234
Resting membrane potential, measurement of, 165
RNA synthesis, in amphibian spinal cord neurons, 141—142
Rodent muscle cells, calcium currents in, 143
Rod outer segments, 116—120
Rohon-Beard neurons, 150
Ryanodine, 75, 247
  binding of, 81
  receptor, 76

S

Saturation kinetics, mechanism of, 183—185
Saturation process, mechanisms of inhibition of, 185—187
Sea urchin eggs, and cortical granule discharge, 100
Second messenger
  calcium as, 365—366
  in neuronal calcium currents, 262
Secretion, 88—89
Secretory cells, calcium triggering in, 104—105

Secretory responses, modes of, 104—105
Seizures, effect of calcium channel blockers on, 238
Selection, by affinity, 18
Selectivity
　ionic, 10
　property of, 4
Selectivity of action, 227—232
Sensing, property of, 7
Sensory receptor cells
　classification of, 108
　signal processing in, 120—122
Sensory transduction, 64, 108
Sequestering mechanisms, 64
Side effects, for calcium channel blockers, 359
Skeletal muscle
　effect on drugs of calcium channels in, 224—225
　excitation-contraction coupling in, 71—77
　L-channel function in, 262
　voltage-sensitive calcium channels in, 199
Skeletal muscle fibers, calcium channel models for, 181
Smooth muscle. See also Vascular smooth muscle
　effect of drugs on calcium channels in, 224
Snail neurons. See also Neurons
　anomalous mole-fraction effect in, 170
　calcium currents in, 13
Sodium channel currents, single, 41
Sodium channels
　effect of BAY K 8644 on, 229
　Mn^{2+}-sensitive, 307
　in myocyte development, 142
　and myocyte differentiation, 144
　TTX resistant, 307
　TTX-sensitive fast, 306
Somatostatin, and calcium channel activation, 31
Spider toxins, 215
Spinal cord neurons, calcium conductions in, 139—142
Squid presynaptic calcium channels, gating of, 92—93
^{85}Sr, K^+-stimulated influxes of, 172
SR-7037, structural formula of, 206
SR 33557
　binding site for, 222
　structural formula of, 206
SR^{2+} ions, effect on calcium channels of, 161
Stereoselectivity, 216
Stochastic shifts, 160
Structure-activity studies, 206—207, 215
Structure-function relationships, 249
Stylonychia nytilus, Ca^{2+}-dependent action potentials in, 177
Subarachnoid hemorrhage
　calcium channel blockers in, 270
　nimodipine in, 383
Swiss 3T3 fibroblasts, 176
Sympathetic neurons, survival of, 145
Synaptic activity, of sensory receptor cells, 109
Synaptosomes, brain, 194
Syrian hamster, cardiomyopathic, disturbed Ca^{2+} metabolism in, 299—301

T

Tacharrhythmias, 346—349, 352—353
　calcium channel blockers for, 296
　mechanisms of paroxysmal, 336
Tail currents, 23—25, 27
Taste cells, transduction in, 108
T channels, 253, 263
　characteristics of, 254
　in antiarrhythmias action, 329
　drugs acting on, 202, 230—232
　and inactivation mechanisms, 42
　inhibitors of, 204, 231, 248
　vs. L channels, 254
　and neurotransmitter secretion, 105
　properties of, 198
　voltage-dependent kinetics of, 172
Teramethrin, antiarrhythmic action of, 329
Terodiline, 200—201
Tetramethrin, in T channels, 209
Tetrandrine
　in L class channels, 202
　structural formula of, 206
TH, see Tyrosine Hydroxylase
Thalamic cells, receptive firing of, 135
Thrombocythemia, 375
Thrombosis, arterial and venous, 374
Thrombotic thrombocytopenic purpura (TTP), 374—375
Thyroid parafollicular cells, and calcium-dependent hormone secretion, 99
Tiapamil, dosage regimens for, 359
Tiger salamander eye, retina isolated form, 110, 112
Torsades De Pointes, calcium channel blockers and, 353
Trans-diclofurime, 202, 206
Transducers, calcium channels as, 252
Transducin, 115
Transduction, 63—64, 108—111. See also Photo-transduction; Sensory transduction
Transmitter secretion, triggered by calcium channels, 90—91
Troponin, blockade of, 310
TTP, see Thrombotic Thrombocytopenic Purpura
TTX-sensitive channels, 144
Twitch activation, 72
Tyrosine hydroxylase (TH)
　electrical activity in, 148
　in sympathetic neurons, 152

U

Uterine smooth muscle, Na^+ permeation through calcium channels in, 176
U-shaped inactivation, 39
U-shaped inactivation curve, 47

Index

V

Vascular diseases, 372—375
Vascular smooth muscle
   excitation-contraction coupling in, 81—85
   pharmacomechanical coupling in, 70
Vasculitis, 374
Vasodilation, effect of calcium channel blockers on, 370—371
Vasodilators, 312. See also Dihydropyridines
"Vasospastic angina", 282, 285
Verpamil, 80, 200, 311. See also Blockers, calcium channel
   for actue myocardial infarction, 289—290, 292
   antianginal potency of, 275
   as antiarrhythmic agent, 328—329, 339, 360
   in atherosclerosis, 325
   binding site for, 222
   as calcium entry blocker, 383
   in cardiomyopathic hamsters, 300—301
   in cardiovascular disease, 293
   in chronic effort angina, 281
   in dilated cardiopathies, 298
   dosage regimens for, 359
   and drug metabolism, 246
   effects on AV node of, 358
   effect on calcium channels of, 161
   effect on heart rate of, 355
   for effort angina, 277—278
   enantiomers of, 216—217
   hepatotoxicity prevented by, 270
   in hypertrophic cardiomyopathy, 297—298
   and left ventricular hypertrophy, 313
   in manic patients, 390
   vs. nifedipine, 280—281
   in peripheral vascular disease, 372
   selectivity of action of, 228
   structural formula for, 201
   vasoselectivity of, 312
   in vasospastic angina, 283
Ventricular heart muscle, calcium channels in, 176
Ventricular myocytes
   protonation and deprotonation rates for, 178
   single channel current amplitudes recorded from, 172
Vertebrate neurons, calcium inactivation in, 53
Vitamin D metabolites, 239
VOC, see Voltage-operated Channel
Voltage, inactivation processes controlled by, 37—38
Voltage-activated calcium currents, 126—129
Voltage clamp technique, 165—166
   currents recorded by, 176
   objective of, 189
Voltage dependence, of G-protein coupling, 31
Voltage-dependent block, 15—17
Voltage-dependent calcium channels, types of, 253—256
Voltage-gated calcium channels, 104
   diversity of, 105
   and transmitter-gated channels, 132—133
   transmitter modulation of, 128—129
Voltage-operated-channel (VOC), 36
Voltage-sensitive calcium channels, 197—199
Voltage sensor, and ligand binding data, 218, 220

W

Warfarin, 246
Wash-out, 58
Withdrawal syndrome, effect of calcium channel blockers on, 238

X

Xenopus laveis, 194
Xenopus neurons, 140—141, 155
Xylotrupes dichotomus, Ca spikes in larval muscle fibers of, 177

Z

Zn^{2+} ions, as competitive inhibitor, 177

HEALTH SCIENCES LIBRARY
LUTHERAN COLLEGE
3024 FAIRFIELD AVE.
FORT WAYNE IN 46807-1697